0
25
Punta Anegada
Punta Catalina
Bahía
San Sebastian
ATLANTIC OCEAN
ISLA
GRANDE
Río Grande
R. Grande
Blanco
Viamonte
Almirantazgo
R. Azopardo
Sierra Beauvoir
Sierra de Valdivieso
L. Fagnano
Sierra Alvear
R. Lashifashaj
Ushuaia
CANAL BEAGLE
Harberton
Sierra Lucio Lopez
Peninsula
Mitre
ESTRECHO DE
LE MAIRE
Isla de los Estados
Hoste
Isla Navarino
I. Picton
I. Lennox
I. Nueva
Peninsula Hardy
BAHIA NASSAU
Islas Woollaston
69°
68°
67°
66°
65°
64°
Cabo de Hornos

D. [illegible]

2000.

Flora of Tierra del Fuego

Books written or edited by D. M. Moore

The Vascular Flora of the Falkland Islands. 1968.
British Antarctic Survey.
Plant Cytogenetics. 1976. Chapman & Hall.
Flora Europaea. 5 vols. (with T. G. Tutin *et al.*).
1963–80. Cambridge University Press.
Flora Europaea Check-list and Chromosome Index.
1982. Cambridge University Press.
Green Planet. 1982. Cambridge University Press.

David M. Moore

Professor of Botany
University of Reading, England

Flora of Tierra del Fuego

with line drawings by
R. N. P. Goodall
Flora Patagónica
S. Parkinson

Anthony Nelson England
Missouri Botanical Garden USA

First published in 1983 by Anthony Nelson
PO Box 9, Oswestry, Shropshire SY11 1BY, England
and simultaneously in the USA
by Missouri Botanical Garden
PO Box 299, St Louis, Missouri 63166 USA

ISBN 0 904614 05 0
LC 83–060628

Photoset by Keyspools Ltd, Golborne, Lancs, England
Printed by Livesey Limited, Shrewsbury, England

Contents

List of colour plates

Preface

Tierra del Fuego, the complex archipelago south of the Estrecho de Magallanes, has one of the world's pivotal floras. The plants growing there are those which, out of all the botanical richness of South America, have penetrated furthest towards the icy wastes of Antarctica. Furthermore, many Fuegian species have close relationships with those present on the other side of Antarctica in, most importantly, New Zealand, as well as on the intervening circum-Antarctic islands. Since the modern austral lands result from the fragmentation of the supercontinent of Gondwanaland into which they were united until 80 million years ago, the botanical relationships are of importance for understanding the ancient history of the plants we know today. However, there is also considerable evidence that some plants show affinities across Antarctica because of more recent transoceanic transport of seeds and spores by birds, winds and ocean currents.

Consequently, the plants of Tierra del Fuego are of considerable importance in understanding the evolution and migration of plant species. This book is a small contribution, based on a botanical history of this inclement area involving daring, sacrifice and scientific endeavour, towards providing a firmer foundation for the studies needed to understand more fully the role of these southern lands in the development of temperate floras.

This book aims to be scientifically accurate, given the present state of knowledge. It documents, as far as possible, the reasons for recognising particular species in relation to previous work and an attempt has been made to account for all names used in botanical literature pertaining to Tierra del Fuego. The book also aims to permit reasonably easy identification of the Fuegian plants, and the descriptions, drawings and photographs should be considered as a whole. They do not pretend to be complete but it is hoped that they draw attention to those features of most use in identifying the species. In many instances monographs or other Floras include illustrations and descriptions giving much fuller information; these are referred to in the text.

This account is intended to provide some scientific and historical orientation for an appreciation of the plants of Tierra del Fuego. Although, rightly, still retaining some of its reputation as a remote and inhospitable region, Tierra del Fuego is reached by more and more people each year. It is hoped that, by enabling them to identify the plants they see, this book will help enrich their visit to this fascinating land of fire.

D.M.M.

Acknowledgements

Quien come el calafate vuelve por más
"Whoever eats the calafate returns for more"

This book is based on studies made since 1960, when I first visited Tierra del Fuego. It results from work which has only been made possible by the support of a large number of friends and colleagues, too numerous to acknowledge fully here. Whether this support involved helping my understanding of the plants and their habitats , the provision of a horse, a boat, a bed or a hot bath, an introduction to other friends or, perhaps above all, my acceptance as an adopted 'fueguino' by the kindly inhabitants of Tierra del Fuego, I cannot adequately express my debt – I know that all who have tasted the calafate will understand.

Rae Natalie Prosser Goodall should have been a joint author of this book. That she chose not to be must in no way obscure her substantial contribution in collecting material, documenting field populations and making numerous perceptive observations which are incorporated into the text. The lists of Indian and Spanish names are also the result of her efforts, while her many illustrations of plants which adorn and complement the text need no further comment. The preparation of many of these illustrations was supported by grants from the National Geographic Society to Mrs. Goodall. Many of the other drawings are the work of Sydney Parkinson, botanical artist on Cook's first voyage, and permission to reproduce them was kindly given by the Trustees of the British Museum (Natural History), London, where the unpublished original lithographs are housed. The other illustrations were generously made available by Dra. M. N. Correa (I.N.T.A., Castelar, Argentina), editor of the invaluable *Flora Patagónica*, for which they were prepared.

Over the years various grants from the Royal Society of London, the Natural Environment Research Council (NERC) and the Linnean Society of London's Percy Sladen Fund have permitted my work in South America, while funds from NERC supported my Research Assistants Frances Barnard, Bronwen Yates and Margaret Doggett, whose careful and devoted labours in herbarium and greenhouse so eased my task. The field work in Tierra del Fuego has been facilitated by the Sociedad Explotadora (later Ganadera) de Tierra del Fuego in the years 1960–1968, and, since 1969, by the Instituto de la Patagonia, Punta Arenas, particularly through its Rector, Señor Mateo Martinić B., and its chief botanist, Señor Edmundo Pisano V., friend and companion on many field forays. Ships of both the Armada Argentina and the Armada de Chile have, on occasions, generously provided the maritime transport so essential in the Fuegian archipelago, as did the Empresa Pesquera Bustos, Punta Arenas and the Cooperativo Pesquero, Porvenir. Amongst the many inhabitants of Tierra del Fuego who have furnished hospitality and innumerable other courtesies I should particularly like to thank Mr. Adrian Goodall (Estancia Viamonte), Mr. Thomas D. Goodall (Estancia Harberton) and their families and friends.

In addition to being given access to the extensive private herbarium of Mrs. R. N. P. Goodall, I am indebted to the Directors of the following Institutes and Herbaria for facilitating my study of material held in their care: British Museum (Natural History); Departamento de Botánica Agrícola, I.N.T.A., Castelar; Instituto Botánico Darwinión, San Isidro; Botanical Museum, Göteborg; Royal Botanic Gardens, Kew; Museo de La Plata; Muséum National d'Histoire Naturelle, Paris; Instituto de la Patagonia, Punta Arenas; Museo Nacional de Historia Natural, Santiago de Chile; Naturhistoriska

Riksmuseum, Stockholm; University Institute of Systematic Botany, Uppsala; United States National Arboretum, Washington.

Ing. Agron. O. Boelcke, Buenos Aires (Cruciferae), Dr. A. Burkart, San Isidro (*Lathyrus*), Dr. A. L. Cabrera, San Isidro (Compositae), Dra. M. N. Correa, Castelar (Orchidaceae, *Adesmia*), Dr. J. Grau, Munich (*Myosotis*), Dra. E. G. Nicora, San Isidro (Gramineae) and Mr. T. M. Pedersen, Copenhagen (*Stellaria*), freely shared their taxonomic expertise, while Dr. J. C. Solomon, St. Louis, generously provided the account of *Epilobium* whilst he was still writing up his fine revision of the South American members of this genus. I should also like to acknowledge Mrs. J. Birkey and Miss V. J. Norris for typing much of the text, and Mr. V. Morris and Mr. I. Maclean for photographic assistance with many of the illustrations. Angela Lauder and Richard Gozney, of the British Embassy in Buenos Aires, enlisted the services of several colleagues to bring Mrs. Goodall's original drawings to England; their good offices were indeed greatly appreciated.

Altogether, I have spent well over two of the last twenty one years in Tierra del Fuego and southern Patagonia. Since field studies of austral plants must be undertaken during the British winter, my absences from home have at times included part or all of the family Christmas period. It is, therefore, a particular pleasure to conclude by thanking my wife, Ida, for cheerfully managing house and our sons during my sojourns in the Southern Hemisphere, for her forebearance over the many hours I was closeted with the developing manuscript, and for her assistance in checking the typescript and proofs – as some recompense, she did eventually get to taste the calafate!

I. Introduction

About 2 million years ago the Cordillera de los Andes, which had been pushed ever higher for 50 times as long, finally reached the approximate elevation that it has today. Since then it has acted as a barrier over which the westerly winds prevalent in southern South America were forced to rise, depositing the moisture accumulated as they crossed the Pacific Ocean on the west flanks of the mountains, so leaving the area to the east in a rain-shadow with relatively drier conditions. In Tierra del Fuego the Andes swing eastward before trending south as the islands of the Scotia Arc into Antarctica, so that their Fuegian representatives also interrupt the winds sweeping up from the Southern Ocean.

Some 16,000 years ago the glaciers of the last phase of the most recent (Pleistocene) ice age began to retreat from southernmost South America, which they had intermittently covered for about 2 million years, during the 4 advances of the Pleistocene ice. As the ice retreated it left behind open areas which became available for colonization by plants which had survived the glaciation, either further north in South America or in ice-free areas that extended over large parts of the continental shelf between the Patagonian coast and the Falkland Islands (Islas Malvinas) exposed by "glacioeustatic" lowering of the sea-level, which resulted from the demands of the glaciers (Auer, 1958, 1960). The glaciers are still retreating and, in the form of the Patagonian Ice Cap ('Hielo Patagónico Sur' – *c.*48–51°S. lat), together with the ice mass of Antarctica, some 1,000 km south of Tierra del Fuego, serve to remind us that in these southern regions the last ice age is a recent and even continuing phenomenon.

The final connections across the Estrecho de Magallanes were severed about 10,000 years ago and Tierra del Fuego, as an entity separate from continental South America, was born. Local glaciations persisted well into postglacial times in parts of Tierra del Fuego, such as the Altos del Boquerón and the Cordillera Fuegina (Auer, 1970), but after various fluctuations they have mostly disappeared, apart from the glaciers found today in the Cordillera Darwin and Isla Santa Inés, for example. Since the retreat of the major ice masses, Tierra del Fuego, like other areas, has been subjected to worldwide climatic changes, and generally moister conditions, resulting in the advance of the forests to the eastern fringe of the archipelago, have occurred at least twice (*c.*9,000–7,000 and 4,500–3,000 years ago), interspersed with drier periods when the forest retreated westwards (Auer, 1970). Today the climate apparently favours the retreat of the forests.

Superimposed upon these ponderous fluxes in the vegetation of Tierra del Fuego resulting from climatic changes have been those due to the effect of man. After his long trek from Asia and down the Americas man appeared in Tierra del Fuego about 7,600 years ago (Laming-Emperaire *et al.*, 1972). The Indians, living principally by hunting and scavenging, probably had little major effect on the plant cover of Tierra del Fuego, nor had the earlier European visitors. From 1520, when Magellan landed briefly at Bahía Felipe in northern Isla Grande during his month-long sojourn in the Straits now bearing his name, visitors to Tierra del Fuego gradually increased in proportion to the traffic in the region, which provided the only sea route between the Atlantic and Pacific Oceans until the opening of the Panama Canal in 1914. The activities of whalers and sealers, which expanded considerably after 1787 when Colnett began exploiting the seal fisheries from Isla de los Estados, probably saw the introduction of some European weeds around the shore-encampments periodically established, and the goats introduced by Capt. Piedrabuena to Isla de los Estados in 1873 undoubtedly exerted their traditional catastrophic effect on the vegetation.

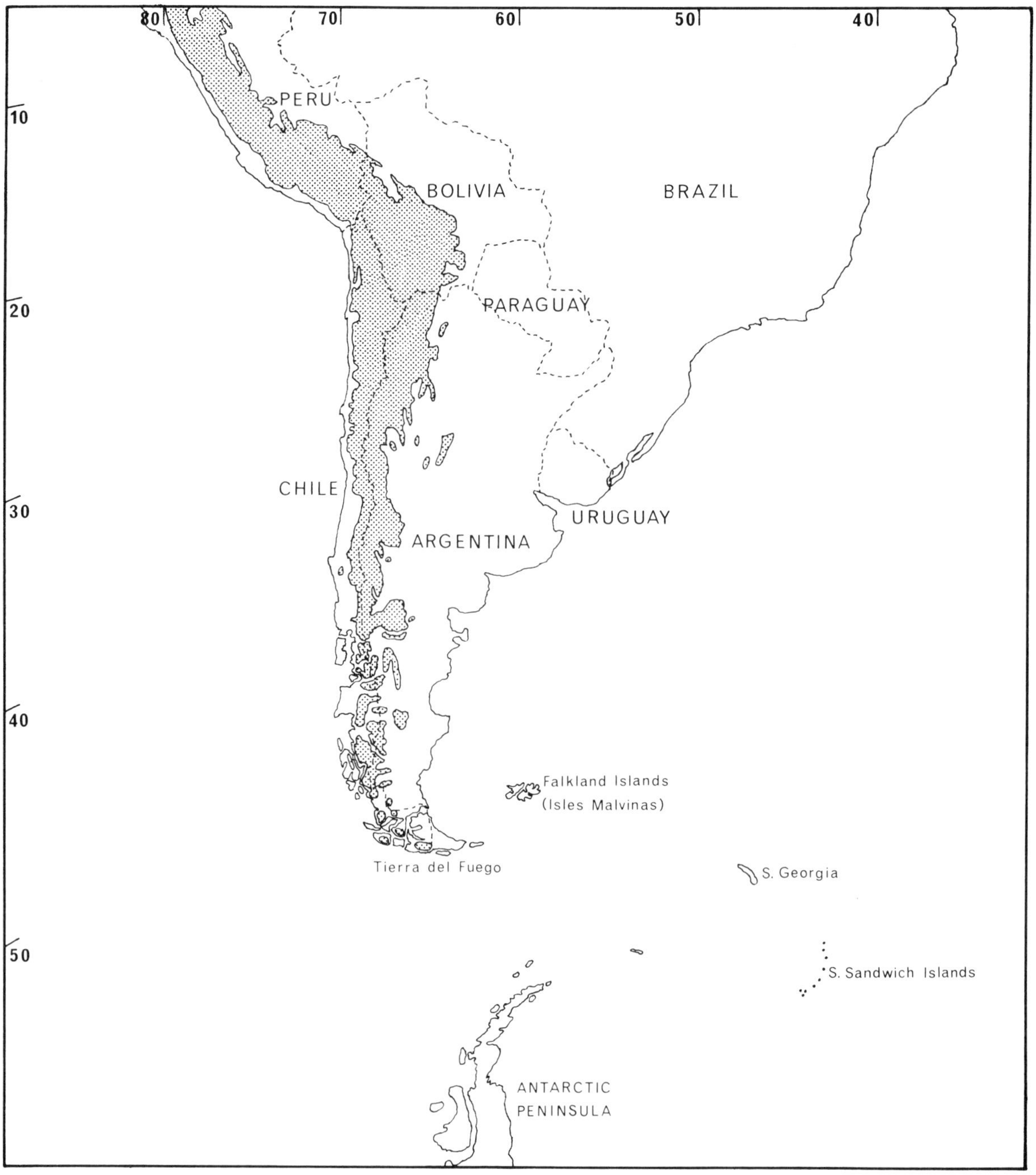

Fig. 1. Map showing Tierra del Fuego in relation to adjacent regions. Mountains over 1000 m are shaded.

However, the first major human impact on the vegetation was initiated by the goldminers. In 1879 Lieut. Ramon Serrano Montaner discovered gold in northern Isla Grande, and thus began the systematic exploitation, from 1881, of the precious metal. One tragic botanical result was the virtual elimination from the Altos del Boquerón of the extensive forests of *Maytenus magellanica*, which provided excellent charcoal for smelting the ore.

Nevertheless, it was the permanent settlement of Tierra del Fuego by Europeans introducing agricultural practices which produced the greatest changes in the native vegetation of the region. The granting of concessions for sheep ranching in Gente Grande (1883), Harberton (1886), Punta Anegada (1890), Bahía Felipe (1891), Caleta Josefina (1893), San Sebastián (1894), Primera Argentina (1896) and Sara (1897) laid the foundations for the pastoral basis of much of modern Fuegian life. The attendant selective grazing, drying of wet areas by sheep-trampling, and clearance of forest to provide further open range, have altered the vegetation and provided habitats for introduced weeds. Oil and tourism, both requiring improved access and bringing their own pressures on the environment, have also helped to change the flora and vegetation in certain areas; even pollution from oil-spillages (e.g. Dollenz, 1978) threatens parts of Isla Grande. Although much of Tierra del Fuego remains virtually unspoiled, and will remain so if the current policies of the Argentinian and Chilean governments can be maintained, this account is inevitably a view of the flora and vegetation as it is now, a transitory vision of a, perhaps regrettably, changing world.

1. TOPOGRAPHY

As defined here, Tierra del Fuego constitutes the archipelago lying south of the Estrecho de Magallanes, between 52° 25′ – 56° S lat. and 63° 47′ – 74° 45′ W. long., separated from the mainland of South America by from 30 km, in Canal Ancho, to 4 km, at the Primera Angostura. Much the largest island is Isla Grande (48,000 km^2), while there are several other islands of significant area – Isla Dawson (1,400 km^2), Isla Santa Inés (3,850 km^2), Isla Hoste (4,800 km^2), Isla Navarino (2,800 km^2) and Isla de los Estados (520 km^2). Associated with these are well over 200 islands ranging down to rocks a few square metres in extent, the whole land surface of Tierra del Fuego totalling some 66,000 km^2.

Most of the north and east of Isla Grande is undulating countryside between sea-level and 200 m, only parts of the Altos del Boquerón exceeding 500 m, and is traversed by the two longest rivers of the archipelago, the Río del Oro, which runs into Bahía Felipe, and the Río Grande, which runs into the Atlantic. Along the north side of the great depression occupied by Seno Almirantazgo, Río Azopardo and Lago Fagnano the land rises gently to a series of peaks (Sierra Beauvoir) averaging 600–700 m (maximum 1,036 m), with deep, densely wooded valleys betwen them.

The main mass of the Fuegian mountains lies south of the Almarantazgo–Fagnano depression. At the west end of Isla Grande the Fuegian extension of the Andes appears as Mte Sarmiento (2,404 m) and the mountains attain their greatest altitude in the Cordillera Darwin (Mte Luis de Savoya, 2,469 m.; Mte Darwin, 2,438 m) before gradually diminishing eastwards along both the north (Mte Vinciguerra, 1,450 m.; Mte Cornu, 1,380 m) and south (Isla Hoste, 1,342 m.; Dientes de Navarino, 1,195 m) shores of the Canal Beagle to the Sierra Lucio Lopez (672 m) at its eastern entrance.

To the south and west of the principal mountain system lies the main part of the Fuegian archipelago. Although substantial areas above 500 m only occur on Isla Santa Inés, the terrain is generally rugged throughout the outer archipelago with a multitude of islands bounded by generally steep, often precipitous, rocky coasts and penetrated by deep fjords. Exposed to the ceaseless battering of the Pacific rollers the west and south sides of the outer islands are particularly rugged as seen, for example, in the cliffs on the south side of the island which gives its name to the world famous Cabo de Hornos.

2. GEOLOGY AND STRUCTURE

The undulating plains of north and east Isla Grande are formed by Tertiary sediments laid down under marine and terrestrial conditions, overlain by sands, gravels and clays deposited during the retreat of the latest, Quaternary, glaciers. In some coastal areas there are also recent marine sands and gravels resulting from local incursions by the sea during the post-glacial period. Underlying the Tertiary and Quaternary deposits are extensive Cretaceous sediments – sandstones, clay-slates, marls and conglomerates – which emerge along the north side of the Almirantazgo–Fagnano depression to form the "Marginal Cordillera" (Kranck, 1932) or "Pre-Cordillera" (Bonarelli, 1917) of the sub-Andean region (Pisano, 1977), where they are strongly folded with the beds often vertical. The softer rocks erode to give the generally more rounded contours of these northern mountains, with wide valleys ploughed out by the Pleistocene glaciations.

The main mass of the Fuegian mountains, lying south of the Almirantazgo depression, is composed of strongly deformed and metamorphosed Palaeozoic to Jurassic rocks. The core of this "Central Cordillera" is composed of strongly tectonised mica- and quartz-schists, and is penetrated in many places by lower Tertiary granites and grano-diorites. The schists are flanked by dark slates (greywackies), marls and quartz-porphyries. South of these the Jurassic-Lower Cretaceous igneous greenstones cross Fuegia from Isla Capitán Aracena, through Península Brecknock (including Mte Sarmiento) and thence across S.W. Isla Hoste to Cabo de Hornos. South and West of the greenstones the so-called Andean diorites of the Patagonian batholith, which is interpreted as the root of a Late Jurassic andesitic volcanic chain (Dalziel *et al.*, 1974), extend to the outermost islands of the Fuegian archipelago.

There are abundant signs of glaciation throughout the Central Cordillera and the archipelagic regions. The acute, 'horn' peaks of, for example, Montes Sarmiento, Cornu and Olivia, the numerous glacial cirques at higher elevations and the U-shaped valleys, with their "hanging" tributary valleys, are examples of widespread land-forms resulting from ice-action by the relatively recent glaciations, of which the glaciers still existing in the region are a forcible reminder.

3 CLIMATE AND SOILS

Although the underlying rocks obviously have an important influence on the soils produced it is clear from the studies of Diaz *et al.* (1960), Etcheverherre (1972) and Pisano (1977), that soil development is strongly correlated with climate in Tierra del Fuego and southern Patagonia. The two are consequently dealt with in this section.

1. *Climate*

Despite the lack of detailed meteorological coverage of Tierra del Fuego, most of the stations being in coastal areas, the main characteristics of the climate can be outlined here thanks to the valuable summaries of, for example, Jerez and Arancibia (1972) and Pisano (1977). The complicated topography, particularly in the south and west, together with the paucity of information from that area, means that the frequent dramatic local variations in climate cannot be described with any confidence.

The climate of Tierra del Fuego and southern Patagonia is controlled by the South Pacific and South Atlantic Polar Fronts (Ojeda, 1966), which result in low pressure areas south and south east of the region throughout the year, giving gradients for westerly winds, though the northward movement of polar air masses, particularly in the winter, often gives rise to cold southerly winds. There are no suitable data from Fuegia but information from Punta Arenas, on the north side of the Estrecho de Magallanes and partially sheltered from the west, shows the features of the winds, which are generally stronger in the summer:–

	Jan.	Feb	Mar.	April	May	June
Direction	W	W	WNW	WNW	WNW	WNW
Speed (km/hr)						
Monthly mean	22	22	21	19	20	18
Monthly max.	138	143	138	123	138	110
Max. gust	150	154	160	130	150	130

	July	Aug.	Sept.	Oct.	Nov.	Dec.
Direction	W	W	W	W	W	WNW
Speed (km/hr)						
Monthly mean	19	19	21	23	22	22
Monthly max.	130	128	148	138	130	120
Max. gust	135	150	200	150	160	150

The winds reach Fuegia from the west laden with moisture picked up over the Pacific Ocean. As they reach the coast and rise over the mountains they deposit their moisture as rain or snow, thus being much drier as they continue eastwards. The consequent dramatic rainfall gradient from south and west Fuegia to the conspicuously more arid rain shadow of north and east Isla Grande is demonstrated in Table 1. The apparent anomaly of the westernmost point, Isla Evangelistas, having a lower rainfall than the more easterly station at Bahía Felix on Isla Desolación, suggests that the low-lying outermost islands do not induce the winds to release their moisture as effectively as those of higher relief and within the influence of the cordillera. In addition to the dramatic rainfall gradient there is a much greater seasonal difference in N.E. Fuegia than in the south and west, where the oceanic conditions are demonstrated by the comparable figures for the wettest and driest months (Table 1).

Mean annual temperature shows a tendency to increase north-eastwards across Fuegia, as does the seasonal difference. The lowest temperatures are recorded from stations immediately north and south (Pampa Guanaco, Ushuaia) of the principal mountain system, which has a considerable cooling effect. The permanent snow-line is at about 700 m, although with many local variations, and in several places glaciers stream down to sea level. On the basis of the climatic data four major climatic zones can be recognised in Fuegia (Pisano, 1977), as indicated in Table 1, and with these are closely correlated the principal vegetation types (Chapter III).

2. *Soils*

Although still not at all well known the soils of Tierra del Fuego clearly correlate closely with the rainfall, age and type of the underlying substrate, and the prevailing relatively cool conditions. In the Cold Steppe zone, developed from recent glacial deposits, the soils are predominantly Meadow-Planosols, the least acid in Fuegia (pH 6–7), with an upper clay-humus usually underlain by a hard clay pan at depths of 30–50 cm. When disturbed they erode readily in the relatively dry, strong winds. In low-lying areas, near the coast and inland depressions (e.g. San Sebastián), the high water table gives rise to saline conditions and the soils are then termed Solonchaks. Further south the soils become more acid (pH 4–5.2) as the rainfall increases and the Submontane zone is characterised by podsols developed over the glacial deposits. In undulating or hilly areas the soils are shallow, have good drainage and are readily penetrated by roots as far as the iron pan at depths up to 30 cm; they have up to 30% organic content and generally support scrub or deciduous forest, which can lead to the development of acidic brown forest soils with the incorporation of leaf-litter and minerals. In areas of poor drainage the soils develop into gley-podsols in which an impermeable clay pan maintains the upper layers saturated with water to give a peaty humus of high (*c.*95%) organic content, over which bogs in which *Sphagnum* is important can develop.

Podsols and acid brown forest soils develop in the southern and western zones of high precipitation only where the relief and/or permeable morainic and other glacial detritus permit reasonable drainage. Much of the region, however, is underlain by diorites and similar igneous rocks which break down slowly, and supports shallow soils poor in minerals and which, since they are acid and almost constantly waterlogged, exhibit slow decomposition of plant remains so that extensive peat deposits accumulate to support a wet tundra-bog vegetation. The colder conditions at higher elevations, especially in the main mountain systems, lead to considerable fracturing of the rocks by frost-action and the resultant accumulations of debris, from coarse sand to large fragments, are low in organic matter. Consequently, such soils as are present are mostly lithosols, though at lower elevations in valleys etc. they incorporate a little superficial humus to constitute immature Regosols. In addition peri-glacial features, such as solifluction and "frost-heave", further reduce the stability of these skeletal soils.

Table 1. Meteorological data from stations representing the climatic diversity of Tierra del Fuego. Since no information is available from northernmost Isla Grande records from Punta Dúngeness, on the north shore of the east end of the Estrecho de Magallanes, are included (After Pisano, 1977).

Station	Position		Precipitation (mm)			Temperature (°C)		
	Lat. °S	Long. °W	Annual Total	Wettest Month	Driest Month	Annual Mean	Max. monthly mean	Min. monthly mean
Cold Steppe								
Punta Dúngeness	52°24′	68°25′	295.0	36.0	14.5	7.4	11.6	3.1
Cerro Sombrero	52°46′	69°22′	—	—	—	6.8	11.3	0.8
Cullen	52°53′	68°27′	376.9	38.1	14.3	—	—	—
Porvenir	53°16′	70°22′	—	—	—	6.4	11.2	1.3
Río Grande	53°48′	67°47′	379.1	52.1	21.0	—	—	—
Submontane								
Pampa Guanaco	54°07′	68°42′	—	—	—	2.7	9.6	−4.2
Ushuaia	54°47′	68°16′	545.4	57.6	33.4	4.4	10.0	−1.7
Puerto Williams	54°55′	67°36′	467.3	51.3	26.5	6.0	9.6	1.9
Oceanic Cold Temperate								
Bahía Harris	53°50′	70°24′	869.0	79.5	45.2	—	—	—
Oceanic Tundra								
Bahía Félix	52°58′	74°04′	4845.8	506.8	339.0	—	—	—
Evangelistas	52°24′	75°06′	2772.8	272.8	215.4	6.5	8.8	4.4

II. History of Botanical Exploration

Four hundred and sixty-two years ago Ferdinand Magellan discovered the Straits which bear his name; the area to the south, which he called Tierra del Fuego, has subsequently been shown to comprise an archipelago of several large and numerous smaller islands. This region, at the southern extremity of South America, has been of considerable interest to scientists of all persuasions and Table 2 summarises information on the approximately 120 botanists concerned, at least in part, with flowering plants and ferns, who have left some trace of their activities in the literature or in herbaria. Although the list is almost certainly incomplete, we are reasonably confident that no-one who made a significant contribution to knowledge of the Fuegian flora has been omitted.

Several accounts have dealt with the developing knowledge of the Fuegian flora as part of the more general scene in temperate South America (Castellanos, 1945; Soriano, 1948; Parodi, 1961) or the circum-Antarctic zone (Godley, 1965, 1970). Here, we shall concentrate solely on providing an outline of the events in Tierra del Fuego and, although somewhat artificial, it is helpful to recognise several periods for which the publications of, for example, Hooker (1847), Spegazzini (1896), Alboff (1896), Dusén (1900), Skottsberg (1916) and Auer (1970), as well as the present work, provide useful summaries.

The first collections from Tierra del Fuego are considered to have been made by George Handisyd, surgeon aboard the British freebooter *Welfare*, as it traversed the Straits of Magellan in 1690 en route to its dubious activities along the coast of the then Vice-Royalty of Peru and Chile. Eight sheets in the Sloane collection at the British Museum are labelled as from Port Gee, Tierra del Fuego. This locality cannot be traced, and it is just possible that it refers to Puerto Galant, a frequent point of call on the north side of the Straits, where the other localities for Handisyd's austral collection are to be found. Similarly, all of Commerson's collections in 1767 were made along the north shore of the Straits of Magellan, apart from his brief visit to Isla Isabel. For about the next 80 years virtually all botanical exploration was centred on southern Fuegia, principally along the Canal Beagle, Isla de los Estados and south through the Islas Wollaston. Only James Anderson, aboard King's *Adventure*, and Darwin on the *Beagle*, made collections further north, in Seno Almirantazgo and Bahía Inútil, and Isla Isabel, respectively. Being ship-based, in an area where shore parties were liable to urgent recall because of the rapidly changing weather, most of the early collections are coastal; only Banks and Solander are recorded as appreciably leaving the shore, of Bahía Buen Suceso, moving "into the Countrey about two miles to see an Indian town which some of our people had given us intelligence of" (Beaglehole, 1962: 224). Despite this, of the 417 species currently considered to be native to Tierra del Fuego, Hooker (1847) was able to list 153, or just over one third, as occurring in the archipelago in the first comprehensive summary of the Fuegian flora and its relationships. In addition, during this period, Parkinson, botanical artist on Cook's first expedition, provided the first, often exquisite, paintings of Fuegian plants based on living material.

The next 40 years saw the botanical exploration of the south side of the Estrecho de Magallanes from its eastern to western entrances by Robert Cunningham, aboard HMS *Nassau* surveying the Straits and the Chilean channels (Cunningham, 1871), and Paul Savatier on *La Magicienne* of the French Naval Division in the Pacific (Savatier, 1880; Franchet, 1889). During this period the Chilean and Argentinian governments extended the exploration of their austral possessions to Tierra del Fuego. Thus, the botanist Pablo Ortega was sent by the Museo Nacional, Santiago de Chile, on the 1879 expedition of Lieut. R. Serrano Montaner, whose discovery of gold in northern Isla Grande led to the gold rush of 1881 and subsequent years (Martinić, 1976). Most notably, however, the Argentinian Carlos Spegazzini accompanied the 1882 expedition of Lieut. Giacomo Bove to southernmost Argentina, collecting extensively along the north, west and south coasts of Isla Grande, as well as on many islands in the Canal Beagle and on Isla de los Estados (Spegazzini, 1883, 1896). When he published his findings in 1896 the flora of Tierra del Fuego was raised above that known by Hooker by a further 160 species, to about three-quarters of those currently recognised. Less well documented are the smaller collections made by botanists Tonini de Turia and Lista, accompanying the fact-finding missions sent to Tierra del Fuego in 1895–97 by Argentinian President Roca (Lenzi, 1967; Belza, 1973). European and North American activities in the area continued, however, notably during the French part of the International Polar Programme of 1882–83, led by Capt. L. -F. Martial, which was largely based on Bahía Orange, Isla Hoste, during which Hyades, Hahn and Hariot were able to collect widely in the islands of southern Fuegia (Francet, 1889). Also of importance during this era are the collections of the Frenchmen H. Rousson and P. Willems who collected 130 specimens in the first serious attempt to investigate the flora of the east coast of Isla Grande, but there is, unfortunately, very poor documentation of their efforts (Hariot, 1900).

The last decade of the nineteenth century saw two major advances in knowledge of Fuegian plants. Firstly, the Swedish expedition to southernmost America, led by the noted geologist Otto Nordenskjold, gave Per Dusén, their botanist, opportunity to collect widely throughout the northern part of Isla Grande, as well as further south along Río Azopardo and elsewhere, for 5–6 months in 1895–97 (Dusén, 1900). Secondly, the emigré Russian botanist Nicolas Alboff, who did so much to develop the herbarium of the Museo de La Plata, spent over 2 months on the Canal Beagle in the autumn of 1896. Based principally on Ushuaia, he made the first serious collections of the Fuegian high montane flora in the Montes Martiales (Alboff, 1896), amassing almost 3,000 numbers belonging to 350 species. In the summer of 1897–98 the Belgian Antarctic Expedition, led by Henry K. Arctowski, spent some time in Tierra del Fuego and the botanist, M. E. Racovitza, aboard the *Belgica*, was able to collect in parts of the western and south-eastern archipelago. Although his collections were neither particularly significant nor extensive, they were written up by Wildeman (1905), who took the opportunity to gather together earlier records and to make the first and so far only attempt at analysing the size of the floras of some constituent islands in the Fuegian archipelago. Despite Wildeman's dubious taxonomic treatment, he listed about 90% of the species currently known to be Fuegian, together with several aliens. His analyses illustrated the undoubtedly rich flora of Isla Grande, which he showed as having 4 times as many species as Isla de los Estados and 5–7 times as many species as Islas Hoste and Navarino, probably the best known at that time.

The Swedish Antarctic Expedition of 1901–1903, led by Otto Nordenskjold, provided the impetus for a further development in the study of the Fuegian flora. Although its botanist, Carl Skottsberg, spent only relatively short periods in Fuegia, principally along the Canal Beagle, and his collections were largely lost when their ship, the *Antarctica*, was crushed in the ice and wrecked in February, 1903, he was able to make the first useful observation on the ecology and breeding systems of some 73 Fuegian species (Skottsberg, 1904, 1908 a, b). Perhaps of more importance, his experiences stimulated him to revisit the area during 1908–09 as part of a scientific expedition which travelled south along the Andes from San Carlos de Bariloche to Tierra del Fuego, and also visited the Falkland Islands (Islas Malvinas). Not only did he collect extensively, but he also expanded his earlier ecological studies to provide valuable descriptions of the plant communities in relation to those encountered elsewhere in southernmost South America. His summary (Skottsberg, 1916), in which he noted 290 species (about two-thirds of those now recognised) and their distribution, and described the vegetation types and the floristic affinities, is basic to all subsequent studies of the Fuegian flora.

With the granting of a concession for sheep farming at Gente Grande in 1883 (Martinić, 1976) and the foundation of Ea Harberton in 1886 (Goodall, 1979) the pastoral development of Tierra del Fuego began (Moore and Goodall, 1977). Whilst the initial agricultural exploitation was entirely dependent upon the initiative and enterprise of the pioneers, the governments in Santiago and Buenos Aires gradually began to assess the resources of their southernmost provinces. Thus, for example, Eduardo Holmberg and Cressenso Calcagnini examined the lumbering potential of Isla Grande in 1902 (Holmberg, 1906), and Guido Bonarelli studied the value of the Fuegian bogs in 1917, as did Guiñazu and Caldenius in the summer of 1927–28 (Guiñazu, 1934). These visitors made useful collections, as did many others during the early decades of the present century, but none added significantly to knowledge of the Fuegian flora and vegetation. Perhaps the most telling comments were those of the great Carlos Spegazzini who, on his second and final visit to Tierra del Fuego, after 41 years, commented on the dramatic changes in the flora and fauna along the Canal Beagle – "La modificación marcada de la vegetación de esas admirables regiones, sea por el incendio de los bosques, en todos los puntos donde asentó el hombre blanco, sea por la invasión de esencias forasteras, que desalojaron o aniquilan, cada día mas, las esencias indígenas" (Spegazzini, 1924).

The Finnish Expedition of 1928–29, led by Väinö Auer, which was followed by similar expeditions in 1937–38, 1947–53 and 1957, allowed H. Roivainen to make detailed ecological and floristic studies of the bogs over much of Isla Grande (Roivainen, 1954). Of greater significance, however, was the work of Auer himself, whose studies on deposits including fossil pollen and soils have provided detailed information on the changing vegetation and climate of Tierra del Fuego during and since the retreat of the last glaciations in the area some 12,000 years ago (Auer, 1956, 1958, 1959, 1965, 1970).

The increasingly good land communications, at least in parts of Isla Grande, resulted in a number of botanists visiting the area. Although few further noteworthy scientific advances were made in Fuegian Botany during the period up to the early 1960's, at least two contributions should be mentioned. The increasingly important alien flora of Tierra del Fuego was clearly noted by Ruiz-Leal along the north shore of the Canal Beagle (Ruiz-Leal, 1954), while Holdgate (1961) provided the first rigorous descriptions of Fuegian soils, on Isla Navarino.

In 1964 the first botanist resident in Tierra del Fuego began a detailed programme of exploration which has considerably increased knowledge of the distribution and performance of Fuegian plants, especially in southern and eastern Isla Grande, but also elsewhere, particularly Isla de los Estados. Natalie Goodall, of Estancia Harberton and Ushuaia, has also prepared the first illustrations of Fuegian plants based on living material since those of Sydney Parkinson, who accompanied Banks on Cook's first voyage over 200 years ago. In 1968, together with D. M. Moore, the only major botanical survey of the mountains along the north side of the Canal Beagle was mounted since that of Alboff, over seventy years previously. The foundation of the Instituto de la Patagonia, Punta Arenas, in 1969, signalled a comprehensive programme studying the ecosystems of the province of Magallanes, Chile. As part of this programme, studies of the flora and vegetation of various parts of Fuegia, especially the west and south, have considerably advanced botanical knowledge of Tierra del Fuego (Pisano, 1971, 1972a, b, 1977a). Isla de los Estados, much visited by earlier voyagers, provided a focus for the botanists, led by Eliza Nicora, associated with the 1967 expedition sponsored by the Argentinian Centro de Investigaciones de Biología Marina and the Naval Hydrographic Service, and this was expanded by the work of Ted Dudley and his associates during the visit of the US National Science Foundation in 1971 to give the most comprehensive botanical coverage of that most eastern part of the Fuegian archipelago (Dudley, 1981).

On the basis of all the earlier work studies of the Fuegian flora are now concerned with a detailed examination of the cytological structure of populations and their variation, the affinities between Fuegian plants and apparently similar species elsewhere in the world, the structure and relationships of the vegetation, the economic importance of the plants and vegetation, (e.g. Moore, 1980; Moore and Chater, 1971 etc.) and the impact of man on them. Similar problems have long been current in other parts of the world, showing that, in at most 400 years, the study of Fuegian plants has moved from initial observation to detailed scientific inquiry.

In a brief summary such as this it is impossible to do justice to the endeavour, fortitude and spirit of the botanists who contributed to the chronicle outlined in Table 2. Nowadays, as one enters Tierra del Fuego by plane, car or luxury cruiser, it is often difficult to envisage the problems and dangers faced by earlier visitors. Yet, once away from the main highways in Isla Grande, and certainly among the other islands of the Fuegian archipelago, one is immediately faced with the problems of yesteryear; it is there that the greatest gaps in our knowledge lie and where the problems faced by the earliest workers in Fuegia must still in large measure be faced.

CHRONOLOGICAL SUMMARY OF VASCULAR PLANT COLLECTORS IN TIERRA DEL FUEGO

Collector	*Dates of visit/ collections*	*Collecting areas*	*Location* and size of collection*	*Notes and references*
G. Handisyd	12.ii.–v. & 5–12 xii.1690	Isla Isabel ? Isla Grande–"Port Gee"	BM 8 sheets	Surgeon aboard British freebooter *Welfare* (Capt. J. Strong) (Gunckel, 1971; Middleton, 1909)
P. Commerson	12.xii.1767	Isla Isabel	P	Naturalist on Bougainville's voyage (Oliver, 1909)
J. Banks, D. Solander	14–20.i.1769	Isla Grande–Bahía Thetis; Bahía Buen Suceso and inland	BM 104 sheets	Naturalists on Cook's first voyage aboard *Endeavour* (Beaglehole, 1962; Groves, 1962; Stearn, 1969)
J. R. & G. Forster & A. Sparrman	20.xii.1774–3.i.1775	Isla Hoste–Seno Christmas Isla de los Estados–Isla Año Nuevo	B, BM, GOET, K, KIEL, L, LE, LIVU, MHA, P, PH, UPS 40 spp.	Naturalists aboard *Resolution* on Cook's second voyage (Forster, 1777, 1787, 1789; Beaglehole, 1961; Carolin, 1963)
W. Anderson	20.xiii.1774–3.i.1775	?	BM	Surgeon's mate aboard *Resolution* (Beaglehole, 1961; Godley, 1965)
A. Menzies	26.i.–12.ii.1787	Isla Grande–Bahía Thetis Isla de los Estados–Bahía Año Nuevo	BM 5 spp.	Surgeon with Capt. Colnett (Cowan, 1954; Godley, 1960, 1965)
J. Anderson	xii.1826–iv.1827 i.–viii.1828 iv.–v.1829; xii.1829–viii.1830	Isla Dawson Isla Grande–Seno Almirantazgo Isla Grande–Bahía Inútil Isla Hermite–Caleta San Martín Canal Beagle	BM 2 sheets probably more	Botanist aboard *Adventure* with Capt. King's surveying voyages (Godley, 1965; Hooker, 1847; King, 1839)
W. H. B. Webster	25.x.–23.xii.1828	Isla de los Estados–Pto Cook Isla Hermite–Caleta San Martín	12 spp. ?	Surgeon aboard *Chanticleer* (Capt. H. Forster). (Webster, 1834; Godley, 1965)
J. Eights	Late 1829	Isla de los Estados–? Pto Cook ? Cabo de Hornos	K, NYS 7 spp.	Naturalist aboard *Annawan* (Capt. Palmer) on 1st US Exploring Expedition
J. W. Reynolds	Late 1829	Isla de los Estados–? Pto Cook	?K	Naturalist aboard *Seraph* on 1st US Exploring Expedition
C. Darwin	17.xii.1832–25.11.1833	Isla Grande–Bahía Buen Suceso Isla Hermite–Caleta San Martin Isla Navarino along Canal Beagle to Isla Stewart	CAM, K 11 spp.	Naturalist aboard *Beagle* (Capt. R. Fitzroy) on survey of southern S. American coasts. (Fitzroy, 1839; Darwin, 1839; Hooker, 1847; Godley, 1965)
	?27–29.i.1834 25–26.ii.1834 5–6.111.1834	Isla Isabel Isla Wollaston Isla Navarino-Wulaia	11 spp. ? 37 spp.	
C. Wilkes	17.ii.–30.iii.1839	Isla Grande–Bahía Buen Suceso Isla Hoste–Penin. Hardy, Bahía Orange	US 59 spp.	Commander of 2nd US Exploring Expedition, which included 3 naturalists (Wilkes, 1845; Gray, 1854; Brackenridge, 1854)

Collector	*Dates of visit/ collections*	*Collecting areas*	*Location* and size of collection*	*Notes and references*
J. D. Hooker	20.ix.–7.xi.1842	Isla Hermite–Caleta San Martin Isla Horn	K 39 spp.	Botanist and Assistant Surgeon aboard HMS *Erebus* (Capt. J. C. Ross) on scientific voyage to the Antarctic regions (Hooker, 1847)
?	1857	?	W	Austrian frigate *Novera* (Soriano, 1948)
Giglioli	1866–67	?In Punta Arenas in 1867 (Martinić, 1977)	W	Naturalist aboard Italian corvette *Magenta* (Capt. Arminjon) (Soriano, 1948)
R. O. Cunningham	10.i.–15.iv.1867	Isla Grande–Pta Catalina; Cabo Espiritu Santo; Cabo San Vicente; Bahía Gente Grande; Pta San Isidro; Bahía Felipe	?K	Naturalist aboard surveying ship HMS *Nassau* (Capt. R. C. Mayne) (Cunningham, 1871)
	4–5.iii.1868	Isla Isabel; Isla Contramaestre Isla Grande–Bahía Gente Grande		
	26–29.xii.1868 22–23.iii.1869	Isla Desolación–Bahía Tuesday; Pto. Angosto; Surgidero Valentine; Pto. Churruca Isla Santa Inez–Bahía Swallow		
P. A. L. Savatier	12–15.11.1877 29–30.1.1879	Isla Desolación–Pto. Churruca	K, P 30 nos.	Chief medical officer aboard French ship *La Magicienne* (Savatier, 1880; Franchet, 1888)
P. Ortega	i.–ii.1879	Isla Grande–Bahía Gente Grande; "Fuegia oriental"	LP, SGO . At least 2 new spp.	Botanical collector sent by Museo Nacional, Santiago de Chile with expedition of Lieut. R. Serrano Montaner. Some of his collections incorrectly attributed to E. Ibar Sierra by Philippi and Muñoz (1961) (Pisano, 1977a)
C. Spegazzini	6.ii.–29.vi.1882	Isla de los Estados–Pto Roca; Pto Hoppner; Pto Cook; San Juan; Bahía Pactolus; Bahía Blossom; Pto Vancouver Isla Capitan Aracena–Pto Hope Isla Grande–Penin. Brecknock, Seno Negri, Seno Chair; Yendegaia; Ushuaia; Remolino; Bahía Sloggett; Caleta Olla Islas Basket, Burnt and Chair Isla Navarino–Canal Murray Isla Picton Isla Hoste–N. coast (Ialambaia) Isla Melville–Pto Tom Isla Gordon–Pto Divide	LP 313 species	Botanist aboard Argentinian corvette *Cabo de Hornos* (Capt. Piedrabuena) and later schooner *San Jose* on the Expedición Austral Argentina led by Lieut. Bove (Bove, 1883; Spegazzini, 1883, 1896; Parodi, 1961)
	9.vii.1882	Isla Isabel Isla Grande–Bahía Gente Grande; Pto San Isidro; Pto Anegada; Bahía Possessión		

Collector	Dates of visit/ collections	Collecting areas	Location* and size of collection	Notes and references
Hochstetter	1882	Not known	LP	No information
P. Hyades, P. Hahn & P. Hariot	31.x.1882–9.ix.1883	Isla Grande–Lapataia; Ushuaia; Bahía Sloggett; Bahía Buen Suceso Isla Dawson–Pto Kelp Isla Capitan Aracena–Pto Sholl; Pto Hope; Pto Ariadne; Cabo Turn Isla Camden Isla Hoste–Penin. Hardy; Bahía Tekenika; Bahía Orange; Isla Burnt; Isla Packsaddle Isla Hermite–Caleta San Martín; Bahía Maxwell Islas Wollaston–I. Grevy, I. Wollaston, I. Otter, I. Mouten, I Saddle. Isla Horn Isla Picton–Caleta Banner Isla Navarino–Pto Goree; Isla Button; Canal Murray Isla de los Estados–Pto Cook; Pto San Juan; Pto Vancouver	P 216 spp.	Doctors (Hyades, Hahn) aboard *Romanche* (Capt. L.-F. Martial) on French expedition as part of International Polar Programme. Hariot, botanical assistant Muséum National d'Histoire Naturelle, Paris, joined the party in May 1883. (Hariot, 1884; Martial, 1888; Franchet, 1889)
F. Tonini de Furia	5.ii.–2.iii.1885	Isla Grande–Ushuaia; Isla Gable; Bahía Aguirre; Bahía Thetis; Bahía Buen Suceso Isla Picton–Caleta Banner	?LP 85 nos.	Naturalist who accompanied F. M. Paz aboard *Villarino* when he visited Fuegia on being appointed governor of the Argentinian territory (Paz, 1886; Belza, 1973)
R. Lista	20.xi.1886–24.i.1887	Isla Grande–Bahía San Sebastian south along coast to Bahía Thetis; Ushuaia	No information	Botanist and governor of Terr. Santa Cruz/Naturalist on expedition sent by Presidente Roca to explore E. Isla Grande. Arrived aboard *Villarino* (Capt. F. Spurr), later transferred to *Bahía Blanca* (Lista, 1887; Soriano, 1948; Lenzi, 1967; Goodall, 1979)
F. Delfin	1887	Isla Grande–Bahía Espiritu Santo	SGO	No information
L. A. Lee	2–8.ii.1888	Isla Desolación–Pto Churruca; Island Harbour ?Isla Isabel	US At least 2 nos., probably more	Botanist aboard USS *Albatross* on voyage to Estrecho de Magellanes and W. Patagonian Channels (Townsend, 1910)
H. Rousson & P. Willems	1890–91	Isla Grande–Bahía Buen Suceso north along coast to Cabo Peñas, then west to Bahía Inútil Isla Dawson	?P 130 spp.	Sent by French Ministry of Instruction Publique et des Beaux-Arts to explore Tierra del Fuego (Rousson & Willems, 1891, 1892; Hariot, 1900)
B. Ansorge	28.ii.1896 ?1897	Isla Grande–Río Cullen; El Páramo	UPS	Subadministrator and later Administrator of the gold-mining camp at El Páramo (Belza, 1973)

Collector	Dates of visit/ collections	Collecting areas	Location* and size of collection	Notes and references
P. Dusén	20.xii.1895–6.iv.1896, 1897	Isla Grande–Cabo Peñas; Río Grande; Cabo Domingo; Cabo San Sebastian; Cabo San Martín; El Páramo; Arroyo Alfa; Barrancas de Carmen Silva; Springhill; Río del Oro; Bahía Gente Grande; Porvenir; Seno Almirantazgo; Río Azopardo; Río Condor, Ushuaia Isla Dawson–Bahía Harris Isla Desolación–Pto Angosto	UPS	Botanist with Swedish Expedition to southernmost South America 1895–1897 (Dusén, 1897, 1900, 1905)
O. Nordenskjold	i.1896	Isla Grande–Río Grande to Lago Fagnano	UPS	Geologist and leader of Swedish Expedition to southernmost South America 1895–97
N. M. Alboff	5.ii.–20.iv.1896	Isla Grande–Lapataia; Ushuaia; Montes Martiales; San Sebastian Isla Navarino–north coast and mountains Isla de los Estados–Pto San Juan	CORD, LP, SI 350 spp. 2734 nos.	Russian botanist instrumental in developing herbarium of Museo de la Plata (Alboff, 1896, 1902). Travelled aboard Argentinian transport *Primero de Mayo* (Capt. Noguera)
M. E. Racovitza	10.xii.1897–13.1.1898	Isla Dawson–Bahía Harris Isla Capitan Aracena–Pto Hope Isla Londonderry–Bahía du Torrent Isla Desolación–Bahía Isthmus Isla de los Estados–Pto San Juan Isla Grande–Seno Ventisquero; Lapataia; Pto Harberton	?BR *c.*200 nos.	Rumanian botanist aboard *Belgica* (Capt. A. de Gerlache de Gomery) on Belgian Antarctic Expedition (Arctowski, 1901; Gerlache, 1902; Wildeman, 1905; Chapman, 1964)
B. Brown	1899–1900	Isla Grande–Canal Beagle; W. coast	NY	American collector (Macloskie, 1906)
C. E. Gallardo	28.i.–26.ii.1901	Isla Grande–Ushuaia; Montes Martiales; Remolino; Ea Harberton; Bahía Sloggett Isla Hoste–Penin. Hardy, Bahía Tekenika Isla de los Estados–Pto Cook Isla Dawson	LP 1 spec. known, perhaps more extant	Leader of Argentinian mission to investigate penal colony on Isla de los Estados (Dabbene, 1904; Gallardo, 1910; Barclay, 1926)
R. Dabbene	,,	,,	BA 200 nos. 80 spp. BA, SI, UPS	Member of above mission; mainly interested in birds and Indians
C. J. F. Skottsberg	6.1.1902 4–29.iii.1902 16.ix.–6.xi.1902 18.xi. 1903	Islas Año Nuevo–Isla Observatorio Isla Grande–Ushuaia; Harberton Isla Navarino Isla Grande–Ushuaia; Lapataia; Harberton Isla Hoste–Penin. Hardy, Bahía Tekenika Isla de los Estados–Pto. Cook; Pto Vancouver	BA, SI, UPS	Botanist on Swedish Antarctic expedition led by O. Nordenskjold. Collections mostly lost when their ship, the *Antarctica* (Capt. Larsen), was wrecked (Skottsberg, 1905, 1906, 1909)
L. Rodriguez	Summer–Autum 1902	Isla de los Estados	BAB	No information

Collector	Dates of visit/ collections	Collecting areas	Location* and size of collection	Notes and references
E. Kunze	1902	Isla Grande–San Sebastian, Bahía Inútil	BAB	No information
E. L. Holmberg & C. Calcagnini	2.iii.–1.iv.1902	Isla Grande–Río Grande to Río Fuego and Cerro Cotamil	BAB, SI	Botanist and agronomist on expedition sent by Argentinian Ministry of Agriculture to assess feasibility of timber extraction in C. Isla Grande (Holmberg, 1906)
M. S. Pennington	28.ii.–10.iv.1903	Isla Grande–Pto Villarino; Ushuaia; Remolino; Lapataia Isla Navarino	BAB, BAF, SI c.215 nos.	Collected material in S. Patagonia and Fuegia for Instituto de Botánica y Farmacología, Buenos Aires (Autran, 1905; Soriano, 1948)
E. L. Holmberg	3.i.1904	Ushuaia	SI	See above
C. M. Hicken	13–16.i.1904	Isla Grande–Ushuaia; Porvenir	SI	Noted Argentinian botanist, founder of Instituto Darwinión, who travelled and collected widely in Argentina (Parodi, 1961)
A. de A. Sierra	10.xi.1904	Isla Navarino	BAF 1 no.+?	No information
A. Snaider	Summer 1906–07	Isla Grande–Ushuaia	BAB, P	No information
C. J. F. Skottsberg	25.ii.–27.iii.1908	Isla Dawson–Bahía Harris Isla Grande–Seno Almirantazgo, Pto Gomez, Bahía Hope; Río Azopardo; W. end Lago Fagnano, Isla Lagrelius; Río Fontaine; Río Betbeder; head of Río Rojas	BA, SI	Botanist on Swedish scientific expedition to S. Patagonia, Tierra del Fuego and Falkland Islands (Islas Malvinas)
	4–19.iii.1909	Isla Grande–Bahía Romanche; Yendegaia; Lapataia; Río Rojas; Ushuaia; Mte Olivia; Harberton; Bahía Sloggett Isla Londonderry–Pto Fortuna Isla Navarino–Río Douglas Isla Hoste–Penin. Hardy, Bahía Tekenika, Bahía Orange, Bahía Packsaddle Isla Wollaston–Isla Otter		Used Chilean boats *Huemul* in 1908 and *Porvenir* in 1909 (Skottsberg, 1911, 1916)
A. Benove	22.x.–23.xii.1910	Isla Dawson–Bahía Lomas; Río de la Misión; Río San Martín	LP, SI 3 nos. +?	Collected algae for Museo Salesiano in Punta Arenas. Resident or visitor at Salesian Mission (Buen Pastore) at Bahía Harris (Chamorro, 1936)
Colegio Salesiano	3.i.1911	Isla Dawson	?	Probably collected by workers from the Salesian College in Punta Arenas
C. M. Hicken	20.i.1912	Isla de los Estados–Pto Cook	SI 1 no. +?	Noted Argentinian botanist (see above)
E. L. Holmberg	i.1912	Isla Grande–Ushuaia Isla Lennox Isla de los Estados–Pto Cook	SI, 4 nos. +?	Argentinian botanist (see above), perhaps collecting jointly with C. M. Hicken
G. B. De Gasperi	? before 1915	Isla Grande ?	F 98 spp.	Italian who collected plants and rocks at least (Dainelli, 1922)

Collector	*Dates of visit/ collections*	*Collecting areas*	*Location* and size of collection*	*Notes and references*
G. Bonarelli	iii.1917	Isla Grande–Porvenir, Río Grande	SI 1 no. +?	Visited Tierra de Fuego to study bogs (Bonarelli, 1917; Guiñazu, 1934; Chamorro, 1936)
F. Pico	27.xii.1918–2.ii.1919	Isla Grande–Río Fuego; Cabo San Vicente, Caleta Falsa	BAB 2 nos. +?	Inspector who visited Tierra del Fuego for Argentinian Dirección de Tierras (C. Goodall, pers. comm.)
M. Gusinde	xii.1919 i.1920 i.–ii.1922 iii. & viii.–xii.1923 ii.1924	Isla Grande–Río Grande; Río Fuego across to Pto Harberton; Remolino; Ushuaia; Penin. Brecknock (Pto Edwards) Isla Navarino–Pto Williams; Río Douglas	GB, S 86 nos. +?	Noted anthropologist who studied Fuegian indians on several visits (Skottsberg, 1926)
F. Pastore	7.ii–7.iii.1921	Isla Dawson–Bahía Lomas; Bahía Harris Isla Grande–Ushuaia	BAB, SI 4 nos. +?	Argentinian geologist later chief of Geological Survey (Guiñazu, 1934)
Herborn	5.i.1923	Isla Grande–Ushuaia	1 no. +?	No information
C. Spegazzini	13.–19.i.1923	Isla Capitan Aracena–Pto Shell Isla O'Brien–Pto Ballenas Isla Grande–Lapataia; Ushuaia; Pto Harberton Isla Hoste–Penin. Hardy; Bahía Orange	LP, BAF 4 spp. +?	Eminent Argentinian botanist (see above) on his second visit to Tierra del Fuego, principally to study Fungi (Spegazzini 1924)
J. R. Guiñazu & C. C. Caldenius	Summer 1927–28	Isla Grande–Cabo Domingo to Lago San Luis; Río Ewan and Lago Fagnano		Commissioned by Dr. J. M. Sobral, ex-Director of Mines and Geology, Buenos Aires, to study bogs and quaternary glacial deposits of Tierra del Fuego (Guiñazu, 1934)
H. Roivainen	5.xii.1928–iii.1929	Isla Grande–Ea Ruby; Sierra Carmen Silva; Cabo San Pablo; Ea Los Cerros; Ea La Marina; Ea Vicuña; Lago Blanco; Lago Lynch; Altos del Boquerón; Río Condor; Pto Yartou; Lago Fagnano; Cabo San Pablo; Lago Deseado; Seno Agostini; Seno Contralmirante Martinez; Bahía Broken; Bahía Sarmiento; Bahía Keta Isla Dawson–Pto Valdéz; Bahía Isla	H RNG 19 nos.	Botanist on expedition to Fuegia sent by Finnish Geographical Society and led by V. Auer (Räsänen, 1932; Roivainen, 1954)
W. B. Spencer	5.v.–29.vi.1929	Isla Capitan Aracena–Bahía Warp Isla Londonderry–Caleta Engaño Isla Grande–Ushuaia Isla Navarino–Bahía Douglas Isla Hoste–E. end Penin. Duma	BM 2 nos. +?	British biologist and anthropologist who visited S. Fuegia to study surviving indians and died there (Marett and Penniman, 1931)
A. Donat	15.xi.–4.xii.1930	Isla Grande–Vicuña; Lago Blanco; Porvenir; Bahía Inútil; Bahía Felipe	BA, BM, GH, K, LIL, M, NY, SI, U 7 nos. +?	Argentinian botanist who visited Fuegia at least twice during his expeditions to prov. Santa Cruz (Soriano, 1948)

Collector	Dates of visit/ collections	Collecting areas	Location* and size of collection	Notes and references
	26.xi.1931	Isla Grande–Bahía Inútil, Pto Nuevo		
M. Auschalon	28.i.1930	Isla Grande–Monte Olivia	?	No information
A. Castellanos	7.xii.1932–8.ii.1933	Isla Grande–Ushuaia to E. of Harberton; north to Lago Fagnano; Río Fuego; Río Grande; San Sebastian; Ea Cullen; Ea San Julio	BAB, LIL, SI ?	Professor of Botany in Buenos Aires who visited Isla Grande with M. P. Gomez to collect plants and insects for Museo Argentino de Ciencias Naturales (Humphrey, 1970)
	28.xii.1933–iii.1934	Isla de los Estados–Bahía Crossley; Cabo Colnett; Isla Observatorio; Pto San Juan; Pto Parry	BAB, LIL, SI ? 4 nos. +?	
Y. Mexia	3.26.iii.1936	Isla Grande–Río Grande; Ea La Esperanza; Cabo Peñas; Ea Viamonte	GB, LIL, UC 65 nos.	Collector attached to 1st University of California at Berkeley Botanic Garden expedition to South America led by Goodspeed, principally to study *Nicotiana*, as well as general collecting (Goodspeed, 1941; Mexia, unpub.)
O. L. Giacobbi & I. A. Taccari	xii.1938	Isla Grande–Lago Fagnano	BAF 1+?	Ingenieros agrónomos from Buenos Aires who seem to have principally collected fungi (Marchionatto, 1940). Giacobbi with Auer's 2nd Finnish Expedition 1937–38
R. Santesson	22.i.–19.iv.1940 27.xii.1940–8.i.1941 28.i.–15.ii.1941	Isla Grande-Porvenir; Río Grande; Viamonte; Lago Fagnano; Yendegaia; Lapataia; Ushuaia; Las Cotorras; Pto Yartou	S SW US	Member of expedition sent by Royal Museum of Natural History, Stockholm (Humphrey, 1970)
A. C. Umana	ii.–iii.1941	Isla Grande–Bahía Thetis; Leticia	LP ?192 nos.	No information
A. Castellanos	29.–31.i.1942	Isla Grande–Río Grande; Ea San Cristobal	BA 2+nos.	See above. No further information
R. A. Pérez-Moreau & S. A. Guarrera	1–4.ii.1942	Isla Grande–Lapataia		Botanists from Faculty of Science, University of Buenos Aires
L. A. Tortorelli	ii.1942	Isla Grande–Ushuaia, Mtes Martiales	BAB 1+?	Forester, Faculty of Agronomy, University of Buenos Aires
S. A. Guarrera	14.–30.ii.1948	Isla Grande–Bahía Aguirre; Ushuaia towards Mte Olivia	BA 2+	Algologist from Faculty of Science, University of Buenos Aires
F. Verwoorst	20.xi.1948–23.i.1949	Isla Grande–Ushuaia; Mtes. Martiales; Ea Harberton Isla Hoste–Penin. Hardy Isla Navarino–Wulaia, Bahía Douglas Islas Wollaston–Isla Grevy, Isla Bayley, Isla Wollaston, Isla Herschel, Isla Jordan, Isla Deceit	LIL 460 nos.	Botanist from Instituto Miguel Lillo, Tucuman.

Collector	Dates of visit/ collections	Collecting areas	Location* and size of collection	Notes and references
D. Cozzo	± 1–15.ii.1949	Isla Grande–Bahía Aguirre	BA 12 spp. 16 nos.	Anatomist from Museo Argentino de Ciencias Naturales who visited Fuegia to study woody plants (Cozzo, 1949; Parodi, 1961)
R. R. Martin	1946 and 1947	Isla Grand–Río Tierra Mayor	BAI 31 spp.	Sent by Dirección de Economia Forestal, Administración Nacional de Bosques, Buenos Aires (Martin, 1949)
R. N. Luti	21–23.xii.1949, 9–15.ii. & 6–12.iii.1950	Isla Grande–Ushuaia; Tierra Mayor; Mtes Martiales; Lapataia; Remolino	CORD 166 nos.	Professor of Ecology University of Cordoba, who collected to and from 2 visits to Antarctica. Algae and lichens also collected (Luti, unpub.)
R. Barrientos	13.ii.1951	Isla Grande–Ea Cameron	2 nos. +?	Chilean; no further information
A. Ruiz-Leal	23.xii.1949 9.i.1950 24.ii.1953	Isla Grande–Ushuaia; Ea Fique; Tierra Mayor	LP, R–L, UC 5 nos. +?	Noted botanist from Mendoza who visited Fuegia on 3 occasions, the second with Sr. Carretero, the third with F. A. Roig (Ruiz-Leal, 1954)
J. H. Hunziker	6.i.–8.ii.1955	Isla Grande–Ushuaia–Mtes Martiales; Ea Nueva Argentina; Lago Argentino; Lago Antuk; Ea María Behety; Ea Los Flamencos; Ea San Julio; Ea San Cristobal; Ea Sara; Ea San Martín; San Sebastian; Río Grande	BAB 331 nos.	Botanist and cytogeneticist from Faculty of Science, University of Buenos Aires (Hunziker, unpub.)
Konopacki	ii.1955	Isla Grande–Río Grande	BAB 1 no. +?	Staff-member of I.N.T.A., Castelar, Bs Aires (Correa, pers. comm.)
E. Grondona	ii.1955.	Isla Grande–Ea Sara; Ea San Martín; San Sebastian; Río Grande to Ea Pirinaica; Lago Fagnano; Tierra Mayor valley; Ushuaia; Mte. Olivia; Pto Harberton	BAB, SI 50 +? nos.	Argentinian botanist who has collected widely in Patagonia and Isla Grande of Tierra del Fuego
A. Soriano	1.1955	Isla Grande–San Sebastian; Ea María Behety; Ea Sara; Río Grande	BAB 4 nos. +?	Botanist from Faculty of Agronomy, University of Buenos Aires
O. Magens	26.xi.–15.xii.1955 9.i.1956, 27.xi.1956–11.i.1957, xxi.1959	Isla Grande–Cerro Manantiales; Pto Percy; Caleta Josefina; Ea Cameron; Ea Viamonte	HAM, B, RNPG 28 +? nos.	German gardener resident in Punta Arenas who, among other things, planned and planted gardens and the large conservatory-garden of Cerro Sombrero
M. J. Molina	30.xi.1957 1–6.xii.1959 22.ii.–1.iii.1962 22–28.i.1968	Isla Grande–Ushuaia; road to Glacier Martial; Pta Maria; Cabo Santa Inés	RGAL	Salesian Professor of Natural Sciences from Comodoro Rivadavia who visited Fuegia on 4 occasions (pers. comm., 1972)

Collector	Dates of visit/ collections	Collecting areas	Location* and size of collection	Notes and references
E. Aubert de la Rue	10–18.xi.1958 probably longer. Also in 1961	Isla Grande–Ea Cameron; Caleta Josefina Isla Navarino–Pto Williams Islas Picton and Nueva Isla Hoste Isla Desolación Isla Santa Inés Islas Diego Ramirez	P ? nos.	French geologist who travelled in west Chile south from Chiloe (Aubert de la Rue, 1959)
E. J. Godley	27.xii.1958–9.ii.1959	Isla Navarino–Pto Williams; Dientes de Navarino; Seno Grandi; Isla Bertrand; Wulaia Isla Lennox Isla Picton Isla Hoste–Penin. Hardy, Bahía Orange Isla Hermite–Caleta San Martín	BM, CHR, K 281 nos.	New Zealand botanist with Royal Society of London Expedition to W. Chile between 42°–56°S lat. (Godley, 1960, 1963, 1964, unpub.)
J. Waldron	1960	Isla Grande–Ea Cullen	?K, HIP	Wife of estancia manager (Waldron, unpub.)
M. N. Correa & R. L. Pérez-Moreau	8.–29.i.1960	Isla Grande–Ushuaia; Río Pipo–Montes Martiales; Laguna Escondida; Lago Fagnano; Pta María; Río Grande; Ea Sara; Cullen	BAB 289 nos.	Taxonomists of Inst. Botánica Agrícola, I.N.T.A., Castelar, Buenos Aires
D. M. Moore	19–21.i.1961	Isla Grande–Yendegaia Isla Navarino–Pto Williams, Río Lauta	LA, RNG 26 nos.	
S. Crespo	1962	Isla Grande–Lapataia	BAB 1 +? nos.	Taxonomist of Inst. Botánica Agricola, I.N.T.A., Castelar, Buenos Aires
H. Correa Luna & Reynosa	9.ii.–15.iii.1962 10.ii.1965	Isla Grande–Lapataia; Lago Roca; Glacier Martiales; Cerro Condor; E. end Lago Fagnano; Ea La Pampa; Pto Harbeton; Cascada Almanza; Pta María; Kosobo; Ea Viamonte Isla de los Estados–Bahía Crossley	BAB, DPN ? nos.	Botanists with Dirección General de Parques Nacionales, Buenos Aires (Correa Luna, 1964; Dimitri, 1972)
J. A. Vallerini & J. Robertson	iii.1962	Isla Grande–Ea Cullen; Ea Sara; Ea Viamonte; Ea María Behety; Ea Flamenco; Ushuaia	BAA, BAB ? nos.	Ecologist with I.N.T.A, San Carlos de Bariloche and pastures agronomist from Reno, Nevada, USA
J. Daciuk	i.1963	Isla Grande–Lapataia; Río Pipo; Valle Carbajal; Mtes Martiales; Ea Viamonte; Río Grande to Ea Cullen	BAB, DPN	Botanist with Dirección General de Parques Nacionales, Buenos Aires
D. M. Moore	19–21.iii.1964	Isla Grande–Ea Cameron; Ea La Escondida; Ea Vicuña	AAS, C, CHR, GH, K, LP, S, SGO, UC, US 67 nos.	
R. N. P. Goodall	28.xi.1964–25.v.1965 21.ix.–29.xii.1965	Isla Grande–Ea Harberton; Bahía Moat; Ea Viamonte; Cabo Peñas; Cabo Domingo; Missión Salesiano (YPF Oil Camp)	DPN, KE, LP, RNG, RNPG, UC 242 nos.	Wife of manager of Ea Harberton and keen student of Fuegian plants and animals.

Collector	*Dates of visit/ collections*	*Collecting areas*	*Location* and size of collection*	*Notes and references*
R. Martínez-Crovetto	i.1965	Isla Grande–Río Grande; Lago Fagnano; Ushuaia	BAA, CTES, LP, SI *c.*350 nos.	Ethnobiologist from Faculty of Agronomy, University of North East, Corrientes (Martínez-Crovetto, 1968, unpub.)
R. N. P. Goodall	3.i.–10.v.1966 14.xi.1966–28.v.1967 13.viii.–12.xii.1967	Isla Grande–Bahía Sloggett; Bahía Moat; Ea Harberton from Bahía Almirante Brown east to Pampa de los Indios and from Isla Gable north to mountains; Paso Garibaldi; Laguna Escondida; Lago Fagnano; Cabo San Pablo; Río Fuego; Viamonte; Pta María; Cabo Domingo; Cabo Peñas; Ea María Behety; Ea San Jose; Laguna Victoria	BAB, KE, LP, MICH, MU, NA, RNG, RNPG, SI, UC, US 856 nos.	See above
E. Nicora, S. Crespo, E. Ancibor & M. A. Torres	xi.1967 (2 weeks)	Isla de los Estados–Pto Cook to Pto Vancouver	BAB, RNG, SI	Members of expedition sponsored by Argentinian Centro de Investigación de Biología Marina and Servicio Hidrografía Naval
R. N. P. Goodall	5.i.–21.iv. and 15.ix.–31.xii.1968	Isla Grande–Ea Harberton from Bahía Almirante Brown east to Pampa de los Indios and from Isla Gable north to Mte. Almanza, Mte. Roja, Mte. "Lincoln Constance"; Bahía Aguirre; Mte. Lucio López; Ea Moat; Ea Punta Segunda, Mte. Fernandez; Río Encajonado; Pta Remolino; Paso Garibaldi; Mte. Atukoyak; Mte. Shenolsh; Lago Yehuin; Cabo San Pablo; Ea La Marina; Cabo Domingo; Ea Sara; Ea Cullen Isla de los Estados–Pto Roca (All above with D. M. Moore) Isla Grande–Pta María; Ea San José; Cabo Peñas; Río San Pablo; Ea Vicuña; Cerro Sombrero; San Sebastian; Porvenir; Río Olivia; Valle Andorra	MICH, MU, NA, RNG, RNPG, UC 879 nos.	See above (Moore, 1969, 1970, 1971; Goodall, unpub.)
L. Constance	3–11.ii.1968	Isla Grande–Mte. "Lincoln Constance"; Río Lapataia	UC 5 nos.	Professor of Botany, University of California at Berkeley. Collected with D. M. Moore and R. N. P. Goodall (see above)
D. M. Moore	5.i.–4.iii.1968	With R. N. Goodall (see above), plus Isla Grande–Río Moat, Cerro Moore; Sierra Lucas Bridges, Mte "Moore"	BAA, BAB, C, CHR, H, HIP, LP, K, MO, NA, RNG, SI, UPS, US 987 nos.	(Moore, 1968, 1970, 1971; unpub.)

Collector	Dates of visit/ collections	Collecting areas	Location* and size of collection	Notes and references
	5.–21.iii.1968	Isla Grande–San Sebastián to Caleta Josefina; Río Grande; Lago Blanco; Ea Vicuña; Cerro Cuchillo; Ea Cameron; Río Mayor; Bahía Inútil; Altos del Boquerón; Bahía Gente Grande; Seno Almirantazgo, La Paciencia, Bahía Parry, Pto Toto, Isla Tres Mogotes; Río Azopardo		
M. & R. Barrett	28.xi.–9.xii.1968	Isla Desolación–Pto Angosto Isla Capitan Aracena–Bahía Morris Isla London Isla Stewart–Pto Fanny Isla Navarino–Wulaia, Bahía Windhond, Seno Grandi, Pto Inútil Isla Hoste–Penin. Hardy, Caleta Misión; Bahía Allen Gardiner Isla Jordan–Pto Maxwell Isla Barnavelt Isla Grey–Bahía Gretton Isla Button–Bahía de Julio Isla Aguirre–N.E. part	NA, RNG, RNPG 414 nos	Aboard R.V. *Hero* sent by US National Science Foundation to S. Chile, principally to study whales and dolphins (Goodall, unpub.)
R. N. P. Goodall	7.i.–23.iv.1969	Isla Grande–Lapataia; Ushuaia; Ea Harberton; Bahía Aguirre; Lago Fagnano; Ea Viamonte; Cabo Peñas; Río Grande; Ea Lago Gaviota; Cerro Sombrero	BAB, MICH, MU, NA, RNG, RNPG, SI, UC, US 401 nos.	(Moore and Goodall, 1973, 1974)
	12.–26.xi.1969	Isla Grande–Bahía Thetis to Policarpo		Accompanied expedition from Argentinian Centro de Investigación de Biología Marina
S. B. Young	3–8.x.1969	Isla Desolación–Bahía Tuesday; Pto Churruca Isla Clarence–Bahia Pond	US 93 nos	Biologist aboard *Hero* sent by US National Science Foundation to S. Chile (Young, 1970, 1973, unpub.)
H. Roivainen	29.xi.1969–27.i.1970	Isla Grande–Pta Espora to Río Grande, south to Laguna Escondida; Paso Garibaldi; Lapataia; Ushuaia, Mtes Martiales, Mte Olivia, Río Pipo, Mte Susana	H, RNG *c.*2,100 nos.	Noted Finnish botanist on return visit to Fuegia (see above)
R. N. P. Goodall	4.i.–17.iii. & 26.x.–31.xii.1970	Isla Grande–Lapataia; Lago Roca; Ushuaia; Ea Harberton; Mtes Martiales; Ea Fique; Paso Garibaldi; Tierra Mayor valley; Ea Viamonte; Río Grande; Ea Los Flamencos; Ea María Behety; Ea Cullen; Bahía San Sebastian; Cabo Domingo	AAH, BAB, HIP, MICH, MU, NA, RNG, RNPG, UC 531 nos.	

Collector	Dates of visit/ collections	Collecting areas	Location* and size of collection	Notes and references
E. Pisano V.	12.–13.xii.1970 14–26.ii.1971	Isla Grande–Fiordo Parry	HIP, RNG 114 spp.	Chief Botanist, Inst. Patagonia, Punta Arenas, aboard fishing boat *Doña Mercedes* in collaboration with members of Taranaki Alpine Club, New Zealand (Pisano, 1971)
R. N. P. Goodall	2.i.–11.v. & 2.ix.–24.xi.1971	Isla Grande–Río Cullen to Cabo Espiritu Santo; Ea Las Violetas; Ea San Martín; Ea María Behety; Río Grande; Cabo Peñas; Ea Viamonte; Río San Pablo; Yendegaia; Río Rojas; Mte Dalla Vedova; Cordon Pirámide; Lapataia; Ushuaia; Mte Olivia; Ea Harberton over Sierra Lucas Bridges to Lago Fagnano; Caleta Josefina; Cabo Domingo; Ea San Julio; Ea Olivia; Ea La Correntina	AAH, BAB, HIP, MICH, NA, RNG, RNPG, UC 1236 nos.	Collections during i.–iii.1971 made in company with J. Wood (Goodall, unpub.) Collections during 17–24.xi.1971 made in company with O. Boelcke, M. N. Correa, D, M. Moore and B. Piccinini (see below)
T. R. Dudley, R. N. P. Goodall & G. Crow	13.x.–12.xi.1971	Isla Grande–Bahía Buen Suceso; Bahía Valentin Isla de los Estados–Bahía Crossley; Bahía Flinders; Pto Hoppner; Pto Parry; Pto Basil Hall; Pto Cook; Pto San Juan; Pto Roca; Pto Vancouver; Pto Celular; Bahía Alejandro; Bahía Capitan Canepa; Bahía Franklin; Isla Observatorio; Isla Alférez Goffre	AAS, BAB, BLFU, E, GH, HIP, LP, MO, MSC, NA, P, RNG, RNPG, SGO, SI, UC 2000 nos. 138 spp.	Botanists aboard R.V. *Hero* sent by US National Science Foundation to Península Mitre, Isla de los Estados and Antarctica (Dudley, 1981)
D. M. Moore	6–14.xi.1971	Isla Grande–Porvenir north to Bahía Lee; Pta Catalina; Pta Espora; Pta Baja; Bahía Felipe; Cerro Sombrero; San Sebastian; Río Marazzi; Caleta Josefina to Río Chico; Bahía Inútil; Pto Yartou; Pto Arturo; Ea Darwin	BAA, BAB, C, H, LP, K, RNG, UPS, US 159 nos.	Collecting in conjunction with E. Pisano V.
E. Pisano V.	,,	,,	HIP ? nos.	Joint collecting trip with D. M. Moore (see above)
D. M. Moore	17–21.xi.1971	Isla Grande–Cerro Sombrero to Cabo Domingo; Ea Jose Menendez; Ea San Jose; Río Rasmussen; Ea San Julio; Bahía San Sebastian; Ea Cullen; Ea Viamonte; Lago Fagnano	C, CHR, H, HIP, LP, K, RNG, UPS, US 143 nos.	Collecting in company with O. Boelcke, M. N. Correa, R. N. P. Goodall and B. Piccinini (see above and below)
O. Boelcke, M. N. Correa and B. Piccinini	,,	,,	BAA, BAB ? nos.	Botanists from Depto. de Botánica Agrícola, Castelar and Faculty of Agronomy, University of Buenos Aires (see above)

Collector	Dates of visit/ collections	Collecting areas	Location* and size of collection	Notes and references
E. Pisano V.	8–13.xii.1971	Isla Capitan Aracena–Bahía Morris	HIP, RNG ? nos.	Chief Botanist, Instituto de la Patagonia, Punta Arenas, accompanied by O. Ortiz, archaeologist. Sailed aboard fishing cutter *Adrián* (Pisano, 1972a)
D. M. Moore	8–9.xii.1971	,,	BAA, BAB, CHR, H, LP, K, RNG, UPS, US 83 nos.	Accompanied Pisano and Ortiz for part of expedition (see above)
D. M. Moore & R. N. P. Goodall	2. & 14–16.xii. 1971 4–11.i.1972	Isla Grande–Porvenir; Altos del Boquerón; Río del Oro; Río Chico valley; Ea Lago Gaviota; Barrancas de Carmen Silva; Cabo Peñas; Ea María Behety; Ea Los Flamencos; Ea Las Violetas; Ea Sara; Ea Cullen north to Río Alfa; El Páramo; Ea San Martín; Río Grande; Ea Harberton	BAA, BAB, C, CHR, H, HIP, LP, K, RNG, RNPG, UPS, UC, US 403 nos.	Also concerned with recording distribution of species without collecting voucher specimens, and obtaining cytological material
D. M. Moore	23–31.xii.1971	Isla Grande–Ea Harberton, Cordón No Top, Sierra Lucas Bridges, Mte Spion Kop, Río Lata, Mte Square, Mte Pink, Isla Gable; Mte Hueheupen	BAA, BAB, C, H, HIP, LP, K, RNG, UPS 65 nos	Also concerned with recording distribution of species without collecting voucher specimens, and obtaining cytological material
E. Pisano V.	12.–17.i.1972	Isla Hoste–Bahía Orange, Caleta Misión Islas Diego Ramirez: Isla Gonzalo Isla Herschel–Caleta Martial Isla Hornos– Isla Navarino–Pto Toro	HIP, RNG 477 nos.	Chief Botanist, Instituto de la Patagonia, Punta Arenas, accompanying Chilean naval vessel *Lientur* servicing navigational lights in S. Fuegia (Pisano, 1972b)
M.-C. Latour	6–10.i.1972	Isla Grande–Ea Cullen; Bahía San Sebastian; Ea Los Flamencos; Río Grande; Lago Fagnano; Paso Garibaldi	BAR *c.*150 nos.	Botanist from Estación Experimental Agropecuaria, I.N.T.A., Bariloche, Argentina, on visit to study artificial pastures in N. Isla Grande (Latour, unpub.)
R. N. P. Goodall	19,i.–7.vii & 14.x.–31.xii.1972	Isla Grande–Ea Harberton, Río San Pablo, Ea Dos Hermanas, Ushuaia, Lapataia, Río Olivia, Mte Portillo, Ea Moat, Tierra Mayor, Rancho Hambre, Lago Fagnano, Lago Kosobo, Ea Margarita, Ea Rivadavia, Ea Río Ewan, Ea Los Flamencos, Ea San Julio, Los Chorrillos, San Sebastián, Río Cullen, Arroyo Alfa, Arroyo Beta, Cerro Cabeza de León, Ea María Behety, Ea Las Hijas, Ea Sarmiento, Río Candelaria, Río MacLellan, Pto Popper, Ea La Criolla, Ea Los Cerros, Ea Carmen, Ea San José	AAH, BAB, MICH, MU, NA, RNG, RNPG 302 nos.	Collections mainly for dye experiments and seeds

Collector	Dates of visit/ collections	Collecting areas	Location* and size of collection	Notes and references
R. N. P. Goodall	2.i.–10.v. & 29.xii.1973	Isla Grande–Ea Santa Ana, Cabo San Pablo, Ea María Behety, Cabo Peñas Lake, Lago Fagnano, Ushuaia, Mte Martial, Tierra Mayor, Mte de las Carpas, Ea Harberton	AAH, BAB, MO, NA, RNG, RNPG 251 nos.	
		Isla de los Estados–Pto San Juan	6 nos.	Collected by W. Brickner aboard R.V. *Hero*
J. & C. Leontsinis	21.iii–6.iv.1973	Isla Grande–Mte Martial, Cerro Bonete, Mte Alvear, Bahía Golondrina, Cerro Guanaco, Cerro Condor, Lapataia, Cerro Portillo, Valle Las Carpas	RNPG, RNG 76 nos.	Assistants employed by R. N. P. Goodall
R. E. G. Pichi-Sermolli	16–27.i.1974	Isla Grande–Pta Espora; San Sebastian; Río Grande; Lapataia; Lago Roca; Ushuaia; Isla Redonda; Mte Martial; Río Olivia; Pto Remolino; Paso Garibaldi; Laguna Escondida Isla de los Estados–Pto Roca	FI, K, P, SI 66 spp. 107 nos.	Noted Italian Pteridologist, accompanied by M. P. Bizzarri, as part of expedition sponsored by AMF Mares-Gruppo Richerche Scientifiche e Tecniche Subacquee. (Pichi-Sermolli & Bizzarri, 1978) T. Ristori collected the plants at Pto. Roca
R. N. P. Goodall	2.i.–4.vii & 27.x.–11.xii.1974	Isla Grande–Isla Casco, Isla Conejo, Mte E. of Garibaldi, Mte Martial, Ushuaia, Río Pipo, Río Olivia, Laguna Escondida, Valle las Carpas, Tierra Mayor, Ea Harberton, Pta María, Río Grande, Laguna San Ricardo, Mte Olivia, Río Hambre, Ea Río Ewan, Mte Paso Tristen	AAH, BAB, MO, NA, RNG, RNPG 123 nos.	
		Isla Navarino–Río Douglas Isla Hornos–Bahía Garibaldi Isla Hoste–Rice Cove Isla Wollaston–Washington Channel	34 nos.	Specimens collected by S. Kester and R. Dott aboard R.V. *Hero*
R. O. Slayter & J. Mulroy	29.i.–4.ii.1974	Isla Grande–Paso Garibaldi; Mte Martial; Viamonte; Pta María	DUKE	Australian and N. American ecologist and taxonomist studying alpine communities
R. N. P. Goodall & E. R. Lutken	12.xii.–30.xii.1974 10.i.–2.iii.1975	Isla Grande–Cabo Domingo, Ea Cauchicol, Ea La Marína, Ea San Jose, Ea Despedida, Cabo Peñas, Pta María, Mtes Martiales, Valle E. of Río Olivia, Ea Harberton, Ushuaia, Porvenir S. to Bahía Inútil, Porvenir N. to Bahía Gente Grande, Caleta Percy, Caleta Clarence, Bahía Felipe, Bahía Azul, Pta Espora, Cerro Sombrero, San Sebastián, Ea Sara, Lago Fagnano	BAB, CHR, ERL, MO, NA, RNG, RNPG 115 nos.	
G. D. McSweeney	4.ii.1975	Isla Grande–Ushuaia	HIP 46 nos.	No information (Pisano, pers. comm.)

Collector	Dates of visit/ collections	Collecting areas	Location* and size of collection	Notes and references
R. N. P. Goodall	5.iv.–21.xii.1975	Isla Grande–Río Olivia, Ushuaia, Isla Conejo, Islas Bridges, Pta María, Lapataia, Cabo Peñas to Río Grande, Ea Las Violetas, Cabo Espiritu Santo, Arroyo Beta, Penin. Páramo	BAB, CHR, MO, NA, RNG, RNPG 61 nos.	
J. P. Hjerting, B. Søegaard & S. Ødum	13–16.iii. & 7–11.v.1975	Isla Grande–First Narrows to Ushuaia, Lago Roca, Paso Garibaldi, Laguna Escondida, Lago Fagnano, Cabo San Pablo, Río Fuego, San Sebastián, Bahía Inútil, Porvenir	*c* 21 nos.	Nordic Arboretum Expedition to Southernmost Argentina and adjacent Chile, 1975. Mainly collecting seed and living material, mainly of woody plants. (Ødum *et al.*, 1977)
P. Samsing	xi.1975	Islas Wollaston–Isla Horn	HIP 93 nos	Member of Expedición al Cabo de Hornos del Club de Deportes Náuticas, Punta Arenas (Pisano, pers. comm.)
A. Atalah	22–25.xi.1975	Isla Navarino–Pto Williams	HIP 14 nos.	No information (Pisano, pers. comm.)
P. Seibert	26.xii.1976–3.i.1977	Isla Grande–Ushuaia, Ea Harberton, Lago Fagnano, Bahía Inútil, Porvenir	BAB 62 nos.	Phytosociologist from University of Munich
W. A. Adams	18.1.–4.iv.1977	Isla Hoste–Seno Ponsonby	RNG 20 nos.	Member of Cambridge University expedition to Isla Hoste (Andrews-Speed, 1979; Moore, 1979)
R. N. P. Goodall	4.i.1976 & 2.ii.–26.vii.1977	Isla Grande–Arroyo Grande, Cabo Peñas, Pta María, Cerro Condor, Páramo, Río Lainez, Lapataia, Ea Cullen, Bahía Inútil, Río Marazzi, Pto Nuevo	BAB, MO, NA, RNG, RNPG 76 nos.	
		Isla Gordon: Romanche Bay, Morning Cove. Canal Beagle, Brazo N.W.; Pta Micalvi, Seno Serrano, Seno Hyatt	5 nos.	Collected by T. D. Goodall aboard R.V. *Hero*
R. N. P. Goodall	28.v.–3.vi.1978	Isla Grande–Porvenir, Gente Grande, Bahía Felipe, Bahía Lomas, Ea Pepita, Pta Espora	BAB, MO, NA, HIP, RNG, RNPG 12 nos.	
R. N. P. Goodall	20.ix.–11.xii.1979 & 7.i.–11.v.1980	Isla Grande–Río Pipo, Ea Harberton, Pta María, Las Violetas, Ea Viamonte, Bahía Brown, Ushuaia, Los Chorrillos	RNG, RNPG 25 nos.	
S. Ødum, S. Laegaard & E. Schmidt Nielsen	i.–ii.1978 iv.1978	Isla Grande–Ushuaia, Lapataia	C	Danish scientific expedition to Patagonia and Tierra del Fuego (Madsen *et al.*, 1980)
E. Pisano V.	15.ii.–3.iii.1980	Islas Wollaston–Isla Wollaston, Isla Bayly, Isla Bandurrias	HIP, RNG 205 nos.	See above. Multi-disciplinary expedition from Instituto de la Patagonia, Punta Arenas aboard Chilean naval vessel *Lientur* (Pisano, 1980)

*Herbarium abbreviations are as recommended by Holmgren and Keuken (1974), except for the following: **BAR** Estacion Experimental Agropecuaria, I.N.T.A., San Carlos de Bariloche, Argentina. **DPN** Dirección General de Parques Nacionales, Buenos Aires, Argentina. **RGAL** Museo Regional Provincial Santa Cruz, Río Gallegos, Argentina. **RNPG** Private herbarium of R.N.P. Goodall, Ea Harberton, Tierra del Fuego, Argentina.

III. Vegetation

As noted previously (p.4) the major climatic gradient across Tierra del Fuego, from oceanic conditions with high rainfall in the south and west to more continental conditions with much lower rainfall and more marked seasonal differences in the north and east, influences the soils and, together, they determine the broad pattern of vegetation throughout the archipelago. In response to this gradient four major vegetation zones (Fig. 2) are evident as one progresses from north and east to south and west Fuegia – Patagonian Steppe, Deciduous Forest, Evergreen Forest and Magellanic Moorland – with a fifth zone, of Alpine Vegetation, at high elevations in the mountains. It should be noted that local edaphic, topographic and microclimatic conditions frequently result in communities characteristic of one of the major vegetation zones being present in another. The more prominent of these are noted in the account.

Within each of these principal vegetation zones numerous associations and communities, differing in their structure and floristic composition in response to local environmental conditions, can be recognised. In this account it is not possible to describe the full diversity of the vegetation in Tierra del Fuego and it aims only to indicate briefly the principal plant communities. Only the more interesting or important plant species present in each community are mentioned. More detailed information may be found in the publications mentioned in the account, the most recent and comprehensive survey being that of Pisano (1977a).

A. PATAGONIAN STEPPE

Within this zone plant communities belonging to grassland heath and shrub formations are recognised, as well as the remnants of an extinct forest described later (p.28).

a. *Grassland*

***Festuca gracillima* grassland.** The rolling grassy plains of northern Isla Grande are dominated by *Festuca gracillima*, typically with *Agropyron fuegianum*, *Agrostis flavidula*, *Festuca magellanica*, *Poa alopecurus* ssp. *alopecurus* and *Trisetum spicatum* as common associates or even, locally, co-dominants among the grasses, while in the moister facies of this grassland species such as *Alopecurus magellanicus*, *Deschampsia antarctica* and *Phleum alpinum* often become important. Apart from the grasses the following species are among the most frequent associates of *Festuca gracillima*:– *Acaena pinnatifida*, *Armeria maritima*, *Calceolaria uniflora*, *Cerastium arvense*, *Draba magellanica*, *Erigeron myosotis*, *Leucheria hahni*, *L. purpurea*, *Luzula alopecurus*, *Phaiophleps biflorus*, *Senecio magellanicus*, *Silene magellanica*, *Sisyrinchium patagonicum*, *Taraxacum gilliesii*, *Vicia bijuga* and *Viola maculata*. The drier facies of this grassland may be recognised by the prominence of *Festuca pyrogea*, *Phacelia secunda* and *Valeriana carnosa*, and the occurrence of *Stipa chrysophylla* and *Rytidosperma virescens*.

Mesic grassland In areas along valleys or watercourses, and in depressions with a clay horizon below the surface which prevents downward, but not slow lateral, movement of water, the moist conditions support a mesic grassland forming the meadows or 'vegas' that grade into the moister facies of the *Festuca gracillima* grassland. A number of communities can be recognised, in which grasses, such as *Alopecurus magellanicus*, *Deschampsia antarctica*, *D. kingii*, *Hierochloë redolens* and *Hordeum comosum*, and graminoid herbs, such as *Carex acaulis*, *C. curta*, *C. darwinii* and *Juncus scheuchzerioides*, are prominent or dominant. Among the species of other life-forms, *Gunnera magellanica* and *Samolus spathulatus* are locally dominant in such mesic grassland,

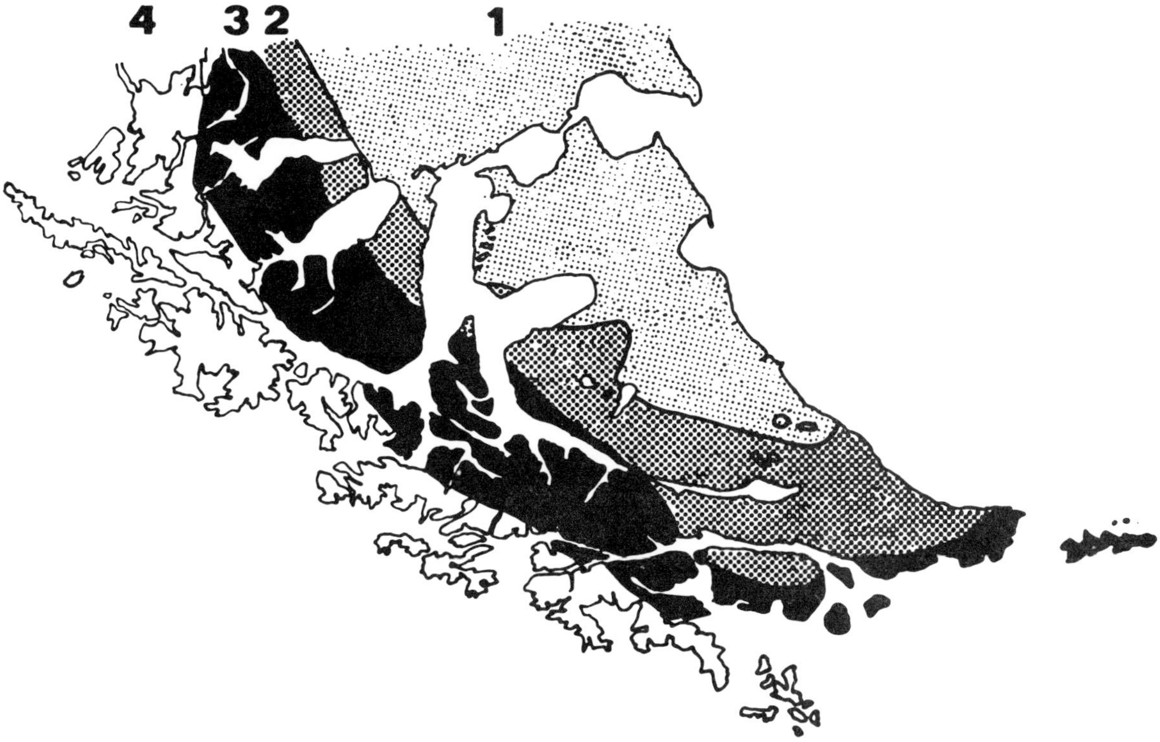

Fig. 2. Principal vegetation zones of Tierra del Fuego. 1 (grey). Fuego-Patagonian steppe. 2 (dark grey). Deciduous forest. 3 (black). Evergreen forest. 4 (white). Magellanic moorland. High montane vegetation not indicated.

while predominant are, for example, *Acaena magellanica, Anagallis alternifolia, Azorella caespitosa, Blechnum penna-marina, Caltha sagittata, Euphrasia antarctica, Gentianella magellanica* and *Pratia longiflora.*

Saline grassland. Depressions in which there is no lateral movement of the subterranean water or, if so, evaporation exceeds the flow, accumulate salts. The resultant vernally wet saline soils support communities in which species of *Puccinellia*, particularly *P. magellanica* and *P. biflora*, are locally dominant. Among the important associates may be included *Arjona pusilla, Chenopodium antarcticum, Myosurus patagonicus, Plantago barbata* and *Salicornia ambigua*. Saline ares which remain moister longer, as around the margins of small pools and lakes, tend to support open communities dominated by *Eriachaenium magellanicum*, with which most of the above species are associated as well as, for example, *Colobanthus quitensis, Suaeda argentinensis* and *Triglochin concinna*.

b. *Scrub*

***Lepidophyllum* scrub.** In the northern third of Isla Grande coastal sands frequently support a community dominated by the shrub *Lepidophyllum cupressiforme*, as do some inland sands, notably between Bahía Inútil and Bahía San Sebastián. Floristically, the community grades into the drier facies of *Festuca gracillima* grassland and also the saline grassland. The dominant, which reaches 0.5–1.4 m high, may be accompanied by *Berberis buxifolia* and *Senecio patagonicus* in the upper layer, while the lower layers include such species as *Acaena pinnatifida, Berberis empetrifolia, Boopis australis, Descurainia antarctica, Jaborosa magellanica, Oxalis enneaphylla, Phacelia secunda*, and *Valeriana carnosa*, as well as the introduced *Leymus arenarius*.

***Chiliotrichum* scrub.** Towards the southern part of the area included in the Patagonian Steppe, where the annual rainfall approaches 350 mm, the *Festuca gracillima* grassland gradually gives way to communities dominated by *Chiliotrichum diffusum*. Since this evergreen shrub is prominent in communities in many open areas of the deciduous and evergreen forests it probably signals the ecotone between the grass steppe and the deciduous forest, but since the communities show strong floristic affinities with the former it is here included within the steppe zone, though it certainly transcends whatever boundary there might be. In addition to many species characteristic of the moister facies of the *Festuca gracillima* grassland, the *Chiliotrichum* scrub contains species indicating affinities with the sheltered, mesic conditions of forest-margins, e.g. *Acaena ovalifolia*, *Adenocaulon chilense*, *Anemone multifida*, *Aster vahlii*, *Baccharis patagonica*, *Calceolaria biflora*, *Cotula scariosa*, *Schizeilema ranunculus* and *Viola magellanica*.

c. *Heath*

***Empetrum* heath.** Throughout the steppe areas of northern Isla Grande the prevailing grassland and shrub communities locally give way to communities dominated by the ericoid dwarf shrub *Empetrum rubrum*. This heath develops over acid, permeable, often rather shallow, soils in which clay is absent at least in the upper horizons. Some of the commonest species associated with the dominant, such as *Arjona patagonica*, *Armeria maritima*, *Festuca gracillima*, *Oxalis enneaphylla* and *Poa alopecurus*, are widespread in the grass steppe but other prominent members of the heath support its distinctiveness, e.g. *Azorella caespitosa*, *Baccharis magellanica*, *Colobanthus subulatus*, *Luzula chilensis*, *Nassauvia darwinii* and *Perezia recurvata*. The *Empetrum* heath seems to have extended somewhat as a result of grazing in certain areas.

***Nardophyllum* heath.** The cushions, lax or compact, of *Nardophyllum bryoides* occur as relatively minor components of both *Empetrum* heath and *Lepidophyllum* scrub, but on some coastal gravels the species assumes local dominance, as in parts of N.W. Isla Grande. Because of the ericoid leaves and habit of the dominant this community is here considered as a heath, with the most characteristic associated species including *Berberis empetrifolia*, *Euphorbia collina*, *Lepidophyllum cupressiforme*, *Oxalis enneaphylla* and *Suaeda argentinensis*.

B. DECIDUOUS FOREST

South of the steppe zone, where the annual rainfall rises from about 400 mm to 800 mm, deciduous forest occurs along both flanks of the mountains, at altitudes from sea level to almost 500 m in sheltered areas. The dominant species are *Nothofagus pumilio* and, to a lesser extent, *N. antarctica*, but shrub, bog and grassland communities occur frequently in the zone. The deciduous forest has been modified by man in many areas, as a result of logging operations and clearing for sheep-grazing, but relatively unaltered communities are still comparatively common.

a. *Forest*

***Nothofagus pumilio* forest.** The best stands of *Nothofagus pumilio*, with trees up to 25 m or so, occur at lower elevations on well-developed soils (brown forest podsols) with good drainage. Here the canopy is dense and relatively few species occur in the shrub and herb layers, the most frequent being *Adenocaulon chilense*, *Asplenium dareoides*, *Berberis ilicifolia*, *Maytenus disticha*, *Blechnum penna-marina*, *Codonorchis lessonii*, *Dysopsis glechomoides*, *Lagenifera hariotii* and *Marcrachaenium gracile*, with *Misodendrum punctulatum* and *M. quadriflorum* parasitising the dominant. Where the forest is more open, as a result of variation in the soil, available water or microclimate, the increased light permits other species to prosper. Thus, *Cardamine glacialis*, *Chrysosplenium macranthum*, *Gunnera magellanica* and *Senecia acanthifolius* flourish where the soils are wet, while under drier conditions are commonly found, for example, *Acaena ovalifolia*, *Cotula scariosa*, *Rubus geoides*, *Schizeilema ranunculus* and *Viola magellanica*.

With increasing altitude and/or exposure the size of the *Nothofagus pumilio* decreases to give a medium height wood (rarely above 5 m) which, at timber-line, is reduced to a scrub having a prostrate, "krumholz" habit. The above trend is generally accompanied by an increase of *N. antarctica* to provide a series of ecotonal communities leading to forest dominated by that species.

***Nothofagus antarctica* forest.** The dominant species has a wider ecological amplitude than *N. pumilio*, with which it cannot compete sucessfully in the most favourable habitats, and so becomes prominent in places where shallower soils, higher water table, greater aridity etc. render them unsuitable for the latter species. The trees of *Nothofagus antarctica* rarely exceed 6 m in height, often have contorted trunks and in many cases are reduced to shrubs; furthermore, the tree layer is frequently rather open so that shrub and herb layers tend to be much richer than in *N. pumilio* forest. In addition to many of the associates listed in the preceding section, the *N. antarctica* communities include, for example, *Berberis buxifolia*, *Chiliotrichum diffusum*, *Embothrium coccineum*, *Empetrum rubrum*, *Galium aparine*, *Luzula alopecurus*, *Osmorhiza chilensis*, *O. depauperata* and *Ranunculus peduncularis*, together with such species as *Anemone multifida* and *Vicia magellanica* in their drier facies, or *Pratia repens* and *Uncinia lechleriana* where the soil is wetter.

b. *Scrub*

Around its margins and in clearings the deciduous *Nothofagus* forest is usually gradually replaced by shrub communities. These contain a mixture of species present in the forest and the adjacent vegetation and probably represent transitional types, as suggested above (p. 25).

***Chiliotrichum-Berberis buxifolia* scrub.** As noted on p. 25, scrub dominated by *Chiliotrichum diffusum* occupies a buffer zone between the grass-steppe and the deciduous forest. The co-dominance of *Berberis buxifolia* indicates the more mesic conditions closer to the forest, as does the local prominence of *Ribes magellanicum* and shrubby examples of *Maytenus magellanica* and *Embothrium coccineum*.

***Pernettya mucronata* scrub.** In many areas the wetter facies of the deciduous forest grades into bog or wet grassland communities. In such places a scrub dominated by *Pernettya mucronata* develops at the forest margin. *Empetrum rubrum* is frequently important in such scrub, while both *Chiliotrichum diffusum* and *Berberis ilicifolia* can be frequent.

c. *Heath*

***Empetrum–Bolax* heath.** Within the deciduous forest zone, particularly where *Nothofagus antarctica* is dominant, a dwarf-shrub heath develops over shallow, well-drained soils, especially if the sites are subjected to cold-air drainage or local "frost-pockets". Such heath is dominated by *Empetrum rubrum* and *Bolax gummifera*, while among the important associates may be included *Abrotanella emarginata*, *Azorella lycopodioides*, *Drapetes muscosus*, *Festuca magellanica*, *Lycopodium magellanicum*, *Myrteola nummularia*, *Nanodea muscosa* and *Pernettya pumila*. A very similar heath frequently occurs at and above timberline as will be noted later (p. 30) when alpine vegetation is considered.

d. *Grassland*

In flatter areas of the deciduous forest zone, where the ground water approaches the surface, as in parts of some river valleys, a mesophytic grassland develops. This has much the same aspect and many of the same species as the similar grassland developed within the steppe areas mentioned earlier (p. 24). However, the occurrence of species such as *Carex magellanica*, *C. microglochin*, *Geum magellanicum*, *Senecio acanthifolius* and *Uncinia lechleriana* gives it a distinctive appearance.

e. *Bogs*

Where the water table reaches the surface various bog communities replace the deciduous forest, frequently grading into the mesophytic grassland.

***Sphagnum* bog.** Communities dominated by the moss *Sphagnum magellanicum* are particularly widespread in the deciduous forest zone, though they do extend into the evergreen forest. The absorption of water by the moss means that the communities are raised above the general ground surface and this, together with the direct incorporation of rainwater, results in a low mineral content of the water. The most frequent associate is *Empetrum rubrum*, which may be locally co-dominant, together with such species as *Carex banksii*, *C. curta*, *C. magellanica*, *Gunnera magellanica*, *Marsippospermum grandiflorum*, *Perezia lactucoides*, *Pernettya pumila* and *Tetroncium magellanicum*, while *Astelia pumila*, *Drosera uniflora* and *Pinguicula antarctica*, for example, are locally important. As the hummocks of this bog increasingly rise above the water table they may dry out to give an *Empetrum*-dominated community resembling facies of the heath described earlier (p. 26).

***Marsippospermum* bog.** Under similar ground-water conditions to those of the above communities but where the general climate is drier, as towards the northern limits of the deciduous forest, *Marsippospermum grandiflorum* becomes dominant in the bog communities. Cushions of mosses, especially *Sphagnum*, and liverworts are important in the lowest layers of this community, while frequent associates such as *Carpha alpina*, *Cortaderia pilosa*, *Festuca contracta*, *Rostkovia magellanica*, *Schoenus antarcticus* and various *Carex* species give it a meadow-like appearance; indeed, these communities frequently intergrade with the mesophytic grassland described on p. 24.

C. EVERGREEN FOREST

As, towards the south and west of Fuegia, the annual precipitation increasingly exceeds 800–850 mm the evergreen *Nothofagus betuloides* becomes important in the forest. Initially, *N. betuloides* is intermingled with *N. pumilio* in an association which has been distinguished as mixed evergreen-deciduous forest, "Bosque Magallánico Mixto" (Pisano 1973a, b, 1974) or evergreen transitional forest (Young, 1973), but which can be better considered ecotonal between the two life-forms (Moore 1978; Pisano 1977a). With increased precipitation *N. betuloides* assumes dominance in a series of communities that extend westwards and southwards into areas of up to at least 4,000 mm annual precipitation. Shrub and bog communities are also prominent in the zone.

a. *Forest*

***Nothofagus betuloides* forest.** The purest stands of *Nothofagus betuloides* develop away from the coasts, usually on rather shallow, peaty soils which, because of sloping parent rock or partly permeable substrates, do not accumulate excessive water. They extend from sea-level to about 350 m, although above *c.*200 m the trees tend to follow glacial valleys, stream-sides and such sheltered areas, decreasing in stature to form a scrub which terminates in a "krumholz" timberline as described (p. 26) for the deciduous forest. In this forest such species as *Empetrum rubrum* and *Philesia magellanica* may become locally sub-dominant in the more open facies, while throughout the commonest associates include *Berberis ilicifolia*, *Lebetanthus myrsinites* and *Blechnum magellanicum* in the shrub layer and, in the herb layer, *Blechnum penna-marina*, *Gunnera magellanica*, *Hymenophyllum* spp. (especially *H. pectinatum*, *H. secundum* and *H. tortuosum*), *Luzuriaga marginata* and *Senecio acanthifolius*.

***Nothofagus betuloides-Drimys* forest.** In coastal areas *Drimys winteri* becomes co-dominant or locally dominant within a mosaic with the *Nothofagus*; in better drained sites, as along watercourses, *Maytenus magellanica* and *Embothrium coccineum* may become locally important. The associated shrubs include all those noted above in the *Nothofagus betuloides* communities, together with the handsome *Desfontainia spinosa*. Because of edaphic and microclimatic factors, as well as the poverty of the soil microfauna, the decomposition of plant remains is slow and there is frequently a deep covering of at most partially decomposed tree-trunks on the forest floor. These conditions provide very suitable conditions for the filmy ferns, the *Hymenophyllum* spp. mentioned in the previous section, but with the prominent addition of *H. ferrugineum*, *H. peltatum*, and *Serpyllopsis caespitosa*; to the other important members of the ground layer already noted should be added *Acaena ovalifolia*, *Adenocaulon chilense* and *Dysopsis glechomoides*.

***Maytenus-Drimys* forest.** In some protected areas of the Altos del Boquerón, along the north side of Bahía Inútil on Isla Grande, occasional specimens of *Drimys winteri* are to be found. There is convincing historical evidence that these are the remnants of a forest dominated by this species and *Maytenus magellanica* until the years 1880–1890 (Pisano, 1977a), when its destruction, and eventual extinction, began with its exploitation by Europeans for firewood and for charcoal production in the goldfields then developing. Little is known of the floristic composition of this forest but, interestingly, *Embothrium coccineum* seems to have been important. Along the west coast of Isla Grande towards the entrance of Seno Almirantazgo scattered stands of *Embothrium* forest, with robust trees up to 5–8 m, are still to be found and it seems that local microclimatic and edaphic factors permitted this westerly finger of evergreen forest to reach northwards into the zone which, climatically, largely supports Patagonian steppe vegetation.

***Pilgerodendron* forest.** The *Nothofagus-Drimys* coastal forest often incorporates specimens of *Pilgerodendron uvifera* when the soil becomes boggy. As these conditions progress to allow the development of *Sphagnum* communities on the ground surface *Pilgerodendron* frequently increases in importance, often forming pure stands, the dominant sometimes reaching 6–10 m but usually being much shorter in Fuegia to give communities which may better be considered as scrub. In addition to the associates of the *Nothofagus-Drimys* communities mentioned earlier the *Pilgerodendron* forest and scrub contains *Acaena pumila*, *Astelia pumila*, *Bolax caespitosa*, *Gleichenia quadripartita*, *Marsippospermum grandiflorum*, *Myrteola nummularia*, *Perezia magellanica* and *Schoenus antarcticus* as important members.

b. *Scrub*

Towards its upper limit and as it approaches exposed or very boggy areas at lower elevations the *Nothofagus betuloides* is reduced to a shrubby habit, as noted earlier (p. 26). However, rather distinctive shrub communities are also present within the zone.

***Fuchsia* scrub.** In coastal areas, often descending virtually to highwater mark, the *Nothofagus-Drimys* forest is fringed by a scrub reaching 1–2.5 m high in which *Fuchsia magellanica* is prominent. Both *Ribes magellanicum* and *Pernettya mucronata* may be locally dominant or co-dominant, while *Chiliotrichum diffusum* is often important, particularly with *P. mucronata*, to give a series of wet shrub communities flanking the major components of the evergreen forest. Many herbs of the coastal mixed forest occur in this scrub, but among the additional noteworthy species are, for example, *Cardamine geraniifolia* and *Ranunculus biternatus*.

***Hebe* scrub.** More exposed coastal areas than those occupied by the *Fuchsia* communities support a dense scrub up to 1 m high, dominated by *Hebe elliptica*. This scrub, which reaches south to Cabo de Hornos, has a rather poor herb layer, though there are various associated shrubs, including *Berberis ilicifolia*, *Empetrum rubrum*, *Escallonia serrata*, *Lebetan-*

thus myrsinites, *Nothofagus betuloides* and *Pernettya mucronata*.

c. *Grassland*

Where there are openings within the forest and scrub communities or along their margins, particularly where there is a degree of lateral water movement, such as where streams enter the sea, small areas of mesic grassland develop. Prominent species in such areas include *Poa alopecurus* ssp. *fuegiana*, *P. robusta*, *Hierochloë redolens*, *Carex banksii*, *Deschampsia kingii*, *Cardamine glacialis* and *Senecio smithii*.

d. *Bogs*

***Sphagnum* bog.** As pointed out previously (p. 27), raised bogs dominated by *Sphagnum magellanicum* occur most characteristically in the deciduous forest zone; nevertheless, they are not at all uncommon in the evergreen forest. The associates of the *Sphagnum* are generally similar to those already noted, but the occurrence of such species as *Donatia fascicularis*, *Oreobolus obtusangulus*, *Schoenus andinus* and *Senecio trifurcatus* indicates a partial transition towards the distinctive bogs of the Magellanic moorland. As the moss hummocks rise high enough above the water table to become somewhat drier they frequently permit the entry of *Pilgerodendron uvifera*, thus providing a gradation to those facies of the coastal forest in which that species is prominent (see p. 28).

D. MAGELLANIC MOORLAND

In the high rainfall (2,000–5,000 mm) areas of westernmost and southernmost Tierra del Fuego is found a series of bog communities that, collectively, have been termed Magellanic moorland (Godley, 1960) or Tundra magallánica (Pisano, 1977a), which extends north through the outer Chilean archipelago to the Golfo de Peñas at 48°S lat. (Moore, 1979). In sheltered areas, even on the outermost islands, fragments of the evergreen forest, dominated by *Nothofagus betuloides*, *Drimys winteri* and *Pilgerodendron uvifera*, develop, but exposure to the fierce westerly gales results in Magellanic moorland, in which prostrate dwarf shrubs, cushion plants, grass-like plants and bryophytes can all be important, giving rise to a blanket peat the prevalence of which is one of the most characteristic features of the zone. The moorland seems to result from a combination of high rainfall, relatively low temperatures, exposure, poor drainage and intractable igneous rock. As pointed out by Crow (1975) and Moore (1979) the moorland consists of a mosaic of communities, that may be small or quite extensive; a full description of them here is neither practicable nor necessary.

a. *Scrub*

In crevices on rocky, better drained prominences in the Magellanic moorland zone a sparse scrub often develops, composed of dwarfed, often prostrate individuals of species such as *Berberis ilicifolia*, *Chiliotrichum diffusum*, *Drimys winteri*, *Embothrium coccineum*, *Empetrum rubrum*, *Escallonia serrata*, *Nothofagus antarctica*, *N. betuloides* and *Philesia magellanica*. Such scrub grades into the patches of evergreen forest which occur in the zone, as noted earlier, and also into the bog communities to be mentioned next.

b. *Bogs*

The characteristic vegetation of the Magellanic moorland zone is provided by a series of bog communities in which prostrate dwarf shrubs, cushion plants, sedges or rushes and bryophytes can all be important.

Cushion bog. Probably the most notable communities of the areas with an annual precipitation of more than about 2,000 mm comprise a dense low covering of cushion plants. The most important of such species are *Astelia pumila*, *Bolax caespitosa*, *Caltha dioneifolia*, *Donatia fascicularis*, *Drapetes muscosus*, *Gaimardia australis* and *Phyllachne uliginosa*.

Graminoid bog. Although quite commonly associated with the principal components of the cushion bog just listed, the rather grass-like (graminoid) *Schoenus antarcticus*, *Tetroncium magellanicum* and *Uncinia kingii* locally become dominant to provide a bog of rather different appearance, superficially resembling rather the *Marsippospermum* bogs of the Deciduous Forest zone.

Amongst the associates of the various dominants noted above in the cushion and graminoid bog communities the following are examples of the most consistently encountered: *Acaena pumila*, *Carpha alpina*, *Drosera uniflora*, *Gaultheria antarctica*, *Gunnera lobata*, *Marsippospermum grandiflorum*, *Myrteola nummularia*, *Nanodea muscosa*, *Oreobolus obtusangulus*, *Perezia magellanica*, *Pernettya pumila*, *Tapeinia pumila* and *Tribeles australis*. Apart from *Sphagnum*, which is virtually absent from the zone, mosses and, particularly, liverworts are prominent and locally dominant in these bog communities of the Magellanic moorland.

c. *Grassland*

In some sheltered coastal localities the evergreen forest found there gives way to small extensions of grassland comparable to that found in sites with good drainage within the zone where such forest is dominant (see p. 29). In addition, a very distinctive grassland, to be described next, occurs within the Magellanic moorland zone.

Maritime tussock grassland. This is the most conspicuous vegetation on the coasts of some of the outermost islands of the southern Fuegian archipelago which have a markedly oceanic climate, extending to some of the islands off the north coast of Isla de los Estados (Crow, 1975). The dominant species is *Poa flabellata*, which forms tussocks up to 3 m high, each comprising a fibrous stock 1–1.5 m in diameter surmounted by a dense crown of leaves. At the maximum development of this grassland, as on the Islas Diego Ramírez (Pisano, 1972b), the leaves of adjoining tussocks interlace to form a canopy that excludes all associates. With reduction in the density of the cover *Crassula moschata* forms sparse mats between the tussocks, while openings resulting from the activities of penguins, albatrosses and seals allow an open community of *Colobanthus quitensis* and *Plantago barbata* to develop.

E. ALPINE VEGETATION

In general the transition from forest to alpine vegetation in Tierra del Fuego is marked by a rather clear timberline, which usually comprises a dense belt of "Krumholz" formed by the *Nothofagus* species. The timberline lies at an average altitude of 500–600 m on Isla Grande, though it is lower on smaller peaks and higher on the larger mountain ranges. Towards the south and west of the archipelago the timberline becomes lower and often indistinct.

The vegetation of the alpine regions seems to be governed by 3 principal factors – exposure to wind, availability of water and the physical nature of the substrate (Moore, 1975).

a. *Heath*.

A series of communities in which heath-like plants, such as *Empetrum rubrum* and *Pernettya pumila*, are prominent are widespread in the Fuegian uplands. They closely resemble similar communities at lower elevations and form a rather complex mosaic within which two intergrading groups may be discerned.

Cushion heath. This is well-developed at and just above timberline where *Bolax gummifera* forms prominent cushions up to 1 m or more high. Close associates usually include cushion-forming species such as *Abrotanella emarginata*, *Azorella lycopodioides*, *Colobanthus subulatus*, *Drapetes muscosus* and, under wetter conditions, *Bolax caespitosa*, *Caltha appendiculata*, *C. dioneifolia* and *Plantago barbata*. Upon and between the cushions *Empetrum rubrum* is conspicuous, together with *Acaena magellanica*, *Festuca contracta*, *Gamochaeta spiciformis*, *Luzula alopecurus*, *Pernettya pumila* and *Stipa rariflora*, for example, as well as *Saxifragodes albowiana* and *Tetrachondra patagonia*, which are virtually restricted to these communities.

With increasing altitude and exposure *Bolax* decreases in importance and the heath becomes more open, with smaller, lower cushions. Here *Armeria maritima*, *Azorella selago*, *Cerastium arvense*, *Draba magellanica*, *Leucheria hahnii*, *Onuris alismatifolia*, *Perezia magellanica* and *Trisetum spicatum* may be prominent. Towards its upper limit the cushion heath becomes increasingly open and impoverished until it is represented only by scattered cushions of *Saxifragella bicuspidata*, sometimes accompanied by *Azorella selago* or *Cerastium arvense*. The heath is also open where seepage water runs over the ground surface, which is frequently gravelly, and here *Viola tridentata* and the rare *Koenigia islandica* occur. In the mountains north of Lago Fagnano, where the cushion heath is developed over a deep clay soil, the substrate is often unstable on steep slopes as a result of wind or trampling by guanaco or sheep and often soil-runs develop. A few species of the open facies of the cushion heath, such as *Calandrinia caespitosa*, *Oxalis enneaphylla*, *Nassauvia pygmaea* and *Senecio alloeophyllus* can tolerate these condition, to which *Phaiophleps biflora* ssp. *lyckholmi* and *Tristagma nivalis* are virtually restricted.

Dwarf shrub heath. Although intergrading considerably with the cushion heath, communities in which *Empetrum rubrum* and, to a lesser degree, *Pernettya pumila* and *Myrteola nummularia* dominate their structure develop along the margins of rock screes, on islands of more stable ground within the screes and on some other well-drained sites. Most of the associated species are also present in the cushion heath but differ in that species such as *Cystopteris fragilis* and *Senecio darwinii* have their principal alpine occurrences here. *Grammitis magellanica*, *Hymenophyllum falklandicum* and *Polystichum andinum*, which occur on rock faces and in crevices in the screes, may also enter the heath.

a. Coastal community, Seno Almirantazgo.
Senecio candidans, with *Acaena magellanica*, *Apium australe* and *Poa patagonica* in foreground.

b. *Lepidophyllum* scrub, Bahía Inútil. (Photos: D. M. Moore).

PLATE I.

a. *Festuca gracillima* grassland, Timaukel.

b. Deciduous forest (*Nothofagus pumilio*) in autumn, Río Grande. (Photos: D. M. Moore).

PLATE 2.

a. *Empetrum-Bolax* heath, Ea Cameron. *Nothofagus antarctica* at margin.

b. Magellanic moorland, dominated by *Donatia fascicularis*, Bahía Moat. (Photos: D. M. Moore).

PLATE 3.

a. Ecotone between grass-steppe and deciduous forest, Ea Vicuña.

b. *Bolax gummifera* cushion-heath, above timberline on Cerro Atukoyak.

c. *Sphagnum* bog, Isla Navarino.

d. *Hippuris vulgaris* freshwater community, Lago Fagnano. (Photos: D. M. Moore).

PLATE 4.

b. *Feldmark*

At higher elevations on many Fuegian mountains there are extensive areas of undulating or gently sloping ground covered by tallus deposits. These tend to be largely devoid of higher plants, hence the appropriate alternative term 'Desierto Andino' (Pisano 1974), but they provide the rather specialised habitat of *Moschopsis rosulata* and *Nassauvia lagascae*. Dense lichen communities, usually dominated by *Usnea*, are also common. The presence of large areas of open, exposed mineral soil allow the occurrence of such species as *Nassauvia latissima*, *N. pygmaea*, *Saxifraga magellanica* and *Senecio humifusus* and others which indicate a transition to the cushion heath, and the demarcation between it and true feldmark is often arbitrary.

c. *Alpine "meadow"*

Over much of the Fuegian high country south of the Azopardo–Fagnano depression there are frequent streams or seepage areas, derived in many cases from permanent glaciers or slow-melting snow patches, which often support a rather distinct and rich vegetation that has somewhat inappropriately been termed "meadow".

In relatively fine soil along stream-margins mats of *Abrotanella linearifolia*, *Caltha appendiculata* and *Plantago barbata* occur, amongst which grow such species as *Acaena antarctica*, *A. tenera*, *Caltha sagittata*, *Lagenifera nudicaulis*, *Ourisia fuegiana*, *Oxalis magellanica*, *Poa alopecurus* ssp. *fuegiana*, *Primula magellanica*, *Tapeinia obscura* and *Viola commersonii*. Where the substrate is rocky or composed of coarser soil, species such as *Cardamine glacialis*, *Epilobium australe*, *Hamadryas magellanica* and *Nassauvia magellanica* are also found.

The sheltered parts of many montane streams are flanked by a broad belt of marshy ground which, together with seepage areas having impeded drainage, can support communities in which grassy leaved plants are prominent. *Agrostis magellanica*, *Carex banksii*, *C. magellanica*, *Carpha alpina*, *Cortaderia pilosa*, *Deschampsia atropurpurea*, *Rostkovia magellanica*, *Schoenus antarcticus*, *Uncinia kingii* and *U. lechleri* can all be found in such areas, together with many of the streamside species noted above. Bryophyte communities, often grading into *Sphagnum* bogs, are also present.

F. LITTORAL VEGETATION

A number of communities at or just above highwater mark can be found around the coasts of Fuegia. These are usually of small extent and are often absent, as on the steep shores frequent in south and west Fuegia where the dominant terrestrial vegetation reaches the sea-margin.

a. *Sandy shores*

In the northern half of Isla Grande communities dominated by *Salicornia ambigua* occur near or below highwater mark on sandy-silty shores. The dominant forms almost pure stands a few centimetres high, but where there is local drainage, as along the sides of small channels, species such as *Plantago barbata*, *P. maritima*, *Puccinellia magellanica* and *Suaeda argentinensis* may be associated.

Sandy beaches in the same area may support an almost pure community of *Senecio candidans*, although *Acaena magellanica*, *Adesmia pumila* and *Rumex magellanicus* can all be associated. Further south in Fuegia these communities can also include, for example, *Apium australe* and *Poa robusta*.

b. *Rocky shores*

In crevices of rock slabs and small areas of wet sand at or just above highwater mark it is usual to find an assemblage of maritime species various combined into several limited communities. The commonest species found in such sites include *Apium australe*, *Colobanthus quitensis*, *C. subulatus*, *Crassula moschata*, *Plantago barbata*, *P. maritima*, and *Polygonum maritimum*, while *Abrotanella submarginata* and the Fuegian endemic *Poa darwiniana* are restricted to such places in the southern parts of the archipelago.

G. FRESHWATER VEGETATION

Many species present in the various bogs of Tierra del Fuego may be found as local emergents in freshwater pools and along the margins of lakes. A relatively small number of species, however, are largely confined to freshwater habitats, though some of them are occasionally members of wetland, terrestrial communities.

In pools and at the margins of lakes and slow-moving streams the freshwater communities are dominated by such species as *Eleocharis melanostachys*, *Hippuris vulgaris*, *Myriophyllum quitense*, *Potamogeton linguatus* and *Ranunculus aquatilis*. All these species, with the exception of the *Eleocharis*, seem to be restricted to the Patagonian Steppe and Deciduous Forest zones, as are such associates as *Littorella australis*, *Potamogeton strictus*, *Ruppia filifolia*, *Scirpus californicus* and *Zannichellia palustris*. *Lilaeopsis hillii* and *Montia fortana* extend, like *Eleocharis melanostachys*, into the Evergreen Forest zone, while *Callitriche antarctica* and *Isoetes savatieri*, which probably grows at greater depths (up to *c.*5 m) than any of the other freshwater species, reach the Magellanic moorland zone.

IV. Geographical affinities of the Flora

In this Flora 545 species are recognised as occurring in Tierra del Fuego. Of these, 417 species are considered to be native to the region, while 128 species are believed to be aliens. Since the alien species have been brought to the archipelago, either deliberately or accidentally, through the activities of the Europeans who inherited the area from the original indian tribes, their occurrence has a significance different to that of the native species, whose distribution results from natural causes, and here they will be considered separately.

I. NATIVE FLORA

It has long been recognised that the flora of Tierra del Fuego represents an extension of that of the temperate mainland of South America to the north (e.g. Hooker, 1847; Skottsberg, 1916; Cabrera, 1971; Hueck and Seibert, 1972). Thus, about 93% of Fuegian species occur on the mainland, as can be seen from the general geographical summaries given later (pp. 46–353) in the systematic account. This is not surprising since, after the Pleistocene ice began to retreat from southernmost South America (including Tierra del Fuego) about 16,000 years ago (p.1), the colonising plants would be expected to come principally from the adjacent ice-free regions to the north. Nevertheless, other factors have been involved in producing the flora of Tierra del Fuego that we see today and it is interesting to look a little more closely at the distribution of the species occurring there.

The four principal climatic and vegetation zones described in passing from north and east to south and west Fuegia (pp. 24–30) parallel rather closely the sequence seen north of the Estrecho de Magallanes in traversing cool temperate Argentina and Chile from the Atlantic to the Pacific Oceans. Since all plant species can only occupy areas with climatic and ecological conditions within their range of tolerance it is not surprising that some species restricted to the drier steppe areas of N.E. Tierra del Fuego extend northwards in the drier parts east of the Andes, while those in the wettest Magellanic moorland zone of Fuegia are found in similar areas of the western Chilean channels. Very many Fuegian species, however, have a rather wider ecological amplitude and extend northwards along the Andes, which provide a rather wide range of cool temperate conditions as a result of their altitude.

In addition to plant-distributions which can be related to the almost continuous climatic belts running northwards from Fuegia into southern South America, are those which appear to be related to the ability of some plants to disperse over rather large distances. Thus, a number of Fuegian species are found across the tropics in various parts of the Northern Hemisphere, while others have traversed the wide expanses of the Southern Ocean to show floristic affinities with the islands surrounding Antarctica and even with New Zealand and, to a lesser degree, Australia or further north.

Some species seem to be restricted to Tierra del Fuego. Such endemics are important because they give an indication of the extent of species-formation since Fuegia became available for colonisation by plants. However, it must be remembered that further exploration in other areas, principally north of the Estrecho de Magallanes, may change the status of some of these endemics.

On the basis of the information presented in this Flora the affinities of the Fuegian plants may be expressed approximately as follows:

a.	North along both flanks of the Andes in W. Argentina and E. Chile	64%
b.	North along the Andes in W. Argentina	2%

c. North only or principally in C. and E. Argentina 4%
d. North along west side of Andes, principally in W. Chile 8%
e. Various distributions in South America but crossing tropics to north temperate areas 12%
f. Various distributions in South America but extending to different degrees into the circum-Antarctic zone 7%
g. Endemics, not known outside Tierra del Fuego 3%

a. The majority (*c.*75%) of Fuegian species occurring along both flanks of the Andes do not reach north of *c.*38°S. lat., indeed 11% of them, including e.g. *Acaena lucida, Agrostis fuegiana, Deschampsia kingii, Nassauvia magellanica, Perezia lactucoides, Ranunculus pseudotrullifolius* and *Viola tridentata*, scarcely occur north of 50°S. lat. This floristic affinity between the southern Andes and Tierra del Fuego is paralleled by the similarity of the vegetation, with deciduous forest and associated communities being prominent in both regions. However, the difference between the generally moister western and drier eastern slopes of the cordillera is reflected in the distribution of species having different environmental requirements. Thus, species such as *Adesmia lotoides, Arjona patagonica, Calceolaria biflora, Draba magellanica, Ephedra frustillata, Hamadryas delfinii* and *Sisyrinchium patagonicum*, which reach significantly further north on the east than the west side of the Andes, tolerate more arid conditions than, for example, *Azorella lycopodioides, Berberis ilicifolia, Carex vallis-pulchrae* and *Fuchsia magellanica*, for which the reverse situation seems to be true.

Of the species included in this section which occur north of *c.*38°S. lat., only a few, such as *Acaena ovalifolia, Anagallis alternifolia, Berberis empetrifolia, Carex fuscula, Perezia pilifera* and *Plantago barbata*, have been found at increasing altitudes in the Andes north of 30°S. lat. Indeed a few species, such as *Eleocharis albibracteata, Juncus chilensis, J. stipulatus* and *Saxifraga magellanica*, reach Colombia and Ecuador, while the extension of *Colobanthus quitensis* and *Daucus montanus* to the mountains of Mexico provides a link with the "bipolar" disjunct species included below in section e.

b. A number of species included in the preceding section, such as *Boopis australis, Gutierrezia baccharoides, Nardophyllum bryoides, Rumex magellanicus* and *Senecio arnottii*, are quite widely distributed in W. Argentina, while their known occurrence in E. Chile is restricted to the drier areas along the north shore of the Estrecho de Magallanes and, occasionally, Ultima Esperanza. They thus form a link with the species in this section which have so far only been found in W. Argentina, e.g. *Benthamiella nordenskjoldii, Botrychium dusenii, Bromus stamineus, Carex sorianoi* and *Tetrachondra patagonica*.

c. Species with their sole or principal extra-Fuegian distribution in W. Argentina, included in the previous section, probably indicate a preference for the drier conditions east of the mountains than those of section a, which also occur on the west side of the Andes. The species included in this section largely occur north of the Estrecho de Magallanes in the arid areas of central and eastern Argentina, e.g. *Androsace pusilla, Boopis patagonica, Draba australis, Frankenia chubutensis, Hypochoeris patagonica* and *Nassauvia fuegiana*. Others, such as *Chenopodium antarcticum, Puccinellia pusilla* and *Suaeda argentinensis*, clearly seem to have distributions governed by the occurrence of relatively low rainfall and the open saline habitats they prefer.

d. A significant proportion of Fuegian species (*c.*8%) occur north of the Estrecho de Magallanes only in the higher rainfall areas west of the Andes, principally in W. Chile. As might be expected most of them are principally or entirely restricted to the evergreen forest and magellanic moorland zones in Tierra del Fuego. Some, such as *Abrotanella linearifolia, A. submarginata, Agropyron elymoides, Caltha dioneifolia, Gunnera lobata, Phyllachne uliginosa* and *Senecio darwinii*, only just extend beyond Fuegia, while others, such as *Acaena pumila, Astelia pumila, Donatia fascicularis, Misodendrum brachystachyum, Pinguicula antarctica* and *Tapeinia pumila*, have been found as far north as *c.*40°S. lat. in W. Chile.

e. Just over 12% of the native species of Tierra del Fuego have disjunct distributions across the tropics to the cool temperate regions of the Northern Hemisphere. Some of these, such as *Carex curta, Cystopteris fragilis, Scirpus cernuus, Trisetum spicatum* and *Zannichellia palustris*, occur in suitable temperate areas throughout the world, and may be considered virtually cosmopolitan. However, most of these "bipolar" disjunct species reflect the geographical proximity of the Americas where they seem to have crossed the tropics, usually from north to south (e.g. Raven, 1963; Moore, 1972), by some form of long-distance dispersal. Species such as *Anemone multifida, Carex gayana, Chenopodium macrospermum, Osmorhiza chilensis, O. depauperata, Polemonium micranthum,*

Salicornia ambigua and *Triglochin concinna* have their only Northern Hemisphere occurrences in North America. A larger number, however, also occur in Europe – e.g. *Armeria maritima*, *Carex capitata*, *C. macloviana*, *C. magellanica*, *C. microglochin*, *Cerastium arvense*, *Deschampsia flexuosa*, *Gentiana prostrata*, *Hippuris vulgaris*, *Koenigia islandica*, *Plantago maritima*, *Ranunculus aquatilis* and *Triglochin palustris*. Of particular interest is a small number of species, including *Calamagrostis stricta*, *Galium aparine*, *Polygonum maritimum* and *Spergularia marina*, which occur in Europe and southern South America, apprently without areas in North America.

Many of these "bipolar" disjuncts have distributions which accord with the view that much of their journey between the Northern and Southern Hemispheres was achieved via the temperate mountain pathways provided by the Andes and Rocky Mountains. However, at least one species, *Koenigia islandica*, has its only South American occurrences in Tierra del Fuego so that the distance traversed from its source in temperate North America and Eurasia is considerable, although it may still be undiscovered along the Andes.

f. About 7% of Fuegian species show important affinities eastwards into the circum-Antarctic zone, occurring on various of the islands scattered across the Southern Ocean. The floristic relationships between the Australian-New Zealand region and South America have long been of interest because of the wide expanses of open ocean now separating these areas and the affinities have been explained in two ways (e.g. Moore, 1972). Since the southern continents were united into Gondwanaland until some 180 million years ago, genera and families shared by them may result from the fragmentation of that "supercontinent" by continental drift. Shared species, such as those considered here, indicate a much more recent situation when the land masses had attained more or less their present positions and the plants must have attained their disjunct distributions by long-distance dispersal over the intervening ocean.

The number of Fuegian species present on various areas in the circum-Antarctic can be summarised as follows:

Island (s)	Falkland Islands	South Georgia	Iles Crozet
Approx. longitude	59°W	37°W	50°E
No. of species	22	21	16
Island (s)	Iles de Kerguelen	Macquarie Island	New Zealand
Approx. longitude	70°E	159°E	170°E
No. of species	15	12	15

Most are present in the Falkland Islands (Islas Malvinas), while the number of species found in other areas generally decreases with distance from Tierra del Fuego, although the available land surface and diversity of habitats on some islands are certainly also important factors. It should also be noted that there is not a gradual "sieve" effect. Thus, whilst *Oreomyrrhis hookeri* and *Valeriana sedifolia* only reach the Falkland Islands, and *Acaena tenera* only South Georgia, *Carpha alpina*, *Geum parviflorum* and *Oxalis magellanica* do not occur between Tierra del Fuego and New Zealand. On the other hand, species such as *Apium australe*, *Blechnum penna-marina*, *Callitriche antarctica*, *Festuca contracta*, *Juncus scheuchzerioides* and *Ranunculus biternatus* occur on most of the islands between those two areas and are truly circum-Antarctic.

g. On present information, not more than 3% of the native species of Tierra del Fuego are known only from the archipelago. This relatively low level of endemism is consistent with the youth of the flora, the area only having become available to plants during the past 16,000 years (p.1). Of the presumed endemics, 11 seem to have their closest relatives in Fuegia north of the Estrecho de Magallanes: *Atriplex reichei*, *Chiliophyllum fuegianum*, *Descurainia antarctica*, *Festuca cirrosa*, *Onuris alismatifolia*, *Ourisia fuegiana*, *Poa darwinii*, *P. yaganica*, *Senecio eightsii*, *S. humifusus* and *S. websteri*. *Nassauvia latissima* seems to be most closely related to *N. magellanica*, of the southern Andes and Fuegia, and the Falkland Islands (Islas Malvinas) endemic, *N. serpens*, while *Epilobium conjungens* has its closest relatives in New Zealand.

2. ALIEN FLORA

About 128 species present in Tierra del Fuego are considered here to have been introduced into the archipelago by man (Moore and Goodall, 1977). They are probably all of European origin, although many certainly arrived via settlements in S. Chile, S. Argentina and the Falkland Islands (Islas Malvinas). The earliest certain records of alien species in Tierra del Fuego are in 1882 (Spegazzini, 1896), thirteen years after the first resident Europeans arrived, but there is evidence that some may have come with sealers and whalers much earlier.

A rather crude classification of the habitats occupied by the adventives suggests that only 9 species (about 7.2%) of the alien flora seem to have entered relatively unmodified natural communities. Interestingly, these species include those, such as *Cerastium fontanum*, *Poa annua*, *P. pratensis*, *Rumex acetosella*, *Sagina procumbens* and *Veronica serpyllifolia*, which seem to have been amongst the earliest arrivals in Tierra del Fuego and which also figure prominently in the alien floras of other southern cool temperate regions. About 60% of the alien species are dependent upon habitats clearly disturbed by human activity, such as road-sides and tracks, a third of the species being restricted to highly modified sites such as gardens. The remaining species occur in naturally open habitats, such as beaches, river-banks etc., or in vegetation which has been modified by domestic animals, principally sheep.

The majority of the alien species seem to have been introduced accidentally by man, in the ballast of ships, among the packing of machinery or attached to animals brought from north of the Estrecho de Magallanes or the Falkland Islands (Islas Malvinas), for example. However, a number of species were certainly introduced deliberately. Seed mixtures containing such species as *Trifolium repens*, *Holcus lanatus* and *Anthoxanthum odoratum* have been used for many years to improve the natural pastures, particularly where forests have been cleared for sheep-rearing. *Mentha × piperita* has long been cultivated in gardens along the Canal Beagle, escaping to semi-natural wet habitats, while even *Taraxacum officinale* seems to have been deliberately introduced to N.E. Isla Grande for use as a salad-vegetable (Moore and Goodall, 1977).

The increasing impact of man in parts of Tierra del Fuego is undoubtedly extending the area of many introduced species. *Cirsium vulgare*, for example, arrived in Isla Grande prior to 1917 but it was not common. During the 1960's, however, it became much more widespread, particularly along the roads. Being of earth, the roads are regularly levelled by big graders and the frequent disturbance of the verges, together with the greatly increased vehicular traffic, has undoubtedly been important in the recent spread of this species. As man's activities continue to expand in Tierra del Fuego the introduced species may be expected to become more prominent in the Fuegian flora and their occurrence should be monitored closely in the future.

V. Systematic Account of the Flora

A. INTRODUCTION

a. *Format*

The order of families in the systematic account follows that given by Stebbins (1974), incorporating as it does many modern views on their relations, while the sequence of genera within each family is largely based on the second edition of Engler's *Syllabus der Pflanzenfamilien* (Melchior and Werdermann, 1954; Melchior, 1964). Within a genus the order of the species follows recent revisions and other taxonomic works wherever possible. Descriptions are provided for all native and introduced species. The species' names accepted in this Flora are given in bold italic type and numbered, the number being prefixed by an asterisk (*) when the species is considered to have been introduced into Fuegia (see Moore and Goodall, 1977), while synonyms are shown in smaller italics. Only the essential diagnostic characters are given for species that have been reported from Tierra del Fuego whose presence or status, for some reason, is regarded as doubtful and their names are not numbered.

Where appropriate, infraspecific categories within a species are mentioned, either formally or in notes following the geographical distribution, usually with some indication of their distinguishing characters and often, when data are available, with a comment on their validity. These categories are retained at the level recognised in the recent literature and there is therefore no uniformity in their circumscription. Thus, although the subspecies is generally a major morphologically and geographically definable node in the species' pattern of variation, the category of variety has been used much less consistently, in some instances being essentially synonymous with the subspecies and in others referring to a much more restricted entity or even to a rather sporadic local variant or "form".

b. *Nomenclature and literature citations*

As far as possible all names appearing in the botanical literature of Tierra del Fuego are included in the Flora. A complete index to all names and synonyms is provided (pp. 383–396). The abbreviations of author names are listed and explained in the appendices, as are those for the titles of books and journals cited. References to citations of synonyms are given chronologically. Details of such references, as well as of those appearing elsewhere throughout the Flora, are given in the list of references (p. 373), which constitutes, therefore, a rather comprehensive guide to the botanical literature of Tierra del Fuego.

Common names, both English and Spanish, used in Tierra del Fuego are not included in the species' accounts but are listed separately (p. 360). A separate list of Indian names and uses is also provided (p. 367).

c. *Descriptions*

All descriptions apply to the species as they occur in Tierra del Fuego, as does the information on flowering period and altitudinal range. Characteristics of the family as represented in Tierra del Fuego are not repeated in the generic descriptions. Similarly, features of the genus common to all the Fuegian representatives are not repeated in the species' descriptions, which follow the general sequence: habit, stems, leaves, inflorescence, flowers, fruit and seed. Unqualified measurements refer to length; measurements connected by the multiplication sign refer to length by breadth; numbers enclosed in brackets indicate dimensions beyond the normal range. Explanations of the abbreviations used are incorporated in the glossary (p. 354), in which the meanings of the technical terms used in the descriptions are given.

d. *Chromosome numbers*

The diploid (2*n*) chromosome number is given for a species when it is known from material from Tierra del Fuego; these numbers are documented by Moore (1981). Where such information is not available from Fuegian material it is given from Falkland Islands (Islas Malvinas) specimens, as published by Moore (1967), and is then marked by an asterisk (*).

e. *Ecology and geographical distribution*

A summary of the habitats occupied by each species is provided by reference to the major vegetation types described in chapter III (pp. 24–32). This information only applies to the species in Tierra del Fuego, since outside the archipelago their ecological preferences may be different.

The distribution of each species within Tierra del Fuego is summarised by a general phrase referring to its occurrence in the major sectors of the archipelago or, where appropriate, Isla Grande. Only the largest islands are named. More detailed data, derived from collections and the literature, are given in the distribution maps (Maps 1–42), in the hope that these, mostly almost certainly incomplete, will provide a basis to which future information can be added as it becomes available, partly at least from users of this Flora.

Where appropriate, the extra-Fuegian distribution of each species is summarised. For geographical areas that are sufficiently conspicuous to be shown in it, the *Times Atlas* (Anon., 1977) has been followed for place-names.* The names of other localities are derived from Goodall (1976, 1979) and the maps of Tierra del Fuego issued by the Instituto Geográfico Militar, Chile and the Servicio de Hidrografía Naval, Argentina.

f. *Illustrations*

All native genera, and a majority of species, are illustrated by line drawings (Figs. 3–282); these are grouped according to the artist. Thus, each group of drawings is accredited as follows – R. N. P. Goodall, Flora Patagónica (without distinguishing individual artists) and Sydney Parkinson; the copyright for the originals of these drawings is retained by the artists or their institutions where the originals are housed (see Acknowledgements). Fig. 1 and the map on the end-papers were drawn by David Angus, Figs. 2 and 283–284 by the author.

*The Fourth Commission of the Committee of 24, in Resolution 2065 of the 20th General Assembly of the United Nations, resolved that Commission documents in Spanish should refer to "Islas Malvinas (Falkland Islands)" while documents in other languages of the United Nations would refer to "Falkland Islands (Islas Malvinas)". This formula has been followed in this book.

B. SEQUENCE OF TAXA

Pteridophyta

LYCOPHYTINA
I. Lycopodiaceae
1. *Huperzia*
2. *Lycopodium*
II. Isoetaceae
1. *Isoetes*
FILIOPHYTINA
III. Ophioglossaceae
1. *Botrychium*
2. *Ophioglossum*
IV. Grammitaceae
1. *Grammitis*
V. Azollaceae
1. *Azolla*
VI. Hymenophyllaceae
1. *Hymenophyllum*
2. *Serpyllopsis*
VII. Aspleniaceae
1. *Asplenium*
VIII. Athyriaceae
1. *Cystopteris*
IX. Aspidiaceae
1. *Polystichum*
Rumohra
X. Blechnaceae
1. *Blechnum*
XI. Gleicheniaceae
1. *Gleichenia*

Spermatophyta

CONIFEROPHYTINA
XII. Podocarpaceae
1. *Dacrydium*
XIII. Cupressaceae
1. *Pilgerodendron*
CHLAMYDOSPERMO-PHYTINA
XIV. Ephedraceae
1. *Ephedra*
ANGIOSPERMOPHYTINA
Dicotyledons
MAGNOLIIDAE
Magnoliales
XV. Winteraceae
1. *Drimys*
Ranunculales
XVI. Berberidaceae
1. *Berberis*
XVII. Ranunculaceae
1. *Caltha*
2. *Hamadryas*
3. *Ranunculus*
4. *Myosurus*
5. *Anemone*
Papaverales
XVIII. Papaveraceae
1. *Papaver*
2. *Fumaria*
HAMAMELIDAE
Fagales
XIX. Fagaceae
1. *Nothofagus*
CARYOPHYLLIDAE
Caryophyllales
XX. Caryophyllaceae
1. *Arenaria*
2. *Stellaria*
3. *Cerastium*
4. *Colobanthus*
5. *Sagina*
6. *Spergularia*
7. *Silene*
8. *Dianthus*
XXI. Chenopodiaceae
1. *Chenopodium*
2. *Beta*
3. *Atriplex*
4. *Salicornia*
5. *Suaeda*
XXII. Portulacaceae
1. *Calandrinia*
2. *Montia*
Polygonales
XXIII. Polygonaceae
1. *Koenigia*
2. *Polygonum*
3. *Bilderdykia*
4. *Rumex*
5. *Rheum*
Plumbaginales
XXIV. Plumbaginaceae
1. *Armeria*
DILLENIIDAE
Urticales
XXV. Urticaceae
1. *Urtica*
Violales
XXVI. Violaceae
1. *Viola*
XXVII. Frankeniaceae
1. *Frankenia*
Capparales
XXVIII. Cruciferae
1. *Sisymbrium*
2. *Descurainia*
3. *Erysimum*
4. *Hesperis*
5. *Cardamine*
6. *Lunaria*
7. *Onuris*
8. *Draba*
9. *Erophila*
10. *Capsella*
11. *Thlaspi*
12. *Lepidium*
13. *Cardaria*
14. *Coronopus*
15. *Brassica*
16. *Rapistrum*
17. *Raphanus*
Ericales
XXIX. Ericaceae
1. *Gaultheria*
2. *Pernettya*
XXX. Epacridaceae
1. *Lebetanthus*
XXXI. Empetraceae
1. *Empetrum*
Primulales
XXXII. Primulaceae
1. *Primula*
2. *Androsace*
3. *Anagallis*
4. *Samolus*
ROSIDAE
Rosales
XXXIII. Droseraceae
1. *Drosera*
XXXIV. Rosaceae
1. *Rubus*
2. *Geum*
3. *Acaena*
XXXV. Crassulaceae
1. *Crassula*
2. *Sedum*
XXXVI. Saxifragaceae
1. *Saxifraga*
2. *Saxifragodes*
3. *Saxifragella*
4. *Chrysosplenium*

5. *Tribeles*
6. *Ribes*
7. *Escallonia*

Fabales

XXXVII. Leguminosae
1. *Adesmia*
2. *Anarthrophyllum*
3. *Ulex*
4. *Astragalus*
5. *Vicia*
6. *Lathyrus*
7. *Medicago*
8. *Trifolium*

Haloragales

XXXVIII. Haloragaceae
1. *Gunnera*
2. *Myriophyllum*

XXXIX. Hippuridaceae
1. *Hippuris*

Myrtales

XL. Thymelaeaceae
1. *Drapetes*

XLI. Myrtaceae
1. *Myrteola*
2. *Tepualia*

XLII. Onagraceae
1. *Fuchsia*
2. *Epilobium*

Proteales

XLIII. Proteaceae
1. *Embothrium*

Santalales

XLIV. Santalaceae
1. *Nanodea*
2. *Myoschilos*
3. *Arjona*

XLV. Misodendraceae
1. *Misodendrum*

Celastrales

XLVI. Celastraceae
1. *Maytenus*

Euphorbiales

XLVII. Euphorbiaceae
1. *Dysopsis*
2. *Euphorbia*

Rhamnales

XLVIII. Rhamnaceae
1. *Discaria*

Geraniales

XLIX. Linaceae
1. *Linum*

L. Geraniaceae
1. *Geranium*
2. *Erodium*

LI. Oxalidaceae
1. *Oxalis*

LII. Tropaeolaceae
1. *Tropaeolum*

Polygalales

LIII. Polygalaceae
1. *Polygala*

Umbellales

LIV. Araliaceae
1. *Pseudopanax*

LV. Umbelliferae
1. *Hydrocotyle*
2. *Azorella*
3. *Bolax*
4. *Schizeilema*
5. *Huanaca*
6. *Osmorhiza*
7. *Daucus*
8. *Conium*
9. *Oreomyrrhis*
10. *Apium*
11. *Lilaeopsis*
12. *Levisticum*

ASTERIDAE

Gentianales

LVI. Desfontainiaceae
1. *Desfontainia*

LVII. Gentianaceae
1. *Gentiana*
2. *Gentianella*

Polemoniales

LVIII. Solanaceae
1. *Jaborosa*
2. *Benthamiella*

LIX. Polemoniaceae
1. *Polemonium*
2. *Microsteris*

LX. Hydrophyllaceae
1. *Phacelia*

LXI. Boraginaceae
1. *Amsinckia*
2. *Myosotis*
3. *Plagiobothrys*

Lamiales

LXII. Labiatae
1. *Scutellaria*
2. *Lamium*
3. *Glechoma*
4. *Satureja*
5. *Prunella*
6. *Mentha*

LXIII. Tetrachondraceae
1. *Tetrachondra*

LXIV. Callitrichaceae
1. *Callitriche*

Plantaginales

LXV. Plantaginaceae
1. *Plantago*
2. *Litorella*

Scrophulariales

LXVI. Scrophulariaceae
1. *Limosella*
2. *Calceolaria*
3. *Digitalis*
4. *Veronica*
5. *Ourisia*
6. *Hebe*
7. *Euphrasia*

LXVII. Lentibulariaceae
1. *Pinguicula*

Campanulales

LXVIII. Campanulaceae
1. *Campanula*

LXIX. Lobeliaceae
1. *Pratia*

LXX. Stylidiaceae
1. *Donatia*
2. *Phyllachne*

Rubiales

LXXI. Rubiaceae
1. *Oreopolus*
2. *Nertera*
3. *Galium*

Dipsacales

LXXII. Valerianaceae
1. *Valerianella*
2. *Valeriana*

LXXIII. Calyceraceae
1. *Moschopsis*
2. *Boopis*

Asterales

LXXIV. Compositae

Astereae
1. *Lagenophora*
2. *Bellis*
3. *Gutierrezia*
4. *Lepidophyllum*
5. *Nardophyllum*
6. *Chiliotrichum*
7. *Chiliophyllum*
8. *Conyza*
9. *Erigeron*
10. *Aster*
11. *Baccharis*

Inuleae

12. *Antennaria*
13. *Gnaphalium*
14. *Gamochaeta*

15. *Adenocaulon*
16. *Eriachaenium*
Heliantheae
17. *Madia*
Helianthus
Anthemideae
18. *Achillea*
19. *Anthemis*
20. *Matricaria*
21. *Chamomilla*
22. *Tanacetum*
23. *Leucanthemum*
24. *Cotula*
25. *Abrotanella*
26. *Artemisia*
Senecioneae
27. *Senecio*
Cardueae
28. *Carduus*
29. *Cirsium*
30. *Silybum*
Mutiseae
31. *Macrachaenium*
32. *Nassauvia*
33. *Leucheria*
34. *Perezia*
Cichorieae
35. *Hypochoeris*
36. *Tragopogon*
37. *Taraxacum*
38. *Sonchus*
39. *Agoseris*
40. *Crepis*
41. *Hieracium*
MONOCOTYLEDONS
ALISMATIDAE
Najadales
LXXV. Juncaginaceae
1. *Tetroncium*
2. *Triglochin*
LXXVI. Potamogetonaceae
1. *Potamogeton*
LXXVII. Zannichelliaceae
1. *Zannichellia*
LXXVIII. Ruppiaceae
1. *Ruppia*
COMMELINIDAE
Restionales
LXXIX. Centrolepidaceae
1. *Gaimardia*
Poales
LXXX. Gramineae
Poeae
1. *Festuca*
2. *Lolium*
3. *Poa*
4. *Puccinellia*
5. *Dactylis*
6. *Cynosurus*
7. *Catabrosa*
Bromeae
8. *Bromus*
Triticeae
9. *Leymus*
10. *Elymus*
11. *Agropyron*
12. *Triticum*
13. *Secale*
14. *Hordeum*
Aveneae
15. *Avena*
16. *Arrhenatherum*
17. *Koeleria*
18. *Trisetum*
19. *Deschampsia*
20. *Aira*
21. *Hierochloë*
22. *Anthoxanthum*
23. *Holcus*
24. *Agrostis*
25. *Ammophila*
26. *Calamagrostis*
27. *Phleum*
28. *Alopecurus*
Phalarideae
29. *Phalaris*
Stipeae
30. *Stipa*
Cortaderieae
31. *Cortaderia*
Danthonieae
32. *Rytidosperma*
Juncales
LXXXI. Juncaceae
1. *Rostkovia*
2. *Marsippospermum*
3. *Juncus*
4. *Luzula*
Cyperales
LXXXII. Cyperaceae
1. *Scirpus*
2. *Eleocharis*
3. *Schoenus*
4. *Oreobolus*
5. *Carpha*
6. *Uncinia*
7. *Carex*
ARECIDAE
Arales
LXXXIII. Lemnaceae
1. *Spirodela*
LILIIDAE
Liliales
LXXXIV. Iridaceae
1. *Tapeinia*
2. *Sisyrinchium*
3. *Phaiophleps*
LXXXV. Liliaceae
1. *Philesia*
2. *Luzuriaga*
3. *Astelia*
4. *Tristagma*
5. *Allium*
LXXXVI. Amaryllidaceae
1. *Alstroemeria*
Orchidales
LXXXVII. Orchidaceae
1. *Codonorchis*
2. *Gavilea*
3. *Chloraea*

C. KEY TO FAMILIES

a. Plant reproducing by spores (Pteridophyta):
b. Plant free-floating on water **V. Azollaceae**
b. Plant rooted in substrate, aquatic or terrestrial:
c. Leaves without distinct lamina and petiole:
d. Submerged aquatic; leaves in basal rosette **II. Isoetaceae**
d. Terrestrial plant; leaves not in basal rosette **I. Lycopodiaceae**

c. Leaves with distinct lamina and petiole:
e. Leaves simple:
f. Sporangia few, embedded in distinct linear fertile spikes **III. Ophioglossaceae**
f. Sporangia numerous, in 2 rows of sori on underside of lamina **IV. Grammitaceae**
e. Leaves variously divided:
g. Leaves delicate, membranous, translucent; plant resembling bryophyte **VI. Hymenophyllaceae**
g. Leaves not as above; plant not resembling brypohyte:
h. Lamina with 3–6 pairs of semi-orbicular pinnae; sporangia on margins of ultimate segments of branched fertile spike **III. Ophioglossaceae**
h. Lamina and sporangia not as above:
i. Leaves dichotomously branched **XI. Gleicheniaceae**
i. Leaves not dichotomously branched:
j. Leaves simply pinnate or pinnatifid; sporangia borne on distinct fertile fronds **X. Blechnaceae**
j. Leaves 2- to 3-pinnate or -partite, if simply pinnate the sporangia not borne on distinct fertile fronds:
k. Sori and indusia elliptical to ovate **VII. Aspleniaceae**
k. Sori orbicular; indusium orbicular or ovate-lanceolate:
l. Indusium orbicular, peltate **IX. Aspidiaceae**
l. Indusium ovate-lanceolate, acuminate, attached to one side of sorus **VIII. Athyriaceae**
a. Plant reproducing by seeds (Spermatophyta):
m. Ovules not enclosed in an ovary, borne either on the upper surface of scales arranged in cones or solitary at apex of lateral stems; perianth absent; trees or shrubs:
n. Leaves brownish, not imbricate; internodes long **XIV. Ephedraceae**
n. Leaves green, densely imbricate; internodes short:
o. Stem prostrate; leaves spirally arranged; seeds solitary **XII. Podocarpaceae**
o. Stem erect; leaves in 4 rows; seeds in woody cone **XIII. Cupressaceae**

m. Ovules completely enclosed in an ovary; perianth usually present; herbs, trees or shrubs (Angiospermophytina):
p. Plant free-floating on surface of water, not rooted **LXXXIII. Lemnaceae**
p. Plant not aquatic or, if so, rooted in mud:
q. Hemiparasite on branches of *Nothofagus* trees **XLV. Misodendraceae**
q. Not parasitic or, if so, not on tree-branches:
r. Trees or shrubs, sometimes dwarf-shrubs:
s. Leaves compound:
t. Flowers in umbels **LIV. Araliaceae**
t. Flowers not in umbels:
u. Flowers zygomorphic; fruit a pod **XXXVII. Leguminosae**
u. Flowers actinomorphic; fruit an achene **XXXIV. Rosaceae**

s. Leaves simple:
v. Leaves opposite, rarely in whorls of 3:
w. Stipules present **XLVIII. Rhamnaceae**
w. Stipules absent:
x. Leaves spiny; flowers orange-yellow **LVI. Desfontainiaceae**
x. Leaves not spiny; flowers not as above:
y. Leaf-lamina at least 10×6 mm; flowers pendent, with prominent hypanthial tube **XLII. Onagraceae**
y. Leaf-lamina not more than 10×6 mm; flowers not pendent; hypanthial tube absent:
z. Corolla zygomorphic, 2-lipped **LXII. Labiatae**
z. Corolla actinomorphic; petals free:
a′. Leaves not more than 1 mm wide, strongly revolute, not punctate **XXVII. Frankeniaceae**
a′. Leaves at least 2 mm wide, not strongly revolute, punctate **XLI. Myrtaceae**
v. Leaves alternate:
b′. Flowers grouped in capitula **LXXIV. Compositae**
b′. Flowers not in capitula:
c′. Perianth tubular, red; fruit a woody follicle **XLIII. Proteaceae**
c′. Perianth not tubular and red; fruit not a woody follicle:
d′. Perianth not differentiated into sepals and petals:
e′. Leaves semi-cylindrical, not more than 3 mm wide **XXI. Chenopodiaceae**
e′. Leaves flat, at least 5 mm wide:
f′. Stipules absent; flowers hermaphrodite; fruit a drupe **XLIV. Santalaceae**

f′. Stipules present, caducous; flowers unisexual; fruit a woody cupule enclosing 1 or 3 nuts **XIX. Fagaceae**

d′. Perianth of calyx and corolla; fruit a capsule, berry or drupe

g′. Leaves not more than 2 mm wide, strongly revolute **XXXI. Empetraceae**

g′. Leaves more than 2 mm, if less then not strongly revolute:

h′. Corolla tubular to urceolate, shortly 5-lobed:

i′. Scandent shrub; sepals free **XXX. Epacridaceae**

i′. Erect or prostrate (tree or) shrub; sepals united **XXIX. Ericaceae**

h′. Petals free:

j′. Plant with spines; petals 3 **XVI. Berberidaceae**

j′. Plant without spines; petals at least 4:

k′. Sepals 2, caducous; petals up to 14; carpels separate **XV. Winteraceae**

k′. Sepals 4–5, persistent; petals 4–5:

l′. Sepals free; ovary inferior **XXXVI. Saxifragaceae**

l′. Sepals united at base; ovary superior **XLVI. Celastraceae**

r. Herbs, sometimes woody at base:

m′. Aquatic plant with submerged or floating leaves; inflorescence sometimes emergent:

n′. Leaves divided into numerous filiform segments:

o′. Leaves pinnately divided; flowers in terminal spike **XXXVIII. Haloragaceae**

o′. Leaves palmately divided; flowers solitary or in pairs **XVII. Ranunculaceae**

n′. Leaves entire or minutely toothed:

p′. Fruitlets with stalks several times their own length **LXXVIII. Ruppiaceae**

p′. Fruitlets sessile:

q′. Flowers in spikes **LXXVI. Potamogetonaceae**

q′. Flowers not in spikes:

r′. Flowers in heads or in branched inflorescences **LXXXI. Juncaceae**

r′. Flowers solitary or few, sessile or shortly pedicellate, usually in leaf axils:

s′. Leaves in whorls of 8 or more **XXXIX. Hippuridaceae**

s′. Leaves not in whorls of 8 or more:

t′. Leaves with basal sheath; ovary not lobed **LXXVII. Zannichelliaceae**

t′. Leaves without basal sheath; ovary deeply 4-lobed **LXIV. Callitrichaceae**

m′. Land plant or aquatic with emergent stems or leaves:

u′. Leaves apparently absent; stem succulent **XXI. Chenopodiaceae**

u′. Leaves obvious:

v′. Leaves opposite or in whorls:

w′. Leaves in whorls:

x′. Leaves pinnately divided; flowers in terminal spike **XXXVIII. Haloragaceae**

x′. Leaves entire or minutely toothed; flowers solitary, in pairs or axillary cymes:

y′. Flowers at least 10 mm in diameter; fruit a capsule **LXXXVII. Orchidaceae**

y′. Flowers not more than 3 mm in diameter; fruit a nut or of 2 mericarps:

z′. Corolla rotate, 3- to 4-lobed; stamens 4–5 **LXXI. Rubiaceae**

z′. Corolla absent; stamen 1 **XXXIX. Hippuridaceae**

w′. Leaves opposite:

aa. Perianth absent; stamen 1 **LXIV. Callitrichaceae**

aa. Perianth present; stamens 2 or more:

bb. Styles 2–10:

cc. Styles 8–10 **XLIX. Linaceae**

cc. Styles 2–6:

dd. Leaves entire **XX. Caryophyllaceae**

dd. Leaves lobed to serrate, or bicuspidate at apex:

ee. Stipules present; leaves palmately lobed; styles 5; fruit a schizocarp **L. Geraniaceae**

ee. Stipules absent; leaves serrate, or bicuspidate at apex; styles 2; fruit a capsule **XXXVI. Saxifragaceae**

bb. Style 1:

ff. Ovary inferior:

gg. Stipules present; stamens 4–5 **LXXI. Rubiaceae**

gg. Stipules absent; stamens 3 or 8:

hh. Stamens 3; calyx an annular ring; corolla 5-lobed **LXXII. Valerianaceae**

hh. Stamens 8; sepals 4, free; petals 4, free **XLII. Onagraceae**

ff. Ovary superior:

ii. Sepals 2 **XXII. Portulacaceae**

ii. Sepals 4–5:

jj. Stipules present **XXV. Urticaceae**

jj. Stipules absent:

kk. Stamens 5:

ll. Style 3-lobed; capsule dehiscing by 3 valves **LIX. Polemoniaceae**

ll. Style 2-fid; capsule dehiscing by 2 valves **LVII. Gentianaceae**

kk. Stamens 2, 4 or 10:

mm. Petals absent; fruit a drupe
XL. Thymelaeaceae

mm. Petals present; fruit of nutlets, follicles or a capsule:

nn. Sepals and petals free; fruit of follicles **XXXV. Crassulaceae**

nn. Sepals and/or petals united; fruit of nutlets or a capsule:

oo. Corolla actinomorphic; leaves not more than 3.5 × 2 mm
LXIII. Tetrachondraceae

oo. Corolla zygomorphic; leaves at least 5 × 3 mm:

pp. Corolla with tube or lower inflated lip; fruit a capsule
LXVI. Scrophulariaceae

pp. Corolla neither with tube nor lower lip inflated; fruit of 4 nutlets **LXII. Labiatae**

v′. Leaves alternate, spirally arranged or in basal rosettes:

qq. Perianth absent, or not differentiated into sepals and petals, being entirely sepaloid or petaloid:

rr. Perianth petaloid:

ss. Stipules sheathing, scarious (ochreae)
XXIII. Polygonaceae

ss. Stipules absent:

tt. Stamens numerous; carpels free
XVII. Ranunculaceae

tt. Stamens not more than 6; carpels united:

uu. Perianth-segments or -lobes 4–5; stamens 4–5 **XLIV. Santalaceae**

uu. Perianth-segments 6; stamens 1, 3 or 6:

vv. Flowers actinomorphic:

ww. Stamens 3; ovary inferior
LXXXIV. Iridaceae

ww. Stamens 6; ovary superior
LXXXV. Liliaceae

vv. Flowers zygomorphic:

xx. Stamen 1; anther and stigma borne on a column
LXXXVII. Orchidaceae

xx. Stamens 6; free, not united to style
LXXXVI. Amaryllidaceae

rr. Perianth sepaloid, scale-like (glumaceous) or absent:

yy. Perianth absent, or represented by scales or bristles; flowers in axils of specialised scale-like bracts which are usually arranged into spikelets:

zz. Leaves filiform, in basal rosette; ripe fruit dehiscent; cushion forming plants
LXXIX. Centrolepidaceae

zz. Leaves ± linear and grass-like; ripe fruit indehiscent; grass-like herbs:

a′a′. Flowers usually with a bract above and below; leaf sheaths usually open; stems usually with hollow internodes, terete or flattened **LXXX. Gramineae**

a′a′. Flowers with a bract below only; leaf-sheaths usually closed; stems usually with solid internodes, often triquetrous
LXXXII. Cyperaceae

yy. Perianth present or, if minute or absent, the flowers not arranged in spikelets and without scale-like bracts:

b′b′. Leaves linear ± grass- or rush-like, or reduced to basal sheaths:

c′c′. Flowers in spikes or racemes; fruit of indehiscent follicles
LXXV. Juncaginaceae

c′c′. Flowers solitary or in cymes; fruit a capsule **LXXXI. Juncaceae**

b′b′. Leaves not linear;

d′d′. Leaves compound **XXXIV. Rosaceae**

d′d′. Leaves simple:

e′e′. Stipules united in scarious sheaths (ochreae) **XXIII. Polygonaceae**

e′e′. Stipules free, or absent:

f′f′. Ovary superior; fruit an achene or schizocarp:

g′g′. Perianth-segments ± connate if present; fruit an achene
XXI. Chenopodiaceae

g′g′. Perianth-segments free; fruit a schizocarp
XLVII. Euphorbiaceae

f′f′. Ovary inferior; fruit a drupe:

h′h′. Leaves ovate to orbicular, palmately lobed **XXXVIII. Haloragaceae**

h′h′. Leaves linear, entire
XLIV. Santalaceae

qq. Perianth with distinct sepals and petals:

i′i′. Petals free:

j′j′. Leaves with conspicuous, red, viscid glandular hairs **XXXIII. Droseraceae**

j′j′. Leaves not as above:

k′k′. Flowers in umbels **LV. Umbelliferae**

k′k′. Flowers not in umbels:

l′l′. Carpels and styles free or almost so:

m′m′. Stipules absent
XVII. Ranunculaceae

m′m′. Stipules present
XXXIV. Rosaceae

l′l′. Carpels or styles, or both, united:

n′n′. Corolla zygomorphic:

o′o′. Corolla with basal spur; stamens 5 **XXVI. Violaceae**

o′o′. Corolla without spur; stamens 8 or 10:

p′p′. Stamens 8, free; leaves simple, peltate; fruit a schizocarp **LII. Tropaeolaceae**

p′p′. Stamens 10, all or 9 united in tube; leaves compound, pinnate; fruit a pod **XXXVII. Leguminosae**

n′n′. Corolla actinomorphic:

q′q′. Petals 10; cushion plant **LXX. Stylidiaceae**

q′q′. Petals 4–5; usually not a cushion plant:

r′r′. Sepals 2, sometimes caducous

s′s′. Petals 5; stamens 3–7; stigma-lobes 3 **XXII. Portulacaceae**

s′s′. Petals 4; stamens numerous; stigmas 4–6 **XVIII. Papaveraceae**

r′r′. Sepals 4 or 5:

t′t′. Sepals 4; petals 4 **XXVIII. Cruciferae**

t′t′. Sepals 5; petals 5:

u′u′. Leaves pinnate **L. Geraniaceae**

u′u′. Leaves not pinnate:

v′v′. Leaves radially palmate; leaflets emarginate **LI. Oxalidaceae**

v′v′. Leaves simple, oblanceolate to peltate, entire to deeply lobed:

w′w′. Leaves peltate, deeply lobed; fruit a schizocarp **LII. Tropaeolaceae**

w′w′. Leaves oblanceolate to sub-orbicular; fruit a capsule **XXXVI. Saxifragaceae**

i′i′. Petals united:

x′x′. Flowers grouped into dense heads (capitula) surrounded by involucral bracts or uppermost leaves:

y′y′. Involucre with tubular, scarious sheath at base enclosing top of scape **XXIV. Plumbaginaceae**

y′y′. Involucre without basal scarious sheath:

z′z′. Anthers united into cylinder; stigma 2-fid **LXXIV. Compositae**

z′z′. Anthers free; stigma simple **LXXIII. Calyceraceae**

x′x′. Flowers not grouped into capitula:

aaa. Ovary inferior:

bbb. Cushion plant; leaves densely imbricate; stamens 2 **LXX. Stylidiaceae**

bbb. Not a cushion plant; leaves not imbricate; stamens 5:

ccc. Stems ascending to erect; corolla actinomorphic; stigma 3-lobed; fruit a capsule **LXVIII. Campanulaceae**

ccc. Stems prostrate; corolla zygomorphic; stigma 2-lobed; fruit a berry **LXIX. Lobeliaceae**

aaa. Ovary superior:

ddd. Flowers zygomorphic:

eee. Stamens 2 or 4, not united:

fff. Calyx 4- to 5-lobed; corolla spurred; leaves pale green, covered with viscid hairs above **LXVII. Lentibulariaceae**

fff. Calyx 2-lipped; corolla not spurred; leaves not as above **LXVI. Scrophulariaceae**

eee. Stamens 8 or 10, most or all united in tube:

ggg. Inner 2 sepals much larger than outer 5, petaloid; petals 3; fruit a capsule **LIII. Polygalaceae**

ggg. Sepals subequal; petals 5; fruit a pod or lomentum **XXXVII. Leguminosae**

ddd. Flowers actinomorphic:

hhh. Sepals 2–4; corolla 2- to 4-lobed, scarious; flowers in terminal spikes or heads **LXV. Plantaginaceae**

hhh. Sepals 5 or calyx 5-lobed; corolla 5-lobed; flowers not in spikes or heads:

iii. Flowers solitary, or in umbellate or racemose inflorescence, or 2–4 in cyme:

jjj. Plant a dense cushion or with leaves pinnatifid **LVIII. Solanaceae**

jjj. Plant not cushion-forming; leaves entire to crenate-serrate **XXXII. Primulaceae**

iii. Flowers numerous, in often scorpioid cymes:

kkk. Style simple; fruit of 2 or 4 nutlets; leaves entire **LXI. Boraginaceae**

kkk. Style 2-fid; fruit a capsule; leaves usually pinnatifid **LX. Hydrophyllaceae**

Pteridophyta

I. LYCOPODIACEAE

Perennials, with stems often somewhat woody at base. Leaves small, without veins or with midvein only. Homosporous; sporangia in axils of leaves or sporophylls, which may be grouped into spikes.

a. Stems ascending, divided dichotomously into equal branches; sporangia axillary **1. *Huperzia***
a. Stems creeping, with short lateral branches; sporangia in terminal spikes **2. *Lycopodium***

Literature: G. Looser, *Rev. Univ.* **46**: 213–244(1961).

1. *Huperzia* Bernh.

Stems short, ascending, divided dichotomously into branches of equal length. Sporangia axillary, reniform, stipitate, borne near the top of the branches; spores foveolate-punctate.

1. *H. selago* (L.) Bernh. ex Schrank & Mart. 1829, *Hort. Monac.* 3.
Lycopodium selago L.1753, *Sp. Pl. 1102 – Skottsberg, 1916.*
L. fuegianum Roivainen 1936, *Ann. Bot. Soc. Zool.-Bot. Fenn. Vanamo* **6(8)**: 13.

Stems 2–5 cm, with suberect branches, with many rows of imbricate leaves. Leaves 4–8 × 1.5–2.5 mm, linear-lanceolate, acute, with numerous minute teeth on margin, gold-green, sessile. (Fig. 3).

Bogs, with *Donatia, Caltha, Marsippospermum* and *Sphagnum*; 100–650 m.

W.C. to S.W. Isla Grande. (Map 1a).
Falkland Islands (Islas Malvinas).
Eurasia.
N. America.

Fuegian material has been separated as a distinct species on the basis of having more fleshy, linear leaves and smaller spores than N. Hemisphere material. However, these characters have not proved satisfactory and there is no clear discontinuity in any morphological features.

2. *Lycopodium* L.

Stems creeping, with short lateral branches. Leaves arranged spirally or in whorls, sessile. Sporangia in terminal spikes; sporophylls unlike the leaves; spores muricate-reticulate.

a. Stems leafless, ± below ground; sporophylls less than 4 times as long as wide **2. *magellanicum***
a. Stems leafy, overground; sporophylls more than 5 times as long as wide **1. *confertum***

1. *L. confertum* Willd. 1810, *Sp. Pl.*, Ed. 4, **5**: 27.

Stems up to 50(–100) cm, on surface of ground, procumbent, rooted at intervals, leafy; branches 5–25 cm, procumbent to ascending, densely leafy. Leaves 4.5–7.5 × 0.7–0.9 mm, linear to linear-lanceolate, acute to acuminate, entire. Spikes (8–)12–18(–20) × *c.*10 mm, erect, sessile; sporophylls 4–6 × 0.8–0.9 mm, ovate-lanceolate, acuminate, denticulate, gold-green, reflexed at maturity. (Fig. 4).

Pebble screes, open areas in montane cushion plant communities; 400–600 m.

S. Isla Grande, Isla de los Estados. (Map 1b).
W. Argentina at *c.*51°35′S.lat.
Chile north to *c.*41°S.lat.
Falkland Islands (Islas Malvinas).

2. *L. magellanicum* (P.Beauv.) Swartz 1806, *Syn. Fil. Lycop.* 180.
L. clavatum L. var. *magellanicum* (P.Beauv.) Hooker f. 1847, *Fl. Antarct.* **1(1)**: 113 – Franchet 1889; Alboff, 1896.

Stems up to 30–50(–100) cm, just below surface of ground, procumbent, rooted at intervals, with white, ovate, cucullate scales; branches (2–)5–30 cm, prostrate to erect, leafy. Leaves (1.7–)3–5(–5.8) × 0.6–1 mm, linear to linear-lanceolate, acute, entire. Spikes 10–40 × 3–5 mm, erect, subsessile or on peduncles 5–20 mm, with distant, linear-subulate to -lanceolate, bract-like leaves; sporophylls 2.3–4.3 × 1.1–2.1 mm, ovate to ovate-lanceolate, acuminate, denticulate, becoming yellowish and with apex ± reflexed at maturity. (Fig. 5).

Shallow soil on rocks, open shrub and grass communities, drier parts of bogs, forest margins and clearings, feldmark; 0–800 m.

Almost throughout Fuegia. (Map 1c).
W. Argentina north to *c.*41°S.lat.
Chile north to *c.*38°S.lat.
Juan Fernandez.
Falkland Islands (Islas Malvinas), South Georgia, Iles Crozet, Iles de Kerguelen.

II. ISOETACEAE

Perennials, with short, ± lobed stem. Leaves in basal

rosette. Heterosporous; sporangia solitary, enclosed by leaf-base.

1. *Isoetes* L.

Aquatic. Leaves subulate, without stomata, with broad base bearing thin ligule above. Megasporangia on outer leaves, the megaspores tetrahedral, white; microsporangia on inner leaves, the microspores bilateral, brownish.

Literature: H. P. Fuchs, *Nova Hedwigia Beih.* **3**: 1–160(1962).

1. ***I. savatieri*** Franchet 1889, *Miss. Sci. Cap Horn* 391.

Stem obscurely 3-lobed. Leaves 28–120(–200) × 2.5–4 mm, subulate, somewhat 4-angled above, erect, the base up to 10 mm, not persistent; ligule 4.5–5 × 1–2 mm, triangular. Sporangia 2–4.3 mm in diameter, incompletely covered with thin tissue (velum); megaspores *c.*0.4 mm in diameter, ridged; microspores ± spinose. (Fig. 6).

Freshwater lakes, usually deeply submerged.

W.C. Isla Grande (53°32′S.lat., 69°09′W.long.). (Map 1d).
W. Argentina north to *c.*41°S.lat.
Chile north to *c.*41°S.lat., also at 37°S. and 32°06′S.lat.

III. OPHIOGLOSSACEAE

Perennials, with short, fleshy rhizome without scales. Leaves not circinate in bud, divided into a sterile lamina and a fertile simple or compound spike. Homosporous; sporangia large, thick-walled, sessile, without annulus, opening by transverse slit.

a. Lamina entire; fertile spike simple, with sunken and coalescent sporangia **2. *Ophioglossum***
a. Lamina pinnate; fertile spike compound, with exposed, free sporangia **1. *Botrychium***

1. *Botrychium* Swartz

Leaves with pinnate lamina having dichotomous free veins; fertile spike 1- to 2- pinnate, with marginal, free sporangia.

1. ***B. dusenii*** (Christ) Alston 1960, *Lilloa* **30**: 107.
B. lunaria (L.) Swartz var. *dusenii* Christ 1906, *Ark. Bot.* **6(3)**: 5 – Spegazzini, 1896.

Rhizome vertical, cylindrical, ± fleshy, producing 1(2) leaves per season, only 1 of which develops a fertile spike. Sterile lamina 10–28 × 7–20 mm, ovate to triangular-ovate in outline, with 3–7 pairs of pinnae, sessile; pinnae semi-orbicular to cuneiform, cuneate at base, entire to shallowly and irregularly crenate. Fertile spike (6–)9–21(–37) × 4–12 mm, the stalk 2–24 mm, overtopping sterile lamina; sporangia 1–1.7 mm in diameter, globose, along margins of ultimate segments, yellowish brown; spores verrucose. (Fig. 8).

Coastal sands and shell deposits, open moist soil; 0–30 m.

C. & S. Isla Grande, Isla de los Estados, Islas Wollaston. (Map 1e).
W. Argentina north to *c.*49°S.lat.
Falkland Islands (Islas Malvinas).

2. *Ophioglossum* L.

Leaves with entire lamina having reticulate veins; fertile spike simple, with 2 rows of sunken, coalescent sporangia.

Literature: R. T. Clausen, *Amer. Fern Jour.* **38**: 176–186(1948).

1. ***O. crotalophoroides*** Walter 1788, *Fl. Carol.* 256.

Rhizome swollen, fleshy, producing 1(2) leaves per season, usually only one of which develops a fertile spike. Sterile lamina 30–40 × 20–30 mm, ovate to oblong, obtuse, concave, tapering at base to point of insertion. Fertile spike with a slender stalk 10–15 mm widening above to ± compressed cylindrical part bearing 10–17 sporangia on each side; spores tuberculate. (Fig. 7).

Moist depressions in cleared open *Nothofagus antarctica* woodland; 250–300 m.

S. side of Isla Grande. (Map 1f).
W. Argentina at *c.*49°S.lat.
E. Chile and north along Andes to C. Colombia (*c.*5°N.lat.).
Guatemala, North America; Falkland Islands (Islas Malvinas), South Georgia, ? Tristan da Cunha.

IV. GRAMMITACEAE

Rhizome short – creeping to erect, covered with scarious scales. Leaves tufted, simple; veins free. Sori in 2 rows, one on each side of midvein. Indusium absent; sporangia with slender stipe; spores tetrahedral.

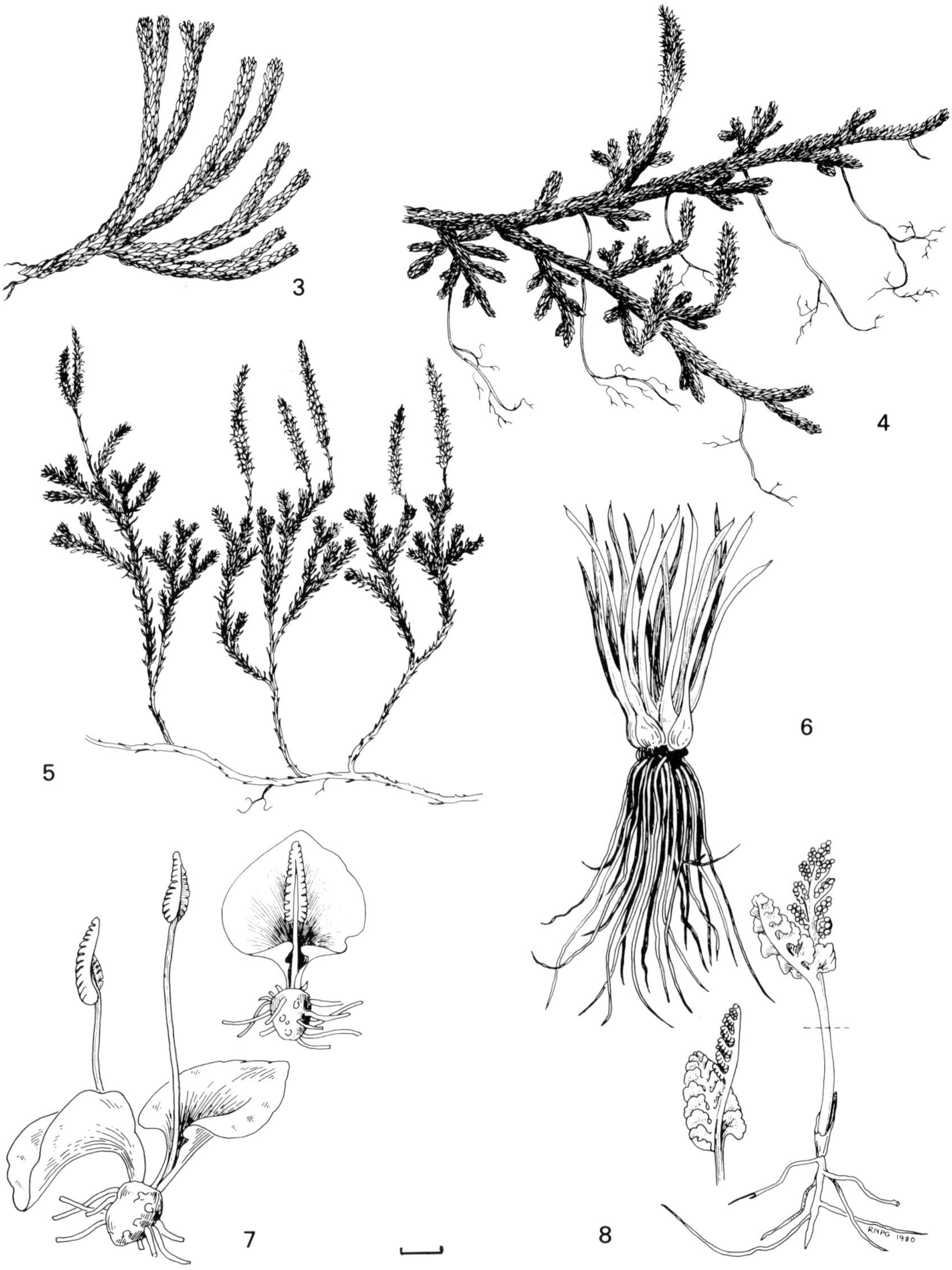

Fig. 3. *Huperzia selago*. Fig. 4. *Lycopodium confertum*. Fig. 5. *Lycopodium magellanicum*. Fig. 6. *Isoetes savatieri*. Fig. 7. *Ophioglossum crotalophoroides*. Fig. 8. *Botrychium duseni*. Scale = 1 cm.

(R. N. P. Goodall)

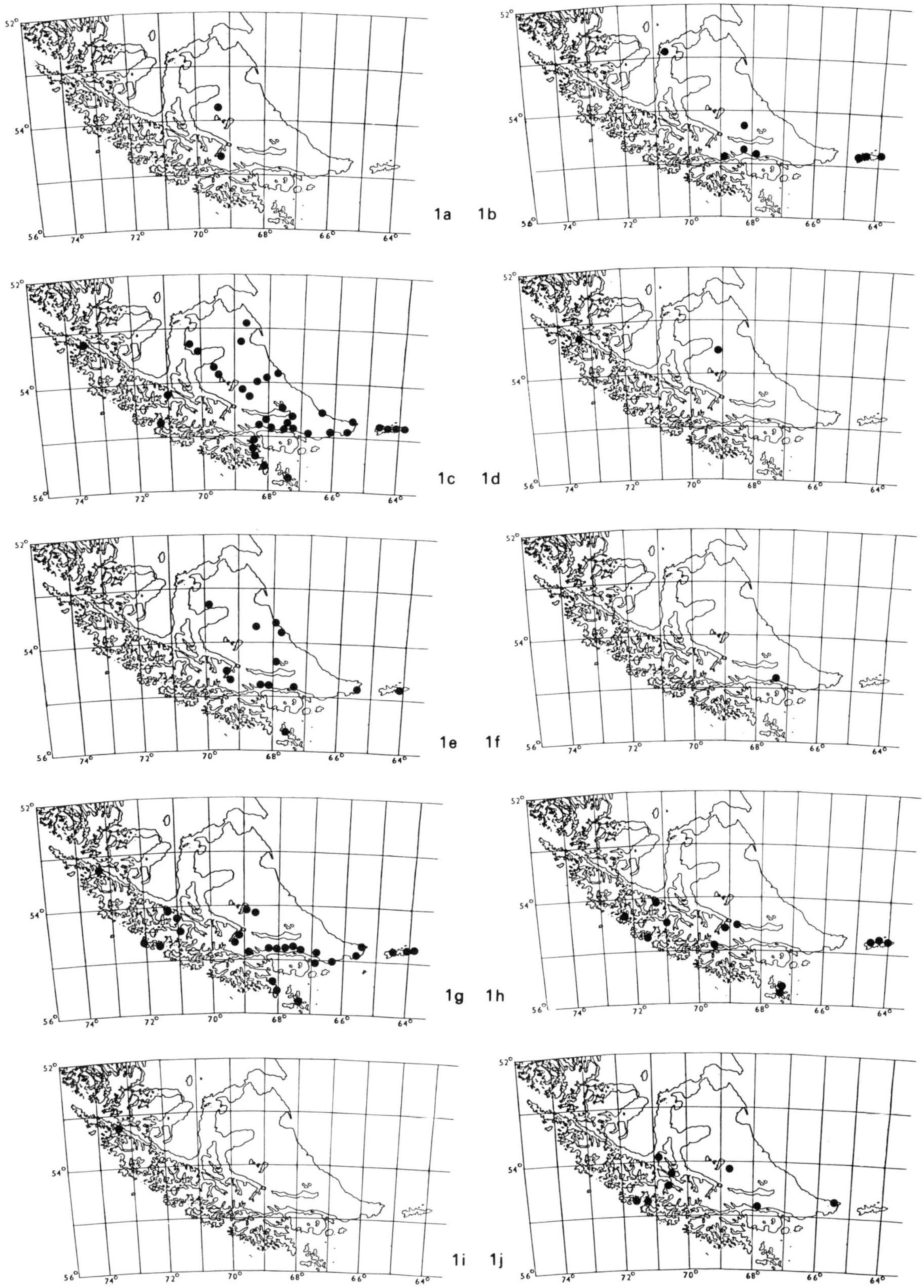

Map. 1. a, *Huperzia selago*. b, *Lycopodium confertum*. c, *Lycopodium magellanicum*. d, *Isoetes savatieri*. e, *Botrychium lunaria*. f, *Ophioglossum crotalophoroides*. g, *Grammitis magellanica*. h, *Hymenophyllum darwinii*. i, *Hymenophyllum pectinatum*. j, *Hymenophyllum magellanicum*.

1. *Grammitis* Swartz

Leaves glabrous, entire, rarely with irregular shallow lobes towards apex. Sori coalescing when mature.

Literature: E. R. de la Sota, *Opera Lilloana* **5**: 1–229(1960); *Darwiniana* **14**: 42–54(1966); B. S. Parris and D. R. Given, *N.Z. Jour. Bot.* **14**: 85–111(1976).

1. ***G. magellanica*** Desv. 1811, *Ges. Nat. Freunde Berl. Mag.* **5**: 313.
   G. nana Brack. 1854, *U.S. Expl. Exped.* **16**: 1.
   G. australis R. Br var. *nana* (Brack.) Franchet 1889, *Miss. Sci. Cap Horn* 397 – Spegazzini, 1896.
   G. billardieri Willd. var. *magellanica* (Desv.) de la Sota 1960, *Opera Lilloana* **5**: 208.
   Polypodium billardieri (Willd.) C. Chr. var. *magellanicum* (Desv.) C. Chr. 1910, *Ark. Bot.* **10(2)**: 15 – Skottsberg, 1916.

Scales on rhizome 1–3 × 0.2–0.5 mm, linear-lanceolate to -ovate, acuminate to obtuse, sometimes lacerate. Leaves (1.2–)2–11.5 × 0.2–0.75 cm, sublinear to subspathulate, obtuse, coriaceous, with indistinct midvein. Sori 2–3.5 mm, subglobose to elliptic-oblong, on distal half of lamina. (Fig. 9).

Damp rocks and rock-crevices, epiphytic on trunks of *Nothofagus* and *Drimys*; 0–670 m.

C. & S. Fuegia. (Map 1g).
W. Argentina north to *c.*41°S.lat.
Chile north to *c.*38°S.lat.
Falkland Islands (Islas Malvinas), South Georgia, Tristan da Cunha, Marion Island, S. South Africa, Iles Kerguelen, Iles Crozet, Macquarie Island, Campbell Island, Auckland Islands, Stewart Island, New Zealand, Tasmania, S.E. Australia.

In South American material of this species de la Sota (1960, 1966) has recognised two forms differing in habitat and spore-size. Plants occurring on rocks, usually at higher altitudes, have larger spores (*c.*40–54 µm in diameter), long, rarely shortly creeping rhizomes and leaves up to 5 cm (*G. magellanica* f. *nana*), while the epiphytes in the forests have spores 32–40 µm in diameter, erect to shortly creeping rhizomes and leaves more than 5 cm, and belong to the type form.

Some authors have separated these two taxa as distinct species, the former thus being called *G. poeppigiana* (Mett.) Pichi Serm. (Pichi Sermolli and Bizarri, 1978) or *G. armstrongii* Tindale (Parris and Given, 1976). However, we find that the distinguishing characters intergrade in Fuegian material, and the correlation of the modal morphological differences with distinct habitats, together with the anatomical similarity of the taxa (de la Sota, 1966), suggests ecotypic variation within the single species recognized here.

V. AZOLLACEAE

Small floating plants. Leaves small, in 2 alternate rows, imbricate, 2-lobed. Heterosporous; sori surrounded by indusium, in pairs on lower, submerged lobe of each leaf, the larger sori with several microsporangia, the smaller with 1 megasporangium.

Literature: G. Looser, *Rev. Univ.* **46**: 257–260(1961).

1. *Azolla* Lam.

Leaves with upper lobe herbaceous, the lower larger and hyaline. Microsporangia producing several groups of microspores (massulae) bearing branched hairs (glochidia); megasporangia producing 1 megaspore with 3 floats.

1. ***A. filiculoides*** Lam. 1783, *Encycl. Méth. Bot.* **1**: 343.

Stems up to 6 cm, slender, branched. Leaves 1–1.5 mm, oblong to ovate, glaucous to reddish, the lobes ovate. Microsporangial sori *c.*1.5 mm in diameter, globose, the megasporangial sori smaller, ovoid. (Fig. 10).

Still freshwater.

S. Fuegia (*c.*55°S.lat.). Widespread in N. & S. America, naturalized in Europe.

Reported once (Wildeman, 1905) from Fuegia at Bahía Orange, Península Hardy, Isla Hoste.

VI. HYMENOPHYLLACEAE

Plants small, with the habit of bryophytes and often growing with them. Rhizomes slender, extensively creeping. Leaves with translucent lamina, one cell thick except on the veins, usually deeply pinnatisect. Sori marginal, with receptacle ± enclosed by 2-valved to tubular indusium; spores tetrahedral.

Literature: J. Diem & J. S. de Lichtenstein, *Darwiniana* **11**: 611–760(1959)

a. Pinnae entire, glabrous, or pinnately divided and with hairs on veins; sori with 2-valved indusium, the valves separate for $\frac{1}{3}$ – $\frac{3}{4}$ of length
   1. *Hymenophyllum*

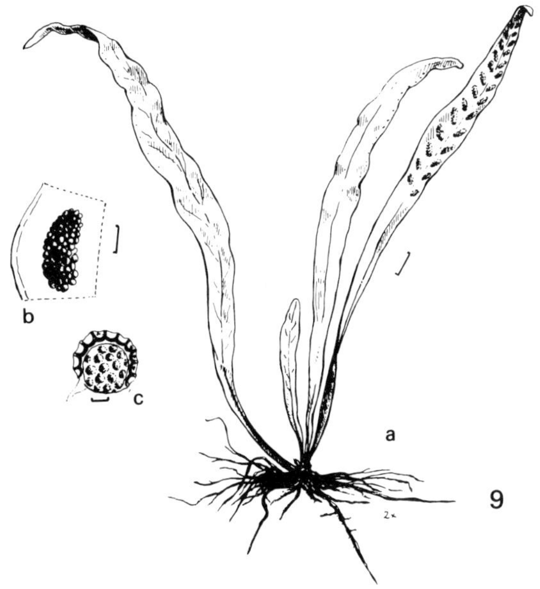

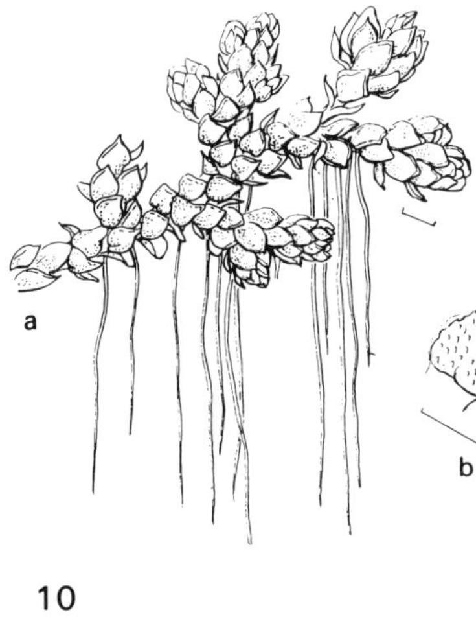

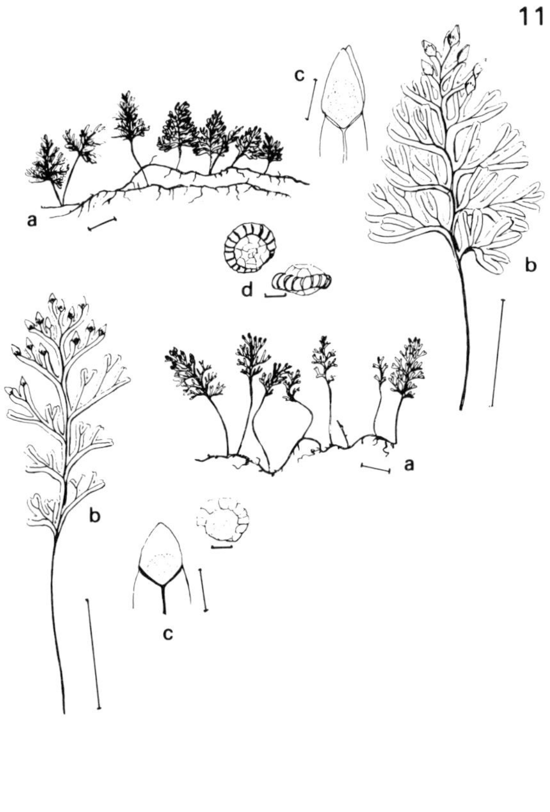

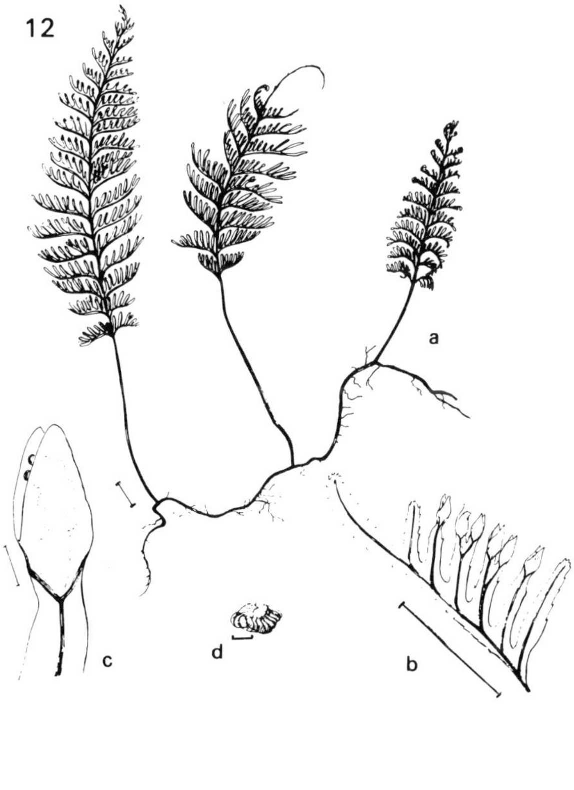

Fig. 9. *Grammitis magellanica*. a, Plant. b, Sorus. c, Spore. Scale = 1 cm (a), 1 mm (b) or 1 μm (c). Fig. 10. *Azolla filiculoides*. a, Plant. b, Sorus. Scale = 1 mm. Fig. 11. *Hymenophyllum darwinii*. a, Plant. b, Fertile pinna. c, Sorus. d, Sporangia. Scale = 1 cm (a–c) or 0.1 mm (d). Fig. 12. *Hymenophyllum pectinatum*. a, Plant. b, Fertile pinna. c, Sorus. d, Sporangium. Scale = 1 cm (a–b), 1 mm (c) or 0.1 mm (d).

(R. N. P. Goodall)

a. Pinnae entire, with hairs on veins; sori with tubular indusium, divided into 2 lobes for $\frac{1}{6}$ of length **2. *Serpyllopsis***

1. *Hymenophyllum* Sm.

Rhizome usually glabrous, smooth. Leaves with pinnae divided, usually glabrous, persistent. Sori with 2-valved indusium, the valves separate for $\frac{1}{3}-\frac{3}{4}$ of their length. Receptacle fertile at the apex.

a. Leaves hairy **10. *ferrugineum***
a. Leaves glabrous:
 b. Sori inserted ± perpendicular to plane of leaf-lamina, arcuate at base; receptacle almost completely covered with sporangia:
  c. Leaf-segments on both sides of pinnae **9. *dentatum***
  c. Segments only on side of pinna directed towards leaf-apex:
   d. Pinnae and segments horizontal, rigid, twisted; segments usually not more than 1 mm wide **6. *secundum***
   d. Pinnae and segments oblique or vertical, flexuous; segments usually more than 1 mm wide:
    e. Sori usually 1 per pinna; lower pinnae with 2–3 segments; petiole not winged **7. *falklandicum***
    e. Sori (1) 2–5 per pinna; lower pinnae with 3–8 segments; petiole usually narrowly winged when young. **8. *peltatum***
 b. Sori inserted in same plane as leaf-lamina, not arcuate at base; receptacle partly covered with sporangia, always bare in lower part:
  f. Leaf-segments with entire margin **1. *darwinii***
  f. Leaf-segments with dentate or ciliate margin:
   g. Segments only on side of pinna directed towards leaf-apex; receptacle filiform, included **2. *pectinatum***
   g. Segments on both sides of pinnae; receptacle cylindrical, exserted:
    h. Leaves flat or slightly undulate **3. *magellanicum***
    h. Leaves crispate:
     i. Leaf-segments undulate-denticulate; sori ovate; indusium plicate, acuminate at apex, 2- to 4-fid, without cilia **5. *plicatum***
     i. Leaf-segments dentate-ciliate; sori broadly ovate; indusium not plicate, truncate at apex, entire, strongly ciliate **4. *tortuosum***

1. *H. darwinii* (Hooker f.) Van den Bosch 1863, *Ned. Kruidh. Arch.* **5(3)**: 157.
H. abruptum Hooker f. 1874, in Hooker & Baker, *Syn. Fil.* **1**: 88 – Franchet, 1889; Spegazzini, 1896.
H. skottsbergii C.Chr. 1910, *Ark. Bot.* **10(2)**: 22.
H. rarum auct., non R.Br. – Hooker, 1847.

Rhizome flexuous, blackish, glabrous. Lamina 7–15 × 8–11 mm, ovate to linear or deltoid in outline, 2- to 3- pinnatisect, with 5–12 usually alternate pinnae per side, sometimes imbricate in depauperate specimens; rhachis winged at least in lower part; segments 0.7–1 mm wide, linear, usually very wrinkled and tortuous, sometimes ± crispate, truncate or emarginate at apex, bright to yellowish-green; petiole 5–25 mm, glabrous, black, shiny, with small tufts of dark red hairs at base. Sori terminal, more frequent near apex and on upper segments; indusium 1–2 × 0.8–1 mm, rhomboid to ovate, flat, entire, divided for up to $\frac{2}{3}$ of length; valves acute to subacute, rarely rounded; receptacle $\frac{1}{2}-\frac{2}{3}$ as long as indusium, filiform, very slender; sporangia 8–15, 0.3–0.4 mm in diameter. (Fig. 11).

Floor of *Drimys – Nothofagus betuloides* forest: 0–40 m.

S. Fuegia. (Map 1h).
W. Argentina north to *c.*40°S.lat.
Chile north to *c.*51°S.lat., and *c.*42 – 39°S.lat.

2. *H. pectinatum* Cav. 1802, *Descr. Pl.* 275.

Rhizome rigid, blackish, with small whitish hairs when young. Lamina 30–130 × 15–40 mm, lanceolate, rarely linear, in outline, 2-pinnatisect, with 8–20 pinnae per side; rhachis with whitish hairs on lower side; segments *c.*1 mm wide, linear, somewhat obtuse, 8–10 on one side of pinna, ± parallel, dentate, especially at apex of infertile segments, green or somewhat yellowish; petiole 10–80 mm, with whitish hairs when young, later glabrous. Sori terminal, borne at apex of segments, most segments fertile on upper pinnae; indusium 1.5–2 × *c.*0.8 mm, obovate, entire, divided to $\frac{2}{3}$ of length; valves obtuse to somewhat acuminate; receptacle up to $\frac{3}{4}$ as long as indusium, slightly thickened in middle, covered with sporangia for distal $\frac{2}{3}$; sporangia 35–45, *c.*0.3 mm, in diameter. (Fig. 12).

Floor of *Nothofagus betuloides* forest, and epiphytic on trees. 0–30 m.

W. Fuegia (Isla Desolación). (Map 1i).
W. Argentina north to *c.*40°S.lat.

Chile north to *c*.39°S.lat.
Juan Fernandez.

3. *H. magellanicum* (Desv.) Willd. ex G. Kunze 1847, *Bot. Zeit.* **5**: 226.

Rhizome rigid, blackish, glabrous. Lamina 50–150 × 40–150 mm, ovate-lanceolate to deltoid in outline, 2- to 3-pinnate to 4-pinnatifid, with (3–)5–12 pinnae per side, the lowest pinnae with long petiolules forming wide angle with the rachis, the veins blackish; segments 0.5–1 mm wide, linear, somewhat obtuse, flat, serrate-dentate, with short, slender, acuminate teeth on both sides of pinna; petiole 40–150 mm, with dentate, undulate wings 1–1.5 mm wide and sparse reddish hairs at base. Sori in upper part of lamina, inserted on subaxillary segments; indusium 1.5–2 × 0.5–0.75 mm, oblong, the upper margin ± serrate-dentate, divided to $\frac{1}{3}(-\frac{1}{2})$ of length; receptacle 3–3.5 mm, exserted $\frac{1}{3}$ or more, slender, somewhat thickened at base; sporangia 20–30, *c*.0.3 mm in diameter.

Wet coastal scrub; 0–40 m.

C. & S. Fuegia except the extreme west. (Map 1j).
W. Argentina north to *c*.41°S.lat.
Chile north to *c*.40°S.lat.
Juan Fernandez.

4. *H. tortuosum* Hooker & Grev. 1831, *Icon. Fil.* **2**: pl.129.

H. nigricans auct., non Colla – Franchet, 1889; Spegazzini, 1896.

Rhizome flexuous, dark brown, with hairs up to 1.5 mm. Lamina (20–)30–100(–140) × 10–60 mm, lanceolate to ovate or deltoid in outline, slightly asymmetrical, 3-pinnate to -pinnatifid, with 5–14 pinnae per side, the lower at least 1 cm, with almost black nerves; segments 0.5–1(–2) mm wide, obtuse, the margin plicate or tortuous, dentate and finely ciliate, dark green, on both sides of pinna; petiole (10–)22–75(–90) mm, with crispate, dentate-ciliate wings. Sori subaxillary and terminal, in middle and upper part of lamina, inserted on somewhat reduced segments; indusium 1.5–2.5 × 1–1.5 mm, narrowly obovate to oblong, obtuse, with swollen base when young, the margin finely ciliate, divided for up to $\frac{1}{3}$ of length, with very narrow mouth between valves; receptacle up to 3 mm, slender, somewhat thickened at base, covered with sporangia in upper $\frac{1}{3}$; sporangia 20–50, 0.3–0.4 mm in diameter. (Fig. 13).

Bogs, on floor and trunks of wet *Nothofagus* forest, wet rock faces; 50–900 m.

C. & S. Fuegia. (Map 2a).
W. Argentina and Chile north to *c*.41°S.lat.
Juan Fernandez, Falkland Islands (Islas Malvinas).

The above refers to var. *tortuosum*. Var. *glomerata* Diem & Lichtenstein, characterized by its smaller, very crispate lamina with all the segments densely clustered, is reported from Isla Dawson.

5. *H. plicatum* Kaulf. 1824, *Enum. Fil.* 268.

H. dichotomum auct., non Cav. – Skottsberg, 1916.

Like *H. tortuosum* but leaf-segments not ciliate; indusium plicate, acuminate and without cilia at apex.

Recorded by Dusén (1900) from W. Fuegia (Isla Desolación).
W. Argentina *c*.42°–41°S.lat.
Chile 51°20′–*c*.37°30′S.lat.

6. *H. secundum* Hooker & Grev. 1831, *Icon. Fil.* **2**: pl.133.

H. tunbridgense auct., non (L.)Sm. – Alboff, 1902.

Rhizome flexuous, blackish, when young covered with small hairs which persist only at the base of petioles, otherwise glabrous. Lamina 20–120 × 5–30 mm, obovate to lanceolate in outline, usually recurved, sometimes truncate at base, 2- to 3-pinnate; pinnae 3–20 per side, rigid, usually markedly arcuate; rhachis slender, with narrow, entire wings in upper half; pinnules and segments ± flabelliform; segments 0.7–0.9 mm wide, linear to lanceolate, obtuse, often bifurcate, usually flat, with acute teeth, bright green with dark bluish metallic sheen, only on side of pinna directed to apex; petiole 5–50 mm, with narrow entire wing above. Sori in upper part of lamina, inserted on very reduced segments, usually those nearest the rhachis, 1–7 per pinna, arcuate at base; indusium 1.5–2 × 0.5–1 mm, oval, convex, flattened in lower part, entire, divided for up to $\frac{2}{3}$ of length, the valves widely separated when mature, dark brown to almost black; receptacle *c*.$\frac{3}{4}$ as long as indusium, cylindrical, somewhat flattened; sporangia 25–45, *c*.0.3 mm in diameter. (Fig. 14).

Bogs, on floor and trunks of wet forest, stream-sides; 0–700 m.

S. & W. Fuegia. (Map 2b).
W. Argentina north to *c*.40°31′S.lat.
Chile north to *c*.42°S.lat.
Juan Fernandez.

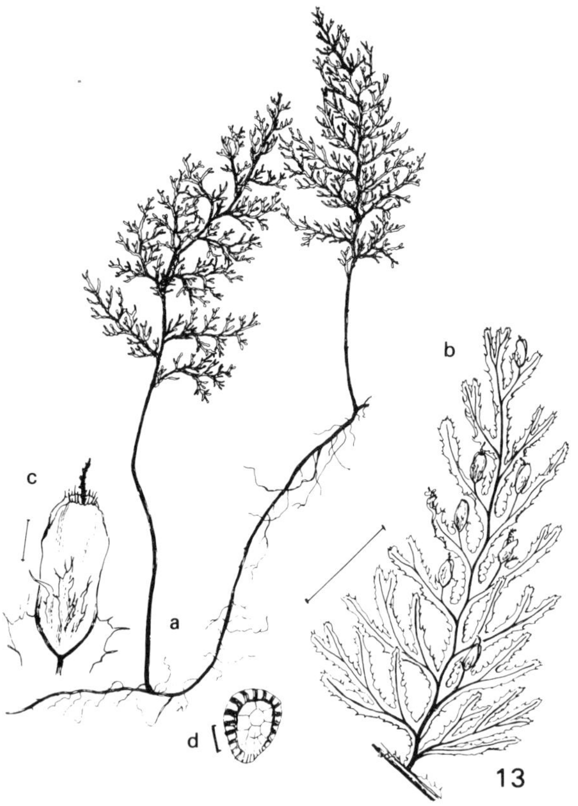

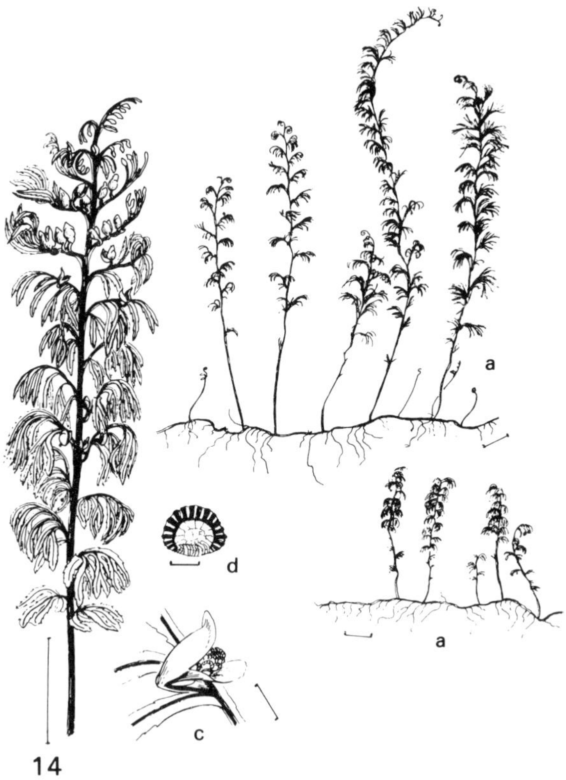

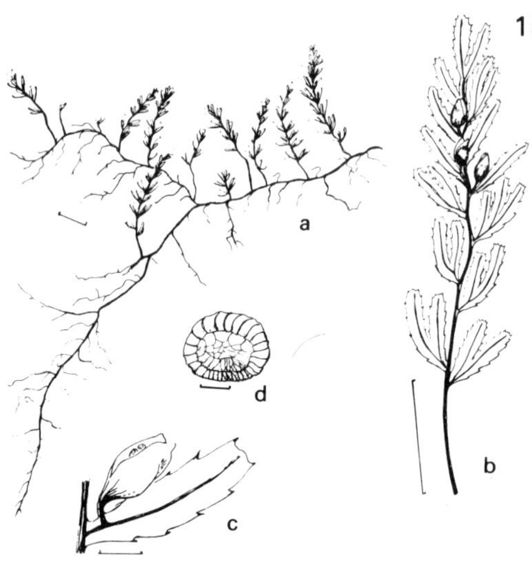

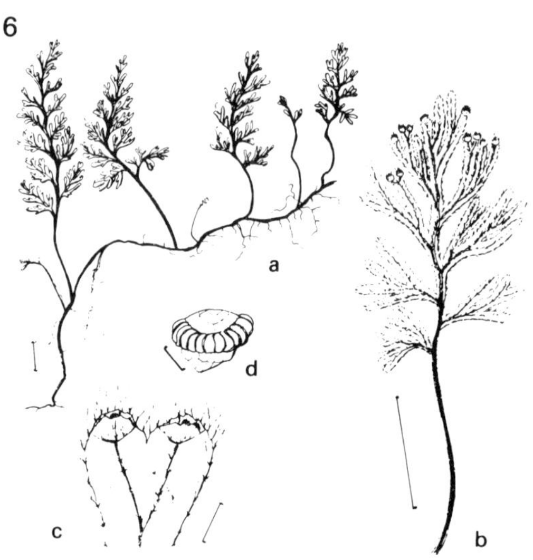

Fig. 13. *Hymenophyllum tortuosum*. a, Plant. b, Fertile pinna. c, Sorus. d, Sporangium. Scale = 1 cm (a–b), 1 mm (c) or 0.1 mm (d). Fig. 14. *Hymenophyllum secundum*. a, Plant. b, Fertile leaf. c, Sorus. d, Sporangium. Scale = 1 cm (a–b), 1 mm (c) or 0.1 mm (d). Fig. 15. *Hymenophyllum falklandicum*. a, Plant. b, Fertile leaf. c, Sorus. d, Sporangium. Scale = 1 cm (a–b), 1 mm (c) or 0.1 mm (d). Fig. 16. *Hymenophyllum ferrugineum*. a, Plant. b, Fertile leaf. c, Sorus. d, Sporangium. Scale = 1 cm (a–b), 1 mm (c) or 0.1 mm (d).

(R. N. P. Goodall)

7. *H. falklandicum* Baker 1874, in Hooker & Baker, *Syn. Fil.*, ed.2, 68.

H. caespitosum auct., non Gaudich. – Dusén, 1900.
H. tunbridgense auct., non (L.) Sm. – Alboff, 1902
H. wilsonii auct., non Hooker – Hooker, 1847.
Trichomanes sibthorpioides auct., non Bory – Hicken, 1921.
T. flabellatum auct., non Van den Bosch – Alboff, 1902.

Rhizome rigid, brown, glabrous or with a few hairs at base of petiole. Leaves thin, dense, forming dense mats; lamina 20–40(–55) × 7–15(–20) mm, linear-lanceolate, 2-pinnatifid; pinnae 3–10 per side, subopposite to alternate, oblique, at least the apical with 2–3 segments, on side towards apex; rhachis flexible, almost black, finely winged towards apex; segments 2–5 × 1–1.6 mm, truncate, acutely dentate, usually flat, brownish to dark green; petiole 4–12 mm, smooth. Sori 1(–3) per pinna, up to 12 per leaf, inserted on very short, narrow segments, strongly arcuate at base; indusium *c.*1.5 × 1 mm, ovate-lanceolate, sometimes acuminate, convex, entire, rarely slightly pinnate, divided for up to $\frac{2}{3}$ of length, dark brown when mature; receptacle up to $\frac{1}{2}$–$\frac{2}{3}$ as long as indusium, cylindrical to slightly clavate; sporangia 20–30, 0.43–0.45 mm in diameter. (Fig. 15).

Floor of *Nothofagus* forest, wet rocks; 0–1000 m.

S. & W. Fuegia. (Map 2c).
W. Argentina north to *c.*40°S.lat.
Chile north to 51°20'S.lat.
Juan Fernandez, Falkland Islands (Islas Malvinas), South Georgia.

H. nahuelhuapiense Diem & Lichtenstein 1959, *Darwiniana* **11**: 722, has been doubtfully reported from Tierra del Fuego (near Ushuaia) on the basis of sterile material. It is distinguished from *H. falklandicum* by having the lower pinnae with less dentate segments on both sides.

8. *H. peltatum* (Poiret) Desv. 1827, *Mém. Soc. Linn. Paris* **6(3)**: 333.

H. menziesii C. Presl. 1843, *Hymenoph.* 51.

Rhizome rigid, dark brown, sometimes sparsely hairy. Lamina 25–120 × 10–30 mm, lanceolate to lanceolate-ovate in outline, 2-pinnate to -pinnatifid; pinnae 5–14 per side, alternate, gradually arcuate, especially the lower, somewhat decurrent; segments 0.5–1.2 mm wide, sometimes slightly imbricate, the margin incurved, dentate, on side of pinna towards apex; petiole 6–24 mm, somewhat villous, sometimes winged. Sori 1–6 per pinna, on the middle and apical pinnae, inserted on segments near the rhachis, strongly arcuate at base; indusium 1.6–2.5 × 0.75–1.3 mm, ± ovoid, arcuate near base, entire, divided to 1/2–2/3 of length; receptacle included, up to $\frac{3}{4}$ as long as indusium, cylindrical, slender; sporangia 15–35, 0.38–0.4 mm in diameter.

Floor of wet forest; 0–40 m.

S. Fuegia, except the extreme west. (Map 2d).
W. Argentina north to *c.*41°S.lat.
Chile north to *c.*40°S.lat. and at *c.*31°S.lat.

9. *H. dentatum* Cav. 1801, *Descr. Pl.* 276.

H. bridgesii Hooker 1844, *Sp. Fil.* **1**: 97 – Spegazzini, 1896; Dusén, 1900.
H. trichochaulon Phil. 1896, *Anal. Univ. Chile* **94**: 360.

Rhizome dark brown, covered with brownish hairs. Lamina 80–100 × 40–60 mm, broadly obovate to deltoid in outline, subacuminate, 3-pinnate; pinnae 8–12 per side; segments 1–3 × 0.75 mm, linear, obtuse, usually with incurved margin, shallowly dentate, on both sides of pinna; petiole up to 25–60 mm, with dense reddish-brown hairs. Sori subaxillary, most frequently on upper pinnae, strongly arcuate at base; indusium 1–1.5 × 0.7–1 mm, ovate to suborbicular, divided almost to base, entire or rarely somewhat denticulate, the outer valve with 2–3 longitudinal crests reaching more than halfway up indusium and ending in a spur; receptacle 0.7–1.3 mm, cylindrical or clavate, usually not reaching upper margin of indusium, totally covered with sporangia; sporangia 18–30, 0.32–0.35 mm in diameter.

Wet rocks and slopes; 0–*c.*80 m.

S. Fuegia. (Map 2e).
W. Argentina north to *c.*40°S.lat.
Chile north to *c.*38°S.lat.

10. *H. ferrugineum* Colla 1835, *Mem. Accad. Sci. Torino* **39**: 30.

H. subtilissimum G. Kunze 1837, *Analecta Pterid.* 49 – Hariot, 1884.
H. aeruginosum (Poiret) Carmichael 1818, *Trans. Linn. Soc. Lond.* **12**: 513 – Dusén, 1900.

Rhizome flexuous, somewhat reddish-brown, with golden hairs, at least when young. Lamina 30–110 × 20–30 mm, oblong to lanceolate in outline, 2- to 3-pinnate, 6–10 pinnae per side, the pinnules and segments on both sides of pinnae, with reddish-brown hairs on both sides of lamina; segments 1–1.4 mm wide, linear, often divided, rounded at apex, with entire margin bearing fine stellate hairs inserted on very small protuberances; petiole up to 6.5 mm, densely hairy when young. Sori small, terminal, on upper part of lamina, inserted on symmetrical segments; indusium 0.6–0.8(–1.6) mm in diameter, suborbicular, submerged in segment, the base included in the branches of the bifurcate nerve and inserted across width of the segment, the margins free, sinuate, with long, fine, stalked, stellate hairs; receptacle included, globose in upper half which is covered with a few sporangia; sporangia 8–16, 0.25–0.28 mm, in diameter. (Fig. 16).

On floor and trunks of *Drimys – Nothofagus betuloides* forest, wet rock-crevices; 0–600 m.

S. Fuegia. (Map 2f).
W. Argentina and Chile north to *c.*40°40′S.lat.

Fuegian collections have been referred to var. *donatii* Looser, but the material now available suggests that it is not useful to maintain this distinct from var. *ferrugineum*.

2. *Serpyllopsis* Van den Bosch

Like *Hymenophyllum* but with long, reddish hairs on rhizome, and veins and apex of lamina, with more elongate, simple, entire pinnae, and with the indusium tubular, having valves separate for up to $\frac{1}{6}$ of length and dentate at apex.

1. ***S. caespitosa*** (Gaudich.) C.Chr. 1910, *Ark. Bot.* **10(2)**: 29.
   H. caespitosum Gaudich. 1825, *Ann. Sci, Nat.* **5**: 99.
   H. dusenii Christ 1900, in Dusén, *Svenska Exped. Magellansl.* **3(5)**: 242.
   Trichomanes caespitosum (Gaudich.) Hooker 1846, *Sp. Fil.* **1**: 132 – Hooker, 1847, Franchet, 1889, Spegazzini, 1896.

Rhizome up to 50 cm, ± flexuous, dark brown. Leaves 7–20(–24) × (2.5–)3–8 mm, ovate-oblong, pinnate, with 4–10 alternate to subopposite, imbricate pinnae on each side; pinnae 4–6 × 1–1.5 mm, oblong to narrowly elliptical, obtuse, with entire, involute margin; petiole 2–6 mm. Sori near apex of leaf, 1–3 terminal on lobes; indusium 0.8–3 × 0.7–1.5 mm, obovate, truncate and dentate at apex, clothed with long, reddish hairs; receptacle clavate, exserted; sporangia 12–20, ± 0.5 mm in diameter. (Fig. 17).

Open, wet rock screes, bogs, feldmark, cliffs, on trunks in wet forests; 20–1000 m

S. Fuegia. (Map 2g).
W. Argentina and Chile north to *c.*40°S.lat.
Juan Fernandez, Falkland Islands (Islas Malvinas).

VII. ASPLENIACEAE

Rhizome short, erect, covered with clathrate scales. Leaves in apical tufts; veins free. Sori borne along one side of fertile veins, covered with indusium when young. Indusium resembling sorus in size and shape, opening towards vein; sporangia long-stipitate; spores bilateral.

1. *Asplenium* L.

Leaves glabrous or sparsely glandular hairy; lamina 2- to 3-pinnate. Sori elliptical to ovate; indusium slightly smaller than sorus.

1. ***A. dareoides*** Desv. 1811, *Ges. Nat. Freunde Berlin Mag.* **5**: 322.
   A. magellanicum Kaulf. 1824, *Enum. Fil.* 175 – Hooker, 1847; Franchet, 1889.

Rhizomes with linear-lanceolate, dark brown scales 2–4 mm. Leaves with lamina 15–60(–75) × (10–)15–25(–35) mm; pinnules with 2–4(5) lobes, sometimes somewhat dentate at apex; petiole (15–)25–60(–115) mm, brownish black at base, green above. Sori 1–2(3) per pinnule, *c.*1.5 × 1 mm; indusium whitish. (Fig. 18).

Shaded rock-crevices and stream banks, moist scrub, floor and fallen logs in moist scrub and most forest communities; 0–500 m.

Throughout C. & S. Fuegia. (Map 2h).
W. Argentina at *c.*36°48′S.
Chile north to *c.*40°S.lat.
Juan Fernandez, Falkland Islands (Islas Malvinas).

VIII. ATHYRIACEAE

Rhizome short, decumbent, covered with soft scales. Leaves delicate, dying in winter; veins free. Sori in 2 rows, one on each side of midvein of pinnule. Indusium differing from sorus in shape, attached by its base below sorus; sporangia with slender stipes; spores bilateral, echinate.

1. *Cystopteris* Bernh.

Leaves glabrous; lamina 2- to 3-pinnate. Sori orbicular; indusium ovate-lanceolate, cucullate, becoming deflexed and shrivelled as sporangia mature.

1. ***C. fragilis*** (L.) Bernh. 1806, *Neues Jour. Bot.* **1(2)**: 27.

Rhizome with lanceolate, acuminate, thin, brown

scales 2–4 mm. Leaves with lamina 40–220 × (20–)30–50(–60) mm, lanceolate, acute, with 15(–17) alternate to subopposite pinnae on each side; pinnae ovate-lanceolate, subsessile, the second pair from the base usually the longest; pinnules ovate to oblong, acute, cuneate at base, dentate to deeply pinnatifid; petiole 15–100 mm. Sori *c.*0.4 mm; indusium 0.5–1 mm, whitish. (Fig. 21).

Moist rock-crevices and mossy ledges, floor and fallen logs in moist scrub and most forest communities, montane cushion plant communities; 0–600 m.

C. & C.S. Fuegia. (Map 2i).
W. Argentina north to *c.*36°48′S.lat.
Chile north to *c.*30°S.lat.
Falkland Islands (Islas Malvinas), South Georgia, Iles de Kerguelen, New Zealand, S.E. Australia, North America, Eurasia, Africa.

Plants in the circum-Antarctic area from Fuegia to Australasia have been distinguished as var. *apiiformis* (Gand.) C.Chr. However, it is not clear how valid this taxon is within the wide range of variation shown by this cosmopolitan species.

IX. ASPIDIACEAE

Rhizome ascending to erect, with opaque scales. Leaves 1- to 3-pinnate; veins free. Sori superficial; indusium attached at base; spores bilateral.

1. *Polystichum* Roth

Rhizome short, branched; scales dense, lacerate. Leaves tufted, coriaceous. Sori orbicular; indusium peltate, orbicular, persistent.

a. Lamina not more than 2 cm wide; pinnae usually lobed **1. *andinum***
a. Lamina at least 5 cm wide; pinnae 1- to 2-pinnate:
 b. Longest pinnae not more than 9 cm, pinnate; pinnules lobed to shallowly pinnatifid **2. *mohrioides***
 b. Longest pinnae at least 11 cm, 2-pinnate; at least basal pinnules on pinnae cut to midvein **3. *multifidum***

1. *P. andinum* Phil. 1858, *Linnaea* **29**: 108.
P. mohrioides var. *plicatum* C.Chr. 1940, *Ark. Bot.* **10(2)**: 17.
P. plicatum auct., non (Poeppig ex G. Kunze) Hicken - Diem, 1958.

Leaves with lamina 6–14(–21.5) × 1–1.5(–2) cm, oblong to oblong-lanceolate, pinnate; pinnae 5–10 × 4–7 mm, ovate to triangular, obtuse, pinnatifid, rarely pinnatisect, to crenate near base, becoming decreasingly dissected towards apex; petiole with dense, pale brown, ovate-lanceolate, long-acuminate, serrate-fimbriate scales 4–7 × 1–2 mm. Sori *c.*0.5 mm in diameter, in 1–2 rows along each side of midribs of pinnae in distal $\frac{2}{3}$–$\frac{1}{2}$ of lamina, often coalescing when mature; sporangia dark brown. (Fig. 19).

Rock screes, cliffs, feldmark; 800–1100 m.

S. part of Isla Grande, Isla Desolación. (Map 3a).
W. Argentina and E. Chile north to *c.*40°S.lat.

2. *P. mohrioides* (Bory ex D'Urv.) C. Presl. 1836, *Tent. Pteridogr.* 83.
Aspidium mohrioides Bory ex D'Urv. 1825, *Fl. Is. Mal.* 26 – Hooker, 1847; Hariot, 1884; Dusén, 1900.
A. plicatum Poeppig ex G. Kunze 1834, *Linnaea* **9**: 94.
A. coriaceum auct., non Swartz – Spegazzini, 1896.
Polystichum plicatum (Poeppig ex G. Kunze) Hicken 1915, in Hosseus, *Trab. Inst. Bot. Farm.* **33**: 9.

Leaves with lamina 17–59 × 5.5–16 cm, elliptic- to ovate-lanceolate in outline, 2-pinnate; pinnae 2.8–8.5 × 1.2–2.3 cm, lanceolate to oblong-lanceolate in outline, pinnate; pinnules 7.5–12 × 4.3–7 mm, ± ovate to subrhomboid, acute, obtusely lobed to shallowly pinnatifid; petiole 3–18 cm, with dense scales, the scales 5–12(–25) × 1–5.5 mm, ovate-lanceolate to ovate, acuminate, dark brown with paler margins. Sori *c.*0.5 mm in diameter, on veins to lobes of pinnules on pinnae in distal half of lamina, usually coalescing when mature; sporangia dark brown. (Fig. 22).

Rock-crevices, cliffs, stream banks, moist slopes in *Nothofagus* forest; 0–450 m.

C & S. Fuegia, except the extreme west. (Map 2j).
W. Argentina and E. Chile north to *c.*37°S.lat.
Falkland Islands (Islas Malvinas), South Georgia.

Fuegian material belongs to var. ***elegans*** (Remy ex C. Gay) C.Chr. 1910, *Ark. Bot.* **10(2)**: 17. Var. *mohrioides* is known only from the Falkland Islands (Islas Malvinas) and South Georgia.

3. *P. multifidum* (Mett.) T. Moore 1857, *Ind. Fil.* **1**: 84.
Aspidium multifidum Mett. 1856, *Fil. Lechl.* **1**: 20; Dusén, 1900.
Polystichum vestitum auct., non (Forster f.) C. Presl. – Hooker 1847; Spegazzini, 1896.
P. orbiculatum auct., non (Desv.) Gay – Dusén, 1900; Skottsberg, 1916.

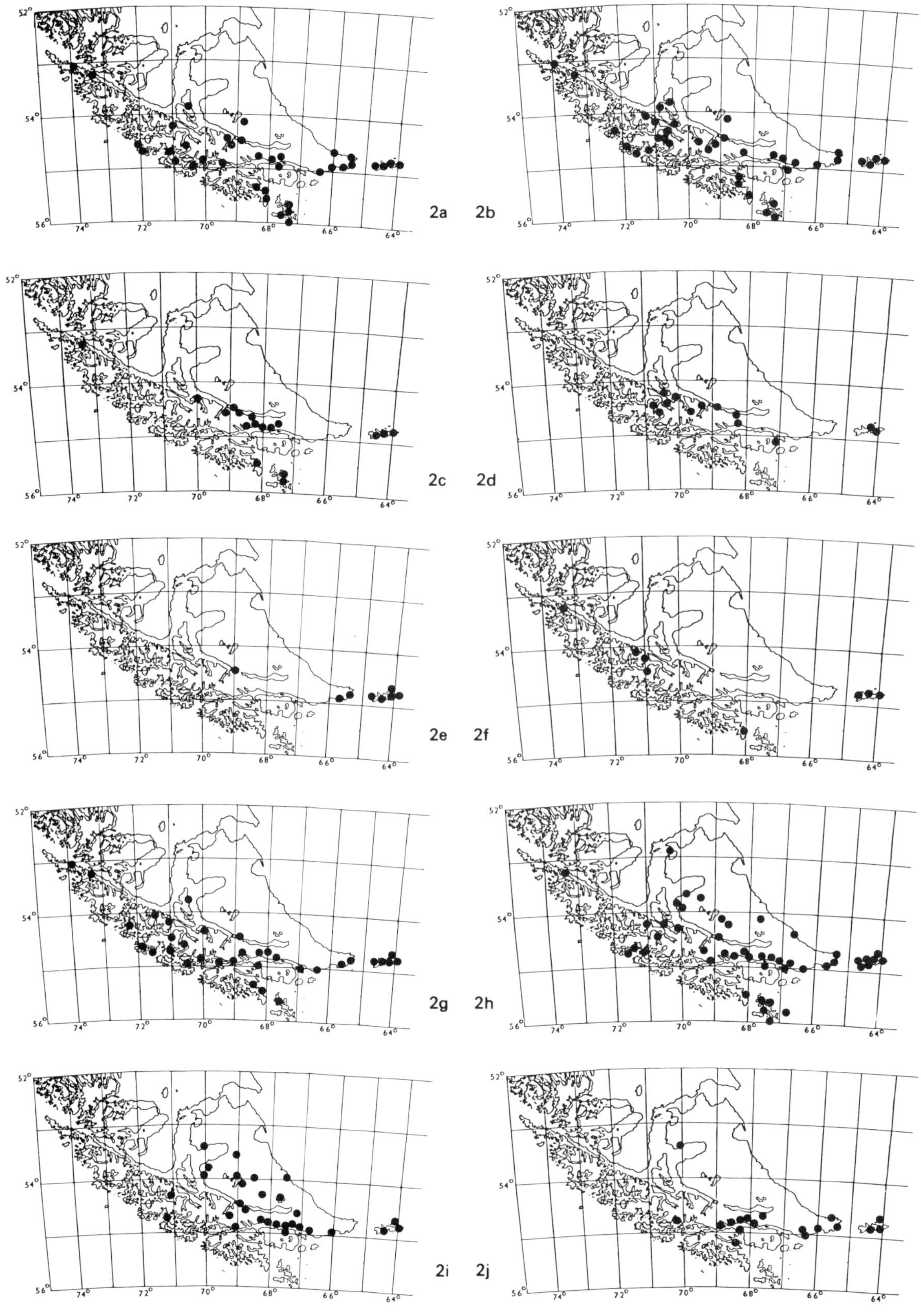

Map 2. a, *Hymenophyllum tortuosum*. b, *Hymenophyllum secundum*. c, *Hymenophyllum falklandicum*. d, *Hymenophyllum peltatum*. e, *Hymenophyllum dentatum*. f, *Hymenophyllum ferrugineum*. g, *Serpyllopsis caespitosa*. h, *Asplenium dareoides*. i, *Cystopteris fragilis*. j, *Polystichum mohrioides*.

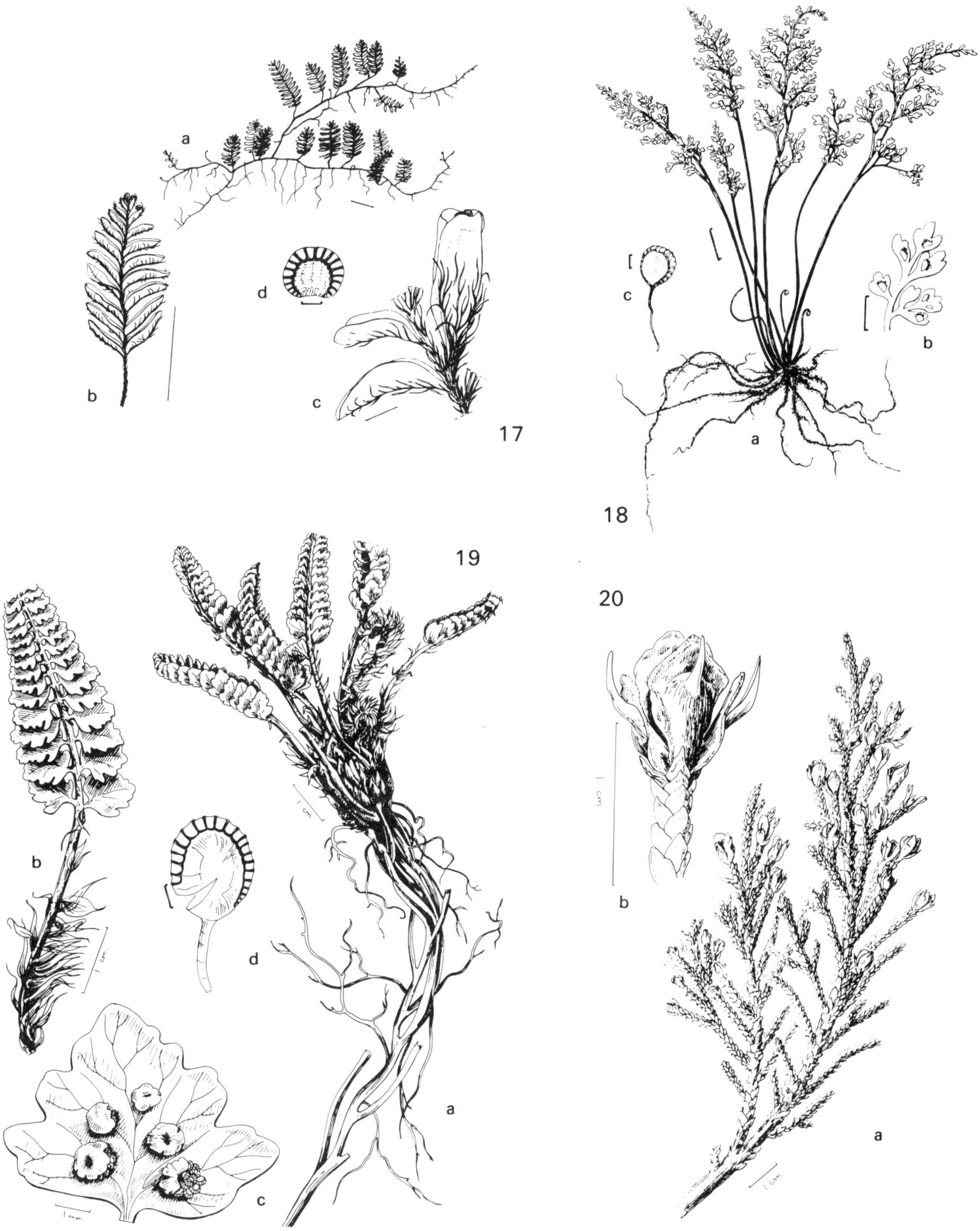

Fig. 17. *Serpyllopsis caespitosa*. a, Plant. b, Fertile leaf. c, Sorus. d, Sporangium. Scale = 1 cm (a–b), 1 mm (c) or 0.1 mm (d). Fig. 18. *Asplenium dareoides*. a, Plant. b, Fertile pinna. c, Sporangium. Scale = 1 cm (a), 5 mm (b) or 0.1 mm (c). Fig. 19. *Polystichum andinum*. a, Plant. b, Leaf. c, Fertile pinna. d, Sporangium. Scale = 1 cm (a–b), 1 mm (c) or 0.1 mm (d). Fig. 20. *Pilgerodendron uviferum*. a, Fertile branch. b, Female cone. Scale = 1 cm.

(R. N. P. Goodall)

Leaves with lamina 57–65 × 16–19 cm, elliptic- to ovate-lanceolate in outline, 3-pinnate; pinnae 11–12 × 2–2.6 cm, lanceolate, acuminate, 2-pinnate; pinnules 13–16 × 6–8 mm. lanceolate, at least the basal cut to the midvein into linear-lanceolate to ovate, obtuse, often lobed segments; petiole 20–23 cm, with scales at base, the scales 4–24 × 1–4 mm, ovate, acuminate, dark brown with transparent margin. Sori 0.4–0.8 mm in diameter, several on ultimate segments in distal half of lamina, usually coalescing when mature; sporangia brown. (Fig. 23).

Coastal slopes and cliffs, usually with shrubs or trees; 0–30 m.

S. Isla Grande and adjacent islands to south; Isla de los Estados. (Map 3b).
W. Argentina and E. Chile north to *c.*26°S.lat.

Rumohra adiantiformis (Forster f.) Ching 1934, *Sinensia* **5**: 70, was recorded (under *Polystichum adiantiforme* (Forster f.) Sm.) as reaching Fuegia by Skottsberg (1916). No material or locality was cited and there are no other reports.

It may be confused with *Polystichum mohrioides* and perhaps *P. multifidum*, but is readily distinguished by its long, creeping rhizome clothed with peltate, not basifixed, scales. It occurs in Chile north of *c.*43°S. lat. extending to C. America, and in Australia, New Zealand, New Guinea, South Africa, Madagascar, Mauritius and the Falkland Islands (Islas Malvinas).

X. BLECHNACEAE

Rhizome erect, often produced into an erect "trunk", or creeping, clothed with scales. Leaves usually coriaceous; veins free, forked. Sori borne on distinct, fertile leaves, in 2 continuous bands throughout the length of each pinna, protected by incurved leaf-margin when young.

1. *Blechnum* L.

Leaves with scales or hairs at least near base; lamina pinnate. Sori coalescing when mature.

a. Rhizome creeping; petiole naked or sparsely covered with scales at base; sterile pinnae ± twice as long as wide **1. *penna-marina***
a. Rhizome erect, often produced into an erect "trunk"; petiole covered with rigid hairs at base; sterile pinnae more than twice as long as wide **2. *magellanica***

1. *B. penna-marina* (Poiret) Kuhn 1868, *Fil. Afr.* 92.
Lomaria alpina Sprengel 1827, *Syst. Veg.*, ed. 16, **4(2)**: 62 – Hooker, 1847; Hariot, 1884; Spegazzini, 1896.

Rhizome up to 14 cm, creeping, wiry, with sparse reddish to pale brown, ovate, ± acuminate to obtuse scales. Leaves along rhizome but often clustered towards apex. Sterile leaves procumbent to ascending; lamina (3–)5–11(–17) × 1–1.7 cm, linear to linear-lanceolate, subcoriaceous to ± herbaceous, green to reddish green; pinnae (5)6–8.5 × (2.5–)3–4.3 mm, oblong to triangular, obtuse to subacute, entire, with broad basal insertion, opposite to alternate; petiole (3–)5–9 cm, naked or with sparse, ovate to ovate-lanceolate, acute to acuminate, brownich scales towards base. Fertile leaves erect, usually longer than the sterile; pinnae revolute, ± arcuate, distant. Spores smooth. (Fig. 25).

In all communities except feldmark and wetter parts of bogs; 0–500 m.

Throughout Fuegia except the extreme north. (Map 3c).
W. Argentina and Chile north to *c.*38°S.lat.
S. Brazil. Falkland Islands (Islas Malvinas), South Georgia, Tristan da Cunha, Gough Island, Prince Edward Islands, Iles Crozet, Iles de Kerguelen, Macquarie Island, Campbell Island, Auckland Islands.

2. *B. magellanicum* (Desv.) Mett. 1856, *Fil. Lechl.* **1**: 14.
Lomaria magellanica Desv. 1811, *Ges. Nat. Freunde Berlin Mag.* **5**: 330 – Hooker, 1847; Hariot, 1884; Spegazzini, 1896.

Rhizome erect, stout, often produced into a short "trunk" up to 10 cm in diameter, clothed with bases of old leaf-petioles bearing numerous leaves in an apical tuft. Sterile leaves ascending to erect; lamina 12.5–31(–50) × 6–16(–20) cm, lanceolate, coriaceous, green; pinnae 32–85(–100) × 7–13 mm, linear to linear-lanceolate, acute, entire, shallowly cordate at base, alternate to subopposite; petiole 7.5–12(–35) cm, with rigid, brown, linear-lanceolate scales 2.5–3.5(–4.2) × *c.*1 mm at the base, becoming shorter, paler and membranous above. Fertile leaves 23–36 cm, erect; pinnae *c.*50 × 2.5–4 mm, sometimes arcuate. Spores smooth. (Fig. 24).

Open *Drimys-Nothofagus betuloides* forest, coastal scrub, open wet peaty montane slopes; 0–550 m.

S. Fuegia. (Map 3d).
W. Argentina north to *c.*40°S.lat.
Chile north to *c.*36°S.lat.
Falkland Islands (Islas Malvinas).

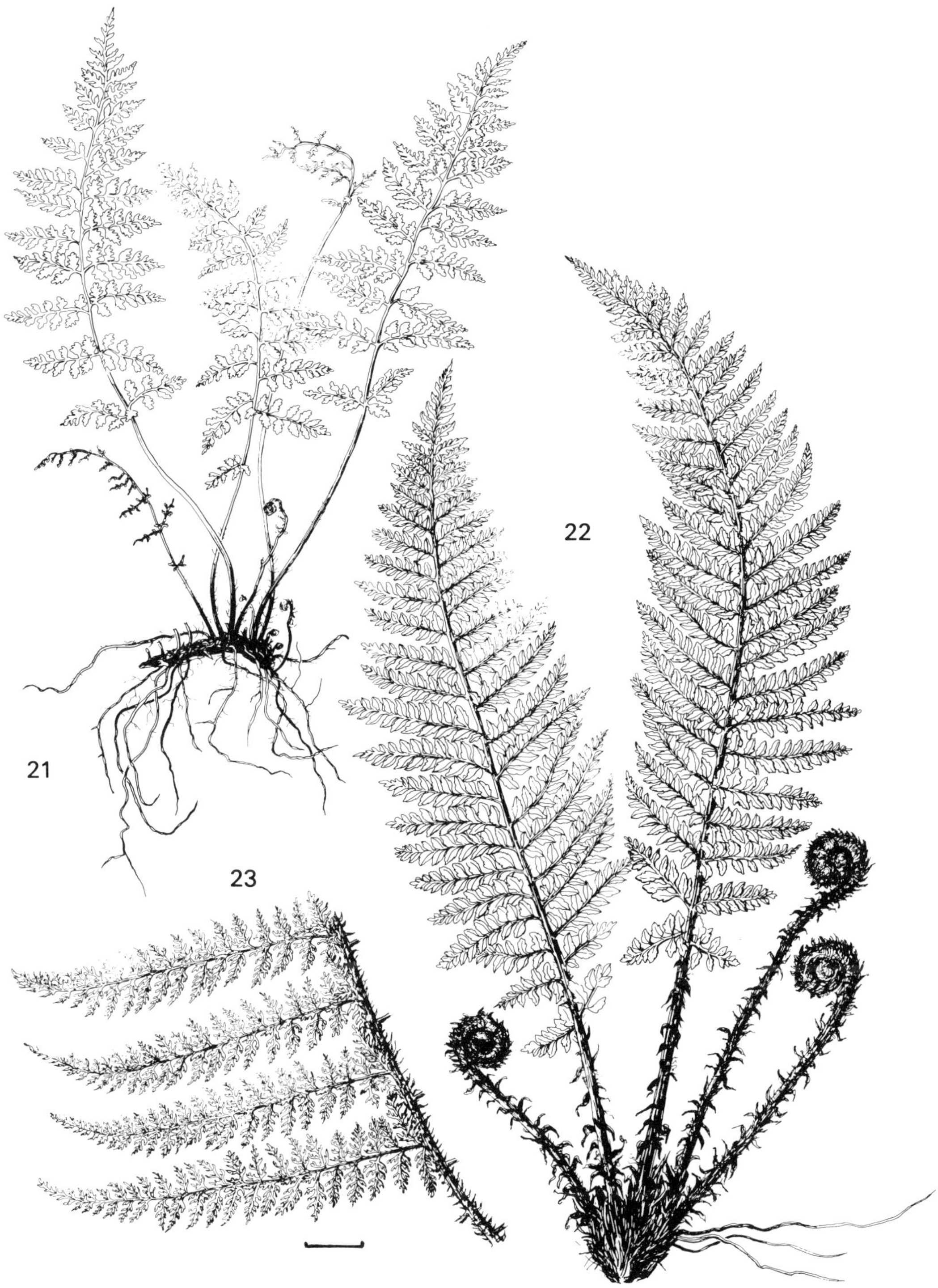

Fig. 21. *Cystopteris fragilis*. Fig. 22. *Polystichum mohrioides*. Fig. 23. *Polystichum multifidum*, pinnae. Scale = 2 cm.
(R. N. P. Goodall)

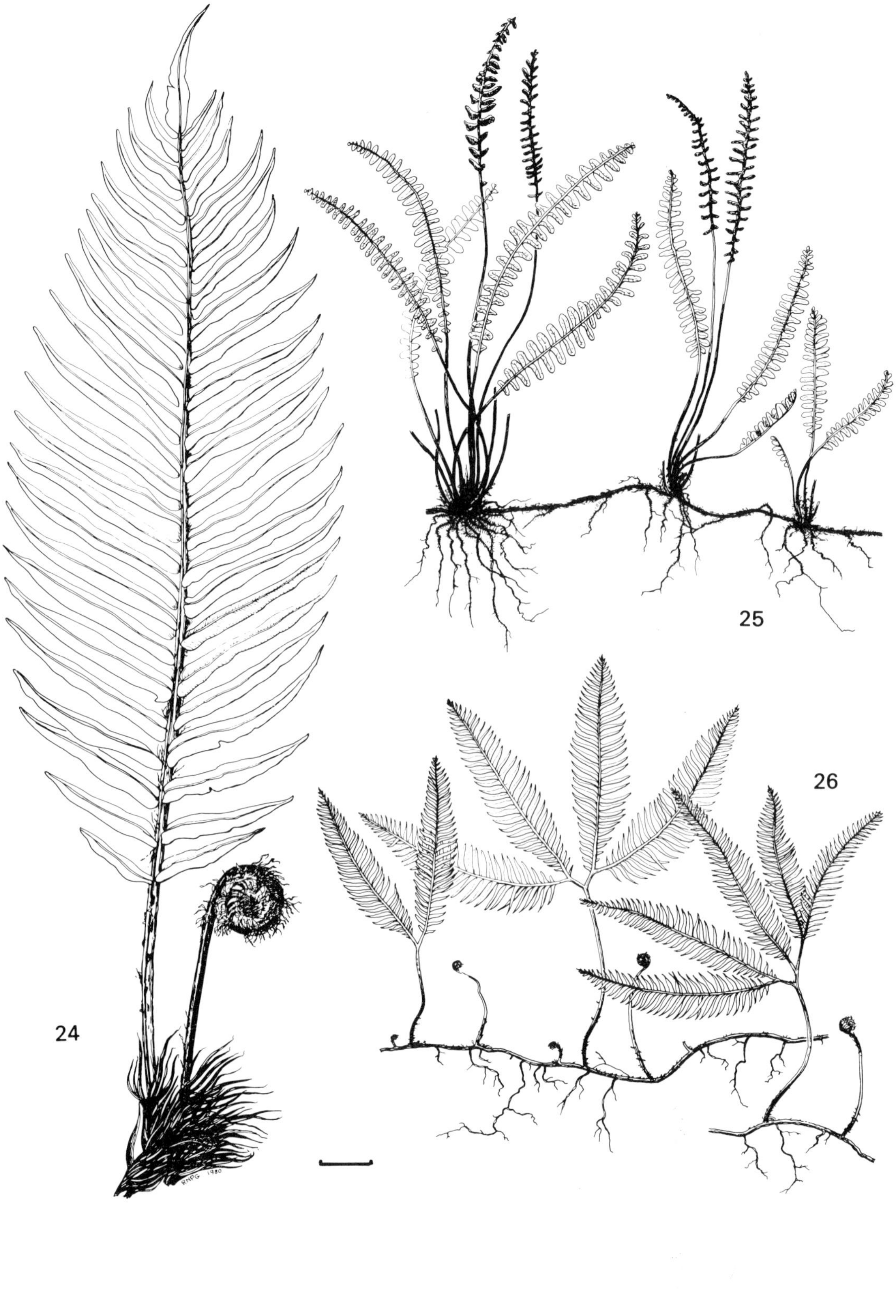

Fig. 24. *Blechnum magellanicum*. Fig. 25. *Blechnum penna-marina*. Fig. 26. *Gleichenia quadripartita*. Scale = 2 cm.
(R. N. P. Goodall)

XI. GLEICHENIACEAE

Rhizome creeping, dichotomously branched, clothed with scales. Leaves coriaceous; veins free. Sori in 1 row on each side of midvein of pinnule, covered by revolute margin of pinnule.

1. *Gleichenia* Sm.

Rhizome bearing leaves at intervals. Leaves 1- to 2-pinnate. Sori with few, usually 3–4, sporangia.

Literature: G. Looser, *Rev. Univ.* **19(5–7)**: 761–775(1934); **47**: 17–31(1962). G. Kunkel, *Nova Hedwigia* **11**: 507–511(1966).

a. Branching of leaves dichotomous **2. *quadripartita***
a. Primary branching of leaves trichotomous, with axis in the middle of each dichotomy **1. *cryptocarpa***

1. *G. cryptocarpa* Hooker 1846, *Sp. Fil.* **1**: 7.
Dicranopteris cryptocarpa (Hooker) Looser 1936, *Notas Helechos Chile* **2**: 8.
Sticherus cryptocarpus (Hooker) Ching 1940, *Sunyatsenia* **5**: 282.

Rhizome 1.5–3 mm in diameter; scales 3–4 mm, suborbicular to ovate, acute to acuminate, peltate, dark brown. Lamina 3–7 × 4.5–7.5 cm, 2-pinnate, yellow to yellow-brown at maturity; pinnae 4–6.5 × 1.5–2.5 cm, linear to oblong-lanceolate in outline, acute; pinnules (3–)8–22 × *c.*1 mm, linear, acute, with undulate, revolute margins, greyish green. Sporangia 1–4 per sorus.

Montane rocks. 300–400 m.

S. Isla Grande (W. end Lago Fagnano), Isla de los Estados. (Map 3e).
W. Argentina north to *c.*40°S.lat.
Chile *c.*37°–40°S.lat.
Falkland Islands (Islas Malvinas).

Perhaps only a developmental variant of *G. quadripartita*.

2. *G. quadripartita* (Poiret) T. Moore 1862, *Ind. Fil.* 382.
G. acutifolia Hooker 1846, *Sp. Fil.* **1**: 7 – Hariot, 1884.
Dicranopteris quadripartita (Poiret) Looser 1936, *Notas Helechos Chile* **3**: 10.
Sticherus quadripartitus (Poiret) Ching 1940, *Sunyatsenia* **5**: 284.

Rhizome *c.*1.3 mm on diameter; scales *c.*2–4 mm, ovate to lanceolate, acute to acuminate, peltate, dark brown. Lamina 9–15 × 11–22 cm, 1- to 2- pinnate, blackish above and reddish brown beneath at maturity; pinnae 8–13 × 2–3 cm, lanceolate in outline, acuminate; pinnules 10–19 × 1.2–2.5 mm, linear, acute, with undulate, revolute margins, dark green. Sporangia 1–3 per sorus. (Fig. 26).

Open, better drained parts of coastal forest; 0–50 m.

C. & S. Fuegia. (Map 3f).
W. Argentina *c.*40°S.lat.
Chile north to *c.*39°S.lat.

Spermatophyta Coniferophytina

XII. PODOCARPACEAE

Dioecious evergreen shrubs. Leaves simple, scale-like. Male cones of densely imbricate scales, the scales with 2 pollen-sacs on lower surface; female scales 1-several, free, with 1 ovule on upper surface, free. Fruit membranous.

1. *Dacrydium* A. B. Lambert

Leaves imbricate, spirally arranged. Male cones with *c.*15 scales. Seed nut-like.

1. *D. fonkii* (Phil.) Bentham 1880, in Bentham & Hooker, *Gen. Pl.* **3**: 433.

Usually prostrate shrub up to 60 cm, with slender branches up to 5 mm in diameter. Leaves 4–5 × 3–4 mm, triangular-ovate, obtuse. Male cones *c.*6 × 2 mm, the scales *c.*1 mm, subtended by basal bracts, terminal on branches. Seed 3–4 × *c.*2 mm, ovoid. *Flowers November to February*. (Fig. 27).

Magellanic moorland, bogs; 2–20 m.

S. Fuegia (N.W. Isla Hoste). (Map 3h).
W. Chile north to *c.*40°S.lat.

Quinn (1982) refers this and 2 New Zealand species to *Lepidothamnus* Phil., the genus in which it was first described.

XIII. CUPRESSACEAE

Monoecious, evergreen trees. Leaves opposite, simple, scale-like. Cones composed of opposite scales; scales of male cones with 3–5 pollen-sacs on lower surface; scales of female cones with 2-

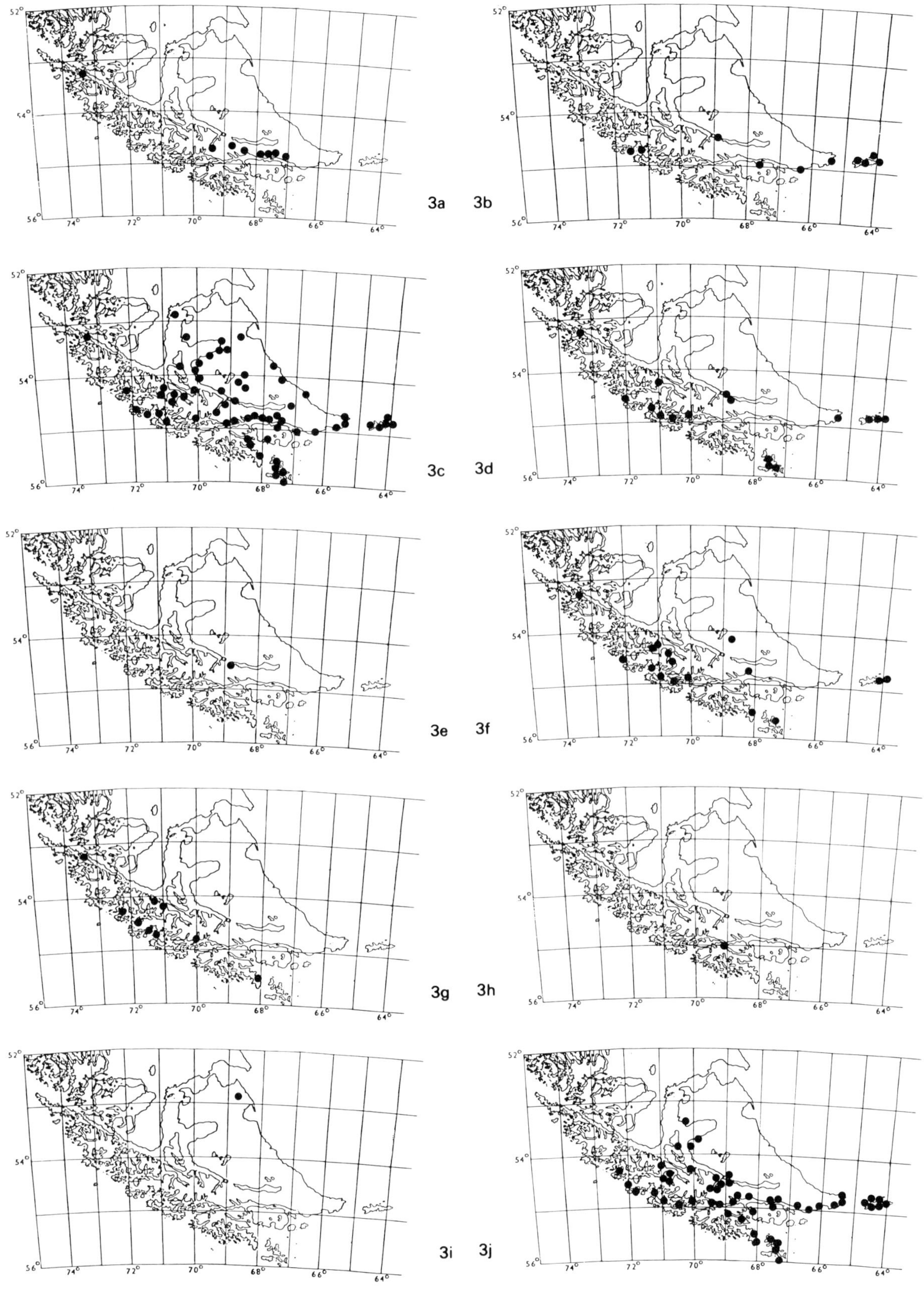

Map 3. a, *Polystichum andinum*. b, *Polystichum multifidum*. c, *Blechnum penna-marina*. d, *Blechnum magellanicum*. e. *Gleichenia cryptocarpa*. f, *Gleichenia quadripartita*. g, *Pilgerodendron uviferum*. h, *Dacrydium fonkii*. i, *Ephedra frustillata*. j, *Drimys winteri*.

numerous, erect ovules on upper surface, completely united to subtending bract. Fruit a woody cone.

1. ***Pilgerodendron*** Florin

Leaves scale-like, imbricate, in 4 rows. Male and female cones on different branches. Cones with 4–6 scales; upper and middle scales with 1–2, 2-winged seeds in fruit.

1. ***P. uviferum*** (D.Don) Florin 1930, *Svensk Bot. Tidskr.* **24**: 133.
   Libocedrus tetragona (Hooker) Endl. 1847, *Syn. Conif.* 44 – Franchet, 1889; Dusén, 1900; Skottsberg, 1916.
   Thuja tetragona Hooker 1844, *Jour. Bot. (London)* **3**: 144 – Hooker, 1847.

Narrowly pyramidal tree 1.5–10 m; trunk up to *c.*35 cm in diameter; bark dark brown, flaking in long strips; shoots tetragonal. Leaves (1.8–)2–4 × 1.5–3 mm, ovate, obtuse, sometimes reflexed. Female cones 8–10 × *c.*8 mm, borne on short shoots; scales oblanceolate, the outer pair with rigid dorsal spine *c.*3 mm, sometimes arcuate. Seeds with wide, oblique wing about twice as long as other, narrower, wing. (Fig. 20).

Drimys – Nothofagus betuloides coastal forest or locally dominant in open stands on sheltered lowland bogs further inland; 0–150 m.

S. & W. Fuegia. (Map 3g).
W. Chile north to *c.*40°S.lat.
W. Argentina *c.*47–41°S.lat.

Chlamydospermo-phytina

XIV. EPHEDRACEAE

Dioecious, dwarf shrubs. Leaves in whorls, reduced to short teeth and ± united sheaths. Inflorescences axillary, on short shoots. Flowers enclosed in 2–4 membranous perianth-segments. Fruit with 1–2 seeds surrounded by 2 pairs of fleshy bracts.

1. ***Ephedra*** L.

Inflorescences with 3–6 pairs of bracts united in basal $\frac{1}{3}$–$\frac{1}{2}$. Male flowers with 2, free perianth-segments and 2–8, 2- to 3- celled anthers, sessile in axils of bracts. Female flowers with 3–4 united perianth-segments, 2–4 borne in axil of upper bracts, the lower bracts sterile; ovule with integument prolonged into a slender tube.

Literature: J. Hunziker, *Lilloa* **17**: 147–174(1949).

1. ***E. frustillata*** Miers 1863, *Ann. Nat. Hist.*, Ser. 3, **11**: 262.
   E. nana Dusén 1900, *Svenska Exped. Magellansl.* **3(5)**: 235 – Skottsberg, 1916.

Stems up to 20 cm, prostrate to ascending, branched, green to yellowish. Leaves 1.2–2.5(–3.5) mm; teeth triangular, subacute to obtuse, concave. Male inflorescence 7–9 mm, oblong-ovoid. Female inflorescence *c.*8 mm, ovoid-globose. Fruit *c.*8 mm, globose, red to purplish; seeds 4–6 mm. (Fig. 28).

Grass steppe and dry rocks; 0–20 m.

N.E. Isla Grande. (Map 3i).
Argentina north to *c.*42°S.lat.
E. Chile north to *c.*51°S.lat.

Not reported from Tierra del Fuego for 80 years and, because of its susceptibility to sheep grazing, perhaps extinct there.

XV. WINTERACEAE

Trees or shrubs. Leaves alternate, simple; stipules absent. Flowers hermaphrodite, actinomorphic, the parts free and spirally arranged. Sepals valvate, slightly differentiated from imbricate petals. Stamens short and wide with lateral anther-locules. Carpels stipitate, folded longitudinally with partly free margins papillate to form stigmatic crests. Fruit an indehiscent berry; seeds numerous, with endosperm.

1. ***Drimys*** Forster & Forster f.

Leaves entire, evergreen. Flowers in cymes. Sepals 2, caducous. Petals up to 14. Carpels up to 10.

Literature: A. C. Smith, *Jour. Arnold Arboret.* **24**: 1–33(1943)

1. ***D. winteri*** Forster & Forster f. 1775, *Char. Gen. Pl.* 84.

Tree up to 20 m; trunk up to 1 m in diameter, with smooth, brown bark. Leaves (2.5–)4–10(–11) × (1–)2–3(–5) cm, oblanceolate to elliptical, obtuse, rarely emarginate, cuneate at base, coriaceous, shiny dark green above, glaucous or white punctate beneath; petiole up to 1.5 cm. Peduncles up to 2.5(–4) cm, 1- to 3-flowered, divergent from axis. Sepals *c.*5(–6) mm, reddish. Petals 1.6–2.5 × (4–)5–7 mm, lanceolate, subacute, shiny, white. Berry 5–9 mm in diameter; seeds 3–4 mm, lunate, black, shiny. *Flowers October to March.* (Fig. 29, Pl. 5a).

Locally dominant or co-dominant with *Nothofagus*, especially *N. betuloides*, in usually lowland sites with abundant water available. 0–300 m.

S. & W. Fuegia. (Map 3j).
W. Argentina north to *c.*38°S.lat.
Chile north to 36°S.lat.

XVI. BERBERIDACEAE

Shrubs. Leaves alternate, simple, those on long shoots transformed into spines; stipules absent. Flowers hermaphrodite, actinomorphic. Perianth-segments free, in whorls of 3. Stamens 6, dehiscing by apical cap. Carpel solitary, superior, 1-celled; style short or peltate; stigma sessile. Fruit a 4- to 8-seeded berry.

1. *Berberis* L.

Wood yellow. Leaves clustered on short shoots in axils of spines. Flowers solitary, axillary, and in axillary corymbs or clusters. Outer perianth-segments small; the innermost with basal nectaries; usually all petaloid.

Literature: L. W. A. Ahrendt, *Bot. Jour. Linn. Soc. London* **57**: 1–410(1961); M. C. Orsi, *Bol. Soc. Argent. Bot.* **17**: 127–149(1976).

a. Leaves linear, needle-like **1. *empetrifolia***
a. Leaves elliptical to oblanceolate or obovate:
 b. Flowers solitary **3. *buxifolia***
 b. Flowers 4 to 8 in corymbs **2. *ilicifolia***

1. *B. empetrifolia* Lam. 1792, *Tabl. Encycl. Méth. Bot.* **1**: 391.

Prostrate shrub with slender branches up to 40 cm and often rooting at nodes. Spines 3-fid, the lateral segments 2.5–7 mm, the median (3.5–)5–12 mm. Leaves 6–14 × 0.7–1.5 mm, linear, often recurved, with sharp, rigid apical spine, entire, with revolute margin, glabrous, with short, red petiole. Flowers solitary or in clusters of 2–3; peduncles up to 10 mm. Perianth-segments in 5 whorls, petaloid, orange, 3.5–5.5 × 1.6–2.5 mm, ovate, obtuse, the outer smaller, the inner concave. Style short. Berry *c.*5.5 × 4–5 mm, ellipsoid, pruinose; seeds 3.5 mm, lunate, finely reticulate. *Flowers October to January.* (Fig. 30).

Open sandy or rocky soils, especially near coast and by rivers and lakes. 0–200 m.

Throughout Isla Grande, Isla de los Estados and islands along Canal Beagle. (Map 4a).
W. Argentina north to 28°S.lat.
Chile north to 30°S.lat.

2. *B. ilicifolia* L.f. 1781, *Suppl.* 210.

Erect shrub or small tree up to 7 m; stems up to 20 cm in diameter, with greyish brown, fissured bark. Spines 3- to 5- fid, 3–11 mm. Leaves 10–45(–50) × 8–15(–35) mm, oblanceolate to obovate or elliptical, obtuse, with apical spine, attenuate at base into short petiole, the margin revolute with 1–5 pairs of spines directed towards apex, coriaceous, glabrous, dark green. Corymbs 4- to 8- flowered; peduncles 4–15 mm; pedicels 5–25 mm. Perianth-segments in 5 whorls, petaloid, orange to golden yellow, 4–13 × 2–8.8 mm, elliptical, obtuse to subacute, the outer smaller, the inner concave. Style short. Berry 5–8 × 3.5–7 mm, ellipsoid, pruinose, 4-seeded; seeds *c.*4.5 × 2 mm, finely reticulate, brown to blackish. $2n = 28$. *Flowers September to December.* (Pl. 5c).

On moist soils, especially by coast and along rivers, in forest with *Nothofagus* and *Drimys*. 0–250 m.

S. & W. Fuegia. (Map 4b).
W. Argentina north to 54°S.lat.
Chile north to 39°S.lat.

3. *B. buxifolia* Lam. 1792, *Tabl. Encycl. Méth. Bot.* **1**: 391.
B. heterophylla auct., non Juss. – Spegazzini, 1896.
B. microphylla Forster f. 1787, *Comment. Gotting.* **9**: 29.

Erect shrub up to 4 m; stems up to 3 cm in diameter, with pale, greyish brown bark. Spines simple or 2- to 3(4)-fid, 7–20 mm. Leaves 7–20(–26) × 3–12 mm, oblanceolate or obovate to ovate or elliptical, obtuse to subacute, with sharp rigid apical spine, attenuate at base into short petiole, the margin thickened, entire, coriaceous, glabrous, bright green above, pale beneath, with prominent yellowish veins. Flowers solitary; peduncles 6–20 mm, pendent. Perianth-segments in 4 whorls, petaloid, yellow, 2.8–8 × 1.8–6 mm, obovate to ovate, the outer smaller, the inner concave. Stigma sessile. Berry 7–10 × 7.5–8.5 mm, subglobose, pruinose, 8-seeded; seeds *c.*5 × 2 mm, lunate, smooth, dark brown. $2n = 28$. *Flowers October to January.* (Fig. 83).

Coastal scrub, forest margins and clearings, moister areas in grass steppe, along streams and rivers. 0–450 m.

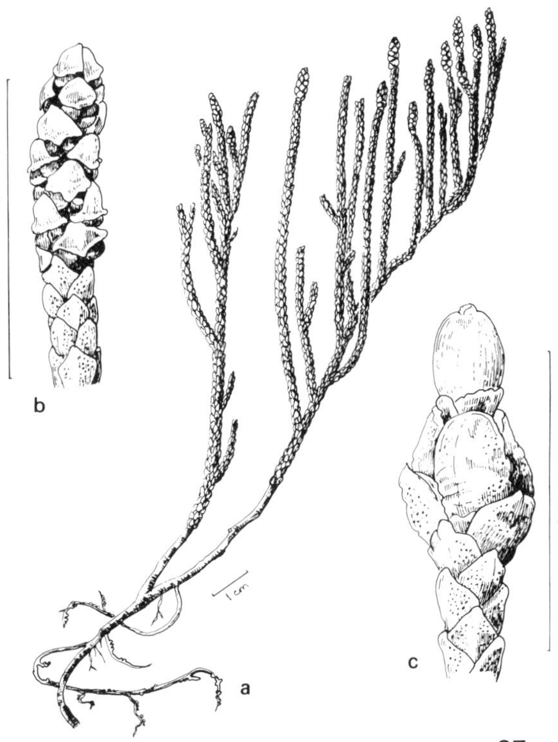

27

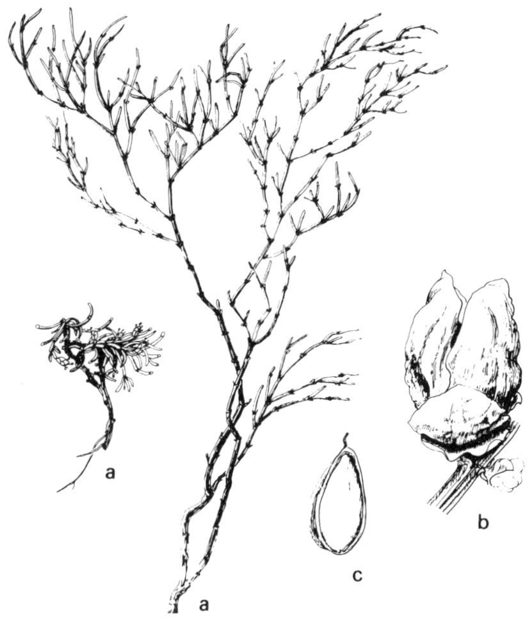

28

29

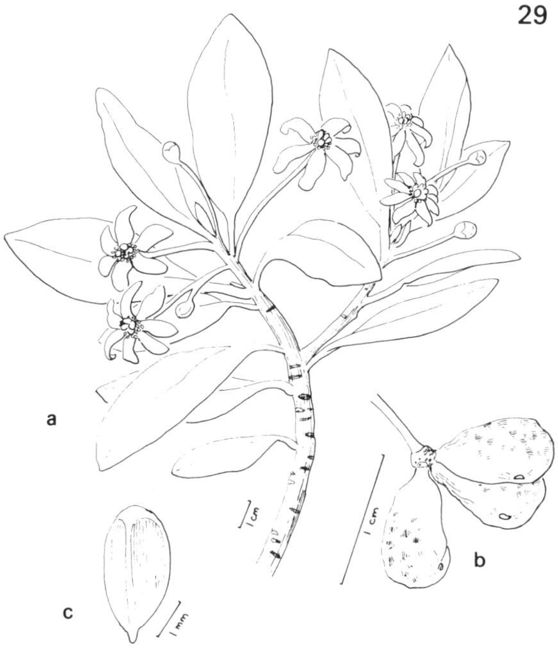

30

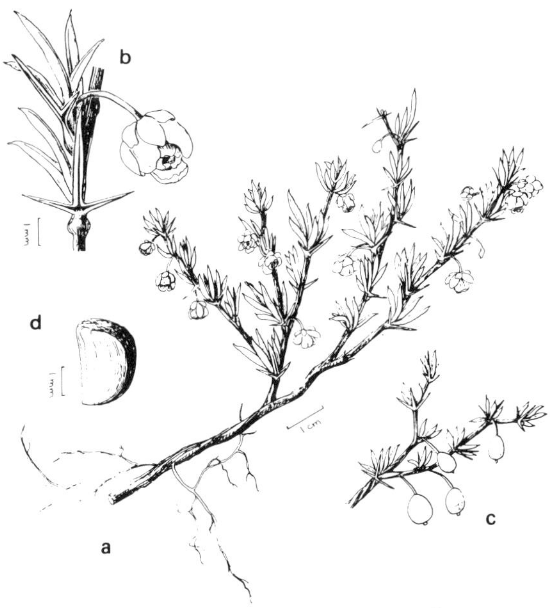

Fig. 27. *Dacrydium fonkii*. a, Plant. b, Male cone. c, Female cone. Scale = 1 cm. Fig. 28. *Ephedra frustillata*. a, Plant. b, Female flower. c, Seed. Scale = 1 cm. Fig. 29. *Drimys winteri*. a, Flowering branch. b, Fruits. c, Seed. Scale = 1 cm (a–b) or 1 mm (c). Fig. 30. *Berberis empetrifolia*. a, Plant. b, Flower. c, Fruiting branch. d, Seed. Scale = 1 cm (a, c) or 1 mm (b, d).
(R. N. P. Goodall)

Throughout Fuegia but only local in N.E. and extreme south and west. (Map 4c).
W. Argentina north to 37°S.lat.
Chile north to 36°S.lat.

B. microphylla Forster f., which has been accorded specific status by Orsi (1976), comprises populations from southernmost Chile and Argentina differentiated from *B. buxifolia* by their stamens having emarginate (not rounded) apices. Within the variation shown by *B. buxifolia* the constancy of this character is not clear and further work is required.

XVII. RANUNCULACEAE

Herbs, sometimes forming dense cushions. Leaves alternate or in basal rosettes; stipules usually absent. Flowers usually hermaphrodite, actinomorphic. Sepals 3–5, free, often petaloid. Petals 3 – numerous. Stamens numerous, spirally arranged. Carpels 1 – numerous, usually free and spirally arranged. Fruit of 1 or more follicles or a head of achenes.

Literature: A. Lourteig, *Darwiniana* **9**: 397–608(1952).

a. Plants dioecious **2. *Hamadryas***
a. Plants monoecious or hermaphrodite:
 b. Fruit of follicles, each with several seeds; leaf-lamina with upturned appressed basal appendages **1. *Caltha***
 b. Fruit of achenes; leaf-lamina without upturned lobes:
  c. Achenes on an elongated receptacle; plant not exceeding 4 cm; leaves up to 1.5 mm wide; sepals with minute spur **4. *Myosurus***
  c. Achenes clustered in heads; plant usually exceeding 4 cm; leaves more than 2 mm wide; sepals without spur:
   d. Perianth-segments all creamy white to yellowish **5. *Anemone***
   d. Sepals green to brown; petals white to yellow **3. *Ranunculus***

1. *Caltha* L.

Perennial herbs, sometimes forming cushions. Leaves alternate, simple, with upturned basal appendages, sometimes fused to upper surface. Flowers hermaphrodite, solitary. Perianth-segments 5–8, petaloid, yellow; nectaries absent. Fruit a 3- to 10-seeded follicle, in heads of 3–90.

Literature: P. G. Smit, *Blumea* **21**: 119–150(1973).

a. Leaf-lamina not more than 5 mm, ovate or suborbicular, with 2 ciliate appressed lobes enclosing 2 basal appendages **3. *dioneifolia***
a. Leaf-lamina more than 5 mm, ovate to triangular, the lobes not ciliate or conspicuously appressed:
 b. Leaf-lamina cordate at base, with appendages infolded from base; stamens 30 or more; follicles 30 or more **1. *sagittata***
 b. Leaf-lamina cuneate at base, with appendages arising from each side of midrib above; stamens not more than 12; follicles not more than 8 **2. *appendiculata***

1. *C. sagittata* Cav. 1799, *Icon. Descr.* **5**: 8.
C. multicapsularis Sol. ex Forster f. 1807, *Trans. Linn. Soc. London*, **8**: 324.

Rhizome horizontal to ascending, up to 15(–30) mm in diameter, often branched distally, covered with leaf-sheaths, bearing leaves towards apex. Leaves (8–)10–35(–45) × 6–29(–50) mm, ovate to triangular, obtuse or retuse at apex, cordate at base, irregularly undulate-serrate; appendages as triangular-ovate basal folds up to ½ as long as lamina; petiole 1–4.5(–15) cm, with prominent, amplexicaul, triangular-ovate, scarious basal sheaths. Peduncles 0.5–2 cm, longer in fruit. Perianth-segments 5–8, 4.5–16 × 1.5–8 mm, ovate to oblong, obtuse, subequal. Stamens 30–80. Carpels (30–)50–85. Follicles 4.5–7 × 1.5–3 mm, asymmetrically oblong, compressed, with straight or slightly curved beak 1.5 mm, sessile in globose head; seeds 5–10, ovoid-ellipsoid, smooth, black. $2n = 48$. Self-compatible. *Flowers October to April.* (Fig. 32, Pl. 5d).

Moist open places near fresh or brackish water. 0–1000 m.

Throughout Fuegia. (Map 4d).
W. Argentina north to *c.*34°S.lat.
Chile north to *c.*30°S.lat.
North along Andes to 16°S.lat.
Falkland Islands (Islas Malvinas).

2. *C. appendiculata* Pers. 1806, *Syn. Pl.* **2**: 107.
C. paradoxa Sol. ex. Forster f. 1807, *Trans. Linn. Soc. London*, **8**: 324.

Rhizome ascending to erect, up to 8 mm in diameter, often branched distally, covered with leaf-sheaths, bearing leaves towards apex. Leaves 6–9(–35) × 2–7(–10) mm, ovate to oblong or elliptical, entire or 3-fid, retuse or rarely obtuse at apex, cuneate at base, irregularly undulate or crenate; appendages 2(–3), oblong to obovate, arising from upper surface on either side of midrib towards base, up to as long as lamina, often absent in upper leaves; petiole 0.4–2 cm, with prominent, amplexicaul, ovate to triangular-ovate, scarious basal sheaths. Peduncles up to c.4 mm. Perianth-segments 5, 5–8(–12) × 1.5–4 mm, oblong to lanceolate, acute to obtuse, subequal. Stamens 8–10(–12). Carpels 3–5(–8). Follicles 4–6(–7) × c.3 mm, asymmetrically ovoid, compressed, with short curved beak, sessile; seeds 3–4, ovoid-ellipsoid, smooth, black. $2n = 48$. Self-compatible. *Flowers October to December.* (Fig. 31).

Wet open ground beside fresh water, bogs, Magellanic moorland. 0–1000 m.

Throughout Fuegia, but more common in S. & W. (Map 4e).
W. Argentina north to c.38°S.lat.
Chile north to 36°S.lat.
Falkland Islands (Islas Malvinas).

3. ***C. dioneifolia*** Hooker f. 1843, *Jour. Bot. (Hooker)* **2**: 307.

Plant usually forming cushions up to 5 cm high. Rhizome erect, up to 5 mm in diameter, branched, covered with leaf-sheaths, bearing appressed leaves distally. Leaves 2–4(–4.5) × 2.5–4(–6) mm, ovate to suborbicular, concave, 2-lobed; lobes semi-ovate to asymmetrically oblong, obtuse, appressed, with hyaline, ciliate margins, enclosing 2 basal, semi-ovate appendages arising above on either side of midrib; petiole 0.7–0.9(–1.6) cm, with scarious, 2-lobed, amplexicaul basal sheath. Peduncles c.4 mm. Perianth-segments 5–7, 3.5–6 × 1.5–2 mm, oblong to subspathulate, obtuse. Stamens 6–9. Carpels 2–5. Follicles 2.5–3 × c.1 mm, asymmetrically ovoid, compressed, with short beak, sessile; seeds c.6, ± ovoid, smooth. $2n = 48$. *Flowers October to December.* (Fig. 34).

Bogs and other open areas with abundant surface fresh water. 0–800 m.

S. & W. Fuegia. (Map 4f).
Chile north to 50°S.lat.

Hybrids between this species and *C. appendiculata* (Moore and Goodall, 1973) have been formally named *C.* × *goodalliana* T. Dudley 1981, *Rhodora* **83**: 290. They may be quite widespread, since the parents frequently occur intermingled in upland bogs.

2. ***Hamadryas*** Comm. ex Juss.

Dioecious, perennial herbs. Leaves basal, variously divided. Flowers unisexual, solitary or in few-flowered cymes or panicles. Sepals 5–8. Petals (7–)12–16(–20), with basal nectaries. Fruit an achene, numerous in heads.

a. Flowers solitary, rarely in pairs; achenes rugose dorsally **2. *delfinii***
a. Flowers in cymes or panicles, rarely solitary; achenes longitudinally striate dorsally **1. *magellanica***

1. ***H. magellanica*** Lam. 1789, *Encycl. Méth. Bot.* **3**: 67.
H. tomentosa DC. 1817, *Reg. Veg. Syst. Nat.* **1**: 227 – Spegazzini, 1896; Alboff, 1897.
H. paniculata Hooker f. 1847, *Fl. Antarct.* **1(2)**: 548.

Rhizome short, vertical or ascending, covered with dried leaf-bases, with leaves at apex. Leaves with lamina 20–70 × 20–80(–110) mm, subtriangular to suborbicular in outline, 3-fid to 3-partite, pubescent, especially beneath, sometimes subglabrous; lateral segments usually 2-fid, all with more or less prominent, acute to obtuse lobules; petiole up to 20 cm, with scarious basal sheath up to 35 × 7 mm. Peduncle as long as or longer than petioles, with 1–5 shortly pedicellate or subsessile flowers; bracts 1.5–2 mm, ovate, pubescent. Sepals 5–7, in male flowers 8–10 × 1–3 mm, elliptical to oblong, obtuse to subacute, pubescent; similar in female flowers but smaller, c.6 × 1.5–2 mm, ovate to elliptical, acute to subacute. Petals (7–)12–16, in male flowers 8–10 × 0.5–1.7 mm, linear to lanceolate, acute to acuminate, yellow; similar in female flowers but fewer and smaller, 6–8 × 0.5–1 mm. Stamens 35–50. Achenes 35–50, 2–2.5 mm, elongate, compressed, glabrous, longitudinally striate, with short, straight or slightly curved beak. *Flowers October to February.* (Fig. 33).

Open soil, among rocks and sometimes open grassland; (0–40) 450–1100 m.

Throughout Isla Grande but commoner in south, islands along Canal Beagle and Isla de los Estados. (Map 4g).
Chile north to 51°50′S.lat.

The larger plants, with deeply cordate leaves, have been referred to f. *paniculata* (Hooker f.) Lourteig, described from Isla de los Estados.

2. ***H. delfinii*** Phil. ex Reiche 1894, *Anal. Univ. Chile* **88**: 77.

Rhizome short, stout, covered with dried leaf-bases, with leaves at apex. Leaves with lamina 15–65 × 20–70(–85) mm, subtriangular in outline, 3-sect, pubescent; lateral segments up to 2- to 3- pinnate, all

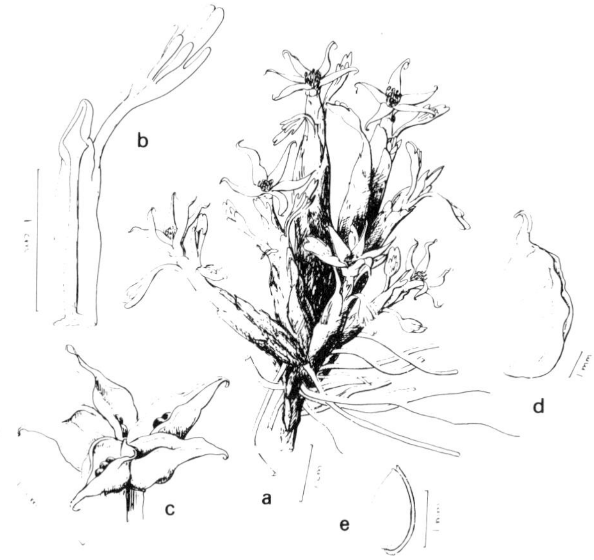

31

32

33

34

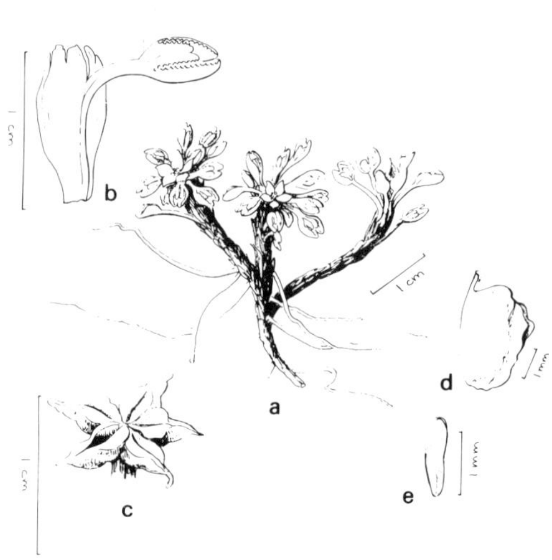

Fig. 31. *Caltha appendiculata*. a, Plant. b, Leaf. c, Fruit. d, Follicle. e, Seed. Scale = 1 cm (a–c) or 1 mm (d–e). Fig. 32. *Caltha sagittata*. a, Plant, with flowers and fruits. b, Follicle. c, Seed. Scale = 1 cm (a) or 1 mm (b–c). Fig. 33. *Hamadryas magellanica*. a, Plant. b, Fruiting head. c, Achene. Scale = 5 cm (a), 1 cm (b) or 1 mm (c). Fig. 34. *Caltha dioneifolia*. a, Plant. b, Leaf. c, Fruit. d, Follicle. e, Seed. Scale = 1 cm (a–c) or 1 mm (d–e).

(R. N. P. Goodall)

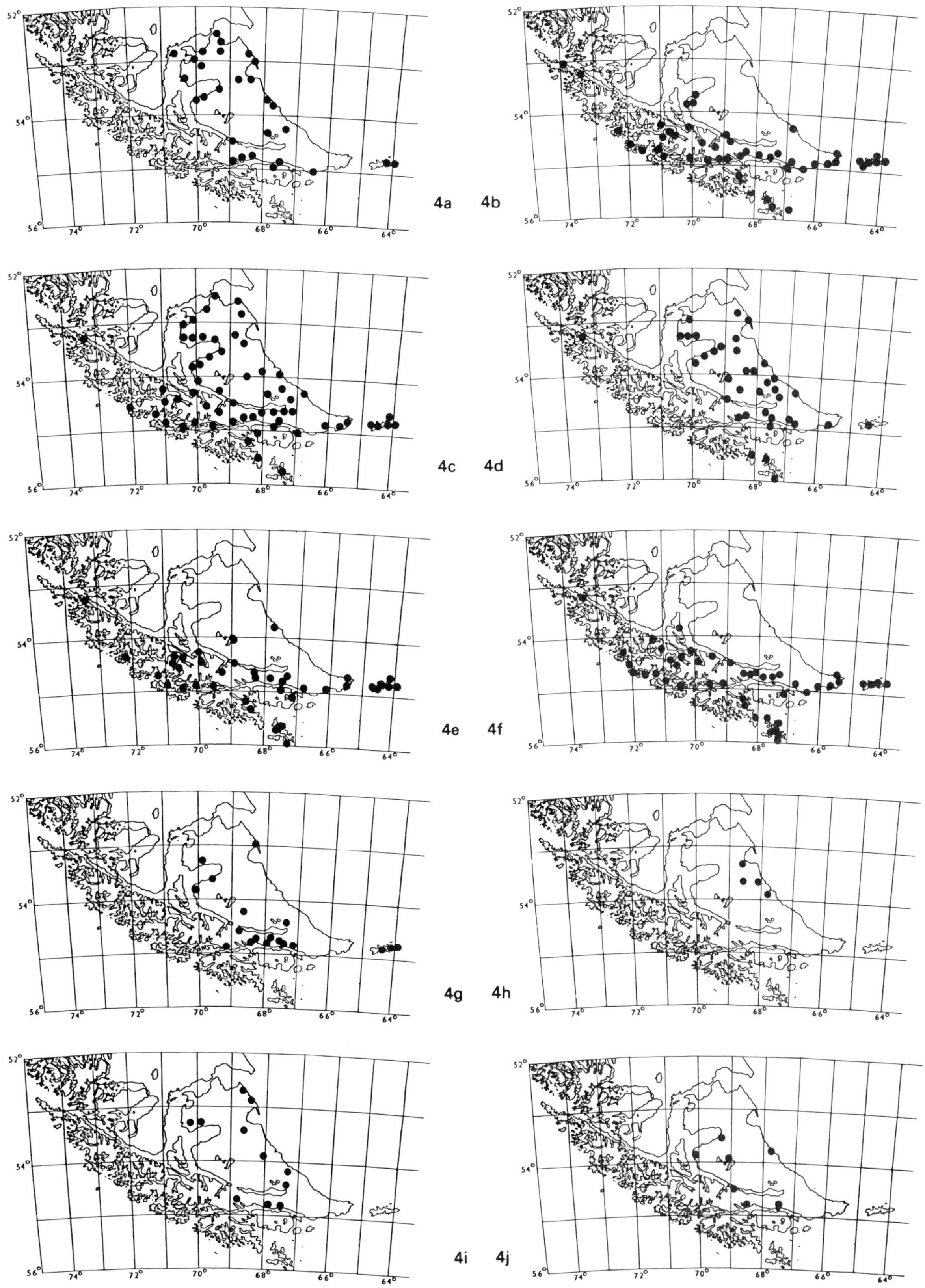

Map 4. a, *Berberis empetrifolia*. b, *Berberis ilicifolia*. c, *Berberis buxifolia*. d, *Caltha sagittata*. e, *Caltha appendiculata*. f, *Caltha dioneifolia*. g, *Hamadryas magellanica*. h, *Hamadryas delfinii*. i, *Ranunculus aquatilis*. j, *Ranunculus pseudotrullifolius*.

segments subacute to subacuminate, variously lobed, with petiolules up to 25 mm; petiole 2–10 cm, with basal sheath up to 35 × 9 mm and scarious at margin. Peduncle shorter to longer than petioles, with 1(2) flowers. Sepals 5–7, 6.5–14 × 2–6 mm, oblong to obovate, obtuse, pubescent at least at base, yellowish, somewhat unequal in female flowers. Petals (8–)11–18, in male flowers 6–12 × 1–2 mm, linear, acute to acuminate, reddish; similar in female flowers but 7–9 mm, linear to sublanceolate. Stamens 30–60. Achenes 30–65, 2–2.5(–4) mm, asymmetrically semi-ovate, compressed, rugose on back, with short, slightly curved beak. *Flowers October to December*. (Fig. 35).

Open grassland and bare soil or rock crevices; 100–250 m.

E.C. Isla Grande. (Map 4h).
W. Argentina north to *c*.32°S.lat.
Chile north to *c*.51°S.lat.

3. *Ranunculus* L.

Perennial, rarely annual, herbs. Leaves basal and usually also cauline, simple or variously divided. Flowers solitary or in cymose panicles. Perianth of 2 whorls; outer segments (sepals) 3–5(6), free, green, yellowish or brown; inner segments (petals) 3–16, with basal nectaries. Fruit an achene, with persistent style (beak), clustered in heads.

a. Plant submerged — **1. *aquatilis***
a. Plant terrestrial or, if aquatic, leaves floating:
  b. Sepals and petals different in number:
    c. Leaves entire or somewhat lobed:
      d. Petals more than 5 mm; achenes nerved — **9. *uniflorus***
      d. Petals 3 mm or less; achenes minutely foveolate — **10. *hydrophilus***
    c. Leaves variously divided:
      e. Plant pubescent; stems erect — **11. *peduncularis***
      e. Plant glabrous; stems creeping, stoloniferous — **12. *fuegianus***
  b. Sepals and petals equal in number:
    f. Sepals and petals 3:
      g. Floral organs glabrous — **2. *pseudotrullifolius***
      g. Sepals and gynophore pilose — **3. *minutiflorus***
    f. Sepals and petals 5 (6):
      h. Sepals and petals 5–6; achene nerved — **8. *cymbalaria***
      h. Sepals and petals 5; achene not nerved:
        i. Petals more than 7 mm — **4. *repens***
        i. Petals less than 7 mm:
          j. Plant glabrous; leaf-lamina biternate; gynophore glabrous — **5. *biternatus***
          j. Plant pubescent; gynophore hairy:
            k. Petals, nectariferous scales and achenes pubescent — **7. *sericocephalus***
            k. Petals and nectariferous scales glabrous; achenes slightly pubescent, ± glabrescent — **6. *maclovianus***

1. *R. aquatilis* L. 1753, *Sp. Pl.* 556.
R. trichophyllus auct., non Chaix – Lourteig, 1952.
R. fluitans auct., non Lam. – Dusén, 1900.

Plant submerged, sometimes with some apical leaves floating; stems slender. Submerged leaves up to *c*.6 cm, 3-partite; segments variously divided, with filiform lacinii, usually glabrous, with some short hairs at apex and in the angles of the division, or with short, whitish or yellowish appressed hairs; petiole up to 27 mm, slender, with ovate, amplexicaul basal sheath; floating leaves 7–20 × 5–20 mm, reniform to orbicular in outline, peltate, 3-partite, glabrous, the segments subdivided laterally, all somewhat lobed, sometimes with narrow lacinii. Flowers solitary or in pairs, axillary, floating or aerial. Sepals 5, *c*.3 × 1.5–2.5 mm, somewhat unequal, oblong or elliptical to slightly asymmetrical, obtuse to subacute. Petals 5, 4.5–6 × 2–2.5 mm, somewhat unequal, elliptical to oblong-subspathulate, obtuse, white. Stamens 10–15. Carpels 15–40. Fruiting head subglobose; achenes 2–2.25 mm, asymmetrically ovate, compressed, somewhat marginate, transversely striate, glabrous or hairy on back, the beak short. *Flowers December to February*. (Fig. 36).

Lakes, ponds, slow moving freshwater streams and rivers.

Canal Beagle northwards. (Map 4i).
W. Argentina north to *c*.32°S.lat.
Chile north to *c*.38°S.lat.
North along Andes to *c*.10°S.lat.
N. America, Eurasia, N. Africa.

As Cooke (1966) has shown, *R. trichophyllus* Chaix, in which Lourteig (1952) included Fuegian material, does not occur in South America.

2. *R. pseudotrullifolius* Skottsb. 1913, *Kungl. Svenska Vet. Akad. Handl.* **50(3)**: 32.
R. trullifolius auct., non Hooker f. – Dusén, 1900; Macloskie, 1905.

R. caespitosus Dusén 1900, *Svensk. Exped. Magellansl.* **3**: 185 – Wildeman, 1905; Macloskie, 1905.
R. dusenii Lourteig 1951, *Darwiniana* **9**: 468.

Plant glabrous; stems up to 13 cm, prostrate, rooting at nodes. Leaves with lamina 5–35 × 1–37 mm, linear to oblong, ovate or obovate, obtuse, cuneate to cordate at base, entire to sometimes deeply 3-lobed, the lateral segments entire or lobed; lobes obtuse, with apical gland; petiole up to 14 cm (to 24 cm when submerged), with ovate, amplexicaul basal sheath with scarious margins. Flowers solitary or in 2- to 3-flowered cymes, axillary, subsessile or with peduncle up to 10 mm. Sepals 3–4, 2–5 × 1–3.5 mm, ovate to suborbicular, obtuse, concave or cucullate. Petals 3, *c.*3–5 × 0.75–2.5 mm, linear-to oblong-spathulate, obtuse, yellow. Stamens 5–13. Carpels 20–65. Fruiting head subglobose; achenes 1.5–2 mm, asymmetrically ovoid to obovoid, somewhat compressed, the beak *c.*1 mm, slender, curved. Self-compatible and sometimes self-pollinated. $2n = 48$. *Flowers December to February*. (Fig. 38).

Wet mud, sand and pebbles by lakes and streams. 0–50 m.

Canal Beagle northwards in Isla Grande. (Map 4j).
Argentina and Chile north to *c.*50°S.lat.
Falkland Islands (Islas Malvinas); Iles de Kerguelen.

3. *R. minutiflorus* Bert. ex. Phil. 1856, *Linnaea* **28**: 661.

R. chilensis auct., non DC. – Skottsberg, 1906.
R. savatieri Franchet 1889, *Miss. Sci. Cap Horn* **5**: 320 – Macloskie, 1905; Alboff, 1902.

Stems up to more than 1 m, erect to ascending or prostrate, sometimes rooting, yellow- or whitish pubescent to sericeous. Basal leaves with lamina up to 60 × 65 mm, suborbicular in outline, cordate at base, 3-lobed to -fid, the segments dentate or serrate, appressed pubescent, especially beneath; petiole up to 21 cm, with 1- to 2- auriculate, acute to obtuse basal sheath; cauline leaves smaller, with petioles up to 4 cm. Flowers solitary, leaf-opposed; peduncle 7–20 mm. Sepals 3(4), 2.5–3.5 × 1.6–2 mm, concave to cucullate, pilose on back. Petals 3–4.5 × 1.5–3 mm, oblong to elliptic-spathulate, yellow. Stamens usually 8–9. Carpels 5–27. Fruiting head ellipsoid; achenes 1.5–2 mm, asymmetrically suborbicular to obovoid, compressed, finely foveolate, marginate, the beak somewhat curved. $2n = 32$. Self-compatible. *Flowers November to March.*

Moist scrub, grassland, forest margins and swamp margins. 0–800 m.

S. & W. Isla Grande and adjacent islands. (Map 5a).
W. Argentina north to *c.*39°S.lat.
Chile north to *c.*37°S.lat.

4. *R. repens L. 1753, *Sp. Pl.* 554.

Plant glabrous or somewhat pubescent; stems up to 60 cm, creeping or ascending to erect, rooting at nodes. Basal leaves with lamina (10–)20–50 × 30–50 mm or more, triangular-ovate in outline, 3-fid to 3-foliolate, cordate at base; segments usually deeply cut into 3 serrate-dentate lobes, the middle segment long stalked; petiole up to 20 cm, with entire or 2-auriculate, amplexicaul basal sheath; cauline leaves smaller, with less divided to entire segments. Flowers solitary or in cymose panicles; peduncle up to 7.5 cm, Sepals 5, 6.5–8 × 3–4.5 mm, ovate, acute, concave, with hyaline margins, hairy especially towards base. Petals 5–6, 11–15 × 7–14 mm, obovate to suborbicular, entire or subemarginate, the margin sometimes irregular, bright yellow. Stamens 60–75. Carpels 35–55. Achenes *c.*2 × 3 mm, asymmetrically globose, compressed, glabrous, finely foveolate, the beak *c.*1.5 mm. stout, curved. *Flowers November to February.*

Disturbed ground near settlements; 0–200 m.

C. & S. parts of Isla Grande.
Introduced from Eurasia and widely naturalized in temperate regions of the world.

5. *R. biternatus* Sm. 1814, in Rees, *Cyclop.*, **29**: No.48.

Plant glabrous; stems up to 40 cm, or more, prostrate, sometimes branched, rooting at the nodes from which arise one or more leaves and sometimes flowers. Leaves with lamina 3–25 × 4–20 mm, triangular-ovate to reniform in outline, 3-fid to 3-foliolate or biternate, cordate at base; ultimate segments 2- to 5- lobed or dentate; petiole up to 8 cm, with ovate-oblong, scarious basal sheath. Flowers solitary, axillary, rarely in 2-flowered cymes; peduncle 1–3 cm, usually erect. Sepals 5, 3.5–4 × 1.5–2.5 mm., ovate to obovate, obtuse, brownish green, reflexed. Petals 5, *c.*4 × 1–2 mm., oblong to obovate, spathulate or linear-spathulate, obtuse, yellow. Stamens 7–12(–15). Carpels 10-numerous. Achenes 2–3 × *c.*2 mm, asymmetrically ovoid to subglobose, weakly marginate, often reddish, the beak *c.1* mm, somewhat curved. $2n = 48$. Self-compatible, frequently self-pollinated. *Flowers November to April.* (Fig. 37).

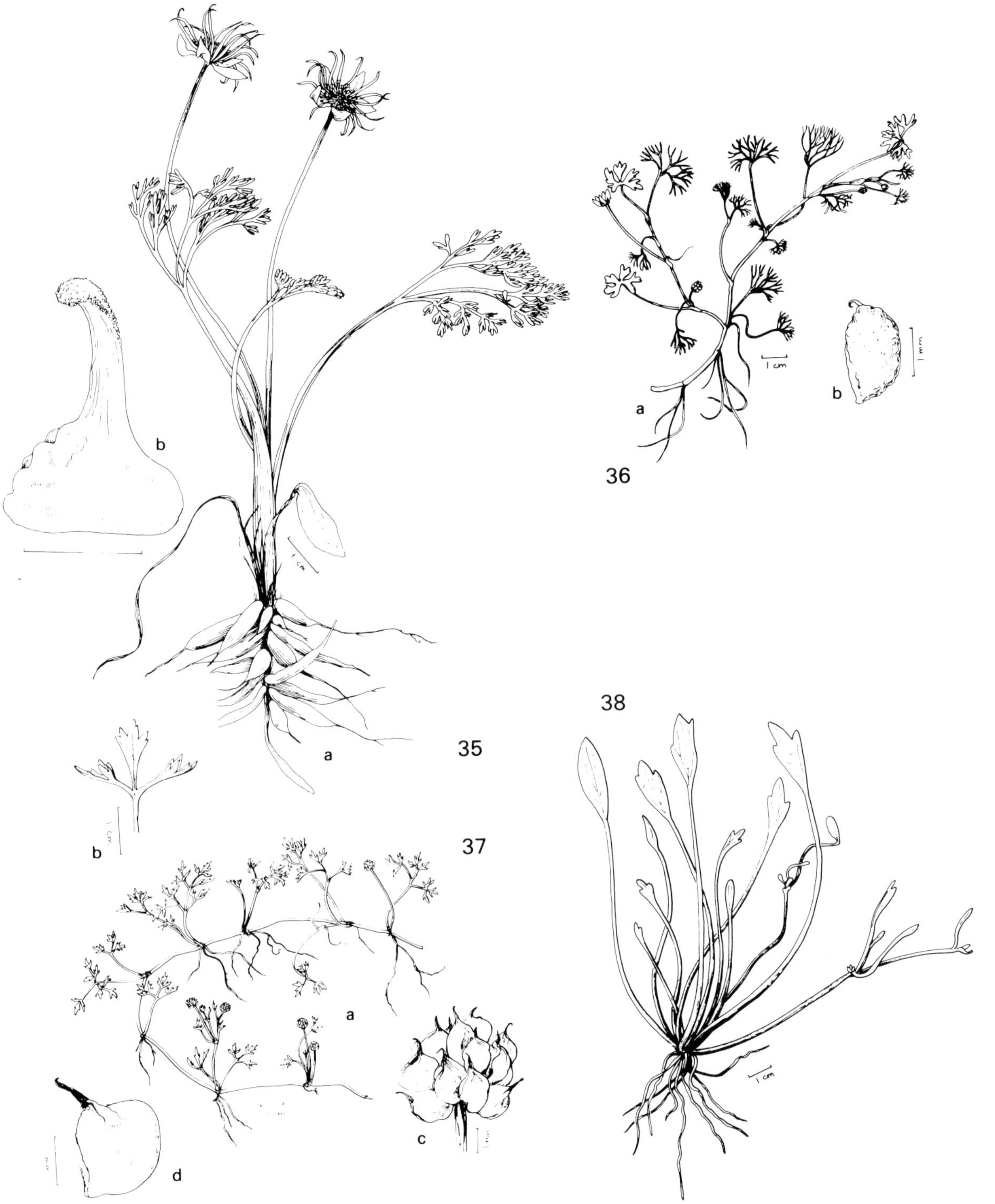

Fig. 35. *Hamadryas delfinii*. a, Plant. b, Achene. Scale = 1 cm (a) or 1 mm (b). Fig. 36. *Ranunculus aquatilis*. a, Plant. b, Achene. Scale = 1 cm (a) or 1 mm (b). Fig. 37. *Ranunculus biternatus*. a, Plant. b, Leaf. c, Fruiting head. d, Achene. Scale = 1 cm (a–b) or 1 mm (c–d). Fig. 38. *Ranunculus pseudotrullifolius*. Plant. Scale = 1 cm.

(R. N. P. Goodall)

Open moist places beside fresh and seawater; 0–300 m.

Throughout Fuegia. (Map 5b).
Argentina north to *c.*50°S.lat.
Chile north to *c.*41°30′S.
Falkland Islands (Islas Malvinas); South Georgia; Islas Diego Ramirez; Iles Crozet; Iles de Kerguelen; Ile Amsterdam; Macquarie Island; Marion Island.

6. *R. maclovianus* D'Urv. 1825, *Fl. Is. Mal.* 48.

Plant white- or yellowish hairy; stems 0.5–*c.*2 cm, erect or ascending, sometimes divided, with dense basal rosette of leaves. Leaves with lamina 3–25 × 5–25 mm, suborbicular to somewhat reniform in outline, usually deeply 3-fid, cordate at base; segments irregularly 2- to 3(–5)-dentate or -lobed, sometimes entire when young, obtuse to acute; petiole 0.8–7 cm, with ovate, dark brownish, scarious basal sheath. Flowers solitary or in 2- to 3-flowered cymes; peduncle up to 3 cm, erect. Sepals 5, 2–3 × *c.*1 mm, unequal, ovate to ovate-lanceolate, obtuse to acute, membranous, brownish, hairy, reflexed. Petals 5, *c.*3 × 1–1.5 mm, spathulate, the limb ovate to suborbicular, obtuse, glabrous or sometimes hairy on back, yellow. Stamens 7–11. Carpels numerous. Achenes 2–2.5 × 1–1.5 mm, asymmetrically ovoid to oblong-ovoid, somewhat compressed, marginate, often narrowly winged on one side, glabrous or sparsely hairy on keel, the beak *c.*0.75 mm. triangular. $2n = 48^{\star}$ Self-compatible. *Flowers December to January.* (Fig. 39).

Open moist places by coast and inland streams. 0–200 m.

Both sides of Canal Beagle. (Map 5c).
S.C. Chile (*c.*40°S.lat.).
Falkland Islands (Islas Malvinas).

7. *R. sericocephalus* Hooker f. 1847, *Fl. Antarct.*, **1(2)**: 225.

Similar to *R. maclovianus* but usually smaller and more densely hairy in all parts, the leaves often hispid; flowers always solitary; peduncles up to 1 cm; petals densely hairy on back; achenes densely hairy on keel and dorsal part, marginate but not winged. $2n = 48^{\star}$. Self-compatible, at least sometimes self-pollinated. *Flowers November to February.* (Fig. 40).

Moist open peat and sand; 0–1000 m.

C. & S. parts of Isla Grande. (Map 5d).
Chile north to 50°55′S.lat.
Falkland Islands (Islas Malvinas).

8. *R. cymbalaria* Pursh 1814, *Fl. Amer. Sept.* **3**: 392.

Plant slightly pubescent, becoming glabrous; stems up to 25 cm, creeping, rooting at the nodes and sometimes twining at the apex. Basal leaves with lamina 3–40 × 2–46 mm, suborbicular, rarely ovate or oblong to reniform in outline, cordate or truncate at base, 3-lobed, the lateral lobes divided or crenate to entire; segments obtuse, rarely acute or emarginate, irregularly crenate, sparsely hairy at base and margins, rarely all pubescent; petiole 2–23 cm, pubescent, with ovate to linear-triangular, subauriculate or decurrent, amplexicaul, scarious basal sheath. Flowers solitary or in few-flowered cymes; peduncles 4–30 cm, pubescent. Sepals 5–6, 3.5–4.5 × 2–2.5 mm, somewhat asymmetrically ovate to suborbicular or obovate, obtuse, concave (especially the 3 outer), yellow, with entire, somewhat undulate margin. Petals 5–6(–7), 2.5–4 × 1–1.5 mm, obovate to elliptic-subspathulate, obtuse, entire, yellow. Stamens 20–50. Carpels numerous. Achenes asymmetrically oblong to obovoid, subtetrahedral, marginate, slightly reticulate, the beak short. *Flowers January to February.* (Fig. 42).

Moist open ground by streams and lakes; 0–50 m.

Local in C. & S. parts of Isla Grande. (Map 5e).
Northwards at increasing altitudes to *c.*22°S.lat. in Argentina and Chile; Bolivia; Uruguay; N. America.

Perhaps under-recorded in Fuegia because of confusion with the more common *R. fuegianus*, which is glabrous and has 3–5 sepals, 5 petals and 14–20 stamens.

9. *R. uniflorus* Phil. ex. Reiche 1894, *Anal. Univ. Chile* **88**: 58.

R. bovei Speg. 1896, *Anal. Mus. Ci. Nat. Bs. Aires* **5**: 44 – Alboff, 1897; Wildeman, 1905; Macloskie, 1905; Dusén, 1900.

Plant fleshy; stems long, glabrous, with roots and clusters of leaves at nodes. Leaves with lamina 11–25 × 7–24 mm, ovate to suborbicular, obtuse or rarely emarginate, truncate to cordate at base, entire, sometimes with undulate margin, glabrous; petiole 6–30 cm, with long, usually appressed, hairs, with scarious basal sheath 2–23 × 1.5–7 mm. Flowers solitary; peduncle 1–7.5 cm, subglabrous, erect.

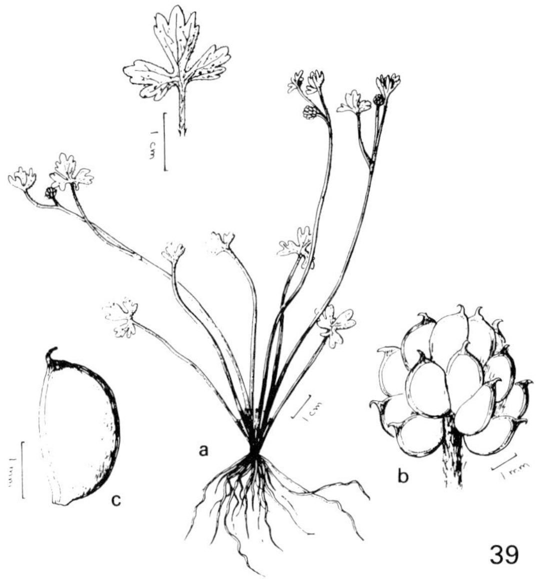

39

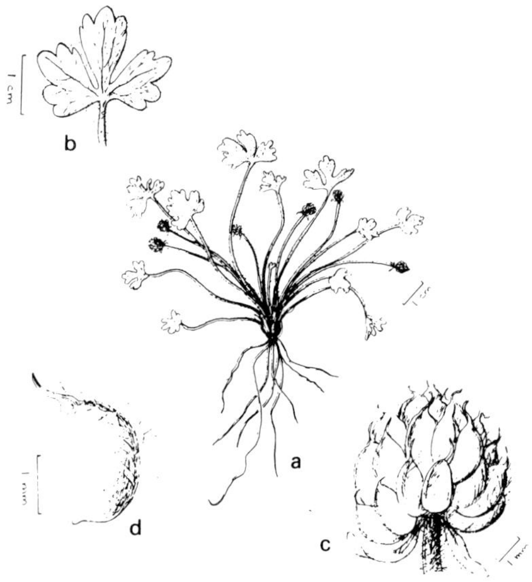

40

41

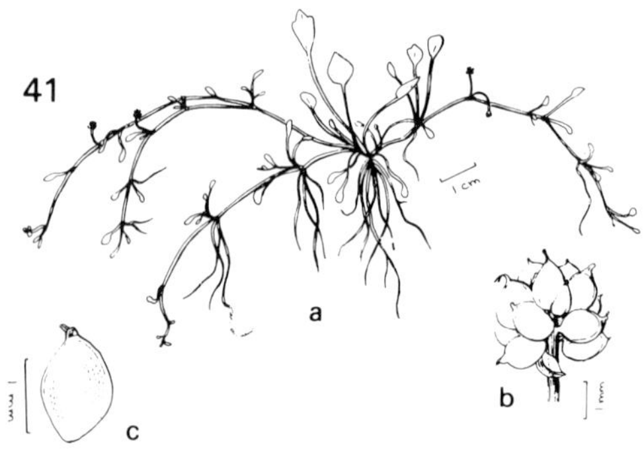

42

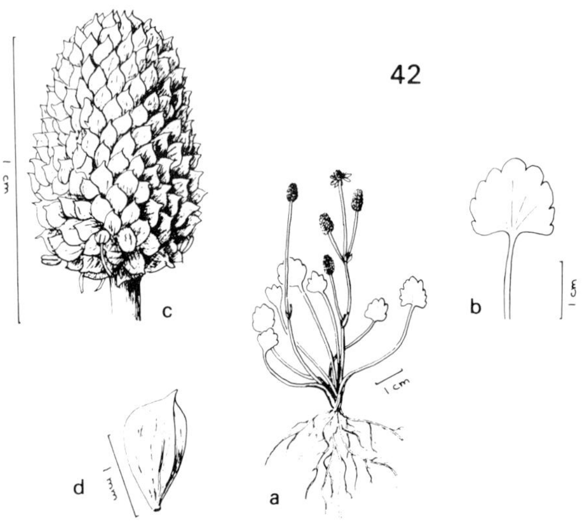

43

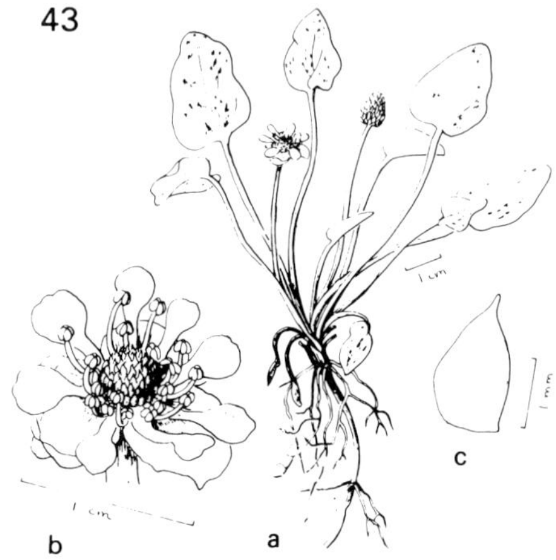

Fig. 39. *Ranunculus maclovianus*. a, Plant. b, Fruiting head. c, Achene. Scale = 1 cm (a) or 1 mm (b–c). Fig. 40. *Ranunculus sericocephalus*. a, Plant. b, Leaf. c, Fruiting head. d, Achene. Scale = 1 cm (a–b) or 1 mm (c–d). Fig. 41. *Ranunculus hydrophilus*. a, Plant. b, Fruiting head. c, Achene. Scale = 1 cm (a) or 1 mm (b–c). Fig. 42. *Ranunculus cymbalaria*. a, Plant. b, Leaf. c, Fruiting head. d, Achene. Scale = 1 cm (a–c) or 1 mm (d). Fig. 43. *Ranunculus uniflorus*. a, Plant. b, Flower. c, Achene. Scale = 1 cm (a–b) or 1 mm (c).

(R. N. P. Goodall)

Sepals 3, 4.5–7 × 3.2–5.5 mm., ovate to suborbicular, obtuse, concave, glabrous, yellowish. Petals (6–)7, 5.75–6 × 1.2–3 mm, spathulate to subspathulate, obtuse, slightly unequal, with elliptical to oblong limb, yellow. Stamens 25–35. Carpels numerous. Achenes *c.*1.5 mm, subprismatic, thick, longitudinally nerved, the beak short, suberect. $2n = 32$. Self-compatible, at least facultatively self-pollinated. *Flowers November to March.* (Fig. 43).

Seasonal pools, moist open areas by streams, lakes and in marshes or wet grassland. 0–100 m.

Throughout most of Isla Grande and adjacent islands in Estrecho de Magallanes. (Map 5f).
W. Argentina north to *c.*28°S.lat.
Chile north to *c.*35°S.lat.

10 ***R. hydrophilus*** Gaudich. 1825, *Ann. Sc. Nat.* **5**: 105.

Glabrous perennial, perhaps sometimes annual; stems up to 10 cm, prostrate, sometimes branched, with roots, rosettes of leaves and often flowers at most nodes. Leaves with lamina 3–10 × 2–8 mm, suborbicular to ovate or elliptical, obtuse to subacute, cuneate at base, entire or weakly undulate to crenate, rarely weakly serrate-crenate; petiole 0.6–4(–6) cm, usually narrowly winged, with ovate, scarious basal sheath. Flowers solitary; peduncles 2–9 mm., ascending or erect. Sepals 3, 1.2–2 × 0.5–1.5 mm, ovate, obtuse to subacute, cucullate, greenish-gold, tinged with pink, glabrous, reflexed. Petals 3–7, 1.2–2 × *c.*0.5–1 mm, oblong to oblanceolate or spathulate, obtuse. Stamens 5–8(–10). Carpels 10–*c.*20. Achenes *c.*1.5 × 1 mm, asymmetrically ovoid, slightly compressed; the beak *c.*0.1 mm. $2n = 32$. *Flowers November to March.* (Fig. 41).

In shallow, temporary pools, on wet soil and sand, rarely in *Juncus* and *Marsippospermum* bogs. 0–40 m.

S. & W. Fuegia. (Map 5g).
W. Argentina north to *c.*37°S.lat.
Chile north to *c.*40°S.lat.
Falkland Islands (Islas Malvinas).

11. ***R. peduncularis*** Sm. 1814, in Rees, *Cyclop.* **29**: No.49.

R. glandulifer Poeppig 1833, *Fragm. Syn. Pl.* 28 – Franchet, 1889; Alboff, 1897; Spegazzini, 1902; Wildeman, 1905; Autran, 1905.

R. alboffii Macloskie 1905, *Rep. Princeton Univ. Exped. Patag.* **8(5,2)**: 405.

Stems up to 55 cm, erect, appressed-pubescent or sericeous. Leaves with lamina up to 65 × 80 mm., subtriangular to subrhomboid in outline, 3-partite, densely appressed pubescent, sometimes sericeous; segments 2- to 3-fid, obtuse to subacute, the lateral more or less deeply divided; petiole up to 21 cm, with subamplexicaul, sometimes subauriculate basal sheath with scarious margin. Flowers solitary or in 2- to 3-flowered cymes. Sepals 5, *c.*4 × 2 mm, ovate to obovate, obtuse to subacute, concave, green, with hyaline margin, densely appressed-pubescent on back. Petals 10–16, 6–11 mm, oblong – to obovate – subspathulate, bright yellow. Stamens 25–75. Carpels 20–70. Achenes 25–70, 4 × 2.5 mm, asymmetrically suborbicular, compressed, marginate, finely foveolate, the beak somewhat curved. $2n = 32$. *Flowers November to February.* (Fig. 44).

Open moist grassland and, more usually, scrub and forest margins.

Throughout Isla Grande and islands along Canal Beagle. (Map 5h).
W. Argentina north to *c.*34°39′S.lat.
Chile north to *c.*33°S.lat.

Plants with large leaves (up to 120 × 110 mm) and narrow, profusely lobed segments, occur in Fuegia and elsewhere within the range of the species, and have been called var. *erodifolius* (C. Gay) Reiche.

Fuegian records (Skottsberg, 1906) of *R. chilensis* DC. were incorrect determinations of *R. minutiflorus*. *R. chilensis* is very similar to *R. peduncularis*, being separated by having 3, not 5, sepals and fewer (8–10) petals than the latter species, which has 10–16. However, some Fuegian specimens have some flowers with variable numbers of sepals and petals which blur these distinctions. A thorough examination of these two taxa throughout their ranges is clearly necessary.

12. ***R. fuegianus*** Speg. 1896, *Anal. Mus. Nac. Bs. Aires.* **5**: 43.

Plant glabrous; stems creeping, sometimes floating, rooting and with leaves and flowers at nodes. Leaves with lamina 6–30 × 8–40 mm., subpentagonal to suborbicular in outline, deeply 3-partite, truncate to broadly and deeply cordate at base, the segments usually 2- to 3-lobed, obtuse, the lateral asymmetrically divided; petiole 1–10 cm, with wide, biauriculate, scarious, slightly pubescent basal sheaths. Flowers in few-flowered cymes or solitary,

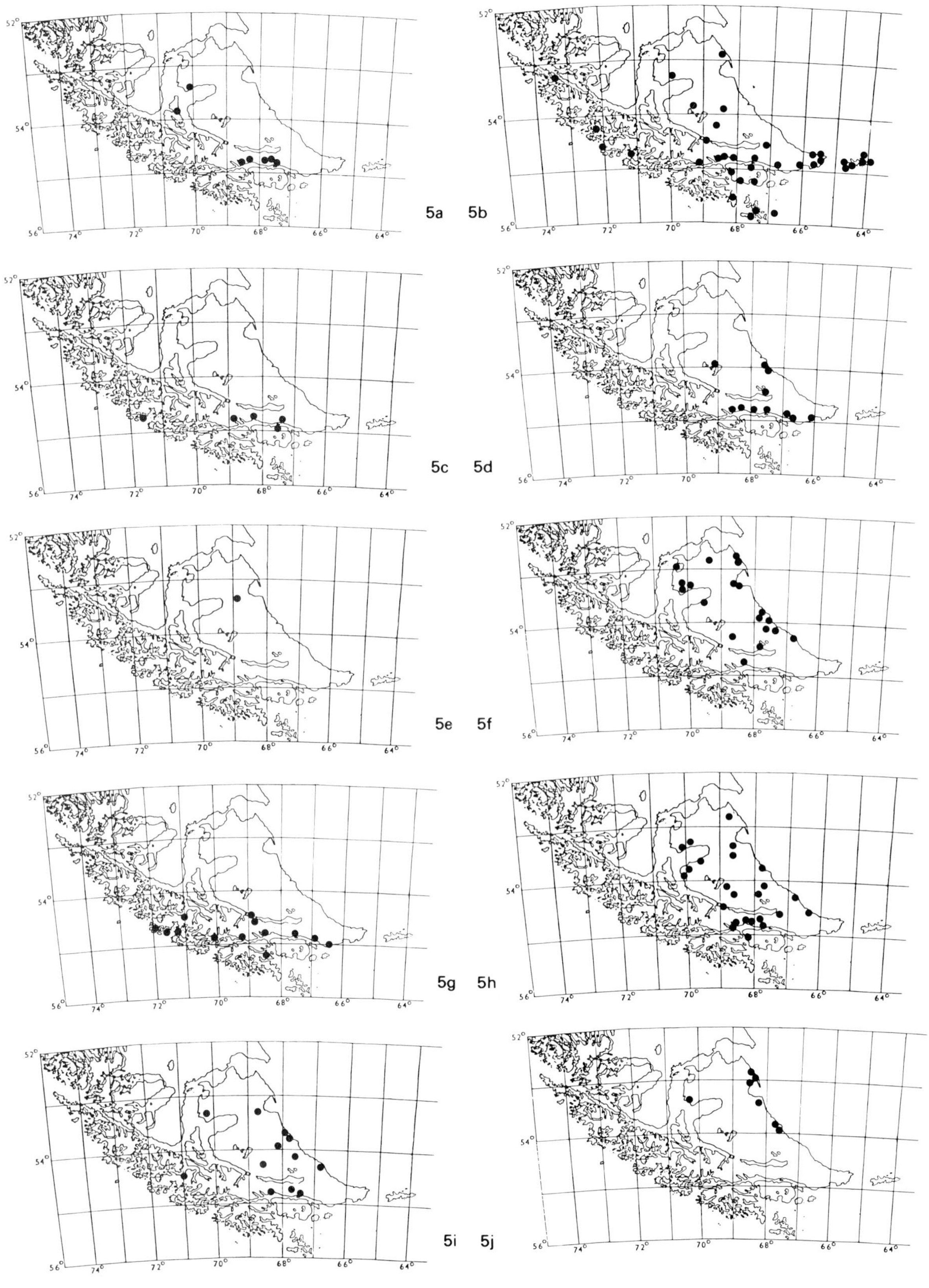

Map 5. a, *Ranunculus minutiflorus*. b, *Ranunculus biternatus*. c, *Ranunculus maclovianus*. d, *Ranunculus sericocephalus*. e, *Ranunculus cymbalaria*. f, *Ranunculus uniflorus*. g, *Ranunculus hydrophilus*. h, *Ranunculus peduncularis*. i, *Ranunculus fuegianus*. j, *Myosurus patagonicus*.

leaf-opposed. Sepals 3–5, 3.5–4 × 2–2.5 mm, concave or cucullate, obtuse, glabrous. Petals 5, 3.5–6×2.3–3 mm, oblong– to suborbicular-spathulate, obtuse to subemarginate, yellow. Stamens 14–20. Carpels 40–65. Achenes *c.*1.5 mm., suborbicular to subtetrahedral, compressed, finely foveolate to almost smooth, the beak short, somewhat curved. $2n = 48$. *Flowers December to February*. (Fig. 46).

Moist places by freshwater, swamps, seasonal pools. 0–330 m.

Throughout Isla Grande westwards to 71°W. long. (Map 5i).
Argentina north to 49°S.lat.
Chile north to *c.*46°40′S.lat.

4. *Myosurus* L.

Small annuals. Leaves entire, in basal rosette. Flowers regular, hermaphrodite, solitary. Sepals 5, free, with oblong-ovate basal spur. Petals 5, with basal nectariferous scale. Stamens 5–10. Carpels numerous, inserted on central axis which elongates in fruit Achenes numerous, aristate, densely imbricate.

1. ***M. patagonicus*** Speg. 1897, *Rev. Fac. Agric. Vet. La Plata* **3**: 488.
M. aristatus auct., non Bentham ex. Hooker – Alboff, 1896; Dusén, 1900.

Caespitose annual 1.2–6 cm. Leaves 6–30 × 0.5–1.5 mm, filiform to linear or narrowly oblanceolate, obtuse to subacute, with ovate to subtriangular, somewaht scarious basal sheath. Sepals 2–3 × 0.5–1 mm, oblong, obtuse, slightly concave, yellowish brown; spur ovate, acute. Petals 3–4 × *c.*0.25–0.5 mm, sublinear, greenish. Fruiting head 4–35 × 1.5–3 mm, cylindrical, tapering distally; achene 1.4–2 mm, irregularly prismatic to ovoid, laterally compressed, pubescent at least distally; dorsal arista prolonged in straight or slightly curved beak. Self-compatible and at least facultatively self-pollinated. *Flowers September to January*. (Fig. 50).

Vernally moist open soil and sand in grassland and *Lepidophyllum* scrub, especially near the coast; 0–20 m.

N. & N.E. parts of Isla Grande. (Map 5j).
Argentina north to 46°S.lat. and at *c.*36°S.lat.
Chile north to 51°30′S.lat. and at *c.*34°S.lat.

According to Lourteig (1952) *M. patagonicus* and *M. apetalus* C. Gay occur immediately north of the Straits of Magellan, the latter having been recorded in Fuegia by Dusén (1900) under the synonym *M. aristatus*. *M. apetalus* is distinguished from *M. patagonicus* by its smaller size, in most parts, and by its glabrous (not pubescent) achenes with a long, curved (not short, straight) beak. The Fuegian plants are variable in size and the achenes, which are all at least to some extent pubescent, have short curved to longer straight beaks. Since the achene-shape, as well as their pubescence, accords more with that described for *M. patagonicus* they are so named here, but a critical revision of the S. American members of the genus is necessary.

5. *Anemone* L.

Pubescent perennial herbs, with woody rhizome. Leaves deeply divided, the basal long-petiolate. Flowering stems with a whorl of 3 partially united, sessile leaves. Flowers hermaphrodite, actinomorphic, solitary or in 2- to 7-flowered cymes. Perianth-segments 5–9, petaloid. Stamens 50–120. Carpels 150 or more. Fruit an achene, clustered in compact head.

1. ***A. multifida*** Poiret 1810, in Lam., *Encycl. Méth. Bot.* **9**: 364.
A. decapetala auct., non Ard. – Franchet, 1889.

Stems up to 45 cm, erect. Basal leaves with lamina up to 10 cm in diameter, subpentagonal in outline, 3-partite; segments 3-partite and unequally, acutely to acuminately lobed; petiole up to 15 cm. Peduncles up to 20 cm. Perianth-segments 10–15 × 4–6 mm, ovate to oblong or, rarely, subspathulate, obtuse to acute, somewhat unequal, pubescent on back, cream to yellowish. Fruiting head 1–2 cm, subglobose to shortly cylindrical; achenes *c.*4 mm, asymmetrically ovate, compressed, covered with long, whitish hairs, with curved lateral beak. $2n = 32$. Self-compatible and at least facultatively self-pollinated. *Flowers November to January*. (Fig. 47).

Grassland, especially among rocks, *Empetrum* heath, open *Chiliotrichum* and dwarf *Nothofagus* scrub; 0–250 m.

Throughout Isla Grande and adjacent islands, except extreme S.W.; Isla de los Estados. (Map 6a).
W. Argentina north to 33°S.lat.
Chile north to *c.*37°20′S.lat.
North America.

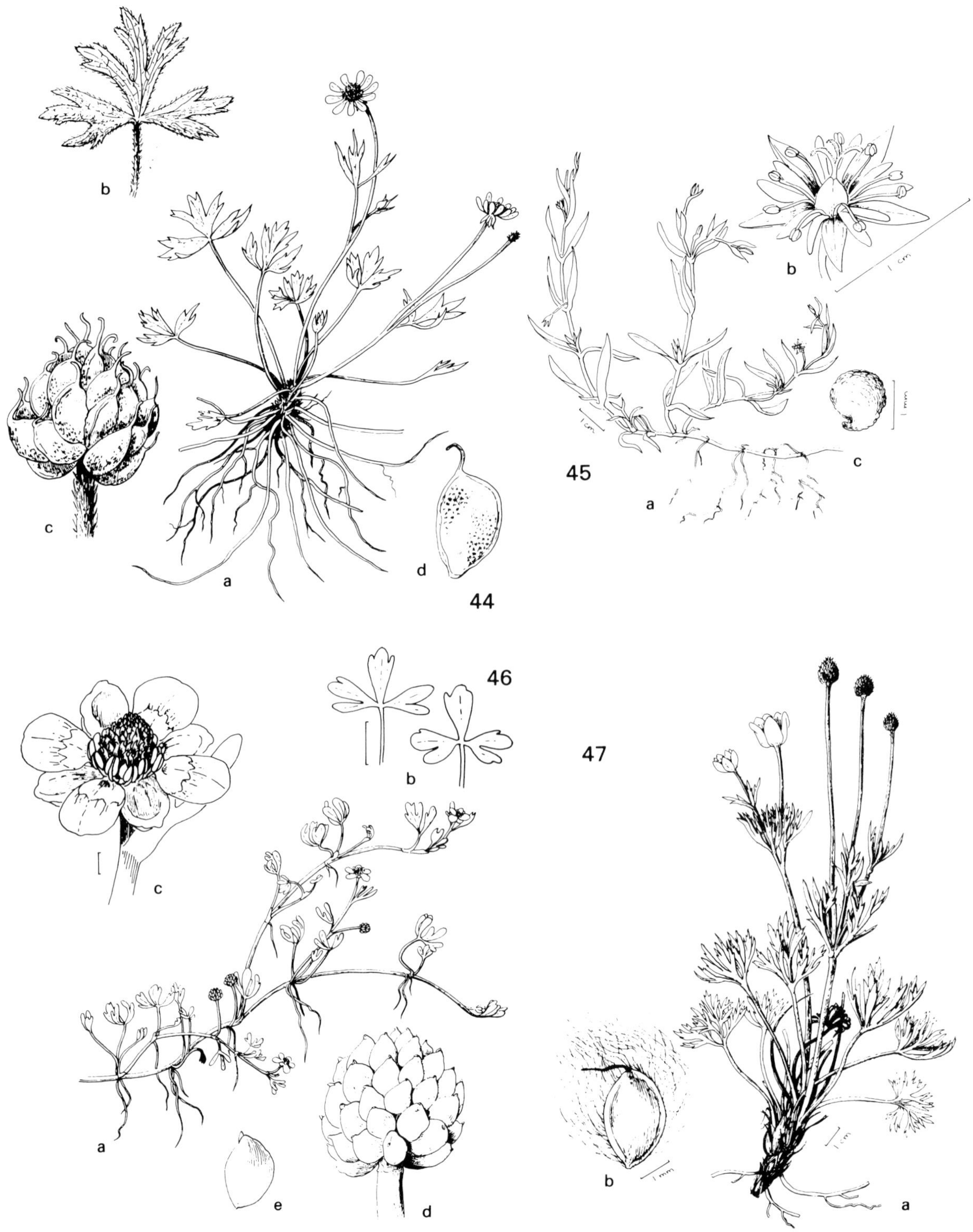

Fig. 44. *Ranunculus peduncularis*. a, Plant. b, Leaf. c, Fruiting head. d, Achene. Scale = 1 cm (a–b) or 1 mm (c–d). Fig. 45. *Stellaria debilis*. a, Plant. b, Flower. c, Seed. Scale = 1 cm (a–b) or 1 mm (c). Fig. 46. *Ranunculus fuegianus*. a, Plant. b, Leaves. c, Flower. d, Fruiting head. e, Achene. Scale = 1 cm (a–b) or 1 mm (c–e). Fig. 47. *Anemone multifida*. a, Plant. b, Achene. Scale = 1 cm (a) or 1 mm (b).

(R. N. P. Goodall)

XVIII. PAPAVERACEAE

Herbs with latex. Leaves deeply dissected, without stipules. Flowers hermaphrodite, actinomorphic, hypogynous. Sepals 2, caducous. Petals 4, caducous. Stamens numerous. Ovary superior, 1-celled; stigmas 4–6, sessile. Fruit a capsule; seeds with endosperm.

1. *Papaver* L.

Leaves in basal rosette. Flowers solitary. Sepals free. Petals free, crumpled in bud. Stigmas in a flat, 4- to 6-rayed disc. Capsule dehiscing by pores below stigmatic disc; seeds numerous.

*1. ***P. nudicaule*** L. 1753, *Sp. Pl* 507.

Robust perennial. Leaves with lamina 2–7 × 1–4 cm, ovate in outline, broadly lobed, cuneate at base; petiole 2–6 cm. Flowering stems up 50 cm. Flowers up to 6 cm in diameter, yellow.

Cultivated in gardens and locally naturalized or adventive on beaches and open ground; 0–5 m.

S. Isla Grande (near Ushuaia).
Native of Asia, widely cultivated and locally naturalized.

****Fumaria officinalis*** L. 1753, *Sp. Pl.* 700, which is included in the Papavercaceae or in a separate family, the Fumariaceae, is introduced from Europe and persists as a weed in gardens and disturbed ground near settlements. It is a glabrous, glaucous scrambling annual herb with watery sap, alternate pinnately compound leaves having flat, lanceolate to linear segments, and zygomorphic flowers in (10–)20 – to 40– flowered, leaf-opposed, bracteate racemes 3–5 cm. The corolla is 6.5–9 mm, with 4, pink, dark-tipped petals, the upper spurred, there are 2, 3-partite stamens and the fruit is a 1-seeded, indehiscent nutlet with 2 apical pits when dry.

XIX. FAGACEAE

Monoecious trees or shrubs. Leaves evergreen or deciduous, alternate, simple, with caducous stipules. Male flowers 1–3 in short spikes; perianth 4- to 6-lobed; stamens few to numerous. Female flowers 1–3, surrounded by 2- or 4-lobed involucre of scales (cupule); perianth 4- to 6-lobed; ovary inferior, 3- or 6-celled, with as many styles; ovules pendent, 2 per cell, 1 usually aborting. Fruit a 1-seeded nut, in groups of 1–3, surrounded by accrescent cupule.

1. *Nothofagus* Blume

Protandrous. Male flowers solitary, in pairs or threes; perianth campanulate; stamens 8–30. Female flowers solitary or in threes; perianth minute, dentate. Cupule hardened in fruit, enclosing 1 or 3 nuts.

a. Leaves evergreen, serrate **3. *betuloides***
a. Leaves deciduous, somewhat lobed or double dentate:
 b. Leaves somewhat lobed, crenate; nuts in groups of 3 **1. *antarctica***
 b. Leaves double dentate; nuts solitary **2. *pumilio***

1. *N. antarctica* (Forster f.) Oersted 1871, *Danske Vid. Selsk.*, Ser. 5, **9**: 354.
Fagus antarctica Forster f. 1789, *Comment. Gotting.* **9**: 24.

Tree 3–15 m, rarely shrub 0.5–2 m; trunk up to 0.8 m in diameter, the bark greyish, ± rough; twigs puberulent to glabrous. Leaves deciduous; lamina 8–40 × 7–18(–22) mm, broadly ovate to oblong or suborbicular in outline, obtuse, truncate to cordate, rarely cuneate, at base, somewhat lobed and crenate, glabrous but puberulent on veins beneath; petiole 2–7 mm. Male flowers solitary, in pairs or threes; stamens 8–13. Female flowers in groups of 3. Nuts in threes, *c.*6 × 5 mm, triquetrous, glabrous; capule 4-partite. *Flowers October to February*. (Fig. 84, Pl. 5b).

Dominant or locally dominant in deciduous forest; forming scrub or isolated bushes extending into swampy or dry areas. 0–550 m.

Throughout Fuegia except the extreme north. (Map 6b).
W. Argentina north to *c.*36°S.lat.
Chile north to *c.*36°S.lat.

2. *N. pumilio* (Poeppig & Endl.) Krasser 1896, *Ann. Naturh. Mus. (Wien)* **9**: 161.

Tree up to 30 m, or small shrub; trunk up to 1.3 m in diameter, the bark brownish grey, ± rough; twigs puberulent to glabrous. Leaves deciduous; lamina 16–31 × 10–27 mm, ovate-oblong to broadly elliptical, obtuse, cuneate at base, obtusely double dentate, subglabrous; petiole 2–7 mm. Male flowers solitary; stamens 20–30. Female flowers solitary. Nuts solitary, *c.*7 × 3–4 mm, triquetrous, puberulent; cupule 2-partite. $2n = 26$. *Flowers October to January*. (Fig. 48).

Dominant or co-dominant in deciduous forest on well-developed soils with good drainage; 0–70 m.

C. Fuegia. (Map 6c).
W. Argentina north to *c*.36°S.lat.
Chile north to *c*.37°S.lat.

3. ***N. betuloides*** (Mirbel) Oersted 1871, *Danske Vid. Selsk.*, Ser. 5, **9**: 354.
N. nitida auct., non Phil. – Skottsberg, 1906.
Fagus betuloides Mirbel 1827, *Mém. Mus. Hist. Nat. (Paris)* **14**: 469 – Franchet, 1889; Hariot, 1900.

Tree up to 20(–30) m or shrub 1–2 m; trunk up to 2 m in diameter, the bark grey or reddish, ± smooth; twigs puberulent, rarely glabrous, somewhat viscid. Leaves evergreen; lamina 12–28 × 8–18 mm, ovate-elliptical to elliptic – suborbicular, acute to obtuse, cuneate at base, serrate, subcoriaceous, glabrous; petiole 2–5 mm. Male flowers solitary; stamens 10–16. Female flowers in groups of 3. Nuts in threes, 5–6 × 4–4.5 mm, triquetrous, glabrous; cupule 4-partite. *Flowers November to February*.

Dominant or locally dominant in evergreen forest, on better-drained sites in higher rainfall areas; 0–500 m.

C. & S. Fuegia. (Map 6d).
W. Argentina north to *c*.46°40′S.lat.
Chile north to *c*.41°S.lat.

XX. CARYOPHYLLACEAE

Herbs, sometimes forming dense cushions. Leaves opposite, entire, usually exstipulate. Flowers regular. Sepals 4–5, free or fused. Petals 4–5, or 0, free. Stamens (3)4–10. Ovary superior, 1-celled; styles 2–6. Fruit a capsule, dehiscing with as many or twice as many teeth as styles.

Literature: R. Martinez-Crovetto, *Bonplandia* **2**: 187–264(1967).

a. Petals absent:
  b. Leaves linear-lanceolate to ovate; styles 3 — **2. *Stellaria***
  b. Leaves linear, often rigid; styles 4–5:
    c. Stamens opposite sepals; sepals widely spreading in ripe fruit — **5. *Sagina***
    c. Stamens alternating with sepals; sepals not widely spreading — **4. *Colobanthus***
a. Petals present:
  d. Petals ± deeply bifid:
    e. Sepals fused to form a distinct calyx-tube — **7. *Silene***
    e. Sepals free:
      f. Styles 5 — **3. *Cerastium***
      f. Styles 3 — **2. *Stellaria***
  d. Petals entire or emarginate:
    g. Sepals more than 10 mm — (see **7.**) ***Dianthus***
    g. Sepals less than 10 mm:
      h. Stipules present — **6. *Spergularia***
      h. Stipules absent:
        i. Capsule-teeth as many as styles — **5. *Sagina***
        i. Capsule-teeth twice as many as styles — **1. *Arenaria***

1. *Arenaria* L.

Perennial herbs. Leaves simple; stipules absent. Flowers hermaphrodite, solitary. Sepals 5, free. Petals 5, entire, white. Stamens 10. Styles 3. Capsule dehiscing by 6 teeth; seeds numerous.

1. ***A. serpens*** Humb., Bonpl. & Kunth. 1823, *Nov. Gen. Sp.* **6**: 32.
A. serpylloides Naudin var. *andicola* (Gillies) Rohrbach 1873, *Linnaea* **37**: 270 – Skottsberg, 1906.

Plant often somewhat caespitose, branched below. Stems (3–)5–12(–15) cm, slender, procumbent, becoming ascending to erect at apices, glabrous or puberulent. Leaves (2.6–)4.5–11 × (0.6–)1–1.9 mm, linear to oblanceolate or subspathulate, obtuse, sometimes ciliate near base, sessile, connate at base. Flowers axillary; peduncles 1.5–6.5 mm. Sepals (1.8–)2.3–3.5 × 1.3–2 mm, ovate to elliptical, obtuse, entire, green with scarious margin. Petals 2–4.8 × 1–2.5 mm, broadly spathulate, rounded at apex. Capsule 4–5 × (3–)3.5–4.5 mm. ovoid to subglobose; seeds 0.9–1.3 × 0.6–1 mm, broadly reniform, shiny, black. Self-compatible and capable of self-pollination. *Flowers December to April*. (Fig. 49).

Open pebbles and sand beside streams and lakes, less commonly on damp open soil; 0–400 m.

C. & S. Isla Grande. (Map 6e).
W. Argentina and E. Chile north to *c*.27°S.lat.

The Fuegian plants belong to var. *andicola* (Gillies) Rohrbach.

2. *Stellaria* L.

Annual or perennial herbs. Leaves simple; stipules absent. Flowers hermaphrodite. Sepals 5, free, entire. Petals 5, bifid, white, rarely absent. Stamens (3–)5–10. Styles 3. Capsule dehiscing by 3 or 6 teeth; seeds numerous.

a. Leaves linear-lanceolate, sessile — **1. *debilis***
a. At least lower leaves petiolate, ovate to suborbicular:

b. Plant glabrous; inflorescence with small, scarious bracteoles 1–1.5 mm **3. *parviflora***
b. Plant usually with 1(2) lines of hairs distally on stems; inflorescence with herbaceous bracteoles more than 2 mm **2. *media***

1. *S. debilis* D'Urv. 1825, *Fl. Is. Mal.*: 52.
S. stenopetala Phil. var. *magellanica* Phil. 1872, *Anal. Univ. Chile* **41**: 678 – Dusén, 1900.
S. axillaris Phil. 1892, *Anal. Univ. Chile* **81**: 762 – Alboff, 1896; Wildeman, 1905.
S. rotundifolia auct., non Poiret – Hariot, 1900.

Lax to caespitose glabrous perennial. Stems 3–20 cm, angled, slender, decumbent. Lower leaves (5–)6.5–20 × 8–3(–15) mm, linear-lanceolate, acute, sessile. Flowers solitary in axils of upper leaves, pedunculate. Sepals 2–4 × 1.1–6 mm, triangular- to ovate-lanceolate, acute, brownish-green. Petals slightly shorter than sepals. Stamens 10. Capsule 3–4 mm, slightly exceeding calyx, ovoid; seeds 0.5–0.9 mm. Self-compatible. *Flowers December to February.* (Fig. 45).

Damp grassland, marshy ground, mud and among pebbles by freshwater, and above highwater mark on beaches; 0–50 m.

N., C. & S.E. Isla Grande; Isla de los Estados. (Map 6f).
Argentina north to *c.*46°S.lat.
Chile north to *c.*45°S.lat.
Falkland Islands (Islas Malvinas).

*****2. *S. media*** (L.) Vill. 1789, *Hist. Pl. Dauph.* **3**: 615.
S. nemorum auct., non L. – Wildeman, 1905.

Annual. Stems 4–35 cm, angled, usually with 1(2) longitudinal lines of hairs distally, decumbent to ascending, much-branched. Lower leaves 5–30 mm, ovate, acute to acuminate, long-petiolate; upper leaves usually larger, ovate to broadly elliptical, sessile. Flowers few or many in axillary inflorescences at or near branch-apices; with herbaceous bracteoles at least 2 mm; pedicels slender, erect or spreading in fruit. Sepals 3.5–6 mm, lanceolate to ovate-lanceolate, acute to subobtuse, glandular-hairy, green, usually with narrow, scarious margin. Petals slightly shorter to slightly longer than sepals, sometimes absent, divided almost to base. Stamens (3–)4–8(–10). Capsule about as long as or slightly exceeding sepals, ovoid-oblong; seeds 0.8–1.4 mm, with rounded or conical tubercles. Self-compatible and frequently, perhaps usually, self-pollinated. *Flowers October to May.*

Open and disturbed ground, especially near settlements; 0–700 m.

Throughout Fuegia.
Introduced from Europe and naturalized in temperate regions of the world.

Fuegian material belongs to ssp. *media*.

3. *S. parviflora* Banks & Sol. ex Hooker f. 1852, *Fl. Antarct.* **2(1)**: 25.
S. decipiens auct., ?non Hooker f. – Skottsberg, 1926; Moore, 1974.

Glabrous perennial, perhaps sometimes annual. Stems up to 22 cm, angled, slender, decumbent. Leaves 3–10 × 2–8 mm, ovate-orbicular, obtuse, shortly cuspidate; petiole 2–6 mm. Flowers usually solitary, rarely in 2(3)-flowered axillary cyme; bracteoles 1–1.5 mm, scarious; pedicels slender. Sepals 2–2.7 mm, ovate, subacute, green, the inner with scarious margin. Petals at least half as long as sepals, all or some often absent, divided almost to base. Stamens 10, or some absent. Capsule up to twice as long as sepals, narrowly ovoid; seeds *c.*1 mm, striate and with low tubercles. *Flowers November to January.*

Moist, often shaded, cliffs, soil and coastal rocks; 0–100 m.

S. Fuegia north to S. and S.W. Isla Grande. (Map 6g).
New Zealand.

T. M. Pedersen (pers. comm.) considers the Fuegian material to comprise a separate subspecies, which will be published in a forthcoming volume of *Flora Patagonica*. This little-known taxon requires further study, particularly since there is some doubt that *S. parviflora* and *S. decipiens* are truly specifically distinct in New Zealand.

3. *Cerastium* L.

Perennial herbs. Leaves simple, sessile; stipules absent. Flowers hermaphrodite. Sepals 4–5, free. Petals 4–5, bifid, white. Stamens 5(10). Styles 5. Capsule dehiscing by 10(20) short teeth; seeds numerous.

a. Petals at least $1\frac{1}{4}$ times as long as sepals; capsule straight **1. *arvense***
a. Petals less than $1\frac{1}{4}$ times as long as sepals; capsule curved **2. *fontanum***

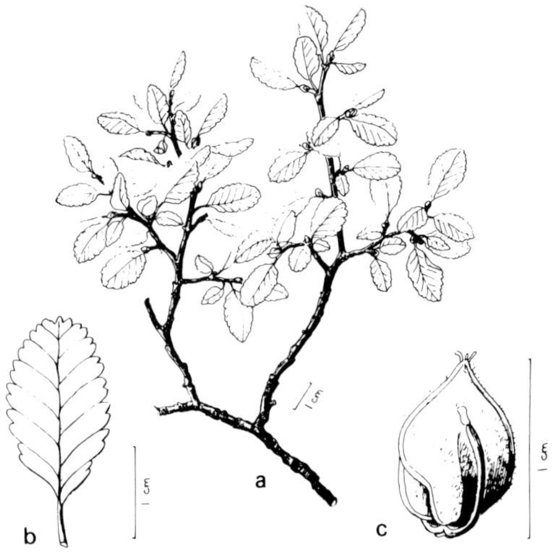

48

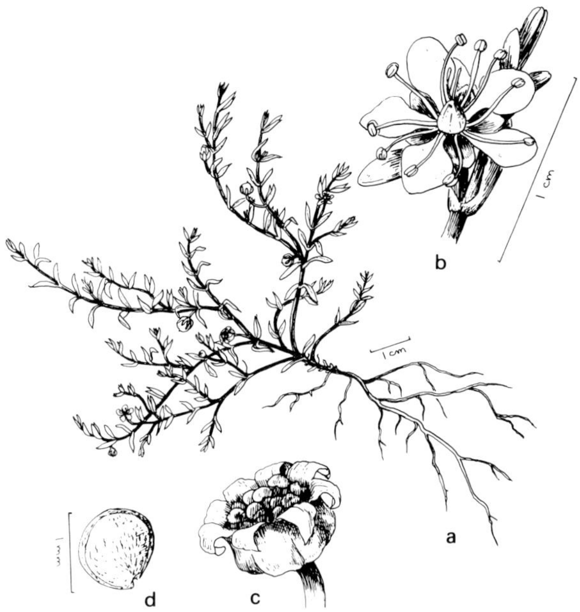

49

50

51

Fig. 48. *Nothofagus pumilio*. a, Twig. b, Leaf. c, Nut. Scale = 1 cm. Fig. 49. *Arenaria serpens*. a, Plant. b, Flower. c, Capsule. d, Seed. Scale = 1 cm (a–c) or 1 mm (d). Fig. 50. *Myosurus patagonicus*. a, Plant. b, Flower. c, Fruiting head. d, Achenes. Scale = 1 cm (a, c) or 1 mm (b, d). Fig. 51. *Cerastium fontanum*. a, Plant. b, Capsule. c, Seed. Scale = 1 cm (a–b) or 1 mm (c). (R. N. P. Goodall)

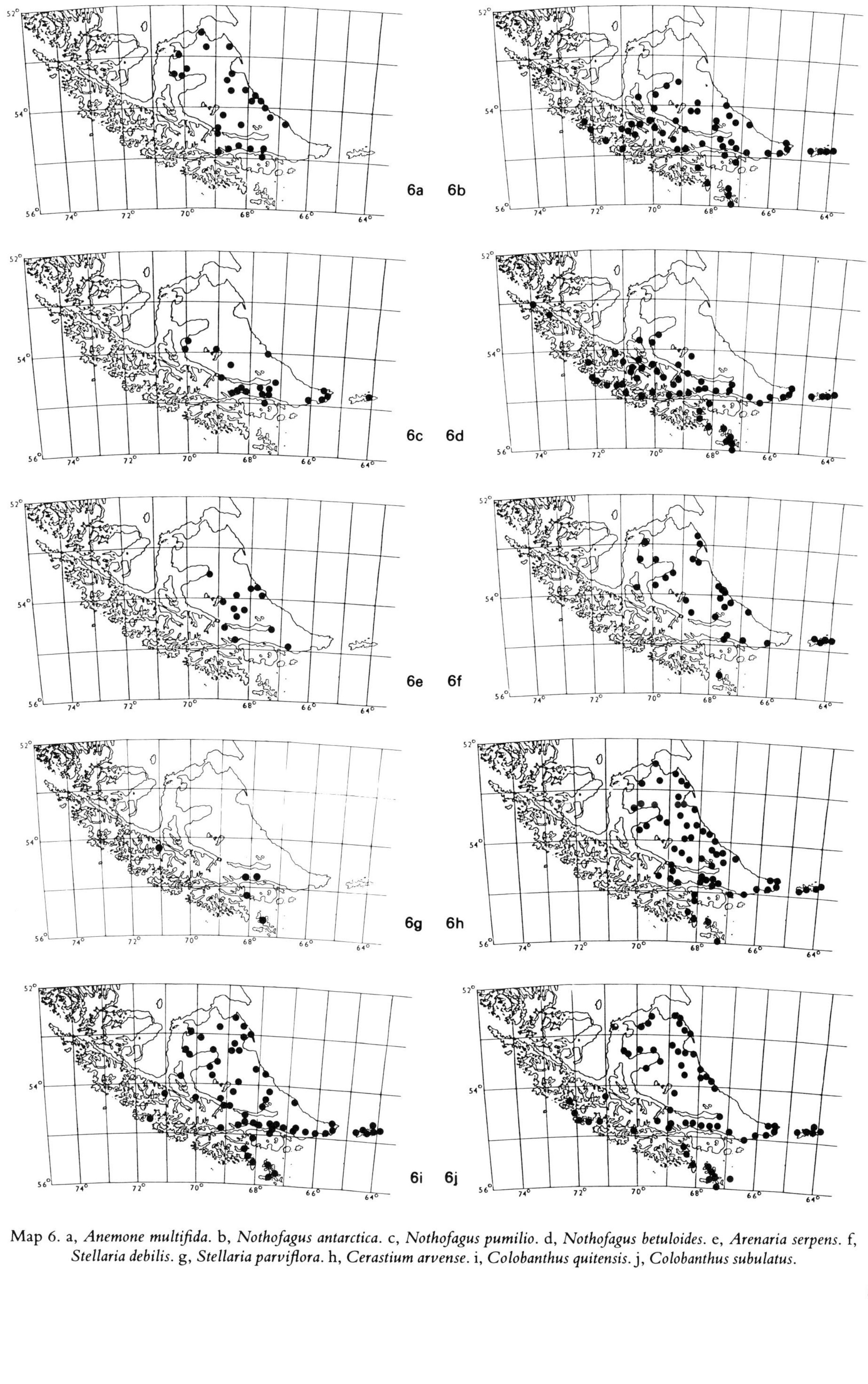

Map 6. a, *Anemone multifida*. b, *Nothofagus antarctica*. c, *Nothofagus pumilio*. d, *Nothofagus betuloides*. e, *Arenaria serpens*. f, *Stellaria debilis*. g, *Stellaria parviflora*. h, *Cerastium arvense*. i, *Colobanthus quitensis*. j, *Colobanthus subulatus*.

1. *C. arvense* L. 1753, *Sp. Pl.* 438.

C. fuegianum Alboff 1896, *Rev. Mus. La Plata* **7**: 362 – Autran, 1905; Wildeman, 1905; Skottsberg, 1906.

C. nervosum Naudin 1845, in C. Gay, *Hist. Chile Bot.* **1**: 277 – Skottsberg, 1916.

C. magellanicum Phil. 1872, *Anal. Univ. Chile* **41**: 678 – Alboff, 1896.

Stems loosely matted to caespitose, glandular-pubescent to hairy; sterile shoots procumbent and rooting at lower nodes; flowering shoots (3–)4–20(–25) cm, ascending. Leaves (4.5–)5–15(–17) × 0.5–2.8(–5) mm, linear to linear- or elliptic-lanceolate, acuminate to subobtuse, glandular-pubescent to hairy with hairs up to 0.4 mm. Flowers 1–8 in lax, terminal dichasia; pedicels 2–20 mm. Sepals 4–7.5(–9) × 0.8–2.2 mm, linear- to ovate-lanceolate, acute or subacute, glandular-pubescent to hairy, green but often purplish towards apex, with scarious margin. Petals 5–12 mm, exceeding sepals, obovate, divided to *c.* quarter of length. Capsule (6–)7.5–9.5 × 2.5–2.8 mm, slightly exceeding sepals, straight; teeth *c.*1 mm; seeds *c.*0.8 mm, tuberculate, dark reddish-brown. Self-compatible and often self-pollinated. *Flowers November to March.*

Grass- and shrub-steppe and heathland, open sand and pebbles, rock and soil screes, feldmark; 0–1,100 m.

Throughout Fuegia, rare in western islands. (Map 6h).
W. Argentina north to *c.*26°S.lat.
Chile north to *c.*25°S.lat.
North along Andes to *c.*0°lat.
Falkland Islands (Islas Malvinas). North America. Eurasia.

Tomentose plants with few, large flowers, which occur at higher elevations, have been described as *C. nervosum* Naudin. They appear to be ecotypic variants of *C. arvense*.

2. *C. fontanum Baumg. 1816, *Enum. Stirp. Transs.* **1**: 425.

C. holosteoides Fries 1828, *Nov. Fl. Suec.*, ed. 2, 158 – Ruiz Leal, 1954.

C. viscosum L. 1753, *Sp. Pl.* 437 – Spegazzini, 1896; Autran, 1905.

C. vulgatum L. 1762, non 1755 – Hariot, 1884; Franchet, 1889; Spegazzini, 1896; Autran, 1905; Skottsberg, 1906, 1916.

Stems rather laxly caespitose, densely hairy; sterile shoots procumbent, basal; flowering shoots 10–32 cm, ascending to erect. Leaves (7–)10–40 × (2.5–)3.5–9 mm, elliptic- to ovate-lanceolate or oblanceolate, obtuse to subacute, with dense hairs 1–2 mm. Flowers 1–18 in rather lax, terminal dichasia; pedicels 1–15 mm. Sepals 5–7 mm triangular- to ovate-lanceolate, acute to subobtuse, densely hairy, green, with scarious margin. Petals 4.5–6 mm, slightly shorter than to slightly exceeding sepals, divided to less than half-way. Capsule 5–9 mm, ± equalling to slightly exceeding sepals, curved; seeds 0.5–0.7 mm, tuberculate, reddish-brown. Self-compatible and usually self-pollinated. *Flowers November to March.* (Fig. 51).

Open and disturbed ground by roads, buildings, and in vegetation modified by man.

Throughout Fuegia.
Introduced from Europe and widely naturalized in temperate regions of the world.

The Fuegian plants belong to ssp. *triviale* (Link) Jalas, the most widespread form in Europe, where it is a common weed.

4. *Colobanthus* Bartl.

Glabrous, often strongly caespitose perennial herbs, usually with strong taproot. Leaves sessile, connate at base; stipules absent. Flowers hermaphrodite, borne singly at apices of branches. Sepals 4–5(6), free, persistent. Petals absent. Stamens as many as sepals and alternating with them. Styles as many as sepals. Capsule dehiscing by as many valves as there are sepals; seeds numerous.

a. Leaves usually longer than branches, usually spreading and herbaceous; peduncles usually much-exceeding leaves; sepals herbaceous, with membranous margin, usually obtuse, sometimes mucronate **1. *quitensis***

a. Leaves usually shorter than branches, imbricate, coriaceous; peduncles equalling or shorter than leaves; sepals rigid and shiny, acuminate **2. *subulatus***

1. *C. quitensis* (Kunth) Bartl. 1831, in C. Presl. *Reliq. Haenk.* **2**: 13.

C. crassifolius (D'Urv.) Hooker f. 1847, *Fl. Antarct.* **1(2)**: 248– Franchet, 1889; Alboff, 1896; Wildeman, 1905; Skottsberg, 1906, 1916.

Stems simple or branched to form mats or lax cushions several centimetres in diameter; branches 0.5–5(–7) cm, bearing 1 to many leafy shoots distally. Leaves (5–)10–25(–35) × (0.3)0.6–1.6 mm, linear to linear-triangular, obtuse to subacute, shortly mucronate, membranous and weakly sheathing at base, often weakly channelled on upper surface, herbaceous, rarely subcoriaceous, usually spreading. Flowers with peduncles equalling or usually much-exceeding leaves. Sepals 4–5(6), (2.2–)2.5–4.5 × 0.6–1.6 mm, ovate to triangular-ovate, usually obtuse, often shortly mucronate and cucullate at apex, with membranous margin, herbaceous. Capsule ± equalling to exceeding sepals, the valves obtuse; seeds 0.5–1 mm, triangular-reniform, smooth, pale reddish-brown with darker, narrow keel. *2*n* = *c*.80. Self-compatible and usually self-pollinated. *Flowers October to April.* (Fig. 72).

Damp sand, gravel and peat, mud, rock crevices; on coasts above high-water mark, by lakes, rivers, and other moist areas inland; common; 0–600 m.

Throughout Fuegia. (Map 6i).
W. Argentina north to *c*.24°S.lat.
Chile north to *c*.25°S.lat.
North along Andes to 0° lat. Mexico.
Falkland Islands (Islas Malvinas); South Georgia; South Shetland Islands; South Orkney Islands; Antarctic Peninsula; Islas Diego Ramirez.

The variation in this widespread species is described by Moore (1970). In exposed habitats it may resemble the next species in habit but can still be readily distinguished by its at most weakly channelled leaves and obtuse, sometimes mucronate sepals.

2. *C. subulatus* (D'Urv.) Hooker f. 1847, *Fl. Antarct.* **1(2)**: 247.

C. lycopodioides Griseb. 1854, *Syst. Bemerk.* 28 – Skottsberg, 1916.
C. patagonicus Hieron. ex. Wildeman 1905, *Result. Voy. S.Y. Belgica* 83 – Spegazzini, 1896.
C. polycnemoides Hieron. 1879, *Bol. Acad. Nac. Ci. Córdoba* **3**: 334 – Spegazzini, 1896; Wildeman, 1905.

Stems much-branched to form compact cushions 3–25 cm in diameter; branches (2–)4–12 cm, leafy along almost whole length. Leaves (2–)3.5–5(–6) × 0.5–1 mm, subulate to narrowly triangular, gradually acuminate, with conspicuously thickened, white margins giving channelled appearance, with midrib prominent beneath, with shiny, white basal sheath, coriaceous, imbricate. Flowers with peduncles not exceeding leaves. Sepals 4, (2–)3–4.5 × 0.5–1.3 mm, narrowly triangular to ovate, acuminate and often somewhat cucullate, usually thickened at margin, coriaceous and shiny. Capsule about half to two-thirds as long as sepals, the valves obtuse; seeds *c*.0.5 mm, subreniform, with minute, acute tubercles on keel, pale reddish-brown. *2*n* = *c*.80. Self-compatible and usually self-pollinated. *Flowers October to March.*

Rock crevices, soil and rock screes, coastal sands, open areas in *Empetrum* heath, saline flats, grass steppe, feldmark; 0–1,100 m.

Throughout Fuegia. (Map 6j).
S. & W. Argentina north to *c*.34°S.lat.
Chile north to *c*.36°S.lat.
Falkland Islands (Islas Malvinas); South Georgia.

Occurs in better drained, usually much drier habitats than *C. quitensis*. Variation in habit and leaf-shape appears to be associated with the degree of dryness of the habitat. Steppe variants with long, lax stems and very narrow leaves, have been described as *C. lycopodioides*.

5. *Sagina* L.

Tufted perennial herbs. Leaves sessile, connate at base; stipules absent. Flowers hermaphrodite, solitary, terminal. Sepals 4(5), free. Petals 4(5), free, usually minute or absent. Stamens as many as sepals and opposite them. Styles as many as sepals. Capsule opening to the base by 4(5) valves; seeds numerous.

1. *S. procumbens L. 1753, *Sp. Pl.*: 128.

Short, non-flowering main stem with a dense central leaf-rosette and numerous slender lateral stems 2.5–20 cm ascending from a rooted, procumbent base, glabrous, rarely ciliolate. Leaves 4–9.5(–11) × 0.4–0.7 mm, linear-subulate, obtuse and mucronate, glabrous. Flowers with peduncles 7–13(–22) mm, slender. Sepals 1–2.5 mm, ovate, obtuse and cucullate, spreading in fruit. Petals sometimes as long as sepals, usually minute or absent. Capsule 2.4–3 mm, exceeding sepals; valves obtuse; seeds *c*.0.3 mm, triangular-reniform, papillose, reddish-brown. Self-compatible and usually self-pollinated. *Flowers October to April, perhaps longer.*

Damp, open soil, rocks, pebbles, coastal sands, disturbed ground; 0–200 m.

Throughout Fuegia.
Introduced; native of Europe but common in temperate regions throughout the world.

The Fuegian and Falkland Islands (Islas Malvinas) plants belong to ssp. *procumbens*.

6. ***Spergularia*** (Pers.) J. & C. Presl

Annual or short-lived perennial herbs. Leaves sessile, decussate; scarious stipules present. Flowers hermaphrodite, in sparingly branched, lax dichasium. Sepals 5, free. Petals 5, free. Stamens (1–)3–6. Styles 3. Capsule dehiscing by 3 valves; seeds numerous.

1. ***S. marina*** (L.) Griseb. 1843, *Spicil. Fl. Rumel.* **1**: 213.

Stems 1–10 cm, arising from slender taproot, decumbent to ascending, flattened, glabrous below, sparsely glandular-hairy above. Leaves 17–35 × 1–2.5 mm, linear, obtuse, mucronulate, flat above, rounded beneath, glabrous, somewhat fleshy; stipules 1–3 mm, broadly triangular-ovate, fimbriate, connate to *c.* halfway to form a sheath around node. Pedicels 2–8 mm. Sepals 3.5–4.5 mm, lanceolate to ovate-lanceolate, obtuse, glabrous or sparsely glandular-pubescent, green, often partly pinkish-purple, with scarious margin. Petals 2.3–3 mm, shorter than sepals, ovate, subacute, pink, white towards base. Capsule 3.5–4.5 mm, slightly exceeding sepals, ovoid-conical; seeds 0.8–1.5 mm, with or without narrow marginal wing, smooth or tuberculate. Self-compatible and usually self-pollinated. *Flowers November to April (or May).* (Fig. 55).

Open mud by seashore; rare; 0–6 m.

N.E. Isla Grande. (Map 7a).
E. Argentina north to *c.*34°S.lat.
Falkland Islands (Islas Malvinas), Europe, N. America.

Fuegian and Falkland Islands plants are morphologically indistinguishable from European material, but they seem to be genetically distinct, hybrids between Northern and Southern hemisphere plants being almost completely sterile (Ratter, 1973).

7. ***Silene*** L.

Perennial herbs. Leaves petiolate; stipules absent. Flowers usually solitary. Calyx tubular, with 5 teeth. Petals 5, free, bifid, with distinct limb and claw, with 2 corona-scales at base of limb. Stamens (5)10. Styles (2–)4–5. Capsule dehiscing by 4–6 teeth; seeds numerous.

1. ***S. magellanica*** (Desr.) Bocquet 1967, *Candollea* **22**: 37.

Melandrium magellanicum (Desr.) Fenzl. ex. Rohrbach 1869, *Linnaea* **36**: 224 – Skottsberg, 1916.

Lychnis magellanicum Desr. 1791, in Lam., *Encycl. Méth. Bot.* **3**: 641 – Spegazzini, 1896.

Flowering stems 5–35 cm, arising with vegetative shoots from stout rootstock, with dense white hairs. Basal leaves (10–)20–50(–60) × 0.5–1(–2) mm, linear, acuminate, with long, winged petiole sheathing at base, coriaceous, densely hairy, imbricate; cauline leaves smaller, sessile, sometimes subconnate at base. Peduncle (6–)20–50(–60) mm, hairy, with leaf-like bracts. Calyx-tube 6–8(–11) mm, pubescent, with 10(–17) veins, green to purplish; teeth 2.5–5 mm, triangular, obtuse, with scarious margin. Petals 10–20 mm, white or cream to purplish; limb *c.* half as long as claw, oblong-obovate, the lobes *c.*4 mm; claw narrowly oblong, auriculate. Capsule ± equalling calyx, ovoid, with recurved teeth; seeds 1.3–1.7 mm, reniform, with 2–3 rows of tubercles on keel, pale brown. *Flowers November to February.* (Fig. 52).

Open scrub, dry heath, grass steppe; 0–250 m.

N. part of Isla Grande, W. part of Canal Beagle. (Map 7b).
W. Argentina north to *c.*41°S.lat.
Chile north to *c.*51°S.lat.

Dianthus L., which is cultivated in may estancia gardens, can be distinguished from *Silene* by the entire or emarginate petals, which lack the corona-scales, epicalyx-scales appressed to the calyx and 2 styles. ***D. deltoides*** L., the maiden Pink, with puberulent flowering stems up to 15 cm, a puberulent calyx 13–15 mm, epicalyx-scales about half as long as calyx, and bearded petals which are pink, rarely white, with a darker basal band, is locally established as a casual in open steppe between Ea Cullen and Bahía San Sebastián. ***D. caryophyllus*** L., the Carnation, which is larger in all parts, glabrous, with unbearded petals and epicalyx-scales about a quarter as long as the calyx, has been found as a casual near Cerro Sombrero.

XXI. CHENOPODIACEAE

Annual or perennial herbs, or shrubs. Leaves alternate, simple, or opposite, scale-like; stipules absent. Flowers inconspicuous, solitary or in spike-like or cymose inflorescences, hermaphrodite or unisexual, actinomorphic. Perianth-segments absent or 2–5(6), more or less connate. Stamens 1–5. Ovary usually superior, l-celled; stigmas (1) 2–3 (–5). Fruit an achene; seeds horizontal or vertical (dorsally or laterally compressed).

a. Leaves scale-like or semi-cylindrical:
b. Stems succulent, jointed; leaves opposite, scalelike **3. *Salicornia***
b. Stems woody, not jointed; leaves alternate, semi-cylindrical **4. *Suaeda***
a. Leaves flat:
c. Leaves linear, entire, not more than 3mm wide **4. *Suaeda***
c. At least lower leaves lanceolate to rhomboid-triangular, entire to pinnatisect, at least 4 mm wide:
d. Flowers all unisexual; fruit tightly invested by 2 bracteoles **2. *Atriplex***
d. Flowers mostly hermaphrodite; fruit not invested by 2 bracteoles:
e. Ovary semi-inferior; fruits adhering together by swollen perianth and receptacle (see **1**) ***Beta***
e. Ovary superior; fruits not adhering, the fruiting perianth membranous **1. *Chenopodium***

1. *Chenopodium* L.

Annual or perennial herbs. Leaves alternate, flat. Flowers hermaphrodite and female, in small cymes (glomerules) variously arranged. Perianth-segments 2–5, unaltered in fruit. Stamens 1–5. Seeds horizontal or vertical, orbicular, compressed.

Literature: P. Aellen, *Feddes Repert.* **26**: 31–64, 119–160(1929); P. Aellen & T. Just, *Amer. Midl. Nat.* **30**: 47–76(1943).

a. Leaves 1- to 2-pinnate or -pinnatifid **3. *antarcticum***
a. Leaves entire to shallowly 3-lobed or sinuate-dentate:
b. Leaves linear-elliptical to lanceolate or elliptical; seeds vertical **2. *glaucum***
b. Leaves triangular to rhombic or ovate; seeds horizontal, at least in terminal flowers of glomerules:
c. All flowers with 5 perianth-segments united to $\frac{1}{2}$-way; seeds horizontal in all flowers, smooth **4. *carnosulum***
c. Terminal flowers in glomerule with 3–5 free perianth-segments and horizontal seeds, the lateral with 3 partly united perianth-segments and vertical seeds; seeds reticulate **1. *macrospermum***

1. *C. macrospermum* Hooker f. 1847, *Fl. Antarct.* **1(2)**: 341.
C. rubrum auct., non L. – Dusén, 1900.

Plant glabrous, fleshy; stems 7–20 cm, procumbent to erect, often branched. Leaves with lamina 10–30 × 7–25 mm, ovate to rhombic or broadly triangular, obtuse, cuneate at base, subentire to sinuate or coarsely serrate; petiole 5–15 mm. Glomerules in a dense, usually branched, bracteate inflorescence. Perianth-segments 1–2 mm, ovate, obtuse to acute, green, 3–5 and free in terminal flowers of glomerule, 3 and united in lateral flowers. Seeds 1.2–2 mm in diameter, shallowly reticulate, horizontal in terminal flowers and vertical in lateral flowers of glomerule. ⋆$2n = 36$. Self-compatible. *Flowers December to January.*

Open moist sand, shingle and clay along coasts and estuaries. 0–10 m.

N.E. coast of Isla Grande. (Map 7c).
Chile north of 35°S.lat.
W. Argentina north to *c.*20°S.lat.
Falkland Islands (Islas Malvinas); Paraguay; Mexico; W. United States.

Fuegian and Falkland Islands (Islas Malvinas) plants have been ascribed to ssp. *macrospermum* (ssp. *crassicaule* (Moq.) Aellen) by Aellen (1929), distinguished principally by their more prostrate habit and axillary glomerules from the more widespread ssp. *halophilum* (Phil.) Aellen, which is erect and with glomerules terminal in bracteate panicles.

⋆2. *C. glaucum* L. 1753, *Sp. Pl.* 220.

Plant glabrous or sublglabrous; stems 2–50 cm, procumbent to erect, much branched. Leaves with lamina 6–15 × 3–8 mm, linear-elliptical to lanceolate or elliptical, obtuse to subacute, cuneate at base, subentire to sinuate or sinuate-serrate; petiole 5–20 mm. Glomerules in spikes or panicles. Perianth-segments 0.75–1 mm, ovate to oblong, obtuse to subacute, green, 4–5 and united at base in terminal flowers of glomerule, 3–5 and free in lateral flowers. Seeds 0.6–1.1 mm in diameter, shallowly reticulate, vertical. *Flowers December to January.*

Stoney and muddy areas by sea.

N.E. coast of Isla Grande.
Introduced from Europe and widely naturalized in temperate regions of the world.

3. *C. antarcticum* (Hooker f.) Bentham & Hooker f. 1880. *Gen. Pl.* **3**: 52.
C. botrys auct., non L. – Spegazzini, 1896.
Blitum antarcticum Hooker f. 1847, *Fl. Antarct.* **1(2)**: 549.

Plant lepidote in many parts, at least when young; stems 2–40 cm, procumbent to ascending, woody

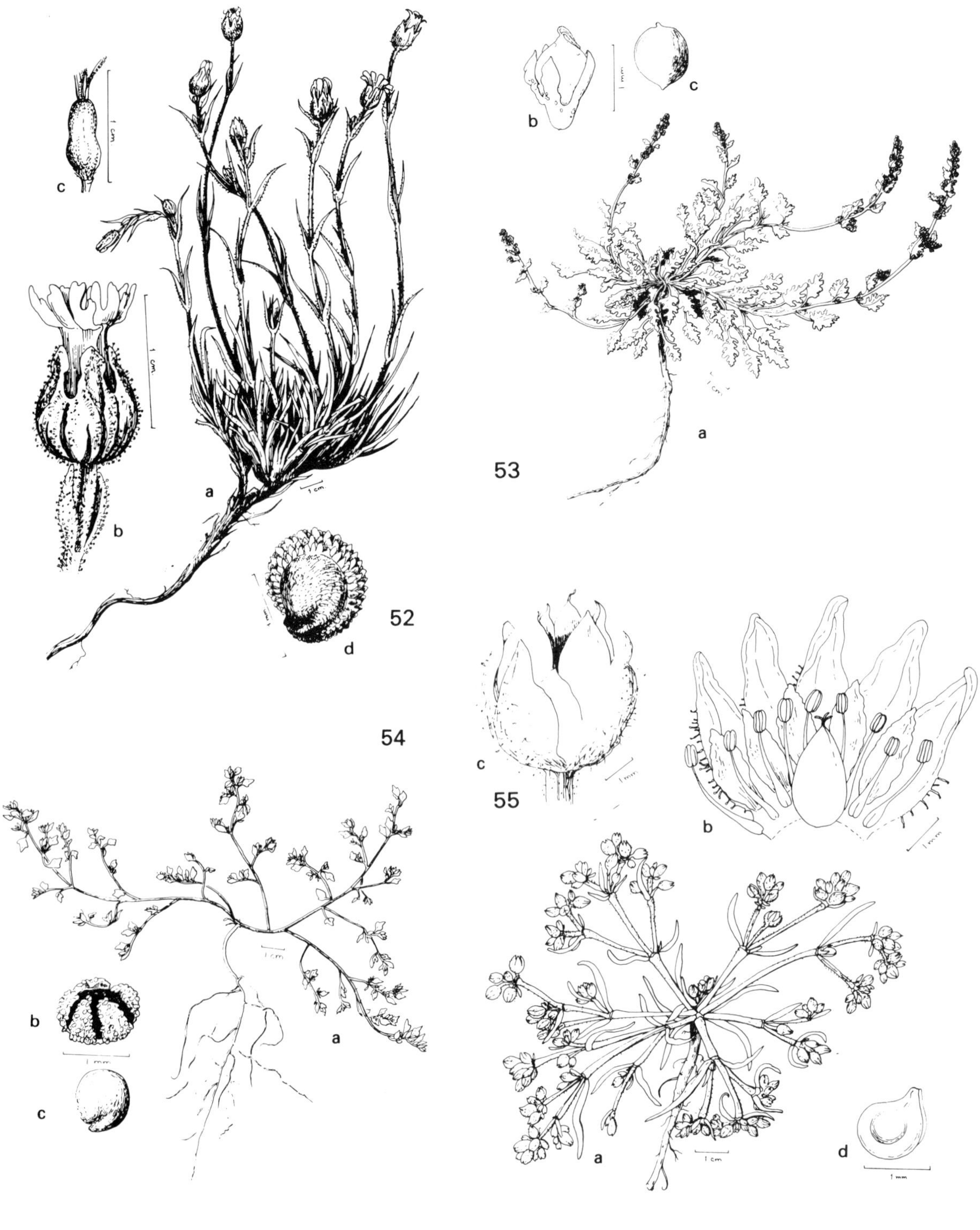

Fig. 52. *Silene magellanica.* a, Plant. b, Flower. c, Ovary. d, Seed. Scale = 1 cm (a–c) or 1 mm (d). Fig. 53. *Chenopodium antarcticum.* a, Plant. b, Young achene. c, Seed. Scale = 1 cm (a) or 1 mm (b–c). Fig. 54. *Chenopodium carnosulum.* a, Plant. b, Young achene. c, Seed. Scale = 1 cm (a) or 1 mm (b–c). Fig. 55. *Spergularia marina.* a, Plant. b, Opened flower. c, Capsule. d, Seed. Scale = 1 cm (a) or 1 mm (b–d).

(R. N. P. Goodall)

below, much branched. Basal leaves with lamina (8–)12–24(–34) × (4–)8–18(–22) mm, lanceolate to triangular-ovate or rhombic in outline, truncate to subcuneate at base, 1- to 2-pinnate or -pinnatifid, the lobes obtuse; petiole 7–40 mm; upper cauline leaves smaller, less divided or subentire with 2 basal teeth, subsessile. Glomerules axillary in terminal simple or, rarely, branched, dense, bracteate spike. Perianth-segments (0.7–)1–2(–3.2) mm, linear-spathulate, obtuse, green, 4–5 and free in terminal flowers of glomerule, (2)3–4 and united for $\frac{1}{2}$–$\frac{3}{4}$ of length in lateral flowers. Seeds (0.7–)0.9–1.1(–1.3) mm in diameter, shallowly reticulate, horizontal in terminal flowers and vertical in lateral flowers of glomerule. $2n = 16$. *Flowers December to March*. (Fig. 53).

Open sand and mud by sea, estuaries and usually brackish lakes.

N. half of Isla Grande; Isla Isabel. (Map 7d).
Argentina north to *c.*50°S.lat.

4. ***C. carnosulum*** Moq. 1849, in DC., *Prodr.* **13(2)**: 64.
C. patagonicum Phil. 1895, *Anal. Univ. Chile* **91**: 421.
C. fuegianum Speg. 1896, *Anal. Mus. Nac. Bs Aires* **5**: 72.
C. vulvaria auct., non L. – Dusén, 1900.

Plant glabrous to farinose; stems 2–20 cm, procumbent to ascending, much branched from base. Leaves with lamina 4–15 × 3–12 mm, triangular to rhombic, 3-lobed, the median lobe often elongate and acute, cuneate to hastate at base; petiole up to 8 mm. Glomerules axillary, in small, axillary and terminal, bracteate spikes. Perianth-segments 5, 1–1.5 mm, ovate, obtuse, green, united up to $\frac{1}{2}$-way. Seeds 1–2 mm in diameter, smooth, horizontal. Self-compatible. *Flowers December to February*. (Fig. 54).

Open sand, pebbles and clay by sea and estuaries. 0–20 m.

Coasts of Isla Grande and adjacent islands. (Map 7e).
Argentina north to *c.*50°S.lat.
W. Chile north to 33°S.lat.
Bolivia; Mexico.

The separation of this species from the widespread European *C. vulvaria* L., which is generally larger and has an unpleasant odour, is not completely satisfactory and further study is needed.

Beta vulgaris L. 1753, *Sp. Pl.* 222, which can superficially resemble *Chenopodium*, is cultivated as a root or forage crop and occurs sporadically as a garden weed in Ushuaia, and perhaps elsewhere. It is an erect, annual to perennial herb with flat, alternate leaves, the lower of which are ovate-cordate to rhombic-cuneate and petiolate, the upper being rhombic to lanceolate and sessile. The hermaphrodite flowers are arranged in glomerules grouped into a spicate infloresence which is interrupted in the lower part. The 5 perianth-segments become swollen in fruit and, together with the swollen receptacle, cause the fruits to adhere together in characteristic clusters.

2. *Atriplex* L.

Annual herbs or small shrubs. Leaves alternate, flat. Flowers unisexual, solitary or in small axillary or terminal cymes (glomerules); male flowers with 5, free perianth-segments and 5 stamens; female flowers without perianth, with 2 filiform stigmas, the ovary semi-inferior. Fruit tightly invested by 2 bracteoles.

a. Small shrub; leaves oblong-elliptical, coriaceous, grey **1. *reichei***
a. Annual herb; at least lower leaves lanceolate to triangular or hastate, membranous, green:
 b. Lower leaves cuneate at base **2. *patula***
 b. Lower leaves truncate or subcordate at base **3. *hastata***

1. ***A. reichei*** Volkens ex. Dusén 1900, *Svensk Exped. Magellansl.* **3(5)**: 190.
Procumbent, grey-farinose shrub; stems up to 40 cm, much branched, prostrate to ascending. Leaves 5–15 × 2–6 mm, oblong-elliptical, obtuse, cuneate at base, entire, coriaceous, grey, shortly petiolate or subsessile. Glomerules in leaf-axils in terminal spicate inflorescences, interrupted towards base. Bracteoles subtending female flowers broadly subdeltoid, entire, smooth, connate at base. *Flowers December to January*. (Fig. 56).

Sand and pebbles by sea and estuaries; 0–20 m.

N. part of Isla Grande. Endemic. (Map 7f).

The relationship of this species to others occurring in S. Patagonia needs further study.

*2. ***A. patula*** L. 1753, *Sp. Pl.* 1053.
Erect annual; stem up to 60 cm, simple or branched, ascending to erect. Lower leaves 1.5–8 × 0.5–5 cm, oblong-hastate to elliptic – triangular or lanceolate,

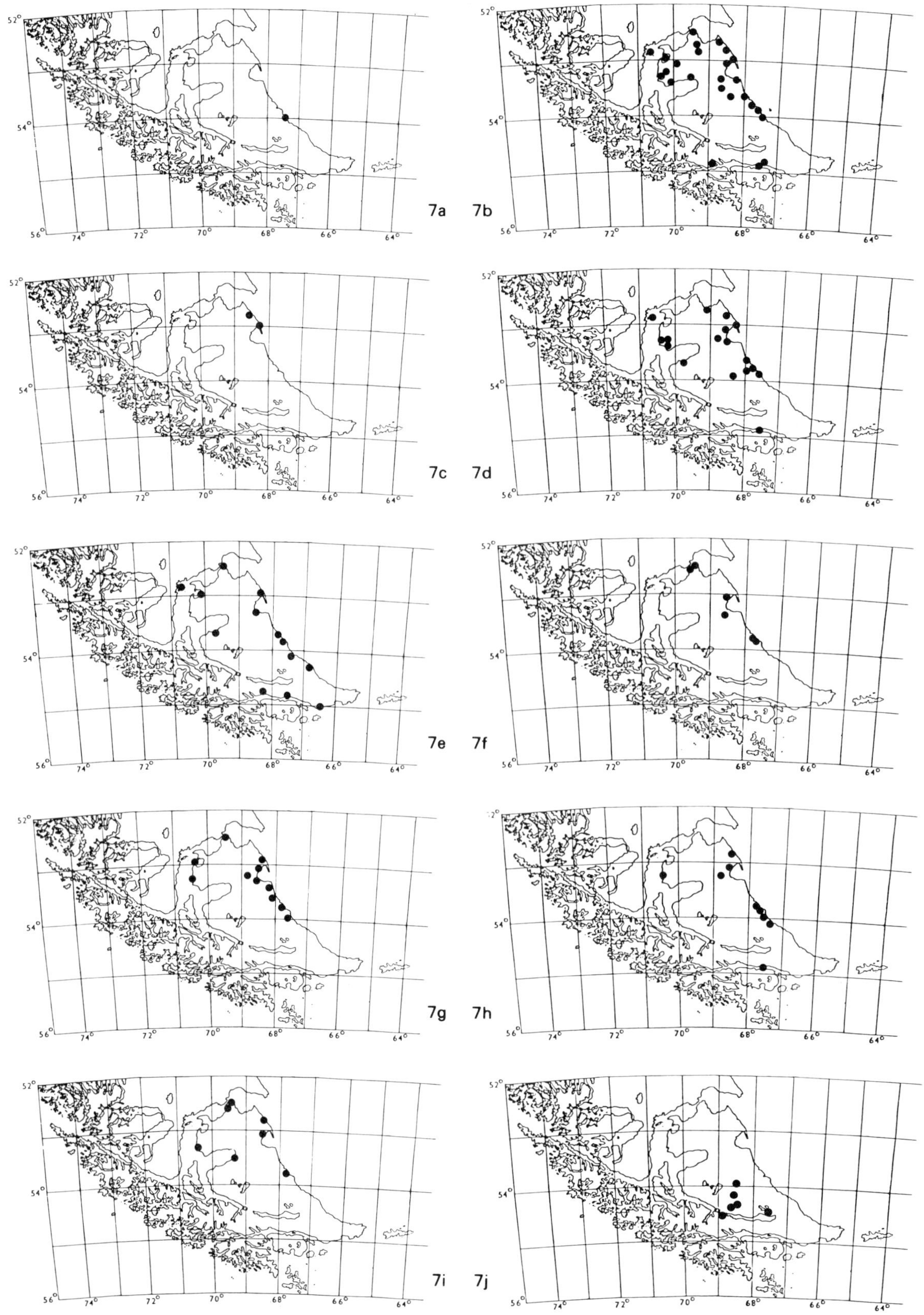

Map 7. a, *Spergularia marina*. b, *Silene magellanica*. c, *Chenopodium macrospermum*. d, *Chenopodium antarcticum*. e, *Chenopodium carnosulum*. f, *Atriplex reichei*. g, *Salicornia ambigua*. h, *Suaeda patagonica*. i, *Suaeda argentinensis*. j, *Calandrinia caespitosa*.

acute to subacute, cuneate at base, entire or shallowly serrate-dentate, the upper smaller, lanceolate to linear, entire. Glomerules in leaf-axils or in long spicate inflorescences; bracteoles subtending female flowers broadly rhomboid, entire or denticulate, smooth or tuberculate on the back. *Flowers January to February.*

Sands and clays by coast and estuaries. 0–20 m.

N. part of Isla Grande.
Introduced from Europe and widely naturalized in temperate regions.

*3. ***A. hastata*** L. 1753, *Sp. Pl.* 1053.
Very similar to *A. patula* but the basal leaves with truncate to subcordate base.

Beaches and disturbed ground near sea. 0–20 m.

S. (Ushuaia) and N.W. (Porvenir) parts of Isla Grande.
Introduced from Europe and widely naturalized in temperate regions.

3. *Salicornia* L.

Perennial herbs; stems much-branched, succulent, jointed, somewhat woody at base, Leaves opposite, scale-like. Inflorescence spike-like, segmented; each fertile segment composed of 2, 3-flowered cymes immersed in and connate with the bract of the segment; flowers hermaphrodite or female, arranged in a triangle. Perianth indistinctly 3-lobed. Stamens 1–2. Style 1; stigmas 2, filiform, plumose. Fruit an ovoid achene; seeds pubescent.

Literature: A. Soriano. *Rev. Arg. Agron.* **14(2)**: 148–172(1947).

1. *S. ambigua* Michx 1803, *Fl. Bor. Amer.* **1**: 2.
S. magellanica Phil. 1895, *Anal. Univ. Chile* **91**: 430.
S. doeringii Lorentz & Niederl. 1881, *Exped. Río Negro* 194 – Dusén, 1900.

Gynodioecious. Stems decumbent, the base 8–16 cm and rooting at nodes, the branches 4–9 cm, ascending to erect; joints 5–20 × 1.8–4 mm. Scale-leaves 0.7–2 mm, lobed. Spikes 8–17 × 2.5–4.5 mm, with 3–7 fertile segments. Anthers *c.*0.5 mm. Stigmas *c.*1 mm. Seeds *c.*1 mm, ovoid, with short, slender, curved hairs, brown. $2n = 18$. *Flowers December to January.* (Fig. 58).

Saline soils and sand, sometimes in open grassland, by sea and brackish lakes.

N. part of Isla Grande. (Map 7g).
Argentina north to *c.*25°S.lat.
Chile north to *c.* 24°S.lat.
Uruguay, Brazil, North America.

4. *Suaeda* Forskål ex Scop.

Glabrous annual herbs or shrubs. Leaves alternate, flat or semi-cylindrical, rather fleshy. Flowers hermaphrodite or female, actinomorphic, sessile, solitary or clustered in axils of leaves. Perianth-segments 5(6), equal, united at base, swollen in fruit. Stamens 1–2 or 5. Stigmas 3–5. Fruit an achene; seed usually horizontal.

Literature: *A. Soriano, Rev. Argent. Agron.* **9**: 343–353(1942).

a. Annual herb; stamens 1–2 — **1. *patagonica***
a. Shrub; stamens 5 — **2. *argentinensis***

1. *S. patagonica* Speg. 1897, *Rev. Fac. Agr. Vet. La Plata* **30–31**: 570.

Annual herb; stems up to 25 cm, procumbent, ascending distally, much branched. Leaves 5–13 × (0.8–)1–2(–2.7) mm, linear, obtuse to subacute, sessile entire, the margin somewhat scarious. Flowers hermaphrodite, sessile, solitary or in small glomerules, in axils of leaves along length of branches. Perianth-segments 1–1.7 × *c.*1 mm, ovate-lanceolate, acute, often prominently winged in middle of back, incurved. Stamens 1–2. Seed *c.*1.5 mm in diameter, lenticular, smooth, horizontal. $2n = 36$. *Flowers October to January.* (Fig. 57).

Mud and pebbles by sea, estuaries and saline lakes. 0–20 m.

N.E. part of Isla Grande; islands of C. Canal Beagle. (Map 7h).
Argentina north to *c.*30°S.lat.

The Fuegian plants belong to var. *patagonica*, in which the perianth-segments become much less swollen in fruit than in var. *crassiuscula* Soriano, from the northern parts of the species' range.

2. *S. argentinensis* Soriano 1942, *Rev. Argent. Agron.* **9**: 348.
S. fruticosa auct., non Forskål – Dusén, 1900.

Gynodioecious shrub; stems up to 24 cm, more or less erect, much branched. Leaves (2–)4–15 × (1.3–)1.5–2.5(–3) mm, linear-oblong to oblanceolate, obtuse, sessile, entire, scarious on margin, semi-cylindrical. Flowers hermaphrodite or female, sessile, solitary or in small glomerules, in axils of

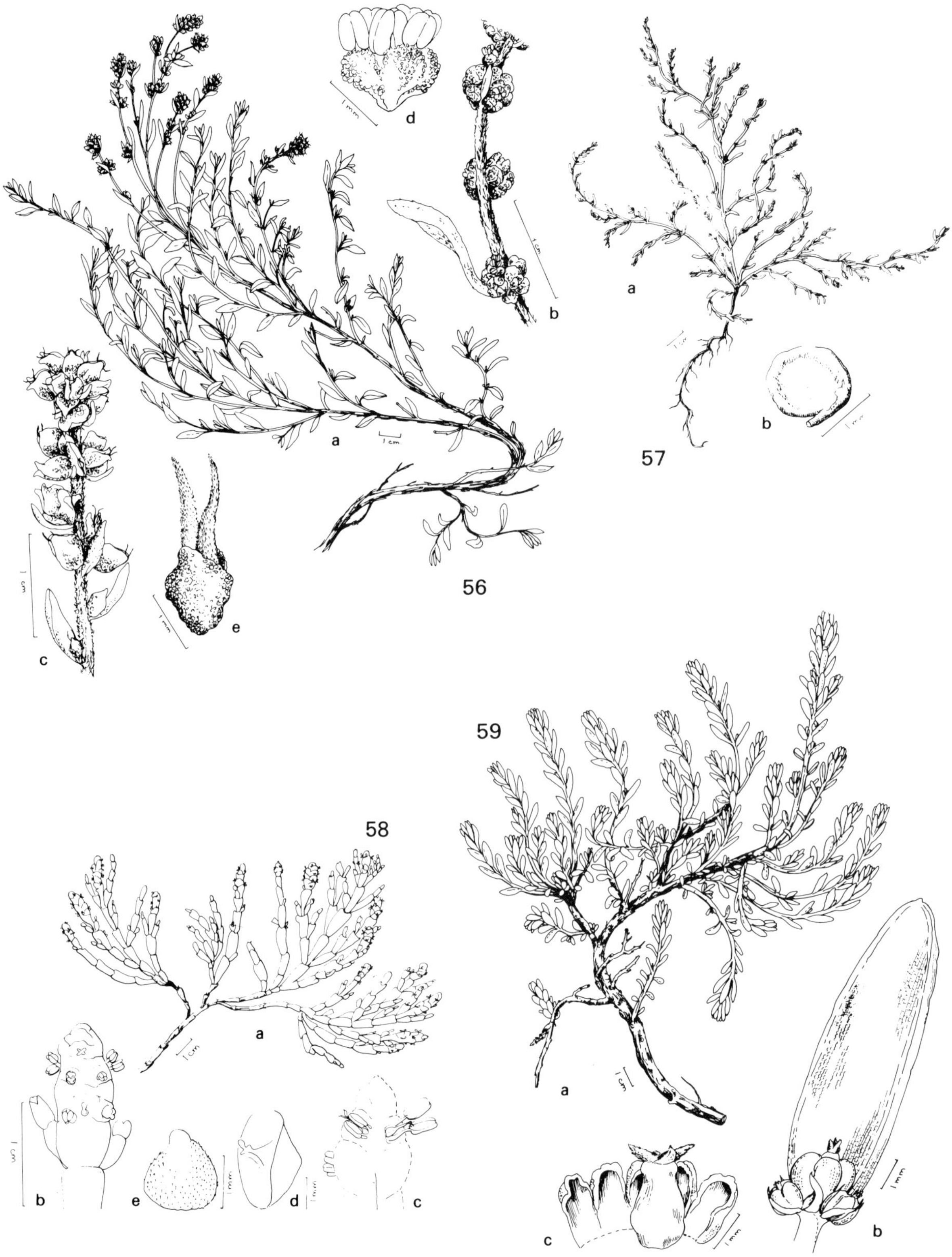

Fig. 56. *Atriplex reichei*. a, Plant. b, Male glomerules. c, Female glomerules. d, Male flower. e, Female flower. Scale = 1 cm (a–c) or 1 mm (d–e). Fig. 57. *Suaeda patagonica*. a, Plant, b, Seed. Scale = 1 cm (a) or 1 mm (b). Fig. 58. *Salicornia ambigua*. a, Plant. b, Fertile segment. c, Flowers. d, Achene. e, Seed. Scale = 1 cm (a–b) or 1 mm (c–e). Fig. 59. *Suaeda argentinensis*. a, Plant. b, Axillary flower-cluster. c, Female flower with perianth opened. Scale = 1 cm (a) or 1 mm (b–c).
(R. N. P. Goodall)

leaves at ends of branches. Perianth-segments 2.5–3 × 1–2.5 mm, ovate-cochleariform, obtuse, incurved. Stamens 5. Seed 1.5–2.5 mm, orbicular-reniform, compressed, smooth, horizontal or vertical. $2n = 36$. *Flowers November to January*. (Fig. 59).

Mud and sand, often in *Lepidophyllum* scrub, by sea and estuaries.

N. part of Isla Grande. (Map 7i).
Argentina north to 38°40′S.lat.
Falkland Islands (Islas Malvinas).

XXII. PORTULACACEAE

Glabrous herbs. Leaves alternate or opposite, often in dense rosettes, simple; stipules absent. Flowers regular, hermaphrodite, solitary or in small terminal cymes. Sepals 2, connate at base, persistent. Petals 5, free or connate at base. Stamens 3–7. Ovary superior, 1-celled. Style 1, with 3 stigmatic lobes. Fruit a capsule; seeds with endosperm.

a. Leaves alternate, in dense rosettes; flowers solitary, at least 10 mm in diameter; capsule with numerous seeds **1. *Calandrinia***
a. Leaves opposite, not in rosettes; flowers in small cymes, less than 5 mm in diameter; capsule with 1–3 seeds **2. *Montia***

1. *Calandrinia* Humb.

Perennial, somewhat woody at base. Leaves alternate, in dense rosettes. Flowers solitary, axillary. Petals free. Stamens 5–7, adnate to base of petals. Capsule opening by 5–7 valves; seeds numerous.

Literature: D. C. A. Suárez de Cullen, *Bol. Soc. Argent. Bot.* **5(1–2)**: 1–29(1953).

1. ***C. caespitosa*** Gillies ex. Arn. 1831, *Edinb. New Philos. Jour.* **3**: 356.
C. fuegiana Gandoger 1912, *Bull. Soc. Bot. Fr.* **59**: 708 – Skottsberg, 1916; Moore 1974, 1981.

Rootstock short, stout, bearing dense rosette of leaves, arising from long, stout, branched, woody, vertical root. Leaves 10–30(–35) × 2–3(–4) mm, linear to linear-spathulate, acute to subobtuse, entire, slightly sheathing at base, somewhat fleshy. Peduncles 20–30 mm, up to 50 mm in fruit. Sepals 5–8(–12) × *c.*5 mm, ovate to triangular-ovate, acute, entire. Petals *c.*8 × 5.5 mm, spathulate to orbicular, white. Capsule 13–19 mm, narrowly ovoid; seeds *c.*1.3 mm in diameter, appressed globose, black, shiny. Self-compatible and usually self-pollinated. $2n = 24$. *Flowers January–March*. (Fig. 60).

Open mud, soil and pebbles, especially along rivers. 0–900 m.

C. & C.S. Isla Grande. (Map 7j).
W. Argentina north to *c.*33°S.lat.
E. Chile north to *c.*31°S.lat.

2. *Montia* L.

Annual to perennial. Leaves opposite and decussate. Flowers in small terminal cymes. Petals united into short tube. Stamens 3–5, united to base of petals. Capsule opening by 3 valves; seeds 1–3.

Literature: D. M. Moore, *Bot. Not.* **116**: 16–30(1963).

*1. ***M. fontana*** L. 1753, *Sp. Pl.* 87.
M. rivularis C. C. Gmelin 1805, *Fl. Bad.* **1**: 301 –

Stems 2–7 cm, up to 20 cm or more in water, erect to decumbent, slender, branched. Leaves 3–20 × (0.8–)1.4–4.5 mm, narrowly spathulate to obovate, entire. Cymes often overtopped by non-flowering shoots. Flowers 2–3 mm in diameter. Petals unequal, white. Capsule 1.5–3 mm, globose, exceeding sepals; seeds 1–1.5 mm, broadly reniform, dark brown to shining black. Self-compatible and usually self-pollinated. *Flowers November to January*. (Fig. 61).

Wet or seasonally moist areas beside streams, lakes etc., usually on mud or wet rocks, often in water 0–230 m.

Throughout Fuegia. (Map 8a).
Widespread in temperate regions of the world but probably native only in the Northern Hemisphere.

The Fuegian plants belong to ssp. *fontana*, distinguished by the shiny, smooth seed which lacks tubercles.

XXIII. POLYGONACEAE

Perennial, rarely annual, herbs. Leaves usually alternate, simple, with stipules united by membranous sheath (ochrea). Flowers actinomorphic, hermaphrodite or unisexual, solitary or in racemose inflorescence. Perianth-segments 3–6, free or united at base, often enlarged and membranous in fruit. Stamens 6–9. Ovary superior, 1-celled; styles 2–4, free. Fruit a nut; seeds with endosperm.

a. Perianth-segments 3 **1. *Koenigia***
a. Perianth-segments 5–6:
b. Leaves palmately lobed; stamens 9 **5. *Rheum***

60

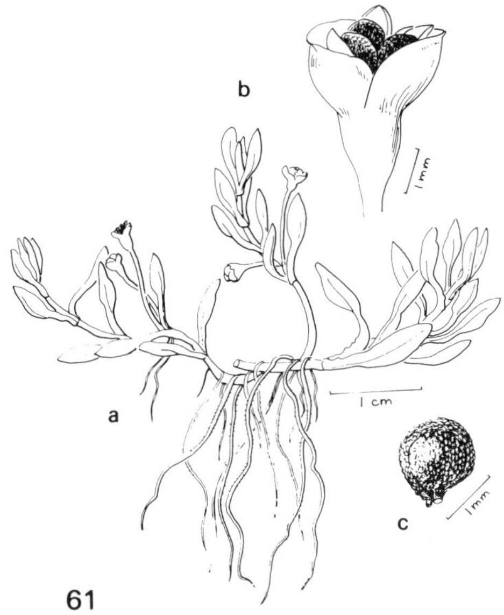

61

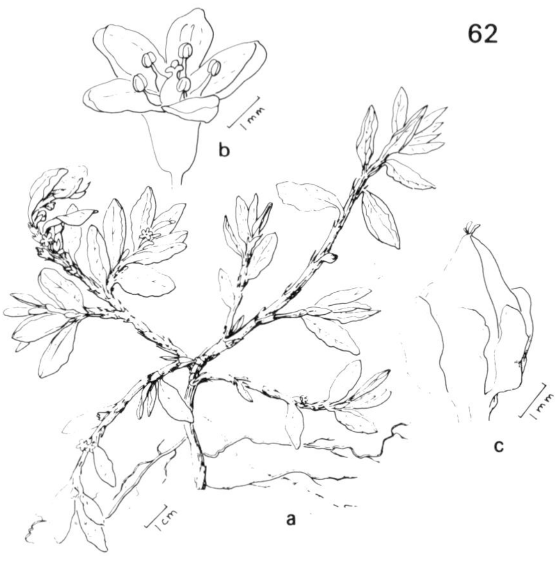

62

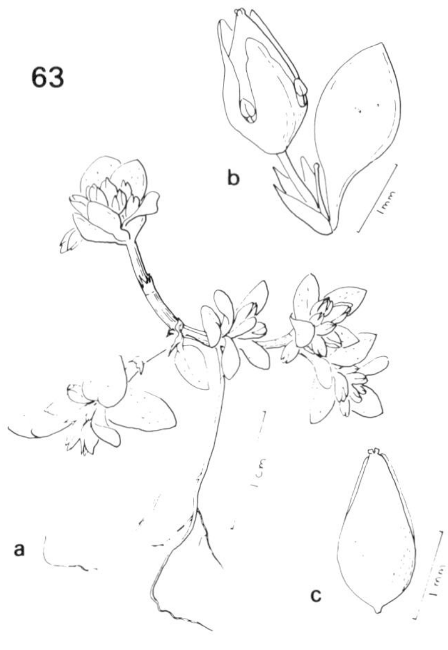

63

Fig. 60. *Calandrinia caespitosa*. a, Plant. b, Part flower. c, Seed. Scale = 1 cm (a) or 1 mm (b–c). Fig. 61. *Montia fontana*. a, Plant. b, Capsule. c, Seed. Scale = 1 cm (a) or 1 mm (b–c). Fig. 62. *Polygonum maritimum*. a, Plant. b, Flower. c, Nut. Scale = 1 cm (a) or 1 mm (b–c). Fig. 63. *Koenigia islandica*. a, Plant. b, Flower. c, Nut. Scale = 1 cm (a) or 1 mm (b–c).

(R. N. P. Goodall)

b. Leaves entire to serrate and crispate; stamens 6–8:
  c. Perianth-segments 6, the inner larger than the outer in fruit **4. *Rumex***
  c. Perianth-segments 5, equal in fruit or the outer larger:
    d. Stems twining; outer perianth-segments keeled or winged in fruit **3. *Bilderdykia***
    d. Stems not twining; outer perianth-segments not keeled or winged in fruit **2. *Polygonum***

1. *Koenigia* L.

Annual herbs. Leaves subopposite. Flowers hermaphrodite, in terminal and axillary clusters. Perianth-segments 3, sepaloid. Stamens 3, alternating with 3, gland-like staminodes. Styles 2.

1. *K. islandica* L. 1767, *Mantissa* 35.
K. fuegiana Dusén 1900, *Svenska Exped. Magellansl.* **3(5)**: 191.

Stems 1–3 cm, erect or ascending, usually unbranched, reddish. Leaves 2.5–3 × 1.5–2 mm, broadly elliptical, rounded at apex, cuneate at base, somewhat fleshy, reddish; petiole 0.7–1 mm, winged; stipules *c.*0.7 × 0.6 mm, scarious, united at base. Flowers solitary or in clusters of 2–5. Perianth-segments *c.*1.2 × 0.6 mm, oblong, acute to obtuse, pale green, reddish distally. Fruit *c.*1.7 × 0.9 mm, ovoid. $2n = 28$. *Flowers February to March.* (Fig. 63).

Wet gravel beside streams and at margins of *Sphagnum* bogs. 30–500 m.

C.S. Isla Grande. (Map 8b).
N. Eurasia and N. America.

Dusén (1900), in describing *K. fuegiana*, differentiated it from the N. Hemisphere species by its more prostrate, much-branched habit. Study of more abundant Fuegian material has revealed no morphological or cytological differences between N. Hemisphere and Fuegian material.

2. *Polygonum* L.

Perennial or annual herbs. Leaves alternate, entire. Flowers hermaphrodite, solitary or in small axillary clusters. Perianth-segments 5–6, usually petaloid, not keeled or winged. Stamens 6–8. Styles (2–)3.

Literature: M. Buchinger, *Bol. Soc. Argent. Bot.* **6(2)**: 98–106(1956).

a. Ochreae brown, ciliate; flowers in dense spikes **4. *persicaria***
a. Ochreae silvery, hyaline at least towards apex, becoming deeply lacerate; flowers in lax spikes:
  b. Perennial; stems woody at base; ochreae in inflorescence longer than internodes, in middle part of stem more than ½ as long as internodes **1. *maritimum***
  b. Annual; stems scarcely woody at base; ochreae all usually less than ½ as long as internodes:
    c. Leaves on main stem much larger than those on branches; perianth-tube very short **2. *aviculare***
    c. Leaves more or less uniform in size; perianth-tube at least ½ as long as segments **3. *arenastrum***

1. *P. maritimum* L. 1753, *Sp. Pl.* 361.
P. chilense Koch 1849, *Linnaea* **22**: 206.

Perennial; stems (5–)10–50(–60) cm, procumbent, branched, stout and woody at base. Leaves 17–27(–31) × 4–8 mm, elliptical to elliptic-lanceolate, acute, with entire, usually revolute margin, glaucous, shortly petiolate; ochreae 5–10 mm, more than ½ as long as internodes in middle part of stem and longer than internodes in inflorescence, with 6–14 conspicuous veins, usually lacerate, reddish brown at base, silvery hyaline distally. Flowers solitary or in axillary clusters of 2–5, with short pedicels forming lax spikes. Perianth-segments 2–2.5 × *c.*1.5 mm, oblong, obtuse, cucullate, green with white or pinkish margin. Nut 4–4.5 × 2–2.5 mm, trigonous, equalling or somewhat exceeding perianth, shiny, dark brown. $2n = 20$. *Flowers November to January.* (Fig. 62).

Sand and shingle by sea and estuaries; 0–10 m.

Coasts of N. and C. Fuegia, less frequent in south. (Map 8c).
Widespread in coastal and saline areas of Chile and Argentina; Falkland Islands (Islas Malvinas); coasts of Europe.

The Southern Hemisphere plants have been distinguished as var. *chilense* (Koch) Skottsb., but no characters can be found to separate them from European specimens.

2. *P. aviculare L. 1753, *Sp. Pl.* 362

Annual; stems 5–40 cm, procumbent to erect, branched. Leaves 20–50 × 5–15 mm, lanceolate to ovate, subacute, much larger on main stem than on branches, green or reddish, shortly petiolate; ochreae 2–7 mm, usually less than ½ as long as internodes, with few faint veins, usually lacerate, silvery hyaline. Flowers solitary or in axillary clusters of 2–6, forming lax, leafy spikes. Perianth with very short tube; segments 2–3 mm, oblong, green with white or pinkish margin. Nut 2.5–3.5 mm, trigonous, included in perianth, dull, punctulate. *Flowers January to April.*

Disturbed ground by roads, beaches, estuarine muds; 0–50 m.

Isla Grande, especially on coasts in northern part. Introduced from Europe and widely naturalized in temperate regions.

3. *P. arenastrum Boreau 1857, *Fl. Centr. Fr.* ed.3, **2**: 559.

Like *P. aviculare* but differs in having leaves of more or less the same size on main stem and branches, and the perianth-tube at least half as long as the segments. *Flowers February to April.*

Disturbed ground by roads and buildings, coastal sands, stream banks. 0–50 m.

C. and N. parts of Isla Grande.
Introduced from Europe and widely naturalized in temperate regions.

Owing to confusion with *P. aviculare* this species is likely to be under-recorded.

4. *P. persicaria L. 1753, *Sp. Pl.* 361.

Annual; stems 20–60 cm, ascending to erect, branched, reddish. Leaves 50–100 × 12–25 mm, lanceolate, acuminate, glabrous or tomentose beneath, often with large blackish spots, shortly petiolate; ochreae 8–12 mm, truncate, ciliate, brown. Flowers in a stout, cylindrical, dense spike, leafless, or with single leaf at base. Perianth-segments 2–2.5 mm, oblong, obtuse, cucullate, pink. Nut *c.*3 mm, usually lenticular, shiny, black. *Flowers February.*

Open shingle and sand at or near beach. 0–10 m.

N.W. Isla Grande.
Introduced from Europe and widely naturalized.

3. *Bilderdykia* Dumort.

Annual herb with twining stems. Leaves alternate, entire. Flowers hermaphrodite, in clusters in spike-like inflorescence. Perianth-segments 5(6), petaloid, the outer larger, keeled or slightly winged in fruit. Stamens 8. Stigmas subsessile, capitate.

1. *B. convolvulus (L.) Dumort. 1827, *Fl. Belg.* 18.

Stems up to 60 cm. Leaves with lamina (4–)5.5–8 × (2–)4.5–5.5 cm, deltate to cordate-sagittate, acute; petiole (2–) 3–7 cm; ochreae 3–3.5 mm, obliquely truncate, pinkish. Pedicels up to 4 mm. Outer perianth-segments *c.*4.2 × 1.5–1.7 mm, cucullate, obtuse, greenish-white. Nut *c.*4 mm, included in perianth, trigonous, punctulate, dull, black. *Flowers February to March.*

Weed of cultivated ground. 0–20 m.

S. coast of Isla Grande.
Introduced from Europe and naturalized in temperate regions.

4. *Rumex* L.

Perennial, rarely annual, herbs. Leaves alternate; ochreae tubular. Flowers hermaphrodite or unisexual, in whorled glomeruli on branched or simple inflorescences. Perianth-segments in 2 whorls of 3, the outer small and thin, the inner (valves) enlarging and usually becoming hard in fruit. Stamens in 2 whorls of 3. Styles 3.

Literature: K. H. Rechinger, *Arkiv. Bot.* **26A(3)**: 1–58(1934).

a. Leaves usually hastate or sagittate, acid-tasting; flowers usually unisexual:
b. Valves about as long as nut — **5. *acetosella***
b. Valves at least $1\frac{1}{2}$ times as long as nut:
c. Inflorescence lax, with simple branches — **6. *acetosa***
c. Inflorescence dense, with most branches repeatedly branched — **7. *rugosus***
a. Leaves not hastate or sagittate, cordate to cuneate at base, not acid-tasting; flowers usually hermaphrodite:
d. Valves with 2 pairs of teeth at least 1 mm — **3. *maritimus***
d. Valves entire or minutely denticulate:
e. Fruiting pedicels at least 4 mm, at least $1\frac{1}{2}$ times as long as valves and nut — **4. *crispus***
e. Fruiting pedicels not more than 1.5 mm, not more than about $\frac{1}{2}$ as long as valves and nut:
f. Leaves up to 0.6 cm wide, sinuate-crispate to midrib; valves lanceolate, with conical tubercle — **2. *crispissimus***
f. Leaves 1.5 cm or more wide, undulate-crispate at margin; valves triangular- to

orbicular – ovate, with rounded-fusiform tubercle **1. magellanicus**

1. *R. magellanicus* Campd. 1819, *Monogr. Rumex* 71.

Perennial; stems 10–30 cm, ascending. Leaves (3–)5–12(–15) × 1.5–2.5 cm, lanceolate to linear-lanceolate, acuminate, cuneate at base, regularly undulate-crispate on margin, the lower long-petiolate; ochreae conspicuous, membranous, pale brown. Inflorescence dense, leafy at base. Flowers hermaphrodite; outer perianth-segments 1.5–2.5 × *c.*1 mm, narrowly elliptical, yellowish green. Valves 3.5–6 × 2–2.5 mm, triangular-to orbicular-ovate, entire or minutely dentate to crenulate, all with rounded-fusiform tubercle; pedicels 0.5–1.5 mm; nut 2.7–3 mm. $2n = 100$. *Flowers November to February*. (Fig. 64).

Open sand and shingle, sometimes mud, on coast and sometimes by streams or lakes. 0–50 m.

Isla Grande west to *c.*71°W.long. (Map 8d).
W. Argentina north to *c.*32°S.lat.
Chile on N. shore of Estrecho de Magallanes.
Falkland Islands (Islas Malvinas) (cf. Moore, 1968).

2. *R. crispissimus* O. Kuntze 1898, *Revis. Gen.* **3**: 269.
R, decumbens Dusén 1900, *Svenska Exped. Magellansl.* **3(5)**: 195.

Perennial; stems 5–20 cm, procumbent, ascending distally. Leaves 2–12(–14) × 0.3–0.6 cm, linear-oblanceolate, obtuse, cuneate at base, alternately sinuate-crispate to midrib, the lower long-petiolate; ochreae membranous, pale brown. Glomerules remote below, contiguous distally in inflorescence, all subtended by leaves. Flowers hermaphrodite; outer perianth-segments *c.*0.5 mm, oblong, rounded. Valves 1.5–2 × 0.75–1 mm, ovate to lanceolate, entire, 1 or all 3 with conical tubercle; pedicels up to *c.*1 mm; nut 1.2–1.5(–1.8) mm. *Flowers December to April*. (Fig. 67).

Open sand and pebbles, rarely mud, usually beside rivers and lakes; 0–50 m.

Throughout Isla Grande except the extreme west. (Map 8e).
W. Argentina and E. Chile north along Andes to *c.*33°S.lat.

3. *R. maritimus* L. 1753, *Sp. Pl.* 335.
R. fueginus Phil. 1895, *Anal. Univ. Chile* **91**: 493-Rechinger, 1934.
R. maritimus var. *fueginus* (Phil). Dusén 1900, *Svenska Exped. Magellansl.* 3(5): 194.

Annual or biennial; stems up to 25 cm, procumbent, ascending distally. Leaves 1–6 × 0.3–1 cm, linear, acute, truncate to auriculate-cordate at base, strongly and irregularly crispate on margin, the lower long-petiolate; ochreae membranous, pale brown. Glomerules remote below, contiguous distally in inflorescence, all subtended by leaves. Flowers hermaphrodite; outer perianth-segments *c.*1 × 0.5 mm, linear-lanceolate. Valves 2.5–4 × 1–1.5 mm, narrowly ovate, with 2 pairs of marginal teeth 1–2 mm, all with fusiform tubercle; pedicels 1–2 mm; nut 1.2–1.5 mm. *Flowers January to February*. (Fig. 66).

Open moist soil by streams and saline lakes; 0–20 m.

C.E. Isla Grande. (Map 8f).
Argentina north to *c.*35°S.lat.
Chile on N. shore of Straits of Magellan.
Europe. N. America.

S. American material is placed in ssp. ***fueginus*** (Phil.) Hultén 1968, *Fl. Alaska* 383, which differs from the European ssp. *maritimus* in having cordate to truncate, not cuneate, bases to the middle cauline leaves and brownish, not yellowish, fruit.

4. *R. crispus L. 1753, *Sp. Pl.* 335.

Perennial; stems up to 65 cm, erect. Leaves up to 25 × 8.2 cm, linear- to oblong-lanceolate, acute, cuneate at base, irregularly undulate and strongly crispate on margin; petiole usually shorter than lamina; ochreae membranous, pale brown. Glomerules remote below, contiguous and dense distally in inflorescence, most subtended by leaves. Flowers hermaphrodite; outer perianth-segments 1.5–2 mm, linear-oblong. Valves 3.5–4.5 × 3.5–5.5 mm, broadly ovate-cordate, subentire or somewhat undulate, usually all with variable tubercles; pedicels 4–5 mm, at least $1\frac{1}{2}$ times as long as valves and nut; nut (1.5–)1.8–2.5 mm. Self-compatible. *Flowers December to Febuary*.

Open sand and pebbles by rivers and beach; 0–20 m.

Coasts of Isla Grande except the extreme west. Introduced from Europe and widely naturalized in temperate regions of the world.

Fig. 64. *Rumex magellanicus*. a, Plant. b, Young nut. c, Fruit. Scale = 1 cm (a) or 1 mm (b–c). Fig. 65. *Armeria maritima*. a, Plant. b, Flower and bract. c, Flower. Scale = 1 cm. Fig. 66. *Rumex maritimus*. a, Plant. b, Flower. c, Valve. d, Nut. Scale = 1 cm (a) or 1 mm (b–d). Fig. 67. *Rumex crispissimus*. a, Plant. b, Fruit. c, Nut. Scale = 1 cm (a) or 1 mm (b–c).
(R. N. P. Goodall)

★5. *R. acetosella* L. 1753, *Sp. Pl.* 338.

Dioecious perennial; stems 3–30 cm, ascending to erect, greenish or often red. Leaves 1–6.5 × (0.25–)0.5–1 cm, linear-lanceolate, acute to mucronate, hastate at base with prominent lobes, usually entire, acid-tasting; petiole shorter to longer than lamina; ochreae lacerate, hyaline, silver to brownish. Glomerules remote, in lax, almost leafless inflorescence. Flowers unisexual; outer perianth-segments 1.5–2 mm, linear-lanceolate. Valves 1–1.5 × *c.*1 mm, broadly ovate, entire or somewhat crenulate, about as long as nut, without tubercles; pedicels usually shorter than valves; nut 1.2–1.5 mm. *Flowers November to March.*

Shingle, sand and open soil, particularly in burned or heavily grazed areas; 0–300 m.

Widespread in Fuegia, except the extreme south and west.
Introduced from Europe and widely naturalized in temperate regions of the world.

★6. *R. acetosa* L. 1753, *Sp. Pl.* 337.

Dioecious perennial; stems up to 80 cm, ascending to erect. Leaves 5–9 × 2–2.5 cm, oblanceolate to oblong, obtuse to subacute, hastate at base with prominent lobes, entire, acid-tasting; petiole up to 15 cm; ochreae short, membranous, pale brown. Glomerules contiguous, in lax, leafless inflorescence with simple branches. Flowers unisexual; outer perianth-segments *c.*0.5 mm, lanceolate, red. Valves 3–3.5 × *c.*1.5 mm, orbicular-cordate, somewhat undulate, at least $1\frac{1}{2}$ times as long as nut, all with small tubercle near base; pedicels shorter than valves; nut 1.8–2.2 mm. *Flowers December to January.*

Moist areas by rivers and swampy ground; 0–100 m.

Locally in S.C. and N.W. Isla Grande.
Introduced from Europe and naturalized in temperate regions of the world.

★7. *R. rugosus* Campd. 1819, *Monogr. Rumex* 113.

Similar to *R. acetosa* but differs in its dense inflorescence with most branches repeatedly branched.

Waste ground in Ushuaia.
Cultivated as a vegetable in Europe and sometimes locally naturalized; origin unknown.

Perhaps the Fuegian material should be included in *R. acetosa.*

5. *Rheum* L.

Robust, perennial herb. Leaves mostly basal, large, palmately lobed; ochreae persistent, not ciliate. Flowers hermaphrodite, in a panicle. Perianth-segments 6, free, equal, not enlarging in fruit. Stamens 9. Stigmas 3, subsessile. Fruit with 3 membranous wings.

★1. *R. rhabarbarum* L. 1753, *Sp. Pl.* 372.

Leaves 15–60 cm in diameter, suborbicular in outline, cordate at base, with stout, fleshy petiole 4–20 cm; ochreae large, membranous. Panicle dense, up to 20 cm. Perianth-segments *c.*3 mm, yellowish.

Cultivated for the edible petioles and locally persisting in gardens abandoned over 20 years ago.

S. side of Isla Grande.
Introduced via Europe from Mongolia.

XXIV. PLUMBAGINACEAE

Perennial herbs. Leaves in basal rosettes, simple; stipules absent. Flowers hermaphrodite, actinomorphic, in bracteate, cymose spikelets. Calyx 5-lobed. Petals 5, united at base. Stamens 5, inserted at base of petals. Ovary superior, 1-celled. Styles 5, united towards base. Fruit dry, membranous, surrounded by calyx with irregular dehiscence; seed 1.

1. *Armeria* Willd.

Spikelets grouped into a terminal, hemispherical capitulum, with an involucre of scarious imbricate bracts and a tubular, scarious sheath at base enclosing top of scape. Calyx infundibuliform, persistent; tube 5-ribbed; limb scarious. Petals persistent. Filaments dilated towards base.

Literature: D. M. Moore and B. Yates, *Bot. Not.* **127**: 183–192(1974).

1. *A. maritima* (Miller) Willd. 1809, *Enum. Pl. Horti. Berol.* 333.
A. elongata (Hoffm.) C. Koch var. *macloviana* (Cham.) Skottsb. 1916, *Kungl. Svenska Vetensk Akad. Handl.* **56(5)**: 285.
A. chilensis Boiss. ssp. *andina* (Poeppig ex. Boiss.) Reiche 1911, *Fl. Chile* **6**: 104.

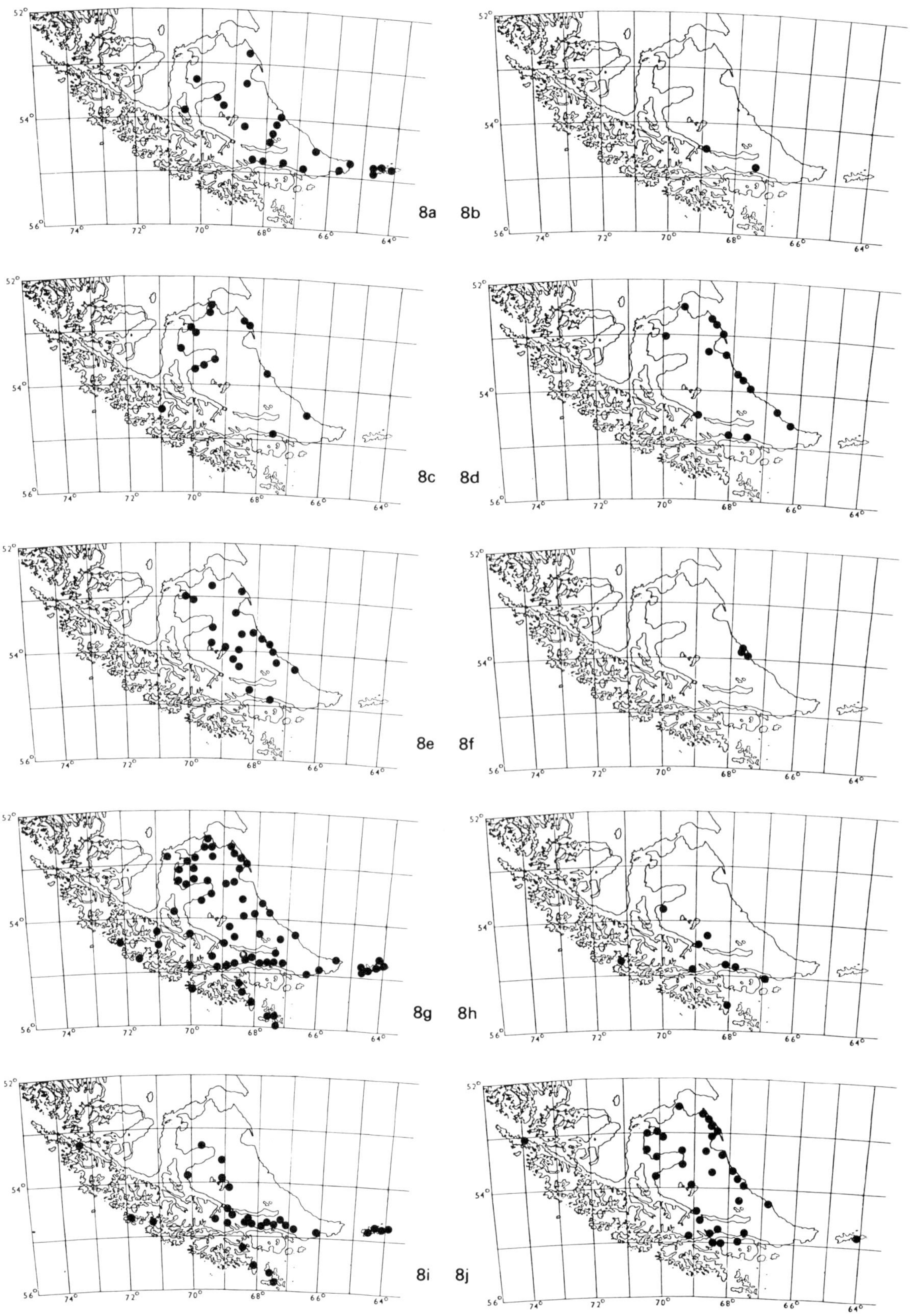

Map 8. a, *Montia fontana*. b, *Koenigia islandica*. c, *Polygonum maritimum*. d, *Rumex magellanicus*. e, *Rumex crispissimus*. f, *Rumex maritimus*. g, *Armeria maritima*. h, *Urtica magellanica*. i, *Viola tridentata*. j, *Viola maculata*.

A. bella Alboff 1896, *Rev. Mus. La Plata* **7**: 385.
A. androsacea Boiss. 1848, in D.C., *Prodr.* **12**: 679.
A. chilensis Boiss. var. *magellanica* Boiss. 1848, in D.C., *Prodr.* **12**: 682- Alboff, 1896.
A. macloviana Cham. 1831, *Linnaea* **6**: 567.
Statice armeria L. 1753, *Sp. Pl.* 394 – Hooker, 1847.

Stock erect, stout, woody, branched. Leaves 6–90 × 0.4–2.5 mm, filiform to linear, subacute, glabrous to pubescent or ciliate. Scapes (0–) 2.5–39 cm, glabrous to villous. Involucral sheath 7–30 mm; capitulum 12–28 mm in diameter; outermost involucral bracts 6–10 × 3–5 mm, oblong-ovate, obtuse, greenish on mid-vein, the inner bracts longer, pinkish distally. Spikelet-bract 5–10 mm, oblong, scarious, pinkish, subequalling to exceeding calyx. Calyx 5–7 mm; tube glabrous to hairy, at least on ribs; lobes with awn 0.5–1 mm, rarely acute, pink. Petals 2.5–3 × *c.*1 mm, oblanceolate to oblong, subacute, pink. Seed *c.*2 mm, fusiform, dark brown. $2n = 18$. Self-compatible with A-type pollen and papillate stigmas, at least facultatively self-pollinating. *Flowers October to February*. (Fig. 65).

Cliffs, coastal rocks and sands, open grassland and scrub, feldmark, screes and peat exposures; 0–900 m.

Throughout Fuegia. (Map 8g).
Southern and W. Argentina north to *c.*31°S.lat.
Chile north to *c.*33°S.lat.
Falkland Islands (Islas Malvinas). N. America. Eurasia.

A very variable species. In Fuegia more southern populations tend to have shorter, more villous scapes and more generally glabrous leaves than those in the northern part of Isla Grande, a trend continued north of the Estrecho de Magallanes. Plants with longer filiform leaves tend to occur inland in drier habitats, while those with shorter, wider leaves tend to be coastal, but the correlation is by no means constant. South American material has been placed in ssp. ***andina*** (Poeppig ex. Boiss.) D. M. Moore and Yates 1974, *Bot. Not.* **127**: 191, but there are no clear morphological distinctions from the N. Hemisphere populations, though the patterns of variation differ. White flowered populations from Isla de los Estados have been formally recognised as var. *goodalliana* T. Dudley 1981, *Rhodora* **83**: 488.

XXV. URTICACEAE

Annual or perennial herbs. Leaves opposite, simple; stipules present, free. Flowers unisexual. Perianth 4-merous. Stamens 4. Ovary superior, 1-celled; style 1, with penicillate stigma. Fruit an achene; seeds with endosperm.

1. *Urtica* L.

Monoecious or dioecious plants with stinging hairs. Inflorescence of axillary, spikelike racemes with clustered cymes. Flowers green, the female with unequal perianth-segments, the 2 larger of which enclose the fruit.

Literature: E. Navas, *Bol. Soc. Argent. Bot.* **9**: 395–413(1961); S. B. Soraru, *Darwiniana* **17**: 246–325(1972).

a. Petiole of lower leaves *c.*$\frac{2}{3}$ as long as lamina; leaves with apical tooth shorter than to as long as the lateral **3. *urens***
a. Petiole of lower leaves not more than $\frac{1}{2}$ as long as lamina; leaves with apical tooth as long as or longer than the lateral;
 b. Plant dioecious or male and female flowers in separate racemes; perianth-segments and peduncle covered with short, rigid hairs **2. *dioica***
 b. Plant with male and female flowers in the same raceme, rarely dioecious; perianth-segments with short, rigid hairs in distal half **1. *magellanica***

1. *U. magellanica* Poiret 1816, in Lam., *Encycl. Méth. Bot.* **12**: 233.

Erect, branched perennial up to 80 cm. Leaves 5–12 × 2–6 cm, ovate-lanceolate, truncate to cordate at base, serrate, the apical tooth linear-lanceolate, much longer than the lateral, the axes of which diverge from the apex; petiole 2–4 cm, not more than $\frac{1}{2}$ as long as lamina in lower leaves; stipules 5–7 × 2–3 mm. Racemes 1–8 cm, with flowers of both sexes, rarely plants dioecious; male flowers 2–3 mm, the perianth-segments with short rigid hairs in distal half, shortly pedunculate; female flowers 1.5–2 mm, sparsely hairy. Achenes 0.5–2 mm, lenticular to ovate, compressed, rugulose. *Flowers November to March*. (Fig. 73).

Open, moist shrub communities and forest-margins. 0–50 m.

S. & W. Isla Grande southwards. (Map 8h).
W. Argentina north to 38°S.lat.
Chile north to 31°S.lat.

The Fuegian plants belong to var. *magellanica*. The other taxon, var. *lilloi* Hauman, from N. W. Argentina, is distinguished by having 1–2 stinging hairs on one of the inner perianth-segments and by the ovoid achenes.

2. *U. dioica L. 1753, *Sp. Pl.* 984.

Like *U. magellanica* but leaves with axes of lateral teeth normally directed towards the apex; plants usually dioecious or, if monoecious, with male and female flowers in separate racemes; perianth-segments and peduncle covered with short, rigid hairs.

Disturbed and cultivated ground near settlements.

Occasionally established in S. & W. parts of Isla Grande.
Introduced from Europe and naturalized in temperate regions of the world.

It can be difficult to separate *U. magellanica* and *U. dioica*, which may indeed be conspecific as suggested by Hooker (1847). However, they differ in leaf anatomy (Navas, 1961) and in the shape of the cystoliths (crystals of calcium carbonate) in the epidermal cells of the leaves. When viewed under the hand lens these are seen to be small dots in *U. dioica* and larger, ellipsoid-cylindrical, in *U. magellanica*.

3. *U. urens L. 2753, *Sp. Pl.* 984.

Erect, branched, monoecious annual up to 35(–64) cm. Leaves 0.8–3.5 × 0.7–2.5 cm, ovate to elliptical, cuneate to cordate at base, serrate, the apical tooth not longer than the lateral; petiole 0.8–1.7 cm, about ⅔ as long as lamina in lower leaves; stipules *c.*1 mm. Racemes up to 1.2 cm, with flowers of both sexes; male flowers 1.2–1.5 mm, the perianth-segments with short, rigid, hairs, sometimes only on the margin; female flowers 1.8–2.1 mm, the perianth-segments ciliate, sometimes hispid dorsally. Achenes 1.2–2.2 mm, ovate, compressed, rugulose. *Flowers October to January*.

Open or disturbed ground, especially near settlements and roads.

Throughout Fuegia.
Introduced from Europe and naturalized in temperate regions of the world.

XXVI. VIOLACEAE

Perennial, rarely annual, herbs. Leaves alternate; stipules present. Flowers hermaphrodite, solitary. Sepals 5, free, persistent. Petals 5, free. Stamens 5, connivent around the ovary. Ovary superior, 1-celled; style 1. Fruit a 3-valved, loculicidal capsule; seeds numerous, with endosperm.

1. *Viola* L.

Leaves simple, with petioles. Peduncle with 2 bracteoles below flower. Sepals prolonged into short appendages below point of insertion. Corolla zygomorphic, the lower petal spurred. Style filiform or thickened distally and then often with stigmatic beak. Seeds smooth, with an elaiosome.

Literature: B. Sparre, *Lilloa* **17**: 377–416(1949), **23**: 347–364 (1950).

a. Annual; stipules leaf-like, coarsely pinnatifid; flowers cream or with some blue-violet on upper and lateral petals **6. *arvensis***
a. Perennial; stipules not leaf-like, entire to fimbriate; flowers white or yellow or bluish and white:
 b. Plant prostrate, forming moss-like patches; leaves 3-lobed at apex; flowers bluish and white **1. *tridentata***
 b. Plant erect; leaves entire at apex; flowers yellow or white:
  c. Petals white, with purplish veins; spur not exceeding 1 mm; lateral petals glabrous **5. *commersonii***
  c. Petals yellow, with brownish veins; spur at least 1 mm; lateral petals at least somewhat bearded inside:
   d. Style long, ± straight, filiform or slightly dilated at apex **4. *magellanica***
   d. Style short, strongly geniculate at base, strongly clavate:
    e. Leaves ciliate and with small black glands on margin, usually cuneate to truncate at base **2. *maculata***
    e. Leaves glabrous and without glands on margin, usually conspicuously cordate at base **3. *reichei***

1. *V. tridentata* Menz. ex. D.C. 1824, *Prodr.* **1**: 300

Glabrous perennial, forming low moss-like mats; stems up to 20 cm, prostrate or procumbent, rooted at intervals, with ascending to erect branches bearing densely imbricate leaves distally. Leaves with lamina 2–7(–9) × 2–4.5(–5.7) mm, obovate to cuneiform, with 3 obtuse to acute lobes at apex, entire; petiole 1.5–4.5(–6.5) mm; stipules *c.* two-thirds as long as petiole, membranous, scarious, adnate except for small, free, apical point. Peduncles 5–15(–20) mm, erect, strongly curved just below flower. Sepals 2.5–3.1 × 1.2–1.5 mm, ovate-lanceolate, obtuse, green or purplish; appendages *c.*0.5 mm, rounded. Petals white with purple veins, 3.5–5.3 mm, the basal longer than the others, obovate to spathulate, obtuse, the lateral glabrous; spur up to 0.5 mm. Style *c.*1.5 mm, geniculate at base, filiform, tapering to simple stigmatic opening. Capsule 3–4 mm in diameter, globose; seeds 1.6–2 mm ovoid, black. $2n = 40$. Self-compatible. *Flowers January to March.* (Fig. 68).

Moist peaty and mineral soils, feldmark, raised parts of *Sphagnum* bogs and in Magellanic moorland. (200–)350–1000 m.

South and west from C. Isla Grande. (Map 8i).
W. Argentina north to 47°52′S.lat.
Chile north to *c.*47°S.lat.
Falkland Islands (Islas Malvinas).

2. *V. maculata* Cav. 1801, *Icon. Descr.* **6**: 20.
V. microphyllos Poiret 1808, *Encycl. Méth. Bot.* **8**: 636 – Skottsberg, 1916.
V. fimbriata Steudel 1856, *Flora (Regensb.)* **39**:423 – Dusén, 1900.

Perennial; stems, up to 11 cm, ascending to erect. Leaves with lamina (6–)9–30(–35) × (3–)5–25(–37) mm, ovate-lanceolate to ovate or suborbicular, obtuse to subacute, cuneate to truncate or, rarely, shallowly cordate at base, crenate to crenate-serrate, with small, black marginal glands, ciliate, sparsely pubescent beneath; petiole (5–)10–50(–70) mm, appressed-pubescent; stipules 3–12 mm, lanceolate, sparsely glandular, fimbriate to fimbriate-dentate, scarious, free. Penduncles 40–160 mm, erect. Sepals 2.5–8 × 0.7–2.5 mm, lanceolate, acute, green; appendages 0.4–0.7 mm, rounded. Petals yellow with brownish veins, 8–17 mm, the basal shorter than to as long as the others, obovate to oblanceolate, obtuse or emarginate, the lateral bearded; spur 1–3 mm. Style *c.*2 mm, geniculate at base, clavate, winged at apex with anterior stigmatic beak. Capsule 5.5–8 mm, ellipsoid; seeds 1.7–2 mm, ovoid, black. Self-compatible. *Flowers December to January.* (Fig. 69).

Dry, open, sandy, pebbly and peaty soils on coast, in grassland and sparse scrub; 0–150 m.

Widespread in Fuegia, but rare or absent in west and extreme south. (Map 8j).
Argentina north to *c.*44°S. lat., in the west to *c.*37°S. lat.
Chile, particularly in the east, north to *c.*34°S.lat.
Falkland Islands (Islas Malvinas).

In Fuegian material Sparre (1949) had distinguished var. *microphyllos* (Poiret) D.C., with narrowly ovate to elliptical leaves having a cuneate base, and sepals glabrous, at least at the apex, and var. *megaphylla* D.C. (correctly var. *maculata* (Moore, 1968: 88)), with ovate leaves having a truncate base, and ciliate sepals. I agree with Sparre that, in view of the considerable variation shown by this species, and the uncertainty as to the relative importance of genetic and environmental components, such infraspecific categories are purely tentative.

3. *V. reichei* Skottsb. 1916, *Kungl. Svenska Vetensk Akad. Handl.* **56(5)**: 265.

Perennial; stems up to 20 cm, procumbent to ascending, rooted at intervals. Leaves with lamina 15–30 × 20–35 mm, orbicular to reniform, obtuse, cordate at base, crenate, somewhat hairy above, glabrous beneath; petiole 15–20 mm, glabrous; stipules 9–13 mm, ovate-lanceolate, ciliate-dentate distally, scarious, free. Peduncles up to 110 mm, erect. Sepals 4.5–6 × 1.5–2 mm, lanceolate, acute, green; appendages *c.*0.75 mm, truncate. Petals yellow, with faint brownish veins, 12–17 mm, the basal longer than the others, obovate, obtuse to acute, the lateral somewhat bearded; spur *c.*3 mm. Style *c.*2 mm, geniculate at base, clavate, somewhat winged at apex with short anterior stigmatic beak. Capsule 6–10 mm, ellipsoid; seeds 1.5–2 mm, ovoid, black. *Flowers November to January.*

Floor of *Nothofagus pumilio* forest; 0–80 m.

C. Fuegia between 53°30 and 55°30′S.lat. (Map 9a).
W. Argentina between 41°30′ and *c.*40°S.lat.
Chile north to *c.*37°S.lat.

4. *V. magellanica* Forster f. 1789, *Comment. Gotting.* **9**: 41.

Perennial; stems up to 16 cm, ascending to erect. Leaves with lamina (6–)10–25(–30) × 10–30(–40) mm, reniform to triangular-ovate, rounded-obtuse, cordate at base, crenate to crenate-serrate, sparsely hairy above; petiole 10–100 cm, glabrous; stipules 3–7 mm, lanceolate, membranous, entire to glandular serrate-fimbriate. Peduncles up to 165 mm, erect. Sepals 4–6.5 × 1–2.5 mm, ovate-lanceolate, obtuse, green; appendages up to 0.5 mm, truncate. Petals yellow, with brown veins, 11–16 mm, the basal longer than the others, obtriangular to obovate, obtuse, the lateral bearded; spur 2–3 mm. Style *c.*2 mm, more or less straight, filiform but slightly dilated at apex, with inconspicuous stigmatic beak. Capsule 6–9 mm, ellipsoid; seeds *c.*1.8 mm, ovoid, black. Self-compatible. *Flowers November to March.* (Fig. 70).

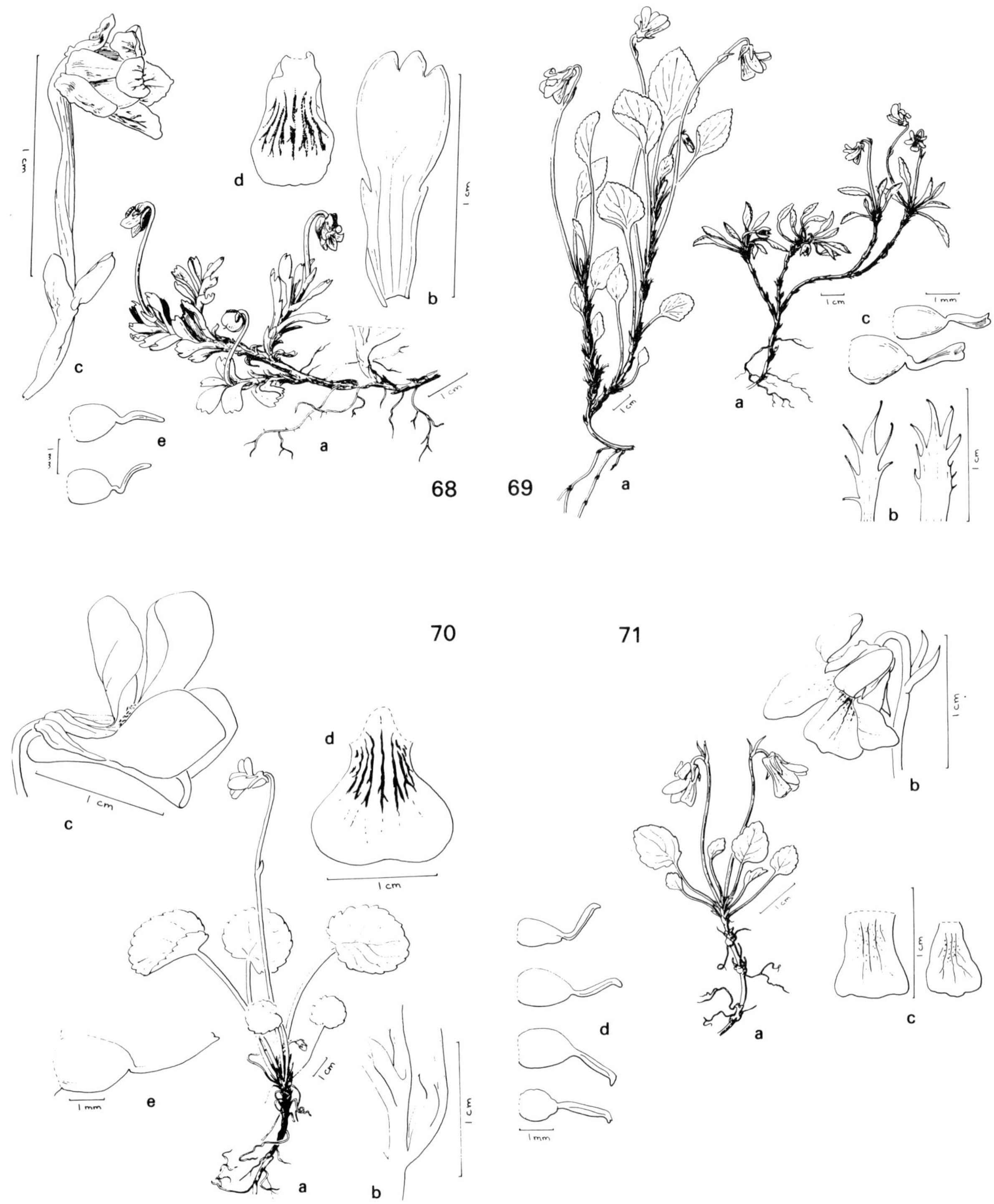

Fig. 68. *Viola tridentata.* a, Plant. b, Leaf. c, Flower. d, Lower petal. e, Ovary and style. Scale = 1 cm (a–d) or 1 mm (e). Fig. 69. *Viola maculata.* a, Plant, two forms. b, Stipules. c, Ovary and style. Scale = 1 cm (a–b) or 1 mm (c). Fig. 70. *Viola magellanica.* a, Plant. b, Stipules. c, Flower. d, Lower petal. e, Ovary and style. Scale = 1 cm (a–d) or 1 mm (e). Fig. 71. *Viola commersonii.* a, Plant. b, Flower. c, Lower petals. d, Ovaries and styles. Scale = 1 cm (a–c) or 1 mm (d).

(R. N. P. Goodall)

Moist places in scrub and, especially, at margins and in clearings of deciduous *Nothofagus* forest; 20–30 m.

Throughout Fuegia except the extreme north-east and south-west. (Map 9b).
Chile north to *c.*52°S.lat. and at *c.*41°S.lat.

5. *V. commersonii* D.C. 1824, *Prodr.* **1**: 297.

Glabrous perennial; stems up to 2 cm, ascending to erect. Leaves with lamina 5–15 × 5–10 mm, ovate, subacute to obtuse, cuneate, rarely truncate, at base, shallowy serrate, with black marginal glands; petiole 4–15 mm; stipules 2.5–4 mm, ovate to oblong, obtuse, entire to denticulate, membranous, free. Peduncles 17–45 mm, erect, curved just below flower. Sepals 3–5 × 1–1.5 mm, linear- to ovate-lanceolate, acute, green; appendages *c.*0.5 mm, truncate. Petals white with purplish veins, 6–10(–13) mm, the basal shorter than the others, obovate-spathulate, obtuse to almost truncate, the lateral glabrous; spur 0.5–1 mm. Style *c.*2 mm, straight, slightly dilated at apex. Capsule 4–5 mm, globose; seeds *c.*1 mm, ovoid. *Flowers December to January.* (Fig. 71).

Sphagnum bogs and wet areas at or above timberline. 300–500m.

S. and W. Fuegia. (Map 9c).
W. Argentina at *c.*50°S.lat.
Chile north to *c.*47°S.lat.

6 *V. arvensis Murray 1770, *Prodr. Stirp. Gotting.* 73.

Annual; stem up to 40 cm, ± erect, branched pubescent. Leaves 12–20 × 4–10 mm, very variable, acute to obtuse, ± cuneate at base, crenate to crenate-serrate; petiole 5–12 mm, pubescent; stipules 15–20 mm, coarsely pinnatifid with terminal segments lanceolate and leaf-like. Penduncles up to 30 mm, erect. Sepals 5–12 × 1–2 mm, lanceolate, acute; appendages 1–2 mm, truncate to emarginate. Petals as long as or shorter than sepals, obovate to oblanceolate, cream, the upper and lateral sometimes bluish; spur 1–4 mm, geniculate near base, with subglobose head having wide stigmatic opening. Capsule 5–10 mm, ellipsoid-oblong; seeds *c.*2 mm, ovoid, brown. Self-compatible. *Flowers December to January.*

Disturbed, open, dry grassland; 0–2 m.

N.E. Isla Grande.

Introduced from Eurasia and locally naturalized in cultivated areas of temperate regions.

XXVII. FRANKENIACEAE

Dwarf shrubs. Leaves opposite, ericoid; stipules absent. Flowers hermaphrodite, actinomorphic, in terminal or axillary cymose inflorescences. Calyx 5–dentate. Petals 6, free. Stamens 6, in 2 whorls. Ovary superior, 1–celled. Style 1, 2– to 3– fid. Fruit a loculicidal capsule; seeds few, with endosperm.

1. *Frankenia* L.

Inflorescences subtended by 4 leaves united at base. Flowers sessile. Calyx tubular. Petals with a scale-like appendage on the claw. Outer whorl of stamens shorter than the inner. Capsule included in persistent calyx.

Literature: M. N. Correa, *Darwiniana* **14**: 68–94(1966).

1. *F. chubutensis* Speg. 1902, *Anal. Mus. Nac. Bs Aires* **7**: 238.
Niederleinia juniperoides auct., non Hieron. – Dusén, 1900.

Densely leafy, prostrate shrub 20–40 cm. Leaves 1–3 × 0.5–0.75 mm, oblong, obtuse, strongly revolute, connate at base, fleshy, glandular above, the 4 subtending inflorescence equal, sometimes incompletely united to give 2 groups. Calyx 5–6 mm, the teeth short, acuminate, ciliolate. Petals 5–6 mm; limb 2.2–3 mm wide, oblong, obtuse, whitish to pink. Style 4–6 mm, the branches 1–1.5 mm; stigmas capitate. *Flowers December to March.* (Fig. 74).

Open, periodically flooded, saline soils on coast; 0–10 m.

Local in northernmost Isla Grande. (Map 9d).
Argentina, especially in the east, north to *c.*47°S.lat.

XXVIII. CRUCIFERAE

Annual to perennial herbs. Leaves alternate; stipules absent. Flowers hermaphrodite, actinomorphic. Sepals 4, free, in 2 decussate pairs, the inner pair often with saccate bases. Petals 4, free, clawed, alternating with sepals, sometimes absent. Stamens 6, an outer pair with short filaments and 2 inner pairs with long filaments, the outer 1–2 pairs sometimes absent. Ovary superior, usually with 2 cells separated by membranous false septum. Fruit usually a capsule with 2 cells separated by false septum, opening from below by 2 valves, rarely indehiscent, called a *siliqua*

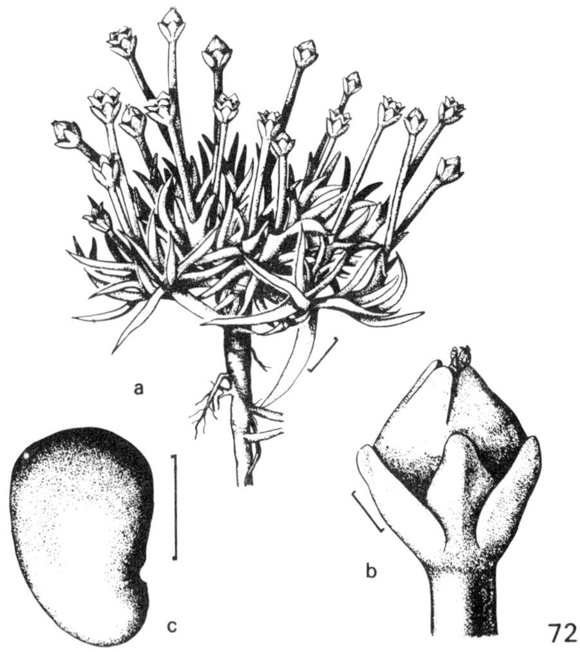

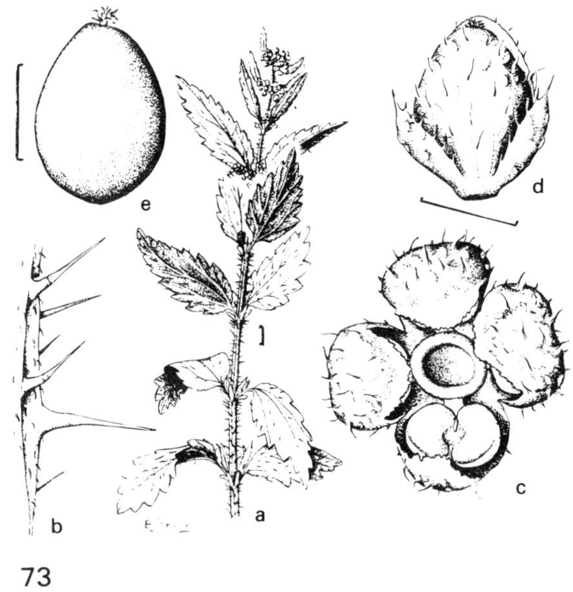

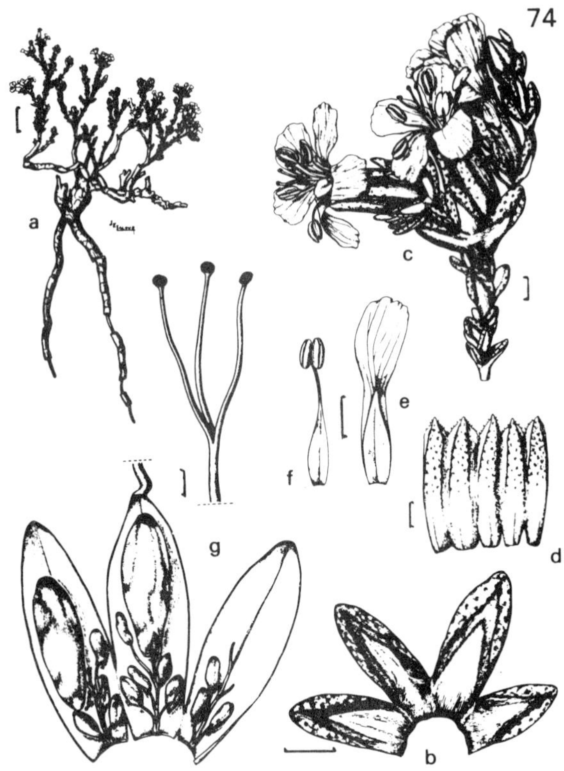

Fig. 72. *Colobanthus quitensis*. a, Plant. b, Capsule. c, Seed. Scale = 1 cm (a), 1 mm (b) or 0.5 mm (c). Fig. 73. *Urtica magellanica*. a, Branch. b, details of hairs. c, Male flower. d, Female flower. e, Achene. Scale = 1 cm (a) or 1 mm (b–e). Fig. 74. *Frankenia chubutensis*. a, Plant. b, Leaves. c, Inflorescence. d, Calyx. e, Petal. f, Stamen. g, Stigmas and ovary. Scale = 5 cm (a), 1 mm (b–f) or 0.5 mm (g). Fig. 75. *Sisymbrium magellanicum*. a, Base of plant. b, Inflorescence. c, Flower. d, Sepal. e, Petal. f, Stamen. g, Nectary. h, Fruit. Scale = 1 cm (a–b), 1 mm (c–f, h) or 0.5 mm (g).

(Flora Patagónica)

if at least 3 times as long as wide and a *silicula* when less than 3 times as long as wide; seeds in 1 or 2 rows in each cell, without endosperm.

a. At least some leaves pinnate or deeply pinnately cut:
  b. Fruit a silicula:
    c. Fruit transversely articulate, the lower joint narrow, with 0–2 seeds, the upper larger, 1–seeded, indehiscent **16. *Rapistrum***
    c. Fruit not transversely articulate, opening by longitudinal valves or indehiscent:
      d. Fruit triangular-obcordate **10. *Capsella***
      d. Fruit didymous or elliptical to suborbicular:
        e. Fruit didymous, indehiscent **14. *Coronopus***
        e. Fruit elliptical to suborbicular, dehiscent **12. *Lepidium***
  b. Fruit a siliqua:
    f. Fruit transversely articulate, the lower joint narrow, seedless, the upper fertile, ± inflated, indehiscent **17. *Raphanus***
    f. Fruit not transversely articulate, opening by longitudinal valves:
      g. Stems with branched hairs or a mixture of branched and simple hairs:
        h. Petals yellow, up to 2 mm; lower leaves up to 40 mm **2. *Descurainia***
        h. Petals white, at least 5 mm; lower leaves at least 100 mm **1. *Sisymbrium***
      g. Stems glabrous or with simple hairs only:
        i. Fruit with an apical beak 4 mm or more **15. *Brassica***
        i. Fruit without an apical beak:
          j. Fruit with conspicuously 1-veined valves, not rolling up on dehiscence **1. *Sisymbrium***
          j. Fruit with veinless valves, which roll up suddenly into spirals on dehiscence **5. *Cardamine***
a. Leaves simple or variously divided, but not pinnately so:
  k. Fruit a silicula:
    l. Fruit broadly cordate, indehiscent **13. *Cardaria***
    l. Fruit ellipsoid to suborbicular or obovate, dehiscent:
      m. Fruit with valves keeled and at least somewhat winged **11. *Thlaspi***
      m. Fruit not keeled and winged:
        n. Plant at least 25 cm; fruit at least 30 mm **6. *Lunaria***
        n. Plant not more than 12 cm; fruit not more than 12 mm:
          o. Petals present, deeply 2-fid; pedicels longer than fruit **9. *Erophila***
          o. Petals absent; pedicels shorter than fruit **8. *Draba***
  k. Fruit a siliqua:
    p. Plant not more than 12 cm; leaves not more than 30 mm; fruit not more than 12 mm:
      q. Leaves glabrous **7. *Onuris***
      q. Leaves pubescent or with marginal cilia **8. *Draba***
    p. Plant at least 25 cm; leaves at least 40 mm; fruit at least 17 mm:
      r. Leaves glabrous or with simple hairs; fruit with an apical beak 4 mm or more **15. *Brassica***
      r. Leaves with branched often stellate hairs or a mixture of branched and simple hairs; fruit without a beak more than 1 mm:
        s. Petals yellow; stigma capitate or shallowly 2-lobed; fruiting pedicels up to 5 mm **3. *Erysimum***
        s. Petals white or violet; stigma deeply 2-lobed; fruiting pedicels at least 10 mm **4. *Hesperis***

1. *Sisymbrium* L.

Annual to perennial, with simple or branched hairs. Leaves pinnatifid to pinnatisect. Fruit a siliqua; seeds in 1 or 2 rows in each cell.

a. Leaves glabrous or with simple hairs; flowers yellow; seeds in 1 row in each cell **2. *irio***
a. Leaves with branched hairs; flowers white; seeds in 2 rows in each cell **1. *magellanicum***

1. *S. magellanicum* (Pers.) Hooker f. 1847, *Fl. Antarct.* **1(2)**: 243.

S. fuegianum (Speg.) Speg. 1901, *Anal. Soc. Ci. Argent.* **47**: 9.
Schizopetalum fuegianum Speg. 1896, *Anal. Mus. Nac. Bs Aires* **5**: 48.
Arabis magellanica (Pers.) Dusén 1900, *Svenska Exped Magellansl.* **3(5)**: 177.
Brassica magellanica Pers. 1806, *Syn. Pl.* **2**: 207 – Alboff, 1897; Macloskie, 1905.

Perennial; stems 30–70 cm, erect, robust, branched above, ± glabrous or with few simple or branched hairs. Lower leaves 100–200 × 40–90 mm, pinnatisect, with 3–7 pairs of lobes with dense branched hairs beneath, at least on veins; terminal lobe obovate, dentate-lobulate, the lateral oblong, coarsely dentate-lobed, smaller and entire towards base; petioles 30–70 mm; upper cauline leaves pinnately cut, with 2–6 pairs of narrow, acute lobes, sessile. Flowers in corymbose panicle; pedicels 2–5 mm. Sepals 2.3–5 mm, the outer ellipical, the inner ovate. Petals 5–7.5 mm, the limb obovate, attenuate into claw, white. Siliqua 17–29 × 1.2–3 mm, linear, somewhat curved, narrowed to style 0.5–1 mm; seeds 1.4–1.7 × 0.7–1 mm, oblong to ovoid, reticulate-striate, pale brown, in 2 rows in each cell. $2n = 18$. *Flowers November to February*. (Fig. 75).

Moist rock-crevices, especially by coast; 0–50 m.

Isla Grande, except the extreme west, and south side of Canal Beagle. (Map 9e).
Argentina, especially in the west, north to *c.*47°20′S.lat.
Chile north to *c.*51°S.lat.

2. S. irio L. 1753, *Sp. Pl.* 695.
S. officinale auct., non L. – Spegazzini, 1896.

Annual; stem up to 60 cm, erect, branched, glabrous or with short, appressed hairs. Lower leaves *c.*7.5 × 2.2 cm, deeply pinnately lobed, with 1–6 pairs of lobes, glabrous or with simple hairs; terminal lobe hastate, larger than the oblong lateral lobes; petiole 8–12 mm; upper cauline leaves hastate, entire, shortly petiolate. Flowers in corymbose panicle; pedicels 3–6 mm. Sepals 1.9–2.5 mm, elliptic-oblong. Petals 2.5–3.5 mm, oblanceolate-spathulate, pale yellow. Siliqua 32–41 × 0.9–1.2 mm, cylindrical with ± regular constrictions, narrowed at apex into style *c.*0.5 mm, somewhat curved; fruiting pedicels 5–10 mm; seeds *c.*0.8 × 0.4 mm, oblong-ovoid, pale brown, in 1 row in each cell.

Disturbed ground near buildings.

S. side of Isla Grande (Ushuaia).
Introduced from Europe and locally naturalized.

2. *Descurainia* Webb & Berth.

Annual, with simple and branched, stellate hairs. Leaves 2– to 3– pinnatisect. Flowers in ebracteate raceme. Fruit a siliqua; seeds in 1–2 rows in each cell.

a. Leaf-segments oblanceolate to obovate; siliqua more than 1.2 mm wide, with 2 rows of seeds in each cell **1. *antarctica***
a. Upper leaf-segments linear; siliqua not more than 1.2 mm wide, with 1 row of seeds in each cell **2. *sophia***

1. *D. antarctica* (E. Fourn.) O. E. Schulz 1924, *Pflanzenreich* **86(IV.105)**: 334.
D. canescens auct., non (Nutt.) Prantl– Dusén, 1900.
D. pimpinellifolia auct., non (Barneoud) O. E. Schulz– Skottsberg, 1926.

Stem 5–8 cm, erect, simple or branched above, with simple glandular and branched stellate hairs. Leaves 14–33 × 7–11 mm, ±ovate in outline, 1– to 2– pinnatifid, with branched stellate hairs; lower segments oblanceolate to obovate, obtuse, the upper similar but smaller. Sepals 0.8–2.5 mm, oblong, sometimes with stellate hairs. Petals 1.5–2 mm, spathulate, pinkish below, pale yellow above. Siliqua somewhat curved towards inflorescence-axis; pedicels 4–11 mm, patent; seeds 1.1–1.2 × 0.6–0.7 mm, ovoid, orange-brown, in 2 rows in each cell. *Flowers December to March*. (Fig. 77).

Sand and gravel beaches, sandstone cliffs, saline flats in *Lepidophyllum* scrub; 0–20 m.

Coasts of N. & S. (locally) Isla Grande. (Map 9f).
Argentina north to *c.*40°30′S.lat.
Chile along north shore of Estrecho de Magallanes.

2. D. sophia (L.) Webb ex Prantl 1891, in Engler & Prantl, *Nat. Pflanzenfam.* **3(2)**: 192.

Stem (3.5–)18–77 cm, erect, branched above, with stellate hairs. Leaves 17–100 × 8–50 mm, ovate to obovate in outline, 2– to 3– pinnatisect, usually with stellate hairs; lower segments lanceolate, subacute, the upper linear. Sepals 2.1–2.7 mm, oblong. Petals 2–2.8 mm, spathulate, pale yellow. Siliqua 16–28 × 0.8–1 mm, ± cylindrical, erect; pedicels 8–11 mm, patent; seeds 0.9–1 × *c.*0.5 mm, ovoid, orange-brown, in 1 row in each cell. *Flowers December to February*.

Disturbed and open ground, especially along roads and by buildings, but also beaches and open scrub; 0–60 m.

N. Isla Grande, locally in the south. Introduced from Europe and widely naturalized.

3. *Erysimum* L.

Annual, with simple and branched hairs. Leaves simple. Flowers in ebracteate, corymbose inflorescence. Fruit a siliqua; seeds in 1 row in each cell.

1. E. repandum L. 1753, *Demonstr. Pl.* 17.

Stem up to 35 cm, erect, simple or somewhat branched, with simple hairs. Leaves 7–8 × 1.5–2 cm, linear-lanceolate to lanceolate, acute, the lower entire to sinuate, the upper dentate. Sepals *c.*5.5 × 0.9–1 mm, linear-lanceolate to -oblanceolate, subacute to obtuse, the inner saccate. Petals with limb *c.*3.3 × 2 mm, suborbicular, and claw *c.*5.5 mm,

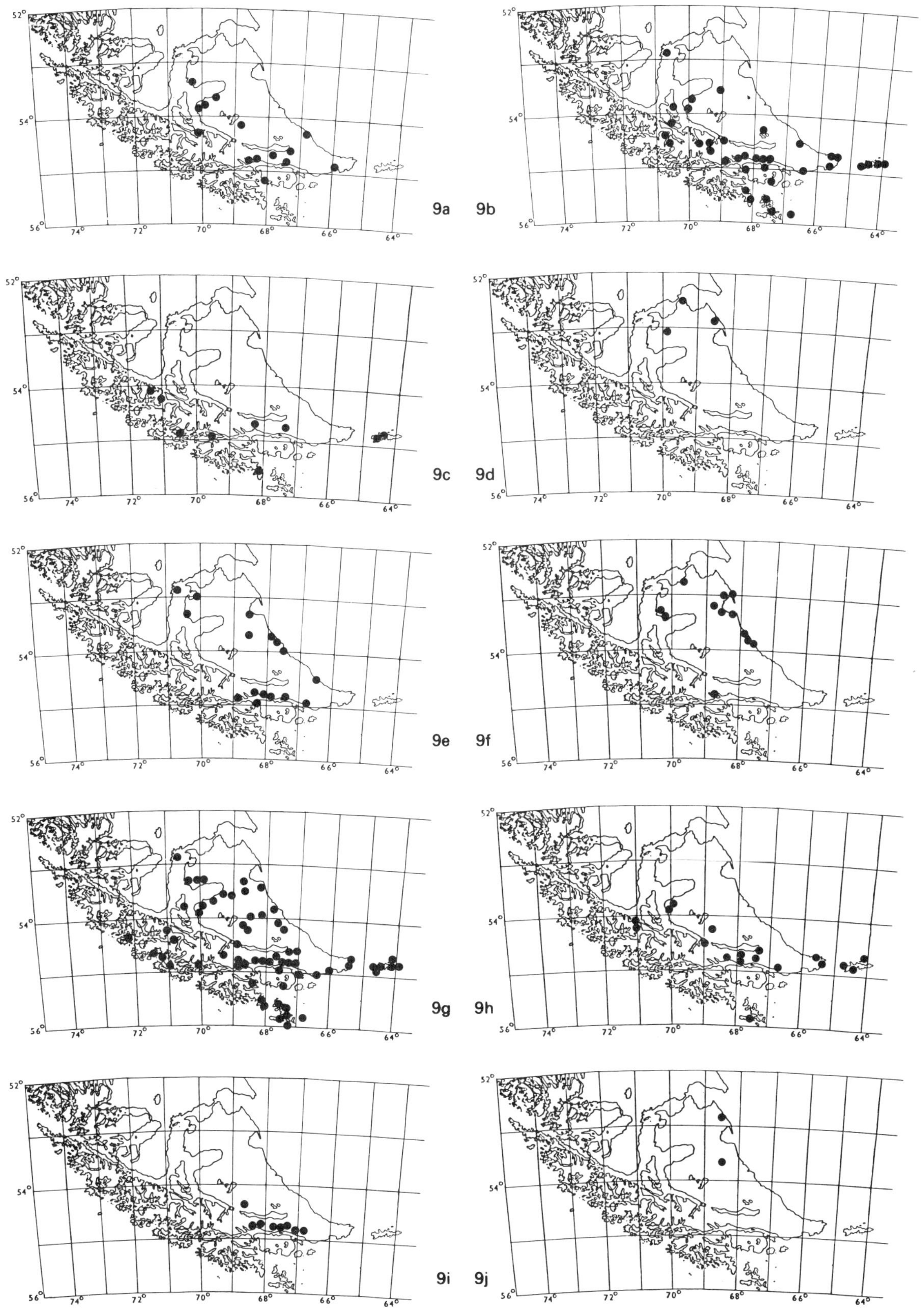

Map 9. a, *Viola reichei*. b, *Viola magellanica*. c, *Viola commersonii*. d, *Frankenia chubutensis*. e, *Sisymbrium magellanicum*. f, *Descurainia antarctica*. g, *Cardamine glacialis*. h, *Cardamine geraniifolia*. i, *Onuris alismatifolia*. j, *Onuris papillosa*.

yellow. Style short; stigma capitate to somewhat 2-lobed. Siliqua 45–70 × 1.3–1.6 mm, ± cylindrical, with simple hairs, patent; pedicels 4–5 mm, patent; seeds *c.*1.4 × 0.5 mm, pale brown. *Flowers January.*

Coastal sands near settlements. 0–10 m.

S. coast of Isla Grande (Ushuaia).
Introduced from Europe, where it is a common weed of disturbed ground.

4. *Hesperis* L.

Biennial or perennial, with simple and stellate hairs. Leaves simple. Flowers in a raceme, ebracteate or the lower bracteate. Fruit a siliqua; seeds in 1 row in each cell.

1. *H. matronalis L. 1753, *Sp. Pl.* 663.

Stem up to 40 cm, erect, usually branched, with short simple and stellate hairs. Lower leaves with lamina 40–70 × 15–20 mm, ovate to elliptical, acute to acuminate, serrate to lyrate, the petiole 20–25 mm; upper leaves smaller, lanceolate, sessile, usually with 2 glands at base. Sepals 6.5–7 × 1.5–1.8 mm, oblong-lanceolate, obtuse, the inner saccate. Petals with limb 7–9 × 4–6 mm, oblong, to obovate, and claw 4–7 mm, white or violet. Style short; stigma deeply 2-lobed. Siliqua 25–90 × 1.5–3 mm, ± cylindrical, glabrous, erecto-patent; pedicels 10–30 mm, ± patent; seeds *c.*3 mm, brown. *Flowers December.*

Disturbed ground by roads, beaches; 0–20 m.

N.W. coast of Isla Grande.
Introduced from Europe as a garden ornamental.

5. *Cardamine* L.

Perennial, glabrous or with simple hairs. Leaves pinnate. Flowers solitary or in corymbose or racemose inflorescence. Fruit a siliqua; seeds in 1 row in each cell.

a. Leaflets deeply divided; sepals at least 4 mm; petals at least 10 mm; siliqua usually at least 60 mm **2. *geraniifolia***
a. Leaflets entire to crenate-dentate; sepals up to 3 mm; petals up to 9 mm; siliqua up to 36 mm **1. *glacialis***

1. *C. glacialis* (Forster F.) DC. 1821, *Reg. Veg. Syst. Nat.* **2**: 264.
C. amara auct., non L. – Wildeman, 1905.
C. antiscorbutica Banks & Sol. ex Hooker f. 1847, *Fl. Antarct.* **1(2)**: 232 – Spegazzini, 1896; Franchet, 1889; Wildeman, 1905.
C. hirsuta auct., non L. – Hooker, 1847.

Stems (2–)5–32(–45) cm, erect or ascending, branched from the base, glabrous or pubescent, often purplish above, with 0–6 cauline leaves. Leaves with lamina 1.5–12 × 0.5–3 cm, oblong in outline, with 2–7 pairs of leaflets; leaflets of basal leaves (2–)4–15(–21) × (3–)7–21(–30) mm, suborbicular, crenate, subsessile, the terminal leaflet often 3-lobed; leaflets of cauline leaves oblong to ovate, entire or rarely crenate, the terminal often 3-lobed. Flowers solitary or in 2-to 16-flowered racemose or corymbose inflorescences; pedicels 2–7(–17) mm. Sepals (1.7–)2–3 mm, broadly ovate, obtuse, green or violet-green, with white margin. Petals (3.5–)4.2–7.5(–8.6) mm, obovate-cuneiform, with long claw, white, sometimes purplish. Siliqua (5.5–)10–30 × 0.5–1.7 mm, compressed, erect, green or purplish, glabrous; pedicels 4–11(–22) mm; seeds 1.1–1.6(–1.9) × 0.6–1 mm, ovoid, dark brown. $2n = 72$. Self-compatible and often, perhaps usually, self-pollinated. *Flowers October to March.* (Fig. 81).

Moist places in most communities; 0–1100 m.

Throughout Fuegia. (Map 9g).
W. Argentina north to *c.*30°S.lat.
Chile north to *c.*40°S.lat.
Falkland Islands (Islas Malvinas). Islas Diego Ramirez.

A very variable species in which indumentum and stature are markedly modified by the environment.

2. *C. geraniifolia* (Poiret) DC. 1821, *Reg. Veg. Syst. Nat.* **2**: 268.
Dentaria geraniifolia (Poiret) Reiche 1896, *Fl. Chile* **1**: 104. – Dusén, 1900.

Stems 16–37(–46) cm, erect, branched, glabrous, succulent, with cauline leaves. Leaves with lamina 5–10 × 3–7 cm, ovate to oblong in outline, with 2–4 pairs of leaflets; petiole 7–12 cm; leaflets 1.5–3 × 2–3 cm, ovate, pinnate to deeply lobed, the lobes dentate-serrate, petiolate; cauline leaves similar to the basal but smaller and only the lower leaflets deeply lobed. Inflorescence a 6- to 15-flowered corymb; pedicels 5–8 mm. Sepals 4–5 mm, ovate to elliptic-oblong, obtuse, cucullate, saccate, green or brownish, with white margin. Petals *c.*11 × 6 mm, obovate-spathulate, white. Siliqua (27–)60–80 × (1.3–)1.8–2.9 mm, linear, compressed, glabrous; pedicels 10–21 mm; seeds 2.5–2.7 × 1.2–1.7 mm, ovoid to ellipsoid, dark brown. $2n = 144$. *Flowers October to January*. (Fig. 85).

Wet rock crevices, wet areas in *Nothofagus pumilio* forest; 0–50 m.

S. Fuegia except the extreme west. (Map 9h).
Chile north to *c*.44°40′S.lat.

6. *Lunaria* L.

Biennial, with simple hairs. Leaves simple. Flowers in a lax raceme. Fruit a large siliqua; seeds in 2 rows in each cell.

*1. ***L. annua*** L. 1753, *Sp. Pl.* 653.

Stem 30–100 cm, erect, simple or branched, with rigid hairs. Leaves up to 7 × 5 cm, broadly cordate to ovate, acuminate, irregularly and coarsely dentate, the lower long-petiolate, the upper subsessile. Pedicels 3–12 mm. Sepals 3–4 mm, oblong, obtuse. Petals *c*.20 × 8 mm, spathulate, reddish-purple, rarely white. Siliqua 30–45 × 20–25 mm, elliptical to oblong or suborbicular; seeds 5–8 mm in diameter, suborbicular, strongly compressed.

Disturbed ground in settlements; 0–30 m.

S. Coast of Isla Grande (Ushuaia).
Introduced from Europe as a garden ornamental, an occasional escape from cultivation.

7. *Onuris* Phil.

Glabrous perennial. Leaves entire. Flowers in a terminal subumbellate raceme. Fruit a siliqua; seeds in 2 rows in each cell.

Literature: O. E. Schulz, *Pflanzenreich* **86(IV.105)**: 199–202(1924).

a. Leaves with marginal papillae, at least towards apex **2. *papillosa***
a. Leaves without marginal papillae, at least towards apex **1. *alismatifolia***

1. ***O. alismatifolia*** Gilg ex Skottsb. 1916, *Kungl. Svenska Vetensk. Akad. Handl.* **56(5)**: 229.

Stem 2–8 cm, much branched, somewhat woody, covered with dead leaf-remains. Leaves in basal clusters, 5–20 × 1.3–2.5 mm, linear-spathulate, obtuse to subacute, gradually narrowed into long petiole, rather fleshy. Flowering stems up to 25 mm, without leaves. Sepals *c*.2.5 × 1.5 mm, elliptical to ovate, obtuse, caducous. Petals 3–4 × 2–3 mm, spathulate, white. Stigma capitate. Siliqua 2–6.5 × 0.7–1.7 mm, ellipsoid; pedicels 2.5–6 mm; seeds *c*.2 × 1.5 mm, ovoid, reticulate. *Flowers November to March*. (Fig. 76).

Feldmark and open dwarf shrub heath; 500–1,000 m.

C. and S. Isla Grande, especially in the south. Endemic. (Map 9i).

2. *O. papillosa* O. E. Schulz 1924, *Pflanzenreich* **86(iv.105)**: 202.

Like *O. alismatifolia* but leaves 0.5–1.5 mm wide, with marginal papillae at least towards apex; flowering stems up to 40 mm. *Flowers November*.

Rock crevices, sand and open grassland; 40–150 m.

N. E. Isla Grande. (Map 9j).
Argentina north to *c*.50°30′S.lat.

8. *Draba* L.

Annual or perennial, with simple or simple and stellate hairs. Leaves entire to dentate. Flowers in a terminal raceme. Fruit a silicula or siliqua; seeds in 2 rows in each cell.

Literature: O. E. Schulz, *Pflanzenreich* **89(IV.105)**: 1–343(1927).

a. Flowering stems with 1–5 sessile leaves **3. *magellanica***
a. Flowering stems without leaves:
 b. Perennial; leaves with rigid cilia, otherwise glabrous or with simple hairs; petals present; fruit a siliqua **1. *funiculosa***
 b. Annual; leaves with stellate hairs; petals absent; fruit a silicula **2. *australis***

1. *D. funiculosa* Hooker f. 1847, *Fl. Antarct.* **1(2)**: 238.

Caespitose perennial; stem 5–100 mm, ascending to erect, simple or sparingly branched, somewhat woody, covered with leaf-remains. Leaves usually imbricate, in basal clusters, 7–12 × 0.7–1.5 mm, linear-lanceolate to -subspathulate, obtuse to subacute, sessile with somewhat sheathing base, dentate with rigid marginal cilia, otherwise glabrous or with setose hairs towards apex, rather coriaceous. Flowering stems 10–30(–63) mm, without leaves; raceme (1)2- to 4-flowered; pedicels 0.7–2(–4) mm. Sepals 1–2 × 0.5–1.3 mm, ovate to oblong, obtuse to acute, with sparse branched or stellate hairs. Petals 3–4(–4.8) × 1.6–2 mm, spathulate, obtuse to ± retuse, white. Siliqua 7–9.5(–11) × 1.5–2.3 mm,

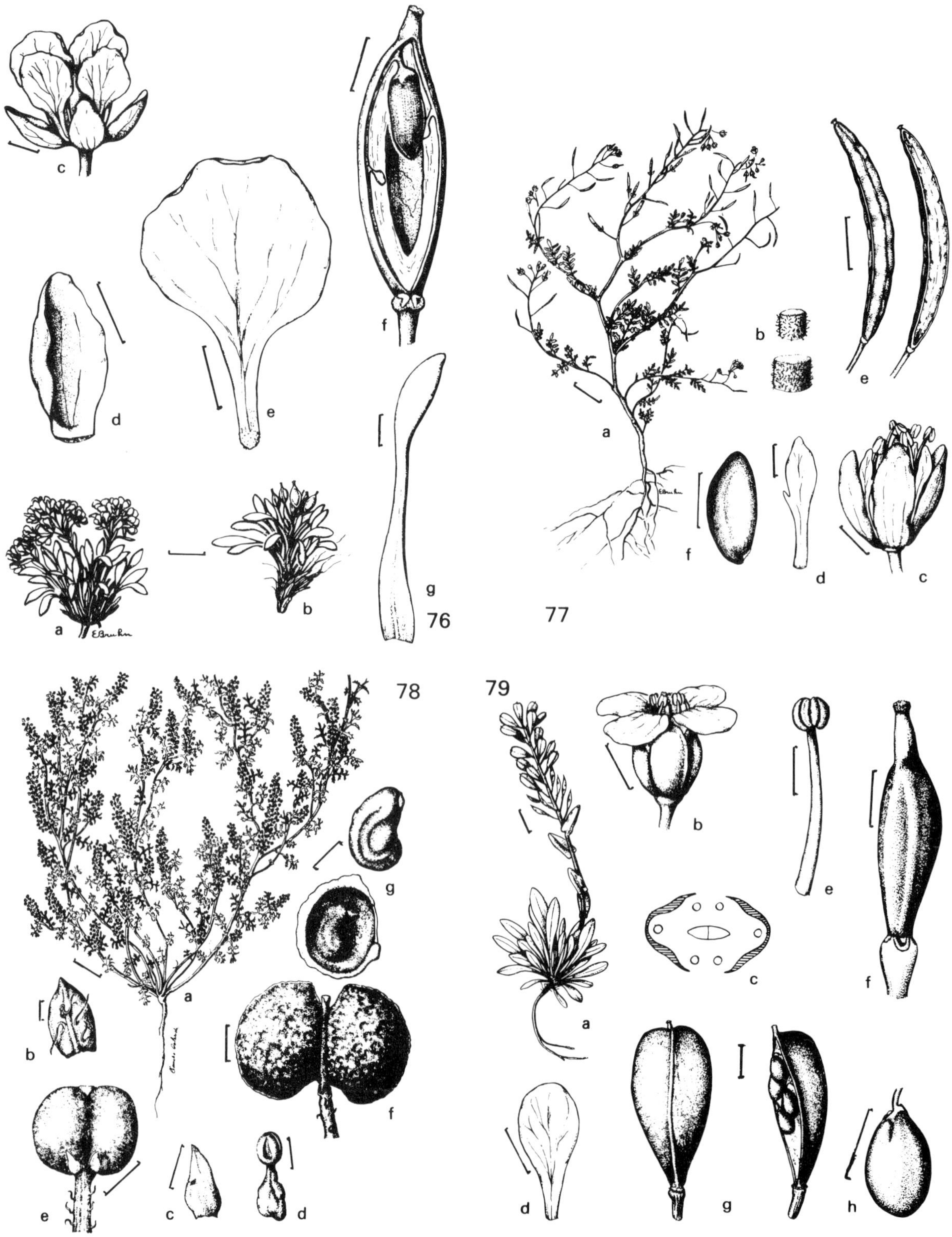

Fig. 76. *Onuris alismatifolia*. a, Plant in flower. b, Plant in fruit. c, Flower. d, Sepal. e, Petal. f, Fruit. g, Leaf. Scale = 1 cm (a–b), 2 mm (g) or 1 mm (c–f). Fig. 77. *Descurainia antarctica*. a, Plant. b, Detail of hairs. c, Flower. d, Petal. e, Fruit. f, Seed. Scale = 1 cm (a), 5 mm (e), 1 mm (c) or 0.5 mm (d, f). Fig. 78. *Coronopus didymus*. a, Plant. b, Leaf-lobe. c, Sepal. d, Stamen. e, Young fruit. f, Mature fruit. g, Fruit-lobe opened to expose seed. Scale = 1 cm (a), 1 mm (b–c) or 0.5 mm (d–g). Fig. 79. *Thlaspi magellanicum*. a, Plant. b, Flower. c, Floral diagram. d, Petal. e, Stamen. f, Ovary with nectary. g, Fruit. h, Seed. Scale = 1 cm (a), 1 mm (b, d, f–h) or 0.5 mm (e).

(Flora Patagónica)

linear, obtuse, compressed, glabrous, erecto-patent; seeds 1–1.1 mm, ovoid, smooth, orange-brown. $2n = 16$*. Self-compatible, usually self-pollinated. *Flowers November to January*. (Fig. 80).

Short grassland, sometimes in open *Chiliotrichum* scrub, rock outcrops; 0–900 m.

Throughout Isla Grande except the west; N. Isla Navarino. (Map 10a).
Argentina north to *c*.52°S.lat. and, in the west, to *c*.50°40′S.lat.
E. Chile north to 50°45′S.lat.
Falkland Islands (Islas Malvinas).

2. *D. australis* R. Br. ex Hooker f. 1847, *Fl. Antarct.* **1(2)**: 235.

Annual. Leaves in basal cluster, 3–7 × 1.5–4 mm, spathulate to lanceolate, obtuse, entire to distantly denticulate with stellate hairs above and simple hairs beneath. Flowering stems 20–50 mm, without leaves, with simple and stellate hairs below; raceme 4- to 12-flowered; pedicels 3–6 mm. Sepals 1–1.2 mm, linear-oblong, obtuse. Petals absent. Silicula 4–11 × 1.5–2.5 mm, oblong to obovate, compressed, sparsely hairy; pedicels 1–6 mm, shorter than fruit; seeds *c*.0.7 × 0.5 mm, ± ovoid, tuberculate. *Flowers September to October*.

Open gravel and soil, beaches and coastal flats; 0–30 m.

N. part of Isla Grande. (Map 10b).
Argentina north to *c*.32°S.lat.

Fuegian material belongs to var. ***ameghinoi*** (Speg.) F. Kurtz 1900, *Collect*. 5, distinguished by its sparsely hairy silicula. May be confused with *Erophila verna*, for differences see p. 115.

3. *D. magellanica* Lam. 1786, *Encycl. Méth. Bot.* **2**: 328.

D. falklandica Hooker f. 1847, *Fl. Antarct.* **1(2)**: 238 – Macloskie, 1905; Skottsberg, 1916.

Caespitose perennial, usually densely covered with stellate hairs, some simple or branched hairs also present; stems (7–)10–40(–50) mm, erect, usually sparingly branched, somewhat woody, covered with leaf-remains. Leaves in dense basal clusters, (5.5–)10–20(–30) × (1.9–)2.3–4.7(–14) mm, oblanceolate to spathulate, obtuse, sessile or with short, winged petiole, the leaves on flowering stems 1–5, usually smaller, ovate to elliptical or obovate, sessile. Flowering stems (15–)35–220(–310) mm, erect, simple or sparingly branched. Raceme 8- to 21-flowered; pedicels (0–)3–5 mm. Sepals 2–3(–3.7) × (1–)1.3–1.5(–1.8) mm, ovate to elliptic-oblong, obtuse, with scarious margin, ± saccate. Petals 3.7–4.7 × 1–2 mm, spathulate, usually emarginate, white. Siliqua 5–9 × 2–3.3 mm, oblong-ellipsoid, subotuse, compressed; pedicel *c*.9 mm; seeds 0.8–1.1 × 0.6–0.9 mm, ovoid, smooth, orange-brown. $2n = 48$*. Self-compatible, usually self-pollinated. *Flowers October to January*.

Open ground in all communities and on all substrates except wet peat; 0–900 m.

Throughout Fuegia except the extreme south and west. (Map 10c).
Argentina north to *c*.52°18′S.lat., then in the west to *c*.33°S.lat.
Chile north to 50°40′S.lat.
Falkland Islands (Islas Malvinas).

A very variable species in which habit, stature and indumentum are readily modified by the environment.

9. *Erophila* DC.

Annual, with branched and stellate, sometimes simple, hairs. Leaves ± entire. Flowers in a terminal raceme. Fruit a silicula; seeds in 2 rows in each cell.

1. *E. verna (L.) Chevall. 1827, *Fl. Gén. Env. Paris* **2**: 898.

Leaves in a basal rosette, 3–20 × 1.5–5 mm, spathulate to lanceolate, obtuse to acute, entire or distantly dentate, with stellate pubescence. Flowering stems 20–120 mm, without leaves, stellate-pubescent below; raceme 2- to 8-flowered; pedicels 1.5–6 mm. Sepals 1.5–2.5 mm, ovate, obtuse. Petals 1.5–6 × 1–2 mm, obovate, deeply 2-fid, white. Silicula 3–9 × 1.5–3 mm, ellliptical to obovate or suborbicular, compressed, glabrous; pedicels 8–23 mm, longer than fruit; seeds 0.3–0.5 mm, ovoid, brown. Self-compatible and self-pollinated. *Flowers October to November or earlier*.

Open and disturbed ground by roads, and in grassland; 0–40 m.

Northernmost Isla Grande. Introduced from Europe and widely naturalized in temperate regions.

Superficially very similar to *Draba australis*, which differs in having no petals or, if present, they are not

Fig. 80. *Draba funiculosa*. a, Plant. b, Inflorescence. c, Fruit. Scales = 1 cm. Fig. 81. *Cardamine glacialis*. a, Plant. 2 forms. b, Fruit. c, Seed. Scale = 1 cm (a–b) or 1 mm (c). Fig. 82. *Lepidium pseudodidymus*. a, Plant. b, Fruit. Scale = 1 cm (a) or 1 mm (b).

(R. N. P. Goodall)

deeply 2-fid, and the fruiting pedicels shorter than the silicula.

10. *Capsella* Medicus

Annual or perennial, glabrous or with simple hairs. At least some leaves pinnatifid. Flowers in a lax ebracteate raceme. Fruit a silicula; seeds in 2 rows in each cell.

*1. ***C. bursa-pastoris*** (L.) Medicus 1792, *Pflanzengatt.* 85.

Stem 3–55 cm, erect, with long, white hairs, rarely glabrous. Basal leaves in a rosette, up to 10 × 2 cm, oblanceolate to narrowly elliptical in outline, obtuse, narrowed to petiole up to 5 cm, entire to pinnatifid, hairy; cauline leaves sagittate-amplexicaul, entire to ± lobed. Sepals 1.5–2 mm, ovate to elliptical, pubescent. Petals (2–)2.5–3 mm, obovate, clawed, white. Silicula 5–8 × 4–8 mm, triangular-obcordate, emarginate at apex, glabrous, green or purplish; pedicel 5–20 mm; seeds 0.8–1(1.2) mm, oblong to ovoid-oblong, pale brown. *Flowers November to April.*

Open disturbed ground, especially near roads and settlements; 0–60 m.

Throughout Isla Grande except extreme west and east.
Introduced from Europe and widely naturalized.

11. *Thlaspi* L.

Annual or perennial, glabrous. Leaves entire to denticulate. Flowers in ebracteate, subumbellate raceme. Fruit a silicula, keeled and winged; seeds 1–5 in each cell.

1. ***T. magellanicum*** Comm. ex Poiret 1806, *Encycl. Méth. Bot.* **7**: 541.

Stems 3–23 cm, ascending to erect. Lower leaves usually in basal cluster, (8–)10–35(–45) × 2–8 mm, oblanceolate to subspathulate, entire to denticulate, green to violet; cauline leaves narrowly elliptical, obtuse, sessile. Pedicels up to 3.5 mm. Sepals 1.5–2 mm, elliptical to oblong, obtuse, green-purple with white margin. Petals (2.5–)3–4 × 1.2–2 mm, spathulate-oblanceolate, obtuse, shortly clawed, white. Silicula 3.5–6 × 2–3.5 mm, obovate-obcordate, emarginate; pedicels 2–8 mm; seeds *c.*1.3 mm, ellipsoid, smooth, brown. *Flowers October to November.* (Fig. 79).

Open sand and soil by coast and rivers, rock-crevices, open grassland, scrub and wood margins; 0–900 m.

Almost throughout Fuegia except the extreme south and west. (Map 10d).
W. Argentina north to *c.*46°S.lat.
Chile north to 50°30′S.lat.

12. *Lepidium* L.

Annual or perennial, with simple hairs. Leaves 1- to 2-pinnate. Flowers in leaf-opposed or terminal, ebracteate racemes. Fruit a silicula; seeds 1 in each cell.

Literature: L. C. Hitchcock, *Lilloa* **11**: 75–134(1945).

a. Ascending to erect annual; silicula as long as to longer than wide, smooth or indistinctly reticulate **1. *racemosum***
a. Prostrate to procumbent perennial; silicula usually wider than long, quite distinctly reticulate **2. *pseudodiymum***

1. ***L. spicatum*** Desv. 1814, *Jour. Bot. Appl.* **3**: 164.
L. racemosum Griseb. 1854, *Syst. Bemerk.*, 116.

Annual. Stem (3)4–17 cm, ascending to erect, branched, with appressed hairs. Basal leaves 17–45 × 3–12 mm, oblanceolate in outline, 1- to 2-pinnate, the cauline leaves smaller, pinnate; ultimate segments oblanceolate to elliptical or linear, acute. Raceme 8- to 50-flowered; pedicels *c.*1 mm. Sepals 0.7–1 × 0.2–0.3 mm, oblong-ovate, obtuse. Petals about ½ as long as sepals, linear, white. Silicula 2.2–2.7 × 1.8–2.7 mm, suborbicular, shallowly emarginate, smooth or indistinctly reticulate, glabrous; pedicels 1.2–2.1 mm; seeds 1.3–1.4 × *c.*0.7 mm, ovoid, ± compressed, orange-brown. *Flowers November to January.*

Coastal sands and gravels, cliffs, open *Lepidophyllum* and *Chiliotrichum* scrub; 0–40 m.

N. coasts of Isla Grande, locally on C. S. coast. (Map 10e).
Argentina and Chile north to *c.*50°S.lat.

2. ***L. pseudodidymum*** Thellung ex Druce 1914, *Rep. Bot. Exch. Club Brit. Is.* **3**: 308.
L. bipinnatifidum auct., non Desv. – Dusén, 1900; Skottsberg, 1906.
L. auriculatum auct., non Regel & Koern. – Skottsberg, 1916.
Senebiera australis Hooker f. 1847, *Fl. Antarct.* **1(2)**: 241.

Perennial. Stems 4–33 cm, prostrate to procumbent, much branched, with appressed hairs. Basal and lower stem leaves 10–20(–40) × 3–15(–20) mm, ovate to oblong or obovate in outline, deeply pinnatisect, petiolate; segments 1–5 pairs, oblanceolate, acute, often with slender acute lobes; upper leaves sessile, with narrow, ± entire segments. Raceme 6- to 60-flowered; pedicels 1–1.5(–3) mm. Sepals 0.8–1 mm, ovate, acute, cucullate. Petals shorter than sepals, linear, white, or absent. Silicula 2–2.5 × 2–3.2 mm, emarginate at apex, shallowly so or entire at base, reticulate; pedicel up to 4 mm; seeds *c.*1.5 mm, ovoid-reniform, reticulate, pale brown. (Fig. 82).

Open sands, gravels and clays, usually near the coast; 0–80 m.

Coasts of Isla Grande except the extreme west. (Map 10f).
Argentina north to *c.*50°S.lat.
Chile north to *c.*45°S.lat.

May be confused with *Coronopus didymus*. For differences see below.

13. *Cardaria* Desv.

Perennial, with simple hairs. Leaves sinuate-dentate. Flowers in ebracteate corymbose panicle. Fruit a silicula, indehiscent, or breaking into 2, 1-seeded nutlets.

*1. ***C. draba*** (L.) Desv. 1814, *Jour. Bot. Appl.* **3**: 163. Plant with branched woody stock; flowering stems up to 40 cm, erect, simple or branched above, ± glabrous to appressed-pubescent. Leaves 3–6 × 1.5–2 cm, obovate, obtuse to subacute, sinuate-dentate, the basal cuneate, with petiole up to *c.*3 cm, the cauline sessile, amplexicaul. Pedicels 3–6 mm. Sepals 1.5–1.8 mm, elliptical to suborbicular, obtuse. Petals 2.8–4.2 × 1–1.5 mm, spathulate, white. Silicula (2)2.5–3.5 × 1.5–3.7 mm, cordate, emarginate, inflated; seeds 1.6–2 × 1–1.2 mm, ovoid. *Flowers April to May.*

Coastal sands and disturbed ground near settlements; 0–20 m.

Local on E. & S. coasts of Isla Grande.
Introduced from Europe where it is a widespread weed.

14. *Coronopus* Haller

Annual or biennial, with simple hairs. Leaves deeply pinnatisect. Flowers in terminal or leaf-opposed racemes. Fruit a silicula, indehiscent or breaking into 2, 1-seeded nutlets.

*1. ***C. didymus*** (L.) Sm. 1800, *Fl. Brit.* **2**: 691.

Senebiera pinnatifida DC. 1799, *Mém. Soc. Hist. Nat. Paris* **1799**: 144 – Spegazzini, 1896.

Stems 2–25 cm, procumbent or ascending, branched, leafy, pubescent. Lower leaves with lamina 10–25 × 5–10 mm, oblong to oblong-ovate in outline, deeply pinnatisect; segments (1)2–5 pairs, oblanceolate, acute, usually pinnatifid with acute lobes having 1–3 short apical hairs; petiole ½ as long to as long as lamina. Pedicels 1–1.5 mm. Sepals 0.5–1 mm, ovate, cucullate, acute. Petals shorter than sepals, linear, white, or absent. Silicula 0.8–1.75 × 1–2.5 mm, didymous, constricted at septum, tuberculate-rugose; seeds *c.*1.5 mm, reniform-ovoid, glabrous, pale brown. Self compatible, usually self-pollinated. *Flowers November to December.* (Fig. 78).

Coastal sands and gravels; 0–10 m.

N. E. Isla Grande.
Native of temperate S. America but widely naturalized in Europe, whence it was probably introduced into Fuegia.

May be confused with *Lepidium pseudodidymum* but most readily distinguished by the fruit being deeply divided at the insertion of the pedicel and with rounded, rugose nutlets; in the *Lepidium* the fruit is entire at the base, flattened in the middle and with reticulate valves.

15. *Brassica* L.

Annual to perennial, glabrous or with simple hairs. Leaves entire to pinnatipartite. Flowers in an ebracteate raceme. Fruit a siliqua, with prominent apical beak; seeds in 1 row in each cell.

*1. ***B. rapa*** L. 1753, *Sp. Pl.* 666.

B. campestris L. 1753, Sp. Pl. 666.

Annual; stems up to 60 cm, erect, herbaceous, simple or branched above. Lower leaves 6–13 × 2–6 cm, lyrate-pinnatifid, obtuse, setose-ciliate, green; petiole up to *c.*3 cm; cauline leaves 4–7(–9) × 1.5–2.5(–4) cm, triangular-ovate to elliptical, obtuse to acute, amplexicaul, sinuate or ± lobed towards base. Inflorescence with open flowers overtopping the buds; pedicels up to 10 mm. Sepals 4–5.5 mm, elliptical, obtuse, erecto-patent. Petals 6–10 × (3–)4.5–5.5 mm, spathulate, yellow. Siliqua 40–70 ×

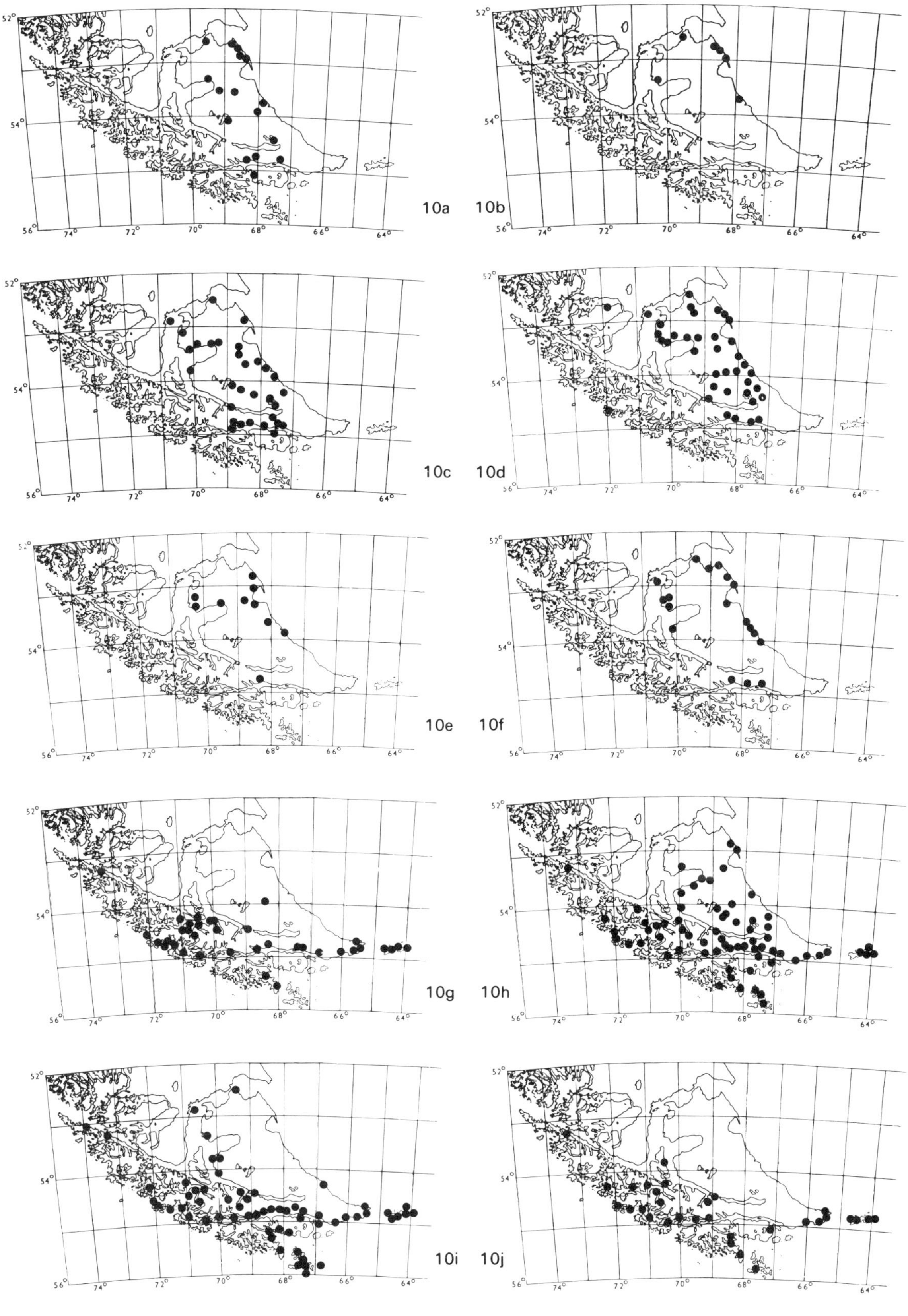

Map 10. a, *Draba funiculosa*. b, *Draba australis*. c, *Draba magellanica*. d, *Thlaspi magellanicum*. e, *Lepidium racemosum*. f, *Lepidium pseudodidymum*. g, *Gaultheria antarctica*. h, *Pernettya pumila*. i, *Pernettya mucronata*. j, *Lebetanthus myrsinites*.

1–1.7 mm, with beak up to 20 mm, suberect; pedicel 9–20 mm; seeds 1.5–2.5 mm in diameter, ovoid to globose, brown. *Flowers December to March.*

Disturbed ground near settlements, coastal sands and gravels.

Sporadically in N. & S. Isla Grande.
Introduced from Europe, where it is a common weed.

B. napus L. 1753, *Sp. Pl.* 666, which was reported by Dusén (1900) from N. E. Isla Grande (Río San Martín), differs from *B. rapa* in having glabrous or sparsely hairy, glaucous leaves and an inflorescence in which the open flowers do not overtop the buds. It is cultivated as a fodder crop or for its edible tuberous tap-root.

B. oleracea L. 1753, *Sp. Pl.* 667, which is widely cultivated as a vegetable (cabbage, brussels sprouts etc.), has been found as a casual on beaches and coastal slopes in N. W. Isla Grande (near Porvenir). It differs from the above two species in being biennial or perennial with woody stems and in having connivent, not erecto-patent, sepals. The upper cauline leaves are at most semi-amplexicaul.

16. *Rapistrum* Crantz

Annual, with simple hairs. Lower leaves pinnate. Flowers in terminal ebracteate racemes. Fruit a silicula, transversely articulate, the narrow lower segment with 0–2 seeds, the larger upper segment 1-seeded and indehiscent.

1. *R. rugosum (L.) All. 1785, *Fl. Pedem.* **1**: 257.

Stem 15–78 cm, erect, simple or branched, hispid at least below. Leaves dark green, hispid at least beneath; lower leaves 40–100 × 10–35 mm, pinnate, with 2–3 pairs of coarsely dentate lateral lobes and a larger terminal lobe, the petiole 20–30 mm; upper leaves 2–2.7 × *c.*0.5 cm, lanceolate, obtuse, lobed or dentate, narrowed at base. Raceme branched; pedicels 2–3 mm. Sepals (2.8–)3–4.2 mm, oblong-elliptical, obtuse. Petals (5.7–)6.2–7.5 × 1.5–2.6 mm, spathulate, pale yellow with darker veins. Silicula erect; lower segment *c.*2 × 1.2–1.7 mm, cylindrical; upper segment 2.5–3 × 2.2–3 mm, subglobose, narrowed to slender beak *c.*2 mm, rugose. *Flowers February.*

Open coastal muds.

S. coast of Isla Grande (Ushuaia).
Introduced from Europe where it is widely naturalized.

Fuegian plants belong to *ssp. orientale* (L.) Arcangeli, in which the fruiting pedicel (*c.*4 mm) is about twice as long as the lower fruit segment.

17. *Raphanus* L.

Annual or biennial, with simple hairs. Leaves lyrate-pinnatifid. Flowers in a terminal, ebracteate raceme. Fruit a siliqua, transversely articulate, the narrow lower segment without seeds, the larger upper segment with 5–12 seeds, indehiscent.

1. *R. sativus L. 1753, *Sp. Pl.* 669.

Tap-root tuberous, white or purplish; stems up to 60 cm, erect, simple or branched, hispid. Lower leaves up to 100 × 40 mm, lyrate-pinnatifid, the lobes coarsely dentate. Sepals *c.*10 mm, oblong-elliptical, obtuse. Petals 16.5–20 × *c.*5 mm, spathulate, yellow and violet or white to violet, usually with darker veins. Siliqua 20–90 × 8–15 mm, the upper segment cylindrical, inflated, with conical beak; pedicel 15–20 mm; seeds *c.*3 mm in diameter, globose, usually black. *Flowers January to February.*

Cultivated and disturbed ground near settlements.

Grown in gardens and occasionally reported as a casual weed. Introduced from Europe.

XXIX. ERICACEAE

Evergreen shrubs. Leaves alternate, simple, coriaceous; stipules absent. Flowers actinomorphic, hermaphrodite or female; bracteoles present. Sepals 5, fused, persistent. Corolla urceolate, shortly 5-lobed. Stamens 10, inserted on disc. Ovary superior, 5-celled; style 1; stigma capitate. Fruit a berry or capsule; seeds numerous.

a. Fruit a capsule or apparently so, ± surrounded by fleshy calyx **1. *Gaultheria***
a. Fruit a berry, with herbaceous calyx at base **2. *Pernettya***

1. *Gaultheria* Kalm ex L.

Calyx 5-lobed, fleshy in fruit. Fruit a loculicidal capsule surrounded by fleshy calyx to appear berry-like; seeds minute.

83

84

85

86

Fig. 83. *Berberis buxifolia.* Fig. 84. *Nothofagus antarctica.* Fig. 85. *Cardamine geraniifolia.* Fig. 86. *Pernettya mucronata.*
(Sydney Parkinson)

1. *G. antarctica* Hooker f. 1847, *Fl. Antarct.* **1(2)**: pl.116.

G. serpyllifolia (Lam.) Skottsb. 1916, *Kungl. Svensk Vetensk.-Akad. Handl.* **56(5)**: 282 – Skottsberg, 1929.
G. procumbens auct., non L. – Autran, 1905.
G. microphylla Hooker f. 1847, *Fl. Antarct.* **1(2)**: 327.

Stems up to 21 (–27) cm, 0.5–1.5 mm in diameter, prostrate to ascending, rooted at intervals, branched. Leaves (2.8–)3.1–7.3 × 2–5.4 mm, ovate to elliptical, obtuse to, rarely, subacute, dentate–serrate, sometimes obscurely so, with margin thickened beneath, glabrous, paler green beneath; petiole up to 1 mm. Flowers 1–2(3) in upper leaf-axils; peduncle 0.5–2 mm, sparsely puberulent, with up to 7 bracteoles; bracteoles 0.7–1 × 0.6–0.7 mm, triangular. Calyx-lobes 0.8–1.6 × 1–2 mm, triangular, acute, somewhat cucullate, entire, glabrous, pinkish to purplish. Corolla 2.5–5 mm, white or pinkish, the lobes 1–1.8 × 1–1.7 mm, ovate, obtuse, recurved. Fruit 4–11 mm in diameter, depressed globose to pyriform, pink or white. Sometimes gynodioecious. *Flowers November to January*. (Fig. 87).

Empetrum heath; *Astelia* and *Marsippospermum* bog; *Sphagnum* swamp. 0–500 m.

C. & S. Fuegia. (Map 10g).
W. Argentina north to *c.*39°S.lat.
Chile north to *c.*40°S.lat.
Falkland Islands (Islas Malvinas).

2. *Pernettya* Gaudich.

Calyx 5-lobed, dry in fruit. Fruit a berry; seeds minute.

Literature: H. Sleumer, *Notizbl. Bot. Garten. Mus. Berlin* **12**: 626–645(1935).

a. Leaves not more than 6 mm, acute to obtuse, not mucronate, the margin entire to serrulate **1. *pumila***
a. Leaves more than 6 mm, acute, mucronate, the margin thickened, serrate **2. *mucronata***

1. *P. pumila* (L.F.) Hooker 1837, *Icon..Pl.* **1**, pl.9.

P. trinervia Gand.1913, *Bull. Soc. Bot. Fr.* **60**: 24.
P. empetrifolia (Lam.) Gaudich. 1825, *Ann. Sci. Nat.* **5**: 102.
Andromeda empetrifolia Lam. 1783, *Encycl. Méth. Bot.* **1**: 155 – Forster, 1789.
Arbutus pumila L.f. 1781, *Suppl.* 239 – Forster, 1789.

Stems up to 23 cm, prostrate to ascending, rarely erect, rooting at intervals, branched. Leaves 2.5–5.5 × 1.3–3.1 mm, ovate to elliptical, acute to obtuse, usually navicular, entire or serrulate, ciliate, glabrous or sparsely appressed – pubescent beneath, shiny above, sessile. Flowers solitary in leaf-axils; peduncle 1–22 mm, sparsely puberulent, with 2–4 bracteoles near base; bracteoles (0.9–)1.4–2 × c.1.5 mm, ovate. Calyx-lobes 2–3 × 1–1.7 mm, ovate, acute to obtuse, entire, somewhat ciliate, pinkish with white margin. Corolla 3–4.5 mm, white or pinkish, the lobes 1–1.9 mm, ovate, obtuse, recurved. Fruit 6–11 mm in diameter, globose, whitish with bluish to purplish tinge. Sometimes gynodioecious, rarely pseudoviviparous. *Flowers November to May*. (Fig. 88).

Empetrum heath; bogs and swamps, open wet *Nothofagus* forest, open peat and sand, rock crevices. 0–900 m.

Almost throughout Fuegia. (Map 10h).
W. Argentina north to ?*c.*46°S.lat.
Chile north to ?*c.*50°S.lat.
Falkland Islands (Islas Malvinas).

2. *P. mucronata* (L.f.) Gaudich. ex G.Don 1834, *Gen. Syst.* **3**: 836.

P. ovalifolia Hombron & Jacquinot ex Dcne 1852, *Voy. Pôle Sud, Bot. Atlas*, pl. 22 fig. L.
Arbutus mucronata L.f. 1781, *Suppl.*239 – Forster, 1789.
A. rigida Banks & Sol. ex Hooker f. 1847, *Fl. Antarct.* **1(2)**: 326.

Stems up to 2 m, much-branched, erect. Leaves 7–14(–19) × 2.5–7.5(–9) mm, lanceolate to ovate, cuneate at base, acute and mucronate, serrate, thickened at margin, glabrous, shiny above, shortly petiolate. Flowers solitary in leaf-axils; peduncle 5–12 mm, sparsely puberulent, with 1 bracteole near middle; bracteole 1.4–1.8 mm, ovate. Calyx-lobes (1.6–)2–2.7 × 0.9–1.5 mm, triangular-ovate, acute to subobtuse, entire, green. Corolla (2.8–)4–4.5(–6) mm, pinkish or white, the lobes (0.8–)1–1.6 mm, triangular-ovate, obtuse, recurved. Fruit (5–)8–10 mm in diameter, globose. *Flowers December to April*. (Fig. 86).

Open, wet *Nothofagus* and *Drimys* forest; dominant or co-dominant with *Chiliotrichum* in shrub communities fringing forest above highwater mark and in transition to swamps and bogs. 0–700 m.

N.W., C. & S. Fuegia. (Map 10i).
W. Argentina north to *c.*31°S.lat.
Chile north to *c.*37°S.lat.

Fuegian material belongs to var. *mucronata*.

XXX. EPACRIDACEAE

Shrubs. Leaves alternate, simple; stipules absent. Flowers actinomorphic, hermaphrodite. Sepals 5, free. Petals 5, fused. Stamens 5, inserted on basal disc, alternating with corolla-lobes. Ovary superior, 5-celled; style 1; stigma capitate. Fruit a capsule, dehiscing by 5 valves; seeds 1–5 per cell, with endosperm.

1. *Lebetanthus* Endl.

Scandent shrubs. Leaves coriaceous. Flowers solitary, axillary, subtended by 8–10 imbricate bracts. Corolla urceolate. Stamens included. Style included. Seeds fusiform.

Literature: S. Arroyo, *Darwiniana* **19**: 312–330 (1975).

1. ***L. myrsinites*** (Lam.) Dusén 1900, *Svensk Exped. Magellansl.* **3(5)**: 139.
   L. americanus (Hooker) Endl. 1841, *Gen. Pl. Ord.* 1411 – Hooker, 1847.
   Prionotes americana Hooker 1836, *Icon. Pl.* t.30 – Hariot, 1884.
   P. myrsinites (Lam.) Skottsb. 1916, *Kungl. Svenska Vet. Akad. Handl.* **56(5)**: 284.

Plant up to 50 cm when unsupported but stems scramble up to 2 m. Leaves 3–8 × 2.2–4.5 mm, subelliptical to ovate, acute, somewhat truncate at base. Sepals 1.5–3 × 0.6–1 mm, lanceolate, acuminate. Corolla (4–) 4.5–5 mm, white; lobes 0.5–0.7 × *c.*1 mm, recurved. Capsule *c.* 4 mm, subglobose, somewhat woody, brown; seeds *c.* 1.5 mm. *Flowers November to February.* (Figs. 89, 91).

Nothofagus betuloides forest and mixed coastal forest with *Drimys*.

S & W Fuegia. (Map 10j).
W. Chile north to *c.*44°S.lat.

In the past *Lebetanthus* has been confused with the Tasmanian genus *Prionotes* but Arroyo (1975) showed that they are clearly distinguished by, amongst other features, the shorter corolla in relation to the calyx and the fewer pendent ovules of the Fuegian genus.

XXXI. EMPETRACEAE

Small, evergreen, usually dioecious shrubs. Leaves alternate, simple; stipules absent. Flowers actinomorphic; bracts present. Sepals 3, free, petaloid. Petals 3, free. Stamens 3, episepalous; anthers dehiscing longitudinally. Ovary superior, 6 – to 9 – celled; style 1; stigma 6 – to 9 – lobed. Fruit a fleshy drupe, with 6–9 stones.

1. *Empetrum* L.

Leaves coriaceous, imbricate. Flowers unisexual, rarely hermaphrodite, solitary in upper leaf-axils. Stigma-lobes dentate-laciniate.

1. ***E. rubrum*** Vahl ex Willd. 1806, *Sp. Pl.,* Ed. 4, **4(2)**: 713

Stems up to 42 cm, procumbent to erect, much branched, usually densely white– to grey– tomentose on younger branches. Leaves 2–5.2 × 0.9–2 mm, oblong to elliptic-obovate, obtuse, with strongly revolute margin, densely tomentose and scabrid on curved edge, shiny, sessile or very shortly petiolate. Flowers subsessile; bracts 3–6, 0.6–1 × 0.5–1.1 mm, ovate to obovate, obtuse, scarious, lanate on margin. Sepals 1.2–2 × 1.2–1.5 mm, ovate, obtuse to subacute, yellow to orange or pinkish. Petals 1.8–2 × 0.7–1 mm in female flowers, 2.5–3.5 × 0.5–1 mm in male flowers, oblong – to oblanceolate – spathulate, obtuse, reddish. Stamens about twice as long as petals. Stigma dark purple. Drupe 4–7 mm in diameter, globose, red, sometimes black. $2n = 26$.*Flowers September to October.* (Fig. 90).

Dominant or co-dominant in dwarf-shrub heath in N. Fuegia, elsewhere common in feldmark communities, bogs and swamps. common or locally dominant in open *Nothofagus* forest and sometimes also frequent in the understorey; 0–660 m.

Throughout Fuegia. (Map 11a).
W. Argentina north, to *c.*42°S.lat.
Chile north to *c.*36°S.lat.
Falkland Islands (Islas Malvinas). Tristan da Cunha.
Introduced in South Georgia (Smith, 1973).

XXXII. PRIMULACEAE

Annual or perennial herbs. Leaves alternate or in basal rosette; stipules absent. Flowers hermaphrodite, actinomorphic, bracteate, solitary or in umbellate or racemose inflorescence. Calyx 5-lobed. Corolla 5-lobed. Stamens 5, inserted on corolla-tube. Ovary superior or semi-inferior, 1-celled; style 1; stigma capitate. Fruit a capsule, circumscissile or dehiscing by 5 valves; seeds numerous, with endosperm.

Literature: A. Lourteig, *Lilloa* **8**: 231–269(1942); *Darwiniana* **7**: 556–559(1947).

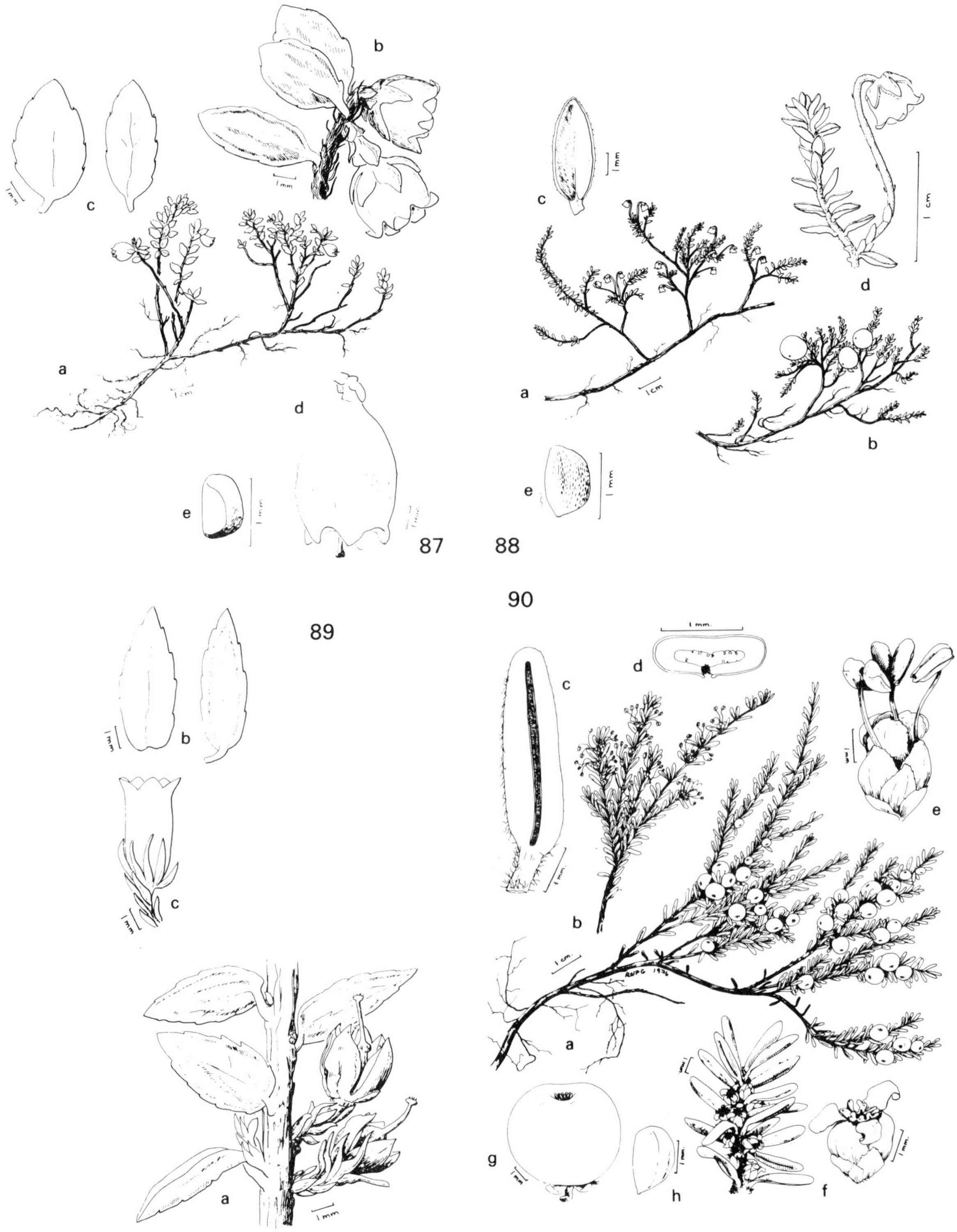

Fig. 87. *Gaultheria antarctica*. a, Plant. b, Flowers. c, Leaves. d, Young fruit. e, Seed. Scale = 1 cm (a) or 1 mm (b–e). Fig. 88. *Pernettya pumila*. a, Plant with flowers. b, Plant with fruit. c, Leaf. d, Flower. e, Seed. Scale = 1 cm (a–b, d) or 1 mm (c, e). Fig. 89. *Lebetanthus myrsinites*. a, Twig with young fruits. b, Leaves. c, Flower. Scale = 1 cm. Fig. 90. *Empetrum rubrum*. a, Female plant. b, Male branch. c, Leaf. d, Section of leaf. e, Male flower. f, Female flowers. g, Berry. h, Seed. Scale = 1 cm (a–b) or 1 mm (c–h).

(R. N. P. Goodall)

a. Leaves alternate; flowers solitary in leaf-axils **3. *Anagallis***
a. Leaves in basal rosette; flowers in terminal inflorescence on leafless scape:
b. Annual, with stellate hairs **2. *Androsace***
b. Perennial, glabrous or with few simple hairs:
c. Leaves farinose; petals bifid **1. *Primula***
c. Leaves glabrous; petals entire **4. *Samolus***

1. *Primula* L.

Perennial herbs. Leaves basal, simple. Flowers in umbels terminal on leafless scape. Corolla infundibuliform, with long tube. Ovary superior. Fruit a capsule dehiscing by 5 valves.

Literature: A. Lourteig, *Bol. Soc. Argent. Bot.* **11**: 140–145(1967).

1. *P. magellanica* Lehm. 1817, *Monogr. Primul.* 62.

P. decipiens Duby 1844, in DC., *Prodr.* **8**: 44.
P. farinosa s p. *magellanica* (Lehm). W. W. Sm. & Forrest 1928, *Notes Roy. Bot. Gard. Edinb.* **16**: 24 – Lourteig, 1947.
P. farinosa var. *magellanica* (Lehm.) Hooker f. 1847, *Fl. Antarct.* **1(2)**: 337 – Lourteig, 1942.

Leaves 1.4–5.5(–8) × 0.3–1.3(–1.8) cm, oblancelolate to oblanceolate-spathulate, obtuse to subacute, obscurely undulate to crenate-serrate and sometimes revolute towards apex, glabrous above, glandular-pubescent near margin and white-farinose beneath. Scape (1.5–)3–27(–32) cm, terete, glandular-pubescent. Umbel 4- to 15- flowered; pedicels 1.5–4 mm, longer in fruit; bracteoles 5.5–9 × 0.5–2 mm, subulate, saccate. Calyx 6–9 × 3–5.5 mm, usually farinose, the lobes 2–3.5 × 1–2.3 mm, triangular to triangular-ovate, acute or subacute. Corolla-tube 5–7 mm, yellowish green in throat; limb 9–19 mm in diameter, the lobes 4–6.5 × 2.5–4.2 mm, cuneate, bifid, white rarely tinged with lilac. Capsule 7–13 × 3–4.5 mm, ovoid-cylindrical; seeds 0.8–1.3 mm in diameter, compressed globose-ovoid, densely round-tuberculate, brown. $2n = 72$. Self-compatible and at least facultatively self-pollinated. *Flowers October to December*. (Fig. 95).

Moist coastal sands and rocks, bogs, moist grassland and open scrub, margins of lakes and rivers, snowmelt areas; 0–850 m.

Throughout Fuegia except extreme south-west. (Map 11b).
Argentina north to *c.*47°55′S.lat.
Chile north to *c.*51°S.lat.
Falkland Islands (Islas Malvinas).

2. *Androsace* L.

Annual herbs. Leaves basal, simple. Flowers in umbels terminal on leafless scape. Corolla urceolate. Ovary superior. Fruit a capsule dehiscing by 5 valves.

1. *A. pusilla* (Dusén) Lourteig 1942, *Lilloa* **8**: 233.

Plant with stellate hairs. Leaves 5–12 × 1–3 mm, linear to oblanceolate, acute, entire or with 1–2 marginal teeth. Scape up to 7 cm, filiform. Umbel 2- to 10- flowered; pedicels up to 13 mm; bracteoles 3–5 × 1–2 mm, lanceolate. Calyx *c.*8 mm, the lobes *c.*4 mm, triangular-ovate, acute. Corolla *c.*2.5 mm, shorter than calyx, white, the lobes *c.*0.8 mm, elliptical, obtuse. Capsule 2.5–3 mm, globose; seeds *c.*1 mm, subtetrahedral, finely pitted. *Flowers December to January*. (Fig. 97).

Open saline grassland; 0–50 m.

C.N. Isla Grande, known only from Ea Lago Gaviota (53°10′S.lat., 68°45′W.long). (Map 11c).
Argentina north to *c.*40°S.lat.

3. *Anagallis* L.

Perennial herbs. Leaves alternate, simple. Flowers solitary, axillary. Bracteoles absent. Corolla rotate, with short tube. Filaments united in tube to *c.*½-way. Ovary superior. Fruit a circumscissile capsule.

1. *A. alternifolia* Cav. 1801, *Icon. Descr.* **6**: 3.

Glabrous; stems up to 16 cm, prostrate, branched, rooting at nodes, leafy towards apex. Leaves (2.2–)3.5–5(–7.5) × (1.3–)2–3(–4.3) mm, ovate-elliptical to -orbicular, acute, rarely apiculate or obtuse, entire, attenuate into short, wide petiole. Peduncles 3–10 mm, slender. Calyx-lobes 2–3.5 × 0.6–1.2 mm, lanceolate, acute to acuminate, green with scarious margin, often pinkish distally. Corolla-lobes 3.5–5.5 × 1.6–2.5 mm, ovate to oblong-lanceolate, obtuse, rarely retuse, pink or white, darker near base. Capsule 2.5–3 mm in diameter, globose, greyish brown; seeds 0.7–0.8 mm, trigonous, tuberculate, black to reddish brown. $2n = 44$*. Self-compatible, facultatively self-pollinated. *Flowers December to April*. (Fig. 96).

Wet rocks, grassland, open scrub and feldmark; 0–675 m.

C. & S. Fuegia. (Map 11d).
W. Argentina north to *c.*26°S.lat.
Chile north to *c.*29°S.lat.
Falkland Islands (Islas Malvinas).

91

92

93

94

Fig. 91. *Lebetanthus myrsinites*. Fig. 92. *Rubus geoides*. Fig. 93. *Acaena pumila*. Fig. 94. *Escallonia serrata*.
(Sydney Parkinson)

4. *Samolus* L.

Perennial herbs. Leaves basal, simple. Flowers in racemes or corymbs terminal on leafless scape. Corolla tubular. Ovary semi-inferior. Fruit a capsule dehiscing by 5 valves.

1. ***S. spathulatus*** (Cav.) Duby 1844, in DC., *Prodr.* **8**: 74.

Leaves 2.5–5.1(–6.5) × 1.7–3.5 cm, spathulate, obtuse to truncate, entire, fleshy, glabrous, shiny. Scape (3.8–)11–29 cm, terete, glabrous. Inflorescence 4- to 8(–10)- flowered; pedicels short; bracteoles 1.4–3 × 0.3–0.5 mm, linear, acute. Calyx 3.5–4.7 mm, the lobes 1.6–2 × 0.7–1.5 mm, triangular, acute to obtuse, green with purple spots. Corolla pink to violet; tube 3–4 mm, the lobes 3–4 × 2.5–2.7(–3) mm, ovate-oblong, obtuse, entire. Capsule 5.5–6(–7) × 4–4.3 mm, ovoid; seeds 0.6–0.8 mm in diameter, angled, reticulate, dark brown. $2n = 78^\star$. *Flowers January to February*. (Fig. 98).

Wet grassland; 0–20 m.

N. W. & S. Isla Grande. (Map 11e).
W. Argentina north to *c.*39°S.lat.
Chile north to *c.*51°30′S.lat.

XXXIII DROSERACEAE

Perennial, insectivorous herbs. Leaves all basal, simple; stipules absent. Flowers hermaphrodite, actinomorphic. Sepals 5, united at base. Petals 5, free. Stamens 5, free. Ovary superior, 1-celled. Styles 3–5, united below. Fruit a loculicidal capsule; seeds numerous, with endosperm.

1. *Drosera* L.

Upper surface and margin of leaf-lamina covered with long, red, gland-tipped, motile hairs which entrap and digest insects. Flowers solitary, or rarely 2–3 in ebracteate cyme, on slender peduncle. Styles clavate, somewhat laciniate. Capsule opening to base by 3–5 lanceolate valves; seeds with ± inflated testa forming a wing.

1. ***D. uniflora*** Willd. 1809, *Enum. Pl. Horti Berol.* 340.

Stock short, vertical, covered with remains of dead leaves. Leaves horizontal, sometimes ascending to suberect; lamina 3–5 × 2–4(–4.5) mm, orbicular or slightly wider than long, tapering to petiole; petiole 4–11 × 0.5–1.4 mm, glabrous, widening to scarious basal sheath with sparsely fimbriate-glandular margin. Peduncle (8–) 10–20 mm, erect, Sepals 3–4 × 2.5–3 mm, unequal, obovate-spathulate, obtuse, rarely emarginate, dark purplish. Petals 5–6 × 2–3 mm, obovate to spathulate, obtuse, white. Capsule 2–3 mm, ovoid-globose, smooth; seeds *c.*0.8 mm, obovoid, black. Self-compatible. *Flowers December to February*. (Fig. 99).

Wet open peat and gravel, *Sphagnum* bogs, wet communities dominated by *Astelia, Gaimardia* and *Rostkovia*. 0–650 m.

S. and W. Fuegia. (Map 11f).
Chile north to *c.*40°S.lat.
Falkland Islands (Islas Malvinas).

XXXIV ROSACEAE

Perennial herbs, often woody at base. Leaves alternate, simple or compound; stipules present. Flowers hermaphrodite or unisexual, actinomorphic. Sepals (3–)5(–7), free or united at base, often with an epicalyx. Petals 5, free, or absent. Stamens 3–5 or numerous, in whorls. Carpels 1–2 or numerous, free or united, receptacle convex to deeply concave. Fruit fleshy or dry; seeds without endosperm.

a. Flowers congested in dense heads; petals absent; carpels 1–2 — **3. *Acaena***
a. Flowers not congested in dense heads; petals present; carpels numerous:
  b. Plant creeping; flowers solitary; fruit a group of 1-seeded drupelets — **1. *Rubus***
  b. Plant erect; flowers in small panicles; fruit a cluster of achenes with persistent styles — **2. *Geum***

1. *Rubus* L.

Leaves usually simple. Flowers hermaphrodite, solitary, axillary. Sepals 5, free. Petals 5, alternating with sepals. Stamens numerous. Carpels numerous, free, on convex receptacle. Fruit a group of numerous. 1-seeded drupelets.

1. ***R. geoides*** Sm. 1789, *Pl. Icon. Ined.* **1** pl. 19.

Stems 6–72 cm, slender, creeping, somewhat woody, branched, rooted at several nodes. Leaves with lamina (1.7–)2.4–5.7 × 1–2.5 cm, oblong or orbicular- to oblong- ovate, obtuse, sometimes emarginate, cordate at base, simple or rarely with 2 unequal leaflets at base, irregularly serrate-denatate,

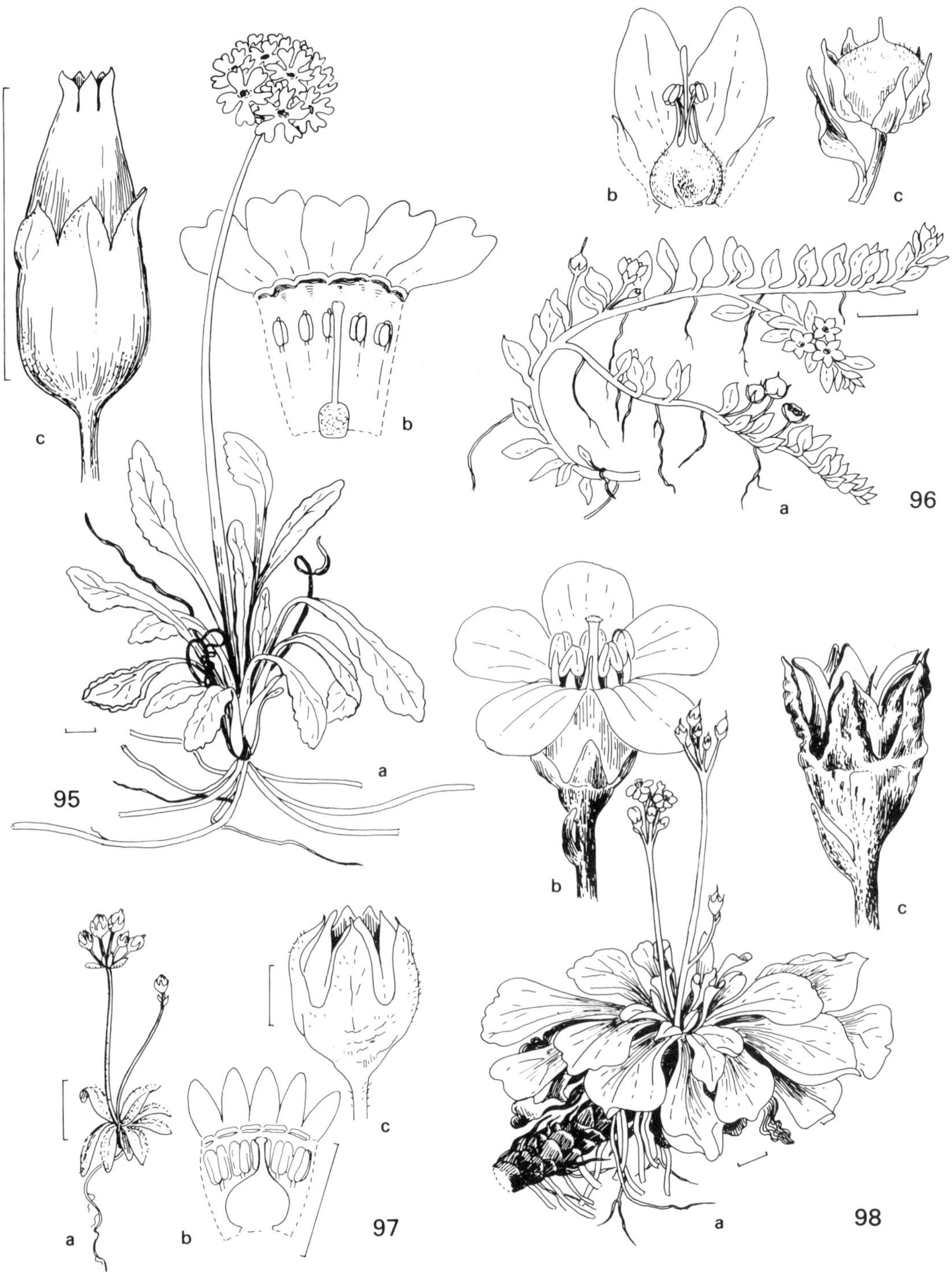

Fig. 95. *Primula magellanica.* a, Plant. b, Opened corolla. c, Fruit. Scale = 1 cm. Fig. 96. *Anagallis alternifolia.* a, Plant. b, Half-flower. c, Fruit. Scale = 1 cm. Fig. 97. *Androsace pusilla.* a, Plant. b, Opened corolla. c, Fruit. Scale = 1 cm (a) or 1 mm (b–c). Fig. 98. *Samolus spathulatus.* a, Plant. b, Flower. c, Fruit. Scale = 1 cm.

(R. N. P. Goodall)

frequently lobed, appressed hairy on veins beneath; petiole 1–4 cm, slender, with linear-lanceolate, brown, scarious stipule 4–5 mm, partly united to base. Peduncles 0.5–2 cm, appressed-hairy. Sepals 5–7 × 2.5–3.5(–4.5) mm, ovate-triangular, acute to acuminate, sparsely hairy. Petals 5–11 × 4–6.5 mm, obovate, obtuse, usually emarginate, white or cream. Fruit 1.3–2 (–2.5) cm in diameter, subglobose, orange or red. $2n = 28^{\star}$. Self-compatible, sometimes self pollinated. *Flowers November to January*. (Figs. 92, 100).

Moist open to partly shaded areas among rocks, grassland, shrub communities, forest clearings and margins; 0–550 m.

C. & S. Fuegia. (Map 11g).
W. Argentina north to *c.*48°S.lat. and at 36°48′S.lat.
Chile north to *c.*37°S.lat.
Juan Fernandez.
Falkland Islands (Islas Malvinas).

2. *Geum* L.

Leaves pinnate. Flowers hermaphrodite, in small panicles. Sepals 5, united at base, alternating with 5 epicalyx-lobes. Petals 5, free. Stamens numerous. Carpels numerous, free, on convex, hairy receptacle. Fruit a cluster of achenes with persistent styles.

a. Petals yellow; flower 20 mm or more in diameter; calyx-lobes reflexed in mature flower and fruit **1. *magellanicum***
a. Petals white; flower up to 10 mm in diameter; calyx-lobes not reflexed at maturity **2. *parviflorum***

1. *G. magellanicum* Comm. ex Pers. 1806, *Syn. Pl.* **2**: 57.
G. chiloense auct., non Balbis ex Seringe – Spegazzini, 1896.

Stock stout; stems 12.5–48(–82) cm, erect, simple or branched above, hairy. Basal leaves 50–140(–320) mm, ± oblanceolate in outline, with 3–7 pairs of leaflets, hairy; terminal leaflet 18–28(–48) × 21–35(–60) mm, cordate to suborbicular, 3-lobed, crenate to crenate-dentate; lateral leaflets up to 9–20(–27) × 7–14(–22) mm, the larger distal, suborbicular to ovate; cauline leaves smaller, often with 3 leaflets. Racemes 3– to 6– flowered; peduncles and pedicels 7–20 mm, up to 65 mm in fruit; bracts up to 30 mm. Flowers 20–25 mm in diameter. Calyx *c.*7 mm; lobes 5.6–6.5 × 3.4–4 mm, triangular-ovate, obtuse to subacute, reflexed in mature flower and fruit; epicalyx-lobes *c.*1 mm. Petals 9–10.5 × 9.7–10.5 mm, spathulate, somewhat retuse, yellow. Achenes 2.5–3 × 0.6–1.3 mm, compressed, with style *c.*4.7 mm, hooked, the terminal part deciduous. $2n = 84$. Self-compatible, facultatively self-pollinated. *Flowers December to February*. (Fig. 101).

Moist areas, usually in shrub communities and forest openings; 0–50 m.

Throughout Fuegia except the extreme south and west. (Map 11h).
W. Argentina north to *c.*50°S.lat.
Chile north to *c.*46°S.lat.

2. *G. parviflorum* Comm. ex Sm. 1805, in Rees, *Cyclop.* **5**: 12

Stock stout; stems 5.5–33 cm, erect, simple or branched above, hairy. Basal leaves 52–120(–150) mm, ± oblanceolate in outline, with 3–6 pairs of leaflets, hairy; terminal leaflet 17–40 × 21–44 mm, cordate to suborbicular, ± 3-lobed, crenate; lateral leaflets up to 10–17 × 10.5–16 mm, the larger distal, suborbicular; cauline leaves 15–28 mm, sometimes 3-lobed, or bract-like. Raceme 3– to 6– flowered; peduncles and pedicels up to 8 mm, up to 27 mm in fruit; bracts *c.*6 mm. Flowers *c.*10 mm in diameter. Calyx 4.5–5.5 mm; lobes 2–2.5 × *c.*1 mm, ovate, obtuse, not reflexed at maturity; epicalyx-lobes *c.*1 mm. Petals *c.*3 × 1.3(–2) mm, ovate to obovate, the claw sometimes ciliate, white. Achenes 1.2–1.5 × 0.6–0.7 mm, compressed, with style 1.7–2 mm, hooked, the terminal part persistent. $2n = 56$. *Flowers December to March*. (Fig. 102).

Moist montane forests and scrub of *Nothofagus pumilio* and *N. antarctica*, extending above timberline in *Bolax* communities; 200–900 m.

S. & W. Isla Grande. (Map 11i).
W. Argentina north to 43°29′S.lat.
Chile north to *c.*40°S.lat.
New Zealand.

3. *Acaena* L.

Stems usually woody at base. Leaves pinnate. Flowers hermaphrodite or unisexual, in globose to cylindrical cymose or racemose inflorescences, with or without basal bracts. Sepals 3–7, free. Petals 0. Stamens 3–5, rarely more. Carpels 1–2, united, with prominent plumose stigmas; receptacle deeply concave. enclosing the carpels. Fruit of 1–2 achenes, enclosed in hard, persistent receptacle (cupule); cupule with 2–4 or numerous, usually barbed, spines, rarely spines rudimentary.

Literature: E. Grondona, *Darwiniana* **13**: 208–342(1964); D. W. H. Walton, *Darwiniana* **19**: 500–509(1975).

a. Main stems usually with long internodes, not covered by leaf-bases; stipules with free herbaceous apex; cupule with 2–4 spines on upper margin:
 b. Fruiting inflorescence less than 10 mm in diameter (including spines); spines usually less than 4 mm:
  c. Leaves usually glabrous; cupules with golden-yellow subsessile glands (manna) at base **4. *tenera***
  c. Leaves densely hairy; cupules without manna **1. *antarctica***
 b. Fruiting inflorescence 10 mm or more in diameter (including spines); spines usually 5 mm or more:
  d. Cupule with 2(3) spines, strigose; anthers white or yellowish before anthesis **2. *ovalifolia***
  d. Cupule with 3–4 spines, glabrous, pubescent or pilose; anthers purple or blackish before anthesis, rarely yellow and purple-tinged **3. *magellanica***
a. Main stems with short internodes, covered by leaf-bases; stipules with apex not or scarcely free, scarious; cupule with numerous spines over whole surface, or spines rudimentary:
 e. Cupules with rudimentary spines **5. *lucida***
 e. Cupules with well-developed spines:
  f. Leaves glabrous; leaflets lobed **7. *pumila***
  f. Leaves hairy, at least when young; leaflets deeply pinnately divided or entire to emarginate, dentate or laciniate:
   g. Leaves with lowest pair of leaflets emarginate to dentate or laciniate **8. *sericea*** group
   g. Leaves with lowest pair of leaflets deeply pinnately divided or entire **6. *pinnatifida*** group

1. *A. antarctica* Hooker f. 1847, *Fl. Antarct.* **1(2)**: 269.

A. microcephala Schlecht. 1856, *Linnaea* **28**: 463–Grondona, 1964.

Stems up to 15 cm, prostrate, rooting at nodes, the internodes not covered by leaf-bases; branches prostrate or ascending, rarely erect, appressed-hairy. Leaves with lamina 0.8–2 × 0.6–1 cm, oblong to elliptic-oblong, with 2–4 pairs of leaflets; leaflets 3–5 × 1.5–3 mm, elliptical to obovate, obtuse to acute, sometimes emarginate, cuneate to truncate at base, serrate-dentate, densely silvery sericeous, shortly petiolulate or sessile; petiole 4–80 mm, densely sericeous; stipules with acute, free apex *c.*2 mm, scarious to subherbaceous, glabrous or sparsely hairy. Peduncles 5–20 mm, sericeous, with globose terminal inflorescence *c.*4 mm in diameter; bracts 1.2–1.5 mm, oblong-spathulate, usually emarginate. Sepals 4–5, *c.*1 mm, elliptical to ovate, glabrous or with sparse marginal hairs. Stamens 2(3); anthers purplish, held somewhat above stigma. Stigmas reddish purple. Cupule *c.*1 mm, obovoid, glabrous; spines 4, *c.*0.5 mm, acicular, glabrous. $2n = 42$. *Flowers January to March.* (Fig. 105).

Moist areas with dry facies of *Astelia* and *Sphagnum* bogs, stone screes, gravel by streams, open soil in *Nothofagus antarctica* scrub; (50–)200–800 m.

C. & S. Fuegia. (Map 11j).
W. Argentina north to *c.*40°S.lat.
Chile north to *c.*51°S.lat.
Falkland Islands (Islas Malvinas).

2. *A. ovalifolia* Ruiz & Pavón 1798, *Fl. Peruv.* **1**: 67.

Stems up to 80 cm, prostrate or ascending, rooting at lower nodes, the internodes not covered by leaf-bases; branches ascending to erect, sericeous. Leaves with lamina 1.5–6 × 1–3.5 cm, oblong to elliptic- or obovate- oblong in outline, with 2–4 pairs of leaflets; leaflets 7–25 × 4–15 mm, oblong to elliptic- or obovate- oblong, subacute or obtuse, cuneate or truncate at base, serrate-dentate, hairy beneath, glabrous or sparsely hairy on veins above, the lateral sessile, the terminal with petiolule 1–4 mm; petiole 5–40 mm, hairy; stipules with acute free apex *c.*5 mm, scarious to herbaceous, sparsely sericeous. Peduncles 30–80 mm, sericeous, with terminal globose inflorescence 5–15 mm in diameter, up to 40 mm in fruit; bracts 2–3 mm, linear to oblong, acute, entire or laciniate, hairy. Sepals 4–5, *c.*1.5 mm, oblong-oblanceolate, pubescent to shortly sericeous. Stamens 2(3); anthers purplish, at same level as stigma. Stigmas pale yellow. Cupule 2.5–3.5 mm, obconical, with rigid white hairs; spines 2(3), 10–14 mm, acicular, glabrous. $2n = 42$. Self-compatible, usually self-pollinated. *Flowers November to February.*

Moist places, usually in forest margins or clearings, but also on shady cliffs and scrub; 0–500 m.

C. to S. Isla Grande and adjacent islands along Canal Beagle; Isla de los Estados. (Map 12a).
Argentina, especially in the west, north to *c.*35°S.lat.
E. Chile and then north along Andes to *c.*12°S.lat.; Juan Fernandez.
Falkland Islands (Islas Malvinas).

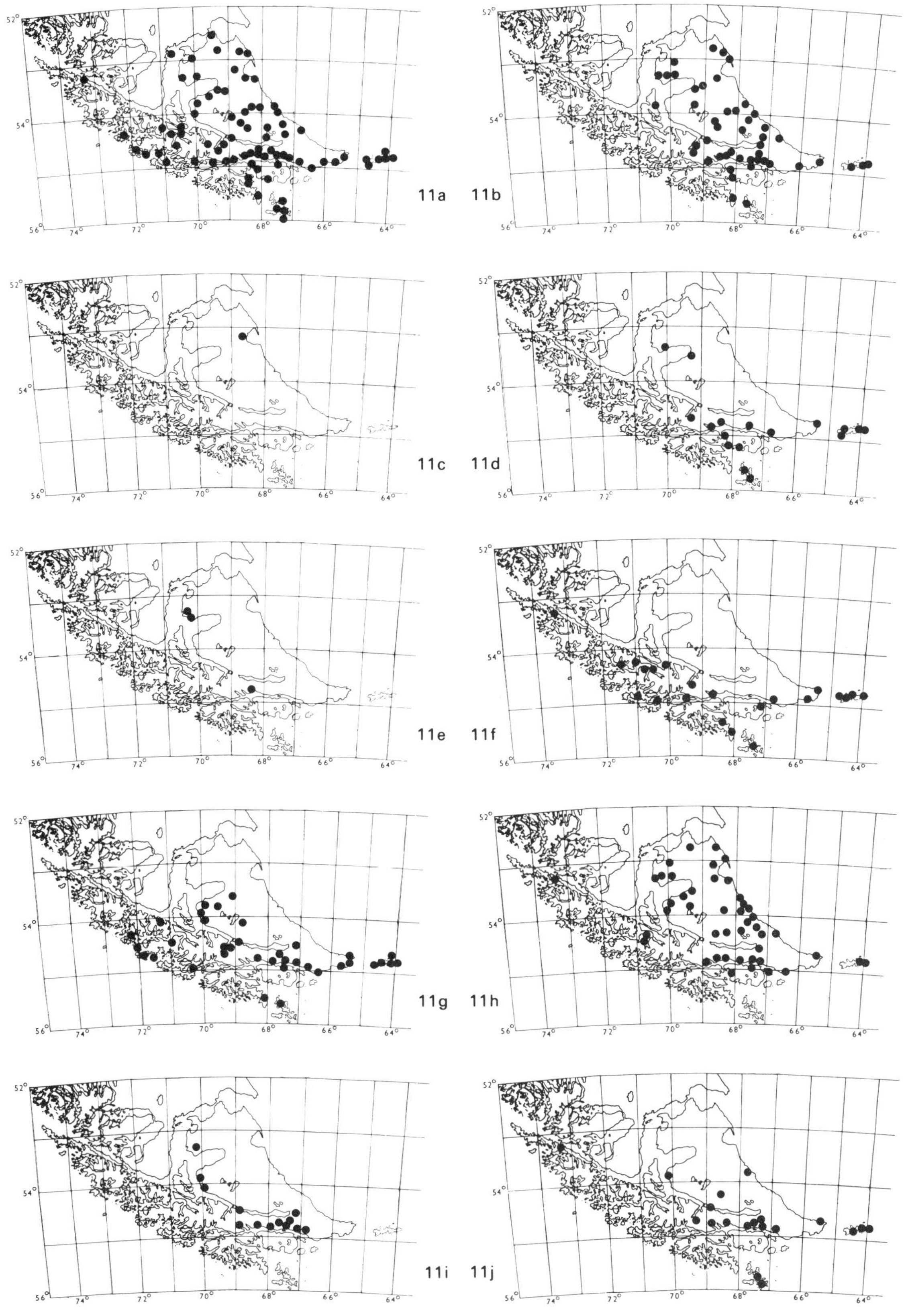

Map 11. a, *Empetrum rubrum*. b, *Primula magellanica*. c, *Androsace pusilla*. d, *Anagallis alternifolia*. e, *Samolus spathulatus*. f, *Drosera uniflora*. g, *Rubus geoides*. h, *Geum magellanicum*. i, *Geum parviflorum*. j, *Acaena antarctica*.

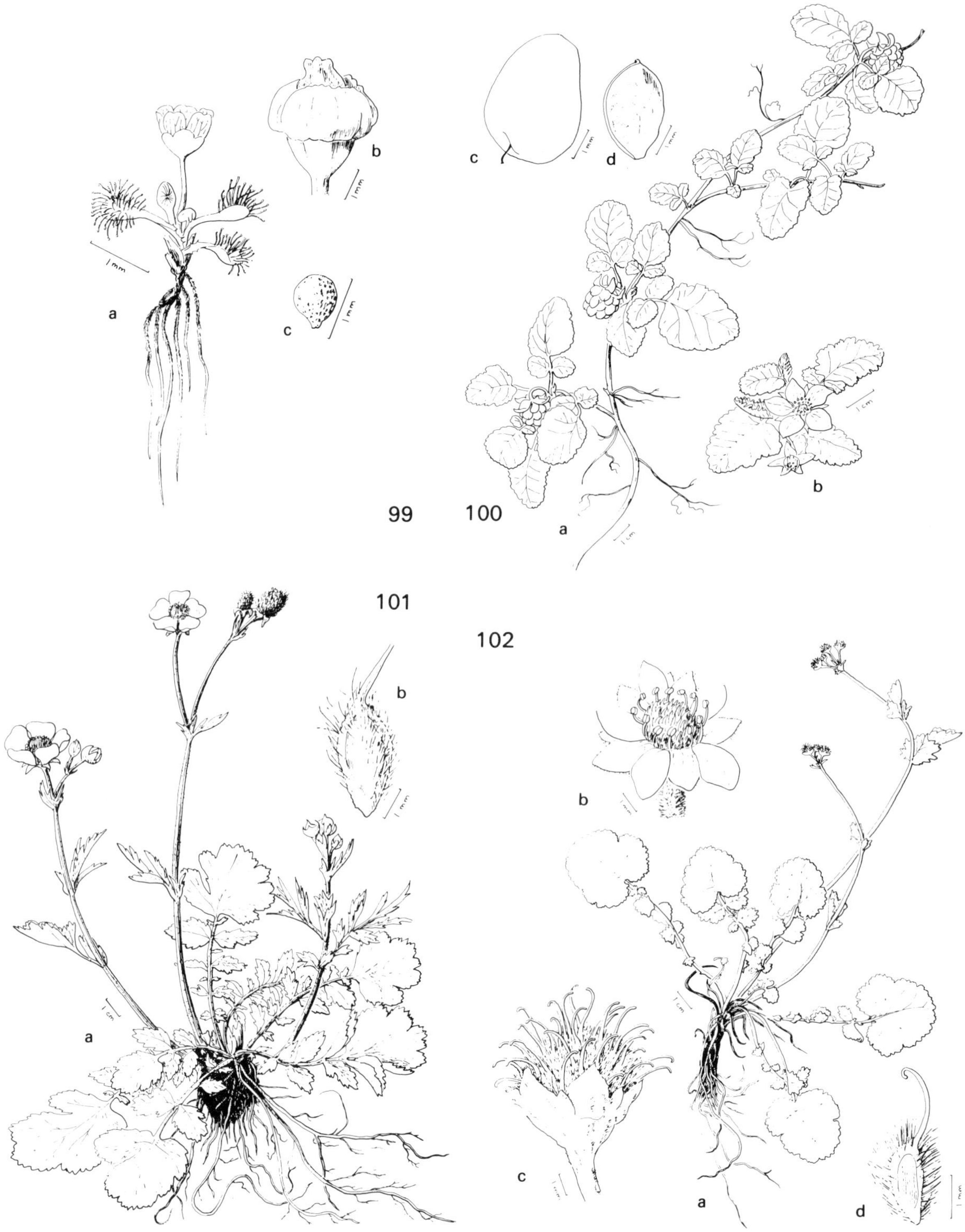

Fig. 99. *Drosera uniflora*. a, Plant. b, Fruit. c, Seed. Scale = 1 mm. Fig. 100. *Rubus geoides*. a, Plant in fruit. b, Flowering shoot. c, Drupelet. d, Seed. Scale = 1 cm (a–b) or 1 mm (c–d). Fig. 101. *Geum magellanicum*. a, Plant. b, Achene. Scale = 1 cm (a) or 1 mm (b). Fig. 102. *Geum parviflorum*. a, Plant. b, Flower. c, Fruit. d, Achene. Scale = 1 cm (a) or 1 mm (b–d).
(R. N. P. Goodall)

3. *A. magellanica* (Lam.) Vahl 1804, *Enum. Pl.* **1**: 297.

A. decumbens (Gaertner) D. W. H. Walton 1971, *Bull. Brit. Antarct. Surv.* **25**: 30.
A. basibullata Bitter 1911, *Bibliotech. Bot. (Stuttgart)* **17**: 228 – Skottsberg, 1916.
A. nudicaulis Alboff 1896, *Rev. Mus. La Plata Bot.* **7**: 367
A. laevigata (Lam.) Aiton 1810, *Hort. Kew.* ed.2, **1**: 67 – Hooker, 1847; Dusén, 1900.
A. adscendens Vahl 1804, *Enum. Pl.* **1**: 297 – Hooker, 1847; Franchet, 1889; Spegazzini, 1896; Hariot, 1900.

Stems up to 40 cm or more, prostrate to ascending, rooted, the internodes not covered by leaf-bases; branches prostrate to erect, glabrous to sericeous. Leaves with lamina 2–8 × 1.3–2.5 cm, ± oblong in outline, with 5–8 pairs of leaflets; leaflets 7–20 × 4–12 mm, oblong– to elliptic-obovate, obtuse, cuneate to truncate at base, serrate to serrate-dentate, glabrous to sericeous, especially beneath, usually glaucous beneath, sessile or shortly petiolulate; petiole 6–60mm; stipules scarious or herbaceous, with acute, entire or divided, glabrous to sericeous, green apex 2–7 mm. Peduncles 7–14 cm, glabrous to sericeous, with terminal globose inflorescence 10–30 mm in diameter (including spines), rarely with small lateral inflorescences below; bracts 1–2 mm, linear to oblong-lanceolate, fimbriate, sericeous. Sepals 4–5 (6), 1.5–1.8 × 0.6–1 mm, elliptical to elliptic-oblong, obtuse, usually sparsely hairy, green or purplish. Stamens 2–4(5); anthers longer than wide, usually purple. Cupule 2–7 mm, obconical to cylindrical, glabrous to pubescent, especially distally; spines 3–4, (1–)5–10 mm, unequal, acicular, glabrous. $2n = 42$, 84. Self-compatible. *Flowers November to March.* (Fig. 104).

Coastal sands and rocks, open shrub communities, open forest, bog-margins, moist grassland, feldmark; 0–1100 m.

Throughout Fuegia. (Map 12b).
W. Argentina and E. Chile north to *c.*27°S.lat.
Falkland Islands (Islas Malvinas), South Georgia, Iles de Kerguelen.

4. *A. tenera* Alboff 1896, *Rev. Mus. La Plata* **7**: 367.

Similar to smaller examples of *A. magellanica* but distinguished by its usually glabrous leaves, with usually purplish-tipped marginal teeth on the leaflets, glabrous peduncle, smaller globose inflorescence less than 10 mm in diameter (including spines), and cupules with spines *c.*2 mm and golden-yellow subsessile glands (manna) at base. $2n = 42$★. *Flowers December to January.* (Fig. 103).

Feldmark and montane cushion plant communities, often in moss-mats; 550–1000 m.

S. part of Isla Grande. (Map 12c).
South Georgia.
?W. Argentina at *c.*41°S.lat.

5. *A. lucida* (Lam.) Vahl 1804, *Enum. Pl.* **1**: 296.

A. parvifolia Phil. 1894, *Anal. Univ. Chile* **84**: 622.

Stems up to 20 cm, procumbent or ascending, rarely suberect, much branched, rooted, the internodes covered by dead leaf-bases; branches ascending to suberect. Leaves with lamina 1–3.8 × 0.3–1 cm, oblong to linear-oblong in outline, with 5–10 pairs of leaflets; leaflets 1.5–4 × 1–2 mm, oblong to elliptical or ovate-elliptical, entire or 2– to 3–pinnatisect, acute, cuneate at base, sessile, pilose or sericeous, especially on margins and beneath; petiole 0.8–2 mm; stipules scarious, laciniate, sometimes with free apex *c.*1 mm. Peduncles 15–100 mm, pilose, with terminal globose inflorescence 5–10 mm in diameter and often lateral flowers or small inflorescences below; bracts 2–4 mm, oblong to ovate-orbicular, acute to obtuse or lobed, pilose. Sepals 4–5, *c.*1 mm, elliptical to ovate, acute to obtuse, pilose. Stamens 2(3); anthers wider than long, purple. Stigmas irregularly laciniate. Cupule 2–4 mm, ovoid to ellipsoid, glabrous or sparsely pilose; spines rudimentary, *c.*1 mm near apex. $2n = 42$★. Selt-compatible. *Flowers November to February.*

Open sand and gravel in grass and shrub communities, rock-crevices, feldmark; 45–660 m.

Isla Grande except the extreme west and east. (Map 12d).
W. Argentina and E. Chile north to *c.*50°S.lat.
Falkland Islands (Islas Malvinas).

6. *A. pinnatifida* group.

This comprises 3 species recorded from Tierra del Fuego by Grondona (1964), which it has proved impossible to separate satisfactorily on the basis of the extensive Fuegian material now available. Further study is required, as an aid to which the characters used to separate the species are indicated.

A. pinnatifida Ruiz & Pavón 1789, *Fl. Peruv.* **1**: 68.

A. multifida Hooker f. 1847, *Fl. Antarct.* **1(2)**: 265.

Stems 3–15 cm, procumbent to suberect, branched, the internodes covered by leaf-bases. Leaves with lamina 20–80 × 5–20 mm, linear-oblanceolate in outline, with 4–13 pairs of leaflets, hairy, more

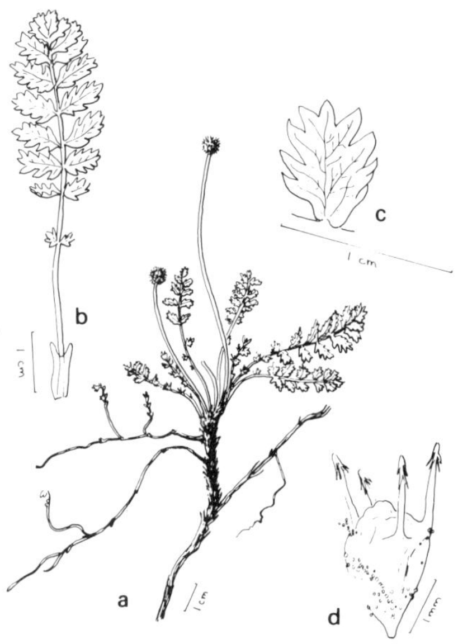

103

104

105

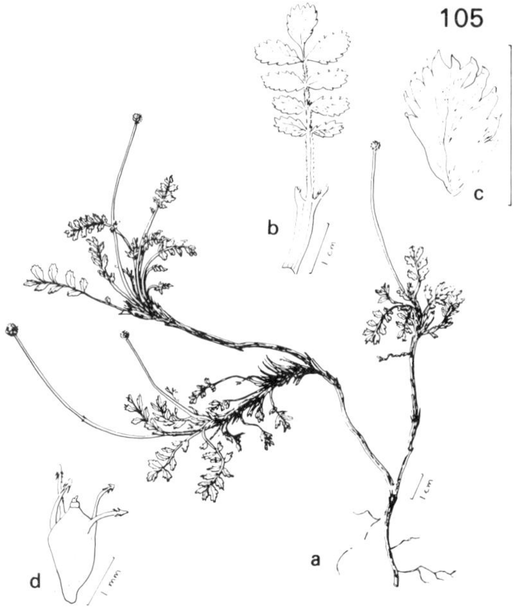

106

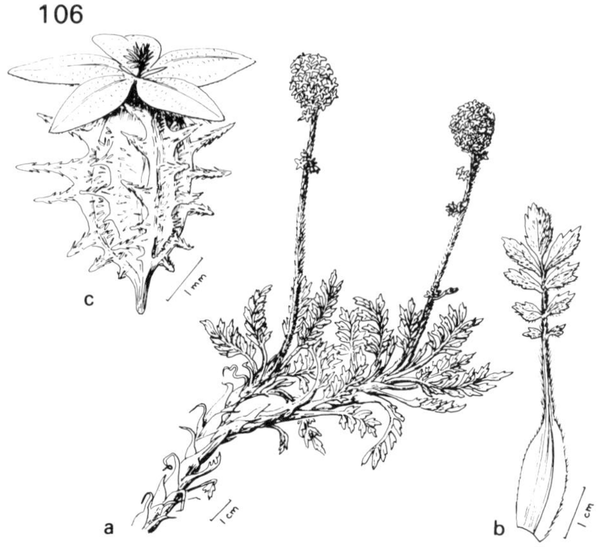

Fig. 103. *Acaena tenera*. a, Plant. b, Leaf. c, Leaflet. d, Cupule. Scale = 1 cm (a–c) or 1 mm (d). Fig. 104. *Acaena magellanica*. a, Plant. b, Flower. c, Immature cupule. d, Spine-apex. Scale = 5 cm (a), 1 cm (b–c) or 1 mm (d). Fig. 105. *Acaena antarctica*. a, Plant. b, Leaf. c, Leaflet. d, Cupule. Scale = 1 cm (a–c) or 1 mm (d). Fig. 106. *Acaena sericea*. a, Plant. b, Leaf. c, Cupule. Scale = 1 cm (a–b) or 1 mm (c).

(R. N. P. Goodall)

a. *Drimys winteri.*

b. *Nothofagus antarctica.*

c. *Berberis ilicifolia.*

d. *Caltha sagittata.* (Sydney Parkinson).

PLATE 5.

a. *Escallonia serrata.*

b. *Ribes magellanica.*

c. *Gunnera magellanica.*

d. *Perezia magellanica.* (Sydney Parkinson).

PLATE 6.

densely so beneath; leaflets 3–10 × 3–9 mm, oblong to orbicular-elliptical or ovate, subacute to obtuse, pinnately divided into 3–13 linear to elliptic-ovate segments; petiole 8–50 mm, hairy, with sericeous stipules completely united. Peduncles 6–30 cm, erect, hairy, with terminal, continuous or interrupted, cylindrical to cylindrical-globose inflorescence 10–150 mm, sometimes with small clusters of flowers below; bracts 3–5 mm, elliptic-orbicular to ovate or linear-oblong, acute, sericeous. Sepals 4–6, 1–2.5 mm, oblong to elliptical, obtuse to subacute, sericeous. Stamens 2–5; anthers orbicular to reniform, purple to blackish. Stigma irregularly lobed, purplish. Cupule 3–10 × 1.5–4 mm, ellipsoid to obovoid, with 3–5 winged angles, pubescent, with unequal, broad-based spines over whole surface or restricted to apical part. $2n = 42$. Self-compatible, often self-pollinated. *Flowers November to April.*

Sands and shingle, open *Lepidophyllum*, *Chiliotrichum* and *Empetrum* communities, clearings and margins of *Nothofagus antarctica* woodland, open grassland; 0–250 m.

Throughout Isla Grande except the extreme east and west. (Map 12e).
Argentina north to *c.*46°S. lat. and, in the west, to *c.*28°S.lat.
Chile north to *c.*52°S.lat.

A. poeppigiana C. Gay 1846, *Hist. Chile Bot.* **2**: 284.

Said to differ from *A. pinnatifida* in having narrower leaves, generally less than 7 mm wide, with the lowest pair of leaflets generally less than 2.5 mm apart. Both characters show continuous variation, not associated with other features, in Fuegian material. Reported from N. E. Isla Grande.

A. platyacantha Speg. 1897, *Rev. Fac. Agron. La Plata* **20–21**: 515.

Said to differ from *A. pinnatifida* in having more globose inflorescences, leaflets with fewer segments (1–3, rarely more), and cupules with more broadly based spines. None of these characters show correlated discontinuities in Fuegian material here ascribed to *A. pinnatifida*. Reported from N. E. Isla Grande.

7. *A. pumila* Vahl 1804, *Enum. Pl.* **1**: 298.

Stems 2–5 cm, prostrate, branched, rooted, the internodes covered by leaf-bases. Leaves with lamina 1.2–60 × 4–16 mm, oblong or linear-oblong in outline, obtuse, with 7–12 pairs of leaflets; leaflets 2.5–10 × 2–5 mm, asymmetrically elliptic-ovate to ovate, obtuse, cuneate to truncate at base, obtusely serrate-dentate towards apex, the margin revolute, comose on teeth, glabrous elsewhere, sessile; petiole 2–15 mm, glabrous; stipules scarious, purplish, completely united to petiole. Peduncles 2–40 mm, pilose, with terminal spike of few, remote, shortly pedicellate flowers; bracts 3–4 mm, oblong, entire or irregularly lobed, glabrous. Sepals 4–5, *c.*1 × 0.5 mm, oblong or elliptic-oblong, obtuse, glabrous, purplish. Stamens 2(–4); anthers rather wider than long, purple. Stigma ± orbicular. Cupule 2–3.5 × 2–2.5 mm, ellipsoid, glabrous, with spines 0.5–0.8 mm, over whole surface. *Flowers November to February.* (Fig. 93).

Sphagnum and *Astelia* bogs, wet bryophyte carpets over peat; 0–500 m.

S. Fuegia. (Map 12f).
Chile north to *c.*49°10′S.lat. and at *c.*39°S.lat.
Falkland Islands (Islas Malvinas).

8. *A. sericea* group.

Includes 2 species reported from Tierra del Fuego by Grondona (1954).

A. sericea Jacquin f. 1813, *Eclogae Plant. Rar.* **1(5–6)**: 81–82.
A. fuegina Phil. 1893, *Anal. Univ. Chile* **84**: 620.
A. philippii Dusén 1900, *Svenska Exped. Magellansl.* **3(5)**: 167.
A. cuneata Hooker & Arn. 1833, *Bot. Misc.* **Suppl.** 307 – Hariot, 1900.

Stems up to 50 cm or more, procumbent to suberect, branched, the internodes covered by leaf-bases. Leaves with lamina 15–80 × 10–30 mm, oblong in outline, with 3–7 pairs of leaflets, sericeous to pilose-sericeous; leaflets 5–15 × 4–7 mm, oblong to elliptical or rhombic, obtuse, cuneate at base, dentate-serrate, sessile; petiole 15–70 mm, pilose-sericeous, with hairy stipules completely united. Peduncles 6–25 cm, erect, hairy, with terminal globose inflorescence up to 8 mm in diameter, sometimes with flowers or clusters of flowers below; bracts 4–5 mm, linear to elliptical or orbicular, acute to obtuse, sericeous. Sepals 4–6, 0.6–1.2 mm, oblong to elliptical, obtuse to subacute, sericeous to pilose-sericeous. Stamens 2–5; anthers elliptic– to broadly orbicular, purple. Stigma ± orbicular, laciniate, purple. Cupule 5–11 × 2.5–4 mm, ellipsoid to obovoid, with 3–5 ± winged angles, pubescent to ± lanate, with short acicular spines over whole surface, or less frequent at base. Self-compatible, at

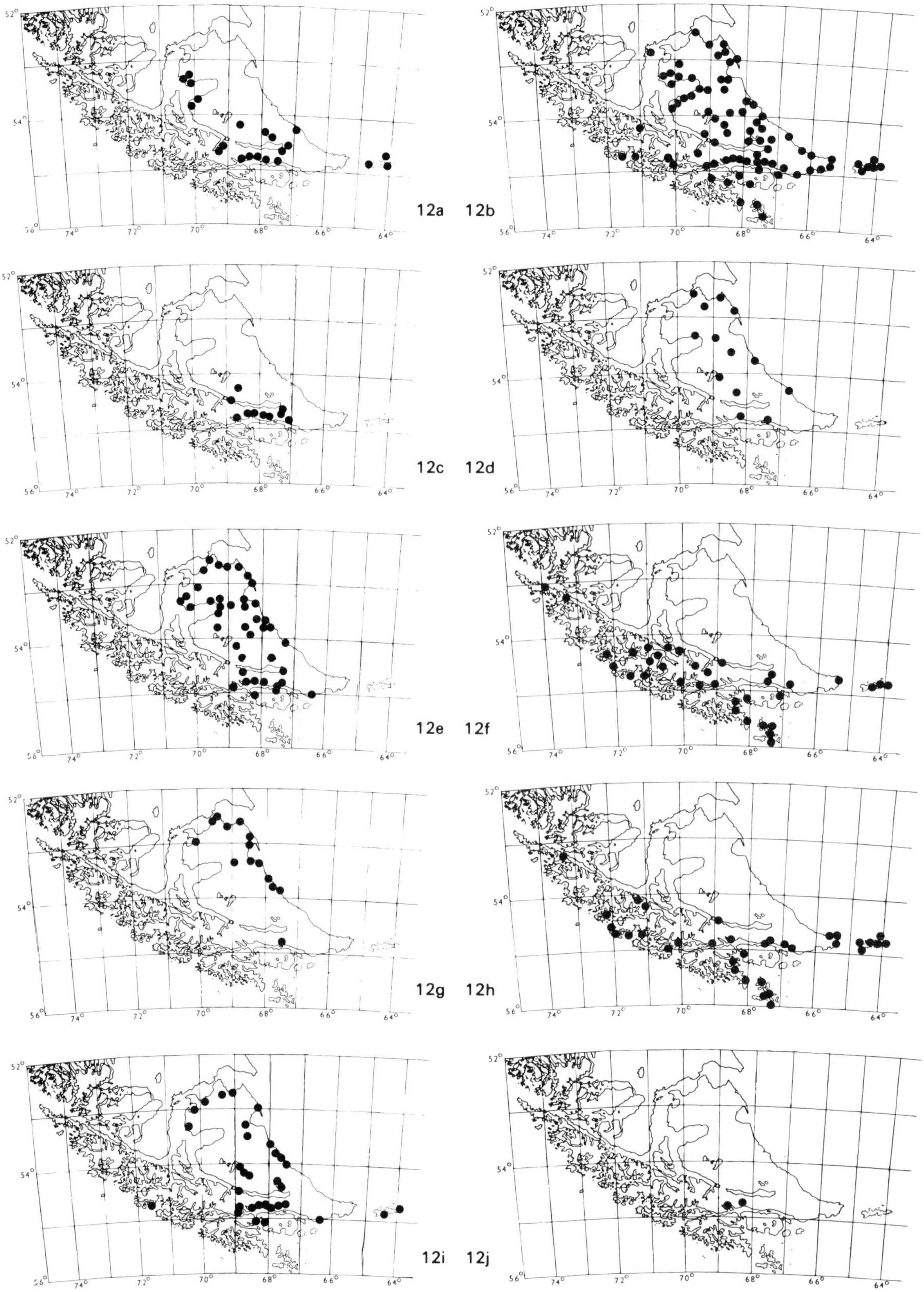

Map 12. a, *Acaena ovalifolia*. b, *Acaena magellanica*. c, *Acaena tenera*. d, *Acaena lucida*. e, *Acaena pinnatifida*. f, *Acaena pumila*. g, *Acaena sericea*. h, *Crassula moschata*. i, *Saxifraga magellanica*. j, *Saxifragodes albowiana*.

least facultatively self-pollinated. *Flowers December to January*. (Fig. 106).

Open coastal sands and gravels, often in *Lepidophyllum* scrub; 0–40 m.

N. & N. E. Isla Grande. (Map 12g).
Argentina north to *c*.48°S.lat., and in the west to *c*.30°S.lat.
Chile north to *c*.28°S.lat.

A. splendens Hooker & Arn. 1833, in Hooker, *Bot. Misc., Suppl.* 306.
A. integerrima Gillies ex Hooker & Arn. 1833, in Hooker, *Bot. Misc., Suppl.* 306 – Skottsberg, 1916.
A. adenocarpa Bitter 1911, *Biblioth. Bot. (Stuttgart)* **74**: 86 – Dusén, 1900; Skottsberg, 1916.

Said to differ from *A. sericea* in having most leaflets obovate with crenate-dentate margins, and a globose to cylindrical inflorescence. These characters do not permit the recognition of discontinuities in Fuegian material. Reported from N. E. Isla Grande.

XXXV. CRASSULACEAE

Glabrous perennial herbs. Leaves opposite or alternate, simple, succulent; stipules absent. Flowers hermaphrodite, actinomorphic. Sepals 4–5, united at base. Petals 4–5, free. Stamens 4 or 10. Ovary superior; carpels 4–5, free or slightly united at base, each with basal nectariferous scale. Fruit a group of follicles; seeds minute, with little or no endosperm.

a. Leaves opposite; sepals and petals 4 **1. *Crassula***
a. Leaves alternate; sepals and petals 5 **2. *Sedum***

1. *Crassula* L.

Leaves opposite, connate at base. Flowers solitary in leaf-axils. Sepals 4. Petals 4. Stamens 4, alternating with petals. Carpels 4.

1. *C. moschata* Forster f. 1789, *Comment. Gotting., Ser. 2*, **9**: 26.
Tillaea moschata (Forster f.) DC. 1828, *Prodr.* **3**: 382– Franchet, 1889; Spegazzini, 1896.
Bullardia moschata (Forester f.) D'Urv. 1825, *Fl. Is. Mal.* 53 – Hooker, 1847.

Plant forming extensive, lax mats; stems up to 30 cm, prostrate, rooting at nodes, green or reddish, with erect or ascending leafy branches up to *c*.15 cm. Leaves (2–)3.5–5(–7.5) × 1.2–2.3 mm, obovate to oblanceolate, spathulate, obtuse to subacute, entire. Peduncles 1.5–4 mm. Sepals (1–)1.3–1.8 × 0.7–0.9 mm, ovate to oblong, obtuse, unequal, green to purplish. Petals 2.2–3 × 1–1.3 mm, oblong, obtuse, white or pinkish. Nectariferous scales 0.5–1 × 0.2–0.5 mm, oblong to obovate, truncate, purplish red. Follicles 1.5–1.8 mm, suboblong, with curved apical beak *c*.0.3 mm; seeds 6–8, 0.5 mm, ovoid-ellipsoid, smooth. Self-compatible, protandrous. *Flowers November to March*. (Fig. 107).

Moist sandy and rocky shores at or above high-water mark; 0–20 m.

S. & W. Fuegia. (Map 12h).
W. Argentina and Chile north to *c*.41°S.lat.
Falkland Islands (Islas Malvinas). Islas Diego Ramirez.
Iles de Kerguelen, Iles Crozet, Marion Island, New Zealand, Macquarie, Stewart and Chatham Islands.

2. *Sedum* L.

Leaves alternate. Flowers in small cymes. Sepals 5. Petals 5. Stamens 10. Carpels 5.

Literature: E. M. Zardini, *Bol. Soc. Argent. Bot.* **14**: 95–106 (1971).

1. *S. acre L. 1753, *Sp. Pl.* 432.

Laxly caespitose plant with hot, acrid taste; flowering stems 5–9(–12) cm, procumbent to ascending, with short, non-flowering shoots. Leaves 3.2–4(–6) × 2.2–2.7 mm, ovoid-trigonous, obtuse, entire, shortly spurred at base, sessile. Flowers pedicellate or subsessile. Sepals 2–3.5 × 1.5–2.5 mm, ovate-lanceolate, obtuse. Petals 5–7 × 1.7–2 mm, lanceolate, acute to acuminate, yellow; nectariferous scales very small, ± oblong, whitish. Follicles 3–3.5 mm, narrowly conical, with slender apical beak 1–1.5 mm; seeds several. *Flowers January*.

Open disturbed ground.

Northernmost part of Isla Grande.
Introduced from Eurasia as a garden ornamental and locally naturalized in temperate regions.

XXXVI. SAXIFRAGACEAE

Shrubs or perennial herbs, sometimes cushion-forming. Leaves opposite or alternate, simple; stipules absent. Flowers regular, hermaphrodite, solitary or borne in cymes or racemes. Sepals 4–5, free or connate. Petals 0 or 4–5, free. Stamens 5–8 or 10, borne on margin of hypanthium. Ovary inferior to superior, with 2 or 4 carpels connate at least at base, forming 1–3 cells, each with anatropous ovules on axile placentas; style 2 or 4. Fruit a capsule or berry; seeds with endosperm.

a. Shrub:
  b. Leaves glandular-dentate; flowers solitary; fruit a capsule **7. *Escallonia***
  b. Leaves 3-lobed, serrate-dentate; flowers in pendent racemes; fruit a berry **6. *Ribes***
a. Herb, often woody at base:
  c. Plant forming moss-like cushions; leaves deeply emarginate at apex **3. *Saxifragella***
  c. Plant not as above; leaves not emarginate at apex:
    d. Leaves 3(–5)– dentate at apex **5. *Tribeles***
    d. Leaves entire to 3(–5)–lobed, not dentate only at apex:
      e. Flowers without petals **4. *Chrysosplenium***
      e. Flowers with 5 petals:
        f. Fertile stamens 10; plant glandular-hairy **1. *Saxifraga***
        f. Fertile stamens 6; plant mostly glabrous **2. *Saxifragodes***

1. *Saxifraga* L.

Perennial herbs. Leaves alternate. Flowers solitary or up to 5 in lax, terminal cymes. Sepals 5, connate at base. Petals 5. Stamens 10, 5 antipetalous. Ovary semi-inferior, 1-celled; styles 2; stigmas flattened, capitate. Fruit a capsule, shallowly 2–fid; seeds numerous.

1. *S. magellanica* Poiret 1805, in Lam., *Encycl. Méth. Bot.* **6**: 686.
S. cordilleranum Presl 1831, *Reliq. Haenk.* **2**: 55 – Dusén, 1900.

Plant usually somewhat caespitose and rather densely glandular-hairy. Vegetative stems up to 15 cm, erect or ascending, branched, somewhat woody and thickly clothed with dead leaves towards base, with densely imbricate leaves above; flowering shoots up to 13 cm. Leaves (3.5–)5–11(–13) × 2.5–5(–8) mm, oblanceolate to spathulate – oblong, 3(–5)–lobed to 3(–5)–fid, rarely entire, narrowing at base to short, usually wide, petiole; leaves on flowering shoots few, ovate to ovate – lanceolate, obtuse, sessile. Pedicels up to 11 mm. Sepals (1.5–)1.8–3.2 × 1–1.8(–2.2) mm, ovate to oblong, obtuse, green or purplish. Petals 3.5–5 × 1.5–3.8 mm, obovate to ovate, obtuse, white or cream-coloured. Capsule 3–4 mm, ovoid; seeds *c.*0.8 mm, ellipsoid, black. $2n = 76$. Self-compatible. *Flowers November to March.* (Fig. 108).

Rock-crevices, cliff-ledges, screes, open soil among rocks, especially at higher elevations. 0–1000 m.

Almost throughout Fuegia. (Map 12i).
North in W. Argentina and E. Chile, and along Andes to *c.*5°N.lat.
Falkland Islands (Islas Malvinas).

2. *Saxifragodes* D. M. Moore

Like *Saxifraga* but: plant mostly glabrous; flowers always solitary, terminal; sepals free; stamens 6, one of which and the 2 staminodes are antipetalous.

1. *S. albowiana* (F. Kurtz) D. M. Moore 1969, *Bot. Not.* **122**: 324.
Saxifraga albowiana F. Kurtz 1896, in Alboff & F. Kurtz *Rev. Mus. La Plata* **7(2)**: 370.
Saxifragella albowiana (F. Kurtz) Engler 1919, in Engler & Irmscher, *Pflanzenreich* **69 (iv.117)**: 673.

Stems 1.5–10 cm, prostrate, rather slender, branched at base. Leaves (1.2–)2.5–4.5 × (1–)2–5 mm, suborbicular to ovate-elliptical, entire or 3-lobed up to ¼ length of lamina, rarely sparsely hairy near base; middle lobe the largest, rounded, the lateral lobes acute to obtuse; petiole (1–)3–4.5 mm. Peduncles up to 1.5 mm. Sepals 0.8–2.1 × 0.8–1.7 mm, ovate to triangular-ovate, acute, green, with small purple flecks near base. Petals slightly shorter than to slightly exceeding sepals, elliptic – to oblanceolate, subacute to obtuse, cream-coloured with pinkish purple flecks, sometimes purplish near base. Capsule 3–4 × 2.5–3 mm, ovoid, compressed, deeply bifid; seeds 0.5–0.6 mm, subreniform, tuberculate, brown. *Flowers January to February.* (Fig. 110).

Open, damp soil, sand and pebbles above timberline. 500–900 m.

S. and W. Fuegia. (Map 12j).
Chile north to *c.*51°S.lat.

3. *Saxifragella* Engler

Perennial, cushion-forming herbs. Leaves opposite, deeply emarginate at apex. Flowers solitary, axillary. Sepals 5, connate at base. Petals 0. Stamens 5. Ovary 2-celled; styles 2; stigmas capitate. Fruit a capsule, bifid to about halfway, dehiscing by ventral fissure; seeds few.

1. *S. bicuspidata* (Hooker f.) Engler 1891, in Engler & Prantl, *Natürl. Pflanzenfam.* **3(2a)**: 61.
Saxifraga bicuspidata Hooker f. 1847, *Fl. Antarct* **1(2)**: 281– Dusén, 1900.

Plant forming moss-like cushions, glabrous. Stems up to c.10 cm, much-branched, densely covered with reddish-brown leaf remains towards base and

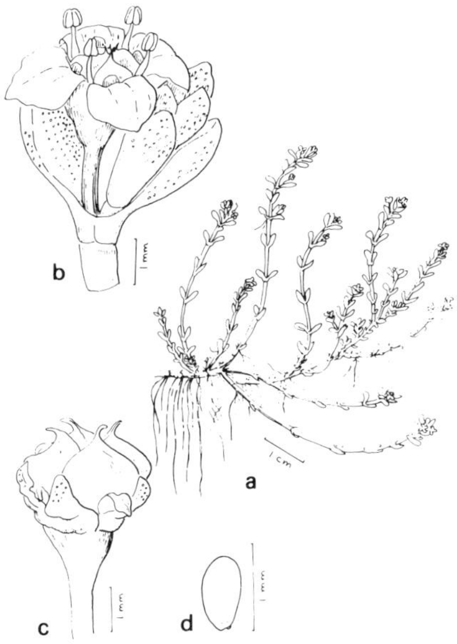

107

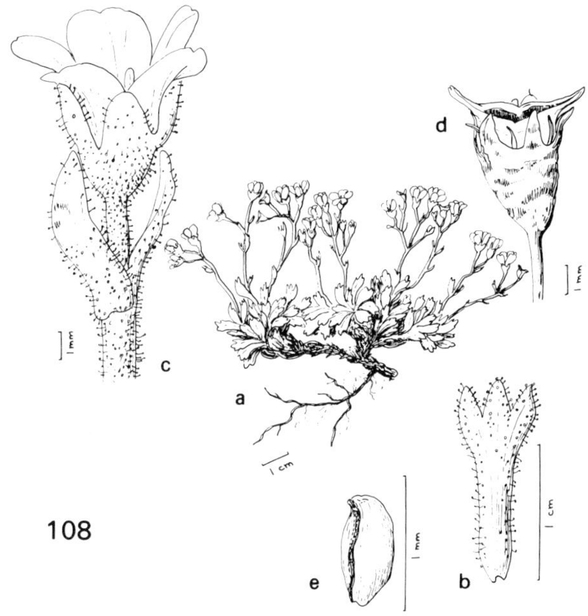

108

109

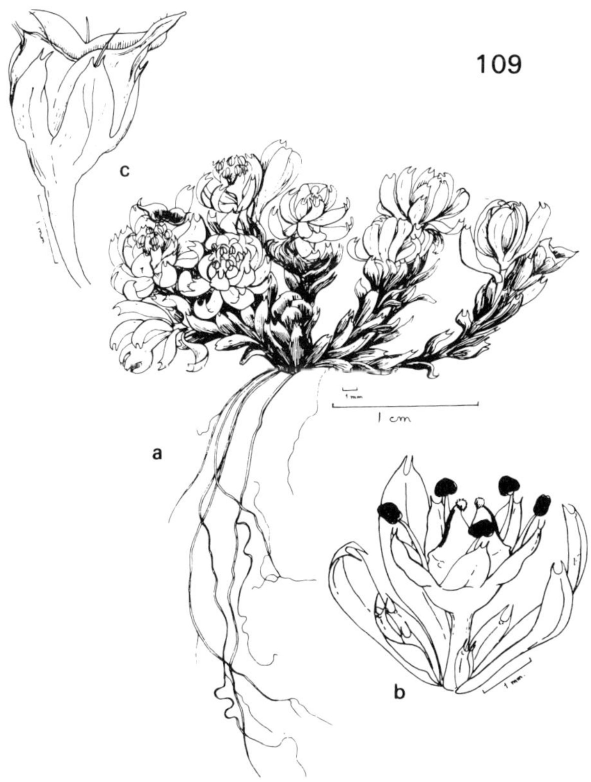

110

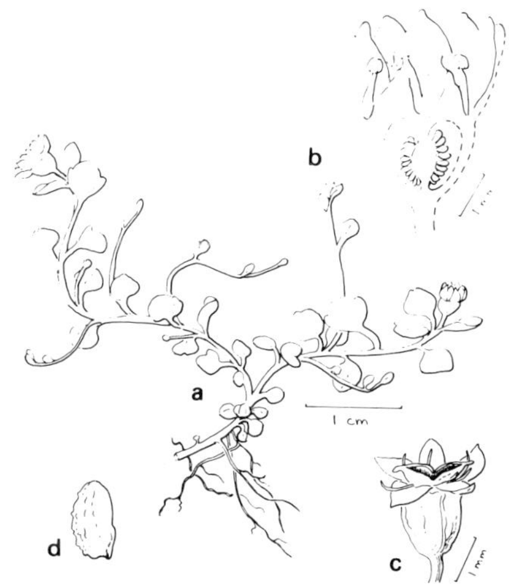

Fig. 107. *Crassula moschata*. a, Plant. b, Flower. c, Fruit. d, Seed. Scale = 1 cm (a) or 1 mm (b–d). Fig. 108. *Saxifraga magellanica*. a, Plant. b, Leaf. c, Flower. d, Capsule. e, Seed. Scale = 1 cm (a–b) or 1 mm (c–e). Fig. 109. *Saxifragella bicuspidata*. a, Plant. b, Flower. c, Capsule. Scale = 1 cm (a) or 1 mm (b–c), Fig. 110. *Saxifragodes albowiana*. a, Plant. b, Half-flower. c, Capsule. d, Seed. Scale = 1 cm (a) or 1 mm (b–d).

(R. N. P. Goodall)

with imbricate green leaves above. Leaves (2.5–)3–5(–7.5) × 0.7–1.4 mm, linear-spathulate, deeply emarginate with acute sinus at apex, somewhat narrowed at base into wide petiole, entire, rather coriaceous. Peduncles shorter than leaves before anthesis, elongating later. Sepals 1–1.5 × 0.4–1.2 mm, triangular, deeply emarginate at apex. Stamens slightly exceeding sepals. Capsule 2.5–3.5 mm, keeled; seeds *c.*0.8 mm, ellipsoid, reticulate, dark brown. *Flowers December to February*. (Fig. 109).

Rock-crevices and screes, feldmark. 660–1100 m.

S. & W. Fuegia. (Map 13a).
W. Argentina and E. Chile north to 50°43′S.lat.

4. *Chrysosplenium* L.

Creeping perennial herbs. Leaves opposite. Flowers in terminal cyme. Sepals 4, free. Petals O. Stamens 8, inserted on margin of hypanthium. Disc lobed. Ovary partly or almost wholly inferior, 1-celled; styles 2, free; stigma punctate. Fruit a capsule, 2(–4)-lobed, dehiscing by ventral fissure; seeds numerous.

Literature: H. Hara, *Jour. Fac. Sci. Univ. Tokyo, III* **7(1)**: 1–90 (1957).

1. *C. macranthum* Hooker 1842, *Jour. Bot. (London)* **1**: 458.
C. valdivicum auct., non Hooker – Hooker, 1847.
C. elevatum Banks & Sol. ex Hooker f. 1847, *Fl. Antarct.* **1(2)**: 281.

Plant rather succulent, glabrous. Stems up to 30 cm, thick, procumbent below, ascending above. Leaves 1–3.5(–6) × 1–2.5(–4) cm, ovate to obovate, obtuse, cuneate to subtruncate at base, coarsely sinuate-crenate, dark green; petiole up to 2 cm, often winged. Cyme few-flowered, shorter than leaves. Sepals in unequal opposite pairs, 1.5–3.0 × 4.3–5.5 or 1.8–3.4 mm, reniform, patent. Stamens much shorter than sepals; filaments *c.*0.5 mm. Capsule 4–7 mm; seeds *c.*1.2 mm, ovoid-globose, tuberculate, brown. *Flowers October to March*. (Fig. 112).

Wet, usually shaded sites in most associations. 0–600 m.

S. & W. Fuegia. (Map 13b).
Chile north to *c.*52°S.lat.

5. *Tribeles* Phil.

Glabrous, creeping perennial herbs, woody at base, rooted at intervals. Leaves alternate. Flowers solitary, axillary. Sepals 5, connate at base into short tube. Petals 5. Stamens 5. Disc absent. Ovary superior, 3-celled; style 1; stigma 3-lobed. Fruit a capsule, 3-grooved, dehiscing in 3 valves; seeds numerous.

1. *T. australis* Phil. 1864–65, *Linnaea* **33**: 307.
Chalepoa magellanica Hooker f. 1871, *Icon. Pl.* t.1082 – Hariot, 1884.

Stems up to 40 cm, prostrate to ascending. Leaves 6–12 × 2.6–4.5 mm, oblanceolate to obovate-spathulate, 3(–5)-dentate at apex, attenuate at base, pale green, sessile or with wide petiole 2–3 mm. Flowers sessile. Calyx *c.*1.7 mm, the lobes *c.*1.2 × 1.1 mm, triangular to broadly ovate, obtuse. Petals 4.5–5.5 mm, ovate-oblong, obtuse, white, yellowish at base. Stamens ½ as long as petals; anthers small. Capsule *c.*3.5 × 2.7–3 mm, suborbicular, reddish brown; seeds *c.*1 mm, obovoid-globose, dark brown. *Flowers November to January*. (Fig. 111).

Swamps and bogs, wet open soils in other communities; 0–650 m.

S. & W. Fuegia. (Map 13c).
Chile north to 40°S.lat.

6. *Ribes* L.

Deciduous shrubs. Leaves alternate. Flowers in pendent racemes. Sepals 5, free. Petals 5. Stamens 5, antisepalous. Ovary inferior, 1-celled; styles 2, connate at base; stigma simple. Fruit a berry; seeds numerous.

1. *R. magellanicum* Poiret 1812, *Encycl. Méth. Bot.* **10**: 856.

Erect shrub up to 4 m, with reddish brown fissured bark on older branches, the twigs pubescent. Leaves (1–)1.8–5(–6.7) × (1.5–)2–5(–9) cm, 3-lobed, truncate to cordate at base, the lobes triangular-ovate, obtuse to subacute; irregularly serrate-dentate, glabrous or sparsely hairy above, pubescent on veins beneath; petiole 8–30(–40) mm, pubescent. Raceme 3.5–15 cm, with 20 or more yellow to reddish flowers; pedicels *c.*0.5–2 mm, pubescent; bracts alternate, 2.5–6 mm, lanceolate. Receptacle campanulate. Sepals 1.2–2 mm, triangular-ovate, reflexed, rarely with sessile, brown glands, Petals shorter than sepals, suborbicular. Berry 3.5–4.5 mm in diameter, globose, purple at maturity; seeds *c.*2 mm, trigonous. $2n = 16$. *Flowers October to December*. (Pl. 6b).

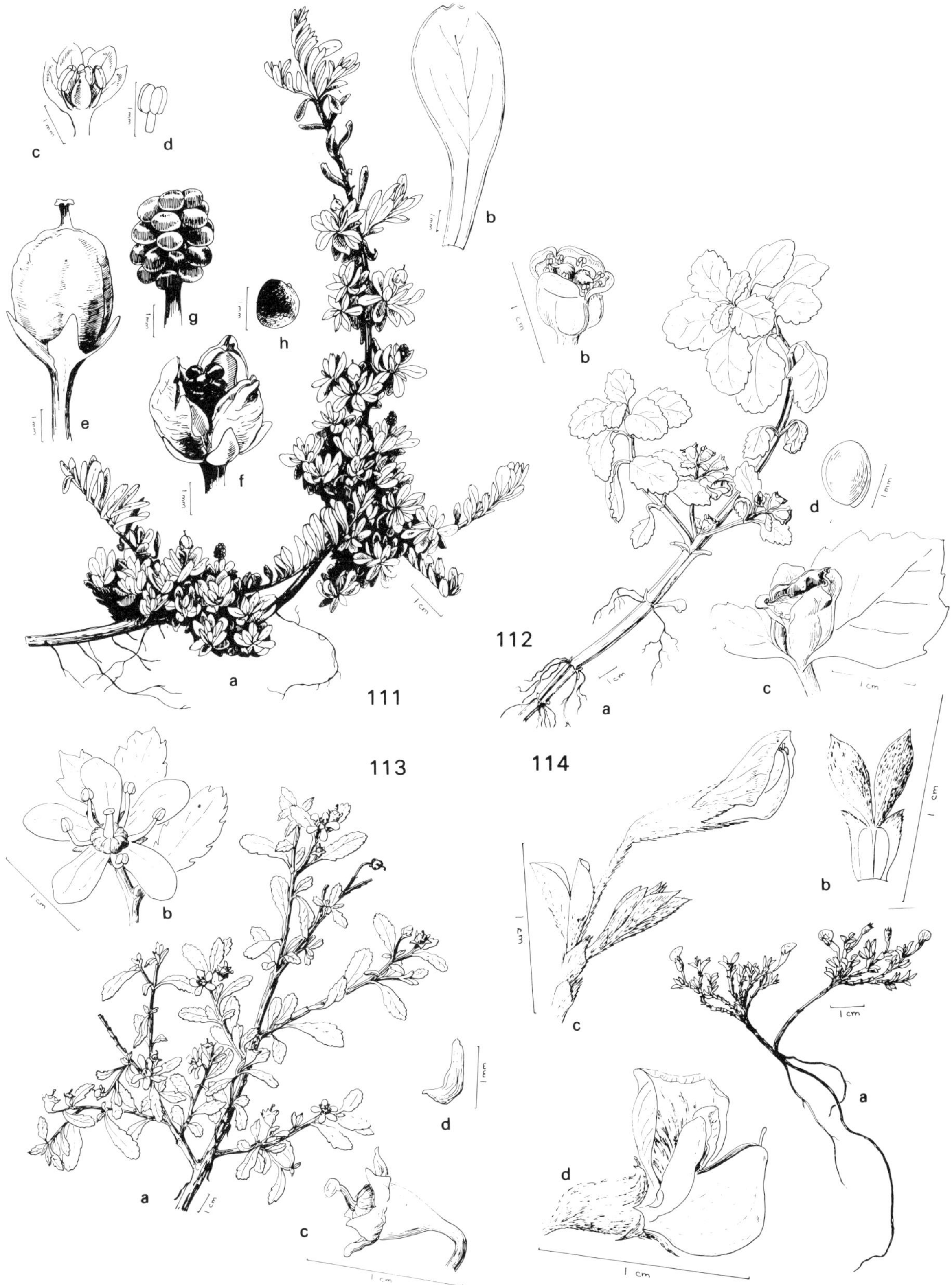

Fig. 111. *Tribeles australis*. a, Plant. b, Leaf. c, Flower. d, Stamen. e, f, Capsule. g, Seeds on placenta. h, Seed. Scale = 1 cm (a) or 1 mm (b–h). Fig. 112. *Chrysosplenium macranthum*. a, Plant. b, Flower. c, Capsule. d, Seed. Scale = 1 cm (a–c) or 1 mm (d). Fig. 113. *Escallonia serrata*. a, Plant. b, Flower. c, Capsule. d, Seed. Scale = 1 cm (a–c) or 1 mm (d). Fig. 114. *Adesmia lotoides*. a, Plant. b, Leaf. c, d, Flower. Scale = 1 cm.

(R. N. P. Goodall)

Open scrub, often with *Berberis buxifolia*, in forest clearings and wood margins. 0–300 m.

Throughout Fuegia except the extreme north and west. (Map 13d).
W. Argentina and Chile north to *c.*40°S.lat.

7. *Escallonia* Mutis ex L. f.

Evergreen shrubs. Leaves alternate. Flowers solitary and axillary or in short terminal racemes. Sepals 5. Petals 5. Stamens 5. Ovary inferior, 1-celled; style 1; stigmas capitate. Fruit a capsule; seeds numerous.

Literature: H. Sleumer, *Verh. K. Ned. Akad. Wet. Afd., Natuurk* **58**: 1–146 (1968).

1. ***E. serrata*** Sm. 1790, *Pl. Icon. Ined.* **3**: 31.

Celastrus venustus Banks & Sol. ex Hooker f. 1847, *Fl. Antarct.* **1(2)**: 279.

Erect shrub 30–50(–80) cm, often appressed to vertical rocks; branches contorted, greyish or whitish, somewhat terete, the twigs angular, glabrous, reddish-brown. Leaves (6–)8–17(–23) × 3–9 mm, obovate-cuneate to -spathulate, obtuse to subacute, deeply 5– to 13– dentate on each side, subsessile, somewhat glaucous, usually glabrous. Flowers solitary, axillary or terminal; pedicels 1–2 mm, glabrous. Hypanthium *c.*2 mm, turbinate. Sepals 1.5–2 mm, triangular, acute, patent or reflexed. Petals 5–7 × 2.5–3 mm, oblong-obovate, white. Disc shallowly 5-lobed. Capsule 3–4 mm, turbinate, 5-ribbed. *Flowers November to February*. (Figs. 94, 113, Pl. 6a).

Wet cliffs, rocks and on shallow peats, often in coastal scrub, sometimes in *Marsippospermum-Empetrum* swamp-heath. 0–120 m.

S. Fuegia except the extreme west. (Map 13e).
W. Argentina north to *c.*40°S.lat.
W. Chile north to *c.*44°S.lat.

E. virgata (Ruiz & Pavón) Pers. 1805, *Syn. Pl.* **1**: 234, has been recorded once from Isla Grande (Skottsberg, 1926), probably between Lago Fagnano and Ea Remolino (Moore, 1974), It is readily distinguished from *E. serrata* by its leaves, which are entire or serrulate towards the apex, and its 6– to 15– flowered terminal racemes in which at least the lower flowers are bracteolate.

XXXVII. LEGUMINOSAE

Shrubs or herbs. Leaves alternate, usually pinnate; stipules usually present. Flowers hermaphrodite, zygomorphic. Sepals 5, ± united at base, subequal. Petals 5, free or somewhat united, the dorsal upright (standard), enclosing the 2 lateral (wings) in bud, the 2 lower ± united along their adjacent margins to form a keel. Stamens 10, all (monadelphous) or 9 (diadelphous) with filaments united into a tube around ovary, sometimes free. Ovary superior, 1-celled; style 1; stigma simple or ± capitate. Fruit a 2-valved, dehiscent pod or a ± indehiscent lomentum; seeds 2– numerous, usually without endosperm.

a. Shrub:
  b. Flowers yellow; leaves glabrous, glandular, not spiny — 1. ***Adesmia***
  b. Flowers orange-reddish; leaves silver-sericeous, spiny — 2. ***Anarthrophyllum***
a. Herb, often woody at base:
  c. Stamens free; fruit a lomentum, with constrictions between seeds — 1. ***Adesmia***
  c. Stamens diadelphous, 9 with filaments united into a tube around ovary, the other stamen free; fruit a dehiscent pod, not lomentaceous:
    d. Leaves with 3 leaflets:
      e. Fruit coiled 2–3 times in a spiral — 6. ***Medicago***
      e. Fruit straight or almost so — 7. ***Trifolium***
    d. Leaves with 1 or more pairs of leaflets:
      f. Leaves without tendrils:
        g. Racemes 3– to 6– flowered; calyx at least 4.5 mm — 3. ***Astragalus***
        g. Flowers solitary; calyx not more than 3.5 mm — 4. ***Vicia***
      f. Leaves with tendrils:
        h. Style pubescent at apex on upper side only — 5. ***Lathyrus***
        h. Style pubescent all round towards apex — 4. ***Vicia***

1. *Adesmia* DC.

Shrubs or perennial herbs, rhizomatous or forming dense cushions. Leaves 3-foliolate or pinnate, sometimes with terminal leaflet. Flowers solitary or almost so, axillary. Calyx campanulate, with 5 subequal teeth. Petals yellow or purplish. Stamens free. Stigma capitate or extrorse. Fruit a lomentum, with constrictions between seeds, laterally compressed, not or occasionally dehiscent; seeds 2–8, ± globose.

Literature: A. Burkart, *Darwiniana* **13**: 9–66(1964); **14**; 463–568(1967).

a. Shrub; leaves with at least 7 pairs of leaflets — 4. ***boronioides***

a. Herb; leaves with not more than 6 pairs of leaflets:
 b. Plant forming cushions; leaves fleshy **3. *salicornioides***
 b. Plant rhizomatous; leaves not fleshy:
  c. Leaves with 3–6 pairs of leaflets, sometimes also with terminal leaflet **1. *pumila***
  c. Leaves with 1 pair of leaflets, sometimes also with terminal leaflet **2. *lotoides***

1. *A. pumila* Hooker f. 1847, *Fl. Antarct.* **1(2)**: 255.

Rhizomes up to 12 cm, slender, much branched, procumbent, covered with dead stipules and leaf-remains above. Leaves 10–30(–45) × 5–10(–13) mm, with 3–5(–6) pairs of leaflets and sometimes a terminal leaflet; leaflets obovate to obtriangular, obtuse or mucronate, cuneate at base, sparsely hairy, ciliate-dentate; petiole *c.*5 mm, white appressed-hairy, with basal amplexicaul stipules 5–8 mm. Peduncles up to 20 mm, solitary, 1-flowered. Calyx 5.5–8 mm, the teeth 2.5–4 × 0.8–1.5 mm, subulate, hairy. Standard 8.5–11 × *c.*8 mm, the limb suborbicular, the claw *c.*3 mm, yellow, purple dorsally, glabrous; wings *c.*6.5 mm, oblong; keel 9–10 mm, glabrous. Filaments up to 10 mm. Fruit 8–14 mm, with 6–8 constructions, linear, pubescent; seeds *c.* 2.5 mm in diameter, brownish black. *Flowers November to February.*

Open soil and, especially, sand, particularly by coasts, open grassland and *Lepidophyllum* scrub; 0–60 m.

N., E. C. & locally, C. S. parts of Isla Grande. (Map 13f).
Argentina north to *c.*47°S.lat.
E. Chile north to *c.*51°S.lat.

2. *A. lotoides* Hooker f. 1847, *Fl. Antarct.* **1(2)**: 255.

Rhizomes up to 7.5 cm, slender, much branched, ascending, often covered with dead stipules and leaf-remains above. Leaves with 2(3) leaflets, sessile, with basal amplexicaul stipules 2.5–5 mm; leaflets 2–6(–9.2) × 1.5–3.5(–4) mm, elliptical to obovate, cuneate at base, acute to mucronate, entire, sericeous above and beneath. Peduncles 5–20 mm, solitary, 1–flowered. Calyx 5.5–7(–8) mm, the teeth 2–4(–5) × 1–1.5 mm, subulate, sericeous. Standard 8–11 mm, the limb orbicular, yellow with brown veins, glabrous; wings 6–8 mm, ± spathulate; keel 6.5–8 mm, somewhat ciliate. Filaments up to 10 mm. Fruit 10–16 mm, with 2–6 constrictions, sericeous; seeds *c.*2 mm in diameter. *Flowers November to February.* (Fig. 114).

Open soil and sand, coastal scrub; 0–40 m.

N. and, locally, C. S. Isla Grande. (Map 13g).
W. Argentina north to c.41°S.lat.
E. Chile north to 50°40′S.lat.

3. *A. salicornioides* Speg, 1899, *Anal. Soc. Ci. Argent.* **47**: 274.
A. carnosa Dusén 1900, *Svenska Exped. Magellansl.* **3(5)**: 159.

Stems thick, woody, much branched above, covered with dead stipules and leaves above, forming dense cushions or mats up to 23 cm in diameter. Leaves with 3 (the uppermost sometimes with 1) leaflets, fleshy, glabrous, the petiole *c.* 1 mm, with basal fleshy stipules having sessile apical glands on lobes; leaflets 2–5 × 0.8–1.2 mm, obovate-cylindrical to -navicular, obtuse, with sessile apical gland. Peduncles *c.*2 mm, solitary, 1(–2)– flowered. Calyx 4–5 mm, the teeth *c.*2 mm, subacute, ± glabrous. Standard 8–11 mm, the limb orbicular, emarginate, yellow, somewhat strigulose dorsally; wings and keel shorter, glabrous. Filaments up to 7 mm. Fruit 6–12 mm, with 2–4 constrictions, glabrous or with whitish glandular bristles; seeds *c.*2.5 mm in diameter. *Flowers November to January.*

Sands and shingles, rock outcrops, grass-steppe, *Empetrum* heath; 0–150 m.

C. & N. Isla Grande. (Map 13h).
W. Argentina north to *c.*46°S.lat.
E. Chile north to *c.*50°41′S.lat.

4. *A. boronioides* Hooker f. 1847, *Fl. Antarct.* **1(2)**: 257.

Erect shrub, up to 1 m; stems branched. Leaves with 15–40 leaflets, glabrous, punctate-glandular, the petiole up to 10 mm, with inconspicuous stipules; leaflets 4–6 × 4–5.5 mm, obcordate, crenate, sessile, the lobes sometimes with apical gland. Inflorescence of 10– to 30–flowered, lateral or terminal racemes; pedicels 3–3.5 mm. Calyx 4–4.5 mm, the teeth *c.* 1 mm, triangular, sparsely pubescent. Standard 6–8 mm, the limb obcordate, yellow; wings and keel *c.*5 mm. Filaments up to 6 mm. Fruit 15–18 mm, with 1–5 constrictions, with black glands; seeds 2–2.5 mm in diameter. *Flowers November to January.*

Coastal steppe, with *Chiliophyllum*: 10–30 m.

N. Isla Grande. (Map 13i).
Argentina north to *c.* 48°S.lat.
E. Chile north to *c.*51°S.lat.

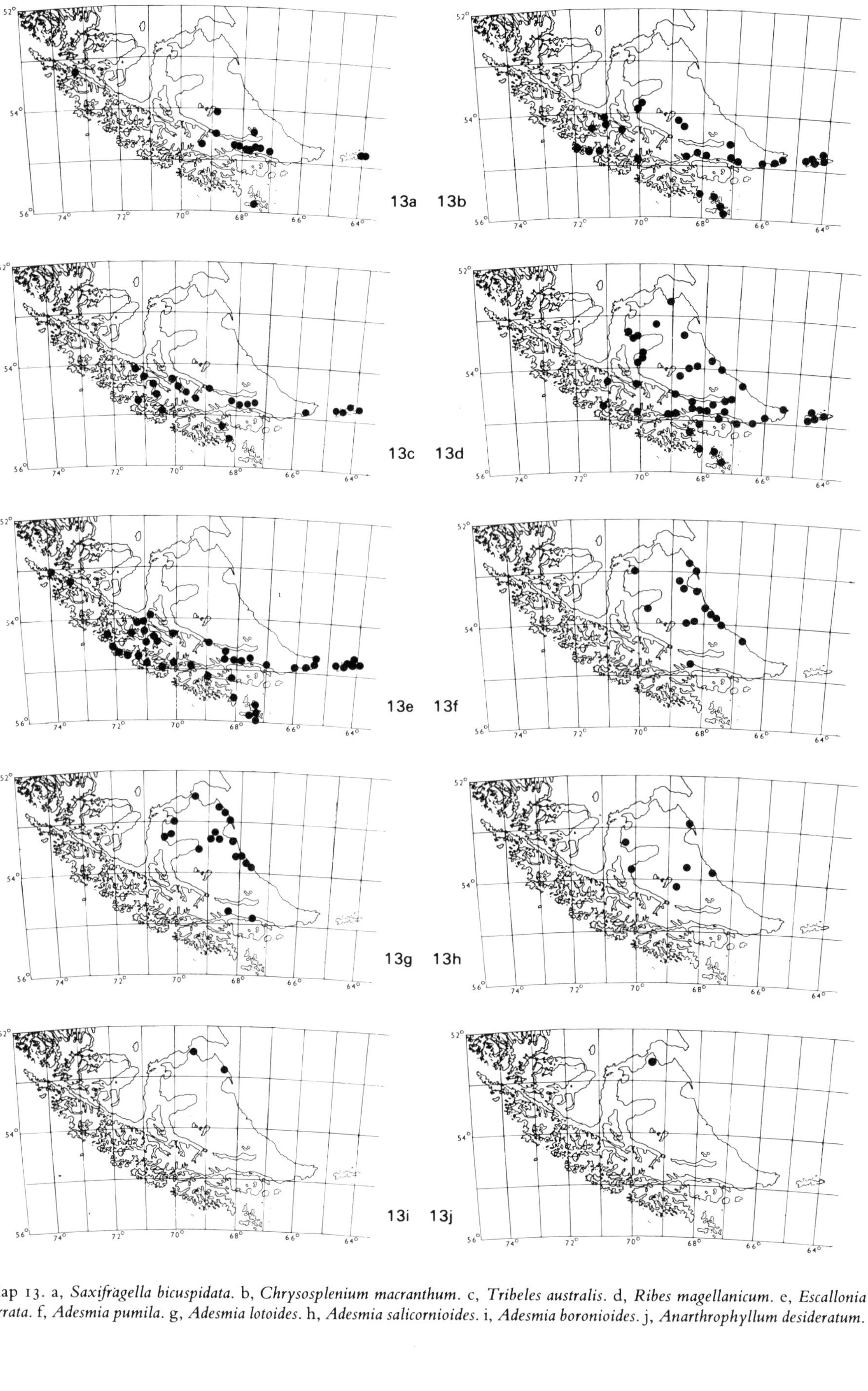

Map 13. a, *Saxifragella bicuspidata*. b, *Chrysosplenium macranthum*. c, *Tribeles australis*. d, *Ribes magellanicum*. e, *Escallonia serrata*. f, *Adesmia pumila*. g, *Adesmia lotoides*. h, *Adesmia salicornioides*. i, *Adesmia boronioides*. j, *Anarthrophyllum desideratum*.

2. *Anarthrophyllum* Bentham ex Bentham & Hooker

Spiny shrubs. Leaves 3-foliolate; stipules present. Flowers solitary, axillary or terminal. Calyx 2-lipped; lower lip conspicuously 2-dentate; upper lip conspicuously 3-dentate. Petals orange-reddish. Stamens monadelphous. Stigma filiform. Fruit a 2-valved pod; seeds numerous.

1. *A. desideratum* (DC.) Bentham ex Jackson 1893, *Index Kew.* **1**: 117.

Plant forming ± lax cushions up to 40 cm high; stems procumbent to erect, much branched, sericeous above. Leaves glaucous, silver-sericeous; leaflets 4–14 × *c.*1 mm, linear, acute, with apical spine *c.*1 mm, rigid; petiole 1–2 mm; stipules *c.*5 × 0.8 mm, subulate, brownish. Peduncle 4–6 mm. Calyx 8–14 mm, brown, silver-sericeous; lower lip 7–9 mm, the teeth 4.5–6 mm, ovate, acuminate; upper lip 8–14 mm, the teeth 3–5 mm, narrowly triangular, acute to acuminate. Standard 5–6 × 4–6 mm, obovate to suborbicular, obtuse; wings 10–15 mm, oblong-oblanceolate; keel 9–12 mm. Filaments up to 13 mm. Fruit 18–25 × 5–8 mm. oblong, acuminate, sericeous; seeds 1.5–2 × *c.*1 mm, dark brown, smooth. (Fig. 115).

Grass steppe and dwarf shrub heath, especially over rocks; 0–20 m.

Northernmost Isla Grande; recorded once from Punta Anegada (Spegazzini, 1896). (Map 13j).
Argentina north to *c.*45°30′S.lat.
E. Chile north to *c.*50°42′S.lat.

Ulex europaea L. 1753, *Sp. Pl.* 741, introduced from Europe as a garden ornamental, may escape and be locally naturalized. It is a shrub with the 3 leaflets of the glabrous leaves reduced to spines on mature plants; the petals are yellow and the hairy, 2-valved pod dehisces explosively.

3. *Astragalus* L.

Perennial herbs. Leaves imparipinnate; stipules present. Flowers in axillary racemes. Calyx tubular, with 5 subequal teeth. Petals pale purple. Stamens diadelphous. Stigma filiform. Fruit a 2-valved pod, inflated at maturity; seeds numerous.

Literature: I. M. Johnston, *Journ. Arn. Arbor.* **28**: 336–409(1947); E. Gomez-Sosa, *Darwiniana* **22**: 313–376(1979).

1. *A. palenae* (Phil.) Reiche 1897, *Anal. Univ. Chile* **97**: 542.
A. brevicaulis Dusén 1900, *Svenska Exped. Magellansl.* **3(5)**: 158.

Plant with woody rootstock *c.*6 mm in diameter, bearing numerous leafy branches at apex; branches up to 25 cm, hairy. Leaves up to 70 × 12 mm, oblong in outline, densely hairy, with 5–8 pairs of opposite or subopposite leaflets; stipules amplexicaul, membranous; leaflets 2.5–7.5 × 1–4 mm, elliptic-oblong to obcordate, retuse, shortly petiolulate, glabrous above, hairy beneath. Racemes 3– to 6– flowered; peduncles (10–) 25–85 mm; pedicels 2–3 mm. hairy; bracteoles 2–3.7 mm, narrowly ovate, acute. Calyx 4.5–7 mm, with appressed black hairs; lobes 2.7–3.5 mm, triangular, acute. Standard *c.*10×4 mm, spathulate, retuse; lateral petals *c.*9.5 mm, obtuse, with spur *c.* 1 mm; keel *c.*9 mm, emarginate. Fruit *c.*24 × 12–15 mm, broadly ellipsoid, cordate; seeds 2–2.4 mm, reniform. $2n = 26$. *Flowers November to January*. (Fig. 116).

Open sand and soil, especially near coast; 0–60 m.

N. E. Isla Grande. (Map 14a).
W. Argentina north to *c.*40°S.lat.
Chile north to *c.*39°S.lat.

4. *Vicia* L.

Climbing or scrambling, rhizomatous, perennial herbs. Leaves pinnate, usually with apical simple or branched tendril; stipules present. Flowers solitary or in small axillary racemes. Calyx campanulate, with 5 unequal teeth, the lower longest. Petals purplish blue to whitish. Stamens diadelphous. Style pubescent all round towards apex; stigma somewhat capitate. Fruit a 2-valved pod; seeds numerous.

Literature: D. M. Moore and B. Scotter, *Darwiniana* **20**: 371–386(1976); R. N. Giangualani, *Parodiana* **1(2)**: 323–339 (1982).

a. Leaf-tendrils usually long, simple to 3-fid; peduncles usually at least 5 mm; racemes (1) 2– to 4– flowered **2. *magellanica***
a. Leaf-tendrils usually short and spike-like, or absent, simple; peduncles not more than 3.1 mm; flowers solitary **1. *bijuga***

1. *V. bijuga* Gillies ex Hooker & Arn. 1833, *Bot. Misc.* **3**: 197.
V. solisii Phil. 1865, *Anal. Univ. Chile* **2**: 326 – Dusén, 1900.
V. saffordii Phil. 1895, *Anal. Univ. Chile* **84**: 266 – Skottsberg, 1926.

Stems up to 60 cm, glabrous or pubescent. Leaves with 1–5 pairs of leaflets, sparsely to densely hairy;

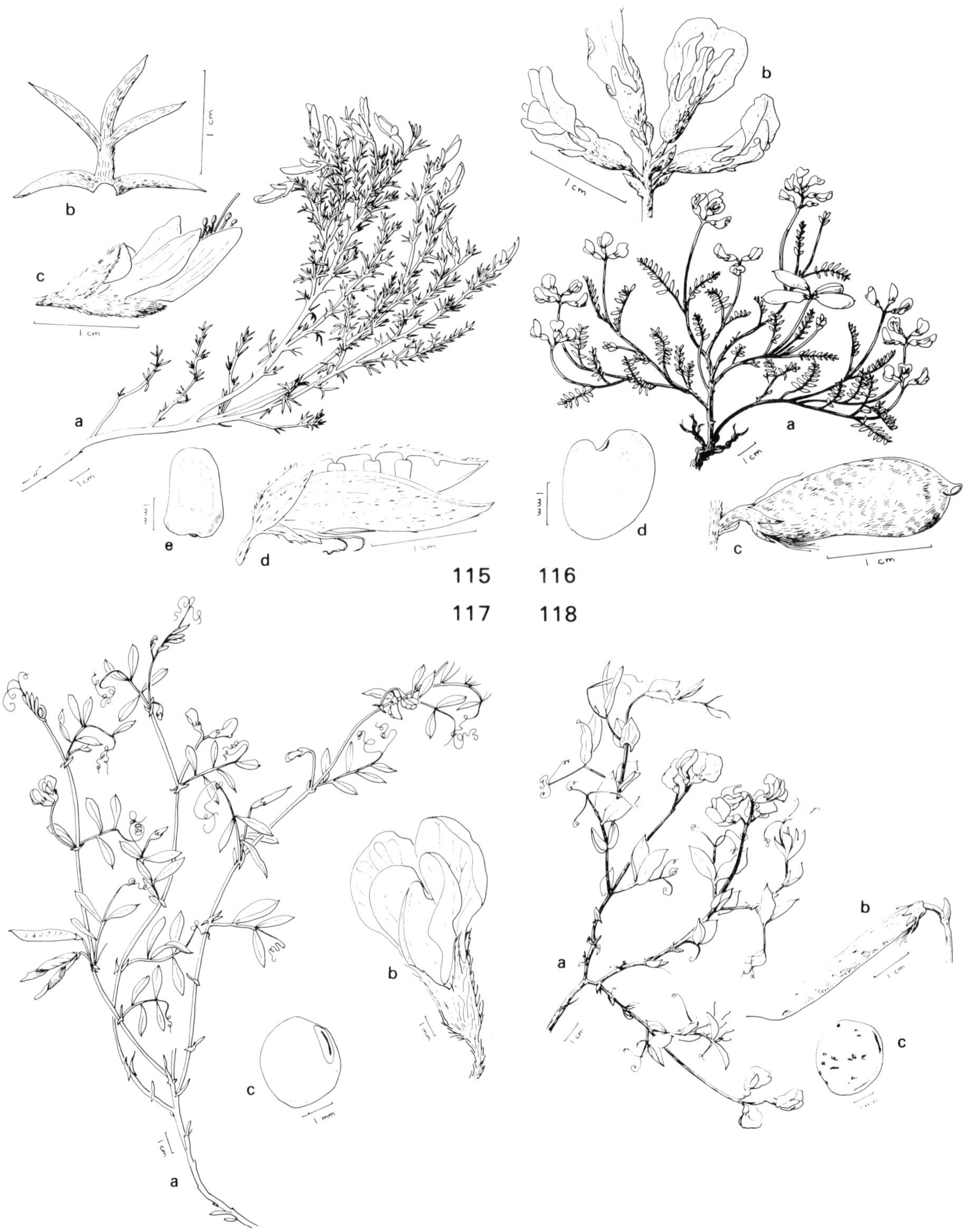

Fig. 115. *Anarthrophyllum desideratum*. a, Plant. b, Leaf. c, Flower. d, Fruit. e, Seed. Scale = 1 cm (a–d) or 1 mm (e). Fig. 116. *Astragalus palenae*. a, Plant. b, Inflorescence. c, Fruit. d, Seed. Scale = 1 cm (a–c) or 1 mm (d). Fig. 117. *Vicia magellanica*. a, Plant. b, Flower. c, Seed. Scale = 1 cm (a) or 1 mm (b–c). Fig. 118. *Lathyrus magellanicus*. a, Plant. b, Fruit. c, Seed. Scale = 1 cm (a–b) or 1 mm (c).

(R. N. P. Goodall)

petiole 2.4–10.1 mm; stipules 1.6–4.6 mm, semi-sagittate, often with lateral lobes; leaflets 3.5–8 (–11) × 0.5–2.1 mm, linear to oblong-elliptical or oblanceolate, acuminate to subobtuse and mucronate, entire; tendrils simple, short and spike-like or absent, rarely rather long. Flowers solitary; peduncle 0.5–3.1 mm. Calyx 2.2–3.5 mm; teeth 0.8–1.2 mm, triangular. Standard 4.9–9 mm, obovate, emarginate; lateral petals and keel 5–7 mm. Fruit 12–25 × 3–6 mm, glabrous to hairy, seeds *c.*2 × 1.5 mm, oblong-globose, smooth, black. $2n = 14$. Self-compatible, at least facultatively self-pollinated. *Flowers November to January.*

Coastal sands and gravels, open grassland, open *Chiliotrichum* and *Lepidophllum* scrub; 0–50 m.

C. & N. Isla Grande. (Map 14b).
W. Argentina north to *c.* 34°S.lat.
E. Chile north to *c.*36°S.lat.

2. *V. magellanica* Hooker f. 1847 *Fl. Antarct.* **1(2)**: 257.
V. kingii Hooker f. 1847, *Fl. Antarct.* **1(2)**: 258 – Franchet, 1889; Alboff, 1896; Skottsberg, 1916; Roivainen, 1954.
V. patagonica Hooker f. 1847, *Fl. Antarct.* **1(2)**: 259 – Spegazzini, 1896; Dusén, 1900; Skottsberg, 1916.
V. graminea auct., non Sm. – Franchet, 1889; Spegazzini, 1896.
V. bijuga auct., non Gillies ex Hooker & Arn. – Alboff, 1896.

Stems up to 60 cm, sparsely hairy. Leaves with 1–5 pairs of leaflets, sparsely hairy; petiole 5.3–33 mm; stipules 2.3–8.2 mm, semi-sagittate, often with lateral lobes; leaflets (6.75–) 10–31.6 × (0.9–) 1.25–3.6 mm, linear to obovate-oblong, acuminate to retuse or mucronate, sometimes emarginate with 2 teeth, entire; tendrils simple or 2– to 3– fid, long, rarely rather short. Flowers 2–4 in racemes, or solitary; peduncle (1.5–) 5.5–28(–41) mm. Calyx 2.7–5 mm; teeth 1–2 mm, linear-triangular. Standard 6.2–12.8 mm, obovate, emarginate; lateral petals 5–10 mm; keel 5–9 mm. Fruit 25–40 × 4–6 mm, glabrous; seeds *c.*2 × 1.8 mm, reniform-globose. $2n = 28$. Self-compatible, at least facultatively self-pollinated. *Flowers November to February.* (Fig. 117).

Open scrub, forest margins and clearings, grassland, coastal sands and gravels; 0–250 m.

Isla Grande except extreme west and east; S shore of Canal Beagle. (Map 14c).
W. Argentina north to 35°S.lat.
Chile north to *c.*30°S.lat.

5. *Lathyrus* L.

Scrambling, rhizomatous perennial herbs. Leaves with 1 pair of leaflets and apical 3– to 5– fid tendril; stipules present. Flowers in axillary racemes, rarely solitary. Calyx campanulate, with 5 unequal teeth, the lower longest. Petals bluish to purple. Stamens diadelphous. Style pubescent on upper side only; stigma capitate or somewhat 2-lobed. Fruit a 2-valved pod; seeds numerous.

Literature: A. Burkart, *Rev. Fac. Agron. Vet. Bs Aires* **8**: 41–127 (1935).

a. Leaflets and stipules with thin, membranous margin **1. *magellanicus***
a. Leaflets and stipules with thickened margin **2. *nervosus***

1. *L. magellanicus* Lam. 1788, *Encycl. Méth. Bot.* **2**: 708.
L. patagonicus Hauman 1913, *Anal. Mus. Nac. Bs Aires* **24**: 399, *nom. nud.*

Stems up to 45 cm, procumbent to erect, tetragonous, glabrous or sparsely hairy when young. Leaves with tetragonous petiole 1.5–10(–17) mm; stipules (2.5–)7–14(–22) × (1.5–)2–8(–10) mm, ovate to suborbicular, acute to subobtuse, with membranous margin, semi-sagittate, sparsely hairy beneath; leaflets (6–)10–30(–40) × 3–10 mm, elliptical, mucronate, with entire and membranous margin, glabrous or sparsely hairy at base; tendril 15–50 mm. Racemes (1) 2– to 7– flowered; peduncle 3–7 cm, tetragonous, glabrous. Calyx-tube 4–6 mm; teeth 1.5–2.7 mm, subulate. Standard (8.5–)10–14 mm, the limb orbicular, emarginate; lateral petals 6.5–10 mm; keel 8.5–10 mm, obtuse, incurved. Staminal tube 6–6.5 mm; free filaments *c.*4 mm, upcurved. Fruit 32–41 × *c.*5 mm, glabrous; seeds 2.5–3 mm, globose, smooth, black. $2n = 14$. *Flowers December to February.* (Fig. 118).

Coastal sands and gravels, open grassland, *Lepidophyllum* scrub; 0–50m.

C. & N. Isla Grande. (Map 14d).
W. Argentina north to *c.* 40°S.lat.
Chile north to *c.*45°S.lat.

A very variable species. Var. *magellanicus*, which turns black on drying, and the smaller var. *glaucescens* Speg., which frequently remains green on drying, have both been recorded from Fuegia and have the distribution given above, other varieties extend to N. W. and N. E. Argentina.

2. *L. nervosus* Lam. 1788, *Encycl. Méth. Bot.* **2**: 708.

Like *L. magellanicus* but plant completely glabrous, never turning black on drying, the leaflets and stipules with margin thickened by a continuous nerve. *Flowers December.* (Pl. 7c).

Open soil in *Chiliotrichum* scrub; *c.*100 m.

N. W. Isla Grande (53°21′S., 70°04′W.). (Map 14e).
Argentina north to *c.*37°S.lat.
Chile north to *c.*46°S.lat.
Uruguay. S. Brazil.

6. *Medicago* L.

Perennial herbs. Leaves 3-foliolate; stipules present. Flowers in axillary racemes. Calyx campanulate, with 5 equal teeth. Petals purple. Stamens diadelphous. Stigma subcapitate. Fruit a spiral pod, tardily dehiscent; seeds 1–10.

1. *M. sativa L. 1753, *Sp. Pl.* 778.

Stem up to 75 cm, erect or ascending, much branched, glabrous to hairy. Leaves with petiole up to 28 mm; stipules 10–19 mm, narrowly triangular, acuminate, dentate, united with petiole; leaflets 11–25 × 4–11 mm, elliptic-oblanceolate to obovate, truncate, cuneate at base, dentate towards apex, with petiolule up to *c.* 5 mm. Racemes 28–40 mm, 10– to 30-flowered; peduncles 15–35 mm. Calyx-tube 6–7.5 mm; teeth 3.5–4 mm, subulate-triangular. Corolla 5–11 mm. Fruit in a spiral of 2–3 turns, glabrous to pubescent; seeds 2–3.2 mm, reniform, greenish brown.

Disturbed ground near buildings, open sand and soil by sea.

Coastal sites of Isla Grande.
Introduced from Europe as a forage legume and locally naturalized in open or disturbed ground.

7. *Trifolium* L.

Annual or perennial herbs. Leaves 3–foliolate; stipules present. Flowers usually numerous in dense racemose heads. Calyx campanulate, with 5 subequal teeth. Petals white, yellow or purplish. Stamens diadelphous. Stigma capitate or subcapitate. Fruit an oblong to subglobose pod, indehiscent or opening by 2 valves or by the top falling off, ± enclosed in calyx; seeds 1–6.

a. Flowers yellow, not more than 7 mm:
  b. Upper leaves with all leaflets subsessile, their petiolules subequal; corolla 6–7 mm **3. *aureum***
  b. Upper leaves with terminal leaflet having distinct petiolule longer than that of lateral leaflets; corolla 2–3.5 mm **2. *dubium***
a. Flowers white, pink or purple, at least 8 mm:
  c. Stems creeping and rooted at nodes; heads all axillary; flowers white or pinkish **1. *repens***
  c. Stems not rooted at nodes; heads all terminal; flowers purple **4. *pratense***

1. *T. repens L. 1753, *Sp. Pl.* 767.

Glabrous perennial; stems up to 50 cm, creeping, rooted at nodes. Leaves palmately 3-foliolate; petiole 35–90 (–130) mm, erect; stipules 8–12 mm, ovate to oblong, shortly subulate, amplexicaul, membranous; leaflets 10–23 (–28) × 7–19(–24) mm, obovate to obcordate or suborbicular, retuse to truncate, denticulate. Heads 13–25 mm in diameter, globose, many-flowered, axillary; peduncles 6.5–20(–31) cm. Flowers white or pinkish; pedicels 0.8–3 mm. Calyx 3.8–6 mm, white with green veins; teeth 1.5–2.8 mm, narrowly triangular. Standard 9–11 mm, folded over pod. Fruit 4–5 mm, oblong, 3– to 6– seeded. *Flowers December to January.*

Grassland, open scrub, disturbed ground; 0–60 m.

Locally throughout Isla Grande and S. shore of Canal Beagle.
Introduced from Europe for forage and naturalized in many localities, especially near settlements.

2. *T. dubium Sibth. 1794. *Fl. Oxon.* 231.

Annual; stems 7–40 cm, procumbent or ascending to erect, slender, hairy. Leaves pinnately 3-foliolate; petiole 1–5.5 (–8) mm; stipules 2–5.2 mm, broadly ovate, acuminate; leaflets 4.2–9 × 3.2–6 mm, obcordate to obovate, the terminal with petiolule distinctly longer than that of lateral leaflets. Heads 4–6 (–8) mm in diameter, globose, (4–) 10– to 20–flowered, axillary; peduncles 1–1.5 cm. Flowers yellow, turning brown; pedicels 0.2–0.4 mm. Calyx 1.3–2 mm, green; teeth 0.6–1 mm, triangular. Standard 2–3.5 mm, folded over pod. Fruit 2.5–3 mm, ovoid, usually 1-seeded. *Flowers December to March (to May).*

Grassland, open scrub, disturbed ground; 0–40 m.

Locally in N.W. and C.S. Isla Grande.

Introduced from Europe, probably with grass seed, and naturalized, especially near settlements.

*3. ***T. aureum*** Pollich 1777, *Hist. Pl. Palat.* **2**: 344.

Similar to *T. dubium* but biennial and more robust; also distinguished by the upper leaves having all leaflets subsessile and with subequal petiolules, the heads (7–)12–17 mm in diameter, the standard 6–7 mm and the fruit usually 2-seeded. *Flowers February to March.*

Open, disturbed *Nothofagus* forests; 0–100 m.

Locally on S. coast of Isla Grande.
Introduced from Europe, probably with grass seed, and locally naturalized near settlements.

*4. ***T. pratense*** L. 1753, *Sp. Pl.* 768.

Perennial, ± pubescent; stems up to 32 cm, decumbent to erect. Leaves pinnately 3-foliolate; petiole 3–13(–23) cm; stipules 12–20 mm, ovate to oblong, acute, setaceous, whitish with green veins; leaflets 20–30(–45) × 12–19(–22) mm, elliptical to obovate, obtuse to emarginate. Heads 21–30 mm in diameter, globose to ovoid, many-flowered, terminal, sessile, subtended by a pair of shortly petiolate leaves. Flowers pinkish purple, sessile. Calyx 6.2–10 mm, green; teeth 3–6 mm, subulate. Standard 12.5–15 mm, folded over pod. Fruit *c.*2 mm, obovoid, with obtuse projections at one side, 1-seeded. *Flowers January to April.*

Open grassland and disturbed ground; 0–50 m.

N. Isla Grande.
Introduced from Europe, probably with grass seed, and locally naturalized near settlements.

XXXVIII. HALORAGACEAE

Aquatic or terrestrial perennial herbs. Leaves alternate or in whorls; stipules absent. Flowers hermaphrodite or, usually, unisexual, actinomorphic. Sepals 2 or 4. Petals 0 or 4. Stamens 2 or 8. Ovary inferior, 1- or 4-celled; stigmas sessile or subsessile. Fruit a nut or drupe.

a. Terrestrial plant; leaves alternate, ovate to orbicular, palmately lobed — **1. *Gunnera***
a. Submerged aquatic plant; leaves whorled, usually pinnate with capillary segments — **2. *Myriophyllum***

1. *Gunnera* L.

Dioecious terrestrial herbs. Leaves alternate, ovate to orbicular, palmately lobed. Flowers in a racemose panicle. Sepals 2, united at base. Petals 0. Stamens 2. Ovary 1-celled; stigmas 2, subsessile. Fruit a 1-seeded drupe.

Literature: A. M. Molina, *Darwiniana*, **21**: 473–489(1978).

a. Leaves shallowly lobed, crenate-dentate — **1. *magellanica***
a. Leaves deeply 5- to 10- lobed, the lobes entire or distantly crenate — **2. *lobata***

1. ***G. magellanica*** Lam. 1789, *Encycl. Méth. Bot.* **3**: 61.

Stems up to 36 cm, stout, prostrate, branched, rooting at nodes, with leaves and stolons produced at apex of stems and branches, and at nodes. Leaves with lamina 0.7–5.5 × 1.5–9 cm, reniform, cordate at base, crenate-dentate, with emarginate to mucronate teeth, sparsely hairy, more densely so on veins beneath and margin; petiole 1.5–20(–30) cm, hairy. Inflorescence (1–)1.5–12 cm; peduncle 1.6–8.5 cm, with reddish gold hairs. Male flowers with pedicels 1–2 mm, the female subsessile or with pedicels up to 0.5 mm. Sepals 1–3mm, united for *c.*$\frac{1}{2}$ of length, the lobes deltate. Filaments *c.*6 mm; anthers 1–1.5 mm, ellipsoid. Stigmas 2.8–6 mm, filiform. Drupe 3.3–4 × 2–3 mm, subglobose, scarlet. $2n = 34$*. *Flowers October to February.* (Pl. 6c).

Moist sheltered places in communities; 0–1000 m.

Throughout Fuegia. (Map 14f).
W. Argentina north to *c.*37°S.lat.
Chile north to *c.*35°30′S.
North along Andes to *c.*1°N.lat.
Falkland Islands (Islas Malvinas).

2. ***G. lobata*** Hooker f. 1847, *Fl. Antarct.* **1(2)**: 274.

Stems up to 21 cm, slender, prostrate, branched, rooting at nodes, with leaves and stolons produced at apex of stems and branches. Leaves with lamina 7–18(–24) × 7–20(–32) mm, suborbicular to reniform, deeply 5- to 10-lobed, the lobes emarginate, entire or distantly crenate, glabrous or with hairs on veins beneath; petiole 8–35 mm, densely reddish brown hairy. Inflorescence 7–10 mm; peduncle up to 30 mm. Male flowers with pedicels *c.*0.5 mm, the female 0.5–0.7 mm. Sepals 2.2–2.4 mm, united for *c.*2/3 of length, the lobes deltate. Filaments *c.*5 mm; anthers *c.*1.3 mm, ellipsoid. Stigmas 3.5–5 mm, linear-subulate. Drupe 3–4 mm in diameter, subglobose. *Flowers November to February.* (Fig. 120).

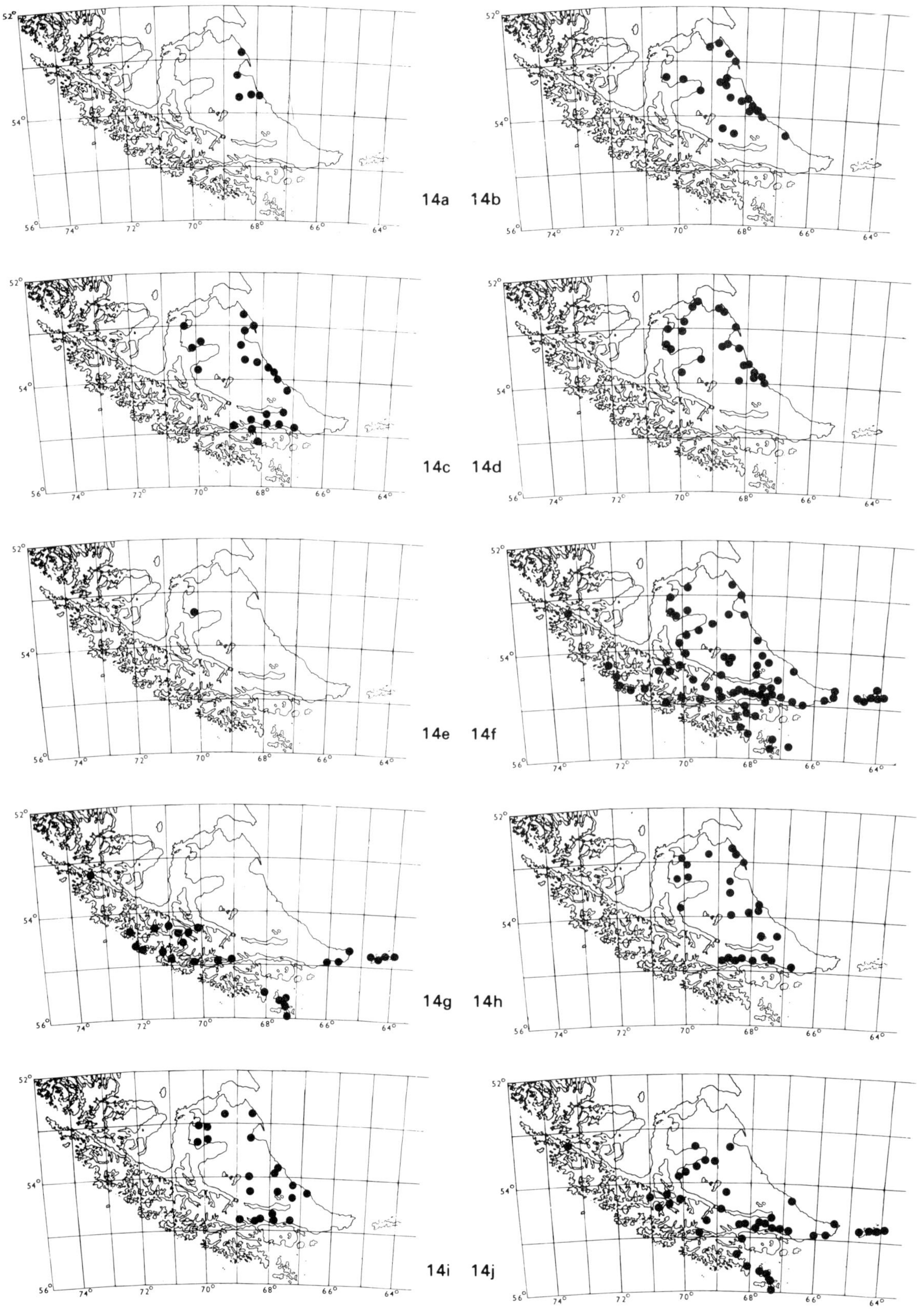

Map 14. a, *Astragalus palenae*. b, *Vicia bijuga*. c, *Vicia magellanica*. d, *Lathyrus magellanicus*. e, *Lathyrus nervosus*. f, *Gunnera magellanica*. g, *Gunnera lobata*. h, *Myriophyllum quitense*. i, *Hippuris vulgaris*. j, *Drapetes muscosus*.

a. *Misodendrum punctulatum.*

b. *Oxalis enneaphylla.*

c. *Lathyrus nervosus.*

d. *Calceolaria uniflora.* (Photos: D. M. Moore).

Plate 7.

a. *Nassauvia magellanica.*

b. *Nassauvia latissima.*

c. *Philesia magellanica.*

d. *Chloraea magellanica.* (Photos: D. M. Moore).

PLATE 8.

Sphagnum bogs, wet open bryophyte communities, Magellanic moorland; 0–300 m.

S. & W. Fuegia. (Map 14g).
Chile north to *c*.51°27′S.lat.

2. Myriophyllum L.

Glabrous, aquatic herbs. Leaves in whorls of (3)4, heteromorphic, pinnatisect and simple. Flowers in bracteate, terminal spikes, hermaphrodite, or usually, unisexual, male above, female below. Sepals 4, free. Petals 4, free. Stamens 8. Ovary 4-celled; stigmas 4, sessile or subsessile. Fruit a nut, separating into 4, 1-seeded nutlets (mericarps).

1. ***M. quitense*** Humb., Bonpl. & Kunth 1823, *Nov. Gen. Sp.* **6**: 89.
   M. elatinoides Gaudich. 1825, *Ann. Sci. Nat.* **5**: 105.

Stems 10–50 cm or more, rooting at lower nodes, branched. Submerged leaves 8–26 × 4–15 mm, pinnatisect, with 8–14 capillary segments, sessile or subsessile; uppermost (aerial) leaves 5.5–10.6 × 1.9–6 mm, oblong to elliptical or ovate to obovate, acute to obtuse, entire, sessile, often reddish; leaves serrate near water surface. Spike 3–5 cm. Bracts 3–8 × 1.5–2.5 mm, oblong-lanceolate, denticulate, reddish. Sepals *c*.0.5 mm, ovate-triangular, remotely denticulate. Petals *c*.2.5 × 1.3 mm, lanceolate, acute, cucullate, purplish, green towards base. Anthers *c*.2 mm, *c*.5 times as long as filaments. Fruit 1.2–1.3 × 1.5–1.8 mm; mericarps oblong, smooth, dark green or reddish, 1–2 often abortive. *Flowers November to January.*

Freshwater streams, ponds and lakes; 0–300 m.

Throughout Isla Grande. (Map 14h).
Argentina, especially in the west, north to *c*.35°S.lat.
Chile north to *c*.49°S.lat.
Andes north to 0°lat. Mexico.
Falkland Islands (Islas Malvinas).

XXXIX. HIPPURIDACEAE

Glabrous, perennial, aquatic herbs. Leaves whorled, simple; stipules absent. Flowers unisexual, hermaphrodite or sterile. Perianth reduced to a small rim around top of ovary. Stamen 1. Ovary inferior, 1-celled; style 1, simple, with stigmatic papillae along its length. Fruit a 1-seeded nut.

1. *Hippuris* L.

Flowers in leaf-axils on shoots emerging above surface of the water, the upper sessile, the lower usually shortly pedicellate.

1. ***H. vulgaris*** L. 1753, *Sp. Pl.* 4.

Plant with stout creeping rhizome, bearing erect shoots up to 43 cm. Leaves in whorls of 8–10, those on emergent shoots 10–20 × 0.9–2.5 mm, linear to linear-oblanceolate, acute, entire, sessile, the submerged leaves 28–40 mm, linear, thinner. Anther *c*.0.6 mm, reddish; filament *c*.0.8 mm. Style *c*.1.4 mm, slender. Nut 2–3 × 1–1.2 mm, ovoid, smooth, brown. $2n = 32$. *Flowers December to January.* (Fig. 122).

Freshwater lakes, ponds and slow-moving rivers; 0–300 m.

Throughout Isla Grande except the extreme west. (Map 14i).
Argentina north to *c*.47°30′S.lat.
Chile north to *c*.40°S.lat.
N. America. Eurasia.

XL. THYMELAEACEAE

Perennial herbs. Leaves opposite and usually connate at base, simple; stipules absent. Flowers hermaphrodite, actinomorphic, in small terminal capitula enclosed by 2 involucral bracts. Calyx tubular, 4-lobed. Petals 0. Stamens 4. Ovary superior, 1-celled. Style 1. Fruit a drupe with a thin, fleshy pericarp; seed 1, closely enveloped by endocarp, without endosperm.

1. *Drapetes* Banks ex Lam.

Plants usually forming lax cushions. Leaves sessile, imbricate. Calyx-lobes slightly spreading. Stamens inserted at mouth of calyx-tube, alternating with lobes. Stigma capitate, with long cylindrical papillae on surface and margins.

1. ***D. muscosus*** Banks ex Lam. 1792, *Jour. Hist. Nat. Paris* **1**: 188.

Cushions or mats 3–14 (–23) cm in diameter and 3–12(–13) cm high, sometimes compact; stems 3–7(–11) cm, prostrate or ascending, much branched above, rather woody, bearing roots towards base, densely covered with leaf-remains. Leaves 2.6–5.5 × 0.7–2.5 mm, elliptical to oblanceolate, obtuse, entire, somewhat cucullate, usually densely sericeous beneath, glabrous above. Involucral bracts 3–5.5 × 2–3 mm, ovate to spathulate, obtuse, densely sericeous above, green with purplish centre and margin. Inflorescence 3- to 6- flowered; peduncle short, elongating to 2.6–8.5 mm in fruit; pedicels 0.5–0.75 mm. Calyx-tube 1.5–2 × 0.2–0.3 mm, glabrous below, pubescent distally, white, the lobes c.1 mm, elliptic-lanceolate, pubescent, purplish. Fruit 1–2.1 × 0.4–1 mm, ovoid, brown, white-pubescent distally. $2n = 18$. Protandrous. *Flowers October to February*. (Fig. 130).

Empetrum heath, open scrub, *Sphagnum* bogs, Magellanic moorland, feldmark; (0–)40–800 m.

Throughout Fuegia, but less common in extreme west. (Map 14j).
W. Argentina and Chile north to *c.*47°50′S.lat.
Falkland Islands (Islas Malvinas).

XLI. MYRTACEAE

Shrubs. Leaves opposite, simple, evergreen; stipules absent. Flowers actinomorphic, hermaphrodite, in racemose inflorescences or solitary, with persistent, small, leaf-like bracteoles. Calyx with 4–5 lobes. Petals 4–5, free. Stamens 8–15, free. Ovary inferior, 3-celled, surmounted by a disc; style slender; stigma capitate. Fruit a capsule or berry; seeds numerous, with little or no endosperm.

a. Erect shrub; fruit a capsule **2. *Tepualia***
b. Prostrate, dwarf shrub; fruit a berry **1. *Myrteola***

1. *Myrteola* Berg

Dwarf shrubs. Flowers solitary. Calyx with 4 lobes. Petals 4. Stamens 8–12. Fruit a berry.

1. *M. nummularia* (Poiret) Berg 1856, *Linnaea* **27**: 397.
Myrtus nummularia Poiret 1798, in Lam., *Encycl. Méth. Bot.* **4**: 407 – Hooker, 1847; Franchet, 1889; Spegazzini, 1896.

Stems up to 40 cm or more, up to 4 mm in diameter, woody, prostrate, branched, rooted at intervals and intertwined to form often extensive patches, glabrous. Leaves with lamina 3–7 × 2–5 mm, elliptical to suborbicular, obtuse, cuneate at base, with revolute margin, glabrous, coriaceous, punctate; petiole 0.5–1 mm. Flowers axillary, towards ends of branches; peduncles 1–3 mm. Calyx-lobes 1.2–1.8 × 0.9–1.2 mm, oblong to ovate, obtuse, green or reddish, coriaceous, persistent in fruit. Petals 2.3–8.3 × 1.6–2.5 mm, obovate to oblanceolate, obtuse, white or creamy. Style 1.8–2.5 mm; stigma at same level as anthers. Berry 4–10 × 4–9 mm, globose to ellipsoid, pinkish to red; seeds 1.2–1.5 mm, reniform, somewhat compressed laterally, smooth, white. $2n = 44^{\star}$. Self-compatible. *Flowers November to February*. (Fig. 119).

Raised parts of bogs, especially with *Sphagnum*, wet *Empetrum*, *Pernettya* and *Nothofagus* scrub, feldmark, especially in bryophyte mats; 0–625 m.

S. Fuegia. (Map 15a).
W. Argentina north to *c.41*°S.lat.
Chile north to 39°S.lat.
Falkland Islands (Islas Malvinas).

2. *Tepualia* Griseb.

Erect shrubs. Flowers solitary or 2–3 in racemose clusters. Calyx with 5 lobes. Petals 5. Stamens 10–15. Fruit a loculicidal capsule.

1. *T. stipularis* (Hooker f.) Griseb. 1854, *Syst. Bermerk.* 31.
Metrosideros stipularis Hooker f. 1847, *Fl. Antarct.* **1(2)**: 275.

Stems up to 3 m, up to 18 mm in diameter, erect, much branched, glabrous. Leaves (3–)5–10 × (1.2–)3–6 mm, elliptical to ovate, subacute, cuneate at base, glabrous, coriaceous, punctate, with somewhat revolute margin; petiole up to 2 mm. Flowers axillary, towards ends of branches; peduncles 3–5 mm. Calyx-lobes 1–2 × 1–1.2 mm, triangular to triangular-ovate, acute, reddish, coriaceous, persistent in fruit. Petals 2–3 mm, obovate, obtuse, white with reddish veins. Style 5–5.5 mm; stigma at same level as anthers. Capsule 4–7 × 3–5 mm, obovoid-turbinate, glabrous; seeds small, linear, reticulate. *Flowers December to February*. (Fig. 121).

Margin and clearings of mixed *Drimys* – *Nothofagus betuloides* forest, among rocks in Magellanic moorland.

W. Fuegia (Isla Desolación). (Map 15b).
W. Argentina at c.41°S.lat.
Chile north to *c.*35°S.lat.

XLII. ONAGRACEAE

Herbs or shrubs. Leaves alternate, opposite or in whorls, simple; stipules absent. Flowers actinomorphic, hermaphrodite. Hypanthial tube present. Sepals 4, free. Petals 4, free. Stamens 8, in 2 whorls of

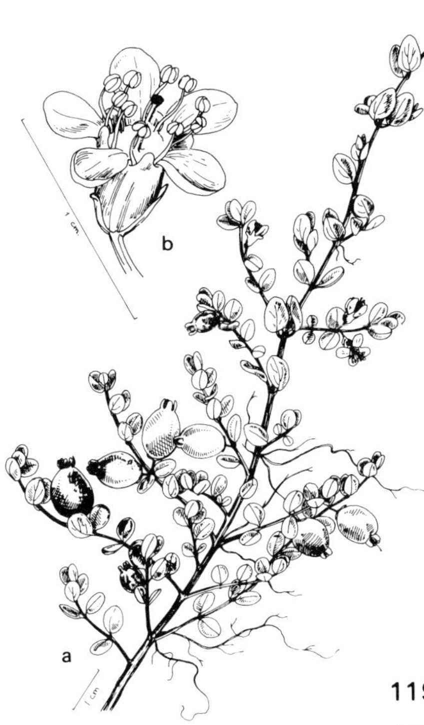

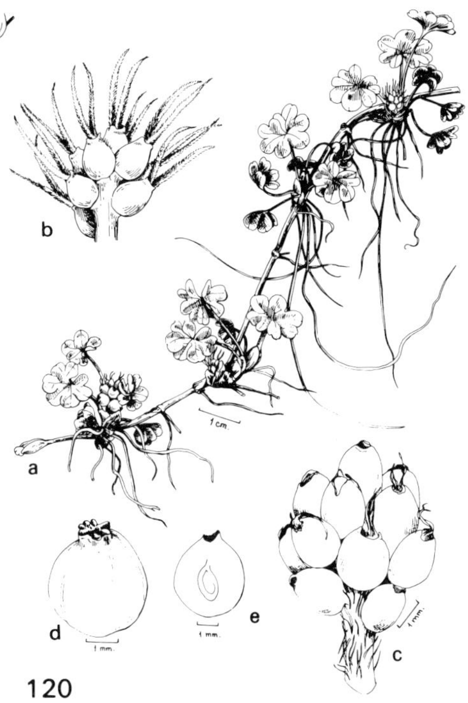

119 120

121 122

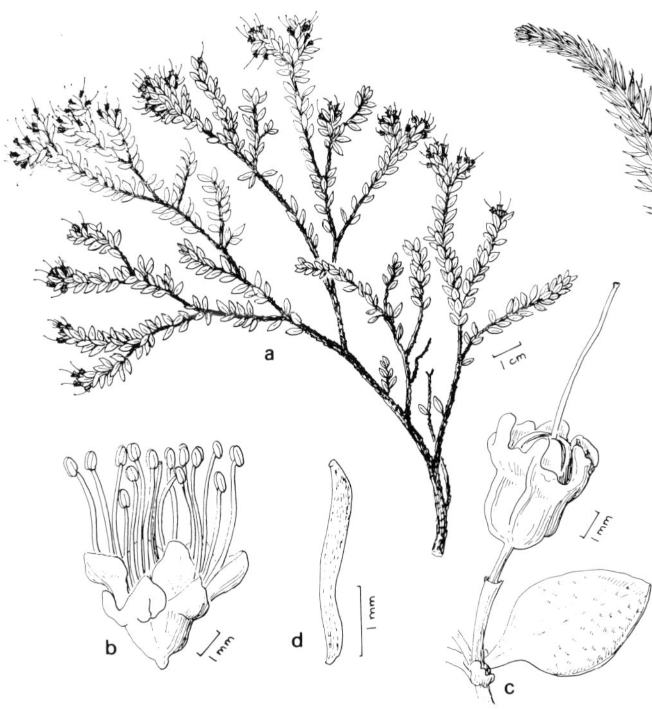

Fig. 119. *Myrteola nummularia*. a, Plant. b, Flower. Scale = 1 cm. Fig. 120. *Gunnera lobata*. a, Plant. b, Female inflorescence. c, Immature fruits. d, Fruit. e, Section of Fruit. Scale = 1 cm (a) or 1 mm (b–e). Fig. 121. *Tepualia stipularis*. a, Plant. b, Flower. c, Fruit. d, Seed. Scale = 1 cm (a) or 1 mm (b–d). Fig. 122. *Hippuris vulgaris*. Scale = 1 cm.
(R. N. P. Goodall)

4. Ovary inferior, 4-celled; style 1; stigma entire or 4-lobed. Fruit a berry or a loculicidal capsule; seeds numerous, without endosperm.

a. Shrub; fruit a berry **1. *Fuchsia***
a. Herb; fruit a capsule **2. *Epilobium***

1. *Fuchsia* L.

Shrubs. Leaves opposite or in whorls of 3, sometimes alternate near ends of branches. Flowers axillary, pendent; hypanthial tube long. Stigma entire. Fruit a berry; seeds glabrous.

1. *F. magellanica* Lam. 1788, *Encycl. Méth. Bot.* **2**: 565.
F. coccinea auct., non Aiton – Franchet, 1889; Spegazzini, 1896.

Stems up to 4 m, up to 25 cm in diameter near base, much branched. Leaves with lamina 10–55 × 6–18 mm, elliptic-ovate, acuminate, shallowly crenate-dentate. Flowers solitary; peduncles 2–5.5 cm. Hypanthial tube 5–10 mm, red. Sepals 12–20 × 4–8 mm, ovate-lanceolate, acuminate, red. Petals 6–20 × 4–7 mm, obovate, obtuse, violet. Berry 1–2 cm, black, $2n = 44$. Pollinated by hummingbirds. *Flowers December to April.* (Fig. 123).

Forest clearings and margins, especially in mixed evergreen deciduous *Nothofagus* forest and coastal *Drimys-Nothofagus* forest; 0–40 m.

S. & W. Fuegia except the outermost islands. (Map 15c).
W. Argentina north to *c.*40°S.lat.
Chile north to *c.*31°S.lat.

2. *Epilobium* L.

Perennial herbs. Leaves opposite, usually alternate above, simple. Flowers axillary, usually grouped into a ± discrete inflorescence; hypanthial tube short. Stigma capitate to clavate. Fruit a loculicidal capsule; seeds with a chalazal plume of hairs (coma).

a. Stems prostrate, rooted above and below flowers; flowers solitary, not in a discrete inflorescence **3. *conjungens***
a. Stems ascending to erect, rooted below flowers; inflorescence ± discrete, usually with numerous flowers:
b. Plant without rhizomes, with fleshy turions or rosettes of leaves produced at or below ground-surface; leaves lanceolate, distantly serrate; seeds with a distinct ridges of fused papillae **2. *ciliatum***
b. Plant with rhizomes, without turions or leaf-rosettes; leaves ovate, distantly denticulate; seeds papillose **1. *australe***

1. *E. australe* Poeppig & Hausskn. ex Hausskn. 1884, *Monogr. Epilobium* 269.
E. lechleri Phil. & Hausskn. ex Hausskn. 1884, *Monogr. Epilobium* 270 – Skottsberg, 1916.
E. antarcticum (Hooker f.) O. Kuntze 1898, *Revis. Gen.* **3**: 97.

Plant with rhizomes, with overwintering leafy shoots produced from rhizome or stem-base; stems (5–)15–35(–50) cm, ascending, with branches near base, sometimes above, pubescent. Leaves 10–30(–40) × (4–)6–14(–22) mm, ovate, acute to acuminate, rarely obtuse, cuneate to rarely sub-cordate at base, distantly and irregularly denticulate, with 4–8(–12) teeth on each side, glabrous beneath, pubescent at least on veins and margin above; petiole 0–2 mm. Inflorescence erect; flowers erect, subtended by leaves; pedicels up to 3 mm, up to 30 mm in fruit. Sepals 2.5–4 × 1.2–1.7 mm, lanceolate, sparsely ciliate, green or reddish. Petals 5–6 (–8.2) × 2.8–3.5 mm, obovate, emarginate, pale pink, the sinus 1–1.6 mm. Hypanthial tube 0.4–0.8 × 1.5–2.6 mm, with ring of erect hairs inside near base. Style 1.4–2 mm, equalling longer stamens. Capsule 33–50 mm, erect, rarely deflexed before dehiscence, sparsely pubescent to glabrous; seeds 1.1–1.4 mm, narrowly obovoid, papillose, the coma 4–7 mm, white to pale brownish. $2n = 36$. Self-compatible; self-pollinated. *Flowers December to March.*

Moist open areas, usually with running water; 0–800 m.

Almost throughout Isla Grande; S. side Canal Beagle; Isla de los Estados. (Map 15d).
W. Argentina north to *c.*33°30′S.lat.
E. Chile north to *c.*34°S.lat.

2. *E. ciliatum* Rafin. 1808, *Med. Repos.* **5**: 361.
E. tetragonum auct., non L. – Hooker, 1847.
E. cunninghamii Hausskn. 1879, *Österr. Bot. Zeitschr.* **29**: 118 – Samuelsson, 1923.
E. valdiviense Hausskn. 1879, *Österr. Bot. Zeitschr.* **29**: 118 – Alboff, 1896; Skottsberg, 1916; Samuelsson, 1923.
E. magellanicum Phil. & Hausskn. ex Hausskn. 1884, *Monogr. Epilobium* 271 – Franchet, 1889; Spegazzini, 1896; Autran, 1905; Skottsberg, 1916; Samuelsson, 1923.

Plant with overwintering fleshy turions or leaf-rosettes produced at or below ground surface; stems (10–)20–60(–90) cm, erect, simple or branched, pubescent, becoming glabrous below, often reddish or purplish above. Leaves 13–65 × 6–18 mm, lanceolate to narrowly ovate, acute to acuminate, obtuse to subcordate, rarely cordate, at base, distantly and irregularly serrate, with (8–)10–18 teeth on each side, glabrous or sparsely ciliate; petiole 0–3 mm. Inflorescence erect, rarely nodding; flowers erect, subtended by leaves which become smaller upwards; pedicels 2–5 mm, up to 12(–22) mm in fruit. Sepals 2.3–2.8 × 0.8–1.2 mm, lanceolate, green or reddish, pubescent, the hairs glandular and eglandular. Petals 3.5–4.5 × 1.8–2.5 mm, obovate, emarginate, pale pink, the sinus 1–1.5 mm. Hypanthial tube 0.9–1.5 × 1.5–2.2 mm, with ring of erect hairs inside. Style 1.5–2.5 mm, equalling or longer than stamens. Capsule 30–60 mm, erect, sparsely glandular- and eglandular-pubescent; seeds 1–1.4 mm, narrowly obovoid, with conspicuous rows of flattened, fused papillae, the coma 5–8 mm, white. $2n = 36$. Self-compatible; self-pollinated. *Flowers January to March.*

Moist disturbed areas, bogs; 0–200 m.

Almost throughout Isla Grande; S. side Canal Beagle. (Map 15e).
W. Argentina and Chile north to *c.*30°S.lat.
Falkland Islands (Islas Malvinas).
North America.

All South American material belongs to ssp. *ciliatum*, which has the natural distribution given above; it is also introduced in Eurasia and Australia-New Zealand.

3. *E. conjungens* Skottsb. 1906. *Wiss. Ergebn. Schwed. Südpolarexped.* **4(4)**: 24.

Stems up to 12 cm, prostrate, rooted above and below flowers, simple or branched, pubescent, especially at nodes. Leaves opposite; lamina 2.5–8.7 × 1.7–6.4 mm, broadly elliptical to orbicular, obtuse, rarely subacute, cuneate to truncate at base, entire or distantly and obscurely dentate, glabrous; petiole 1.3–2.3 mm, winged. Flowers solitary in leaf-axils, erect; pedicels (0–)2.5–7 mm, up to 22 mm in fruit, glabrous. Sepals 3.2–4 × 0.9–1.3 mm, lanceolate, glabrous. Petals 2.8–4.4 × 2.3–2.8 mm, obovate, emarginate, pale pink to white, the sinus *c.* 0.7 mm. Hypanthial tube 0.35–0.6 × 0.8–1.2 mm. Style *c.*1.2 mm, equalling stamens. Capsule 13–35 mm, glabrous; seeds 0.7–1.1 mm, narrowly obovoid, minutely papillose, the coma 4–6 mm, pale yellow-brown. Probably self-pollinated. *Flowers January to March.* (Fig. 124).

Moist areas at and above timberline; 450–1100 m.

S. & W. Fuegia. Endemic. (Map 15f).

XLIII. PROTEACEAE

Shrubs or trees. Leaves alternate, simple; stipules absent. Flowers hermaphrodite, zygomorphic. Perianth tubular, with 4 lobes having reflexed apices, and 2–4 scales alternating with the lobes. Stamens 4, inserted on perianth-lobes. Ovary superior, 1-celled. Style 1; stigma small, clavate. Fruit a follicle; seeds without endosperm.

1. *Embothrium* Forster & Forster f.

Flowers in pairs in terminal racemes having basal bracts, with cylindrical perianth-tube which splits at maturity. Anthers sessile in concave perianth-lobes. Ovary long stipitate. Fruit woody, stalked, with 2 rows of imbricate seeds.

1. *E. coccineum* Forster & Forster f. 1775, *Char. Gen. Plant.* 16.

Shrub or tree up to 12 m; bark reddish brown with numerous grey lenticels; young branches brown-pubescent, becoming grey and glabrous. Leaves (2–)2.5–5(–10) × (0.8–)1.8–3.8 cm, elliptical to subovate, cuneate at base, usually mucronate at apex, entire, coriaceous, glaucous beneath, shiny dark green above; petiole up to 14 mm, compressed. Racemes 4–7 cm, the peduncle 1–4 cm; pedicels (5–)8–14(–17) mm, orange; bracts 4–13 mm, oblong, acute, revolute, red, caducous. Perianth 18–30 mm, red; tube narrow; lobes ovoid, obtuse, keeled, concave. Anthers *c.*4 mm, lanceolate, purple. Ovary with short, truncate, nectariferous scale adnate to base. Follicles 18–35 mm, oblong, terete, tapering to persistent style 16–22 mm, yellowish green, later greyish brown; seeds compressed, with apical, oblong membranous wing. Cross-pollinated, normally by humming birds. $2n = 22$. *Flowers October to April.* (Fig. 126).

Chiliotrichum, *Pernettya mucronata* and *Berberis buxifolia* scrub, with dwarfed *Nothofagus* on better drained sites in bogs, forest-margins and clearings, locally dominant as open forest which was probably more widespread formerly; 0–150 m.

Throughout Fuegia, but rare or absent in N. and E. parts of Isla Grande. (Map 15g).
W. Argentina north to *c.*38°S.lat.
Chile north to 35°S.lat.

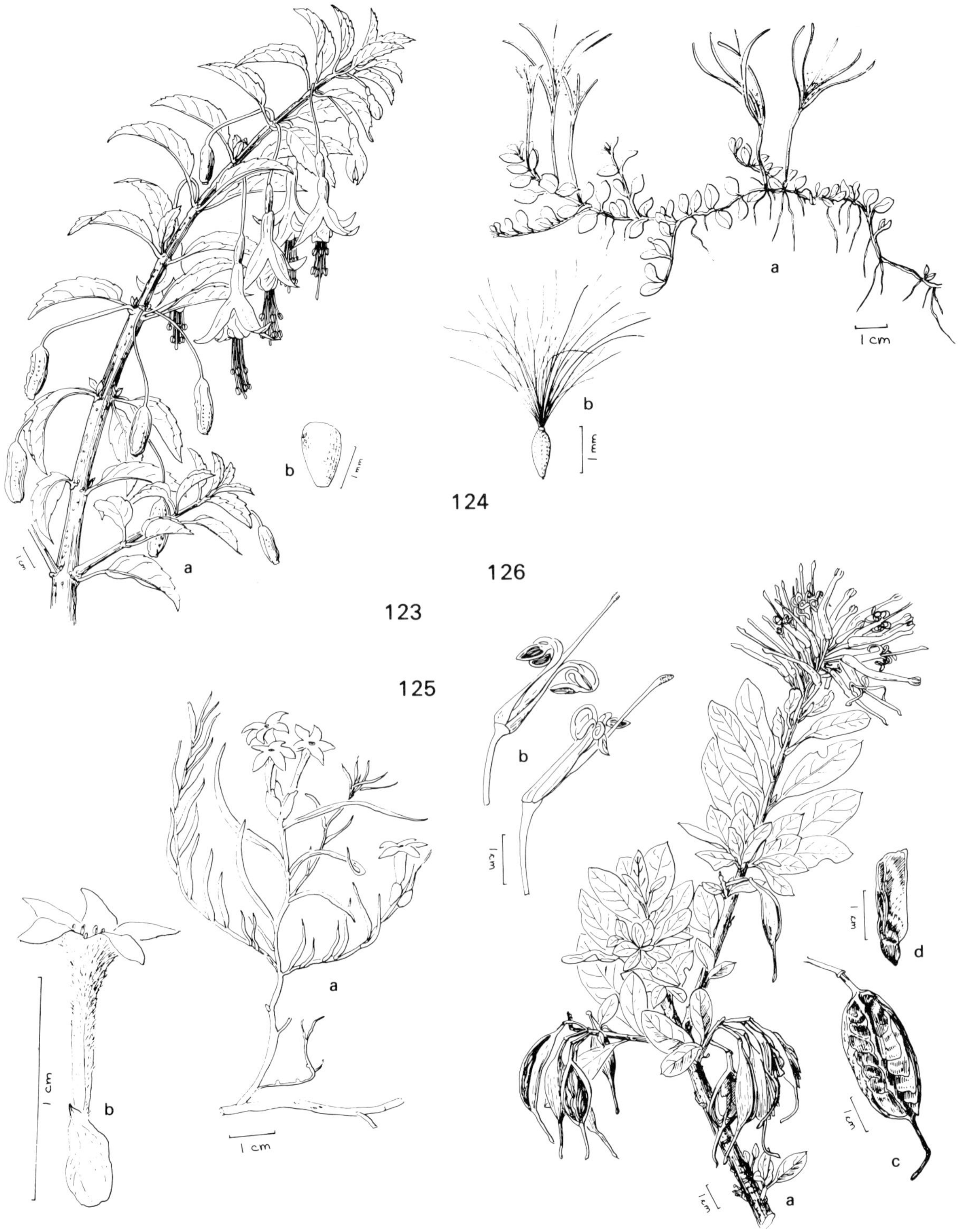

Fig. 123. *Fuchsia magellanica*. a, Plant. b, Seed. Scale = 1 cm (a) or 1 mm (b). Fig. 124. *Epilobium conjungens*. a, Plant. b, Seed. Scale = 1 cm (a) or 1 mm (b). Fig. 125. *Arjona pusilla*. a, Plant. b, Flower. Scale = 1 cm. Fig. 126. *Embothrium coccineum*. a, Branch. b, Flowers. c, Fruit. d, Seed. Scale = 1 cm.

(R. N. P. Goodall)

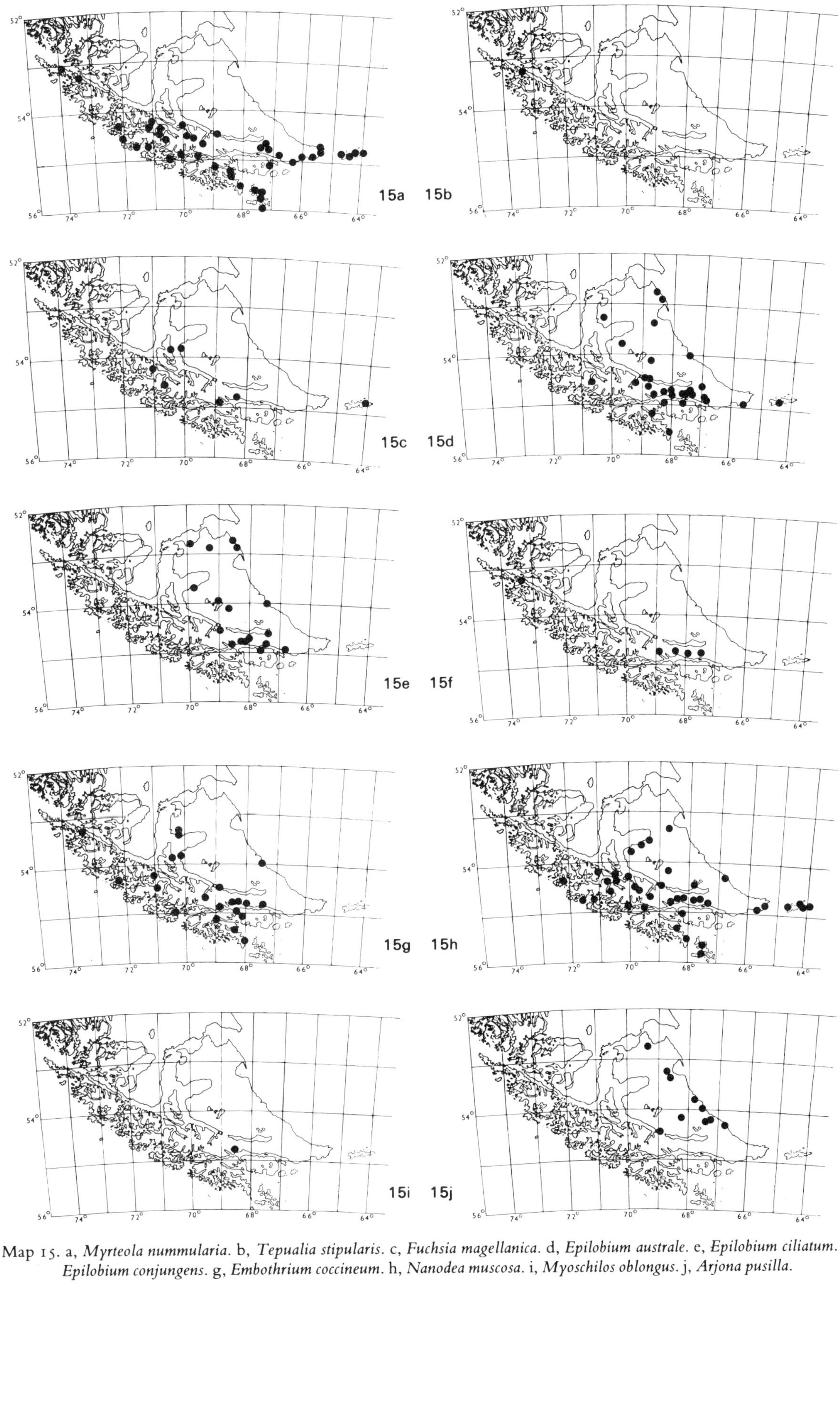

Map 15. a, *Myrteola nummularia*. b, *Tepualia stipularis*. c, *Fuchsia magellanica*. d, *Epilobium australe*. e, *Epilobium ciliatum*. f, *Epilobium conjungens*. g, *Embothrium coccineum*. h, *Nanodea muscosa*. i, *Myoschilos oblongus*. j, *Arjona pusilla*.

XLIV. SANTALACEAE

Hemiparastic herbs or shrubs. Leaves alternate to spirally arranged, simple, entire; stipules absent. Flowers actinomorphic, hermaphrodite, solitary or in spikes. Perianth-segments 4–5, united at base, with or without tube. Stamens 4–5, adnate to perianth opposite lobes. Ovary inferior, 1-celled; style 1, simple; stigma 2- to 3-lobed. Fruit a nut or drupe; seed 1, with endosperm.

Literature: G. Dawson, *Rev. Mus. La Plata Bot. n.ser.* **6**: 5–80(1944).

a. Shrub; leaves at least 5 mm wide; fruit a black drupe **2. *Myoschilos***
a. Herb; leaves not more than 3mm wide; fruit a nut or a red drupe:
 b. Perianth with tube and 5 lobes; fruit a nut **3. *Arjona***
 b. Perianth without tube and with 4 segments; fruit a red drupe **1. *Nanodea***

1. *Nanodea* Banks ex C. F. Gaertner

Perennial herbs. Leaves alternate, sessile, persistent. Flowers solitary in axils of upper leaves. Perianth adnate to ovary, forming a concave disc with 4 ovate, ± acute marginal lobes. Stamens 4. Stigma shortly 2-lobed. Fruit a drupe.

1. *N. muscosa* Banks ex C. F. Gaertner 1807, *Fruct. Sem. Pl.* **3**: 251.

Stems 2–12 cm, erect, branched, reddish. Leaves (3–)4.5–10.5(–19) × 0.5–1.5 mm, linear, acute to acuminate, olive-green, often coppery, usually grouped towards ends of branches. Perianth 3–4 mm in diameter, reddish to yellowish; lobes 1–1.5 mm, ± rhombic, acute. Style stout, reddish. Fruit 5–9 mm, globose, reddish. *Flowers October to December.* (Fig. 131).

Moist *Empetrum-Marsippospermum* heath, *Sphagnum* bogs, moist cushion plant communities below and above timberline; 0–650 m.

Almost throughout Fuegia except the extreme north. (Map 15h).
W. Argentina north to *c.*40°S. lat.
Chile north to 42°S.lat.
Falkland Islands (Islas Malvinas).

2. *Myoschilos* Ruiz & Pavón

Shrubs. Leaves alternate. Flowers in subsessile spikes. Perianth adnate to ovary, forming a thick disc with 5, valvate lobes. Stamens 5. Stigma 3-lobed. Fruit a drupe.

1. *M. oblongus* Ruiz & Pavón 1789, *Syst. Veg.* 73.

Stems up to 1 m, erect, branched, purplish, reddish when young. Leaves 12–30 × 5–12 mm, ovate to ovate-oblong, obtuse to subacute, apiculate, cuneate at base, shortly petiolate. Spikes 5–8 × 3–5 mm, oblong-ovoid. Bracts *c.*2 × 1 mm, ovate, acute, scarious, brownish. Perianth 3–4 mm in diameter, brownish; lobes *c.*1 mm, ± ovate, obtuse to subacute. Fruit 4–7 × 3–5 mm, oblong-ovoid, with thin flesh, black at maturity. *Flowers November to January*. (Fig. 134).

Openings in *Nothofagus betuloides* forest; ?200–400 m.

S. Isla Grande, recently discovered in S. Isla Grande (Cordón del Toro) (Lagaar, unpub.). (Map 15i).
W. Argentina north to *c.*41°S.lat.
Chile north to *c.*32°S.lat.

3. *Arjona* Cav.

Perennial, stoloniferous herbs. Leaves spirally arranged, sessile. Flowers in bracteate terminal spikes. Perianth with long tube and 5 lobes. Stamens 5. Stigma 3-lobed. Fruit a nut.

a. Leaves linear, l-veined; outer bracts glabrous on back, ciliate **1. *pusilla***
a. Leaves ovate to linear-lanceolate, 3- to 5-veined; outer bracts sericeous on back **2. patagonica**

1. *A. pusilla* Hooker f. 1847, *Fl. Antarct.* **1(2)**: 342.

Stems (2–)2.5–5(–9) cm, erect to ascending, usually branched from base. Leaves 10–23 × (0.3–)0.8–1.2(–1.4) mm, the lower shorter, linear, acuminate, erect to arcuate, l-veined, glabrous. Flowers 1–4 in short, terminal spike. Outer bracts 4.5–5.5 mm, ovate to elliptical, mucronate, flat to somewhat keeled, glabrous on back, ciliate. Perianth whitish with purple veins; tube 9–14 × 0.5–1 mm, sericeous, with clusters of short yellow hairs at mouth; lobes 3–4 mm, ovate to lanceolate, acute, recurved, sericeous on back. Stamens with very short filaments, inserted at or near mouth of perianth-tube. Style included. Nut *c.*2.5 × 2 mm, oblong, smooth, brown. *Flowers December to January*. (Fig. 125).

Vernally wet open grassland, especially by freshwater and brackish lakes; 0–50 m.

C. to N. & E. Isla Grande. (Map 15j).
Argentina north to *c.*43°S.lat.

2. *A. patagonica* Dcne 1853, in Hombron & Jacquinot, *Voy. Pôle Sud* 22.

A. tuberosa auct., non Cav. – Dusén, 1900.

Stems 1–14(–20) cm, erect, branched from base. Leaves 6–11(–13) × 1–3 mm, the lower smaller or absent, ovate to linear-lanceolate, apiculate and incurved at apex, 3– to 5– veined, scabridulous. Flowers numerous in terminal spike. Outer bracts 6.5–8 mm, ovate, shortly aristate, keeled, sericeous on back. Perianth white to pinkish or lilac; tube 10–13.5 × *c.*1 mm, sericeous, somewhat hairy inside; lobes 3–5 mm, ovate to oblong, acute. Stamens with short filaments, inserted just inside mouth of perianth-tube. Style included or somewhat exserted. Nut 2–3.5 × 1.8–2.4 mm, oblong, smooth, dark brown. $2n = 28$. *Flowers November to January.* (Fig. 127).

Sands and shingle, open grassland and scrub; 0–250 m.

S. C. to N. Isla Grande. (Map 16a).
W. Argentina north to *c.*32°S.lat.
Chile north to 50°42′S.lat.

XLV. MISODENDRACEAE

Dioecious, semi-paristic shrubs growing on *Nothofagus*, the primary root transformed into an absorbing organ (haustorium) which penetrates the bark of the host. Leaves alternate, deciduous; stipules absent. Flowers unisexual, actinomorphic, borne in spikelets or clusters arranged in a racemose inflorescence. Stamens 2–3. Ovary superior, 3– later 1-celled; style 1, with 3 stigmas. Fruit an achene, with 3 bârbed setae; seed 1, with endosperm.

Literature: C. J. F. Skottsberg, *Pflanzenreich* **62 (IV.68)**: 1–16(1958); R. A. Rossow, *Parodiana* **1(2)**: 245–270 (1982).

1. *Misodendrum* Banks ex DC.

Branches apparently dichotomously or trichotomously divided. Leaves simple, entire or reduced to scarious bract-like scales. Perianth absent in male flowers, adnate to ovary and 3-angled in female flowers. Stigmas short. Seed bearing minute sticky disc at one end.

a. Leaves scale-like, not more than 2.5 mm; stamens 2 **1. *punctulatum***
a. Leaves herbaceous, at least 6 mm; stamens 3:
  b. Leaves of inflorescence linear, ciliate; female spikelets (1) 2-flowered; ovary hairy **2. *linearifolium***
  b. Leaves of female inflorescence ovate to suborbicular, glabrous; female spikelets with 3 or more flowers; ovary glabrous:
    c. Male inflorescence with scale-like leaves; setae not more than 25 mm, thickened at apex **4. *quadriflorum***
    c. Male inflorescence with ovate to ovate-rhomboid leaves; setae at least 30 mm, barbed to apex **3. *brachystachyum***

1. *M. punctulatum* Banks ex D.C. 1830, *Coll. Mém.* **6**: 12.

Branches up to 26 cm, the primary up to 4.5 mm in diameter, spreading, easily breaking at nodes; bark yellowish at first, later reddish brown, prominently verrucose. Leaves (1)1.4–2(–2.3) × (1.4–)2–3 mm, broadly cordate, rounded to truncate, apiculate, entire, scale-like, greenish yellow to pale brown. Inflorescence with 2-flowered spikelets subtended by leaves similar to, but smaller than, those on sterile branches. Male flowers 1–1.25 mm, the female 1–1.5 mm, fusiform. Stigmas 0.2–0.3 mm; ovary glabrous. Achene 2.5–3 mm, narrowly fusiform; setae up to 11 mm, barbed. $2n = 12$. *Flowers July to February.* (Fig. 132, Pl. 7a).

On *Nothofagus antarctica, N. betuloides* and *N. pumilio*: 0–300 m.

From C. Isla Grande south and west. (Map 16b).
W. Argentina north to 39°S.lat.
Chile north to *c.*36°30′S.lat.

2. *M. linearifolium* DC, 1830, *Prodr.* **4**: 671.

Branches up to 25 cm, the primary 1–1.5 mm in diameter, somewhat swollen at nodes; bark dark brown at first, later greyish, smooth, sparsely hairy. Leaves 6–10 × 1–1.5 mm, linear, acute, entire, glabrous or sparsely hairy, dark green. Inflorescence with (1)2-flowered spikelets subtended by leaves similar to those on sterile branches, but yellowish. Male flowers 1–1.5 mm, the female *c.*1 mm, fusiform. Stigmas *c.*0.2 mm; ovary hairy. Achene 2–2.5 mm, ovoid; setae up to 60 mm, barbed. *Flowers September to February.*

On *Nothofagus antarctica* and, sometimes, *N. betuloides*; 0–300 m.

S. & E. Isla Grande.
W. Argentina north to 36°30′S.lat.

Chile north to *c.*36°S.lat.
Fuegian material belongs to var. *contractum* Skottsb., which is usually smaller than the typical variety and may be monoecious or polygamous.

3. ***M. brachystachyum*** DC. 1830, *Coll. Mém.* **6**: 14.
M. oblongifolium auct., non DC. – Dusén, 1900.

Branches up to 15 cm, the primary up to 10 mm in diameter, irregularly angled or ribbed, swollen at nodes; bark dark brown, smooth, glabrous. Leaves (15–)20–35(–40) × 2–5(6) mm, linear to linear-spathulate, obtuse, entire, glabrous, yellow- to dark green. Inflorescence with 7- to 10-flowered spikelets subtended by leaves dissimilar to those on sterile branches. Male inflorescence 2–8 cm; spikelete 6–9 mm; leaves 6–9 × 3–6 mm, ovate to ovate-rhomboid, obtuse to subacute. Female inflorescence 2–8(–15) cm; spikelets 4–20 mm; leaves (4–)10–20(–24) × (2–)2.5–6.5(–7.5) mm, ovate to ovate-spathulate or, rarely, – rhomboid, obtuse to subacute, pubescent at base. Stigmas *c.*0.3 mm; ovary glabrous. Achene 2–2.6 mm, ovoid; setae (25–)30–40 mm, barbed to apex. *Flowers October to January.* (Fig. 133).

On *Nothofagus antarctica, N. betuloides* and *N. pumilio.* 0–500 m.

S. Fuegia. (Map 16d).
W. Argentina north to 39°50′S.lat.
Chile north to *c.*37°S.lat.

4. ***M. quadriflorum*** DC. 1830, *Coll. Mém.* **6**: 2
M. "grandiflorum" – Autran, 1905.

Branches up to 20 cm, the primary up to 10 mm in diameter, smooth; bark dark brown, striate, glabrous. Leaves (8.5–) 10–15 × (1.5–)2–4 mm, obtriangular to obovate, truncate to retuse and erose to denticulate, entire, glabrous, bright green. Male inflorescence 16–25 mm; spikelets (2) 3-flowered, *c.*2 mm; leaves 1.5–2.5 × 1.3–1.5 mm, obovate to obtriangular, truncate, irregularly sinuate-dentate, scale-like, yellow. Female inflorescence up to 9 cm; spikelets 6–11 mm, (3)4(–6)-flowered; leaves (4.7–)6–7(–7.6) × 2.4–3.5(–4.2) mm, ovate to rhomboid or suborbicular, truncate to retuse, serrate, glabrous, green. Stigmas *c.*0.25 mm; ovary glabrous. Achene 2.5–3 mm, subcylindrical; setae up to 25 mm, barbed except at thickened apex. *Flowers October to February.*

On *Nothfagus antarctica* and *N. pumilio*; 0–500 m.

C. & S. Isla Grande and S. side of Canal Beagle. (Map 16e).
W. Argentina north to 39°40′S.lat.
Chile north to *c.*37°20′S.lat.

XLVI. CELASTRACEAE

Trees or shrubs. Leaves alternate, simple; stipules present or absent. Flowers hermaphrodite, actinomorphic, with glandular disc. Sepals 4–5, united at base. Petals 4–5, free. Stamens 4–5, alternating with petals. Ovary superior, 2-celled; style short or absent; stigma lobed. Fruit a loculicidal capsule; seeds 1–2, with endosperm.

1. ***Maytenus*** Molina

Flowers solitary or in axillary cymose inflorescences. Sepals and petals inserted on or below margin of disc. Stamens inserted on disc.

a. Leaves at least 10 mm wide; sepals and petals 5; capsule 2-seeded **1. *magellanica***
a. Leaves not more than 7 mm wide; sepals and petals 4; capsule 1-seeded **2. *disticha***

1. ***M magellanica*** (Lam.) Hooker f. 1847, *Fl. Antarct.* **1(2)**: 254.

Tree or shrub up to 7 m; main stem up to 25 cm in diameter; bark brown to greyish; twigs with lunate to semicircular leaf-scars, glabrous. Leaves with lamina 15–95 × 10–25(–35) mm, ovate to lanceolate, acute to acuminate, rarely obtuse, attenuate at base, serrate and thickened at margin, coriaceous, glabrous, pale green; petiole 3–10 mm. Stipules minute, brown, caducous. Flowers few to numerous in fascicles, on pedicels 2–4 mm; bracteoles 2, *c.*0.5 mm, fimbriate, brown. Calyx-lobes 5, *c.*1 × 1 mm, rounded, convex, patent, glabrous. Petals 5, *c.*2 mm, ovate, obtuse, erose, red with white margin, punctate. Disc orbicular, undulate. Style *c.*0.5 mm; stigma 4-lobed. Capsule 4–6 × 4–4.5 mm, orbicular-obcordate; seeds 2, 4–4.5 × 2–2.5 mm, ellipsoid, with fleshy aril at base. *Flowers September to November.* (Fig. 129).

In *Nothofagus* forest, often with *Drimys*: sometimes locally co-dominant; 0–100 m.

S. Fuegia, except the extreme west. (Map 16f).
W. Argentina north to *c.*40°S.lat.
Chile north to *c.*38°S.lat.

2. ***M. disticha*** (Hooker f.) Urban 1904, in Urban & Graebner, *Festschr. Ascherson* 58
Myginda disticha Hooker f. 1847, *Fl. Antarct.* **1(2)**: 254– Alboff, 1896; Spegazzini, 1896.

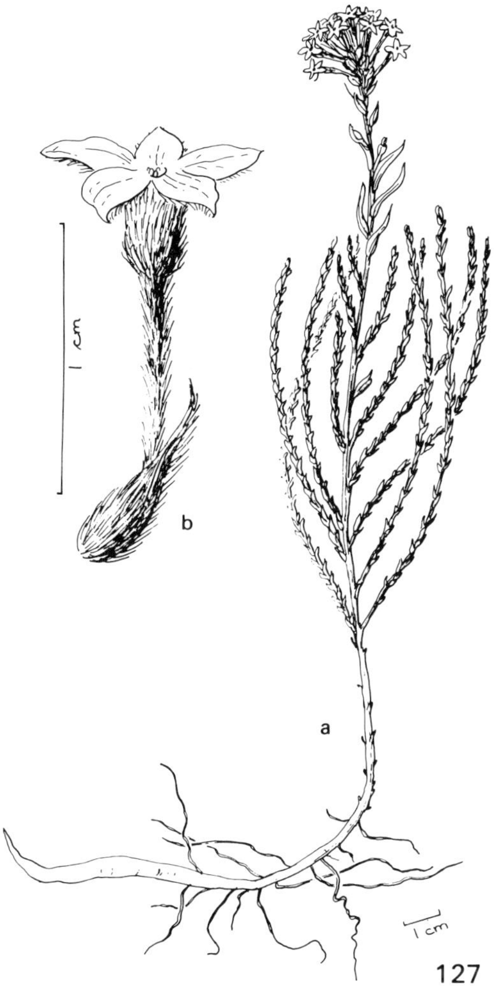

127

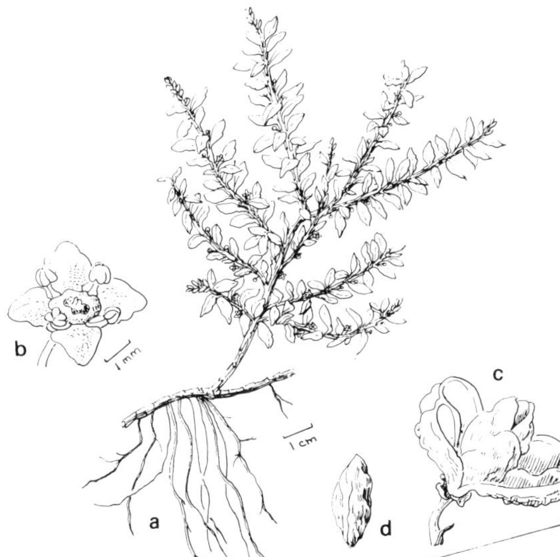

128

129

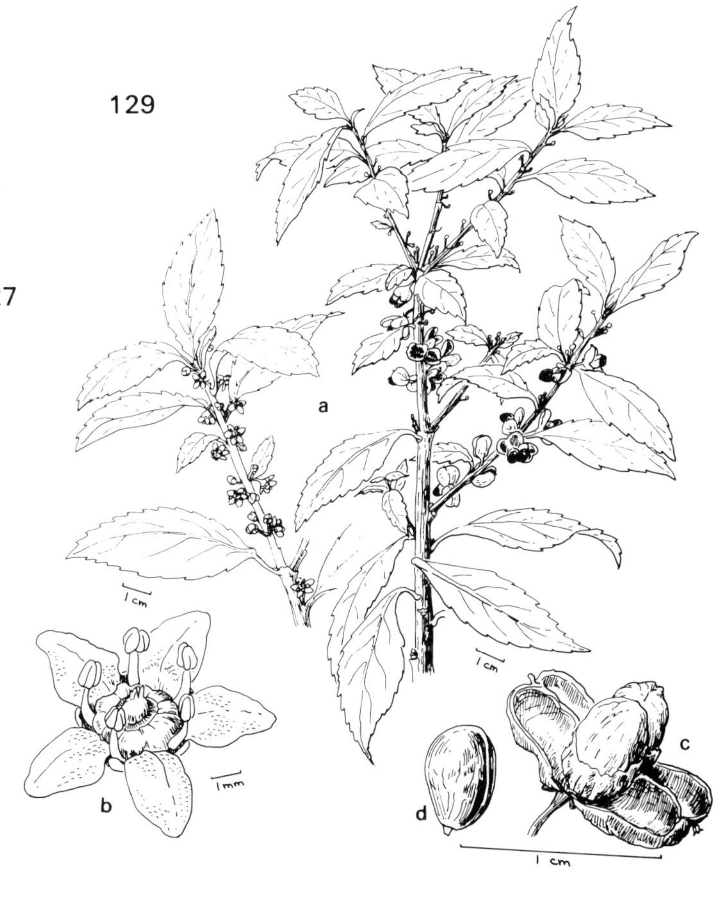

Fig. 127. *Arjona patagonica.* a, Plant. b, Flower. Scale = 1 cm. Fig. 128. *Maytenus disticha.* a, Plant. b, Flower. c, Capsule. d, Seed. Scale = 1 cm (a, c–d) or 1 mm (b). Fig. 129. *Maytenus magellanica.* a Flowering and fruiting branches. b, Flower. c, Open capsule. d, Seed. Scale = 1 cm (a, c–d) or 1 mm (b).

(R. N. P. Goodall)

Prostrate shrub up to 20 cm, much branched, main stems up to 1 cm in diameter; bark light brown; twigs with semicircular to elliptical leaf-scars, puberulent. Leaves with lamina 8.4–14 × 3.9–6.2 mm, ovate-lanceolate to oblanceolate, acute to obtuse or emarginate, often apiculate, cuneate, entire to serrate and somewhat revolute, coriaceous, glabrous, greyish to olive green above; petiole 0.5–1 mm. Stipules absent. Flowers solitary or in small fascicles, on pedicels *c.*1 mm; bracteoles 2, *c.*2 mm, ciliate, brown. Calyx-lobes 4, *c.*0.9 × 0.7 mm, ovate, acute to obtuse, glabrous. Petals 4, *c.*1.3 mm, ovate to suborbicular-rhomboid, obtuse, reddish. Disc surrounding and partly united with ovary. Style absent; stigma 2-lobed. Capsule *c.*6 × 3.5 mm, subglobose-ellipsoid; seed 1, 3–3.5 × *c.*2 mm, ellipsoid, reddish brown, almost covered by membranous aril. (Fig. 128).

On floor of *Nothofagus* forest or thick scrub; 0–600 m.

S. parts of Isla Grande and adjacent islands. (Map 16g).
W. Argentina and E. Chile north to *c.*39°S.lat.

XLVII. EUPHORBIACEAE

Monoecious, perennial herbs. Leaves alternate, simple; stipules present or absent. Flowers unisexual, actinomorphic. Sepals 3 or 0. Petals 0. Stamens 1 or 6. Ovary superior, 3-celled; styles 3. Fruit a schizocarp, of 3 dehiscent 1-seeded mericarps which separate when mature.

a. Plant glabrous, with white latex; flowers in small clusters; stamen 1 — **2. *Euphorbia***
a. Plant pubescent, without latex; flowers solitary, axillary; stamens 6 — **1. *Dysopsis***

1. *Dysopsis* Baillon

Plants pubescent, without latex, Stipules present. Flowers solitary in leaf-axils. Sepals 3, united at least at base. Stamens 6, in 2 whorls, the inner longer; filaments united in basal half.

Literature: A. Lourteig & C. A. O'Donnell, *Lilloa* **8**: 292–296 (1942).

1. *D. glechomoides* (A. Richard) Müller Arg. 1866, in DC., *Prodr.* **15**: 2.

Stems up to 25 cm, creeping to procumbent, rooting at nodes, sparingly branched, slender. Leaves 5–20 × 5–21 mm, suborbicular to ovate, obtuse, truncate to cuneate at base, sinuate-crenate; petiole (6–)9–20 mm; stipules 1–1.5 mm, triangular, acuminate. Male flowers with peduncle (13–) 25–35 mm; sepals 3–3.5 mm, united in basal 1/3, acute; inner stamens 7–10 mm, the outer 5–7 mm. Female flowers with peduncle *c.*5 mm; sepals 1.5–2 mm, united at extreme base, ovate; styles very short. Schizocarp 2.2–3.5 × 2.5–3 mm, subglobose; seeds 1.3–1.5 mm, subglobose, dark brown. $2n = c.28$. *Flowers October to February*. (Fig. 141).

Moist shady places, especially in forest and scrub, but also stream banks and tall bog communities; 0–500 m.

S. Fuegia, except the extreme west. (Map 16h).
W. Argentina north to *c.*40°30′S.lat.
Chile north to 40°S.lat. and at *c.*30°30′S.lat.

2. *Euphorbia* L.

Plants glabrous, with white latex. Stipules absent. Flowers in small groups, including 1 female and several male, surrounded by a lobed herbaceous involucre having 4–5 marginal glands, to form a cyathium; inflorescence of 1–several cyathia in umbellate cyme. Sepals absent. Male flower of a single stamen jointed to the pedicel. Female flowers solitary, pedicellate, surrounded by male flowers.

Literature: R. Subils, *Kurtziana* **10**: 83–248 (1977).

1. *E. collina* Phil. 1857, *Linneae* **29**: 41.
E. portulacoides auct., non Sprengel - Moore & Goodall, 1974.

Stems 5–8 cm, ascending, branched, with remains of leaf-bases below. Leaves 5–10 × 2–3 mm, the upper slightly the larger, elliptical to obovate-spathulate, rounded to obtuse, attenuate at base, with translucent margin. Cyathia solitary or in pairs, surrounded by 3 upper leaves, the peduncles 0.5–1 mm. Involucre 2–3 × 1.5–2.5 mm, turbinate; glands 5, deltate, revolute, green; lobes *c.*0.6 mm, laciniate, alternating with glands. Male flowers 18–20; filaments *c.*0.5 mm. Female flowers with ovary *c.*0.5 mm; styles *c.*0.3 mm. Schizocarp *c.*3 mm, globose-ellipsoid. *Flowers December*. (Fig. 136).

Coastal grass steppe; 0–20 m.

N. W. Isla Grande, known only S. W. of Punta Catalina (52°39′S.lat., 68°59′W. long). (Map 16i).
Argentina and Chile north to *c.*30°S.lat.

A complex and polymorphic species within which many taxa have been described. Following the

130

131

132

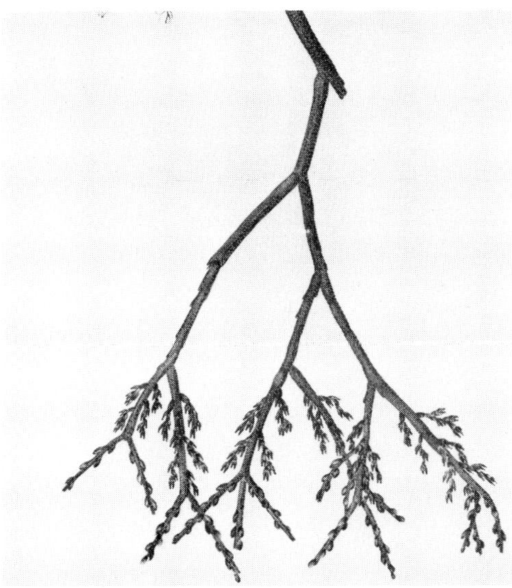

133

Fig. 130. *Drapetes muscosus.* Fig. 131. *Nanodea muscosa.* Fig. 132. *Misodendrum punctulatum.* Fig. 133. *Misodendrum brachystachyum.*

(Sydney Parkinson)

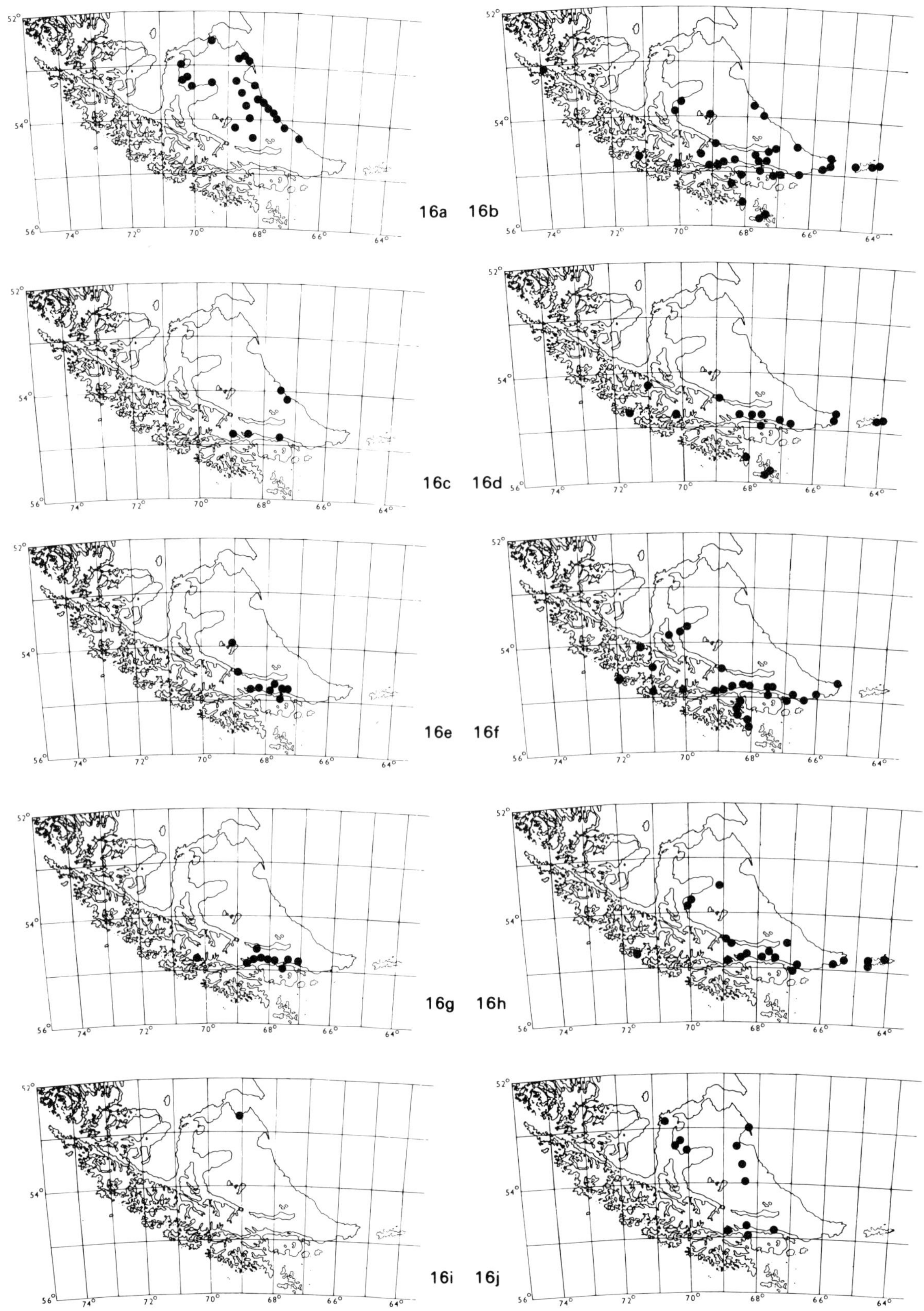

Map 16. a, *Arjona patagonica*. b, *Misodendrum punctulatum*. c, *Misodendrum linearifolium*. d, *Misodendrum brachystachyum*. e, *Misodendrum quadriflorum*. f, *Maytenus magellanica*. g, *Maytenus disticha*. h, *Dysopsis glechomoides*. i, *Euphorbia collina*. j, *Discaria chacaye*.

recent treatment of Subils (1977) the Fuegian plants should be referred to var. *collina*.

XLVIII. RHAMNACEAE

Shrubs. Leaves opposite, simple; stipules present. Flowers hermaphrodite, actinomorphic. Calyx 4-lobed. Petals 0. Stamens 4, attached to base of calyx. Ovary superior, 3-celled; style simple. Fruit dry, nutlike, separating into 3 1-seeded nutlets (cocci), with endosperm.

1. *Discaria* Hooker

Stems spiny. Flowers in axillary fascicles. Stamens alternating with calyx-lobes. Stigma 3-lobed.

1. ***D. chacaye*** (G. Don) Tort, **comb. nov.**
Colletia chacaye G. Don 1832, *Gen. Syst.* **2**: 35.
Discaria serratifolia (Vent.) Bentham & Hooker 1876, in Masters *Gard. Chron.* **2**: 324. – Moore, 1974.
D. discolor (Hooker) Dusén 1900, Svenska Exped. Magellansl. **3(5)**: 153 – Skottsberg, 1916.
Colletia discolor Hooker 1843, *Icon. Pl.* t.538 – Franchet, 1889; Hariot, 1884; Alboff, 1896.

Glabrous shrub up to 1 m, much branched, the main stems with brown or greyish bark, bearing decussate pairs of woody spines, the twigs arising in pairs below spines; spines 10–30 mm, articulate, the older leafless, the younger bearing stipules and often 2 small leaves on articulation. Leaves (5–)10–14 × (2.5–)4–8(–10) mm, elliptical to obovate or oblanceolate, obtuse to emarginate, cuneate at base, rather deeply glandular-dentate, dark green above, coriaceous; petiole 1–4 mm; stipules *c.*2 mm, narrowly triangular, acuminate, ciliate, brown. Pedicels 2–5 mm, up to 9 mm in fruit. Calyx 2–3 mm, campanulate, united to ovary below, persistent; lobes triangular, subacute, cucullate. Style *c.*1 mm, erect. Fruit *c.*3 mm, 3-lobed; nutlets 2.5–3 × *c.*2 mm, ellipsoid, dark brown, shining. *Flowers November to January*. (Fig. 138).

Open *Berberis, Chiliotrichum* and *Pernettya mucronata* scrub, between rocks, margins of *Nothofagus* forest; 0–60 m.

Isla Grande except extreme west, and islands of Canal Beagle. (Map 16j).
W. Argentina north to *c.*36°S.lat.
Chile north to *c.*31°S.lat.

XLIX. LINACEAE

Annual herb. Leaves opposite, simple; stipules absent. Flowers hermaphrodite, actinomorphic, in a lax cyme. Sepals 5, free. Petals 5, free. Stamens 5. Ovary superior, 8- to 10- celled; styles 8–10, free. Fruit a loculicidal capsule; seeds 1 in each cell.

1. *Linum* L.

Leaves sessile, entire. Sepals entire. Petals clawed. Filaments united at base. Stigmas capitate. Capsule dehiscing by 10 valves; seeds flat.

*1. ***L. catharticum*** L. 1753, *Sp. Pl.* 281.

Stems 7–24 cm, erect, slender, glabrous, simple. Lower leaves 3.1–3.6 × 1.8–2.4 mm, obovate, obtuse, crowded; upper leaves 5–11.5 × 1.6–3.8 mm, elliptical, obtuse, distant. Inflorescence much branched; bracts *c.*1. × 0.2 mm, linear-lanceolate, grading into leaves below; pedicels 4–16 mm. Sepals 2.8–3.1 × 0.6–1.1 mm, ovate-lanceolate, acuminate, the margin denticulate, scarious. Petals 4.5–5.5 × 1.3–1.4 mm, obovate to spathulate, white. Capsule 2–2.7 mm, globose; seeds *c.*1.5 × 0.9 mm, ovoid, light brown. *Flowers February to March.*

Disturbed ground near settlements and along paths; 0–30 m.

N. shore of Canal Beagle from Ushuaia to Yendegaia.
Introduced from Europe and locally naturalized in temperate regions.

L. GERANIACEAE

Annual or perennial herbs. Leaves usually opposite, or in basal rosettes, palmately lobed or divided to pinnate; stipules present. Flowers actinomorphic or slightly zygomorphic, hermaphrodite, in cymes or umbels. Sepals 5, free. Petals 5, free. Stamens in 2 whorls of 5, obdiplostemonous, some reduced to staminodes. Ovary superior, of 5 carpels united around a central axis (carpophore); styles 5, free; stigmas filiform. Fruit a schizocarp, separating elastically at the base of the carpophore into 5, 1-seeded mericarps with persistent styles (awns); seeds with little or no endosperm.

a. Leaves pinnate; fertile stamens 5; mature mericarp with awn spirally twisted **2. *Erodium***
a. Leaves palmately divided or lobed; fertile stamens 10; mature mericarp with straight or curved awn. **1. *Geranium***

1. *Geranium* L.

Annual or perennial herbs. Leaves in basal rosettes,

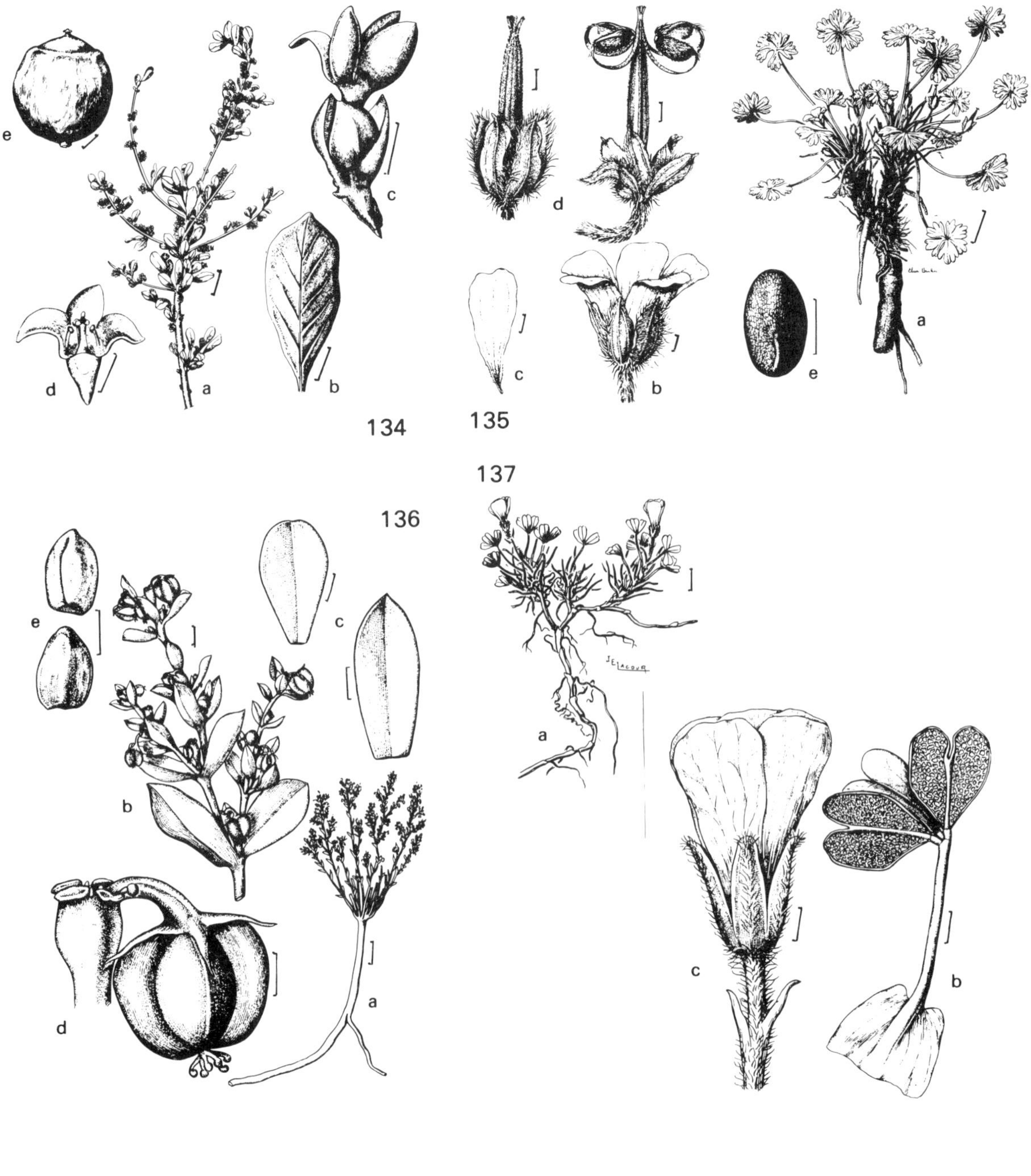

Fig. 134. *Myoschilos oblongus*. a, Plant. b, Leaf. c, Part of inflorescence. d, Flower. e, Fruit. Scale = 1 cm (a) or 1 mm (b–e). Fig. 135. *Geranium sessiliflorum*. a, Plant. b, Flower. c, Petal. d, Fruit, entire and dehisced. e, Seed. Scale = 1 cm (a) or 1 mm (b–e). Fig. 136. *Euphorbia collina*. a, Base of plant. b, Terminal branch. c, Leaves. d, Cyathium and fruit. e, Seeds. Scale = 1 cm (a) or 1 mm (b–e). Fig. 137. *Oxalis magellanica*. a, Plant. b, Leaf. c, Flower. Scale = 1 cm (a) or 1 mm (b–c).
(Flora Patagónica)

when cauline opposite or the uppermost alternate, palmately lobed or divided. Inflorescence cymose. Flowers actinomorphic. Fertile stamens 10. Mericarps usually dehiscent, with straight or curved awn.

a. Annual; petals about equalling or slightly exceeding sepals **3. molle**
a. Perennial; petals at least 1½ times as long as sepals:
b. Plant with basal leaves only; flowers usually solitary; peduncles less than 1 cm **1. sessiliflorum**
b. Plant with basal and cauline leaves; flowers usually in pairs; peduncles at least 1 cm **2. magellanicum**

1. *G. sessiliflorum* Cav. 1787, *Monad. Class. Diss. Dec.* 198.

Perennial, with short, stout, branched, woody stock covered with leaf-remains. Leaves in dense basal cluster; lamina (6–)9–15 × 15–25 mm, suborbicular to reniform in outline, divided almost to base into 5(–7) lobes, pubescent to sericeous; lobes 6–14 × 5–8 mm, obovate to obtriangular, 3- to 7- fid, the segments obtuse to mucronate; petiole 2–8 cm, with appressed often retrorse hairs. Peduncles 1–3(–8) mm, borne on very short annual branches, 1(2)-flowered, densely retrorse-appressed-hairy; pedicels up to 10 mm; bracteoles 4–6 mm, linear-lanceolate, acute. Sepals 3.5–5.5(–7) × 2–2.5 mm, ovate-oblong, subacute, densely hairy, often with hyaline margin. Petals 4.5–7.5 × 3–4 mm, 1½–2 times as long as calyx, obcordate, emarginate, pinkish purple to whitish. Mericarps 2.5–3 × *c.*1.5 mm, oblong to ellipsoid, pubescent; awns 10–15 mm, hispid; seeds 2–2.5 mm, oblong, reticulate, dark brown. *Flowers October to December.* (Fig. 135).

Open sands, gravels, muds and pebbles, *Empetrum* heath, grassland; 0–500 m.

Isla Grande, especially the north and centre, absent from extreme west and east. (Map 17a).
W. Argentina north to *c.*26°S.lat.
Chile north to *c.*34°S.lat.
Bolivia, Peru.
New Zealand, Tasmania, S. E. Australia.

S. American plants belong to ssp. *sessiliflorum*; the other subspecies occur in New Zealand and Australia.

2. *G. magellanicum* Hooker f. 1847, *Fl. Antarct.* **1(2)**: 251.
G. patagonicum Hooker f. 1847, *Fl. Antarct.* **1(2)**: 252.

Perennial, with short, sparingly branched woody stock covered with leaf-remains, bearing annual flowering stems at apex; stems up to 35 cm, procumbent to ascending, often rooted at nodes, usually pubescent. Basal leaves not persistent through summer, larger than but otherwise similar to the cauline; lamina of cauline leaves 9–50 × 11–60 mm, orbicular to oblong-ovate in outline, divided almost to base into 5(–7) lobes, pubescent; 5–27 × 3–20 mm, obovate to ± oblong, 3(–5)- fid, the segments obtuse to mucronate; petiole 1.5–15 cm, retrorse-appressed-pubescent. Peduncles 10–50 mm, borne on long annual stems, usually 2-flowered, retrorse-appressed-pubescent; pedicels 10–40 mm; bracteoles 3–8 mm, linear to linear-lanceolate, acute. Sepals 4–7 × 2–3.5 mm, lanceolate to elliptical, acuminate, pubescent. Petals 6–9 × 3–4.5 mm, 1½–2½ times as long as calyx, obovate, emarginate, pinkish purple, rarely whitish. Mericarps 3.5–4 × *c.*1.5 mm, oblong, hispid and with glandular hairs; awns 9–12 mm, hispid; seeds *c.*2 mm, oblong, punctulate, dark brown. *Flowers December to February.*

Open sands and gravels, open *Chiliotrichum* scrub, grassland, rock outcrops; 0–200 m.

Almost throughout Isla Grande except extreme west and east. (Map 17b).
Argentina north to *c.*46°S.lat.
Chile north to *c.*45°S.lat.

A rather variable species in habit, leaf-shape and indumentum, much of the variation apparently the result of habitat differences.

3. *G. molle L. 1753, *Sp. Pl.* 682.

Similar to some forms of *G. magellanicum* but annual, perhaps rarely perennial, the petals 4–5 mm, equalling or slightly exceeding sepals, the mericarps *c.*2 mm, glabrous, with awns 6–9 mm, the seeds almost smooth. *Flowers November to February.*

Coastal sands and gravels; 0–30 m.

S. coast of Isla Grande.
Introduced from Europe and widely introduced in disturbed and open habitats throughout temperate regions.

2. *Erodium* L'Hérit.

Annual herbs. Leaves alternate, pinnate. Inflorescence umbellate. Flowers slightly zygomorphic. Fertile stamens 5, alternating with staminodes opposite the petals. Mericarps indehiscent, with awn spirally twisted at maturity.

*1. ***E. cicutarium*** (L.) L'Hérit. 1789, in Aiton, *Hort. Kew*, Ed. 1, **2**: 414.

Stemless at first, later with stems 7–30 cm, prostrate to ascending or erect, branched from base, subglabrous to white-hairy. Leaves 2.5–8 × 0.9–2.2 cm; leaflets 3–7 pairs, 5–15 × 3–4.5 mm, ovate in outline, 1– to 2– pinnatifid or pinnatisect; stipules 4–6 × 1.5–2.5 mm, ovate, acute to acuminate, scarious. Umbel 3- to 8-flowered; peduncles 1.8–7(–10) cm; bracts 1.3–2.5 mm, ovate, acuminate, scarious; pedicels 7–18 mm. Sepals 3–7 × 1.2–2.4 mm, elliptic-oblanceolate to ovate, acuminate, accrescent. Petals 5–9 × 1.5–3.5 mm, oblanceolate-spathulate, obtuse, rose-purple, the 2 upper often with blackish basal spot. Mericarps 5–6.5 mm, ± hairy; awns 15–40 mm. *Flowers December to January*.

Coastal sands and gravels, open grassland and *Chiliotrichum* scrub, disturbed ground; 0–80 m.

N. W. Isla Grande.
Introduced from Europe and locally naturalized in modified or naturally open habitats.

LI. OXALIDACEAE

Perennial herbs. Leaves alternate, compound; stipules usually absent. Flowers actinomorphic, hermaphrodite. Sepals 5, free. Petals 5, free. Stamens 10, united at base. Ovary superior, 5-celled; styles 5, free; stigmas capitate. Fruit a loculicidal capsule; seeds numerous, with endosperm.

Literature: R. Knuth, *Pflanzenreich* **95 (IV. 130)**: 1–481(1930).

1. *Oxalis* L.

Plant rhizomatous. Leaves radially palmate. Sepals imbricate. Petals contorted. Seeds projected to some distance by splitting of elastic integument.

a. Leaves with 3 leaflets; rhizome horizontal; petals not more than *c.*10 mm **1. *magellanica***
a. Leaves with 4–20 leaflets; rhizome ± ascending; petals usually at least 14 mm **2. *enneaphylla***

1. *O. magellanica* Forster f. 1789, *Comment. Gotting.* **9**: 23.
O. lactea Hooker 1836, *Comp. Bot. Mag.* **1**: 276. .

Rhizomes up to 20 cm, *c.*2 mm in diameter, simple or branched, densely covered with brown scales towards base, bearing clusters of leaves at intervals. Leaves few, with lamina 4–18 mm in diameter, 3–foliolate; leaflets (2–)3–7(–8.5) × (1–)2–5(–7.5) mm, obcordate, with acute apical sinus $\frac{1}{6}$–$\frac{1}{4}$ of length, cuneate to ± truncate at base, entire, glabrous or sparsely hairy, especially towards base; petiole (0.4–)1–6.5 mm, slender, erect to ascending, with sparse retrorse hairs, often reddish, with wide, ovate, brown, scarious stipules at base. Peduncles 1–10 mm, slender, 1-flowered, usually with patent white hairs, 2-bracteolate about middle; bracteoles 1–2 mm, linear-subulate. Sepals 3–5 mm, ovate-lanceolate to ± oblong, ± hairy. Petals 4–10 mm, obovate-cuneate, white. Capsule *c.*3.5–5 mm in diameter, ± globose; seeds *c.*1.3 mm, ellipsoid, pale brown. *Flowers January to March*. (Fig. 137, Pl. 7b).

Moist grassland, feldmark, streamsides; 300–1000 m.

S. Fuegia. (Map 17c).
Chile north to *c.*39°S.lat.
Bolivia (*c.*19°S.lat.).
New Zealand, Tasmania, S.E. Australia.

2. *O. enneaphylla* Cav. 1799, *Icon. Descr.* **5**: 7.

Rhizomes up to *c.*10 cm, horizontal or ascending, simple or branched, covered in thin, dark brown, linear-lanceolate scales, usually also with clusters of fleshy, pinkish, bulbiliferous scales, bearing leaves and flowers towards apex. Leaves numerous, with lamina 1–3.5 cm in diameter, 4– to 20– foliolate; leaflets 4–13 × 2–14 mm, oblanceolate to obovate, usually prominently 2–lobed at apex, cuneate at base, entire, glabrous to densely hairy, ± glaucous to dark green, often purplish at margins; petiole 3–20 cm., slender, ± erect, glabrous to hairy. Peduncles up to 250 mm, 1-flowered, 2-bracteolate in distal quarter; bracteoles 2–7 mm, ovate to linear-lanceolate. Sepals 6–12 mm, ovate to oblong-ovate, acute to subobtuse or slightly bifid, sericeous, pale green to pinkish, with scarious margins. Petals 14–25 mm, obovate, obtuse, usually emarginate, white to pinkish-purple, often darker at base. Capsule ± equalling the sepals, oblong-ellipsoid; seeds 2–2.5 mm, ovoid-ellipsoid, pale brown. Self-incompatible, trimorphically heterostylous, reproducing vegetatively by means of bulbils. *Flowers November to March*. (Fig. 139).

Coastal sands, open grasslands, open scrub, soil and rock screes, rock-crevices, feldmark; 0–1100 m.

Throughout Isla Grande except the extreme west

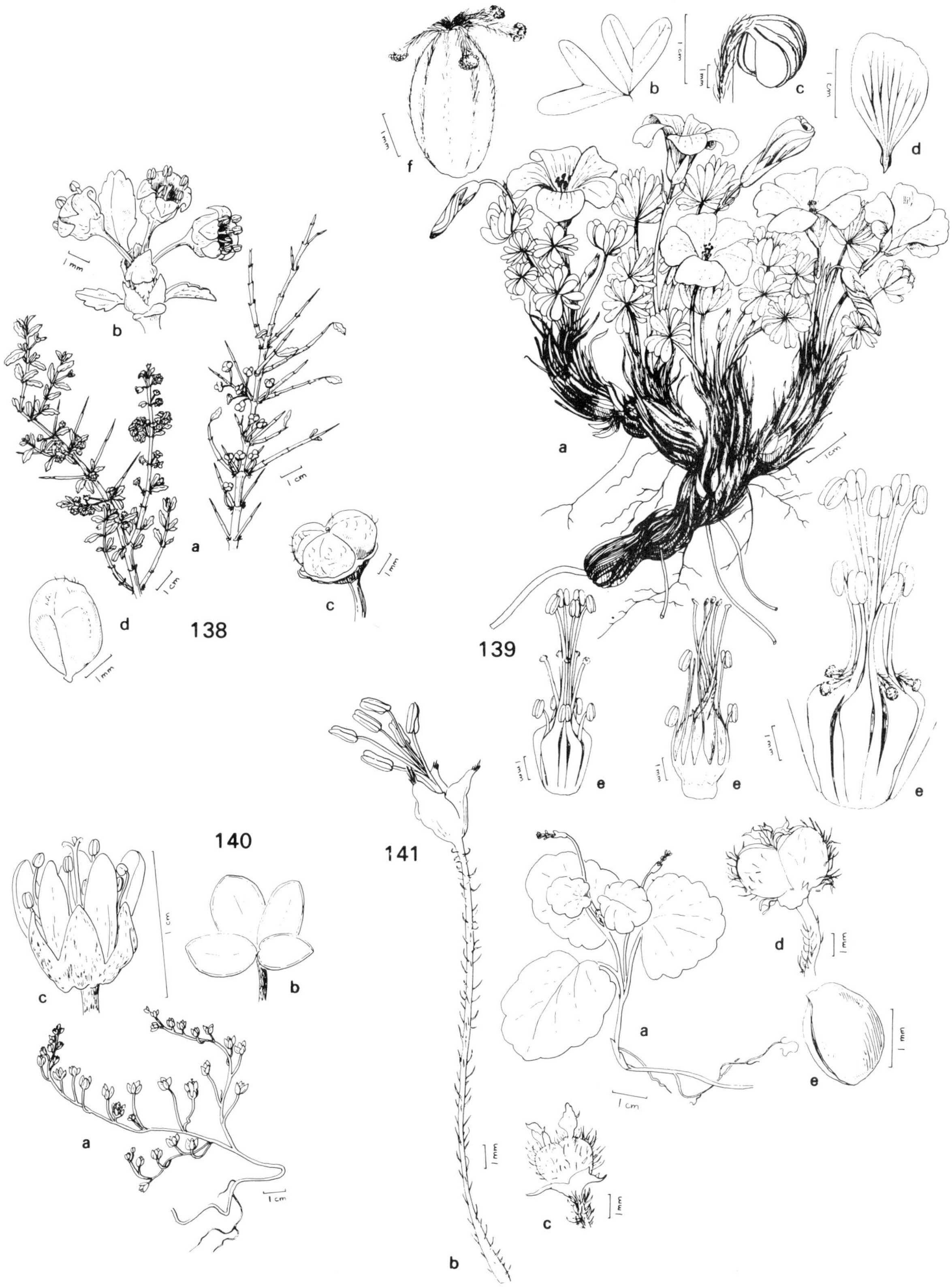

Fig. 138. *Discaria chacaye*. a, Flowering and fruiting branches. b, Flowers. c, Fruit. d, Nutlet. Scale = 1 cm (a) or 1 mm (b–d). Fig. 139. *Oxalis enneaphylla*. a, Plant. b, Leaflets. c, Leaf. d, Petal. e, Stamens, styles and ovary. f, Young capsule. Scale = 1 cm (a–b, d) or 1 mm (c, e–f). Fig. 140. *Tropaeolum patagonicum*. a, Plant. b, Leaf. c, Flower. Scale = 1 cm. Fig. 141. *Dysopsis glechomoides*. a, Plant. b, Inflorescence. c, Young fruit. d, Fruit. e, Seed. Scale = 1 cm (a) or 1 mm (b–e).
(R. N. P. Goodall)

and east; islands of Canal Beagle. (Map 17d).
W. Argentina north to *c.*39°S.lat.
E. Chile north to *c.*50°S.lat.
Falkland Islands (Islas Malvinas).

A very variable species in size, pubescence, leaflet-shape and flower-colour. Much of the variation is apparently environmentally related but detailed experimental studies are needed to assess to what extent genotypic differences and breeding patterns may be involved.

O. fuegensis R. Knuth 1913, *Feddes Repert.* **12**: 36, known only from the non-flowering type specimen collected in an undetermined Fuegian locality, is said to differ from *O. enneaphylla* by its linear scales on the rhizome and conspicuously widened petiole-bases. In view of the variation shown by *O. enneaphylla* such distinctions seem of dubious value and *O. fuegensis* should probably be reduced to a synonym of the widespread species.

LII. TROPAEOLACEAE

Perennial herbs. Leaves alternate, peltate, deeply lobed; stipules absent. Flowers hermaphrodite, indistinctly zygomorphic. Sepals 5, free, the dorsal shortly spurred. Petals 5, free, clawed. Stamens 8, free. Ovary superior, 3-celled; style 1; stigmas 3. Fruit a schizocarp, of 3 indehiscent, 1-seeded mericarps which separate from the central axis when mature, without endosperm.

1. *Tropaeolum* L.

Plant glabrous, somewhat succulent. Flowers solitary in leaf-axils. Sepals valvate.

1. ***T. patagonicum*** Speg. 1897, *Rev. Agron. Vet. La Plata* **3**: 599.

Stems 2–12 cm, prostrate to procumbent, branched. Leaves 4–14 mm in diameter, 3- to 4- foliolate, peltate, the leaflets elliptical to ovate-lanceolate; petioles 3–15 mm, filiform. Peduncles 8–12 mm. Sepals 3–4.5 × 1.5–2 mm, ovate-lanceolate, acute, pale green with purplish-red lines and spots; spur 0.25–0.5 mm, obtuse. Petals 5.5–7 × 2–2.5 mm, oblanceolate to subspathulate, obtuse, white with pale purple stripes near centre. Style 2–4 mm, exserted, the stigmas *c.* 1 mm. *Flowers December to February*. (Fig. 140).

Mud or open wet grassland, especially in seasonally dry shallow lakes; 0–20 m.

N. E. Isla Grande, only known near Cabo Peñas (53°52′S.lat., 67°37′W. long). (Map 17e).
Argentina north to *c.*42°S.lat.

LIII. POLYGALACEAE

Perennial herbs. Leaves alternate, simple; stipules absent. Flowers zygomorphic, hermaphrodite, in racemes. Sepals 5, free. Petals 3, more or less united. Stamens 8, with filaments partly united; anthers opening by pores. Ovary superior, on a small disc, 2-celled; style 1; stigma 2-lobed, only posterior lobe receptive. Fruit a capsule, 2-celled; seeds 1 per cell, with endosperm.

1. *Polygala* L.

Inflorescence of 3- to 12-flowered terminal and lateral racemes; flowers subtended by 1–2 bracts. Inner 2 sepals petaloid, much larger than outer. Petals united at base into corolla-tube, the lower (keel) of different form to other 2 and bearing a fimbriate crest. Filament-tube partly adnate to corolla-tube. Style geniculate, clavate, distally. Capsule compressed, somewhat winged; seeds hairy, with conspicuous strophiole.

Literature: E. M. Grondona, *Darwiniana* **8**: 279–405(1948).

1. ***P. salasiana*** C. Gay 1845, *Hist. Chile. Bot.* **1**: 327.

Roots 1–2.5 mm in diameter, bearing creeping rhizomes with yellow scale-leaves and branching to give slender, glabrous overground stems. Upper leaves 5–9 × 0.8–1.3(–1.7) mm, linear to linear-spathulate, obtuse to subacute, sessile, the lower smaller, elliptic-ovate, shortly petiolate. Racemes sessile or shortly pedunculate; bracts up to 1.5 × 0.4 mm, orbicular to triangular, concave, whitish, persistent. Pedicels 0.7–1.5 mm. Sepals white with green or purple midrib and, often, veins, the outer (1.5–)1.7–2.1 × 0.8–1 mm, elliptic-lanceolate to ovate, obtuse to subacute, the inner 3.2–4.2 × 1.5–1.8 mm, ovate, obtuse, concave. Petals bluish, the keel 1.8–2.6 mm, obtuse or shallowly emarginate, with 5–8 fimbriae, the upper petals 2.4–2.9 mm, narrowly oblong, obtuse or shallowly emarginate. Capsule 2.3–3.2 × 1.5–2.5 mm, elliptical or orbicular; seeds 1.8–2.5 × 0.8–1.1 mm, ellipsoid, black. *Flowers November to January*. (Fig. 142).

Open coastal sands. 0–50 m.

N. & E. C. Isla Grande. (Map 17f).
W. Argentina north to *c.*30°S.lat.
E. Chile north to *c.*31°S.lat.

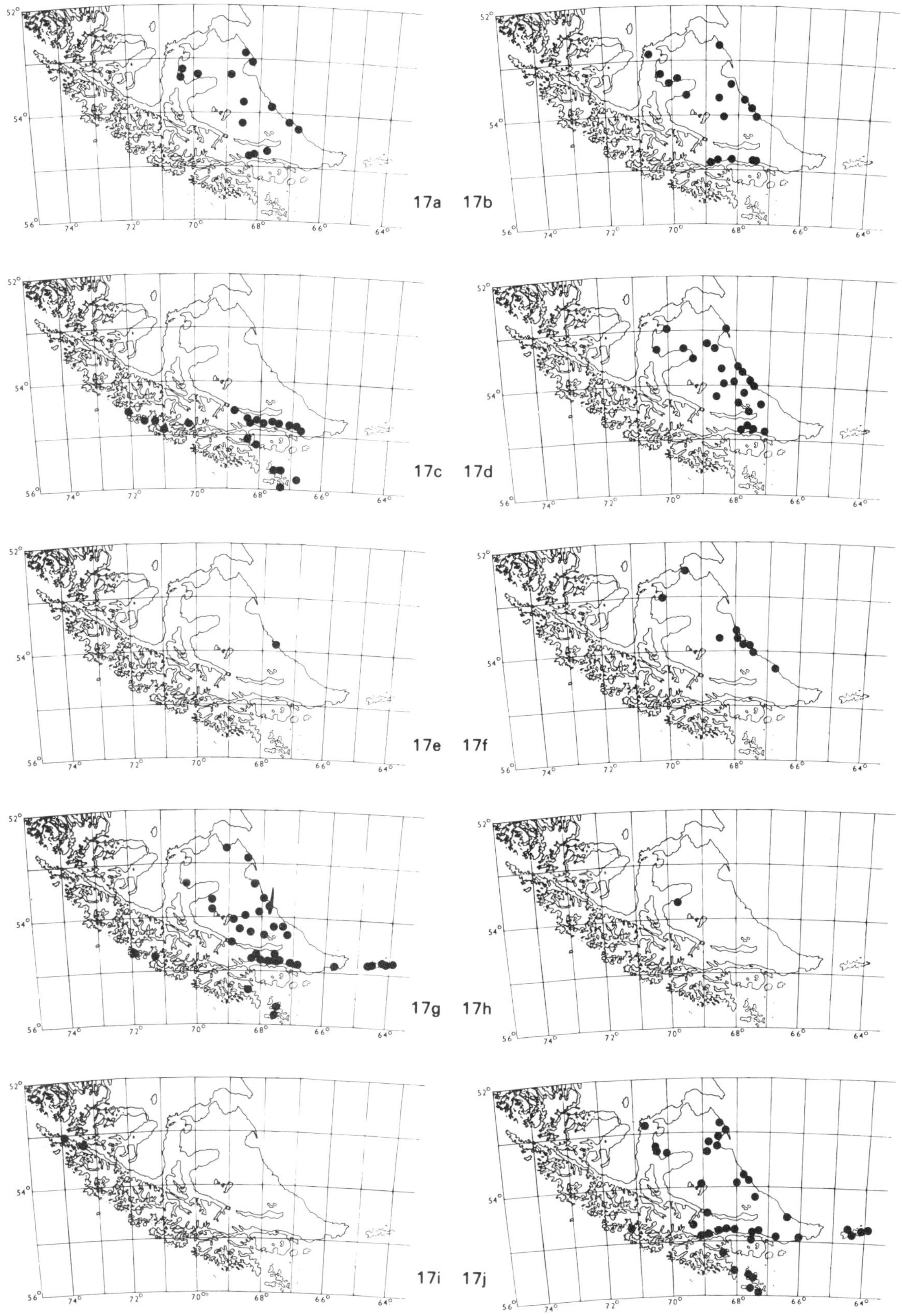

Map 17. a, *Geranium sessiliflorum*. b, *Geranium magellanicum*. c, *Oxalis magellanica*. d, *Oxalis enneaphylla*. e, *Tropaeolum patagonicum*. f, *Polygala salasiana*. g, *Azorella lycopodioides*. h, *Hydrocotyle chamaemorus*. i, *Pseudopanax laetevirens*. j, *Azorella filamentosa*.

LIV. ARALIACEAE

Glabrous, dioecious shrubs. Leaves alternate, compound; stipules absent. Flowers in simple umbels in a racemose inflorescence. Calyx 5-lobed. Petals 5. Stamens 5, arising from nectariferous disc. Ovary inferior, 5-celled; styles 5. Fruit a ± fleshy berry; seeds with endosperm.

1. *Pseudopanax* C. Koch

Leaves digitately (3) 4– to 5– foliolate. Calyx-lobes shallow, truncate. Petals valvate, greenish. Styles united into short column with 5 branches. Seeds 5.

1. *P. laetevirens* (C. Gay) Seem. 1894, in Engler & Prantl, *Natürl. Pflanzenfam.* **3(8)**: 46.

Plant up to 3 m, erect, branched, with greyish bark. Leaflets 2–6 × 0.7–1.8 cm. linear– to elliptic-lanceolate, acute, cuneate at base, shallowly serrate, sessile; petiole 1.5–6 cm. Inflorescence terminal, branched. Umbels 2– to 5– flowered. Petals 1–1.5 × *c.*0.5 mm, oblong-ovate, subacute to obtuse. Stylar branches 0.5–0.75 mm, patent. Fruit 5–7 mm in diameter, oblong-globose, dark bluish. (Fig. 145).

Coastal mixed *Drimys – Nothofagus betuloides* forest.

W. Fuegia (Isla Desolación). (Map 17i).
W. Argentina *c.*50°–40°S.lat.
Chile north to *c.*35°S.lat.

LV. UMBELLIFERAE

Herbs, often forming dense cushions or carpets. Leaves alternate; stipules present or absent. Inflorescence a simple or compound umbel. Flowers actinomorphic, hermaphrodite or unisexual. Calyx indistinctly 5-lobed or absent. Petals 5, free. Stamens 5, alternating with petals. Ovary inferior, 2-celled; styles 2, often with enlarged base (stylopodium); stigma simple. Fruit dry, of 2 indehiscent, 1-seeded mericarps united by their faces (commissure), the fruit terete, or compressed dorsally (parallel to commissure) or laterally (at right-angles to commissure); each mericarp has 5 longitudinal ribs, 1 dorsal, 2 lateral (near edges of commissure) and 2 intermediate (between dorsal and lateral ribs), which may be obscure or prominent, sometimes also with secondary intermediate ribs; at maturity the mericarps normally separate and are usually suspended from the tip of a slender prolongation of the axis (carpophore).

a. Leaves hollow and cylindrical (fistulose) **11. *Lilaeopsis***
a. Leaves not fistulose:
 b. Plant forming dense cushions:
  c. Plant with stellate hairs **3. *Bolax***
  c. Plant glabrous or with simple hairs **2. *Azorella***
 b. Plant not forming dense cushions:
  d. Leaves entire to palmately lobed:
   e. Leaves orbicular or orbicular-reniform in outline:
    f. Leaves shallowly 7– to 8– lobed **1. *Hydrocotyle***
    f. Leaves deeply 3– to 5– lobed or -foliolate **4. *Schizeilema***
   e. Leaves not orbicular-reniform in outline:
    g. Leaf-segments linear to linear-lanceolate, not rigid **5. *Huanaca***
    g. Leaf-segments broadly elliptical to obovate, or subulate and rigid, or leaves entire **2. *Azorella***
  d. Leaves pinnate or deeply pinnate cut:
   h. Umbels simple **9. *Oreomyrrhis***
   h. Umbels compound:
    i. Fruit with spines; at least some bracts 3-fid or pinnatisect **7. *Daucus***
    i. Fruit without spines; bracts entire or absent:
     j. Fruit at least 3 times as long as wide, with slender basal appendages **6. *Osmorhiza***
     j. Fruit about as long as wide, not appendiculate:
      k. Petals yellow; fruit ± dorsally compressed, at least 5 mm **12. *Levisticum***
      k. Petals white; fruit ± laterally compressed, not more than 3 mm:
       l. Stem with purple spots; bracts and bracteoles present **8. *Conium***
       l. Stem without purple spots; bracts and bracteoles absent **10. *Apium***

1. *Hydrocotyle* L.

Perennial, creeping herbs. Leaves simple; stipules present. Umbels simple; flowers hermaphrodite or unisexual. Calyx indistinctly 5-lobed. Stylopodium depressed. Fruit strongly compressed laterally; commissure narrow; ribs obscure.

Literature: R. A. Pérez-Moreau, *Lilloa* **2(2)**: 413–463(1938).

1. *H. chamaemorus* Cham. & Schlecht. 1826, *Linnaea* **1**: 363.

Stems up to 15 cm or more, prostrate, slender, rooted at nodes, glabrous, with leaves borne singly at nodes. Leaves with lamina 5–25 × 7–30 mm, orbicular-reniform, cordate at base, shallowly 7- to 8-lobed, glabrous or sparsely hairy; lobes usually with 3 crenate, rarely serrate or crenulate, teeth; petioles 1–15 cm, slender, sparsely deflexed hairy but more so distally; stipules 1.5–2 × *c.*1.5 mm, oblong-ovate, obtuse, entire, brown, scarious. Umbels 8- to 12-flowered, 2–5 mm in diameter; peduncles 3–36 mm, slender, glabrous or sparsely deflexed-hairy, solitary in leaf-axils; pedicels 0.5–1 mm; bracteoles 0.4–0.8 mm, triangular, acute to acuminate, scarious, with green midvein. Fruit 1–1.5 × 2–2.3 mm, subglobose, glabrous, yellowish brown. $2n = 72^{\star}$. Self-compatible and at least facultatively self-pollinated. *Flowers November to January*. (Fig. 152).

Wet mud by ponds and streams; 0.30 m.

C. W. coast of Isla Grande (53°39′S., 69°47′W.). (Map 17h).
Argentina north to *c.*41°S.lat.
Chile north to *c.*37°S.lat.
Falkland Islands (Islas Malvinas).

2. *Azorella* Lam.

Perennial herbs, often forming dense cushions, glabrous or with simple hairs. Leaves simple; petioles with basal sheath; stipules absent. Umbels simple; flowers hermaphrodite or unisexual. Calyx 5-lobed. Petals brownish to yellowish. Stylopodium usually conical. Fruit terete to compressed dorsally; commissure $\frac{1}{6}$ – $\frac{1}{2}$ as wide as mericarps; ribs prominent to obscure.

a. Petiolar sheath conspicuously fimbriate to laciniate:
  b. Leaves entire **2. filamentosa**
  b. Leaves (2) 3- to 4-fid:
    c. Umbels 2-flowered; flowers sessile; fruit with obscure ribs **1. *lycopodioides***
    c. Umbels 4- to 6- flowered; flowers shortly pedicellate; fruit with ± prominent lateral ribs **3. *fuegiana***
a. Petiolar sheath not conspicuously fimbriate to laciniate:
  d. Leaves entire **5. *caespitosa***
  d. Leaves 3- to 5-fid:
    e. Umbels at least 6-flowered; bracteoles at least 4 **4. *trifurcata***
    e. Umbels not more than 3-flowered; bracteoles not more than 3:
      f. Leaf-lobes triangular-ovate, at least 1 mm wide; commissure narrow **6. *selago***
      f. Leaf-lobes subulate, not more than 1 mm wide; commissure $\frac{1}{2}$ as wide as mericarps **1. *lycopodioides***

1. *A. lycopodioides* Gaudich. 1825, *Ann. Sci. Nat.* **5**: 105.

Stems 3–30 cm, prostrate to ascending, much branched, lax or intertwined to form firm cushions up to 10 cm high, glabrous, covered with remains of leaves and sheaths. Leaves imbricate, often densely so, strict; lamina (2.3–)3–6(–7.5) × 1–2 mm, deeply 3-fid; lobes 2–6.5 × 0.4–0.8 mm, subulate, pungent, rigid, incurved, glabrous, smooth, shiny, with somewhat involute, minutely denticulate margin; petiole 1–5 mm, fimbrio-dentate to – laciniate, sheathing in basal half. Umbels usually with 2 sessile flowers; peduncle up to 2.5 mm, usually shorter than terminal leaves; bracteoles 2, 2–3 × *c.*2 mm, ovate, cymbiform, apiculate, fimbrio-dentate. Fruit 1.5–2 × 1.3–2.5 mm, ± ovoid, ± terete, glabrous, yellow-brown; commissure half as wide as mericarp; ribs obscure. $2n = 16$. *Flowers November to January*.

Gravels, rocks and peats in Magellanic moorland, feldmark, dwarf shrub heath and open scrub; 0–1000 m.

Almost throughout Fuegia. (Map 17g).
W. Argentina north to *c.*45°S.lat.
Chile north to *c.*36°30′S.lat.
Falkland Islands (Islas Malvinas).

2. *A. filamentosa* Lam. 1783, *Encycl. Méth. Bot.* **1**: 344.

Stems 2–16 cm, procumbent to suberect, branched, lax to densely clustered, glabrous, covered with remains of leaves and sheaths. Leaves imbricate, usually ± patent; lamina 3–11.5 × 0.7–3.3 mm, narrowly elliptical, acute to apiculate, cuneate at base, entire with ± involute margin, with up to 8 fimbriae on upper surface, glabrous beneath; petiole 3–17(–26) mm, prominently fimbrio-laciniate, sheathing in basal $\frac{1}{3}$–$\frac{2}{3}$. Umbels (4)5- to 13-flowered; peduncle 3.5–9.5 mm, about equalling terminal leaves; bracteoles 4–8, 2.3–2.7 × 0.5–0.8 mm, lanceolate–triangular, acute to obtuse, fimbriate near base; pedicels 1–2.6 mm. Fruit 1.7–2.5 × 0.8–1.5 mm, ovoid, compressed dorsally, glabrous, green; commissure narrow; ribs prominent. $2n = 16$. *Flowers November to February*.

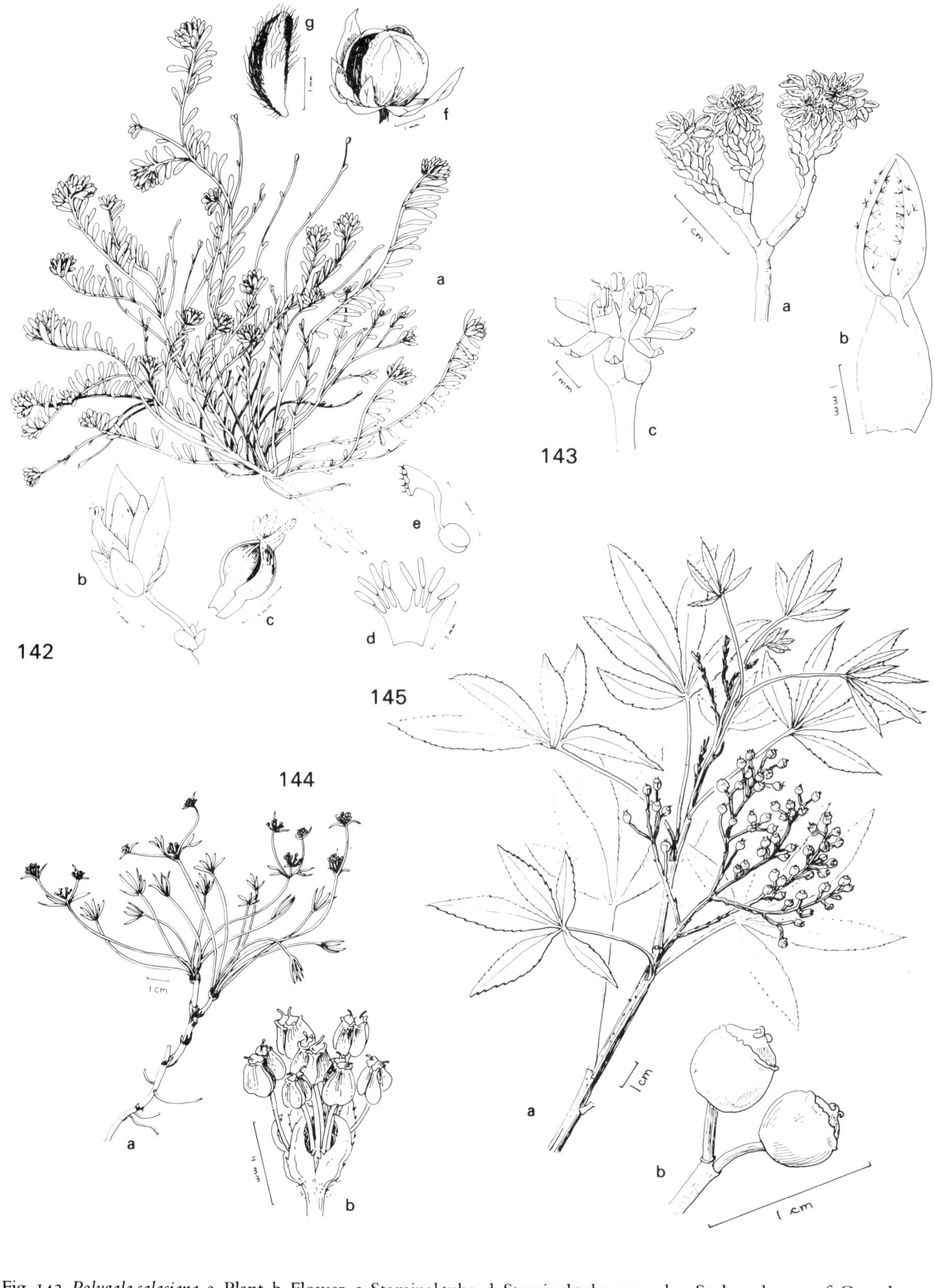

Fig. 142. *Polygala salasiana.* a, Plant. b, Flower. c, Staminal tube. d, Staminal tube opened. e, Style and ovary. f, Capsule. g, Seed. Scale = 1 cm (a) or 1 mm (b–g). Fig. 143. *Bolax caespitosa.* a, Branch. b, Leaf. c, Flower. Scale = 1 cm (a) or 1 mm (b–c). Fig. 144. *Huanaca acaulis.* a, Plant. b, Fruiting umbel. Scale = 1 cm (a) or 1 mm (b). Fig. 145. *Pseudopanax laetevirens.* a, Branch. b, Fruit. Scale = 1 cm.

(R. N. P. Goodall)

Gravels, rocks and peat, scrub, grassland; 0–100 m.

Almost throughout Fuegia except the extreme west. (Map 17j).
W. Argentina north to *c*.48°S.lat.
Chile north to *c*.51°S.lat.
Falkland Islands (Islas Malvinas).

3. ***A. fuegiana*** Speg. 1896, *Anal. Mus. Nac. Bs Aires*, Ser. 2, **5**: 58.

Stems 2–6 cm, procumbent to suberect, branched, lax to densely clustered, glabrous, covered with remains of leaves and sheaths. Leaves imbricate, usually ± patent; lamina 2.5–5 × 1.7–4 mm, broadly elliptical to obovate, truncate 2– to 3– fid, with scabrid fimbriae at base; lobes 0.7–1.8 × 0.6–1 mm, elliptical to oblong or triangular, acute to apiculate, with scabrid fimbriae on margin; petiole 6–16 mm, with short, triangular, fimbriate basal sheath. Umbels 4– to 6– flowered; peduncle 1.4–8 mm, about equalling terminal leaves; bracteoles 4–8, 2.4–3 × 0.6–1 mm, triangular, apiculate, fimbriate near base; pedicels 1–3 mm. Fruit 1.5–1.7 × 1–1.2 mm, ovoid, ± compressed dorsally, glabrous, yellow-green; commissure narrow; lateral ribs ± prominent. *Flowers November to January*.

Empetrum heath, open *Chiliotrichum* scrub, open grassland, sands, rock-crevices; 0–250 m.

Isla Grande, except the extreme west; Isla de los Estados. (Map 18a).
Argentina north to ?50°S.lat.
E. Chile north to *c*.50°S.lat.

4. ***A. trifurcata*** (Gaertner) Hooker f. 1847, *Fl. Antarct.* **1(2)**: 283.

Plant forming dense mats or cushions up to 11 cm high and 45 cm in diameter; stems prostrate to ascending, much branched, somewhat woody, covered with remains of leaves below. Leaves closely imbricate, patent, coriaceous; lamina 6–12(–16) × 2–5(–7) mm, cuneate, 3(–5) – fid, cuneate at base, glabrous; lobes 1.7–4.5(–8) × 0.7–1.7 mm, narrowly triangular, acute, with pungent apex; petiole (2.5–)5–10(–22) mm, with short, fimbriate sheathing base. Umbels 6– to 9 (–12) – flowered; peduncle 2–18(–23) mm, equalling or exceeding terminal leaves; bracteoles 4–8, 1.5–3.5 × 0.6–1 mm, triangular, acuminate, rigid, fimbriate and ± connate at base; pedicels 1.5–1.8 mm. Fruit 2–3 × 1.3–2.6 mm, ovoid, compressed dorsally, glabrous, yellow-green to reddish; commissure narrow; dorsal, lateral and sometimes intermediate ribs ± prominent. $2n = 16$. *Flowers December to February*. (Fig. 146).

Coastal sands and gravels, stream banks, dry *Empetrum* heath, forest margins; 0–200 m.

Isla Grande, except the extreme west; Isla de los Estados. (Map 18b).
Argentina north to *c*.46°S.lat.
Chile north to *c*.37°S.lat.

5. ***A. caespitosa*** Cav. 1799, *Icon. Descr.* **5**: 57.
A. hookeriana Clos. 1847, in Gay, *Hist. Chile Bot.* **3**: 81–Alboff, 1896.

Plant forming dense mats or cushions up to 17 cm high and 60 cm in diameter; stems ascending to erect, much branched, woody, covered with remains of leaves. Leaves closely imbricate, patent; lamina 3–10(–14) × 1–2.5(–3) mm, lanceolate to ovate-lanceolate, apiculate, narrowed slightly at base, entire or rarely distantly serrate, glabrous, rarely with scattered fimbriae, shiny; petiole 2–6.5(–8) mm, with white marginal fimbriae, sheathing at base. Umbels (1–)4– to 8(–11) – flowered; peduncle 1.5–2.2 mm, not exceeding terminal leaves; bracteoles usually 4, 5–6 × *c*.1 mm, acicular, with long, white marginal fimbriae near base; pedicels 1.7–2 mm. Fruit 2.5–3 × *c*.3 mm, ovoid, compressed dorsally, glabrous, greenish brown; commissure narrow; dorsal rib prominent, the lateral and intermediate obscure to prominent. $2n = 16$. *Flowers December to February*.

Sands, gravels and rocky areas, open grassland, dry *Empetrum* heath; 0–50 m.

Isla Grande, except the extreme east and west, islands of Canal Beagle. (Map 18c).
W. Argentina north to *c*.41°S.lat.
E. Chile north to *c*.51°S.lat.
Falkland Islands (Islas Malvinas).

6. ***A. selago*** Hooker f. 1847, *Fl. Antarct.* **1(2)**: 284.

Plant forming dense cushions up to 60 cm high and up to 100 cm in diameter; stems ascending to erect, much branched, woody, covered with remains of leaves. Leaves closely imbricate, appressed; lamina 2.2–4 × (2–)4–7 mm, obovate, deeply 3– to 5– fid, abruptly cuneate at base, convex and glabrous beneath, concave and setose above; lobes 1.5–2.5 × (0.7)1–15 mm, triangular-ovate, apiculate, entire, thickened on margin; petiole *c*.5 mm, glabrous or sparsely fimbrio-dentate above, with partially sheathing base. Umbels 1– to 4– flowered; peduncle 2–2.5 mm, not exceeding terminal leaves; bracteoles 2–3, 2–4 × 0.7–1.3 mm, linear-lanceolate, acute, rigid, dentate towards base; pedicels 0.5–2 mm. Fruit 1–1.5 × 0.7–1.4 mm, oblong to obovoid, strongly compressed dorsally, glabrous, red – to yellow-brown; commissure narrow; ribs prominent. $2n = 16$. *Flowers January to February*.

Feldmark, often with other cushion plants; 450–1100 m.

S. Fuegia. (Map 18d).
Falkland Islands (Islas Malvinas), Iles de Kerguelen, Iles Crozet, Marion Island, Heard Island, Macquarie Island.

3. *Bolax* Comm.

Gynodioecious, perennial herbs forming dense cushions, with stellate hairs. Leaves closely imbricate, simple; petioles with basal sheath; stipules absent. Umbels simple. Sepals 5, petaloid. Petals greenish white. Stylopodium conical. Fruit compressed dorsally; commissure narrow; ribs mostly prominent.

a. Leaves entire **2. *caespitosa***
a. Leaves 3-lobed to 3-fid **1. *gummifera***

1. *B. gummifera* (Lam.) Sprengel 1818, *Spec. Umbellif.* 10.
B. glebaria Comm. ex Gaudich. 1825, *Ann. Sci. Nat.* **5**: 104 – Hooker, 1847;
Azorella gummifera (Lam.) Franchet 1899, *Miss. Sci. Cap Horn* 337 – Alboff, 1896.
A. glebaria (Comm. ex Gaudich.) A. Gray 1854–56, *U.S. Expl. Exped.* 701. – Spegazzini, 1896.

Plant forming hard cushions up to 75 cm high or deep mats up to 135 cm in diameter; stems ascending to erect, branched, woody, covered with remains of leaves, exuding a white latex which becomes reddish brown on drying. Leaves erecto-patent; lamina 2–6.5 × 2–5.5 mm, 3-lobed to 3-fid, cuneate at base, stellate-hairy, or glabrous beneath; lobes (0.3–)1–4 × (0.3–)0.8–2 mm, ovate to linear-lanceolate, obtuse to subacute, entire, the margin often somewhat involute; petiole 4–6 (–12) mm, glabrous to stellate-hairy, sheathing in basal ½. Umbels 3– to 20–flowered; peduncle 2–4 mm, not exceeding terminal leaves; pedicels 2–3 mm. Fruit 2–3.5 × 1–2.5 mm, ovoid, tetragonous, stellate-hairy, dark brown. $2n = 32$. *Flowers October to February*. (Fig. 147).

Cliff and rocks, open scrub and heath, forest-margins and clearings, feldmark; 0–900 m.

Throughout Fuegia except much of the west. (Map 18e).
W. Argentina north to *c.*41°S.lat.
Chile north to *c.*44°30′S.lat.
Falkland Islands (Islas Malvinas).

Plants growing in partial shade tend to be much laxer, with larger leaves.

2. *B. caespitosa* Hombron & Jacquinot ex Dcne 1853, *Voy. Pôle Sud* 62.
B. bovei (Speg.) Dusén 1914, in Macloskie & Dusén, *Rep. Princeton Univ. Exped. Patag* **8, Suppl.**, 195.
Azorella bovei Speg. 1896, *Anal. Mus. Nac. Bs. Aires*, Ser.2, **5**: 58.

Plant forming hard cushions or mats up to 20 cm high and up to 50 cm in diameter; stems ascending to erect, branched, woody, covered with remains of leaves. Leaves patent; lamina (1.5–)2.5–3 × 1.2–1.8 mm, elliptical, obtuse, navicular, ± cucullate, entire, stellate-hairy above; petiole (2–)3–4 mm, glabrous, sheathing in basal. Umbels 3– to 4– flowered; peduncle very short or absent; pedicels 1–3 mm. Fruit *c.*2–3 × 1–2 mm, ovoid, tetragonous, stellate-hairy, brown. *Flowers November to January*. (Fig. 143).

Magellanic moorland, wet facies of feldmark; 0–525 m.

S. Fuegia. (Map 18f).
Chile north to 52°S.lat.

4. *Schizeilema* Domin

Glabrous, perennial, creeping herbs. Leaves deeply lobed to compound; stipules present, ⅔–¾ united to petiole. Umbels usually compound; flowers hermaphrodite. Calyx indistinctly 5-lobed. Petals whitish to purplish. Stylopodium depressed – conical to conical. Fruit somewhat compressed dorsally; commissure narrow; ribs mostly prominent.

1. *S. ranunculus* (D'Urv.) Domin 1908, *Bot. Jahrb.* **40**: 576.
Azorella ranunculus D'Urv. 1825, *Fl. Is. Mal* 46 – Hooker, 1847; Hariot, 1884; Dusén, 1900.

Stems up to 35 cm, prostrate, slender, branched, with leaves, flowers and roots at nodes. Leaves with lamina 4–10(–25) mm in diameter, suborbicular in outline, cordate at base, palmately 3– to 5– lobed to

146

147

149

148

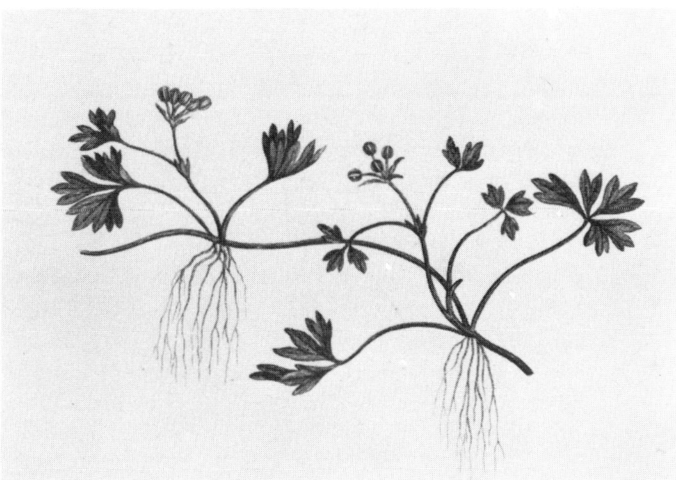

Fig. 146. *Azorella trifurcata*. Fig. 147. *Bolax gummifera*. Fig. 148. *Schizeilema ranunculus*. Fig. 149. *Apium australe*.
(Sydney Parkinson)

– foliolate; segments or leaflets 3–10(–14) × 2–7(–12.5) mm, cuneate, with 2–3(–5), rounded to subacute, shortly apiculate, entire apical lobes; petiole 5–45 mm, slender; stipules 2–4 × 0.3–1.5 mm, ovate-lanceolate, acute to acuminate, sparsely fimbriate to fimbrio-dentate, membranous. Partial umbels 1–2, (3)4– to 6(–8) – flowered; peduncle 4–15 mm, up to 50 mm in fruit; bracts 1–2, 2–3 × *c*.2 mm, ovate, acute, fimbriate; pedicels 1.5–3(–4) mm, patent; bracteoles 4–5, 3–4 × 0.5–1 mm, linear-lanceolate, acute, dentate towards base. Fruit 1.7–2.3 × 1.5–2 mm, ovoid, tetragonous; ribs prominent or the intermediate obscure. $2n = 16$*. *Flowers December to February*. (Fig. 148).

Damp grassland, forest floors, scrub, with cushion plants in feldmark; 20–1000 m.

Throughout Fuegia except the extreme west and north. (Map 18g).
W. Argentina north to *c*.49°S.lat.
Chile north to *c*.41°30′S.lat.
Falkland Islands (Islas Malvinas).

5. *Huanaca* Cav.

Perennial, creeping to erect, often scandent, herbs. Leaves deeply 3– to 7–partite; stipules present, united to petiole. Umbels simple; flowers androdioecious. Calyx 5-dentate. Petals greenish yellow or purple. Stylopodium conical. Fruit compressed dorsally; commissure narrow; ribs prominent.

Literature: M. E. Mathias and L. Constance, Kurtziana **6**: 7–23 (1971).

1. *H. acaulis* Cav. 1800, *Icon. Descr.* **6**: 18.

Stems up to 30 cm, slender, with leaves ± clustered in basal rosette, glabrous. Leaves with lamina palmately divided, glabrous to somewhat hairy; segments 4.5–35 × 0.4–1 mm, linear to linear-lanceolate, acute, entire to 2– to 3– fid; petiole up to 20 cm, slender; stipules up to 7 mm, oblong to ovate, fimbriate, often pinkish. Umbels 3, terminal on stem; median 15– to 30– flowered, with peduncle 0–30 mm, the flowers hermaphrodite; lateral with fewer flowers, the peduncles 8–75 mm, with some or all flowers male; bracteoles 6–10, 6–13 × 0.4–1 mm, linear, entire or lobed, glabrous or hairy. Fruit 1.5–2.5 × 1.2–1.8 mm, ovoid, the mericarps ± concave. $2n = 18$. *Flowers December to February*. (Fig. 144).

Coastal sands, grassland, open scrub; 0–150 m.

N. Isla Grande. (Map 18h).
Argentina north to *c*.46°S.lat.
E. Chile north to *c*.51°S.lat.

6. *Osmorhiza* Rafin.

Perennial herbs. Leaves 2-pinnate; stipules absent. Umbels compound, terminal and lateral; bracts and bracteoles minute or absent. Calyx obscurely lobed. Petals white or greenish. Stylopodium conical. Fruit terete, with short apical beak, tapering at base into slender 'tails'; commissure narrow; ribs filiform, acute, setose.

Literature: L. Constance and R. H. Shan, *Univ. Calif. Publ. Bot.* **23**: 111–156(1948).

a. Pedicels widely diverging; fruit clavate **2. *depauperata***
a. Pedicels erecto-patent; fruit linear-oblong **1. *chilensis***

1. *O. chilensis* Hooker & Arn. 1833, *Bot. Misc.* **3**: 355.
O. berteroi DC. 1830, *Prodr.* **4**: 232 – Skottsberg, 1906.

Stems up to 65 cm, ascending to erect, branched above, terete, striate, glabrous to ± hairy. Leaves with lamina 3–14.5 × (2.5–)4.5–11 cm, triangular-ovate in outline, with 3–5 leaflets; petiole up to 15 cm, sparsely hairy with reddish sheathing base; leaflets 20–70 × 20–70 mm, triangular-ovate in outline, the lower pinnate and with petiolules 4–25(–45) mm, at least the terminal pinnatifid, the ultimate segments prominently serrate. Umbels 10– to 35– flowered; rays 2–6, 3–10 cm; pedicels 6–20 mm, erecto-patent, glabrous. Fruit (12–)14–18 × 1.5–2.5 mm, linear-oblong, densely retrorse-hispid towards base, more sparsely so above. *Flowers November to January*. (Fig. 151).

Deciduous *Nothofagus* forest, moist shaded cliffs; 0–200 m.

Isla Grande, except the extreme north and west, islands on S. side of Canal Beagle. (Map 18i).
W. Argentina north to *c*.40°S.lat.
Chile north to *c*.31°S.lat.
North America.

2. *O. depauperata* Phil. 1894, *Anal. Univ. Chile* **75**: 926.

Similar to *O. chilensis* but with pedicels widely diverging and somewhat smaller ((6.5–)8.5–15 mm), clavate fruits. $2n = 22$. *Flowers November to January*.

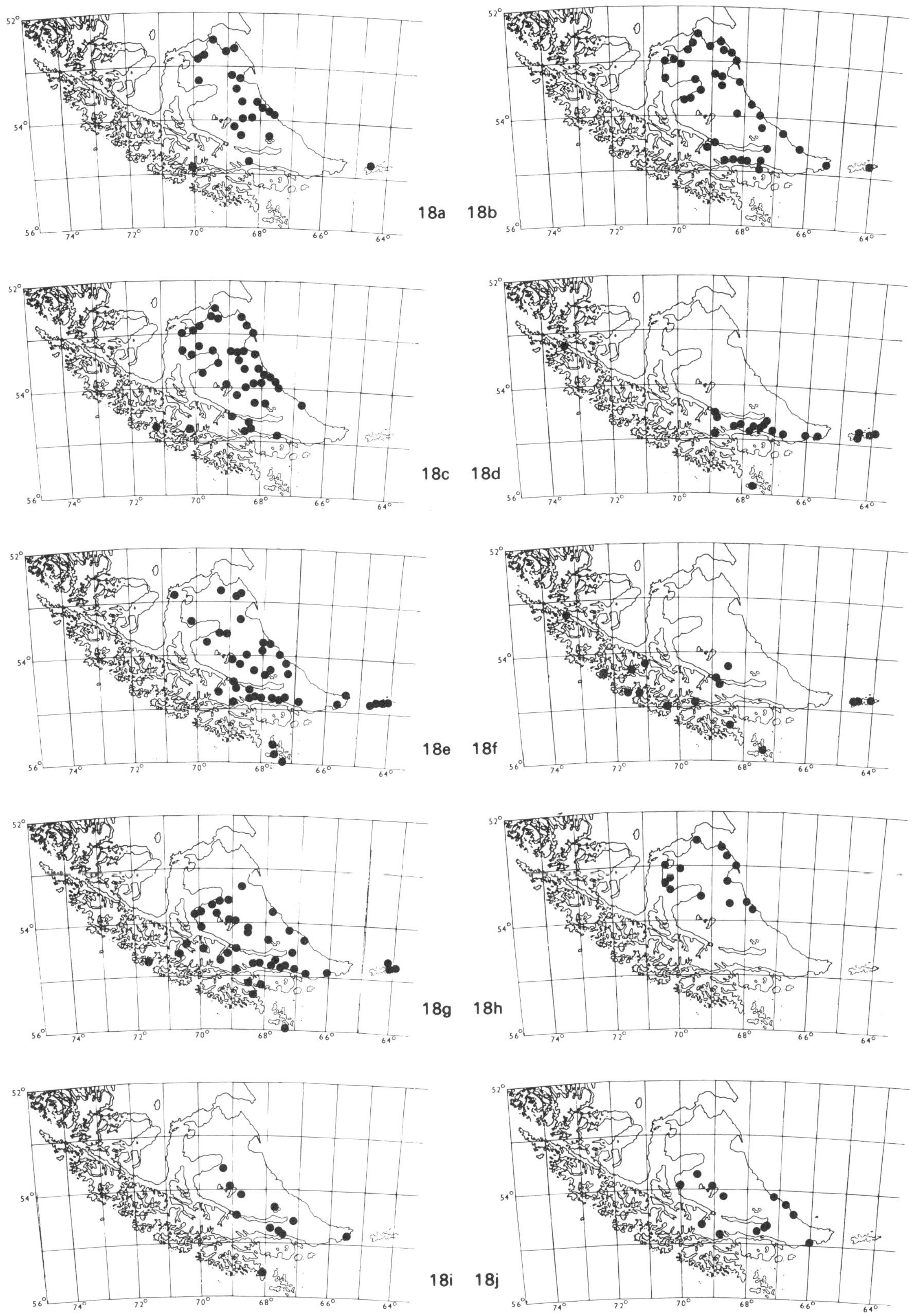

Map 18. a, *Azorella fuegiana*. b, *Azorella trifurcata*. c, *Azorella caespitosa*. d, *Azorella selago*. e, *Bolax gummifera*. f, *Bolax caespitosa*. g, *Schizeilema ranunculus*. h, *Huanaca acaulis*. i, *Osmorhiza chilensis*. j, *Osmorhiza depauperata*.

Deciduous *Nothofagus* forest, moist scrub; 0–300 m.

Isla Grande, except the extreme north and west, islands on S. side of Canal Beagle. (Map 18j).
W. Argentina north to 42°S.lat.
Chile north to *c*.32°S.lat.
North America.

7. *Daucus* L.

Annuals herbs. Leaves pinnate; stipules absent. Umbels compound. Calyx-lobes very small or absent. Petals white to pinkish. Stylopodium conical. Fruit dorsally compressed; commissure narrow; primary ribs filiform, setose, the secondary ribs with barbed spines.

1. *D. montanus* Humb. & Bonpl. ex Sprengel 1820, in Roemer & Schultes, *Syst. Veg.* **6**: 482.
D. australis Poeppig ex DC. 1830, *Prodr.* **4**: 214.

Stems 11–40 cm, erect, simple or with a few branches, subglabrous to retrorse-hispid. Leaves with lamina 20–60 × 13–40 mm, oblong in outline; leaflets 5–12 × 4–10 mm, oblong-ovate in outline, deeply pinnately divided into linear or pinnatisect lobes, the ultimate segments 2–5 × *c*.1 mm. linear; petioles 30–70 mm, with purplish basal sheath. Umbels exceeding leaves at maturity, subtended by leaflike, pinnately divided bracts with linear segments; rays 4–15, 5–60 mm, unequal; partial umbels 4– to 15– flowered, subtended by small, linear bracteoles; pedicels 3–10 mm, unequal. Fruit 3–6 × 1.5–2 mm, oblong, dark brown, the spines 1–1.5 mm. Self-compatible, at least facultatively self-pollinated. $2n = 66$. *Flowers November to January.* (Fig. 150).

Coastal sands, open grassland, clearings in forest and scrub; 0–60 m.

S. & W. C. coasts of Isla Grande. (Map 19a).
W. Argentina and E. Chile, north along Andes to Colombia.
Mexico.

8. *Conium* L.

Subglabrous overwintering annuals or biennials. Leaves pinnate; stipules absent. Umbels compound. Calyx absent. Petals white. Stylopodium depressed – conical. Fruit laterally compressed; commissure narrow; ridges prominent, often undulate-crispate.

1. *C. maculatum L. 1753, *Sp. Pl.* 243.

Stems up to 45 cm, erect, branched, furrowed, with purplish spots. Leaves 6.7–18(–36) × 3–22 cm, ovate to deltoid in outline, 2– to 3– pinnate; leaflets 2.3–25 × 1.2–9 cm, ovate, 1– to 2– pinnate, ultimate segments coarsely serrate. Umbels terminal and axillary; peduncles 3–6 cm; bracts 8–10.5 × 0.4–3 mm, linear-lanceolate to ovate, acute; rays 9–16, 13–23 mm; bracteoles 2–4 × 0.6–1.2 mm, ovate, acuminate; partial umbels 7– to 21– flowered. Fruit 1.5–2.7 × 1.5–2.7 mm, subglobose. *Flowers December to February.*

Disturbed ground by roads and buildings.

N. W. Isla Grande.
Introduced from Europe and locally naturalized.

9. *Oreomyrrhis* Endl.

Perennial herbs. Leaves pinnate; stipules absent. Umbels simple. Calyx minutely 5-lobed. Petals white. Stylopodium conical. Fruit slightly laterally compressed; commissure as wide as mericarps; ribs prominent.

Literature: M. E. Mathias and L. Constance, *Univ. Calif. Publ. Bot.* **27**: 347–416(1955).

1. *O. hookeri* Mathias & Const. 1955, *Univ. Calif. Publ. Bot.* **27**: 369.
O. daucoides auct., non Urban – Franchet, 1889.
O. andicola auct., non (Kunth) Hooker f. – Dusén, 1900.

Stems up to 10 cm, branched at or near base, or absent, sparsely hairy. Leaves with lamina 1.5–7 × 0.6–3 cm, oblong-ovate in outline, with 4–6 pairs of leaflets, sparsely hairy; leaflets 3–15 × 1–8 mm, ovate to oblong, acute to obtuse, pinnatifid, with linear-lanceolate, apiculate segments; petiole 1–5 cm, slender, with membranous, ciliate sheath below. Umbels 9– to 15– flowered; peduncles 2.5–8.5 cm, stout, densely retrorse-hairy; bracteoles 5–8, 2–3 × 1–1.5 mm, ovate, acute, entire to 2– to 5– lobed, appressed hairy, united towards base; pedicels 4.5–10 mm, unequal. Fruit 3–4 × 1–1.5(–2) mm, ± oblong, tapering towards apex, ± glabrous, dark brown. $2n = 12$. *Flowers November to January.* (Fig. 154).

Sands and gravels, open grassland and scrub; 0–250 m.

Almost throughout Fuegia except the extreme east, south and west. (Map 19b).
Argentina north to 51°45′S.lat.
Chile north to 51°25′S.lat.
Falkland Islands (Islas Malvinas).

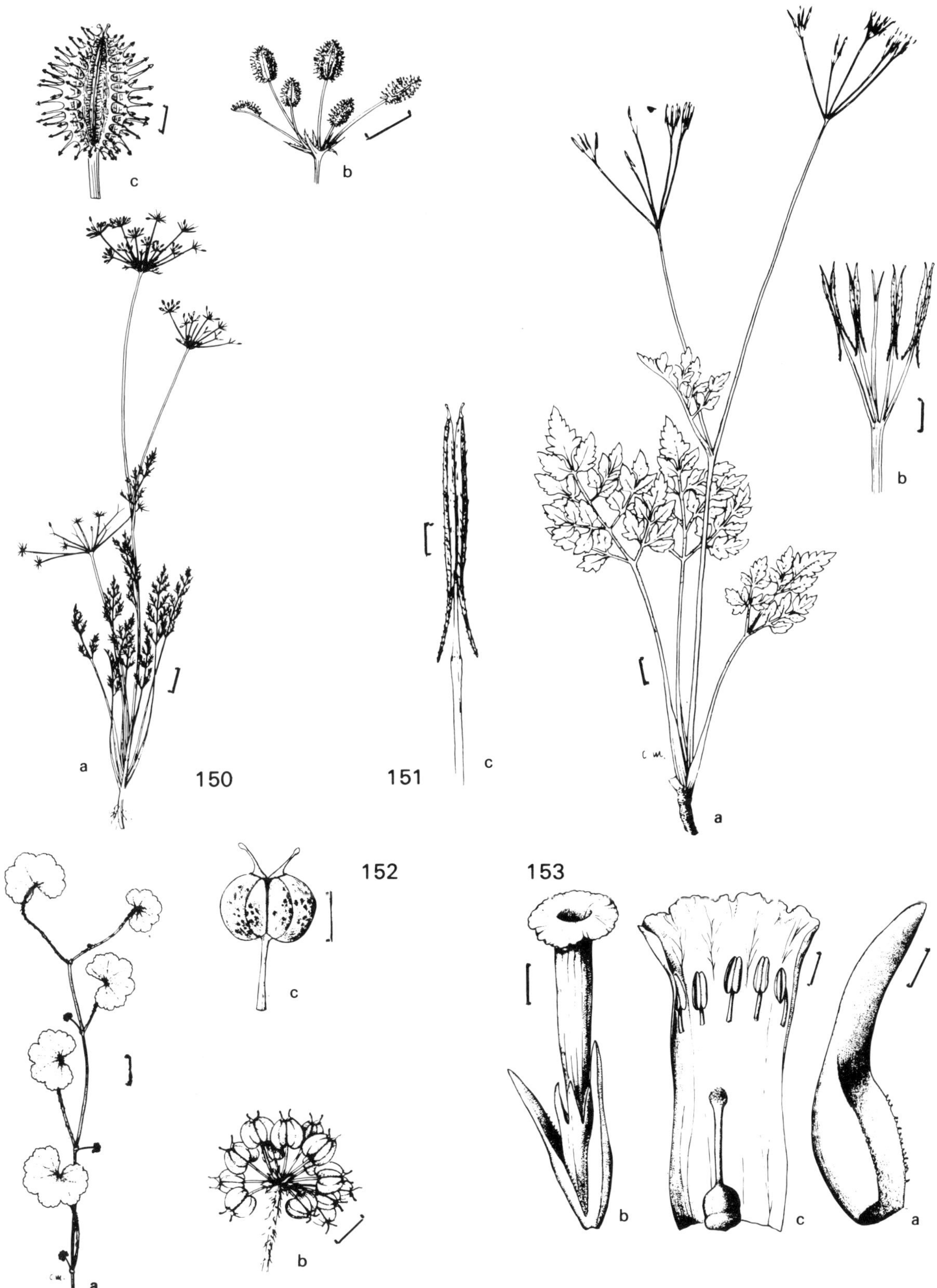

Fig. 150. *Daucus montanus*. a, Plant. b, Partial umbel. c, Fruit. Scale = 1 cm (a) or 1 mm (b–c). Fig. 151. *Osmorhiza chilensis*. a, Plant. b, Partial umbel. c, Fruit. Scale = 2 cm (a), 5 mm (b) or 1 mm (c). Fig. 152. *Hydrocotyle chamaemorus*. a, Plant. b, Umbel. c, Fruit. Scale = 1 cm (a), 2 mm (b) or 1 mm (c). Fig. 153. *Benthamiella nordenskjoldi*. a, Leaf. b, Flower. c, Opened corolla. Scale = 2 mm (b) or 1 mm (a, c).

(Flora Patagónica)

10. *Apium* L.

Glabrous, perennial herbs. Leaves pinnate; stipules absent. Umbels compound. Calyx-teeth absent. Petals white. Stylopodium depressed-subconical. Fruit ± laterally compressed; commissure narrow; ribs very prominent.

1. *A. australe* Thouars 1808, *Fl. Tristan d'Acugna* 43.

A. graveolens auct., non L. – Hooker, 1847; Franchet, 1889; Spegazzini, 1896; Dusén, 1900.

Stems 7.5–54 cm, ascending to erect, branched, striate. Leaves with lamina (3–)4–12 × 2–7 cm, lanceolate to ovate-oblong in outline, with 2–5 pairs of leaflets; leaflets 10–42 × 12–40 mm, the lower deltoid, 3-fid to 3-foliolate, petiolulate, becoming entire and sessile towards apex, the ultimate segments serrate; petiole 4.5–32(–50) cm, with purplish basal sheath. Partial umbels (5–)8– to 24–flowered; rays 5–13, 6–35(–90) mm; bracts and bracteoles absent; pedicels 1–6 mm. Fruit 1.5–2 × 1.5–2 mm, ovoid-globose, glabrous, pale brown; ribs obtuse. $2n = 22$. *Flowers October to March.* (Fig. 149).

Damp places in all ± open coastal habitats, less commonly in wet places inland; 0–75 m.

Throughout Fuegia. (Map 19c).
W. Argentina north to c.43°S.lat.
Chile north to *c.*35°S.lat.
Falkland Islands (Islas Malvinas).
Tristan da Cunha, New Zealand, S. E. Australia, Lord Howe Island.

11. *Lilaeopsis* Greene

Glabrous, creeping, perennial herbs. Leaves basal, reduced to fistulose, transversely septate phyllodes, with basal sheath; stipules absent. Umbels simple. Flowers hermaphrodite. Calyx indistinctly 5-lobed. Petals white with reddish tips. Stylopodium depressed. Fruit not compressed; commissure wide; lateral ribs conspicuously thicker than others.

Literature: R. A. Pérez-Moreau, *Lilloa* **1**: 283–306(1937).

1. *L. hilli* Pérez-Moreau 1935, *Physis* **11**: 478.

Crantzia lineata auct., non Nutt. – Dusén, 1900; Skottsberg, 1916.

Stem up to 17(–33) cm, prostrate, branched, rooted at nodes, bearing tufts of leaves and flowers at some nodes. Leaves (1.7–)2.5–11(–17) × 0.07–0.3 cm, filiform, obtuse, with 1–9 septa, the basal sheath 0.4–1.3 cm, entire. Umbels 4– to 9– flowered; peduncles 2–25 mm, slender; bracteoles 0.5–2 × 0.3–0.4 mm, ovate to ovate-lanceolate, obtuse, entire or crenate-serrate; pedicels 3–7 mm, ascending to pendent. Fruit 1.9–2.3 × 1.6–2.3 mm, subglobose to ovoid, yellowish brown. $2n = 44\star$. *Flowers November to January.* (Fig. 155).

Wet sand, mud and rock-crevices by sea, rivers and lakes, often submerged in still or slow-moving freshwater; 0–20 m.

Much of Isla Grande except the extreme west. (Map 19d).
Argentina north to *c.*46°S.lat.
Chile north to *c.*51°S.lat.

12. *Levisticum* Hill

Stout, perennial herbs. Leaves pinnate; stipules absent. Umbels compound. Flowers hermaphrodite. Calyx-lobes absent. Petals greenish-yellow. Stylopodium subconical. Fruit somewhat compressed dorsally; commissure wide; marginal ridges with rather thick wings, the dorsal low or narrowly winged.

1. *L. officinale Koch 1824, *Nova Acta Acad. Leop.-Carol.* **12**: 101.

Stems 10–250 cm, erect, branched, striate. Leaves 2– to 3– pinnate; lower leaves up to 70 × 65 cm, triangular-rhombic in outline, with *c.*5 pairs of leaflets having long cuneate irregular and deeply dentate lobes; uppermost leaves 45–65 × 3–5 mm, lanceolate, acuminate; petioles 3–15 cm. Peduncle 2.2–7.5 cm; bracts 16–32 × 1.7–2.5 mm, lanceolate, acuminate, often absent; rays 10–15 mm; partial umbels 12– to 20– flowered; bracteoles 5.5–10 × 1.5–3.5 mm, ovate, subulate; pedicels 2–2.5 mm. Fruit 5–7 × 3–5 mm, ovoid-oblong, yellow or brown.

Disturbed ground near buildings; 0–50 m.

S. coast of Isla Grande.
Introduced from Europe, and cultivated for its edible stems, locally escaped and naturalized or persisting in abandoned gardens.

LVI. DESFONTAINIACEAE

Evergreen shrubs. Leaves opposite, simple; stipules absent. Flowers solitary, terminal. Sepals 5, united at base. Corolla shallowly 5-lobed. Stamens 5, inserted at base of corolla-lobes. Ovary superior, (3–)5-celled; style 1; stigma subcapitate. Fruit a berry.

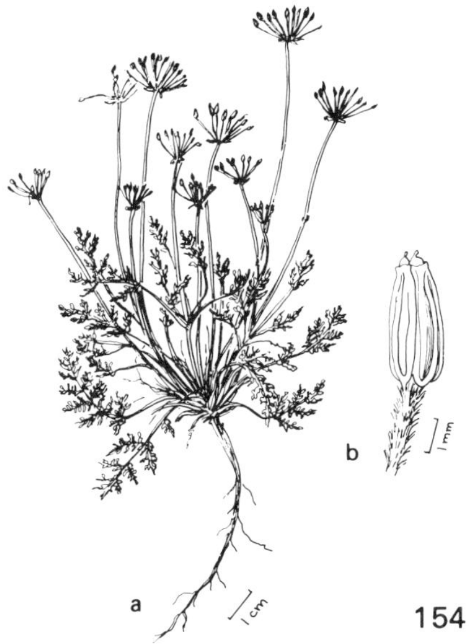

154

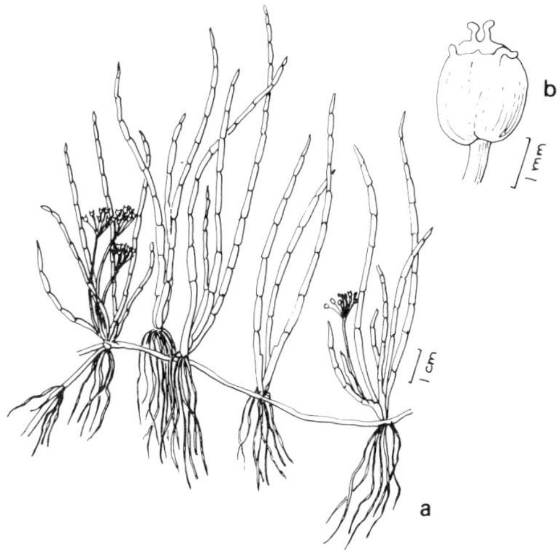

155

156

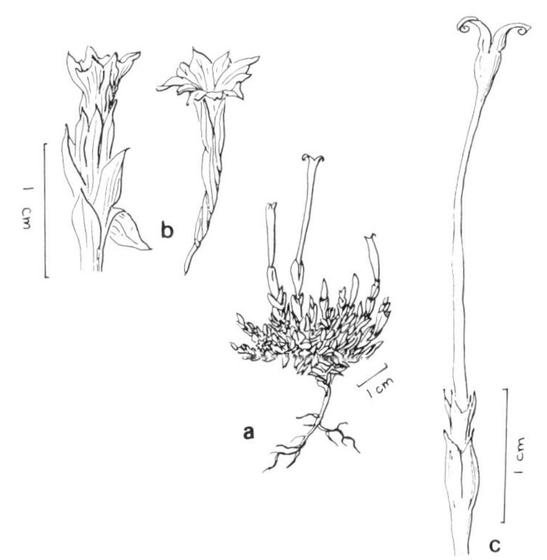

157

Fig. 154. *Oreomyrrhis hookeri*. a, Plant. b, Fruit. Scale = 1 cm (a) or 1 mm (b). Fig. 155. *Lilaeopsis hillii*. a, Plant. b, Fruit. Scale = 1 cm (a) or 1 mm (b). Fig. 156. *Gentiana prostrata*. a, Plant. b, Flowers. c, Fruit. Scale = 1 cm. Fig. 157. *Desfontainia spinosa*. Scale = 1 cm.

(R. N. P. Goodall)

1. *Desfontainia* Ruiz & Pavón

Leaves coriaceous, spiny. Peduncles with 2 basal bracts. Corolla infundibuliform. Anthers subsessile. Fruit many-seeded.

1. ***D. spinosa*** Ruiz & Pavón 1799, *Fl. Peruv.* **2**: 47.

Plant 1.5–3 m, branched, with pale brownish, shiny bark, the twigs greyish. Leaves with lamina 2–4 × 1–2.3 cm, elliptical to ovate or somewhat obovate, acute and pungent, cuneate at base, with 2–8 pairs of spiny, broadly triangular teeth, the spines 2–2.5 mm; petiole 8–10 mm. Peduncles 9–10 mm; bracts *c.*6 × 1–1.2 mm, ± linear. Calyx-segments 7–8 × 3–3.5 mm, elliptic-oblong, obtuse, ciliate, sparsely pubescent. Corolla 4.5–6 cm, with rotund lobes, the tube red, the lobes yellow-orange. Stigma included or slightly exserted. Berry 12–15 × 9–11 mm, oblong-subglobose, greenish-purple; seeds 2–2.5 × 1–1.2 mm, oblong-obovoid, smooth. *Flowers December to February*. (Fig. 157).

Coastal *Drimys – Nothofagus betuloides* forest; 0–30 m.

Isla Desolación and Isla de los Estados (perhaps extinct here). (Map 19e).
W. Argentina *c.*42°–40°S.lat.
Chile north to *c.*40°S.lat.
North along Andes to 4°25′N.lat.

Pollinated by Green-backed Firecrown hummingbird.

LVII. GENTIANACEAE

Glabrous annual or biennial herbs. Leaves opposite, entire; stipules absent. Flowers hermaphrodite, actinomorphic, solitary or in lax cymes. Calyx 5-lobed. Corolla 5-lobed. Stamens 5, inserted on corolla-tube. Nectaries at base of ovary. Ovary superior, 1-celled; style 1, bifid; stigmas capitate. Fruit a septicidal capsule, dehiscing by 2 valves; seeds numerous, with endosperm.

a. Flowers solitary; corolla with appendages between the lobes, not ciliate on lobes and throat **1. *Gentiana***
a. Flowers usually in lax dichasial cymes; corolla without appendages between the lobes, usually ciliate on lobes or throat **2. *Gentianella***

1. *Gentiana* L.

Annual. Flowers solitary, terminal. Calyx-teeth partly joined by a membrane which forms the upper part of the tube. Corolla-lobes with a small appendage between each pair.

1. ***G. prostrata*** Haenke 1789, in Jacq., *Collect. Bot.* **2**: 66.
   G. sedifolia Humb., Bonpl. & Kunth var. *micrantha* Weddell 1855, *Chloris Andina* 73– Alboff, 1896.

Stem 1.5–3 cm, procumbent or ascending, simple or branched from base. Leaves 5–8 × 1.7–2.3 mm, elliptic-obovate, acuminate to obtuse, with membranous margin. Calyx 6–7 mm, the teeth 1–1.2 × 0.8–1 mm, triangular, acute. Corolla 9–10 mm, obconical, bluish, green towards base. Peduncle elongating in fruit. Capsule 7–9 × 2.3–2.5 mm, cylindrical-ellipsoid; seeds c.0.2 mm. *Flowers November to December*. (Fig. 156).

Wet grassland. *Carex* communities and open gravel; 0–600 m.

E. Isla Grande. (Map 19f).
E. Chile north to *c.*30°S.lat.
W. Argentina north to *c.*33°S.lat.
Peru north to *c.*8°S.lat.
N. America, Europe.

2. *Gentianella* Moench

Annual or biennial. Flowers in lax dichasial cymes, rarely solitary. Calyx-lobes not joined by inner membrane. Corolla ± cylindrical, with spreading lobes, without appendages in the sinuses.

1. ***G. magellanica*** (Gaudich.) Fabris ex D. M. Moore 1968, *Brit. Antarct. Survey Sci. Rep.* **60**: 103.
   Gentiana magellanica Gaudich. 1825, *Ann. Sci. Nat.* **5**: 102 – Hariot, 1900.
   G. patagonica Griseb. 1839, *Gen. Sp. Gent.* 237 – Spegazzini, 1896; Alboff, 1896; Dusén, 1900; Skottsberg, 1906.

Stems 2–25 cm, erect, simple or branched, usually above middle. Basal leaves 5.5–15(–25) × 1–5 mm, oblong to oblanceolate, the cauline (7.5–)10–20(–28) × 1.7–9.5 mm, ovate to ovate- or oblong-lanceolate, all obtuse to subacute, entire, sessile. Calyx (5.6–)7–12 mm, tubular, the lobes 4–7.3 × 1–3.7 mm, oblong to oblong-lanceolate, acute to apiculate, green to dark purple. Corolla 10–18 mm, white to bluish purple, the lobes (3–)6–10 × 2.8–6 mm, ovate-lanceolate, obtuse to subacute. Capsule 15–23 × 2–3.5 mm, fusiform-ovoid; seeds *c.*0.8 mm, ovoid, reticulate. $2n = 36$. Self-compatible. *Flowers January to April*. (Fig. 158).

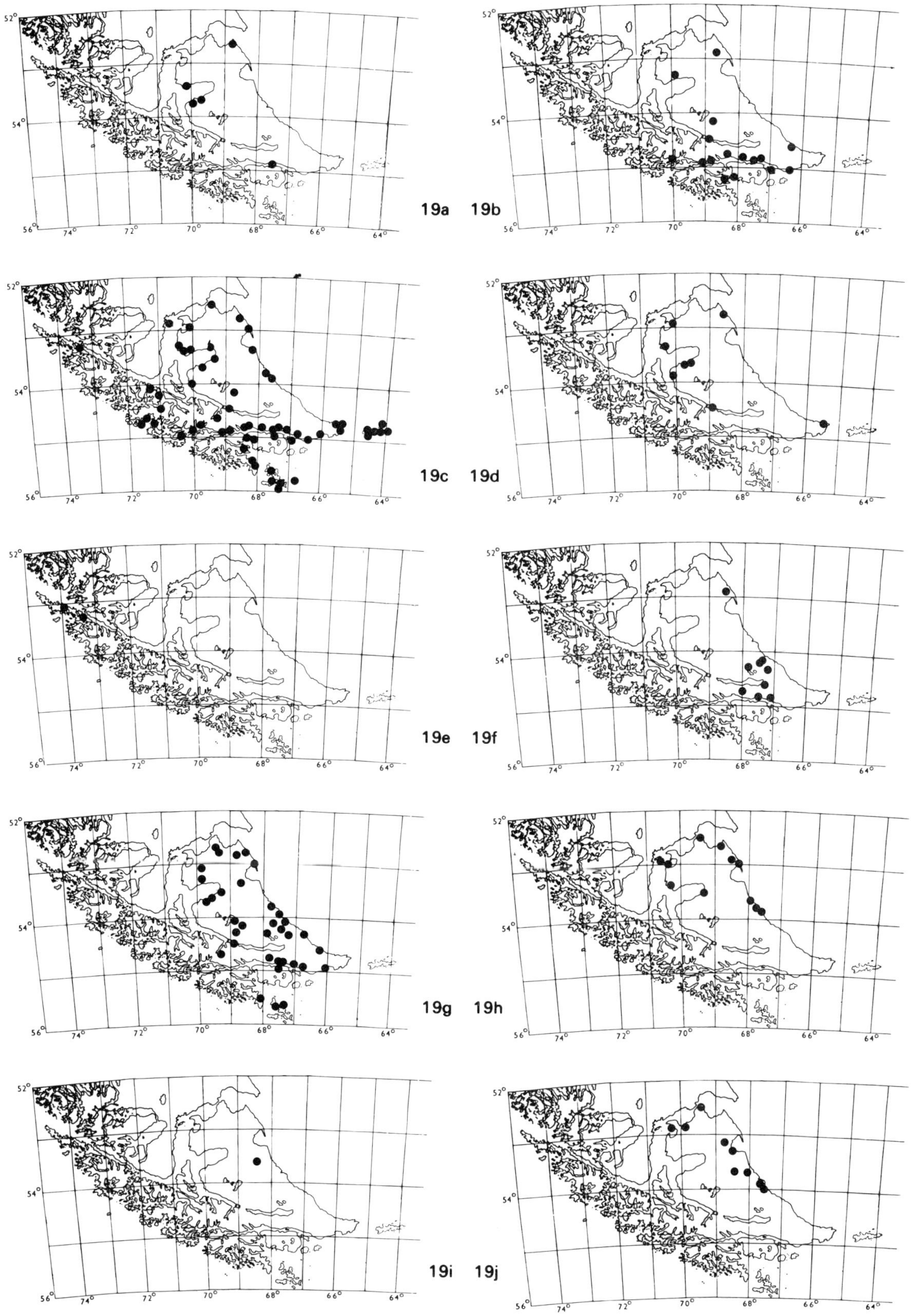

Map 19. a, *Daucus montanus*. b, *Oreomyrrhis hookeri*. c, *Apium australe*. d, *Lilaeopsis hillii*. e, *Desfontainia spinosa*. f, *Gentiana prostrata*. g, *Gentianella magellanica*. h, *Jaborosa magellanica*. i, *Benthamiella nordenskjoldii*. j, *Polemonium micranthum*.

Damp areas in grassland, open shrub communities, stream margins; 0–600 m.

Throughout Isla Grande and islands to the south. (Map 19g).
W. Argentina north to *c.*37°S.lat.
Chile north to 36°40′S.lat.
Falkland Islands (Islas Malvinas).

LVIII. SOLANACEAE

Herbs, sometimes forming dense cushions. Leaves alternate or in fascicles, entire or divided; stipules absent. Flowers actinomorphic, hermaphrodite, solitary or in cymes. Calyx 5-lobed. Corolla 5-dentate to -lobed. Stamens 5, attached to corolla-tube and alternating with lobes; anthers dorsifixed. Ovary superior, 2-celled; style simple. Fruit a capsule or berry; seeds 1-numerous, with endosperm.

a. Plant cushion-forming; leaves linear, imbricate; fruit a capsule **2. *Benthamiella***
a. Plant not cushion-forming; leaves pinnatifid, not imbricate; fruit a berry **1. *Jaborosa***

1. *Jaborosa* Juss.

Procumbent perennial herbs. Leaves in fascicles, pinnatifid. Flowers axillary, solitary or in few-flowered cymes. Calyx tubular to somewhat campanulate, 5-lobed. Corolla tubular to somewhat campanulate, with spreading 5-lobed limb. Disc indistinct. Stigma 2-lobed. Fruit a berry; seeds numerous.

Literature: A. T. Hunziker, *Kurtziana* **4**: 131–138(1967).

1. *J. magellanica* (Griseb.) Dusén 1900, *Svenska Exped. Magellansl.* **3(5)**: 131.

Plant with stout taproot; stems 5–30 cm, prostrate, branched. Leaves with lamina 2–11 × 1–4 cm, ovate to oblanceolate in outline, pinnatifid almost to midrib, the segments pinnately divided into acute, often serrate, lobes, glabrous or white-hairy on midvein beneath; petiole 1–12 cm. Cymes 1- to 4-flowered; peduncles 0.7–3.2 cm, slender. Calyx 2–3 mm, divided to *c.*$\frac{2}{3}$; lobes ovate-lanceolate, acute to acuminate, hairy along midvein on back. Corolla dark purple, sparsely hairy outside; tube 4–5 mm; limb 4–5 mm, the lobes ovate, obtuse to subacute. Stamens inserted at top of corolla-tube. Style exserted from corolla-tube. Berry 10–15 mm in diameter, globose, grey-brown; seeds 2.5–3 mm, reniform-suborbicular, compressed, rugulose, black. $2n = 24$. *Flowers November to December.* (Fig. 159).

Open coastal sands, sometimes in *Lepidophyllum* scrub; 0–20 m.

N. part of Isla Grande. (Map 19h).
Argentina north to *c.*50°S.lat.
Chile along north shore of Estrecho de Magallanes.

2. *Benthamiella* Speg.

Cushion-forming perennial herbs. Leaves alternate, simple, with basal sheath. Flowers axillary, solitary. Calyx tubular to somewhat campanulate, with 5 teeth. Corolla tubular, 5-lobed. Disc present. Style 1, erect; stigma capitate. Fruit a capsule, dehiscing by bifid valves; seeds 1–3(4), fusiform.

Literature: A. Soriano, *Darwiniana* **8**: 233–262(1948). S. Arroyo, *Bot. Not* **133**: 67–76(1980).

1. *B. nordenskjoldi* Dusén ex N. E. Br. 1900. *Icon. Pl., ser.4.* **7(2)**: 2636.
B. patagonica auct., non Speg. – Moore and Goodall, 1974.

Cushions rather lax. Stems woody at base, much branched, covered with imbricate leaves. Leaves 6–10 × 0.75–2 mm, the lower dead and black; lamina 3–5 × 0.75–1.5 mm, linear, rather fleshy, acute, glabrous or somewhat ciliate; basal sheath about as long as lamina, 1.75–2 mm wide, oblanceolate, hyaline, glabrous. Flowers sessile or shortly pedunculate, exceeding leaves. Calyx 4–5 mm, somewhat pubescent; teeth 1–1.5 mm, acute, glabrous or ciliate. Corolla 9–11 mm, white. Stamens included; filaments subequal or 2 somewhat shorter, usually inserted at same level. Style included, reaching to about middle of corolla-tube. *Flowers November.* (Fig. 153).

Open, dry, eroded sandstone slopes; *c.*150 m.

E. C. Isla Grande, known only from Ea San Julio, Cerro Hongo (53°40′S., 68°30′W.). (Map 19i).
W. Argentina between 51° and 52°S.lat.

LIX. POLEMONIACEAE

Annual herbs. Leaves alternate to opposite, simple or compound; stipules absent. Flowers actinomorphic, hermaphrodite. Sepals 5, fused into a tube. Petals 5, fused into campanulate or salveriform corolla. Stamens 5, fused to corolla-tube and alternating with the lobes. Ovary superior, inserted on basal disc, 3-celled; style 1; stigma 3-lobed. Fruit a capsule,

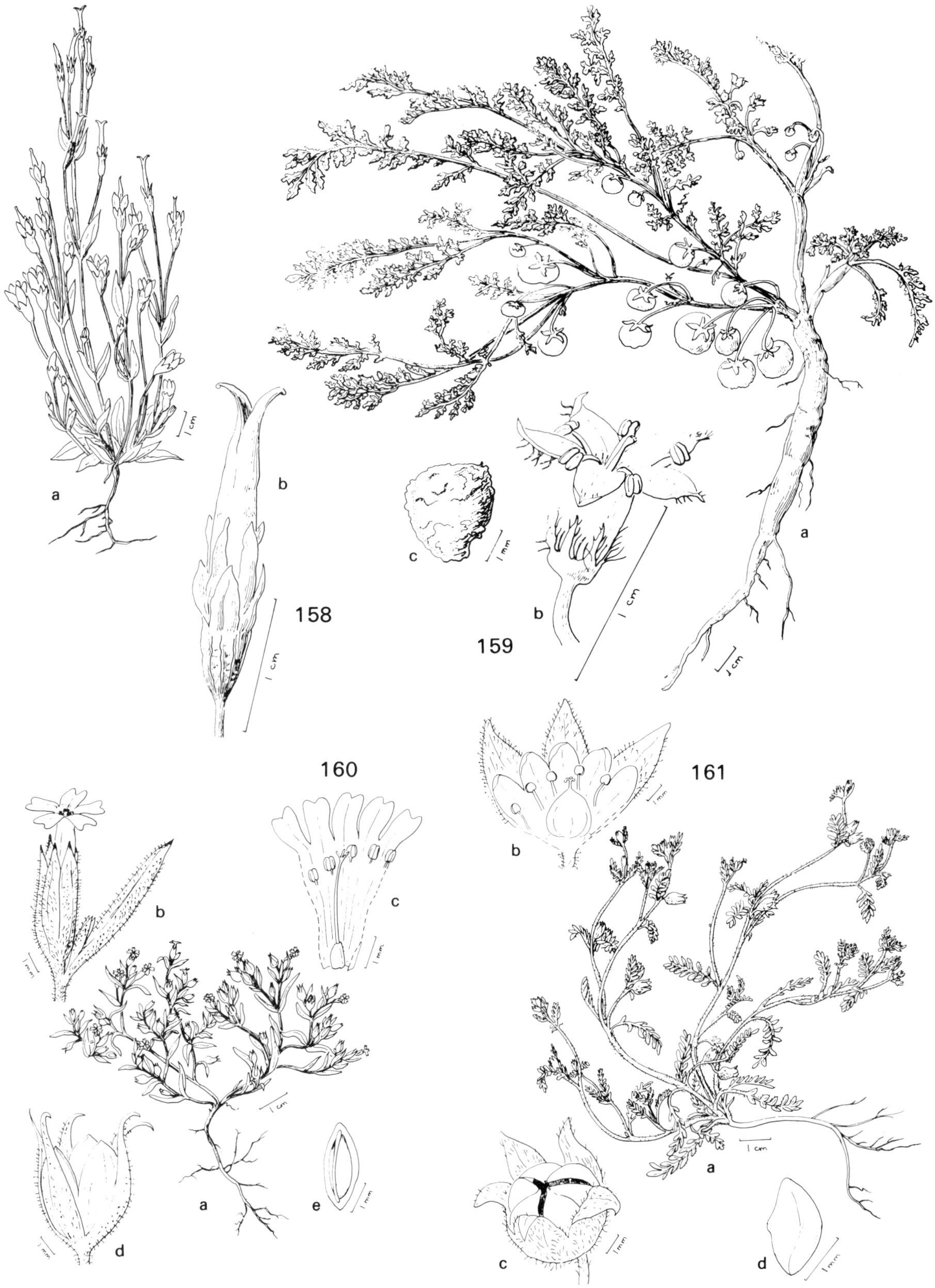

Fig. 158. *Gentianella magellanica*. a, Plant. b, Fruit. Scale = 1 cm. Fig. 159. *Jaborosa magellanica*. a, Plant. b, Flower. c, Seed. Scale = 1 cm (a–b) or 1 mm (c). Fig. 160. *Microsteris gracilis*. a, Plant. b, Flower. c, Opened corolla. d, Fruit. e, Seed. Scale = 1 cm (a) or 1 mm (b–e). Fig. 161. *Polemonium micranthum*. a, Plant. b, Opened flower. c, Fruit. d, Seed. Scale = 1 cm (a) or 1 mm (b–d).

(R. N. P. Goodall)

dehiscing by 3 valves; seeds 3–20, with endosperm, mucilaginous when wet.

Literature: O. E. Borsini, *Lilloa* **8**: 199–230(1942).

a. Leaves simple; corolla exceeding calyx **2. Microsteris**
a. Leaves pinnatisect; corolla shorter than calyx **1. *Polemonium***

1. *Polemonium* (Tourn.) L.

Leaves imparipinnatisect, alternate. Inflorescence a sympodial cyme, each flower produced at apex of 1-leaved shoot and each succeeding flowering shoot developing from axil of uppermost leaf. Calyx campanulate. Corolla broadly campanulate. Seeds 5–8.

Literature: J. F. Davidson, *Univ. Calif. Publ. Bot.* **23**; 209–282(1950).

1. *P. micranthum* Bentham 1845, in DC., *Prodr.* **9**: 318.
P. antarcticum Griseb. 1854, *Goett. Abhandl.* **6**: 131-Borsini, 1942.

Stems 3–25 cm, slender, simple or diffusely branched, glabrous or glandular-hairy. Leaves 20–43 × 7–14 mm; petiole 5–21 mm; leaflets 8–12, 5–15 × 0.5–4.5 mm, narrowly elliptical to spathulate, glandular-pubescent. Flowers with pedicels up to 13 mm. Calyx 3–7 mm, glandular-pubescent; lobes usually 1–1.5 times as long as limb. Corolla 2–6 mm, white or rarely blue; lobes rounded, as long as tube. Stamens included; filaments inserted in basal $\frac{1}{4}$–$\frac{1}{2}$ of corolla-tube, with sparsely pubescent and variably dilated bases. Stigma held at about level of corolla-mouth. Capsule subglobose; seeds *c.*2 mm, oblong, angular, black. $2n = 18$. *Flowers October to December.* (Fig. 161).

Open soil and sand in coastal shrub communities, grassland and *Empetrum* heath, extending in disturbed ground; 0–100 m.

N. & E. C. Isla Grande. (Map 19j).
Argentina north to *c.*40°S.lat.
Chile north to *c.*30°S.lat.
W. North America.

The separation of South American material as *P. antarcticum* and *P. gayanum* Brand was made on the basis of leaf characters and the degree of basal dilation of filaments, which vary continuously throughout the whole distribution of the species and are consequently unreliable.

2. *Microsteris* Greene

Leaves entire, the lower opposite. Flowers usually in pairs in upper leaf-axils. Calyx cylindrical, the lobes more or less united by a hyaline membrane. Corolla salveriform, with slender tube.

Literature: E. T. Wherry, *Brittonia* **5**: 60–63(1943).

1. *M. gracilis* (Douglas ex Hooker) Greene 1898, *Pittonia* **3**: 300.
Gilia gracilis (Douglas ex Hooker) Hooker 1829, *Bot. Mag* **56**: t. 2924 – Borsini, 1942.

Small annual up to 10 cm; stems branched, glabrous to hairy below, glandular-pubescent above. Upper leaves 7–18 × 1.3–3.5 mm, lanceolate, subacute to acute, the lower shorter, spathulate, obtuse; all glandular-pubescent above, glabrous beneath. Flowers with glandular pedicels 2.5–4 mm. Calyx 5–7.5 mm; lobes 2–3 × 0.7–1 mm, acute, subequal. Corolla 7–7.5 mm, creamy tinged with purple or blue distally; lobes 1.6–1.8 × *c.*1.3 mm, obcordate, retuse. Stamens included; filaments inserted unequally in upper $\frac{1}{2}$ of corolla-tube. Capsule 3.2–4 (–4.7) × 2.2–3 mm, ovoid, reticulate; seeds 2–2.3 mm, ovoid, lenticular, pale brown. $2n = 14$. Self-compatible and usually self-pollinated. *Flowers October to March.* (Fig. 160).

Open and sparsely vegetated soil and sand in coastal shrub communities, grassland, *Chiliotrichum* scrub; 0–250 m.

Isla Grande. (Map 20a).
W. Argentina north to *c.*34°S.lat.
Chile, S. Peru.
W. North America.

LX. HYDROPHYLLACEAE

Perennial herbs. Leaves alternate, entire to pinnatifid; stipules absent. Flowers hermaphrodite, actinomorphic, in cymes. Setapls 5, free. Corolla 5-lobed. Stamens 5, inserted at base of corolla-tube. Ovary superior, 1-celled; style 1, 2-fid. Fruit a loculicidal capsule; seeds few, with endosperm.

1. *Phacelia* Juss.

Leaves mostly basal. Stamens long exserted; filaments with 2 scale-like appendages at base. Stigmas long exserted.

Literature: L. R. Heckard, *Univ. Calif. Publ. Bot.* **32**: 1–126(1960); *Quart. Rev. Biol.* **38**: 117–123(1963).

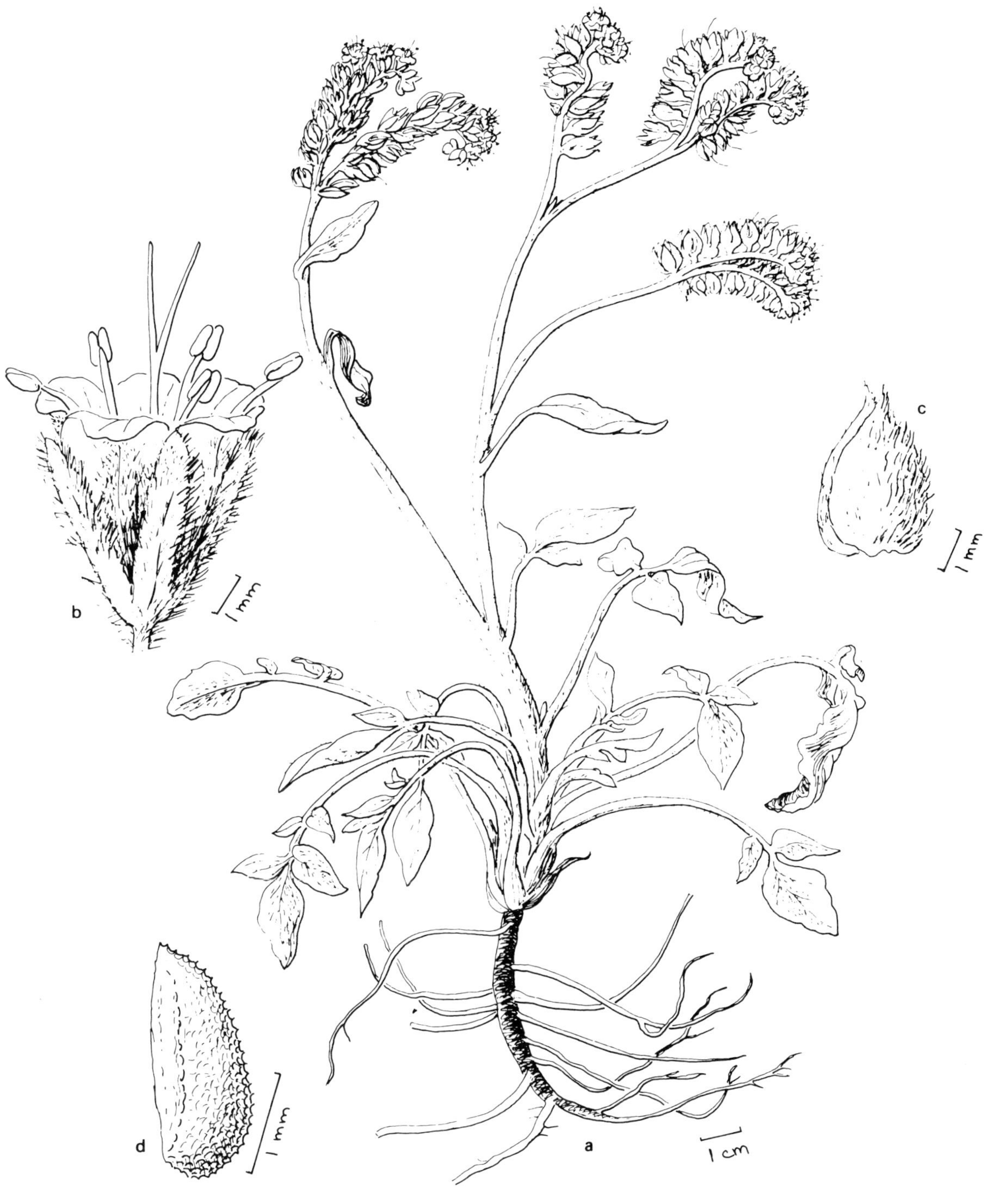

Fig. 162. *Phacelia secunda*. a, Plant. b, Flower. c, Young fruit. d, Seed. Scale = 1 cm (a) or 1 mm (b–d).
(R. N. P. Goodall)

1. *P. secunda* J. F. Gmelin 1791, *Syst. Nat.*, ed, 13, **2**: 330.

P. magellanica (Lam.) Coville 1893, *Contr. U.S. Nat. Herb.* **4**: 159.

P. circinata Jacq. f. 1816, *Eclogae Pl. Rar.* **1**: 135 – Alboff, 1896; Spegazzini, 1896.

Plant pubescent or hispid; stems 5–36 cm, decumbent to ascending or erect, simple or branched. Lower leaves with lamina 20–80 × 10–40 mm, obovate to lanceolate in outline, obtuse, usually deeply divided into ovate-elliptical segments, sometimes entire; petiole as long as lamina. Cymes usually dense; flowers sessile or shortly pedicellate. Sepals 2.5–6 × 0.5–1 mm, oblong, subobtuse. Corolla 4–7 mm, campanulate, usually white, rarely pink or blue; lobes *c.*0.5–1 mm, semi-orbicular. Style about twice as long as calyx, divided for up to distal $\frac{1}{3}$. Capsule 3.5–4.5 mm, oblong-ovoid, brown; seeds 1–4, 2–2.5 mm, oblong-obovoid, pitted, black. $2n = 22$. Self-compatible. *Flowers November to March.* (Fig. 162).

Open sands, gravels and rocks by coast, along rivers, hill-slopes and road-sides; 0–150 m.

C. & N. Isla Grande, locally on S. coast. (Map 20b).
Argentina, especially in the west, and Chile north to *c.*40°S.lat.
Bolivia and Peru north to *c.*10°S.lat.

LXI. BORAGINACEAE

Annual or perennial herbs. Leaves alternate, entire; stipules absent. Flowers hermaphrodite, actinomorphic, usually in scorpioid cymes. Calyx 5-toothed or -lobed. Corolla 5-lobed. Stamens 5, inserted in corolla and alternating with lobes. Ovary superior, 2– or 4– celled; style simple, arising from between 4 lobes of ovary (gynobasic). Fruit of 2 or 4 nutlets.

Literature: I. M. Johnston, *Contr. Gray Herb.* **78**: 1–118(1927); A. Brand, *Pflanzenreich* **97** (**IV. 252**): 1–236(1931).

a. Corolla yellow, with prominent cylindrical tube and campanulate limb, without scales in throat — **1. *Amsinckia***
a. Corolla white to pink or blue at least when mature, rotate with short tube and flat or slightly concave limb, with scales in throat:
  b. Nutlets smooth; calyx divided to *c.*$\frac{1}{2}$-way — **2. *Myosotis***
  b. Nutlets rugose; calyx divided almost to base — **3. *Plagiobothrys***

1. *Amsinckia* Lehm.

Annuals. Cymes terminal, usually bracteate at base. Calyx lobed almost to base. Corolla yellow, with prominent cylindrical tube and campanulate limb, without scales in throat. Stamens included, inserted in upper part of corolla-tube. Style included; stigma capitate. Nutlets ovoid-trigonous, strongly muricate.

1. *A. calycina* (Moris) Chater 1971, *Bot. Jour. Linn. Soc.* **64**: 380.

A. angustifolia Lehm. 1831, *Del. Sem. Horto Hamburg.* 7 – Dusén, 1900; Skottsberg, 1916.

A. hispida I. M. Johnston, nom. illegit. – Johnston, 1927.

Stem 10–50 cm, ascending to erect, often branched, with long, patent and short, hooked hairs. Leaves 20–100 × 2–10 (–30) mm, linear-lanceolate, rarely lanceolate-ovate, acute, with erecto-patent hairs above and beneath. Cyme up to 20 cm; pedicels 1–3 mm. Calyx 3–4 mm, up to 5 mm in fruit, ± hispid; lobes linear-lanceolate, acuminate. Corolla 4–8 mm; limb 1–2 mm. Nutlets 2–2.5 mm, weakly transversely rugose. *Flowers December to January.* (Fig. 167).

Coastal sands and gravels, disturbed open ground; 0–30 m.

N. Isla Grande, locally on S. coast. (Map 20c).
Argentina north to *c.* 38°S.lat.
Chile north to *c.* 20°S.lat.
Bolivia, Peru, Ecuador.

2. *Myosotis* L.

Annuals or perennials. Cymes usually paired and ebracteate, sometimes bracteate at base, or with leaf-like bracts throughout. Corolla white to pink or blue when mature, rotate, with short tube and flat or slightly concave limb, with 5 scales in throat. Stamens usually included, inserted about middle of corolla-tube. Style included; stigma capitate. Nutlets ovoid, ± compressed, smooth and shiny.

a. Calyx with straight, appressed hairs only:
  b. Inflorescence with leaf-like bracts throughout; pedicels up to 2 mm in fruit; corolla-limb 2–4 mm in diameter — **1. *albiflora***
  b. Inflorescence ebracteate, rarely with bracts at base; pedicels up to 10 mm in fruit; corolla-limb up to 8 mm in diameter — **6. *scorpioides***
a. Calyx with hooked hairs, sometimes also with straight hairs:
  c. Calyx open in fruit; limb of corolla flat — **5. *sylvatica***

c. Calyx closed in fruit; limb of corolla saucer-shaped:

d. Leaves with hooked hairs beneath, especially on veins; inflorescence-axis with some patent or erecto-patent hairs **3. *stricta***

d. Leaves without hooked hairs beneath; inflorescence-axis with all hairs appressed:

e. Corolla pale yellow when young, the limb *c.*1.3 mm in diameter; calyx with deflexed, hooked hairs **4. *discolor***

e. Corolla always blue, the limb *c.*3 mm in diameter; calyx with patent, hooked hairs **2. *arvensis***

1. *M. albiflora* Banks & Sol. ex Hooker f. 1847, *Fl. Antarct.* **1(2)**: 329.

Eritrichium albiflorum (Banks & Sol, ex Hooker f.) Griseb. 1854, *Syst. Bemerk.* 131 – Dusén, 1900.

Perennial, with ascending to erect rhizome. Stems (1.3–)4–18(–33) cm, procumbent to ascending, often branched, strigose. Basal leaves with lamina 16–25(–49) × 4.5–9(–19) mm, obovate to spathulate, obtuse, sometimes mucronulate, strigose above, sparsely so beneath; petiole 15–27(–49) mm, strigose; cauline leaves few, similar to basal but smaller, the lower shortly petiolate, the upper sessile. Cymes with flowers in axils of leaf-like bracts; pedicels 0.5–0.7 mm, up to 2 mm in fruit. Calyx 2.7–4 mm, with straight, appressed hairs, divided to *c.*½-way; lobes ovate-lanceolate, acute. Corolla-tube ± as long as calyx; limb 2–4 mm in diameter, flat, white. Nutlets 1.6–1.9 mm. *Flowers December to March.* (Fig. 163).

Wet coastal sands and pebbles, moist areas by streams and lakes; 0–50 m.

N. W. and C. S. Isla Grande, Islas Londonderry, Isla Navarino. (Map 20d).
W. Argentina north to *c.* 51°S.lat.
Chile north to *c.*45°S.lat.

2. *M. arvensis (L.) Hill 1764, *Veg. Syst.* **7**: 55.

Biennial. Stem up to 35 cm, often much branched at base, with patent hairs below, and straight appressed hairs above. Basal leaves up to 35 × 7 mm, oblanceolate, obtuse to subacute, subsessile. Cymes ebracteate, the axis with all hairs appressed; lower pedicels up to 10 mm, the upper shorter, erect in fruit. Calyx up to 7 mm in fruit, closed, with patent, hooked hairs, divided to *c.*½-way; lobes linear-lanceolate to -triangular, acute. Corolla-tube ± as long as calyx; limb *c.*3 mm in diameter, saucer-shaped, bright blue. Nutlets 1.5–2.5 mm. *Flowers December to February.*

Disturbed ground near habitations; 20–40 m.

C. S. Isla Grande (Ea Harberton).
Introduced from Europe, perhaps with grass-seed, and locally naturalized.

Fuegian material belongs to ssp. *arvensis.*

3. *M. stricta Link ex Roemer & Schultes 1819, *Syst. Veg.* **4**: 104.

Annual. Stem (2–)5–28 cm, erect, usually much branched at base, with hooked hairs at base. Basal leaves 17–50 × 5–10 mm, lanceolate, obtuse, subsessile, with hooked hairs beneath, especially on veins, the other hairs straight. Cymes usually ebracteate, the axis with ± appressed and patent or erecto-patent hairs; pedicels up to 1.5 mm, erecto-patent. Calyx up to 4 mm in fruit, closed, with deflexed hooked hairs and appressed, straight hairs, divided to ½-way; lobes linear-triangular, acute. Corolla-tube scarcely exceeding calyx; limb *c.* 1 mm in diameter, saucer-shaped, blue. Nutlets *c.*1.5 mm. *Flowers November to December.*

Coastal sands and gravels, open *Lepidophyllum* and *Chiliotrichum* scrub, roadsides; 0–100 m.

N. Isla Grande.
Introduced from Europe and widely naturalized in disturbed and open sites.

4. *M. discolor Pers. 1797, *Syst. Veg.* ed. 15, 190.

Similar to *M. stricta* but leaves without hooked hairs beneath; inflorescence-axis with all hairs appressed; pedicels up to 4 mm; calyx only with soft, deflexed, hooked hairs; corolla-limb *c.*1.3 mm in diameter, pale yellowish at first, becoming pink to violet or blue. *Flowers December to March.*

Coastal sands and gravels, open disturbed scrub and grassland, roadsides; 0–100 m.

Locally on N. W., N. E. and S. coasts of Isla Grande.
Introduced from Europe and widely naturalized.

Fuegian material belongs to ssp. *discolor.*

5. *M. sylvatica Hoffm. 1791, *Deutschl. Fl.* 61.

Biennial to perennial. Stem up to 25 cm, often much branched, with patent, simple hairs. Leaves 20–40 × 5–12 mm, ovate to linear-ovate or elliptical, obtuse, usually subsessile, with appressed, ± straight hairs. Cymes ebracteate, the axis with appressed hairs;

pedicels up to 7(–15) mm, erecto-patent. Calyx up to 5 mm in fruit, open, with short, hooked hairs, divided to $\frac{1}{3}$-$\frac{1}{2}$ way; teeth linear to linear-triangular, acute. Corolla-tube exceeding calyx by *c.*1 mm; limb up to 8 mm in diameter, flat, blue. Nutlets *c.* 1.7 mm. *Flowers December.*

Disturbed, usually cultivated, ground near habitations; 0–30 m.

C. S. coast of Isla Grande (Ushuaia).
Introduced from Europe and locally naturalized or adventive.

6. M. scorpioides L. 1753, *Sp. Pl.* 131.

Perennial, with prostrate rhizomes. Stems up to 20 cm, ascending to erect, glabrous or with patent, simple hairs. Leaves 12–30 x 4–10 mm, oblong to oblong-lanceolate, obtuse, sessile, subglabrous or with appressed, simple hairs. Cymes ebracteate, rarely with bracts at base, the axis with appressed hairs; pedicels up to 10 mm in fruit, patent. Calyx up to 6 mm in fruit, open, with short, straight, appressed hairs, divided to *c.*$\frac{1}{2}$-way; teeth broadly triangular, acute. Corolla-tube ± as long as calyx; limb up to 8 mm in diameter, flat, blue. Nutlets up to 1.8 mm. *Flowers December to January.*

Disturbed ground near settlements; 30–70 m.

C. S. coast of Isla Grande.
Introduced from Europe; apparently planted for ornament and locally naturalized in moist grassland in disturbed deciduous forest.

3. *Plagiobothrys* Fischer & Meyer

Annuals. Cymes with few to many bracts, interrupted. Corolla white, rotate, with short tube and flat limb, with scales in throat. Stamens included, inserted about middle of corolla-tube. Style included; stigma capitate. Nutlets oblong-ovoid, compressed rugose.

1. *P. calandrinioides* (Phil.) I. M. Johnston 1927, *Contr. Grey Herb.* **78**: 91.
P. lechleri I. M. Johnston 1923, *Contr. Gray Herb.* **68**: 79.
Allocarya procumbens auct., non (DC.) Greene - Skottsberg, 1916.

Stems 5–18 cm, procumbent to ascending, strigose. Leaves 10–50 x 1–2.5 mm, linear to ± oblong, ± obtuse, with ± dense, straight, appressed to patent hairs. Pedicels 0.5–1 mm. Calyx 2–3.5 mm in fruit, strigose, divided almost to base; lobes linear-oblong to lanceolate, obtuse. Corolla-tube ± as long as calyx; limb 1–3 mm in diameter, white. Nutlets 1.5–2.2 mm. $2n = 68$. *Flowers November to February.* (Fig. 168).

Open sands, gravels and rocks by coast, rivers and lakes, open grassland; 0–60 m.

C. and N. Isla Grande. N. Isla Navarino. (Map 20e).
Argentina north to *c.*39°S.lat.
E. Chile north to *c.*33°30′S.lat.

LXII. LABIATAE

Herbs or dwarf shrubs. Leaves opposite, simple; stipules absent. Flowers zygomorphic, hermaphrodite, in axils of leaf-like bracts, solitary or in verticillasters (pseudowhorls) grouped into spike-like or cymose inflorescence. Calyx 5-lobed, usually 2-lipped. Corolla (4)5-lobed, usually 2-lipped. Stamens 4, didynamous. Ovary superior, 2-celled, later 4-celled by secondary septation; style 1, usually branched above, gynobasic. Fruit of four, 1-seeded nutlets.

a. Dwarf shrub — **4. *Satureja***
a. Herb:
  b. Plant smelling of peppermint when crushed; corolla with 4 subequal lobes — **6. *Mentha***
  b. Plant not smelling of peppermint; corolla 2-lipped:
    c. Flowers solitary; leaves suborbicular – obovate — **1. *Scutellaria***
    c. Flowers 2 or more in verticillasters; leaves ovate-suborbicular to ovate or reniform:
      d. Calyx with 5 equal or subequal teeth — **2. *Lamium***
      d. Calyx 2-lipped:
        e. Upper corolla-lip flat, shorter than lower lip — **3. *Glechoma***
        e. Upper corolla-lip hooded, longer than lower lip — **5. *Prunella***

1. *Scutellaria* L.

Creeping, perennial herbs. Flowers solitary. Calyx campanulate, 2-lipped, the lips entire. Corolla 2-lipped; upper lip entire; lower lip 3-lobed, with the middle lobe larger. Fruit included in persistent calyx.

Literature: C. Epling, *Rev. Mus. La Plata, Bot.* **2(8)**: 89–178(1938).

1. *S. nummulariifolia* Hooker f. 1847, *Fl. Antarct.* **1(2)**: 336.

Stems 3–9 cm, ascending from long, prostrate,

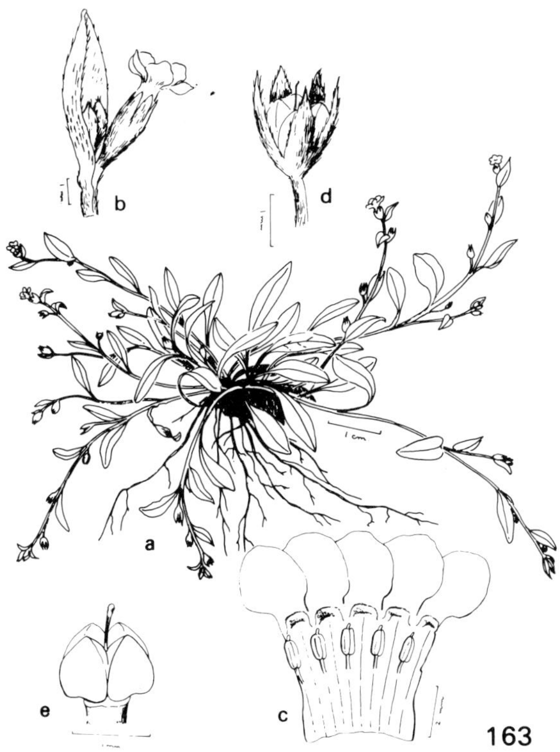

163

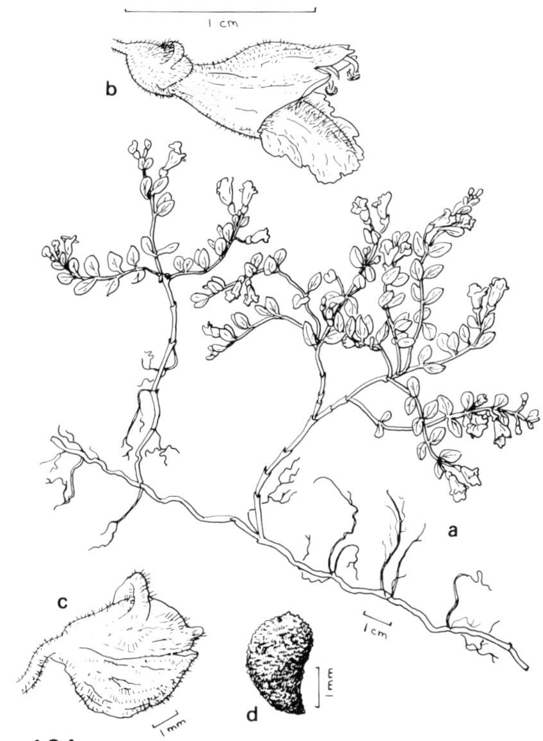

164

165

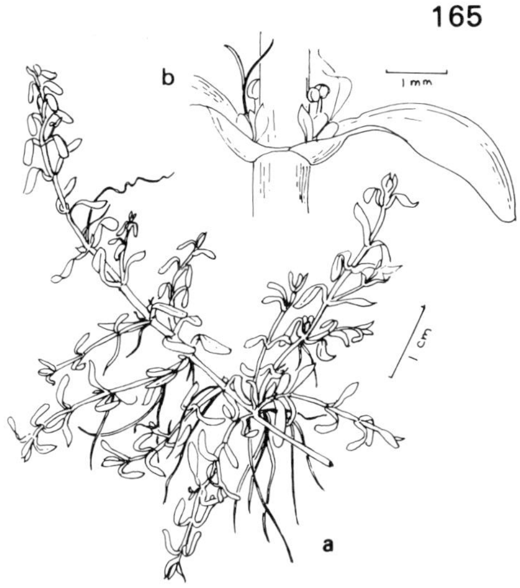

166

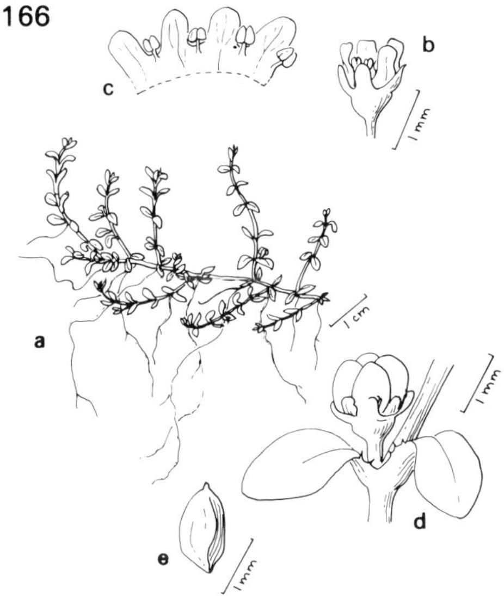

Fig. 163. *Myosotis albiflora*. a, Plant. b, Flower. c, Opened corolla. d, Young fruit. e, Fruit. Scale = 1 cm (a) or 1 mm (b–e). Fig. 164. *Scutellaria nummulariifolia*. a, Plant. b, Flower. c, Fruit. d Seed. Scale = 1 cm (a–b) or 1 mm (c–d). Fig. 165. *Callitriche antarctica*. a, Plant. b, Flowers. Scale = 1 cm (a) or 1 mm (b). Fig. 166. *Tetrachondra patagonica*. a, Plant. b, Flower. c, Opened corolla. d, Fruit. e, Nut. Scale = 1 cm (a) or 1 mm (b–e).

(R. N. P. Goodall)

yellowish rooting base, branched, pubescent, reddish brown to green. Leaves with curved, glandular pubescence; lamina 5–8.5(–10) × 3–8.5 mm, suborbicular-obovate, subcoriaceous; petiole 0.5–3 mm. Pedicels 2–3.5 mm, slender, puberulent. Calyx 2.6–3.5 × 2.3–3 mm, pubescent. Corolla pinkish purple, usually white in throat; tube 6–8 mm; upper lip 1.5–2.5 × *c.*2.5 mm, suborbicular, retuse; lower lip 5–6.5 mm in diameter. Fruit 4.5–5.5 mm, subglobose; nutlets 2.8–3.5 × 1.5–2 mm, ± pyriform, prominently tuberculate, black. *Flowers November to January.* (Fig. 164).

Sands, pebbles and stones, beside sea and rivers; 0–50 m.

N. & N. E. coasts of Isla Grande. (Map 20f).
Argentina north to *c.*45°12′S.lat.
E. Chile north to *c.*51°S.lat.
Falkland Islands (Islas Malvinas).

2. *Lamium* L.

Annual herbs with ± tetragonous stems. Flowers in dense verticillasters. Calyx tubular-campanulate, with 5 subequal mucronate teeth. Corolla 2-lipped; upper lip compressed to form a hood; lower lip with prominent middle, and 2 very small lateral, lobes. Nutlets trigonous.

a. Verticillasters close together; bracts stalked **1. *purpureum***
a. Verticillasters rather distant; at least the upper bracts sessile **2. *amplexicaule***

1. *L. purpureum L. 1753, *Sp. Pl.*579.

Stems 10–31 cm, erect, branched from base, pubescent, often purple tinted. Leaves pubescent; lamina 10–25 × 11–24 mm, ovate, obtuse, cordate to truncate at base, crenate-serrate; petiole 5–30 mm. Bracts like leaves but truncate to subcuneate at base, the upper sometimes subsessile. Inflorescence rather dense, with verticillasters close together. Calyx-tube 3–4.5 mm, pubescent; teeth 3.5–4 mm, patent in fruit. Corolla 10–15 mm, pinkish purple, the tube exceeding the calyx, with a ring of hairs near the base. Nutlets 2–2.2 × 1.1–1.3 mm, concave on 2 sides, black. *Flowers (August to) December to March.*

Disturbed open ground near settlements; 0–40 m.

S. coast of Isla Grande.
Introduced from Europe and locally naturalized.

L. hybridum Vill. 1786, *Hist. Pl. Dauph.* **1**: 251, a native of Europe which may also be introduced in Fuegia, is distingusihed from *L. purpureum* by the irregularly incised leaves and bracts, the upper leaves often being somewhat decurrent along the petioles.

2. *L. amplexicaule L. 1753, *Sp. Pl.* 579

Stems 12–26 cm, ascending, branched from the base, pubescent. Leaves puberulent; lamina 8–12(–20) × 7–15(–21) mm, ovate-suborbicular, obtuse, broadly cuneate to subcordate at base, crenate-lobulate; petiole 7–18(–35) mm. Bracts like leaves but at least the upper sessile, semi-amplexicaul. Verticillasters few, rather distant. Calyx-tube 4–5.5 mm, densely pubescent; teeth 1.5–2 mm, erect in fruit. Corolla 14–17 mm, pinkish purple, the tube shorter than to much exceeding calyx, glabrous inside. Nutlets 2–2.3 × *c.*1.1 mm, flat or convex on sides, black. *Flowers November to February.*

Open grassland, coastal sands and gravels, disturbed ground; 0–30 m.

C. E. & N. coasts of Isla Grande.
Introduced from Europe and locally naturalized.

3. *Glechoma* L.

Perennial herbs. Verticillasters 2– to 5– flowered, secund. Calyx tubular-campanulate, 2-lipped; upper lip with 3 teeth; lower lip with 2 teeth. Corolla 2-lipped; upper lip flat; lower lip 3-lobed. Nutlets obovoid.

1. *G. hederacea L. 1753, *Sp. Pl.* 578.

Non-flowering stems up to 7.5 cm, creeping and rooting; flowering stems 6.5–40 cm, ascending to erect, hairy to subglabrous. Leaves glabrous to pubescent; lamina 5–17(–25) × 12–24(–35) mm, reniform to ovate-cordate, obtuse, crenate; petiole 9–13(–25) mm. Calyx 5.5–6.5 mm, pubescent; teeth 1.2–1.5 × 0.7–1 mm, triangular, apiculate. Corolla 10–20 mm, violet, with purple spots on lower lip; upper lip *c.*3 × 3.5 mm, suborbicular; lower lip 7–15 mm, obovate, with middle lobe larger than the lateral. Nutlets *c.*2 mm, smooth. *Flowers November to December.*

Open disturbed ground.

Local in C. & S. Isla Grande.
Introduced from Europe and occasionally naturalized.

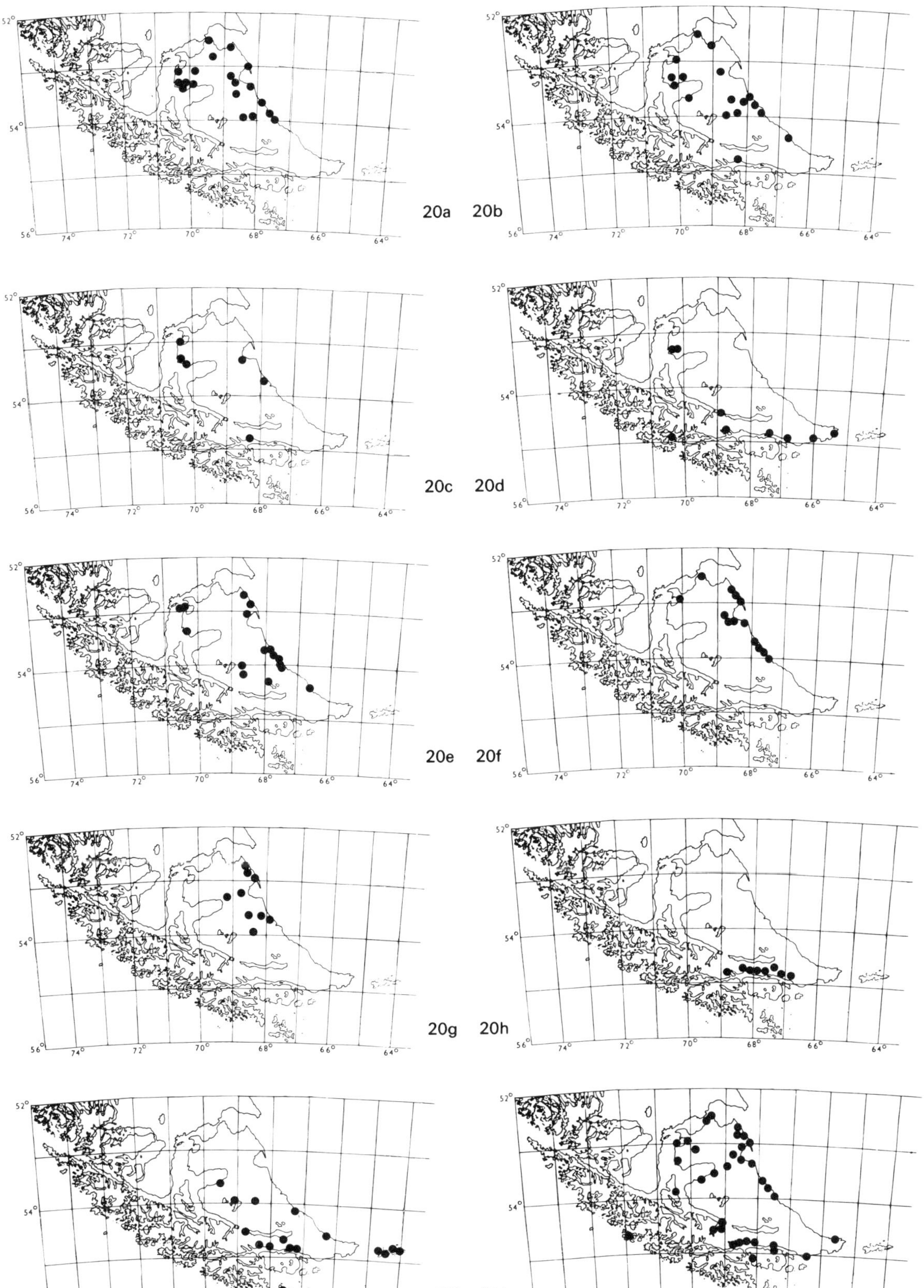

Map 20. a, *Microsteris gracilis*. b, *Phacelia secunda*. c, *Amsinckia calycina*. d, *Myosotis albiflora*. e, *Plagiobothrys calandrinioides*. f, *Scutellaria nummulariifolia*. g, *Satureja darwinii*. h, *Tetrachondra patagonica*. i, *Callitriche antarctica*. j, *Plantago maritima*.

4. *Satureja* L.

Dwarf shrubs. Verticillasters 2– to 4– flowered, or flowers solitary. Calyx tubular-campanulate, 5-lobed. Corolla 2-lipped; upper lip 2-lobed; lower lip 3-lobed. Nutlets ovoid-ellipsoid.

1. ***S. darwinii*** (Bentham) Briquet 1896, *Natürl. Pflanzenfam.* **4(3a)**: 300.
   Micromeria darwinii Bentham 1848, in DC., *Prodr.* **12**: 222 – Spegazzini, 1896.

Stems up to 15 cm, 2–5 mm in diameter at base, procumbent to suberect, with adventitious roots near base, much branched. Leaves glabrous; lamina 2–7 × 1–3 mm, elliptic-oblanceolate to – obovate, obtuse, cuneate at base, entire; petiole *c.*1 mm. Pedicels 1.5–2.5 mm; bracteoles 2, 1.5–2 mm, subulate, at base of flower. Calyx 3–5 × 1.5–2 mm, densely pubescent, purplish green; lobes 0.5–1 mm, narrowly triangular, acute. Corolla 4.5–6.5 mm, pinkish violet; lobes 1–1.5 mm, oblong, obtuse. Nutlets *c.*1.5 × 0.7 mm, smooth, brown. *Flowers November to December.* (Fig. 169).

Dry grassland, open *Chiliotrichum* scrub; 0–150 m.

N. Isla Grande. (Map 20g).
Argentina north to *c.*44°S.lat.
Chile north to *c.*52°S.lat.

5. *Prunella* L.

Perennial herbs. Verticillasters *c.*6- flowered, in dense terminal inflorescence. Calyx tubular-campanulate, 2-lipped, closed in fruit; upper lip ± 3-dentate; lower lip 2-dentate. Corolla 2-lipped; upper lip hooded; lower lip 3-denticulate. Nutlets oblong.

*1. ***P. vulgaris*** L. 1753, *Sp. Pl.* 600

Plant ± pubescent; stems 4–49 cm, ascending to erect, branched, from short rhizome rooting at nodes. Leaves with lamina 10–40(–50) × (5–)7–20(–25) mm, ovate, acute to subobtuse, cuneate at base, entire or crenulate; petiole 3–16(–24) mm. Inflorescence 10–40 × 10–14 mm, oblong to subglobose; bracts 5–8.5 × 7–11.5 mm, suborbicular, with triangular apical tooth, sessile. Calyx 5–8 mm, purple; upper lip with teeth *c.* 0.6 × 1.2 mm, the middle tooth wider than the lateral; lower lip with teeth *c.*3 mm. Corolla 10–14 mm, blue-violet, the tube exceeding calyx, obconical. Nutlets smooth. *Flowers December to March.*

Disturbed ground, coastal cliffs; 0–30 m.

Locally on S. coast of Isla Grande.

Introduced from Europe and occasionally naturalized near settlements.

6. *Mentha* L.

Aromatic, rhizomatous, perennial herbs with ± tetragonous stems. Verticillasters many-flowered, in dense spike-like inflorescence. Calyx tubular, with 5 subequal teeth. Corolla with 4 subequal lobes, but the upper usually wider. Nutlets ovoid.

*1. ***M.** × **piperita*** L. 1753, *Sp. Pl.* 576.

Plant smelling and tasting of peppermint, usually sparsely hairy; stems 20–80 cm, erect, usually branched, reddish or purplish. Leaves with lamina 5–8 × 1.7–3.1 cm, lanceolate to ± ovate, acute, cuneate at base, serrate; petiole 0.6–1 cm. Inflorescence 4–5 × 1.1–1.7 cm, oblong, interrupted at base; bracts 5.5–13 × 2.2–5 mm, ovate to lanceolate, acuminate, with petioles 1–3 mm. Calyx 3–4 mm; tube glabrous; teeth 1–1.4 mm, subulate, long ciliate. Corolla 4–5.3 mm, reddish purple, the tube shorter than calyx. Sterile. *Flowers January to March.*

Damp places, especially near settlements; 0–30 m.

Locally on N. W. & S. coasts of Isla Grande. Introduced from Europe for cultivation as a pot-herb and occasionally naturalized or persisting near abandoned gardens.

LXIII. TETRACHONDRACEAE

Glabrous perennial herbs. Leaves opposite, simple; stipules absent. Flowers hermaphrodite, actinomorphic. Sepals 4, united. Petals 4, united. Stamens 4, inserted on corolla. Ovary superior, 4-lobed to base; style gynobasic. Fruit of 4, 1-seeded nutlets united at base; seeds with endosperm.

1. *Tetrachondra* Petrie

Plant small, creeping. Petioles connate at base. Flowers solitary in leaf-axils. Corolla with short tube, subrotate. Stamens alternating with corolla-lobes. Stigmatic papillae at apex of simple style.

Literature: D. M. Moore, *Bol. Soc. Argent. Bot.* **13**: 2–6(1970).

1. ***T. patagonica*** Skottsb. 1912, *Bot. Jahrb.* **48**, Beibl. **107**: 17.

Plant forming low patches; stems 20–40 mm, slender, branching, rooted at nodes. Leaves with

167

168

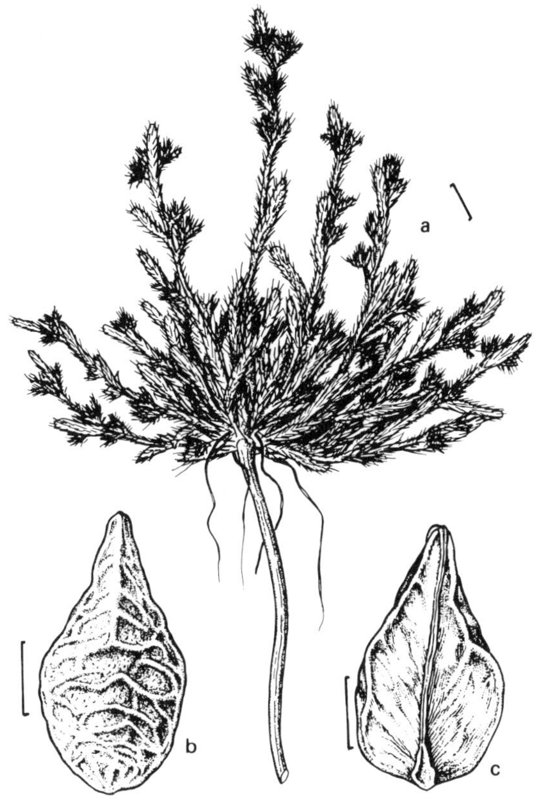

169

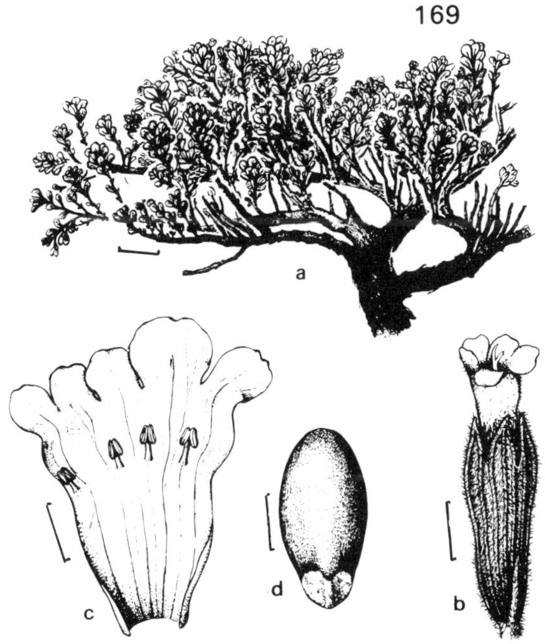

170

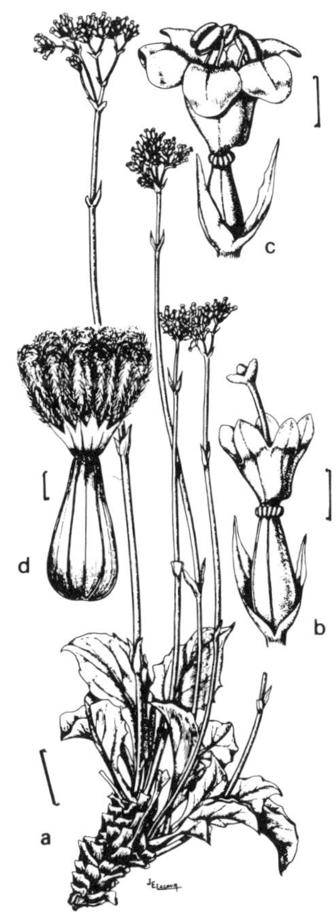

Fig. 167. *Amsinckia calycina*. a, Plant. b, Nut, anterior view. c, Nut, posterior view. Scale = 1 cm (a) or 1 mm (b). Fig. 168. *Plagiobothrys calandrinioides*. a, Plant. b, Nut, anterior view. c, Nut, posterior. Scale = 1 cm (a) or 0.5 mm (b–c). Fig. 169. *Satureja darwinii*. a, Plant. b, Flower. c, Opened corolla. d, Seed. Scale = 1 cm (a), 1 mm (b–c) or 0.5 mm (d). Fig. 170. *Valeriana carnosa*. a, Plant. b, Female flower. c, Male flower. d, Fruit. Scale = 2 cm (a) or 1 mm (b–d).

(Flora Patagónica)

lamina 1.4–3.1 × 0.8–2 mm, elliptical to obovate-oblong, obtuse, entire, rarely distantly denticulate, obscurely gland-dotted, narrowing at base to flat, entire or sparsely ciliate petiole *c.*1–1.5 mm. Calyx-lobes *c.*0.3 mm, triangular, obtuse. Corolla-tube *c.*0.5 mm; lobes *c.*0.6 mm, oblong-ovate, obtuse. Style *c.*0.8 mm. Nutlets 0.9–1.1 × 0.6 mm, oblong-ovoid, smooth to minutely verruculose, glabrous, brown. *Flowers January to February*. (Fig. 166).

Wet places with open mineral soil or bryophyte communities above treeline; 500–950 m.

Mountains along S. side of Isla Grande. (Map 20h).
Argentina between 48°53′S.lat. and 51°08′S.lat.

The Fuegian plants belong to the endemic ssp. *fuegiana* D. M. Moore 1970, *Bol. Soc. Argent. Bot.* **13**: 6. Subspecies *patagonica*, from W. C. and S. E. Sta Cruz province, Argentina, differs in being more erect, in having relatively longer and narrower leaves, which are usually ciliolate-serrulate, and in the nut always being minutely verruculose.

LXIV. CALLITRICHACEAE

Monoecious, aquatic or subaquatic, annual or perennial herbs. Leaves opposite, simple; stipules absent. Flowers solitary, axillary. Sepals and petals absent. Stamen 1. Ovary 4-celled; styles 2, filiform. Fruit 4-lobed, separating into 4 mericarps; seeds with endosperm.

1. *Callitriche* L.

Plant glabrous. Flowers sessile; male of 1 stamen only, in upper leaf-axils; female of ovary only, in lower leaf-axils. Styles papillose.

Literature: N. M. Bacigalupo, *Darwiniana* **22**: 377–396(1979).

1. ***C. antarctica*** Engelm. ex Hegelm. 1867, *Verh. Bot. Ver. Brandenb.* **9**: 20.
   C. obtusangula Le Gall var. *antarctica* (Engelm ex. Hegelm.) Speg. 1896, *Anal. Mus. Nac. Bs Aires*, Ser. 2, **5**: 56.
   C. deflexa auct., non R. Br. ex Hegelm. – Spegazzini, 1896.
   C. verna auct., non L. – Hooker, 1847; Hariot, 1884; Franchet, 1889.

Stems (2–) 4–40 cm, shorter on land than when submerged, procument to suberect, slender, rooted at lower nodes, branched, forming lax mats. Leaves (2.3–)4.2–12(–17) × 0.7–2.7 mm, oblanceolate to spathulate, obtuse to subacute or retuse, entire. Bracteoles 2, 2–7.5 mm, linear to linear-lanceolate, subacute to acuminate, scarious. Filament 3–4.5 mm; anther *c.*0.3 mm. Styles 0.5–1.5 mm. Fruit 0.9–1.2 mm in diameter, brown. $2n = 40$*. *Flowers November to March*. (Fig. 165).

Slow moving or stationary freshwater, wet soil and rocks in most open, lowland communities; 0–100 m.

C. and S. Fuegia except the extreme west. (Map 20i).
W. Argentina and Chile north to *c.*49°S.lat.
Falkland Islands (Islas Malvinas); Islas Diego Ramirez.
South Georgia. Prince Edward Islands, Iles Crozet, Iles de Kerguelen, Heard Island, Macquarie Island, Campbell and Auckland Islands, Tasmania.

C. lechleri (Hegelm.) Fassett 1951 *Rhodora* **53**: 191, which may be distinguished by having the basal leaves narrower than the upper, the fruit laterally compressed and the mericarps with a marginal wing, has been reported from Isla Grande (Bacigalupo, 1979).

LXV. PLANTAGINACEAE

Annual or perennial herbs. Leaves in basal rosettes, simple, sheathing at base; stipules absent. Flowers hermaphrodite or unisexual, in bracteate heads or spikes. Sepals 2–4, united at least at base, persistent. Corolla 2– to 4– lobed, scarious. Stamens 3–4, inserted in corolla-tube, alternating with lobes. Ovary superior, 1– to 2–celled. Style 1, simple. Fruit a circumscissile capsule or indehiscent; seeds 1 – numerous, with endosperm.

Literature: R. Pilger, *Bot. Jahrb.* **62**: 1–113(1928).

a. Terrestrial; stolons absent; flowers hermaphrodite; fruit a circumscissile capsule **1. *Plantago***
a. Aquatic; stolons present; flowers unisexual; fruit indehiscent **2. *Littorella***

1. *Plantago* L.

Terrestrial herbs. Flowers hermaphrodite. Sepals 4. Corolla 4-lobed. Stamens 4. Ovary 2-celled; stigma long. Fruit a circumscissile capsule; seeds often mucilaginous when wet.

a. Corolla-tube hairy, at least near base **1. *maritima***
a. Corolla-tube glabrous:
 b. Leaves ovate to broadly elliptical; capsule with 12 or more seeds **3. *major***
 b. Leaves linear to lanceolate or narrowly elliptical; capsule with not more than 8 seeds:
  c. Scape deeply furrowed; anterior sepals united along almost whole length **2. *lanceolata***

c. Scape not furrowed; all sepals united at base only:
  d. Leaves, bracts and sepals with transparent apical seta **4. *uniglumis***
  d. Leaves, bracts and sepals obtuse to subacute, without apical seta **5. *barbata***

1. *P. maritima* L. 1753, *Sp. Pl.* 114.

Stem up to 9 cm, procumbent or ascending, simple or branched near apex. Leaves 30–120 (–200) × 1–5 mm, linear, obtuse to subacute, not or scarcely narrowed at base, sessile, entire or weakly denticulate, glabrous to strigulose, rather fleshy. Scape 3–20 cm, terete, appressed-hairy; inflorescence 1–7 (–9) cm. Bracts 1.6–4 × 1.2–2 mm, ovate, obtuse to subacute. Sepals 2–3 × *c.*1 mm, oblong-ovate, obtuse. Corolla-tube *c.*1 mm, hairy, at least near base, the lobes 1.2–1.8 mm, ovate-triangular, acute or acuminate. Fruit 4–5 mm, ellipsoid, appressed-pubescent distally; seeds 2–3, *c.*2 × 1 mm, oblong, flat on one side, smooth, dark brown or black. $2n = 12$. Self-compatible and at least facultatively self-pollinated. *Flowers November to December.*

Beaches, coastal rocks, cliffs, salt marshes. *Lepidophyllum* scrub, sometimes on open soil inland; 0–60 m.

Throughout Fuegia, principally in coastal areas. (Map 20j).
Southernmost Argentina and in W. Argentina north to *c.*34°S.lat.
Chile north to *c.*47°S.lat.
Falkland Islands (Islas Malvinas).
N. America, Eurasia.

American material has been separated as ssp. *juncoides* (Lam.) Hultén, but recent studies (Moore et al., 1972) show that South American, North American and European populations of *P. maritima* constitute an interfertile complex with overlapping patterns of morphological and chemical variation.

2. *P. lanceolata L. 1753, *Sp. Pl.* 113.

Stem 1–1.5 cm, erect, covered with dead leaf-bases and, distally, with long silky hairs. Leaves with lamina 52–110 × 8–23(–27) mm, lanceolate to narrowly elliptical, ± acute, slightly narrowed at base, entire or weakly and distantly denticulate, glabrous or somewhat pubescent; petiole 12–60 mm, winged. Scape 9–37 cm, deeply furrowed, appressed-hairy; inflorescence *c.*1–3 cm. Bracts 3–4.2 × 1.5–2 mm, ovate, acuminate. Sepals 2.2–3.8 × *c.*1.5 mm, ovate, acute, the anterior united along almost whole length. Corolla-tube 1.3–2.5 mm, glabrous, the lobes 1.5–2 mm, ovate, acute. Fruit *c.*5 mm, oblong-ellipsoid; seeds (1)2, 2.3–2.8 × *c.*1.1 mm, concavo-convex, smooth, brown. *Flowers November to March.*

Cultivated and disturbed ground, especially by roads and settlements; 0–45 m.

Coasts of Isla Grande, except the S. W.
Introduced from Eurasia and widely naturalized in temperate regions.

3. *P. major L. 1753, *Sp. Pl.* 122

Stem 1–1.5 cm, erect, covered with sheathing leaf-bases. Leaves with lamina (27–)50–155(–180) × (20–)32–80(–130) mm, ovate to broadly elliptical, obtuse, cuneate-truncate at base, entire to irregularly crenate or dentate, pubescent; petiole (10–)22–155(–245) mm. Scape (2–)3.6–10(–13) cm, terete, appressed hairy; inflorescence (1.5–)4.4–22(–26) cm. Bracts 2–3 × 1.4–1.8 mm, ovate, acute. Sepals 1.8–2 × *c.*1 mm, ovate, obtuse. Corolla-tube 1.7–2 mm, glabrous, the lobes 1–1.3 mm, triangular, acute. Fruit 2.5–4 mm, ovoid; seeds 12–15(–30), 1.1–1.6 × 0.6–0.8 mm, compressed, rugulose, dark brown to black. *Flowers January to February.*

Disturbed ground near coastal settlements; 0–10 m.

E. and S. coasts of Isla Grande.
Introduced from Europe and widely naturalized in temperate regions.

4. *P. uniglumis* Wallr. ex Walpers 1843, *Nova Acta Acad. Leop.-Carol.* **19** *Suppl.* **1**: 402.

Stem up to 20 cm, ascending to erect, simple or branched, covered with dead leaf-bases. Leaves 8–38 × 0.9–3 mm, linear, narrowed to transparent apical seta, entire, glabrous or sparsely pubescent, the lamina and petiole not well differentiated. Scape 0.2–2.8 cm, terete, glabrous or appressed hairy; inflorescence 1-flowered. Bracts 2.5–3.2 × 1.5–2.5 mm, ovate to triangular, acuminate with transparent seta at apex. Sepals 3.5–4 × 1.6–2 mm, elliptical, acuminate with transparent seta at apex. Corolla-tube 2.5–3 mm, glabrous, the lobes 1–1.5 mm, ovate-lanceolate, acute. Fruit 3–4 mm, ellipsoid to ovoid, glabrous; seeds 1–4 (–6), 1.8–2.1 × *c.*1 mm, ± ellipsoid, flattened on side, black. $2n = 72$*. *Flowers January.*

Open sand and rock-crevices. 0–30 m.

E. C. Isla Grande. (Map 21a).
W. Argentina and E. Chile north to *c.*30°S.lat.

5. ***P. barbata*** Forster f. 1789, *Comment. Gotting.*, Ser.2, **9**: 25.
P. monanthos D'Urv. 1825, *Fl. Is. Mal.* 37 – Hariot, 1884; Spegazzini, 1896; Alboff, 1896.

Stem 1.5–12 cm, ascending to erect, usually branched, covered with sheathing leaf-bases. Leaves 7–23(–40) × (7–)9–24(–40) mm, linear to linear-lanceolate or ovate to triangular-ovate, obtuse to acute or acuminate, entire to irregularly and distantly dentate, glabrous to pubescent, the lamina and petiole not well differentiated. Scape 0.5–3.5(–5.2) cm, striate, glabrous or appressed hairy; inflorescence 0.3–1.6 cm, (1–)3(–11)-flowered. Bracts 1.5–2.8 × 1.2–3 mm, ovate, obtuse to subacute. Sepals 2.1–3.3 × 1.2–2.5 mm, ovate, obtuse to subacute. Corolla-tube 2-2.8 mm, glabrous, the lobes 1.1–1.5(–1.8) mm, ovate, acute. Fruit 2.7–4.5 mm, ellipsoid, glabrous; seeds 5–8, 2.3–2.7 × c.1.3 mm, ± ellipsoid, flattened on one side, black. $2n = 48$. Self-compatible. *Flowers October to December.* (Fig. 183).

Wet sand, pebbles and rock-crevices by coast, lakes and streams, feldmark, bog-margins and other moist unshaded situations; 0–900 m.

Throughout Fuegia. (Map 21b).
Argentina and Chile north to *c.* 47°S.lat., then north along Andes in both countries to *c.*28°43′S.lat.
Falkland Islands (Islas Malvinas), Islas Diego Ramirez.

A very variable species in habit, leaf-shape, indumentum and inflorescence-size. The variation is partly correlated with habitat and partly with geographical differences. However, because of the independent variation of the characters a satisfactory taxonomic subdivision of the species has yet to be attained.

2. *Littorella* Berg

Monoecious, aquatic stoloniferous herbs. Spike with solitary terminal male flower and 2–7(8) female flowers below. Sepals 3–4 in male and 2–4 in female flowers. Corolla 3– to 4– lobed in male and 2– to 4– lobed in female flowers. Stamens 3–4. Ovary 1-celled; stigma short. Fruit indehiscent, hard; seed 1.

1. ***L. australis*** Griseb. ex Bentham and Hooker f. 1876, *Gen. Pl.* **2**: 1225.

Stolons slender, rooting and with leaf-rosettes at nodes. Leaves 13–45 × 0.8–2.5 mm, linear-subulate and semi-cylindrical, obtuse, with membranous basal sheath 6.5–10 mm. Scape 15–18 mm, slender; bracts 1.1–2.5 × 0.5–1.5 mm, ovate-lanceolate, acuminate. Male flowers with sepals 3–4.5 × 1–1.5 mm, lanceolate, obtuse, united about middle; corolla-tube 3–3.5 mm, the lobes *c.*1.5 mm, ovate. Female flowers irregularly grouped in basal half of axis or above, sessile; sepals 1.5–2 mm, linear, acute, united to about middle; corolla-tube 2.5–4 mm, the lobes *c.*0.1 mm. Fruit 1.5–2 × *c.*0.75 mm, narrowly ovoid, tapering to stout, straight beak, with longitudinal ribs. *Flowers December to January.* (Fig. 171).

Margins of freshwater lakes and ponds; 0–20 m.

Locally in N. E. and S. Isla Grande. (Map 21c).
W. Argentina north to *c.*41°S.lat.
Chile between 40°and 42°S.lat.
Falkland Islands (Islas Malvinas).

LXVI. SCROPHULARIACEAE

Herbs or shrubs. Leaves alternate or opposite, or all basal, simple; stipules absent. Flowers usually zygomorphic (sometimes slightly so), hermaphrodite, in bracteate spikes, panicles or racemes, or solitary in leaf-axils. Calyx 4– to 5– lobed. Corolla 4– to 5– lobed or 2-lipped. Stamens 2 or 4, inserted near base of corolla-tube. Ovary superior, 2-celled; style 1; stigma simple to capitate or 2-lobed. Fruit a septicidal, rarely loculicidal, capsule; seeds numerous, with endosperm.

a. Evergreen shrub **6. *Hebe***
a. Herb:
  b. Lower lip of corolla inflated, deeply concave, saccate **2. *Calceolaria***
  b. Corolla not 2-lipped, or weakly so but with lower lip not inflated and saccate:
    c. Flowers solitary; leaves in basal cluster:
      d. Flowers not more than 3.5 mm; leaves linear to linear-spathulate **1. *Limosella***
      d. Flowers at least 5 mm; leaves ovate to orbicular **5. *Ourisia***
    c. Flowers in spikes, racemes or panicles; leaves alternate or opposite:
      e. Leaves opposite, grey-tomentose beneath; stem usually more than 5 mm in diameter; corolla at least 40 mm **3. *Digitalis***

e. Leaves alternate, if opposite not grey-tomentose beneath; stem not more than 5 mm in diameter; corolla not more than 25 mm:
  f. Leaves deeply 3-fid **7. *Euphrasia***
  f. Leaves entire to crenate:
    g. Calyx and corolla 5-lobed; stamens 4 **5. *Ourisia***
    g. Calyx and corolla 4-lobed; stamens 2 **4. *Veronica***

1. *Limosella* L.

Glabrous annual to perennial herbs. Leaves in basal clusters on slender rhizomes. Flowers weakly zygomorphic, solitary. Calyx 5-lobed. Corolla rotate, irregularly 5-lobed, with short tube. Stamens 4. Stigma 2-lobed.

1. *L. australis* R. Br. 1810, *Prodr. Fl. Nov. Holl.* **1**: 443.
L. aquatica auct., non L. – Dusén, 1900.
L. aquatica var. *tenuifolia* (Nutt.) Hooker f. 1847, *Fl. Antarct.* **1(2)**: 334 – Skottsberg, 1916.

Leaves (6–)7.5–28.5(–95) × 0.3–0.9(–1.4) mm, linear to linear-spathulate, obtuse, entire. Peduncle 1.5–7.5 mm, slender. Calyx 1.9–2.9 mm, subcampanulate, green or purplish; lobes 0.4–0.7(–1.3) mm, triangular-ovate, acute. Corolla (1.5–)2.3–2.6 mm, white or pale lilac; lobes 1.2–1.6 mm, ovate-ligulate, obtuse. Capsule 2–3.2 × 0.9–2 mm, ovoid; seeds *c.*0.6 × 0.3 mm, cylindrical-ellipsoid to reniform, angular, transversely sulcate. $2n = 48$. *Flowers December to March.* (Fig. 174).

Wet mud, sand and pebbles in or beside stream and lake margins; 0–30 m.

Almost throughout Isla Grande except the extreme west. (Map 21d).
Argentina north to *c.*40°S.lat.
Chile north to *c.*40°S.lat.
Falkland Islands (Islas Malvinas).
Perhaps more widespread.

2. *Calceolaria* L.

Perennial herbs, often woody at base. Leaves opposite. Flowers zygomorphic, solitary or 2–4 in lax racemes. Calyx subequally 4-lobed, united with corolla at base. Corolla unequally 2-lipped, with short tube; lips entire, concave-saccate, the lower much larger and inflated. Stamens 2, between lips. Stigma subcapitate. Capsule ovoid-conical.

a. Peduncle with 1 flower; lower lip of corolla (17–)22–25 mm; leaves glabrous, entire to denticulate **1. *uniflora***
a. Peduncle with (1) 2–4 flowers; lower lip of corolla 10–15(–19) mm; leaves white – hairy on margins and veins beneath, serrulate **2. *biflora***

1. *C. uniflora* Lam. 1785, *Encycl. Méth. Bot.* **1**: 556.
C. nana Sm. 1789, *Pl. Icon. Ined.* **1**: t.1.
C. darwinii Bentham 1846, in DC. *Prodr.* **10**: 207.

Stems up to 3 cm, erect to ascending, with leaves clustered towards apex. Leaves glabrous; lamina 10–30 × 6–12 mm, ovate-rhomboid to spathulate, obtuse, cuneate at base, ± entire to denticulate; petiole 3–10 mm, winged. Peduncles 2.4–4.5(–7.5) cm, erect, sparsely puberulent, with shortly stipitate glands above; flower solitary. Calyx 4.5–8 mm; lobes 3–5 mm, triangular-ovate to suborbicular, obtuse, glandular-hairy. Corolla with upper lip 3–5 × 6–10 mm, suborbicular, shorter than to equalling calyx; lower lip (17–)22–25 × 13–17 mm, triangular-obovate, saccate in apical $\frac{1}{3}$–$\frac{1}{2}$, pendent, yellow with red-brown spots or larger markings. Capsule 9–15 × 5–7 mm, brown; seeds *c.*1 × 0.3 mm, navicular, striate, red-brown. $2n = 18$. *Flowers November to January.* (Fig. 172, Pl. 7d).

Coastal sands, *Lepidophyllum* scrub, dry grassland and open *Chiliotrichum* scrub; 0–60 m.

C. & N. Isla Grande, occasional further south. (Map 21e).
Argentina north to *c.*46°S.lat.
E. Chile north to *c.*36°S.lat.

2. *C. biflora* Lam. 1785, *Encycl. Méth. Bot.* **1**: 556.
C. plantaginea Sm. 1789, *Pl. Icon. Ined.* **1**: t.2.

Stems up to 6 cm, procumbent to erect, with leaves clustered towards apex. Leaves white-hairy on margins and veins beneath; lamina (10–)25–100(–150) × 10–40 mm, obovate to rhomboid, obtuse to subacute, cuneate at base, serrulate to crenate-serrulate; petiole 5–25 mm, winged. Peduncles 10–40 cm, erect, sparsely hairy, with 2 small bracts; flowers (1)2–4 in lax raceme; pedicels 10–50 mm, densely stipitate glandular. Calyx 4–6 mm; lobes (1.7–)2–5 mm, triangular-ovate to suborbicular, obtuse, sparsely white-hairy, with shortly stipitate glands. Corolla with upper lip *c.*4 × 5.5 mm, suborbicular, shorter than to equalling calyx; lower lip 10–15(–19) × 12–14 mm, suborbicular, saccate in apical $\frac{1}{2}$–$\frac{3}{4}$, pendent, yellow, sometimes with sparse, small, red-brown spots near throat, with short stipitate glands. Capsule 8–14 ×

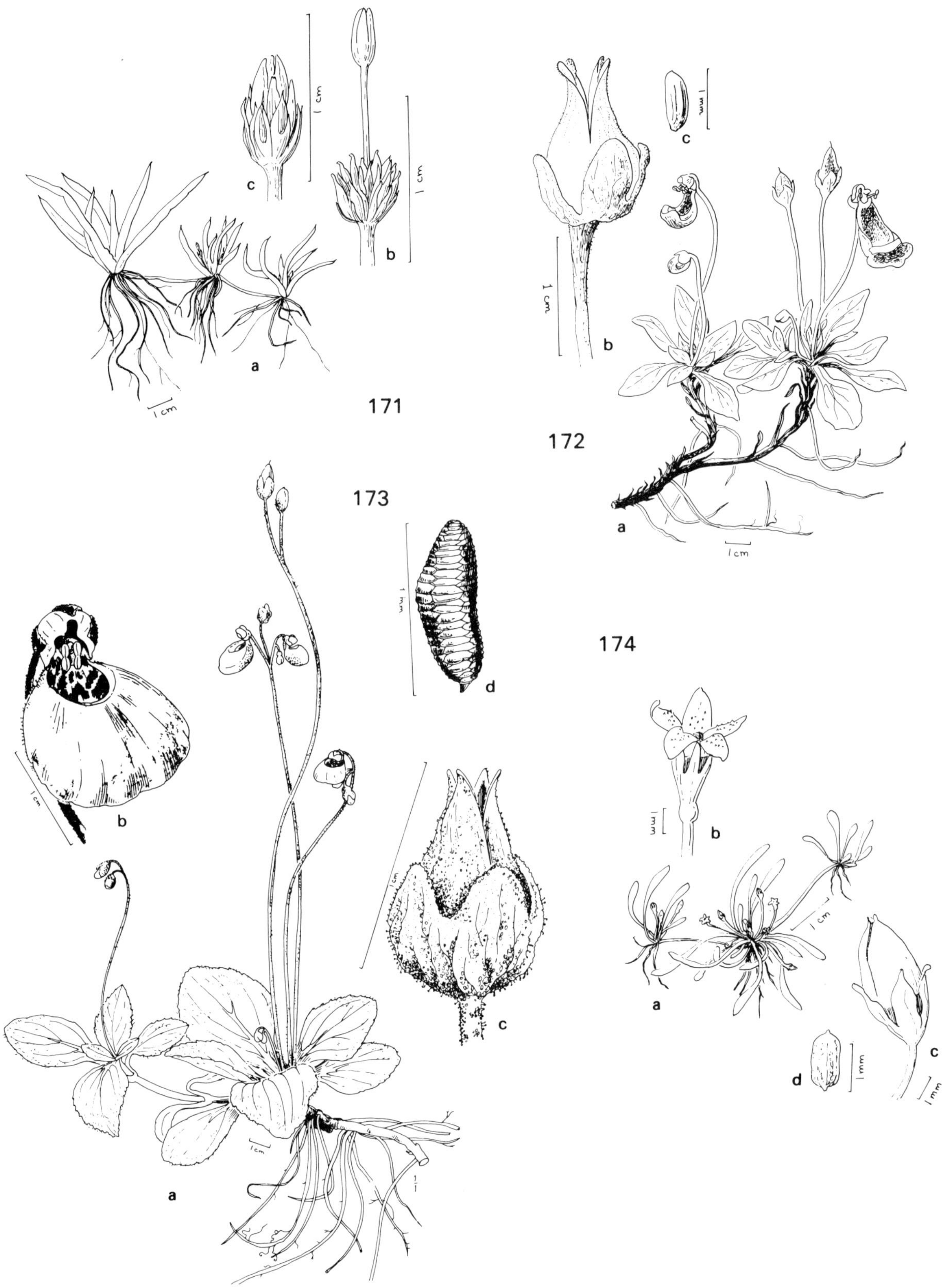

Fig. 171. *Littorella australis*. a, Plant. b, Inflorescence. c, Fruit. Scale = 1 cm. Fig. 172. *Calceolaria uniflora*. a, Plant. b, Fruit. c, Seed. Scale = 1 cm (a–b) or 1 mm (c). Fig. 173. *Calceolaria biflora*. a, Plant. b, Flower. c, Fruit. d, Seed. Scale = 1 cm (a–c) or 1 mm (d). Fig. 174. *Limosella australis*. a, Fruit. b, Flower. c, Fruit. d, Seed. Scale = 1 cm (a) or 1 mm (b–d).
(R. N. P. Goodall)

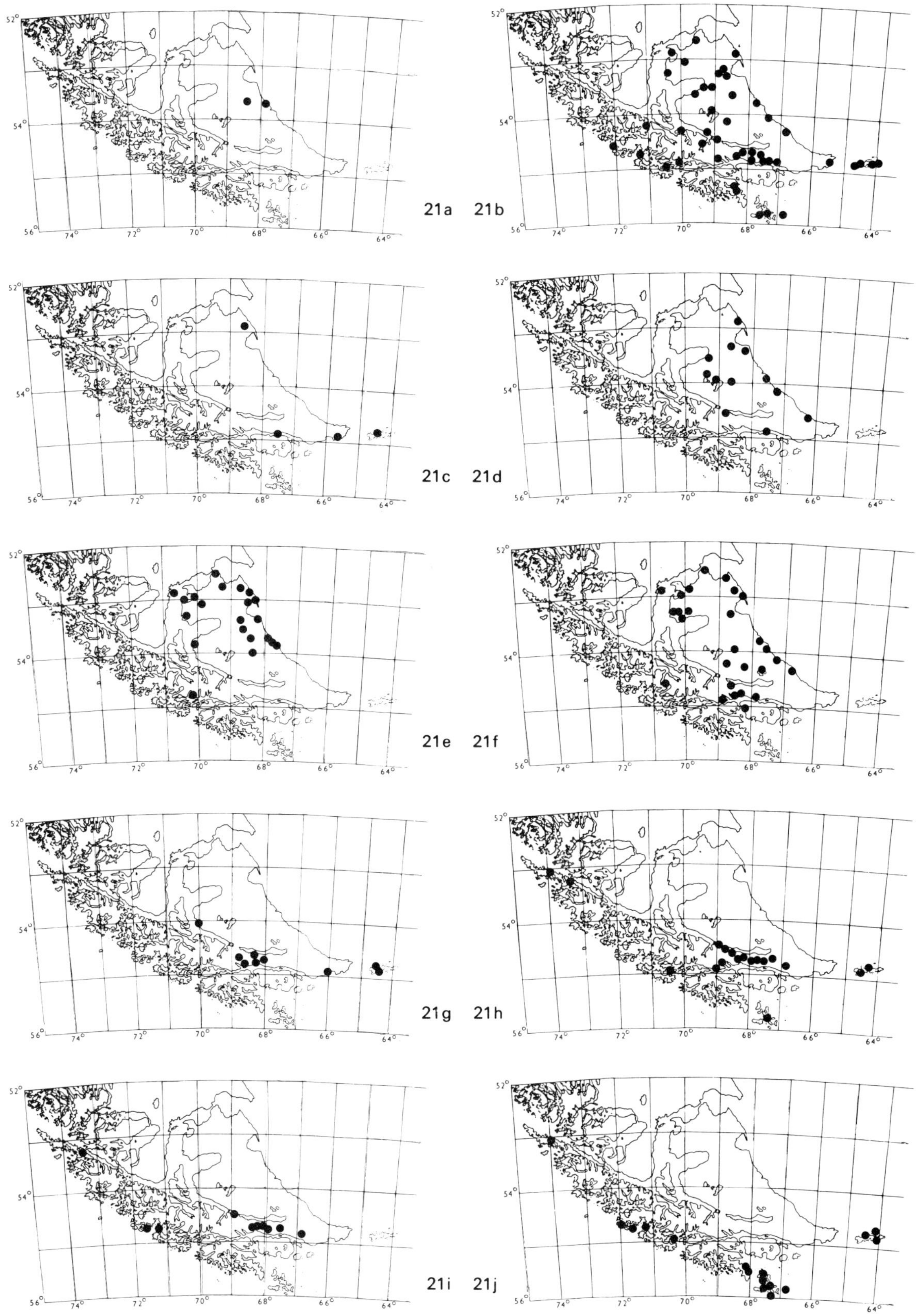

Map 21. a, *Plantago uniglumis*. b, *Plantago barbata*. c, *Littorella australis*. d, *Limosella australis*. e, *Calceolaria uniflora*. f, *Calceolaria biflora*. g, *Ourisia ruelloides*. h, *Ourisia breviflora*. i, *Ourisia fuegiana*. j, *Hebe elliptica*.

5–7 mm, brownish; seeds *c.*0.7 mm, navicular, reticulate. *Flowers December to February (March)* (Fig. 173).

Moist coastal cliffs, open scrub and deciduous forest, sometimes in grassland; 0–250 m.

Almost throughout Isla Grande and south shore of Canal Beagle. (Map 21f).
W. Argentina north to *c.*30°S.lat.
E. Chile north to *c.*51°S.lat.

3. *Digitalis* L.

Biennial or perennial herbs. Leaves alternate, the lowest forming a rosette. Flowers zygomorphic, in terminal, ± secund racemes. Calyx equally and deeply 5-lobed. Corolla 2-lipped, with inflated tube; lips erecto-patent, the lower larger. Stamens 4. Stigma 2-fid.

*1. ***D. purpurea*** L. 1753, *Sp. Pl.* 621.

Stem up to 80 cm, erect, usually simple, glabrous below, grey-tomentose above. Leaves with lamina 3.5–17 × 2.3–8 cm, ovate to lanceolate or elliptical, obtuse, cuneate at base, crenate to dentate, pubescent above, grey-tomentose beneath; petiole 3–11.5 cm, winged. Raceme 9– to 35(–80)– flowered; lower bracts 10–35 × 2–9 mm, lanceolate, acute, entire, sessile, the upper smaller; pedicels 11–20 mm, tomentose. Calyx-lobes 8.5–19 × 2.5–10 mm, acute, the lower ovate, the upper lanceolate. Corolla 40–50 mm, pinkish-purple, with white-margined darker purple spots inside; tube campanulate, constricted at base, exceeding calyx, ciliolate inside; upper lip *c.*5 mm, ± ovate, obscurely 2-lobed; lower lip *c.*12 mm, oblong, with middle lobe more prominent than the lateral. Capsule *c.*11 × 7 mm, brownish; seeds 0.8–0.9 mm, rectangular, tetragonous, reticulate, brown. *Flowers November to March.*

Burned and degraded forest and scrub; 20–90 m.

C. W. & S. coasts of Isla Grande.
Introduced from Europe, perhaps via S. Chile, with grass-seed and locally naturalized, though apparently decreasing.

4. *Veronica* L.

Annual to perennial herbs. Leaves opposite. Flowers weakly zygomorphic, in terminal racemes. Calyx subequally and deeply 4-lobed. Corolla rotate, 4-lobed, the upper lobe largest. Stamens 2. Stigma capitate. Capsule dehiscing loculicidally and also partially septicidally, laterally compressed.

a. Perennial; corolla at least 5 mm in diameter; upper bracts not exceeding pedicels and flowers **1. *serpyllifolia***
a. Annual; corolla not more than 4 mm in diameter; upper bracts exceeding pedicels and flowers;
  b. Leaves oblong to oblong-oblanceolate, entire or weakly dentate **2. *peregrina***
  b. Leaves triangular-ovate, coarsely serrate-dentate **3. *arvensis***

*1. ***V. serpyllifolia*** L. 1753, *Sp. Pl.* 12.

Perennial; stems 5–23(–30) cm, creeping and rooted at nodes when vegetative but flowering stems ascending to erect, pubescent. Leaves 5.6–9.5 × 3.4–7 mm, elliptic-suborbicular to ovate-oblong, obtuse, rounded at base, entire to crenulate, glabrous to puberulent, subsessile or with petiole up to 1.7(–2.5) mm. Raceme (2–)2.5–11.5(–14) cm, up to 36-flowered; upper bracts 2.2–5 × 0.8–1.5 mm, lanceolate to elliptical, obtuse, ciliolate and sometimes sparsely puberulent, not exceeding pedicel and flower; lower bracts larger and becoming leaf-like; pedicels 1.7–3.3 mm, up to 5 mm in fruit, puberulent. Calyx-lobes 1.8–3.8 × 0.7–1.5 mm, ± elliptical, obtuse, sparsely ciliate and puberulent. Corolla 5–10 mm in diameter, pale blue with darker veins; tube 0.5–0.8 mm; lobes 1.5–1.7 × 1.4–2.1 mm, suborbicular. Capsule 2–3 × 3–4 mm, obcordate, ciliate, pale brown; seeds 0.5–0.8 mm, ovoid to suborbicular, pale brown. *Flowers November to May.*

Open, usually moist, soil and sand, in grassland, scrub and open forest; 0–500 m.

Throughout Isla Grande, except the extreme west, and S. side of Canal Beagle.
Introduced from Europe and widely naturalized in areas modified by man and his animals.

2. ***V. peregrina*** L. 1753, *Sp. Pl.* 13.

Annual; stems 3–7(–10) cm, erect, simple or branched, glandular-pubescent. Leaves 8.3–15 × 0.7–2.9 mm, oblong-oblanceolate, obtuse, entire to ± dentate, the lower shortly petiolate, the upper sessile. Raceme 2–6(–8) cm, up to 30– flowered; upper bracts 4.7–8 × 0.6–1.3 mm, oblong-oblanceolate, obtuse, exceeding pedicel and flower; lower bracts larger; pedicels up to 2 mm. Calyx-lobes 2.5–5(–7) × 0.6–1.5 mm, elliptical to linear-lanceolate, acute to obtuse, glandular-pubescent. Corolla 2–4 mm in diameter, whitish; tube *c.*1 mm; lobes 1.3–2 × *c.*1 mm, oblong to oblong-elliptical. Capsule 2.5–4.2 × 3.6–5 mm, obcordate, glabrous, pale brown; seeds 0.5–0.9 mm, ovoid, brown. *Flowers November to January.* (Fig. 178).

Open, usually moist, stones, sand and soil, by streams, lakes and cultivated ground; 0–500 m.

S. to N. E. Isla Grande.
Native throughout western North and South America. Its status in Fuegia, however, is uncertain and it may be introduced here.

Fuegian material belongs to ssp. *xalapensis* (Kunth) Pennell.

*3. ***V. arvensis*** L. 1753, *Sp. Pl.* 13.

Annual; stems 2–10 cm, ascending to erect, simple or branched from base, pubescent. Leaves 4–12 × 3–10 mm, triangular-ovate, obtuse, cordate to truncate or broadly cuneate at base, coarsely serrate-dentate, pubescent, sessile or the lower shortly petiolate. Raceme 2–8 cm, up to 25– flowered; upper bracts 3.5–5 × 1–1.5 mm, linear to linear-lanceolate, entire, ciliate, exceeding pedicel and flower; lower bracts larger; pedicels up to 1 mm. Calyx-lobes 2–2.5 × *c.*1 mm, lanceolate, obtuse, pubescent. Corolla 2–2.5 mm in diameter, blue. Capsule 2–2.5 × 2–2.5 mm, obcordate, ciliate, pale brown; seeds *c.*1 mm, ovoid, brown. *Flowers November to January.*

Coastal sands and gravel, exposed soil in open scrub and grassland; 0–30 m.

Locally on W. and S. coasts of Isla Grande. Introduced from Europe and occasionally naturalized in disturbed and open habitats.

5. *Ourisia* Comm. ex Juss.

Perennial herbs. Leaves opposite, often in basal cluster. Flowers solitary or in terminal panicles. Calyx ± deeply and unequally 5-lobed. Corolla weakly zygomorphic; tube short, slightly oblique; lobes 5, subequal. Stamens 4. Stigma capitate to somewhat 2-lobed.

a. Leaves at least 20 mm; corolla at least 15 mm, red **1. *ruelloides***
a. Leaves not more than 15 mm; corolla not more than 15 mm, white to violet:
b. Leaves at least 4 mm; bracteoles broadly ovate **2. *breviflora***
b. Leaves not more than 4 mm; bracteoles spathulate **3. *fuegiana***

1. *O. ruelloides* (L.f.) O. Kuntze 1898, *Revis. Gen.* **3**: 238.
O. magellanica C. F. Gaertner 1805, *Fruct. Sem. Pl.* **3**: 44– Spegazzini, 1896; Hooker, 1847.
O. poeppigii Bentham 1846, in DC., *Prodr.* **10**: 492.

Rhizome up to 10 cm, horizontal to ascending, rooted at nodes, covered with leaf-remains, bearing leaves clustered towards apex. Leaves with lamina 23–80 (–115) × 14–48 mm, ovate to suborbicular or rarely lanceolate, obtuse, truncate to cuneate at base, crenate, sparsely hairy beneath; petiole 20–45 (–120) mm, sheathing at base. Panicle 2– to 8– flowered; peduncle 40–75 mm; bracts 7–9 mm, suborbicular, semi-amplexicaul, crenate; pedicels 12–35 mm. Calyx 5.4–6 mm, somewhat 2-lipped; lobes 2.5–3.2(–4) × 2–2.8 mm, ovate, obtuse, ciliate. Corolla 16–21 mm, *c.*10 mm in diameter, red; lobes 2–2.7 × 2.8–4(–5) mm, emarginate. Stigma somewhat 2-lobed. Capsule 5–6 × 5.5–7.5 mm, obcordate, compressed; seeds *c.*1.7 mm, cuneate, reticulate, pale brown. *Flowers October to January.* (Fig. 177).

Wet rock crevices, beside and in streams; 0–300 m.

C. & S. Isla Grande except the extreme west; Isla de los Estados. (Map 21g).
W. Argentina north to *c.*41°S.lat.
Chile north to *c.*40°S.lat.

2. *O. breviflora* Bentham 1846, in DC., *Prodr.* **10**: 493.
O. uniflora Phil. 1857, *Linnaea* **29**: 27.

Leaves in basal cluster, villous; lamina (4–) 7–10.5(–14) × 5–12 mm, ovate to orbicular, obtuse, truncate to cordate at base, ± crenate; petiole 0.7–2.5 mm. Flowers usually solitary, or 2–4 in raceme; peduncle (0.7–)2–6 mm, villous; bracts 6–11 mm, broadly ovate, obtuse, crenate, villous above. Calyx 5.5–7.5 mm, divided almost to base; lobes oblong, obtuse, ciliate. Corolla (7–) 10–14 mm, violet with darker veins; lobes 3–3.5 × 2–3 mm, linear to obovate, emarginate. Stigma capitate. Capsule 4–6 mm, suborbicular, glabrous; seeds *c.*1 mm, oblong, reticulate, yellow-brown. *Flowers January to March.* (Fig. 175).

Damp ground by streams, at and above tree-line; 700–1000 m.

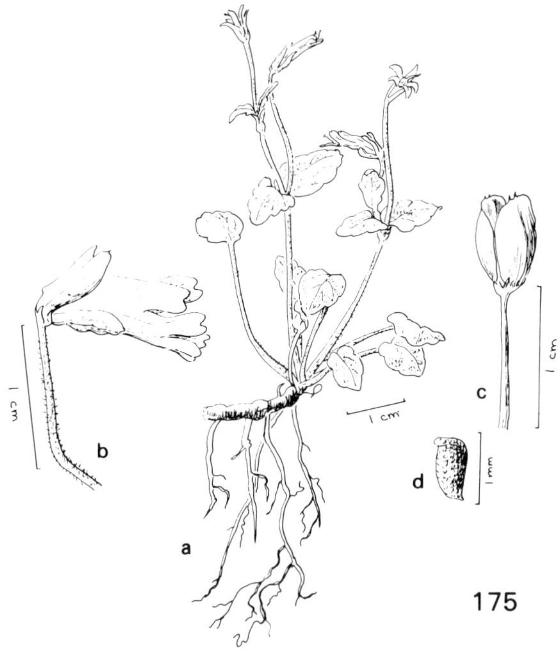

175

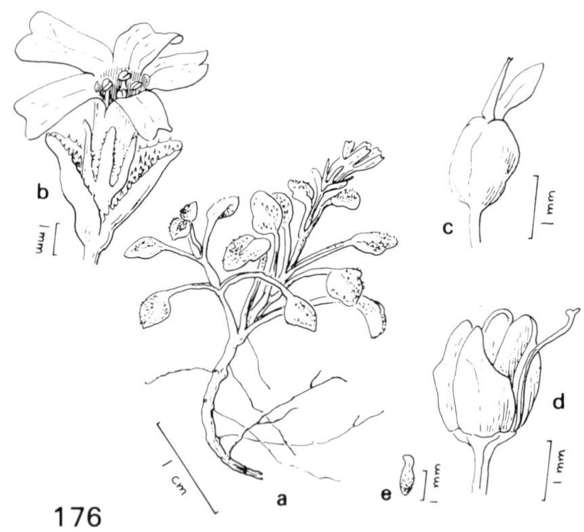

176

177

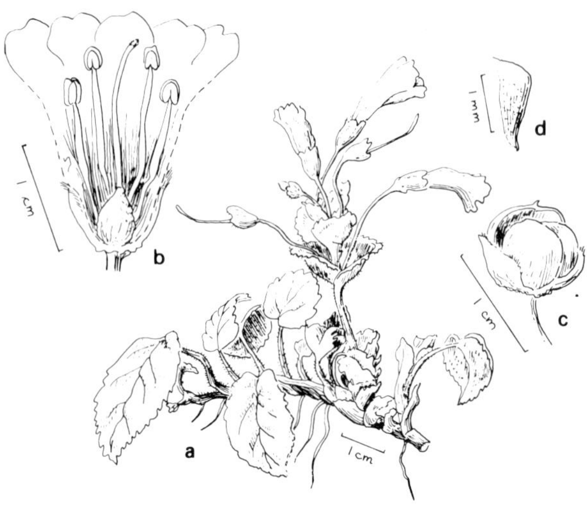

178

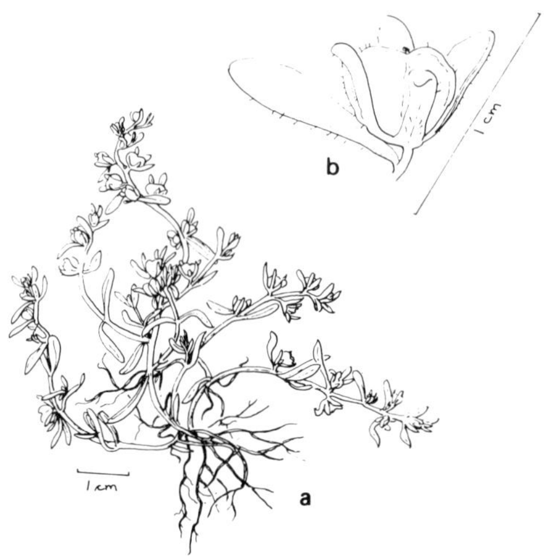

Fig. 175. *Ourisia breviflora*. a, Plant. b, Flower. c, Fruit. d, Seed. Scale = 1 cm (a–c) or 1 mm (d). Fig. 176. *Ourisia fuegiana*. a, Plant. b, Flower. c, Young fruit. d, Fruit. e, Seed. Scale = 1 cm (a) or 1 mm (b–e). Fig. 177. *Ourisia ruelloides*. a, Plant. b, Flower. c, Fruit. d, Seed. Scale = 1 cm (a–c) or 1 mm (d). Fig. 178. *Veronica peregrina*. a, Plant. b, Fruit. Scale = 1 cm.
(R. N. P. Goodall)

C. S. Fuegia. (Map 21h).
Chile north to *c.*51°S.lat.
W. Argentina north to *c.*51°S.lat.

3. ***O. fuegiana*** Skottsb. 1916, *Kungl. Svenska Vet. Akad. Handl.* **56(5)**: 302.

O. muscosus auct., non Bentham – Spegazzini, 1896.
O. nana auct., non Bentham – Dusén, 1900.

Leaves in compact basal cluster; lamina 2–4 × (1.3–) 2–3.5 mm, suborbicular, obtuse, cuneate at base, entire or with few obscure teeth, glabrous beneath, hairy above; petiole 2.5–5 (–9) mm, sparsely hairy. Flowers solitary; peduncle 3–11 mm, sparsely villous; bracts 2.5–4 mm, spathulate, subacute, hairy above. Calyx 2.5–4 mm, divided almost to base; lobes linear, obtuse, ciliate, violet. Corolla 5–5.5 mm, white to pale violet; lobes 2.5–2.8 × 1–1.5 mm, linear-ovate to obovate, emarginate. Stigma capitate. Capsule 2–2.5 mm, suborbicular, glabrous; seeds *c.*0.5 mm, ± oblong, reticulate, red-brown. *Flowers January to March.* (Fig. 176).

Damp ground by streams above tree-line, wet rock-ledges and screes; 600–900 m.

S. Fuegia. (Map 21i).
Endemic.

6. ***Hebe*** Comm. ex Juss.

Evergreen shrubs. Leaves opposite and decussate. Flowers actinomorphic, in bracteate axillary racemes. Calyx deeply 4(5)-lobed. Corolla tubular, with rotate 4-lobed limb. Stamens 2. Stigma capitate. Capsule septicidally, sometimes loculicidally, dehiscent.

1. ***H. elliptica*** (Forster f.) Pennell 1921, *Rhodora* **23**: 39.

Veronica elliptica Forster f. 1786, *Fl. Ins. Austr. Prodr.* 3 – Hooker, 1847; Franchet, 1889; Dusén, 1900; Skottsberg, 1916.

Stems up to 3(–6) m, up to 10 cm in diameter, erect, much branched; twigs usually densely recurved-pubescent. Leaves with lamina 12–35 × 6–15 mm, elliptical to elliptic-obovate, obtuse, apiculate, cuneate to truncate at base, entire to irregularly serrulate and with white-pubescent, somewhat revolute margin; petiole 2–5 mm. Racemes 3– to 10– flowered; peduncle 5–10 mm, sparsely pubescent; bracts *c.*1 × 0.5 mm, triangular, puberulent, pale brown; pedicels 2–4 mm. Calyx-lobes 3–3.7 × 2–3 mm, ovate, acute, ciliolate. Corolla white or lilac, sometimes with bluish streaks; tube 2–3 mm; lobes 5–9 × 4–6 mm, oblanceolate to obovate, obtuse to subacute. Capsule (4–)5–8 × 3.5–5 mm, ovoid, acute to retuse, dorsally compressed; seeds suborbicular, compressed, $2n = 40$. *Flowers November to February.* (Fig. 179).

Coastal cliffs and rocks; 0–40 m.

Extreme S, W. & E. Fuegia. (Map 21j).
Chile north to 45°53′S.lat.
Falkland Islands (Islas Malvinas).
New Zealand.

7. ***Euphrasia*** L.

Annual herbs. Leaves opposite, deeply 3-fid. Flowers zygomorphic, in a terminal spike, sessile in axils of leaf-like bracts. Calyx shortly 4-lobed. Corolla tubular, 2-lipped; upper lip 2-lobed; lower lip 3-lobed. Stamens 4. Stigma capitate.

1. ***E. antarctica*** Bentham 1846, in DC., *Prodr.* **10**: 555.

Stem 2–9(–25) cm, erect, simple or with suberect branches from lower part, deflexed-pubescent, purplish green. Leaves 4.1–7.5(–9) × 4–5.5 mm, cuneiform in outline, sessile; lobes oblong to triangular-ovate, obtuse to subacute, entire, glabrous or sparsely strigulose beneath, often purplish. Bracts 5.5–7.5(–12) × 5.7–10 mm, cuneiform in outline, 3(–6)– fid, green, usually with purple veins; lobes triangular, acute, entire. Calyx (5.5–)7–8 × 3–4 mm, campanulate; lobes 1.5–3.5 × 1.2–2.5 mm, triangular, acute, entire, glabrous or puberulent towards apex, green with purple margin and veins. Corolla white to pale lilac, yellowish in throat; tube 5–5.5 mm; upper lip with lobes 1.5–2 × 2–2.5 mm, oblong; lower lip with lobes *c.*2 × 1.5–2 mm, cuneiform, emarginate. Capsule 4–6 × *c.*3 mm, obovoid, obtuse; seeds 1.5–1.8 mm, cylindrical-ovoid, striate. $2n = 88^{\star}$. *Flowers December to March.* (Fig. 180).

Damp grassland, raised parts of bogs, stream and lake margins; 0–50 m.

Almost throughout Isla Grande except the extreme east and west. (Map 22a).
W. Argentina north to *c.* 49°S.lat.
Chile north to 49° 10′S.lat.
Falkland Islands (Islas Malvinas).

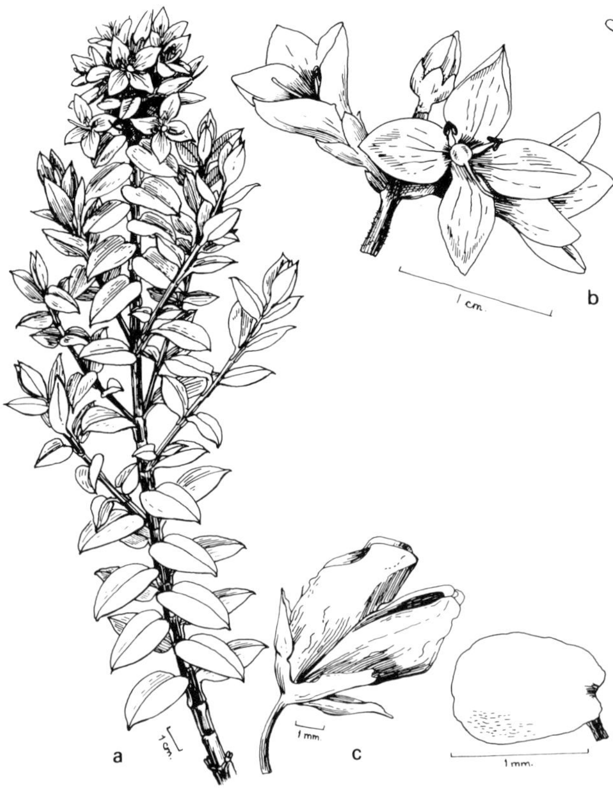

179

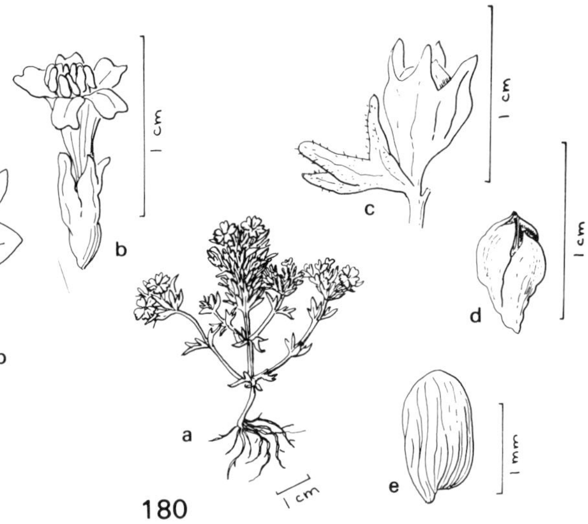

180

182

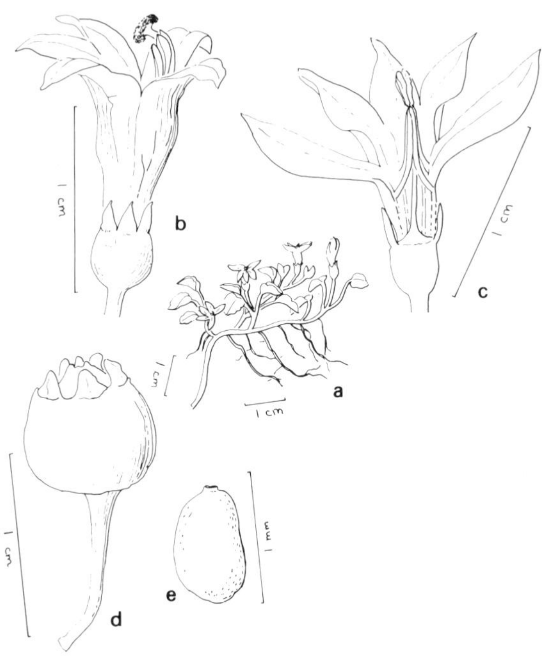

181

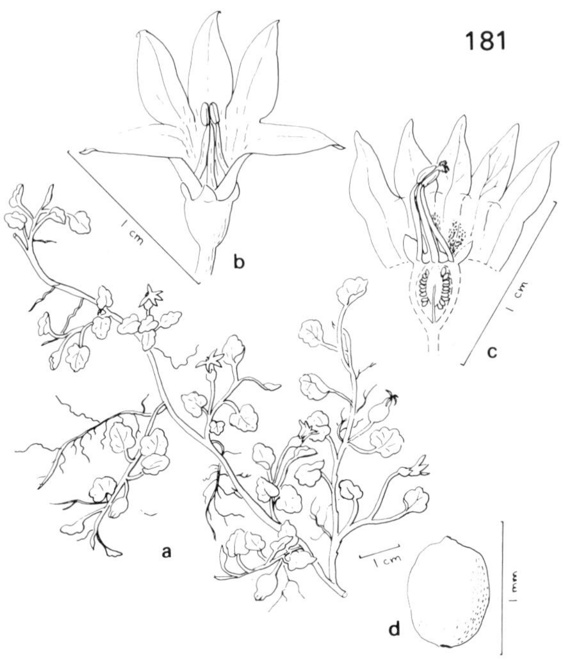

Fig. 179. *Hebe elliptica.* a, Branch. b, Flower. c, Fruit. d, Seed. Scale = 1 cm (a–b) or 1 mm (c–d). Fig. 180. *Euphrasia antarctica.* a, Plant. b, Flower. c, d, Fruit. e, Seed. Scale = 1 cm (a–d) or 1 mm (e). Fig. 181. *Pratia repens.* a, Plant. b, Flower. c, Half-flower. d, Seed. Scale = 1 cm (a–c) or 1 mm (d). Fig. 182. *Pratia longiflora.* a, Plant. b, Flower. c, Half-flower. d, Fruit. e, Seed. Scale = 1 cm (a–d) or 1 mm (e).

(R. N. P. Goodall)

LXVII. LENTIBULARIACEAE

Insectivorous perennial herbs. Leaves in a basal rosette; stipules absent. Flowers solitary. Calyx 2-lipped, persistent. Corolla 2-lipped, spurred. Stamens 2, inserted in corolla. Ovary superior, 1-celled; stigma sessile. Fruit a capsule opening by 2 valves; seeds numerous, without endosperm.

1. *Pinguicula* L.

Leaves soft and fleshy, entire, covered with viscid glands above. Scape slender, naked. Calyx with upper lip 3–, and lower lip 2–, lobed. Corolla with upper lip 2–, and lower lip 3–, lobed, hairy in throat. Stigma 2-lobed.

Literature: S. J. Casper, *Biblioth. Bot. (Stuttgart)* **127/128**: 1–209(1966).

1. *P. antarctica* Vahl 1804, *Enum. Pl.* **1**: 192.
P. magellanica Comm. ex Franchet 1889, *Miss. Sci. Cap Horn* 353.

Leaves 9.5–18 × 7.2–11.6 mm, elliptical to obovate or spathulate, rounded-obtuse, margins strongly involute, yellowish green; petiole 3–4 mm, winged. Scapes 6–12.5 mm. Calyx 3–3.8 mm; lobes 1–3 × 0.8–1.4 mm, ovate to triangular or oblong, obtuse. Corolla 6–7 mm, pale purple; lobes of upper lip 2–2.25 × 1.75–2 mm, oblong, ± emarginate; lobes of lower lip *c*.4 × 3 mm, oblong-obovate, emarginate; spur 1–2(3) mm, ± cylindrical, obtuse, curved downwards. Capsule *c*.5 × 4 mm, subglobose to ovoid; seeds 0.6–0.9 mm, narrowly fusiform. *Flowers November to January*. (Fig. 184).

Sphagnum bogs, wet bryophyte communities, Magellanic moorland. 0–500 m.

S. & W. Fuegia. (Map 22b).
Chile north to *c*.42°S.lat.

LXVIII. CAMPANULACEAE

Perennial herbs. Leaves alternate, simple; stipules absent. Flowers hermaphrodite, actinomorphic. Calyx 5-lobed, the tube united to the ovary. Corolla 5-lobed. Stamens 5, free. Ovary inferior, 3-celled; style 1; stigmas 3. Fruit a capsule, dehiscing by basal pores; seeds numerous, with endosperm.

1. *Campanula* L.

Flowers in a ± branched, lax panicle. Calyx-tube subglobose. Corolla campanulate. Stigmas hairy.

1. *C. rotundifolia L. 1753, *Sp. Pl.* 163.

Stems up to 50 cm, ascending to erect, pubescent below, sparsely leafy up to inflorescence. Basal leaves 12–20 × 12–18 mm, suborbicular to reniform, crenate, with petiole 15–25 mm, the cauline 15–35 × 1–4 mm, narrowly lanceolate to linear, distantly serrate, the uppermost entire and sessile. Calyx-teeth *c*.5 mm, linear to narrowly triangular. Corolla (10–)12–20 mm, blue, the lobes broadly ovate. Capsule (2–)3–5(–12) mm, conical, pendent. Protandrous. *Flowers February to March*.

Disturbed grassland and open coastal pebbles.

Local on S. coast of Isla Grande.
Introduced from Europe with grass seed, apparently diminishing.

LXIX. LOBELIACEAE

Perennial herbs. Leaves alternate, simple; stipules absent. Flowers hermaphrodite, zygomorphic. Calyx 5-lobed. Corolla 2-lipped; upper lip 2-lobed, the lower 3-lobed. Stamens 5. Ovary ± inferior, 2-celled; style 1; stigma 2-lobed. Fruit a berry; seeds numerous, minute, with endosperm.

1. *Pratia* Gaudich.

Flowers solitary in leaf-axils. Calyx-tube adnate to ovary, persistent. Stamens alternating with corolla-lobes; anthers and upper part of filaments united to form a tube around style. Style surrounded by a ring of hairs.

a. Corolla pinkish outside, at least 8mm, the tube at least 4 mm **2. *longiflora***
a. Corolla bluish violet outside, not more than 8 mm, the tube not more than 3.5 mm **1. *repens***

1. *P. repens* Gaudich. 1825, *Ann. Sci. Nat.* **5**: 103.

Stems up to 20 cm, prostrate, branched, rooting at nodes, leafy towards apex. Leaves with lamina 4–10 × 2.5–11 mm, suborbicular- to oblong – ovate, obtuse to rounded, truncate to shallowly cordate or, rarely, cuneate at base, entire or shallowly crenate; petiole 5–25 mm, narrowly winged. Peduncles 5–25 mm, up to 30 mm in fruit, erect. Calyx-lobes 1.4–2.3 × 0.7–1.3 mm, oblong-ovate, obtuse, usually mucronulate, entire, green, often purple-tinged. Corolla 2.5–8 mm, bluish violet outside, white inside, divided to base or lips united in tube up to 3.5 mm; upper lip 3.5–7 mm, the lobes *c*.2.5 × 1–1.6 mm, ovate-triangular, acute; lower lip 4.5–8 mm, the lobes 2.5–3 mm, ovate-triangular, acute. Anthers dark purple. Stigma yellow or white. Berry 5–6 mm, subglobose to cylindrical, green or purplish. $2n = 84$. Self-compatible. *Flowers December to February*. (Fig. 181).

Moist places in grassland, open forest and scrub, coastal rocks and sand; 0–500 m.

Throughout Fuegia except the extreme west. (Map 22c).
W. Argentina north to *c*.46°20′S.lat.
Chile north to *c*.51°S.lat.
Falkland Islands (Islas Malvinas).

2. *P. longiflora* Hooker f. 1847, *Fl. Antarct.* **1(2)**: 325.

Similar to *P. repens* but: stems up to 10 cm; calyx-lobes triangular-ovate, acute; corolla 8–13 mm or more, pinkish with purple veins outside, the lips united in a tube 4–7 mm or more; stigma purple and white. *Flowers December to February*. (Fig. 182).

Moist places in grassland and coastal sands; 0–30 m.

N. & E. Isla Grande. (Map 22d).
W. Argentina north to *c*.49°S.lat.
E. Chile north to *c*.51°S.lat.

Wimmer (1955) included this species in *Hypsela reniformis* (Humb, Bonpl. & Kunth) Presl, which is widespread along the Andes north to *c*.0°S.lat. The separation of *Hypsela*, with an entire corolla-tube, from *Pratia*, with the tube divided to the base, cannot be maintained on the basis of the above two Fuegian species, which are obviously very closely related. Consequently I have preferred to recognise *P. longiflora*, in the present restricted sense, until its relationships with the northern Andean plants are further clarified.

LXX. STYLIDIACEAE

Perennial herbs. Leaves alternate, simple, densely imbricate; stipules absent. Flowers hermaphrodite or unisexual, zygomorphic, solitary, terminal. Calyx 2– to 7– lobed. Petals 5–10, free or united at base. Stamens 2–3, free or united with style. Ovary inferior, 2– to 3– celled; styles 1–3. Fruit a capsule, indehiscent; seeds numerous, with endosperm.

Literature: J. Milbraed, *Pflanzenreich* **35(IV.27–8)**: 19–22(1908).

a. Leaves not more than 5 mm; petals 2.5–3 mm; stamens 2, united with the 1 style **2. Phyllachne**
a. Leaves at least 5 mm; petals 5–6 mm; stamens 3, free; styles 3 **1. *Donatia***

1. *Donatia* Forster & Forster f.

Cushion-forming plants. Flowers hermaphrodite. Calyx-tube united with ovary, irregularly 5– to 7– lobed. Petals 5–10, free. Stamens 3, free. Styles 3; stigmas capitate.

1. *D. fascicularis* Forster & Forster f. 1775, *Char. Gen. Pl.* 5
Polycarpon magellanicum L. f. 1781, *Suppl.* 115–Forster, 1789.

Cushion dense; stems 4–8 cm, ascending to erect, branched, densely covered with leaf-remains below and green leaves distally. Leaves 5–10 × 0.8–1.3 mm, linear to subulate, obtuse, entire, sessile, coriaceous. Calyx-lobes 2–3 mm, linear, unequal. Petals 5–6 mm, linear to lanceolate, obtuse to subacute, white. Styles *c*.5 mm, recurved at apex. Capsule turbinate; seeds obliquely ovoid. $2n = 48$. *Flowers December to February*. (Fig. 185).

Bogs, with *Astelia, Caltha* and *Sphagnum*, often dominant or co-dominant; 0–40 m.

S. & W. Fuegia. (Map 22e).
Chile north to *c*.40°S.lat.

2. *Phyllachne* Forster & Forster f.

Cushion-forming plants. Flowers hermaphrodite or unisexual, the sexes on the same or different plants, with 2–3 bracteoles at base. Calyx-tube united with ovary, minutely lobed. Petals united at base into (4)5(–7)-lobed corolla. Stamens 2, united to apex with style to form a column. Style 1; stigma 2-lobed.

1. *P. uliginosa* Forster & Forster f. 1775, *Char. Gen. Pl.* 115.
Forstera muscoides Willd. 1805, *Sp. Pl.* ed. 4, 148– Hooker, 1847.
F. uliginosa (Forster & Forster f.) Hombron & Jacquinot 1853, *Voy. Pôle Sud*, pl. 16D.

Cushions ± dense; stems up to 5 cm, ascending to erect, dichotomously branched, densely covered with brownish yellow leaves below and green leaves only at apex. Leaves 3–4 × 1–1.3 mm, lanceolate-subulate, acute, mucronate, somewhat sheathing at base, indistinctly serrulate, coriaceous, yellowish green. Bracteoles *c*.2 mm, linear-lanceolate, acute. Calyx-tube 1–1.3 mm, turbinate; lobes indistinct or

183

184

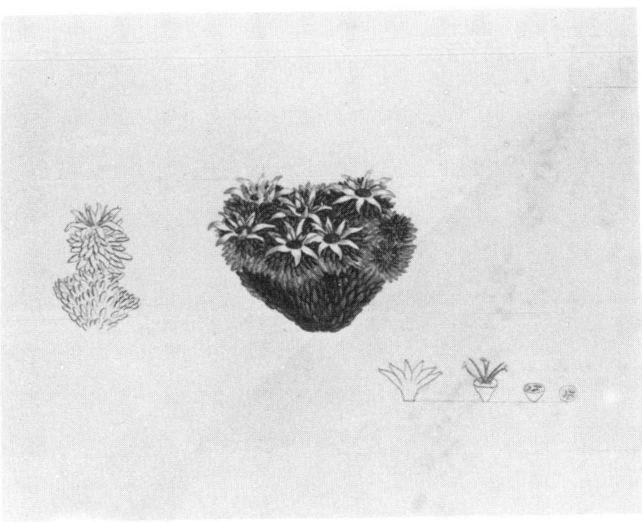

185

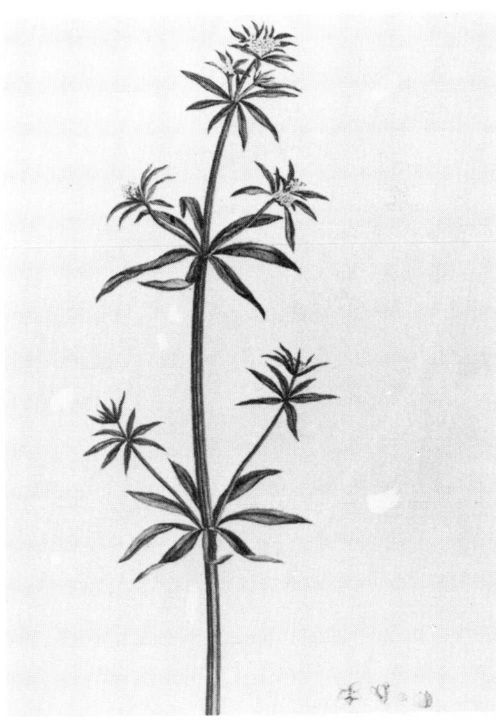

186

Fig. 183. *Plantago barbata*. Fig. 184. *Pinguicula magellanica*. Fig. 185. *Donatia fascicularis*. Fig. 186. *Galium aparine*.
(Sydney Parkinson)

1–2 linear-lanceolate, *c.*1 mm. Corolla 2.5–3 mm, white; lobes *c.*0.6–0.8 mm, ovate-lanceolate. Column 4.5 mm, exserted. Capsule turbinate; seeds obovoid. *Flowers November to April* (Fig. 187).

Bogs and wet rocks; 0–50 m.

S.W. & W. Fuegia. (Map 22f).
W. Chile north to *c.*51°29′S.lat.

LXXI. RUBIACEAE

Perennial herbs, sometimes cushion-forming. Leaves opposite or whorled, simple, entire; stipules present, often leaf-like. Flowers actinomorphic, hermaphrodite, solitary or borne in cymes. Sepals 4–5, free or fused. Petals (3)4–5, fused. Stamens 4–5, inserted on corolla-tube. Ovary inferior, 2-celled; style 1, 2-fid. Fruit a capsule, drupe or schizocarp.

a. Plant forming dense cushions **1. *Oreopolus***
a. Plant with lax creeping or scrambling stems:
 b. Leaves opposite; fruit a drupe **2. *Nertera***
 b. Leaves in whorls; fruit a schizocarp **3. *Galium***

1. *Oreopolus* Schlecht.

Cushion-forming perennial herbs. Leaves opposite, decussate or ternate, fleshy; stipules connate, membranous. Flowers solitary in upper leaf-axils, subsessile. Calyx with short tube, 5-lobed. Corolla hypocrateriform, with long slender tube widening to 5-lobed apex. Stamens inserted at apex of corolla-tube, included or exserted, alternating with corolla-lobes. Stigmas filiform, short. Fruit a septicidal capsule; seeds (1)2–4, with endosperm.

Literature: M. Ricardi, *Gayana, Bot.* **6**: 3–16 (1963).

1. *O. glacialis* (Poeppig & Endl.) Ricardi 1963, *Gayana, Bot.* **6**: 7.
Cruckshanksia glacialis Poepig & Endl. 1847, in Gay, *Hist. Chile Bot.* **3**: 193. – Macloskie 1905; Skottsberg, 1915, 1916.

Plant glabrous except for flowers. Stems arising from thick taproot, 4–12 cm, branched, rather succulent, covered with dead sheathing stipules in lower part. Leaves 7.5–20 × 1.8–5 mm, elliptic – ovate to subspathulate, obtuse, entire, truncate or cuneate at base, with petiole 3–10 mm, the lower ternate, the upper decussate; stipules entire, the free teeth 1–2.5 mm, triangular. Flowers 2–6. Calyx-tube *c.*2 mm, obconical, hairy; lobes 3–3.5 × 2–3 mm, ovate, acute. Corolla yellow, pubescent outside; tube 10–16 mm, filiform, but wider distally; lobes 5–5.5 × *c.*1.5 mm, linear-lanceolate. Stamens included, with filaments almost absent, or exserted, with filaments 1.5–2 mm; anthers *c.*1.5 mm. Style exserted or included. Capsule 5–8 × 3–6 mm, obovoid, hairy, entire or with 1 tooth on each side, bearing persistent calyx at apex; seeds 4–6 × 2–3 mm, oblong-ovoid, somewhat rugose, concave-convex, reddish brown. Heterostylous. *Flowers November to December.* (Fig. 189).

Open rocky and pebbly or sandy areas; 50–250 m.

C. & S.E. parts of Isla Grande. (Map 22g).
W. Argentina north to 38°30′S.lat.
E. Chile north to *c.*33°S.lat.

2. *Nertera* Banks & Sol. ex Gaertner

Glabrous, creeping perennial herbs, slightly foetid when bruised. Leaves opposite. Flowers solitary, axillary or terminal, sessile. Calyx truncate. Corolla subcampanulate, with short tube and 4 lobes. Stamens 4, inserted at base of corolla-tube and alternating with lobes. Stigmas filiform. Fruit a fleshy drupe with 2 stones.

1. *N. depressa* Banks & Sol. ex Gaertner 1788, *Fruct. Sem. Pl.* **1**: 124.

Stems up to 50 cm, *c.*0.8 mm in diameter, prostrate, much branched, rooting at nodes, forming low mats. Leaves with lamina 2.5–6 × 2–5 mm, ovate to orbicular– or triangular-ovate, obtuse, rarely apiculate, truncate or cuneate at base, with entire somewhat thickened margin; petioles 1–3.5 mm, winged above; stipules *c.*0.5 mm, triangular, thick. Corolla pale green, with tube 0.5–0.8 mm; lobes *c.*1.5 × 0.8 mm, ovate-oblong, acute, minutely glandular on margin. Drupe 4–5 mm in diameter, globose, bright red. Strongly protogynous. *Flowers September to November.* (Fig. 188).

Wet rocks and soil beside streams and coast, seepage areas; 0–400 m.

C. S. & S. E. Fuegia. (Map 22h).
W. Argentina north to *c.*41°10′S.lat.
Chile north to *c.*40°S.lat.
Falkland Islands (Islas Malvinas), New Zealand.

3. *Galium* L.

Perennial herbs with tetragonous stems, often scrambling. Leaves and leaf-like stipules in whorls of

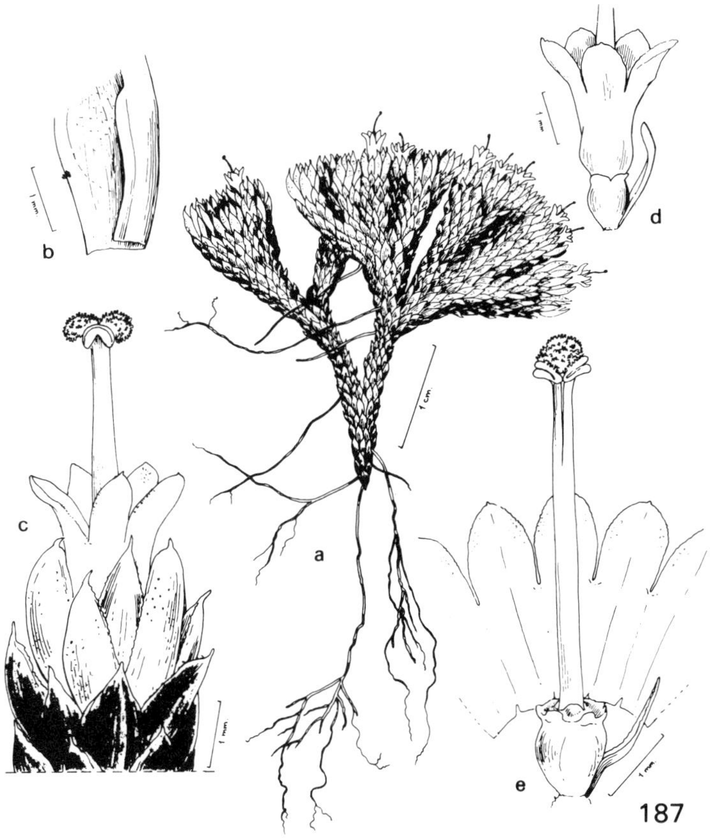

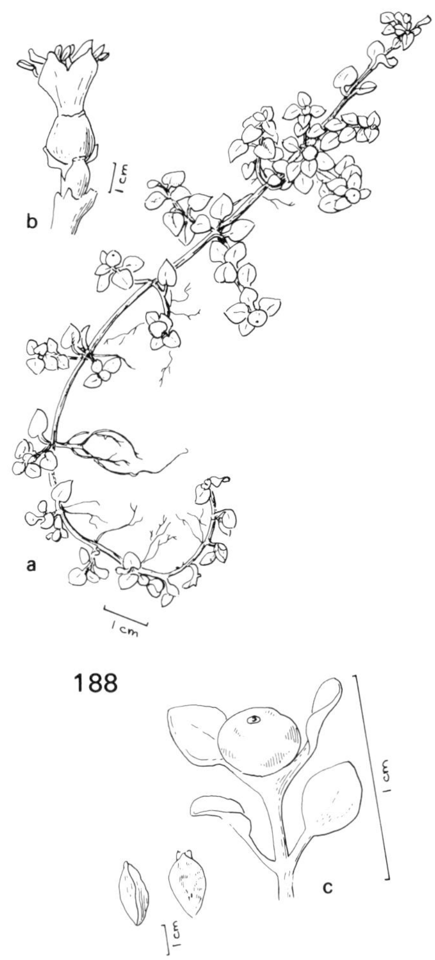

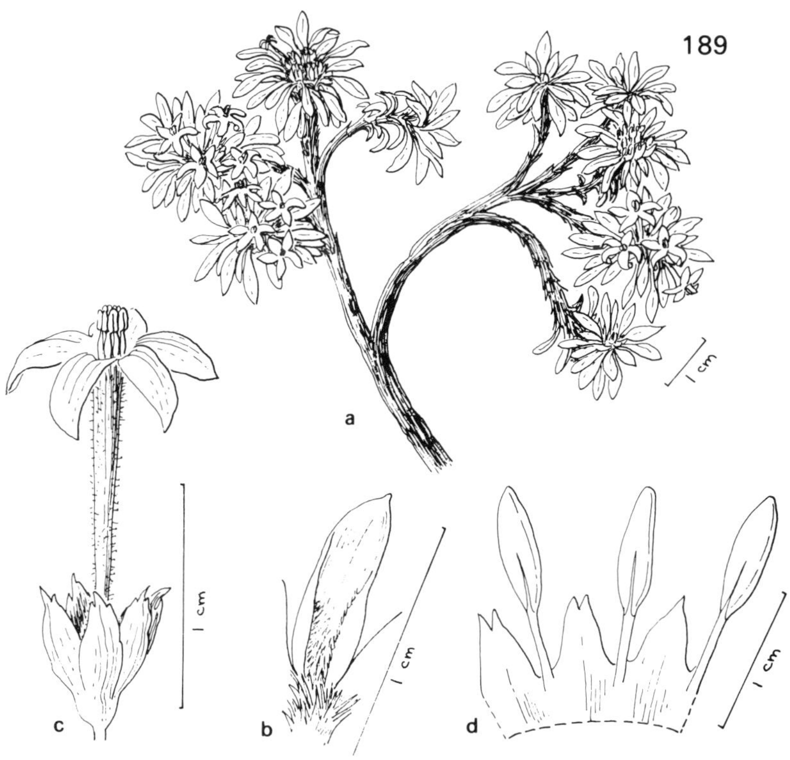

Fig. 187. *Phyllachne uliginosa*. a, Plant. b, Leaf-base. c, Flowering branch apex. d, Corolla, calyx and ovary. e, Opened female flower. Scale = 1 cm (a) or 1 mm (b–e). Fig. 188. *Nertera depressa*. a, Plant. b, Flower. c, Fruit. Scale = 1 cm. Fig. 189. *Oreopolus glacialis*. a, Plant. b, Leaf. c, Flower. d, Part of opened corolla with stamens. Scale = 1 cm.

(R. N. P. Goodall)

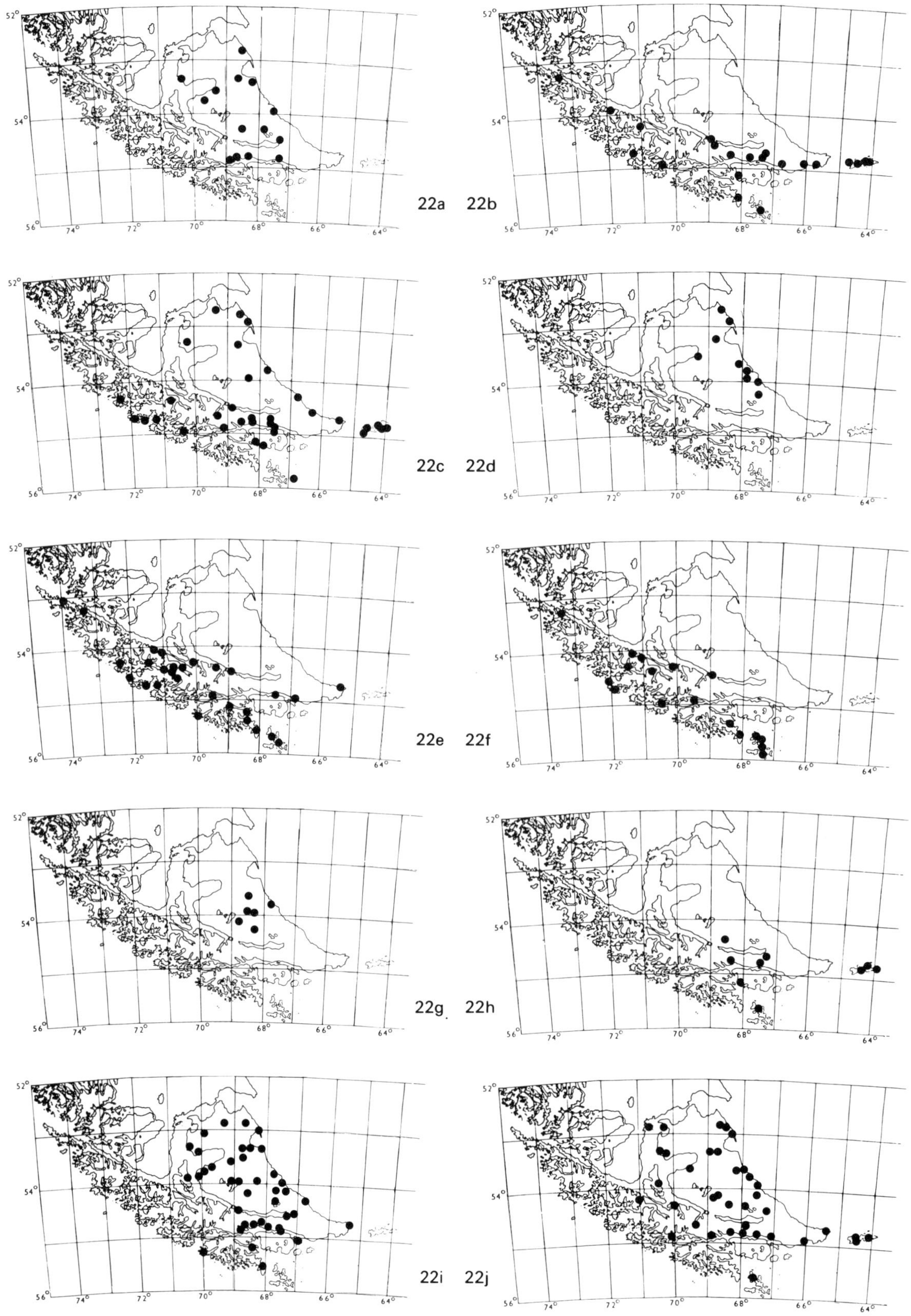

Map 22. a, *Euphrasia antarctica*. b, *Pinguicula magellanica*. c, *Pratia repens*. d, *Pratia longiflora*. e, *Donatia fascicularis*. f, *Phyllachne uliginosa*. g, *Oreopolus glacialis*. h, *Nertera depressa*. i, *Galium aparine*. j, *Galium antarcticum*.

(3)4–9, sessile. Flowers in small axillary cymes or solitary in leaf-axils. Calyx a minute, annular ring. Corolla rotate, with short tube and 3–4 lobes. Stamens 3–4, inserted in corolla-tube and alternating with lobes. Stigmas capitate. Fruit dry, didymous, of two 1-seeded mericarps, with seed-testa adhering to the pericarp.

a. Leaves and stipules in whorls of 5 or more **1. *aparine***
a. Leaves and stipules in whorls of 4 or less:
b. Fruit glabrous; peduncles solitary **2. *antarcticum***
b. Fruit hispid; peduncles 2 or more together **3. *fuegianum***

1. *G. aparine* L. 1753, *Sp. Pl.* 108.
G. chonoense auct., non Hooker f. – Spegazzini, 1896.
G. pseudoaparine Griseb. 1854, *Syst. Bemerk.* 125 – Hariot, 1900.

Stems up to 60 cm, much branched, prostrate to ascending, scrambling, with retrorse prickles on angles. Leaves and stipules in whorls of 5–9, 4–34(–43) × 1–6(–10) mm, linear-oblanceolate to obovate, obtuse to subacute, mucronate, attenuate at base, usually with retrorse marginal prickles, glabrous or upper surface scabrid. Flowers in 2– to 5– flowered axillary cymes; peduncles 1.5–7 cm, tetragonous, prickly, bearing whorls or bracts at base; pedicels up to 1 cm. Corolla-tube *c.*0.5 mm; lobes 4, 1–1.3 × 0.7–0.9 mm, ovate to triangular-ovate, subacute, white. Stamens 4; anthers 0.2 mm. Fruit 2–2.6 × 3.4–3.7 mm, dark green to blackish, covered with white, tuberculate, hooked bristles. $2n = 66$. *Flowers November to February*. (Fig. 186).

Usually in shady places among rocks, shrubs, forest-margins and clearings, but also in more exposed grassland; 0–250 m.

Throughout Fuegia except the extreme west. (Map 22i).
W. Argentina north to *c.*41°S.lat.
Chile north to c.42°S.lat.

2. *G. antarcticum* Hooker f.1847, *Fl. Antarct.* **1(2)**: 303.
G. magellanicum auct., non Hooker f. – Spegazzini, 1896.

Stems up to 30 cm, sparingly branched, decumbent, glabrous or with few small, retrorse prickles on angles. Leaves and stipules in whorls of (3)4, 2–14 × 0.5–2.5(–3.5) mm, elliptic-oblong to obovate or linear-lanceolate, obtuse to acute, attenuate at base, glabrous or scabridulous on margin. Flowers solitary in leaf-axils, subsessile or on arcuate peduncles elongating to 7 mm in fruit. Corolla-tube *c.*1 mm; lobes 3–4, (0.8–)1.5–2 × 0.9–1.5 mm, ovate to triangular-ovate, obtuse, white. Stamens 3–4; anthers *c.*0.4 mm. Fruit 1.7–2.3 × 2.3–3.8 mm, grey-black, glabrous, smooth, $2n = 22$. *Flowers December to February*.

Moist sand, pebbles, rock-crevices, grassland and forest-margins, by coast and inland freshwater; 0–500 m.

Almost throughout Fuegia. (Map 22j).
Chile and W. Argentina north to 48°S.lat.
Falkland Islands (Islas Malvinas), South Georgia, Iles Crozet, Iles de Kerguelen.

3. *G. fuegianum* Hooker f. 1847, *Fl. Antarct.* **1(2)**: 302.

Stems up to 35 cm, branched, ascending, glabrous or with scattered hairs. Leaves and stipules in whorls of (3)4, 3.5–14 × 2–5 mm, elliptic – oblong to broadly elliptical, acute to apiculate, shortly attenuate at base, ciliate, sparsely hairy above and beneath. Flowers in 4– to 6– flowered axillary cymes; peduncles (8–)10–25 mm, bearing whorl of bracts at base; pedicels elongating to 17.5 mm in fruit. Corolla-tube *c.*0.5 mm; lobes 4, 1–1.4 × 0.7–0.9 mm, triangular-ovate, subacute, white to creamy. Stamens 4; anthers *c.*0.2 mm. Fruit 1–1.5 × 1.5–2 mm, blackish, densely covered with hispid, hooked, white– or purple – tipped hairs. $2n = 22$. *Flowers November to January*.

Shrub communities, forest margins and clearings, shady cliffs and slopes; 0–70 m.

Isla Grande and islands immediately south of Canal Beagle. (Map 23a).
Argentina north to *c.*49°S.lat.
Chile north to *c.*37°S.lat.

LXXII. VALERIANACEAE

Annual or perennial herbs, sometimes cushion-forming. Leaves opposite, simple; stipules absent. Flowers actinomorphic, hermaphrodite or uni-sexual. Calyx an annular ring, often with 1–4 minute lobes or teeth. Corolla infundibuliform, 5-lobed. Stamens 3, inserted towards base of corolla-tube, alternating with lobes. Ovary inferior, 3-celled, 1 cell fertile, 2 cells sterile and usually small; style 1; stigma 3-lobed. Fruit dry and indehiscent, 1-seeded with calyx sometimes developing plumed segments; endosperm present.

a. Annual, dichotomously branched; fruiting calyx without plumose setae **1. *Valerianella***
a. Perennial, often cushion-forming, not dichotomously branched; fruiting calyx with plumose setae **2. *Valeriana***

1. *Valerianella* Miller

Erect, dichotomously branched annuals. Flowers hermaphrodite, solitary in the forks of branches and in terminal cymes. Calyx 1-toothed. Style slightly clavate; stigma 3-fid. Fruit without plumose setae.

1. *V. locusta (L.) Laterrade 1821, *Fl. Bordel.* Ed. 2, 93.

Stems 6–30 cm, rather brittle, much branched, sparsely pubescent towards base. Leaves 20–70 × 2–9 mm, entire to denticulate, shortly petiolate or sessile; lower spathulate, obtuse; upper oblong, obtuse to subacute. Cymes capitate; bracts *c.*4 × 1 mm, oblong-spathulate, obtuse, ciliolate. Corolla 1–3 mm, 1–2 mm in diameter, pale lilac; lobes *c.*0.5 mm, patent. Fruit c. 2.5 × 2 mm, subglobose, compressed, glabrous to puberulent; fertile cell with a thickened, spongy outer wall; sterile cells as large as the fertile. *Flowers December to March.*

Disturbed and cultivated ground.

S. coast of Isla Grande (Ushuaia).
Introduced from Europe and locally naturalized.

2. *Valeriana* L.

Erect or cushion-forming perennials. Flowers hermaphrodite or unisexual, in axillary or terminal, usually compound, cymes. Calyx-lobes 2–20, inrolled in flower, accrescent and plumose in fruit. Style clavate; stigma irregularly, shortly 3-lobed. Fruit with plumose setae.

Literature: O. E. Borsini, *Lilloa* **32**: 375–476(1966).

a. Plant erect; lower leaves at least 5 cm **2. *carnosa***
a. Cushion plant; leaves not more than 1 cm **1. *sedifolia***

1. *V. sedifolia* D'Urv. 1825. *Fl. Is. Mal.* 45.
Aretiastrum sedifolium (D'Urv.) Graebner 1906, *Bot. Jahrb.* **37**: 478.

Plant dioecious; stems much branched and closely appressed to form dense cushions 15–60 cm in diameter. Leaves 3–7 × 1–2 mm, oblong- to elliptic-oblanceolate, obtuse or retuse, narrowed slightly to base, entire, thick and fleshy, ciliolate towards base, bright pale green, purplish near base. Inflorescence 3- to 7- flowered; pedicels 0.5–1 mm; bracteoles *c.*2.5 × 1 mm, oblong-lanceolate, obtuse, green, with scarious margins, inserted about middle of pedicel. Corolla yellow; tube 1.5–2.5 mm; lobes 0.5–1 × *c.*0.7 mm, ovate to oblong, obtuse or truncate. Style and stamens slightly exceeding corolla-tube; female flowers with staminodes; male flowers with rudimentary style. Fruit 3–4 × 0.5–1 mm, ellipsoid-pyriform, *c.* 5-ribbed, glabrous, with 2–20 setae united at base into hyaline membrane. $2n = 32$*. *Flowers October to December.*

Rocks and sands, open grassland; 150–200 m.

C. & S. Fuegia. (Map 23b).
Falkland Islands (Islas Malvinas).

V. magellanica Hombron & Jacquinot ex Dcne 1853, in D'Urv., *Voy. Pôle Sud* 54, has been reported once (Borsini, 1966) from N. Isla Grande (between Caleta Clarencia and Cerro Sombrero). It is said to be distinguished from *V. sedifolia* by being gynodioecious and having leaves with an acute, rhombic apex and a widened base. In view of the small amount of available material of both species it is difficult to assess the validity of these characters and the taxa may well be conspecific,(but see Weberling & Uhlarz, 1977).

2. *V. carnosa* Sm. 1791, *Pl. Icon. Ined.* **3**: t.52.

Plant gynodioecious; stems up to 60 cm, erect, simple or branched. Leaves puberulent, especially above; lamina 2–8 × 1–6 cm, obovate, obtuse, cuneate at the base, broadly dentate or with small laciniae, fleshy, glaucous; petiole 1.5–13 cm; upper leaves smaller, often narrower, sessile or shortly petiolate. Inflorescence subglobose to ovoid, the partial cymes 6- to 15- flowered; peduncles 4–15 mm; bracts 3–7 mm, oblong-lanceolate to ovate, acute, entire; pedicels 0.5–1 mm; bracteoles 2.5–3 × *c.*0.5 mm, oblong-lanceolate, acute, entire, ciliate towards base. Corolla pinkish purple; tube 1.5–2.5 mm; lobes 1.5–2.5 × 1.6–2 mm and oblong in hermaphrodite flowers, 0.8–1 × 0.5–1 mm and ovate in female flowers. Style and stamens somewhat exceeding corolla-tube; female flowers with staminodes. Fruit 4–7 × 2–3 mm, pyriform, *c.*6 – ribbed, glabrous, with 12–15 setae 5–8 mm. $2n = 16$. *Flowers November to January.* (Fig. 170).

Coastal sands and shingle, *Lepidophyllum* scrub, open *Empetrum* heath, open grassland; 0–50 m.

C. & N. Isla Grande. (Map 23c).
W. Argentina north to c.41°S.lat.
Chile north to c.36°30′S.lat.

V. lapathifolia Vahl 1805, *Enum. Pl.* **2**: 11, was collected once by Gusinde in "humid woods in the middle of Isla Grande" (Skottsberg, 1926). This species, which occurs in Chile and W. Argentina south from c.40°S.lat., has not subsequently been reported from Fuegia. It may be distinguished from *V. carnosa* by its entire or sinuate, herbaceous leaves and ellipsoid, 3-ribbed fruit, as well as its generally forest habitats.

LXXIII. CALYCERACEAE

Perennial herbs. Leaves alternate, often in rosettes, simple; stipules absent. Flowers hermaphrodite, actinomorphic, grouped into capitula usually surrounded by 1–2 series of involucral bracts, the capitula pedunculate or sessile, solitary or clustered into dense cyme. Calyx of 5 teeth or lobes persisting in fruit. Corolla cylindrical, with 5-lobed limb. Stamens 5, alternating with corolla-lobes; filaments partly united in tube around style, inserted in corolla-tube; anthers free. Ovary inferior, 1-celled; style 1, exserted; stigma capitate. Fruit an achene; seed 1, with endosperm.

Literature: A. Pontiroli, *Rev. Mus. La Plata*, N. Ser. **9**: 175–241(1963)

a. Capitula sessile; involucre absent **1. *Moschopsis***
a. Capitula pedunculate; involucre present **2. *Boopis***

1. *Moschopsis* Phil.

Leaves densely imbricate. Inflorescence sessile, apical, surrounded by upper leaves, without a clearly defined involucre, with clusters of flowers often subtended by 3(–5)-lobed receptacular bracts. Filaments inserted towards base of corolla-tube, united almost throughout, with basal glands.

1. *M. rosulata* (N. E. Br.) Dusén 1907, *Ark. Bot.* **7(2)**: 42.
Acicarpha rosulata N. E. Br. 1900, *Icon. Pl.* 27.
Nastanthus spathulatus auct., non (Phil.) Miers – Moore 1970, 1974.

Plant up to 12 cm in diameter; stem up to 15 cm × 9 mm, usually simple, with scale-leaves towards base. Leaves glabrous, fleshy, with inconspicuous veins; lower distant, subsessile, gradually becoming larger upwards; upper densely imbricate, the lamina 10–30 × 5–10 mm, spathulate, entire or distally 3- to 7-lobed, narrowed to petiole-like base 5–20 mm, often glaucous. Inflorescence 18–50 mm in diameter; receptacular bracts c.5 mm, linear, acute, up to 8 fused in pseudo-involucre, scarious. Calyx-lobes 0.8–2 mm. Corolla 5–6 × 1.1–1.8 mm, yellowish to white; lobes 0.5–1.5 mm, linear. Staminal tube c.4 mm; basal glands green; anthers 1–1.5 mm, brown. Style exserted for 1.5–2 mm. Achene c.2.5 mm, cylindrical-prismatic, 5-ribbed, with lanceolate, apical lobes. *Flowers January to March.* (Fig. 190).

Feldmark and rock/soil screes; 700–1100 m.

C. S. Isla Grande. (Map 23d).
W. Argentina north to c.50°S.lat.
E. Chile north to 51°S.lat.

2. *Boopis* Juss.

Leaves not densely imbricate. Inflorescence pedunculate, lateral or apical, of one or more capitula each surrounded by conspicuous involucre of 5–8 bracts united at base. Filaments inserted at base or towards mouth of corolla-tube, united for at least basal ½ with basal glands; anthers united at base.

a. Stamens inserted in base of corolla-tube; involucral bracts shorter than flowers **1. *patagonica***
a. Stamens inserted towards mouth of corolla-tube; at least some involucral bracts exceeding flowers **2. *australis***

1. *B. patagonica* Speg. 1902, *Anal. Mus. Nac. Bs Aires* **7**: 301.

Plant glabrous; stem 2–6 cm × 2–3 mm, ascending to erect, with short branches up to 2 cm having leaves clustered distally. Leaves with lamina 7–18 × 3–6 mm, oblanceolate-obovate to ovate, obtuse, entire; petiole 5–16 mm. Inflorescence apical, usually solitary, of 3–6 subsessile capitula; capitula 4–7 mm in diameter; involucral bracts 5–8, shorter than flowers, triangular-ovate, obtuse to subacute, united up to ½ way. Calyx-lobes 2–3 mm, obtuse. Corolla 3.5–4 mm, whitish; lobes c.1 mm, oblong-ovate, obtuse. Stamens inserted in base of corolla-tube; filaments united almost to anthers, the tube c.2 mm. Style exserted for up to 2 mm. Achene c.2 mm, cylindrical-prismatic, 5-ribbed, with 5 obtuse apical lobes. *Flowers January to February.* (Fig. 191).

Open moist, probably saline, grassland; 0–50 m.

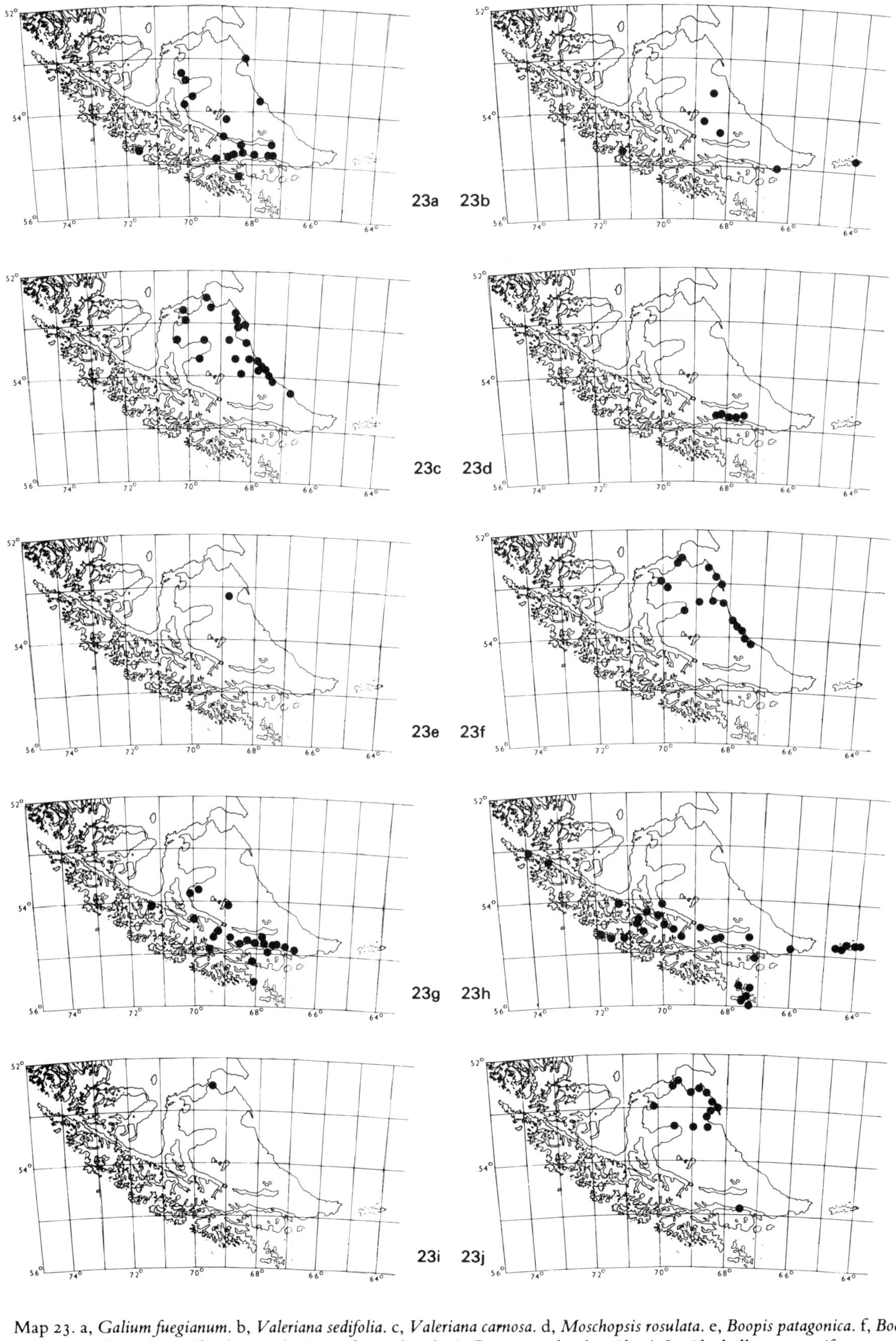

Map 23. a, *Galium fuegianum*. b, *Valeriana sedifolia*. c, *Valeriana carnosa*. d, *Moschopsis rosulata*. e, *Boopis patagonica*. f, *Boopis australis*. g, *Lagenifera hariotii*. h, *Lagenifera nudicaulis*. i, *Gutierrezia baccharoides*. j, *Lepidophyllum cupressiforme*.

N. E. Isla Grande (Ea Lago Gaviota). (Map 23e).
E. Argentina north to *c.*48°S.lat.

2. ***B. australis*** Dcne 1853, in D'Urv., *Voy. Pôle Sud* 87.

Plant glabrous, with stout rootstock up to 2 cm in diameter; stems 5–40 cm × 2–4 mm, procumbent to erect, simple, the lateral free, the central united but free distally, forming rosette up to 18 cm in diameter. Basal leaves in a rosette, 3–14 × 0.2–1 cm, linear, obtuse, without distinct petiole, entire or slightly lobed to distantly pinnatisect in distal half; cauline few, 1–3 cm, linear, entire or somewhat lobed. Capitula apical, 9–15 mm in diameter, with peduncles up to 10 mm; involucral bracts 6–8, 8–15 mm or longer, some exceeding flowers, ovate-lanceolate, obtuse to acuminate. Calyx-lobes *c.*0.5 mm, obtuse. Corolla 3.5–6 mm, whitish or cream; lobes *c.*$\frac{1}{3}$ as long as tube, linear, obtuse. Stamens inserted towards mouth of corolla-tube; filaments united for at least basal $\frac{1}{2}$, the tube *c.*2 mm. Style exserted for 1.5–2.5 mm. Achene 2.5–4 mm, obovoid-prismatic, 5-ribbed, with 5, obovate, apical lobes. $2n = 36$. *Flowers November to April.* (Fig. 192).

Open sands and gravels by coast and estuaries, *Lepidophyllum* scrub; 0–60 m.

Coasts of N. & C. Isla Grande. (Map 23f).
C. & W. Argentina north to *c.*39°S.lat.
Chile along N. shore of Estrecho de Magallanes.

LXXIV. COMPOSITAE

Herbs or shrubs. Leaves alternate, opposite or in basal rosettes; stipules absent. Flowers small (florets), grouped into heads (capitula) simulating a single larger flower which is surrounded by a calyx-like involucre of 1 or more series of involucral bracts. The receptacle of the capitulum is of varied shape, pitted or smooth, with or without receptacular bracts each subtending a floret, sometimes the bracts bristle-like. Florets all similar (monomorphic) or the inner and outer different (dimorphic). Calyx represented by a pappus of numerous simple or plumose hairs in 1 or more series, or of membranous scales, or of teeth or bristles, or of a continuous membranous ring, or absent. Corolla may be (i) filiform or tubular, either actinomorphic, with the corolla-tube having 4–5, ± short, equal lobes, or 2-lipped; (ii) ligulate, with the corolla-tube prolonged only along 1 side as a strap-shaped "ligule" which usually has 3–5 apical teeth. Stamens 5; anthers often tailed at base, united to form closed cylinder around style. Ovary inferior, 1-celled, with 1 ovule; style simple at base, branched above into 2 stigmatic arms. Fruit an achene (cypsela), sometimes with apical beak, surmounted by pappus.

Literature: A. L. Cabrera. Compositae. *In*: M. N. Correa (Ed.) *Fl. Patag.* **7**: 1–451(1971).

a. Plant with milky latex; all florets ligulate:
 b. Leaves all basal:
  c. Pappus of 1 series of plumose hairs, sometimes also with outer series of simple hairs — **35. *Hypochoeris***
  c. Pappus of 1 or more series of simple hairs:
   d. Achene without beak — **41. *Hieracium***
   d. Achene with ± prominent beak:
    e. Achene with prominent ribs without tubercles — **39. *Agoseris***
    e. Achene faintly ribbed; ribs with spiny tubercles — **37. *Taraxacum***

 b. Leaves alternate:
  f. Pappus of 1 series of plumose hairs — **36. *Tragopogon***
  f. Pappus of 1 or more series of simple hairs:
   g. Achenes strongly compressed — **38. *Sonchus***
   g. Achenes not compressed:
    h. Pappus-hairs in 1–2 series; usually at least some part of plant with stellate hairs — **41. *Hieracium***
    h. Pappus-hairs in 3 or more series; plant without stellate hairs — **40. *Crepis***

a. Plant without latex; at least central florets tubular or 2-lipped:
 i. Outer florets ligulate:
  j. Ligulate florets yellow:
   k. Leaves pinnatisect or deeply 3-fid:
    l. Aromatic perennial herb; leaves pinnatisect — **22. *Tanacetum***
    l. Non-aromatic dwarf shrub; leaves deeply 3-fid — **27. *Senecio***
   k. Leaves entire to dentate:
    m. Leaves opposite, ± scale-like, densely imbricate — **4. *Lepidophyllum***
    m. At least upper leaves alternate, neither scale-like nor densely imbricate:
     n. Dwarf shrub; pappus of numerous scales — **3. *Gutierrezia***
     n. Herb; pappus absent or of strigulose hairs:
      o. Annual herb; pappus absent — **17. Madia**
      o. Perennial herb; pappus of strigulose hairs — **27. *Senecio***
  j. Ligulate florets white to pink or purplish:

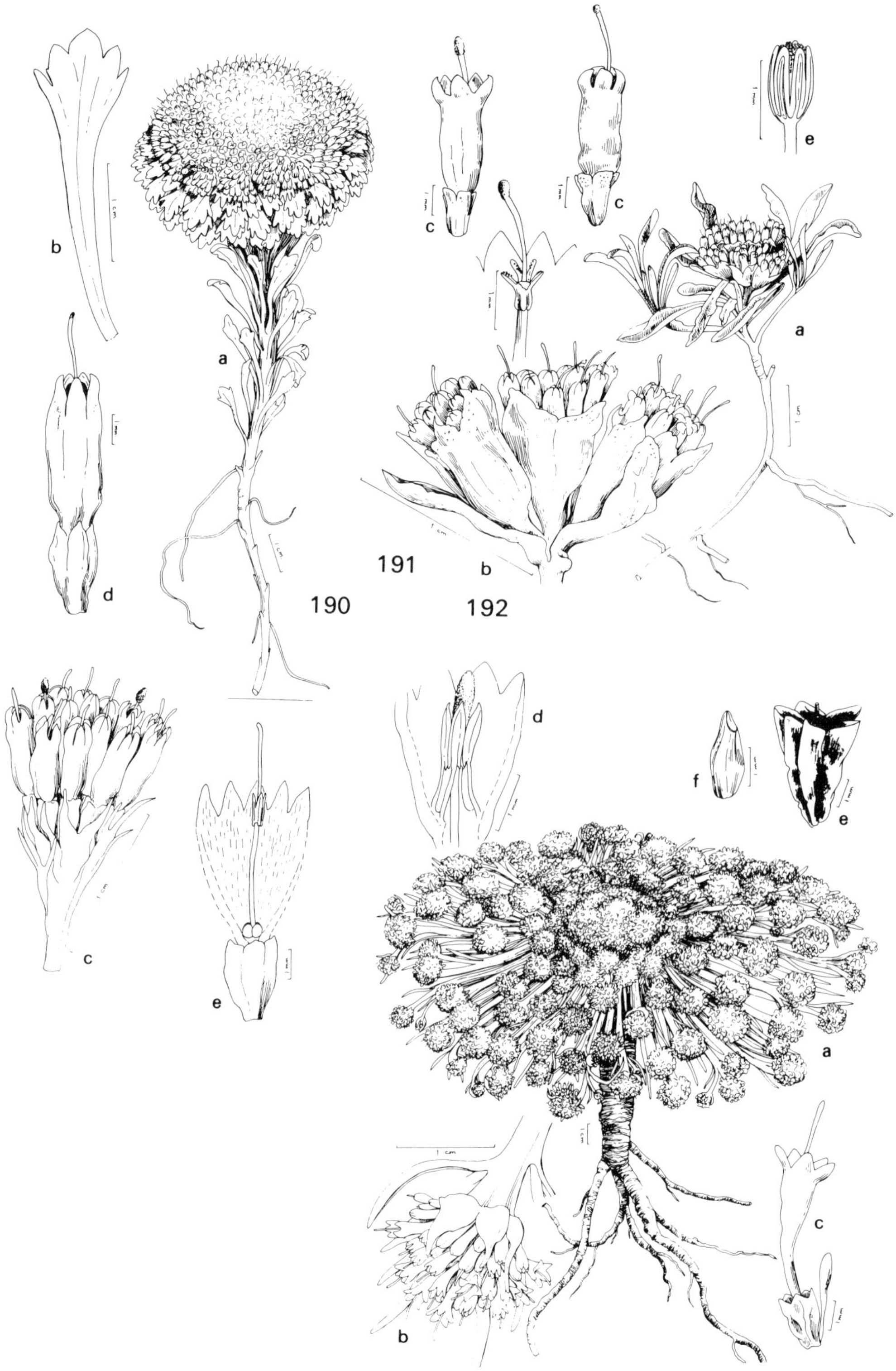

Fig. 190. *Moschopsis rosulata*. a, Plant. b, Leaf. c, Flower. d, Partial inflorescence. e, Half-flower. Scale = 1 cm (a–b, d) or 1 mm (c, e). Fig. 191. *Boopis patagonica*. a, Plant. b, Partial inflorescence. c, Flower. d, Stigma and anthers. e, Anthers. Scale = 1 cm (a–b) or 1 mm (c–e). Fig. 192. *Boopis australis*. a, Plant. b, Partial inflorescence. c, Flower. d, Opened corolla. e, Achene. f, Seed. Scale = 1 cm (a–b) or 1 mm (c–f).

(R. N. P. Goodall)

p. Leaves pinnatisect:
q. Receptacle with bracts; pappus absent:
r. Achene strongly compressed **18. *Achillea***
r. Achene ± terete **19. *Anthemis***
q. Receptacle without bracts; pappus a short, membranous rim:
s. Annual; capitula at least 30 mm in diameter; achene 3-ribbed **20. *Matricaria***
s. Perennial; capitula not more than 20 mm in diameter; achene 5– to 8– ribbed **22. *Tanacetum***
p. Leaves entire to dentate or deeply lobed distally, not pinnatisect:
t. Shrub:
u. Leaves not more than 5 × 2 mm; capitula not more than 13 mm in diameter; achene glandular-pubescent **7. *Chiliophyllum***
u. Leaves at least 8 × 2 mm; capitula at least 22 mm in diameter; achene glandular **6. *Chiliotrichum***
t. Herb:
v. Achene without pappus or pappus a short, membranous rim:
w. Capitulum not more than 9 mm in diameter; achene with apical beak **1. *Lagenifera***
w. Capitulum at least 16 mm in diameter; achene without beak:
y. Leaves in basal rosette; achene compressed, ± appressed– pubescent **2. *Bellis***
y. Leaves alternate; achene ± terete, glabrous **23. *Leucanthemum***
v. Achene with pappus of simple hairs:
z. Ligule short **8. *Conyza***
z. Ligule conspicuous:
a′. Pappus-hairs in 2 or more series; involucral bracts in 1 series, often with shorter supplementary bracts at base of capitulum **27. *Senecio***
a′. Pappus-hairs in 1 series; involucral bracts in 2 or more series:
b′. Involucral bracts in 2–3 series, hairy; achene cylindrical; pappus not more than *c*.4 mm **9. *Erigeron***
b′. Involucral bracts in several series, ciliate; achene 4–angled; pappus at least 7 mm **10. *Aster***
i. Outer florets tubular, filiform or 2-lipped, sometimes with larger outer lip:
c′. Leaves in basal rosette **15. *Adenocaulon***
c′. Leaves alternate:
d′. Leaves deeply lobed to pinnatisect:
e′. Achene without pappus or pappus a short, membranous rim:
f′. Capitula solitary, not grouped into obvious inflorescence:
g′. Dioecious perennial; stems prostrate and rooted at lower nodes; plant not aromatic **24. *Cotula***
g′. Annual, with hermaphrodite florets; stems ascending to erect, not rooted at lower nodes; plant aromatic **21. *Chamomilla***
f′. Capitula in corymb, panicle, or solitary and grouped into spike-like inflorescence:
h′. Capitula in dense, compound corymb; achene with short, membranous pappus **22. *Tanacetum***
h′. Capitula in a panicle or spike-like inflorescence; achene without pappus **26. *Artemisia***
e′. Achene with pappus of hairs or scales:
i′. Leaves with conspicuous white veins above **30. *Silbyum***
i′. Leaves without white veins:
j′. Pappus of 1 series of plumose hairs:
k′. Achene at least 8 mm; leaves at least 40 mm wide **31. *Macrachaenium***
k′. Achene not more than 2.5 mm; leaves not more than 20 mm wide **33. *Leucheria***
j′. Pappus of 2 or more series of plumose or simple hairs or 3–15 caducous scales:
l′. Florets yellow **27. *Senecio***
l′. Florest white, blue or purplish:
m′. Florets 5-lobed, sometimes 1 lobe larger:
n′. Pappus-hairs plumose **29. *Cirsium***
n′. Pappus-hairs simple **28. *Carduus***
m′. Florets 2-lipped, the outer lip often larger:
o′. Pappus of caducous scales or of plumose bristles; involucral bracts in 2 series **32. *Nassauvia***
o′. Pappus of simple, strigose hairs; involucral bracts in 3–9 series **34. *Perezia***
d′. Leaves entire to dentate or serrate:
p′. Leaves not more than 4 × *c*.1.5 mm:
q′. Leaves glabrous; pappus a short, dentate or aristate, membranous rim **25. *Abrotanella***
q′. Leaves white-tomentose beneath; pappus of unequal, scabrid hairs **5. *Nardophyllum***
p′. Leaves more than 5 × *c*.1 mm:
r′. Florets 2-lobed:
s′. Pappus of 1 series of plumose hairs **33. *Leucheria***
s′. Pappus of 2 or more series of plumose or simple hairs or 3–15 caducous scales:

t′. Pappus of caducous scales or of plumose bristles; involucral bracts in 2 series **32. *Nassauvia***

t′. Pappus of simple, strigose hairs; involucral bracts in 3–9 series **34. *Perezia***

r′. Florets tubular or filiform, 4- to 5-lobed:

u′. Capitula with not more than 6 florets; achene without pappus **16. *Eriachaenium***

u′. Capitula with more than 6 florets; achene with pappus of simple hairs:

v′. Involucral bracts in 1 series, often with shorter supplementary bracts at base of capitulum **27. *Senecio***

v′. Involucral bracts in 2 or more series:

w′. Shrub **11. *Baccharis***

w′. Herb:

x′. Pappus of 1 series of hairs:

y′. Biennial or perennial; pappus-hairs united into basal ring and falling as a unit **14. *Gamochaeta***

y′. Annual; pappus-hairs free **13. *Gnaphalium***

x′. Pappus of several series of hairs:

z′. Annual, with hermaphrodite florets; capitula in corymbose cyme **8. *Conyza***

z′. Dioecious, rhizomatous perennial; capitula in terminal glomerule **12. *Antennaria***

Tribe ASTEREAE. Florets tubular or outer with ligules; outer florets female or sterile, the inner hermaphrodite or functionally male. Stigmatic branches flattened, acute to subobtuse.

1. *Lagenifera* Cass.

Perennial herbs with slender rhizomes. Leaves alternate or in basal rosettes, Capitula solitary, on long, usually bracteate, scape-like peduncles; receptacle flat, smooth, without bracts; involucral bracts in 2–3 series. Florets dimorphic; outer ligulate, female; inner tubular, 5-lobed, hermaphrodite or male, forming a disc. Pappus absent.

Literature: A. L. Cabrera, *Blumea* **14**: 285–308 (1966)

a. Leaves entire or mucronately crenate towards apex, glabrous or sparsely appressed-pubescent above **2. *nudicaulis***

a. Leaves with 1–3 pairs of obtuse teeth towards apex, hairy on both surfaces **1. *hariotii***

1. ***L. hariotii*** (Franchet) T. Dudley 1981, *Rhodora* **83**: 481.

Lagenophora hariotti Franchet 1889, *Miss. Sci. Cape Horn* 344 – Cabrera, 1966, 1971; Moore, 1974.
L. hirsuta auct., non Poeppig ex Less. – Skottsberg, 1916.
L. hirsuta var. *gracilis* Alboff 1896, *Rev. Mus. La Plata, Bot.*, **7**: 379.

Stems up to 1 cm, ascending to suberect, hairy. Leaves in ± basal clusters; lamina 6–15 × 6–12 mm, obovate, obtuse and mucronate at apex, cuneate at base, with 1–3 pairs of obtuse, mucronate teeth towards apex, hairy; petiole 5–15 mm, winged, ciliate. Peduncle 1.5–5 cm, terminal, erect, pubescent, with or without 1 short, linear bract Capitulum 6–9 mm in diameter; involucral bracts 2–2.5 × *c.*0.6 mm, oblong-lanceolate, acute to obtuse and laciniate, denticulate, glabrous, green or purplish. Florets reddish-purple to pale lilac or whitish; outer in 2 series, the ligule 2–2.5 × *c.*0.5 mm, lanceolate, retuse; inner *c.*2 mm. Achene *c.*2.5 × 1 mm, oblanceolate, ± compressed, with thickened margin and short, stout, truncate beak. $2n = 18$. *Flowers December to March.* (Fig. 193).

Evergreen and deciduous forests, moist shady slopes; 0–500 m.

C. & S. Fuegia except the extreme south and west. (Map 23g).
W. Argentina north to *c.*38°40′S.lat.
Chile north to 35°30′S.lat.
Juan Fernandez Is.

2. ***L. nudicaulis*** (Comm. ex Lam.) T. Dudley 1981, *Rhodora* **83**: 482.

Lagenophora nudicaulis (Comm. ex Lam.) Dusén 1900, *Svenska Exped. Magellansl.* **3**: 98 – Skottsberg, 1916; Cabrera 1966, 1971; Moore, 1974.
L. commersonii Cass. 1822, *Dict. Sci. Nat.* **25**: 110 – Hooker, 1847.

Similar to *L. hariotii* but leaves 5–10 × 2–6 mm, mucronately crenate, glabrous or slightly appressed – pubescent above; peduncle 1–6.6 cm, glabrous or appressed-pubescent distally, with 1–2 bracts; capitulum 5–9 mm in diameter; involucral bracts acute, entire; ligules 1–1.5 × *c.*0.3 mm. *Flowers January to March.*

Among bryophytes and small herbs on wet mountain slopes and stream banks, usually above timberline; (20–50)200–600 m.

S. W. Fuegia. (Map 23h).
W. Argentina north to *c.* 38°S.lat.
Chile north to c.36°30′S.lat.
Falkland Islands (Islas Malvinas).
Tristan da Cunha, Gough Island.

2. *Bellis* L.

Perennial herbs. Leaves in basal rosette. Capitula solitary, on long ebracteate scapes; receptacle conical, pitted, without bracts; involucral bracts in (1) 2 series. Florets dimorphic; outer ligulate, female; inner tubular, 4- to 5-lobed, hermaphrodite, forming a disc. Pappus absent.

*1. ***B. perennis*** L. 1753, *Sp. Pl.* 886.

Leaves sparsely hairy; lamina 20–55 × 5–18 mm, obovate-spathulate, obtuse, broadly cuneate at base, crenate-serrate; petiole 2–35 mm. Scapes 3–14(–18) cm, hairy. Capitulum 16–25(–30) mm in diameter; involucral bracts 3.5–6.5 × 1.4–2.2 mm, ovate to oblong, obtuse, entire, hairy, green, often black at apex. Outer florets white or pink; ligules 4.5–10 × 1.1–1.4 mm, oblanceolate; inner florets yellow. Achene 1.5–2 × *c.*0.9 mm, obovoid, compressed, with thickened margin, ± appressed – pubescent. *Flowers November to March (to July)*

Moist places in grassland and along streams; 0–500 m.

S. Isla Grande and south shore of Canal Beagle. Introduced from Europe and locally naturalized.

3. *Gutierrezia* Lag.

Dwarf shrubs, glabrous. Leaves alternate. Capitula solitary, terminal on branches; receptacle flat to ± conical, pitted, without bracts; involucral bracts in 3 series. Florets dimorphic; outer ligulate, female; inner tubular, 5-lobed, hermaphrodite. Pappus of numerous scales

Literature: O. T. Solbrig, *Contr. Gray Herb.* **197**: 3–42(1966).

1. ***G. baccharoides*** Sch. Bip. 1855, *Flora (Regensb)* **38**: 115.

Stems branched to form lax cushions up to 5 cm; branches viscid, densely leafy. Leaves 5–13 × 2–5 mm, spathulate, obtuse to subacute, long – cuneate at base, entire, glandular-punctate. Capitula 5–7 mm in diameter, sessile or on peduncles up to 30 mm; involucral bracts 4–5 × 1–2 mm, elliptic-oblong, obtuse, entire, glabrous, yellow-green. Outer florets yellow; ligule *c.*2.5 × 1–1.5 mm, obovate; inner florets 3 mm, yellow. Achene 1.5–2 × *c.*0.8 mm, turbinate, densely velutinous; pappus-scales 1–3 mm, lanceolate, acute. (Fig. 194).

Northernmost Isla Grande (Punta Espora), reported once (Solbrig, 1966). (Map 23j).

W. Argentina north to *c.*33°S.lat.
E. Chile on north shore of Estrecho de Magallanes.

4. *Lepidophyllum* Cass.

Erect, glabrous, evergreen shrubs. Leaves opposite, reduced, densely imbricate. Capitula solitary, terminal on branches; receptacle flat or slightly convex, smooth, without bracts; involucral bracts in 4 series. Florets dimorphic; outer ligulate, female; inner tubular, 5-lobed, hermaphrodite. Pappus of numerous flat, scale-like hairs.

Literature: A. L. Cabrera, *Bol. Soc. Argent. Bot.* **1**: 48–58(1945).

1. ***L. cupressiforme*** (Lam.) Cass. 1816, *Bull. Soc. Philom. Paris* **1816**: 199.

Stems 20–60 cm, much branched. Leaves 2–3 × 1.5–2 mm, elliptic- to ovate-oblong, obtuse, sessile, entire, thick, viscid. Capitula 7–9 × 3–4 mm, sessile; involucral bracts 3–5 × 1–2 mm, elliptic – oblong, obtuse, entire, scarious. Outer florets few, yellow; ligule 3–4 × *c.*1 mm, elliptic-lanceolate; inner florets yellow. Achene 4–4.5 × 1.2–1.5 mm, obovoid, ribbed, hispid towards apex; pappus-scales 3–5 mm, unequal, linear. $2n = 40$. *Flowers November to January*. (Fig. 204).

Saline coastal sands, locally dominant; 0–20 m.

N. Isla Grande, rarely on S. coast. (Map 23i).
E. Argentina north to *c.*48°S.lat.
E. Chile along Estrecho de Magallanes.

5. *Nardophyllum* Hooker & Arn.

Dwarf-shrubs, often cushion-forming. Leaves alternate. Capitula solitary, terminal on branches; receptacle convex, smooth, with or without bracts; involucral bracts in 4 series. Florets monomorphic, tubular, 5-lobed. hermaphrodite. Pappus of unequal, scabrid hairs.

Literature: A. L. Cabrera, *Not. Mus. La Plata, Bot.* **17**: 55–56(1954).

1. ***N. bryoides*** (Lam.) Cabrera 1954, *Not. Mus. La Plata, Bot.* **17**: 61.
N. humile (Hooker f.) A. Gray 1862, *Proc. Amer. Acad. Arts* **5**: 123–Skottsberg, 1916.

Stems much branched to form cushions or mats up to 10 cm; branches procumbent to ascending, the younger densely leafy and tomentose. Leaves 3–5 × *c.*1 mm, linear-oblong, obtuse, sessile, entire, subglabrous above, white-tomentose beneath. Capitula 10–15 × *c.*5 mm, sessile; involucral bracts

193

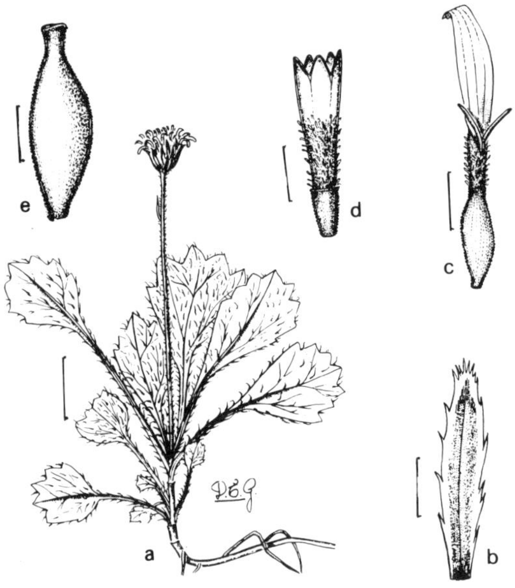

194

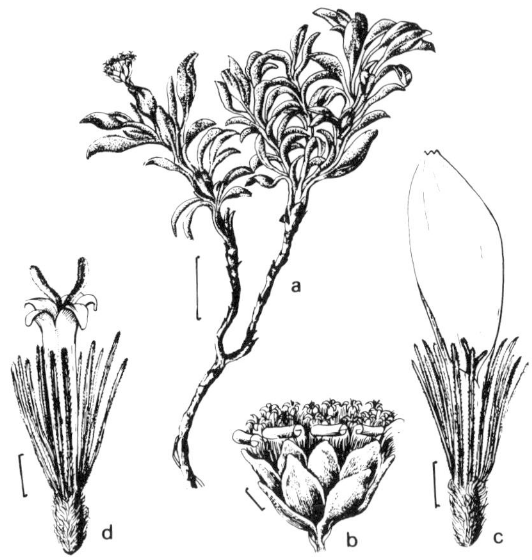

195

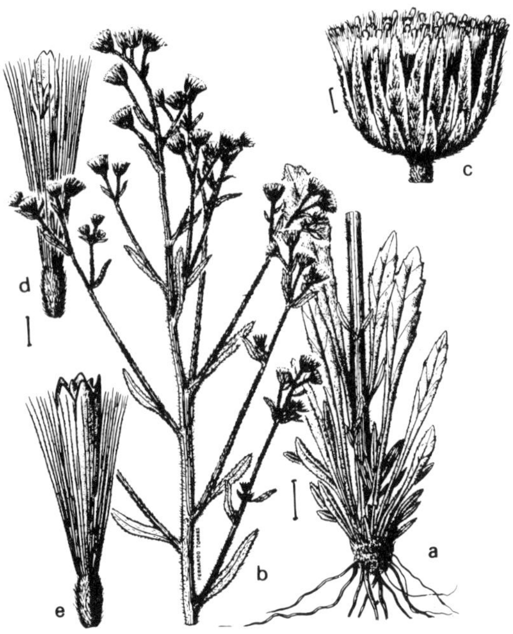

196

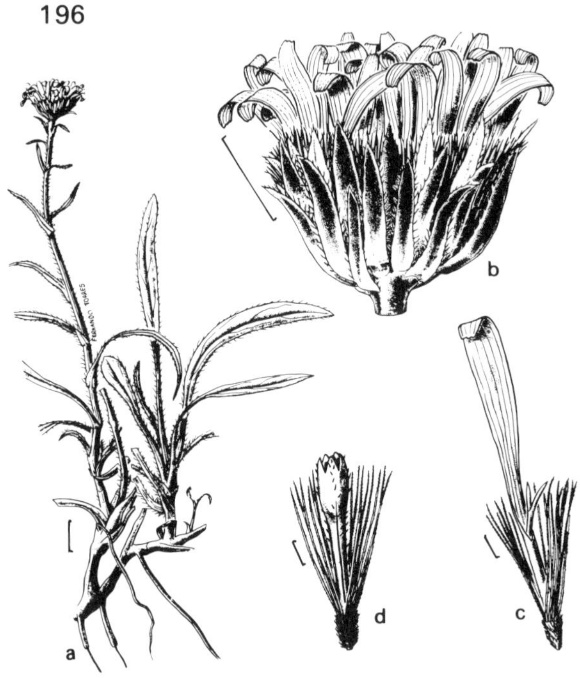

Fig. 193. *Lagenifera hariotii*. a, Plant. b, Involucral bract. c, Marginal floret. d, Disc floret. e, Achene. Scale = 1 cm (a) or 1 mm (b–e). Fig. 194. *Gutierrezia baccharoides*. a, Plant. b, Capitulum. c, Marginal floret. d, Disc floret. Scale = 1 cm (a) or 1 mm (b–d). Fig. 195. *Conyza lechleri*. a, Base of plant. b, Inflorescence. c, Capitulum. d, Marginal floret. e, Disc floret. Scale = 1 cm (a) or 1 mm (b–e). Fig. 196. *Aster vahlii*. a, Plant. b, Capitulum. c, Marginal floret. d, Disc floret. Scale = 1 cm (a), 5 mm (b) or 1 mm (c–d).

(Flora Patagónica)

2.5–5 × 1–1.5 mm, acute, ± lanate, green or purplish, the outer ovate, the inner linear-oblong; receptacle usually with a few linear-lanceolate, scarious scales. Florets *c*.7 mm, golden-yellow. Achene 2–3 × *c*.1 mm, cylindric-obovoid, densely sericeous; pappus-hairs 5–6 mm, subequal, pale brown. $2n = c.50$. *Flowers November to January*. (Fig. 197).

Dry sandy soils, among rocks, *Empetrum* heath; 0–50 m.

N. Isla Grande. (Map 24a).
Argentina north to *c*.51°S.lat.
E. Chile along Estrecho de Magallanes.

This species is very closely related to the more widespread *N. obtusifolium* Hooker & Arn., which Cabrera (1954, 1971) distinguishes by its more lax habit and lack of receptacular bracts. However, Fuegian plants vary from dense cushions to rather lax mats, apparently dependent upon soil-moisture and exposure, and perhaps the intensity of grazing, while receptacular bracts may be present or absent in adjacent plants. Clearly, further intensive study is necessary to determine whether these species should be maintained.

6. ***Chiliotrichum*** Cass.

Erect, evergreen shrubs. Leaves alternate. Capitula solitary, terminal on branches; receptacle flat or somewhat convex, smooth, with bracts; involucral bracts in 2–3 or more series. Florets dimorphic; outer ligulate, female; inner tubular, 5-lobed, hermaphrodite, forming a disc. Pappus of 2–4 series of strigose hairs.

1. *C. diffusum* (Forster f.) O. Kuntze 1898, *Revis. Gen.* **3(2)**: 141.
C. amelloideum Cass. 1817, *Dict. Sci. Nat.* **8**: 577 – Hooker, 1847; Hariot, 1884.
Amellus diffusus Forster f. 1787, *Fasc. Pl. Magell.* 39.

Stems 20–100(–150) cm, ascending to erect, much branched, with flaking, greyish brown bark at maturity, the twigs white-tomentose and leafy. Leaves 8–30 × 2–7.2 mm, oblong-oblanceolate to subelliptical, subacute to obtuse, shortly mucronate, cuneate at base, entire with revolute margins, rather coriaceous, glabrous and shiny dark green above, white-tomentose beneath, sessile or very shortly petiolate. Peduncles 8–42 mm, tomentose. Capitula 22–32 mm in diameter; involucral bracts 3.2–9(–10) × 1.3–2.5 mm, ovate to ovate-lanceolate, acute, entire, ± tomentose, purplish brown; receptacular bracts 6.5–10.7 × 0.5–1.4 mm, linear-lanceolate, subacute, scarious, sericeous distally. Outer florets white; ligule 7–12 × 2.5–4.5 mm, oblong to elliptical or oblanceolate; inner florets yellow. Achene 3–5 × 0.3–1 mm, oblong-oblanceolate, ± triangular in cross-section, 3-ribbed, glandular; pappus 4.5–6.5 mm, yellowish. $2n = 54$. *Flowers November to February*. (Fig. 208).

Dominant or co-dominant in shrub communities at margins and in clearings of evergreen and deciduous forest, drier parts of bogs, rock-crevices, streamsides; 0–400 m.

Throughout Fuegia. (Map 24b).
W. Argentina north to *c*.40°S.lat.
Chile north to *c*.47°S.lat.
Falkland Islands (Islas Malvinas).

C. rosmarinifolium Less. 1831, *Linnaea* **6**: 109, principally differing from *C. diffusum* in its narrower (1–2 mm wide) leaves, has been reported once from N. E. Isla Grande (San Sebastián) by Cabrera (1971). In view of the vegetative variability of austral material, the occurrence of this species in Fuegia requires further confirmation.

7. ***Chiliophyllum*** Phil.

Erect, evergreen shrubs. Leaves alternate. Capitula solitary, terminal on branches; receptacle flat, smooth, with bracts; involucral bracts in 3–4 series. Florets dimorphic; outer ligulate, female, in 1 series; inner tubular, 5-lobed, hermaphrodite, forming a disc. Pappus of 1 or more series of strigose hairs.

1. *C. fuegianum* O. Hoffm. 1900, in Dusén, *Svenska Exped. Magellansl.* **3(5)**: 96.

Stems 90–120 cm, erect, much branched, with grey brown bark at maturity, the twigs shortly tomentose and leafy. Leaves 3–5 × 1–2 mm, oblong to oblong-oblanceolate, obtuse, cuneate at base, entire, rather coriaceous, glabrous above, tomentose to subglabrous beneath, subsessile. Capitula 10–13 mm in diameter, sessile; involucral bracts 3–7 × 1.5–2 mm, acute, ± ciliate, the outer shortest, ovate, tomentose, green, the inner lanceolate, glabrous, brownish; receptacular bracts equalling flowers, linear-lanceolate, laciniate at apex, scarious. Outer florets 5–8, white; ligule 4–7.5 × 2–6 mm, oblong-elliptical; inner florets yellow. Achene 3.5–4 × *c*.0.5 mm, cylindrical, ribbed, glandular-pubescent; pappus 5–6 mm, white to yellowish. $2n = c.54$. *Flowers December to January*. (Fig. 198).

Coastal sands, open sandy clay along estuaries, sometimes in *Lepidophyllum* scrub; 0–10 m.

N. Isla Grande. Endemic. (Map 24c).

8. *Conyza* Less.

Annual herbs. Leaves alternate. Capitula in terminal corymbose cyme; receptacle flat or somewhat convex, without bracts; involucral bracts in 2–3 series. Florets dimorphic; outer shortly ligulate, female; inner tubular, 5-lobed, hermaphrodite, few. Pappus of numerous simple hairs.

1. ***C. lechleri*** (Sch. Bip.) Cabrera 1971, in Correa, *Fl. Patag.* **7**: 46.
   Erigeron bonariensis auct., non L. – Skottsberg, 1916.
   E. spiculosus auct., non Hooker & Arn. – Alboff, 1896; Spegazzini, 1896; Skottsberg, 1916.

Stems 10–40 cm, erect, simple, with long soft and scattered setose hairs, more densely leafy towards base. Lower leaves 60–100 × 5–10 mm, oblanceolate, acute, long-attenuate into petiole-like base, serrate, hairy; upper leaves similar but smaller. Inflorescence-branches 5–14 cm; pedicels 5–20 mm. Capitula 4–7 × 6–12 mm, campanulate; involucral bracts 4–6 × 0.5–0.8 mm, linear, acute, sparsely hairy and usually papillose. Outer florets with ligule 1.2–1.5 × 0.6 mm, elliptic-oblong, retuse, white or pinkish; inner florets yellowish. Achene *c.*1.5 × *c.*0.6 mm, oblong, ± flattened, pubescent; pappus 4–6 mm, tawny brown. *Flowers January to February*. (Fig. 195).

Coastal sands and gravels, cliffs, open ground in grassland and modified forest; 0–280 m.

N. W. & C. to S. Isla Grande; S. side of Canal Beagle. (Map 24d).
W. Argentina north to *c.* 40°S.lat.
Chile north to *c.*51°S.lat.

9. *Erigeron* L.

Perennial herbs, often rhizomatous. Leaves alternate or in basal rosettes. Capitula solitary or in corymbose cyme; receptacle flat or somewhat convex, without bracts; involucral bracts in 2–3 series. Florets dimorphic; outer ligulate, female, in 1–2 series; inner tubular, 5-lobed, hermaphrodite, numerous. Pappus of 1 series of strigulose hairs.

Literature: O. T. Solbrig, *Contr. Gray Herb.* **191**: 3–82(1962).

A difficult genus in southern South America, where most representatives belong to the E. *andicola* complex, which comprises a large number of closely related forms varying in stature, indumentum, leaf-shape, number of capitula etc. There is some evidence (Solbrig, 1962) that this group has an apomictic breeding system, which would perhaps explain the variation pattern. No taxonomic treatment has yet coped satisfactorily with this variation, and the Fuegian representatives are no exception. It seems to me that a detailed experimental study will be necessary before further significant progress can be made. In this account, therefore, I have followed the treatments of Solbrig (1962) and Cabrera (1971) with some modifications and accepting that they do not deal adequately with Fuegian *Erigeron*.

a. Cauline leaves similar to the basal; capitula solitary or 2–6 in cyme; plants up to 30 cm **1. *myosotis***
a. Cauline leaves dissimilar to the basal; capitula solitary; plants not more than 10 cm **2. *patagonicus***

1. ***E. myosotis*** Pers. 1807, *Syn. Pl.* **2**: 431.
   E. philippii Sch. Bip. ex Wedd. 1855, *Chloris Andina* **1**: 192 – Skottsberg, 1916.
   E. sulivani β *magellanicum* Sch. Bip. 1855, *Flora (Regensb.)* **38**: 113 – Spegazzini, 1896.
   E. fernandezi Phil 1894, *Anal. Univ. Chile* **87**: 424 – Skottsberg, 1906.
   E. alpinus auct., non L. – Hooker, 1847.
   E. uniflorus auct., non L. – Skottsberg, 1926.

Rhizome up to 4 cm; stems 3–30 cm, ascending to erect, leafy, hairy. Basal leaves in rosette-like cluster, 20–100 × 1–10 mm, linear-oblanceolate, obtuse to acute, long attenuate in sheathing, petiole-like base, entire, densely hairy to ciliate; cauline leaves similar but gradually smaller upwards. Capitula 7–15 mm in diameter, solitary or up to 6 in cyme; branches of cyme up to 8 cm; involucral bracts 6–8 × 0.6–1 mm, subequal, linear-lanceolate, acute, ± hairy. Outer florets whitish to lilac; ligule 3–6 × 0.7–1.2 mm, linear, ± truncate; inner florets yellow. Achenes *c.*3 × 0.5 mm, cylindrical, velutinous; pappus 1–4 mm, brownish. *Flowers December to March*. (Fig. 199).

Coastal sands and gravels, open *Chiliotrichum* and *Lepidophyllum* scrub, dry grassland and *Empetrum* heath; 0–250(–450) m.

Isla Grande except the extreme west and east. (Map 24e).
W. Argentina north to *c.*40°S.lat.
Chile north to *c.*45°S.lat.

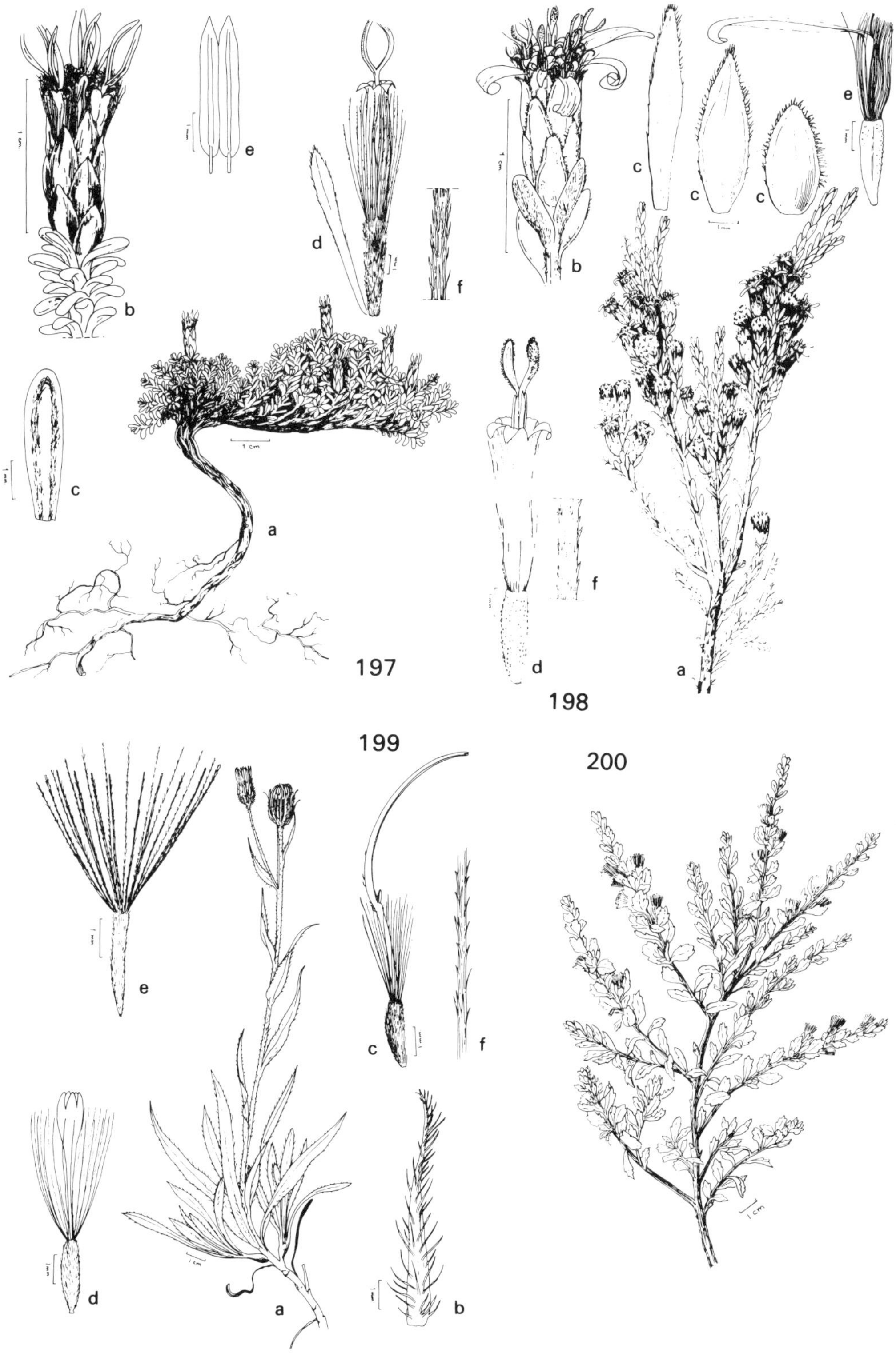

Fig. 197. *Nardophyllum bryoides*. a, Plant. b, Capitulum. c, Leaf. d, Floret. e, Anthers. f, Part of pappus. Scale = 1 cm (a–b) or 1 mm (c–f). Fig. 198. *Chiliophyllum fuegianum*. a, Branch. b, Capitulum. c, Involucral bracts. d, Disc floret. e, Marginal floret. f, Part of Pappus. Scale = 1.5 cm (a), 1 cm (b) or 1 mm (c–e). Fig. 199. *Erigeron myosotis*. a, Plant. b, Involucral bract. c, Marginal floret. d, Disc floret. e, Achene. f, Part of pappus. Scale = 1 cm (a) or 1 mm (b–e). Fig. 200. *Baccharis patagonica*. Scale = 1 cm.

(R. N. P. Goodall)

2. *E. patagonicus* Phil. 1894, *Anal. Univ. Chile* **87**: 414.

E. ciliaris auct., non Phil. – Solbrig, 1962.

Rhizome up to 3 cm; stems 1–11 cm, ascending to erect, sparsely leafy, ± hairy. Basal leaves in rosette-like cluster, 10–60 × 3–6 mm, linear-spathulate, obtuse to acute, long attenuate in sheathing petiole-like base, entire, subglabrous to densely hairy; cauline leaves smaller, linear, acute. Capitula 10–20 mm in diameter, solitary; involucral bracts 6–8 × 0.5–0.7 mm, linear-lanceolate, acute to acuminate, hairy, blackish. Outer florets purplish to whitish; ligule 2–3 × *c.*0.5 mm, linear; inner florets yellowish. Achene *c.*2 × 0.6 mm, cylindrical, pubescent; pappus *c.*4 mm, brownish to white. $2n = 36$. *Flowers November to February.*

Coastal sands, open grassland, mountain rocks, feldmark; 0–600 m.

Almost throughout Isla Grande except the extreme west and east. (Map 24f).
W. Argentina north to *c.*40°S.lat.
E. Chile north to *c.*50°S.lat.

High altitude plants here included in this species have been ascribed to *E. leptopetalus* Phil. by Cabrera (1971). They are of very low stature with somewhat oblanceolate leaves and relatively large capitula. However, they grade into other forms considered here to be *E. patagonicus* and there is evidence that they are ecotypic and environmental variants.

10. *Aster* L.

Perennial herbs. Leaves alternate. Capitula solitary or 2–3, terminal on branches or axillary; receptacle flat, pitted, without bracts; involucral bracts in several imbricate series. Florets dimorphic; outer ligulate, female; inner tubular, 5-lobed, hermaphrodite, forming a disc. Pappus of ciliate-strigulose hairs, in 1 series.

1. *A. vahlii* (Gaudich.) Hooker & Arn. 1836, *Comp. Bot. Mag.* **2**: 49.

Erigeron glabrifolius auct., non DC. – Spegazzini, 1896.
E. scorzonerifolius auct., non Remy – Alboff, 1896; Spegazzini, 1896.

Rhizomes up to *c.*30 cm; stems 5.5–21(–40) cm, ascending to erect, simple or branched near base, striate, sparsely hairy distally, densely leafy below, sparsely so above, reddish. Leaves glabrous or sparsely appressed-pubescent above, ciliate; lower with lamina 15–80 × 2.8–9 mm, obovate to linear-oblanceolate, acute to obtuse and mucronate, cuneate at base, the petiole (7–)11–40(–60) mm, winged, ciliolate, somewhat sheathing at base; upper leaves 9–32 × 1.9–4.4 mm, linear-oblanceolate to linear, acute to acuminate, sessile. Capitulum 12–28 mm; involucral bracts 4.7–10 × 0.8–2 mm, linear-oblanceolate, acuminate, entire, ciliate, green, reddish distally. Outer florets white or pinkish; ligule 5.5–8.5 × 0.6–1.3 mm, linear-oblanceolate, entire or 2– to 3– denticulate at apex; inner florets yellow. Achene 3.2–4 × 0.5 – 0.8 mm, fusiform, tetragonous, appressed-pubescent; pappus 7.5–8 mm, white to yellow-brown. $2n = 24$. Self-compatible. *Flowers December to January.* (Fig. 196).

Damp grassland, drier parts of bogs and marshes, moist coastal sand and rocks, along streams; 0–800 m.

Almost throughout Fuegia. (Map 24g).
W. Argentina north to *c.*36°30′S.lat.
Chile north to *c.*45°S.lat.
Falkland Islands (Islas Malvinas).

11. *Baccharis* L.

Dioecious, evergreen shrubs. Leaves alternate. Capitula solitary or in false spikes; receptacle ± flat, smooth, without bracts; involucral bracts in 3–4 or more imbricate series. Florets monomorphic, yellow or white to purplish; female filiform; male tubular, 5-lobed, with non functional stigma. Pappus of 1 or more series of distally strigulose hairs.

a. Leaves linear to elliptic-oblanceolate, entire; capitula usually with long peduncles; florets white to purplish **3. *nivalis***
a. Leaves subspathulate to oblong-elliptical, crenate-serrate; capitula sessile or with short peduncles; florets yellow:
b. Stems erect; leaves usually at least 9 mm; capitula in false spikes **2. *patagonica***
b. Stems prostrate to erect; leaves not more than 9 mm; capitula solitary **1. *magellanica***

1. *B. magellanica* (Lam.) Pers. 1807, *Syn. Pl.* **2**: 424.

Stems 3.5–33 cm, prostrate to ascending or ± erect, branched, rooted at base, scabridulous, the younger leafy, striate, viscid, green. Leaves (3.7–) 4.6–9 × 1.8–4(–4.5) mm, subspathulate, obtuse, cuneate at base, with 1–2 obtuse, crenate serrations on each side distally, or the upper entire, glabrous or scabridulous on margins, viscid when young, sessile to shortly petiolate. Capitula 4–12 mm in diameter, larger on female than male plants, solitary, terminal on short

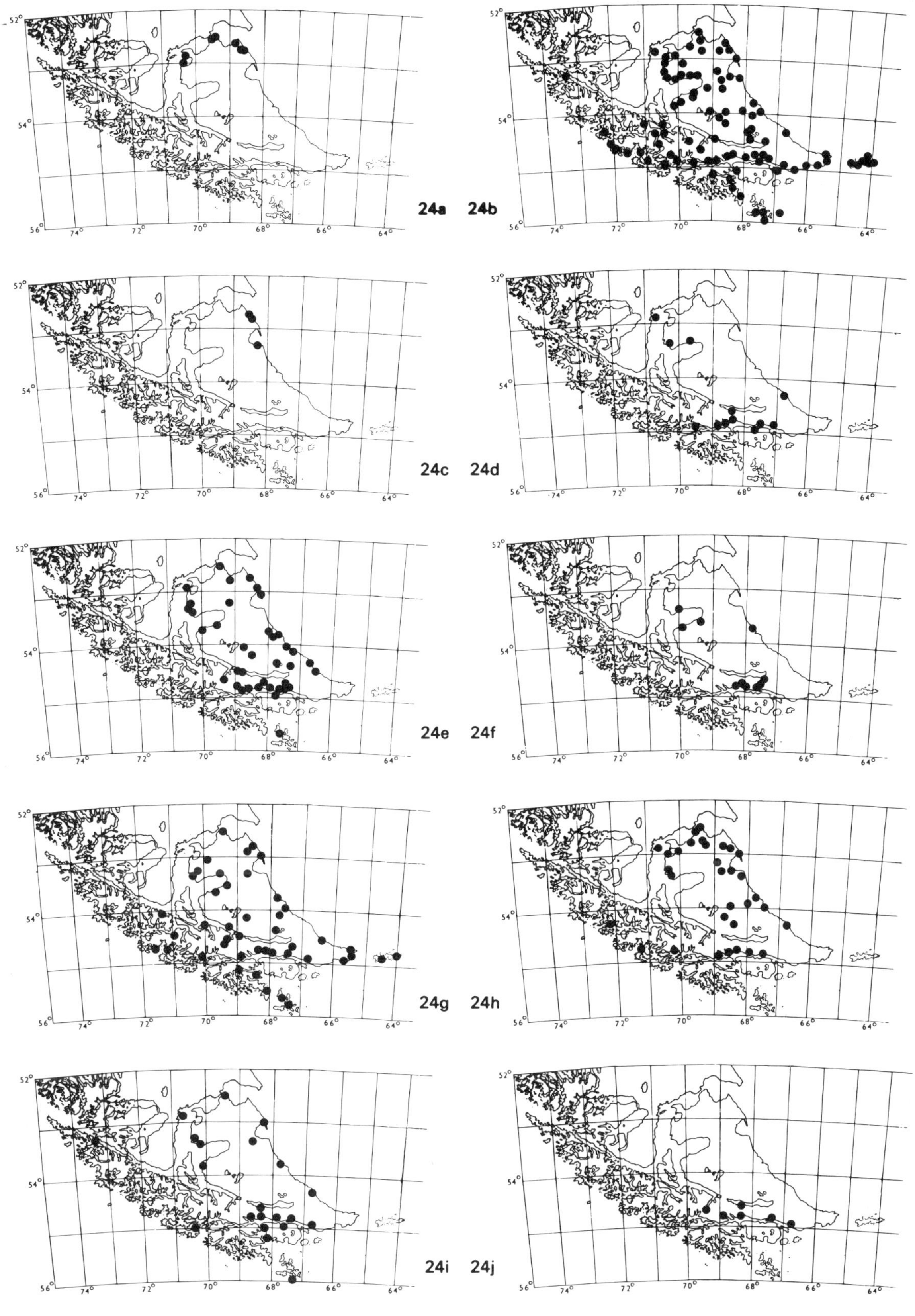

Map 24. a, *Nardophyllum bryoides*. b, *Chiliotrichum diffusum*. c, *Chiliophyllum fuegianum*. d, *Conyza lechleri*. e, *Erigeron myosotis*. f, *Erigeron patagonicus*. g, *Aster vahlii*. h, *Baccharis magellanica*. i, *Baccharis patagonica*. j, *Baccharis nivalis*.

lateral branches, sessile or shortly pedunculate; involucral bracts 2.3–5.5(–6.2) × 0.6–1.5(–2) mm, the outer ovate, obtuse, the inner longer, lanceolate, acute, all entire, ciliate, white and membranous with green or brownish midvein. Florets yellow; female 3–4.5 mm, the pappus in several series; male 3–3.6 mm, the pappus in 1 series. Achene 1.2–2.5 × 0.3–0.5 mm, narrowly obovoid, truncate, pale brown, with *c.*10 whitish ribs; pappus 4–7 mm, white. 2n = 18. *Flowers December to February (to April).* (Fig. 201).

Empetrum heath, *Berberis, Chiliotrichum* and *Pernettya* scrub, open grassland; 0–250 m.

Almost throughout Fuegia except the extreme west, south and east. (Map 24h).
W. Argentina and E. Chile north to *c.*36°S.lat.
Falkland Islands (Islas Malvinas).

2. *B. patagonica* Hooker & Arn. 1841, *Jour. Bot. (Hooker)* **3**: 29.

Stems 30–120 cm, erect, branched, glabrous, sometimes glandular. Leaves (6.3–)9–11(–16) × 2.2–9.5 mm, elliptical to oblong-elliptical, obtuse, broadly cuneate to attenuate at base, with 2–5 obtuse serrations on each side, punctate-glandular above, sessile. Capitula 6–12(–15) mm in diameter, larger on female than male plants, sessile at apices of very short lateral branches to form a false spike; involucral bracts 2.3–6.5 × 0.5–1.3 mm, the outer ovate, obtuse, the inner longer, lanceolate, acute, all entire, ciliolate, white and membranous with greenish brown midvein. Florets yellow; female 2.4–4.3 mm; male 5–7 mm. Achene 1–2 × 0.2–0.5 mm, cylindrical, truncate, *c.*10-ribbed, brownish; pappus 3–7 mm, whitish. *Flowers January to March.* (Fig. 200).

Rock-crevices, *Berberis* and *Chiliotrichum* scrub, forest margins, coastal sands; 0–70 m.

Almost throughout Fuegia. (Map 24i).
W. Argentina north to *c.*40°S.lat.
Chile north to *c.*45°S.lat.

3. *B. nivalis* (Wedd.) Sch. Bip. ex Phil. 1894, *Anal. Univ. Chile* **87**: 705.
Heterothalamus nivalis Wedd. 1856, *Chloris Andina* **1**: 179 – Dusén, 1900.

Plant with slender rhizomes up to 25 cm; stems 2–8 cm, procumbent to ascending, branched, glabrous. Leaves 5–15 × 1–3 mm, linear to elliptic-oblanceolate, acute to subobtuse, entire, glabrous, often purplish distally, sessile. Capitula 6–12 mm in diameter, larger on female than male plants; peduncles 5–20 mm, slender, purplish; involucral bracts 3–6 × 1–2 mm, ovate, acute to subobtuse, ciliate, green, usually with purple margin. Florets white to purplish; female 3–4 mm; male 4–6 mm. Achene 3–4 × *c.*0.8 mm, cylindrical, truncate, brown, white-appressed-pubescent; pappus 8–10 mm, white to brownish. *Flowers January to February.* (Fig. 203).

Pebbles and rocks beside rivers and streams, shingle of glacial moraines; 0–400 m.

S. Isla Grande. (Map 24j).
W. Argentina north to *c.*41°S.lat.
Chile north to *c.*43°S.lat.

Tribe INULEAE Florets tubular or outer with ligules; outer florets usually female, the inner hermaphrodite or male. Stigmatic branches flattened, rounded at apex.

12. *Antennaria* Gaertner

Dioecious, rhizomatous perennial herbs. Leaves alternate. Capitula in terminal glomerule; receptacle ± flat, without bracts; involucral bracts in 3 series. Florets monomorphic; female filiform; male tubular, 5-lobed. Pappus of several series of strigulose hairs, free or united at base.

Literature: A. L. Cabrera, *Not. Mus. La Plata (Bot.)* **19**: 73–79(1957).

1. *A. chilensis* Remy 1849, in Gay, *Hist. Chile Bot.* **4**: 235.
A. magellanica Sch. Bip. 1855. *Flora (Regensb.)* **38**: 117 – Alboff, 1896; Skottsberg, 1916.
Gnaphalium affine var. *parvulum* Speg. 1896, *Anal. Mus. Nac. Hist. Nat. Buenos Aires* **5**: 63.

Rhizomes 2–12 cm; stems 3–13 cm, ascending to erect, tomentose. Basal leaves in rosette, 9–20 × 2–5 mm, oblanceolate-spathulate, obtuse, often apiculate, attenuate at base, entire, densely tomentose, sessile; cauline leaves smaller, linear-oblanceolate. Capitula in glomerules of 4–10, 5–7 × 5–8 mm, sessile; involucral bracts 2.5–3.5 × 1.4–1.7 mm, obtuse, serrate-denticulate, the outer ovate, the inner oblong. Female florets 2–2.5 mm, white or pinkish; male plants not known. Achene 1–1.2 × 0.3–0.5 mm, cylindrical-obovoid, truncate, brown, pubescent; pappus 4–6 mm, white. Apomictic. *Flowers December to February.* (Fig. 207).

Fig. 201. *Baccharis magellanica*. Scale = 1 cm. Fig. 202. *Adenocaulon chilense*. a, Base of plant. b, Inflorescence. c, Capitulum. d, Floret. e, Achene. Scale = 1 cm (a–c) or 1 mm (d–e). Fig. 203. *Baccharis nivalis*. a, Plant. b, Capitulum. c, Floret. d, Pappus. Scale = 1 cm (a) or 1 mm (b–d).

(R. N. P. Goodall)

Open sands and clays by coast and rivers, open grassland, dwarf shrub heath; 0–110 m.

Isla Grande except the extreme west. (Map 25a).
W. Argentine and E. Chile north to *c.*45°S.lat.

Fuegian material belongs to var. ***magellanica*** (Sch. Bip.) Reiche 1903, *Anal. Univ. Chile* **112**: 97, which has the above distribution. Var. *chilensis*, distinguished by its glabrous achenes, reaches further north.

13. *Gnaphalium* L.

Annual herbs. Leaves alternate. Capitula in terminal, leafless corymb; receptacle flat, without bracts; involucral bracts in several imbricate series, scarious, shiny, yellowish. Florets indistinctly dimorphic; outer filiform, female, inner few, tubular, 5-lobed, hermaphrodite. Pappus of 1 series of free, strigulose hairs.

1. *G. luteoalbum L. 1753, *Sp. Pl.* 851.

Stems 13–20 cm, ascending to erect, simple or rarely with short decurrent branches from base, with ± dense, long, appressed whitish hairs. Leaves 8–25 × (1–)2–3 mm, entire, lanuginose; lower oblanceolate to obovate-spathulate, obtuse, attenuate at base; upper smaller, oblong to linear-oblong, acute, often apiculate, amplexicaul. Capitula 4–5 mm, ovoid, in 1 or more dense clusters; involucral bracts 2–3 × 1–1.3 mm, obtuse, entire, glabrous, the outer ovate, the inner oblong-elliptical. Florets 2–3 mm, yellowish. Achenes *c.*0.5 × 0.2 mm, cylindrical, truncate, finely tuberculate; pappus *c.*2.5 mm, white. *Flowers March.*

Open grassland on coastal dunes; 0–10 m.

Local on W. & S. coasts of Isla Grande.
Introduced from Europe and widely naturalized in disturbed ground of temperate regions.

Cabrera (1971) has referred all Patagonian specimens to the perennial *G. montevidense* Sprengel, because they lack a reddish tint to the flowers. However, this feature is not constant in European material, and the habit and other characters of Fuegian and Falkland Islands specimens are identical with N. Hemisphere material of *G. luteoalbum*; this also accords with the habitats occupied by austral examples.

14. *Gamochaeta* Wedd.

Similar to *Gnaphalium* but biennial or perennial; capitula in terminal clusters subtended by leaves, often to form a ± spike-like inflorescence; involucral bracts brownish; pappus-hairs united into a basal ring and falling as a unit.

Literature: A. L. Cabrera, *Bol. Soc. Argent. Bot.* **9**: 359–386(1961).

a. Leaves densely tomentose and of same colour above and beneath; plant usually less than 7 cm **2. *nivalis***
a. Leaves subglabrous to somewhat lanate and darker coloured above, grey-lanate to densely tomentose and paler beneath; plant usually more than 7 cm:
 b. Capitula in terminal cluster surrounded by upper leaves, sometimes with smaller axillary clusters below; outer involucral bracts subobtuse **1. *spiciformis***
 b. Capitula in clusters, each subtended by a leaf, to form a terminal spike-like inflorescence; outer involucral bracts acuminate **3. *americana***

1. *G. spiciformis* (Sch. Bip.) Cabrera 1961, *Bol. Soc. Argent. Bot.* **9**: 381.
Gnaphalium spicatum auct., non Lam. – Franchet, 1889; Alboff, 1896; Dusén 1900.
G. affine f. *medium* Speg. 1896, *Anal. Mus. Nac. Hist. Nat. Bs Aires* **5**: 63.
G. purpureum var. *mucronatum* (Phil.) Skottsb. 1916, *Kungl. Svenska Vetensk. Akad. Handl.* **56(5)**: 317.

Perennial; stems (2–)7–25 cm, erect, procumbent at base, simple, laxly lanate, Leaves 10–40 × (2–)6–11 mm, spathulate to linear-oblanceolate, obtuse, prominently apiculate, attenuate into sometimes long and petiole-like base, entire, sessile, sparsely lanate to subglabrous and green above, lanate-tomentose and paler beneath. Capitula 3.5–4.8 × 1.2–2 mm, numerous, sessile, in ± spherical, apical cluster ± surrounded by upper leaves, often with 1–5 smaller, axillary clusters below; involucral bracts 2–2.5 × *c.*1 mm, the outer ovate, the inner oblong-lanceolate, subobtuse, green or white, usually brown distally, often with purplish tinge. Florets *c.*3 mm, pale yellow, purplish distally; female numerous; hermaphrodite 2–5. Achene *c.*1 × 0.3 mm, oblong, ± compressed, brown, smooth or shortly papillate; pappus 3.5–4 mm, white. $2n = 28$. *Flowers December to March.*

Open sands and soil in most open communities, including forest-clearings and margins, that are not too waterlogged, rock-crevices; 0–950 m.

Almost throughout Fuegia. (Map 25b).
W. Argentina north to *c.*40°S.lat.
Chile north to *c.*45°S.lat.

A rather variable species in which plants from exposed habitats, especially at higher elevation, are of low stature and/or with more densely tomentose leaves showing little difference between upper and lower surfaces. Such examples grade into the larger forms here ascribed to *G. nivalis* and may suggest hybridization. However, the range of environmental modification, which is undoubtedly important, is not known and only detailed experimental studies will show to what extent these two species are truly distinct in Fuegia.

2. *G. nivalis* Cabrera 1961, *Bol. Soc. Argent. Bot.* **9**: 374.

Gnaphalium affine f. *pumilum* Speg. 1896, *Anal. Mus. Nac. Hist. Nat. Bs Aires* **5**: 63.

G. affine auct., non D'Urv. – Alboff, 1896; Autran, 1905; Skottsberg, 1916.

Perennial; stems 3–7(–8) cm, procumbent to ascending, simple, densely tomentose. Leaves 6–18 × 3–5 mm, elliptic-obovate to linear-spathulate, obtuse, often with tiny apiculus, attenuate at base, sessile, entire, densely greyish tomentose above and beneath. Capitula 2.5–3.5 × 1.5–2 mm, 6–15, sessile, in apical cluster ± surrounded by upper leaves, sometimes with 1–2 smaller axillary clusters below; involucral bracts 2.5–3.5 × 0.5–1.1 mm, the outer ovate, tomentose at base, the inner obtuse, whitish, brown distally. Florets *c.*3.5 mm, whitish, purple distally; female 10–20; hermaphrodite fewer. Achene 0.8–1 × *c.*0.3 mm, oblong, ± compressed, brown, smooth; pappus 3–3.5 mm, white. *Flowers January to March.*

Open soil and rock-crevices; (30–)600–900 m.

S. Isla Grande. (Map 25c).
W. Argentina and perhaps E. Chile north to *c.*36°S.lat.

Similar to small forms of *G. spiciformis*, from which it is best distinguished by the leaf-indumentum and scarcely apiculate leaf-apex.

3. *G. americana* (Miller) Wedd. 1856, *Chloris Andina* **1**: 151.

Gnaphalium americanum Miller 1768, *Gard. Dict.*, Ed. 8, No. 17.

G. consanguineum Gaudich. 1825, *Ann. Sci. Nat.* **5**: 103.

G. falcatum auct, non L. – Hooker, 1847.

Biennial or perennial; stems (2–)4–18 cm, ascending to erect, tomentose. Basal leaves in rosette-like cluster, 10–27 × 2.5–7 mm, spathulate, obtuse, with tiny apiculus, attenuate at base, sessile, entire, sparsely lanate and green above, lanate-tomentose and paler beneath. Capitula 4–5 × 3–3.5 mm, numerous, in compact clusters in leaf-axils to form a spike-like inflorescence up to 7 cm; involucral bracts 3–5 × 0.5–1 mm, the outer ovate, acuminate, the inner lanceolate, acute, greenish or white, brown distally. Florets 2.5–3 mm, whitish, purple distally; female numerous; hermaphrodite few. Achene *c.*0.6 × 0.2 mm, oblong, ± compressed, minutely papillate, brown; pappus 3–3.5 mm, white. *Flowers December to February*. (Fig. 209).

Sands, bare soil, open grassland, *Chiliotrichum* scrub, degraded, open deciduous forest, rock-crevices; 0–100 m.

Isla Grande, except the extreme west. (Map 25d).
Much of Argentina and Chile, north through S. America to the Caribbean islands.

Some intermediate forms suggest that hybridization may occur between this species and *G. spiciformis*.

15. *Adenocaulon* Hooker

Perennial herbs. Leaves mostly in basal rosette. Capitula in lax, terminal racemes; receptacle flat, without bracts; involucral bracts few. Florets dimorphic; outer tubular, 4- to 5-lobed or ± 2-lipped, female; inner tubular, 5-lobed, functionally male. Pappus absent.

1. *A. chilense* Less. 1831, *Linnaea* **6**: 107.

Plant with short, stout rhizome; stems 12–50 cm, slender, simple, tomentose, stipitate-glandular above, usually with 1–2 cauline leaves. Basal leaves with lamina (4–)6–20 × 3–7 cm, ovate to elliptical or lanceolate, obtuse, cuneate to truncate at base, entire to crenate-serrate, glabrous above, densely white-tomentose beneath; petiole 2–10 cm, tomentose beneath, winged distally; cauline leaves linear, acuminate, bract-like. Capitula 3–4 × 3–6 mm; peduncles up to 70 mm, slender, stipitate-glandular; involucral bracts *c.*6, 2–2.2 × 0.8–1 mm, ovate, acuminate, green, often purplish distally, white-tomentose towards base. Florets white; female 1.5–2 mm; male 2–3 mm. Achene 5–7 × 1.5–2 mm, obovoid, stipitate-glandular distally. $2n = 46$. *Flowers December to March.* (Fig. 202).

Nothofagus forest; 0–300 m.

C. & S. Fuegia, except the outermost islands. (Map 25e).
W. Argentina north to *c.*40°S.lat.
Chile north to *c.*39°S.lat.

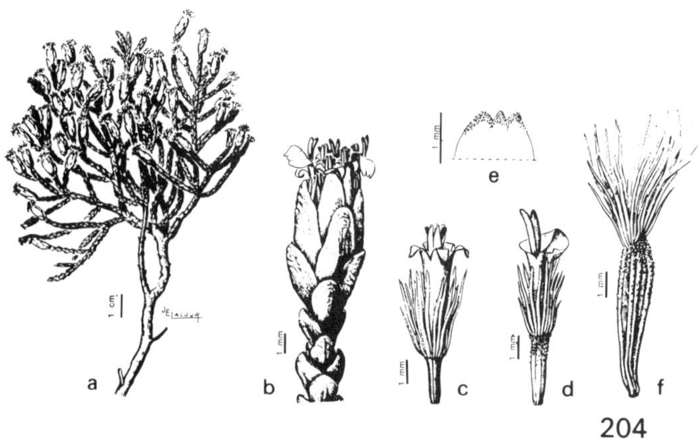

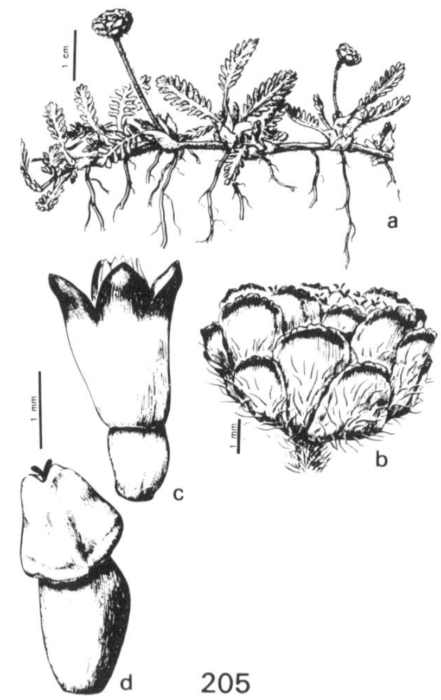

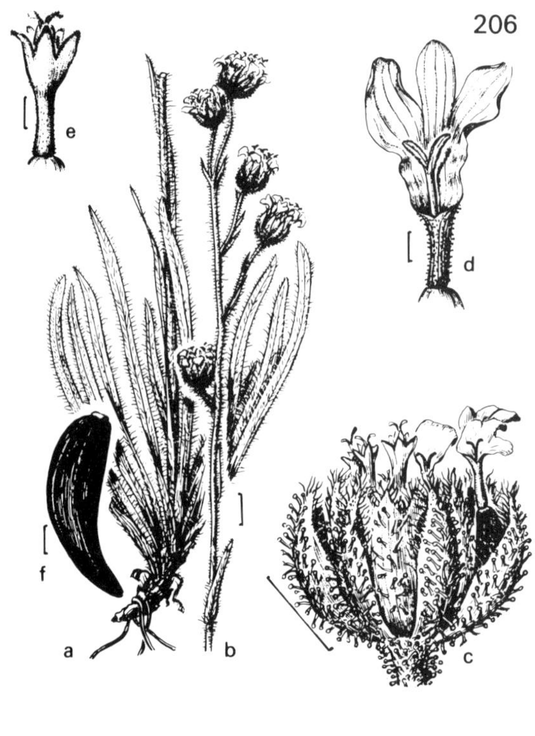

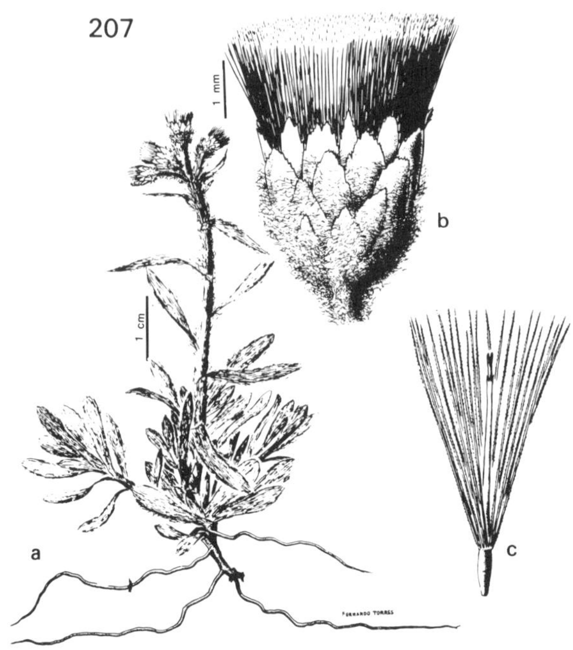

Fig. 204. *Lepidophyllum cupressiforme*. a, Branch. b, Capitulum. c, Disc floret. d, Marginal floret. e, Ligule-apex. f, Achene. Scale = 1 cm (a) or 1 mm (b–f). Fig. 205. *Cotula scariosa*. a, Plant. b, Capitulum. c, Disc floret. d, Marginal floret. Scale = 1 cm (a) or 1 mm (b–d). Fig. 206. *Madia sativa*. a, Base of plant. b, Inflorescence. c, Capitulum. d, Marginal floret. e, Disc floret. f, Achene. Scale = 1 cm (a–b), 5 mm (c) or 1 mm (d–f). Fig. 207. *Antennaria chilensis*. a, Plant. b, Capitulum. c, Achene. Scale = 1 cm (a) or 1 mm (b–c).

(Flora Patagónica)

16. *Eriachaenium* Sch. Bip.

Perennial herbs. Leaves alternate. Capitula small, solitary, axillary; receptacle flat, without bracts; involucral bracts few, in 1–2 series. Florets dimorphic; outer 1–2, tubular, 4-lobed, female; inner *c*.4, tubular, 5-lobed, functionally male. Pappus absent.

1. *E. magellanicum* Sch. Bip. 1855, *Flora (Regensb.)* **38**: 121.

Plant with stout oblique to vertical rhizome; stems 2–8 cm, prostrate to ascending, simple or slightly branched, fleshy, glabrous. Leaves 5–20 × 1–6 mm, linear-oblanceolate to oblanceolate, acute to obtuse, cuneate and ± sheathing at base, subentire to undulate-dentate and somewhat revolute, glabrous above, ± densely white-tomentose beneath, sessile. Capitula 3–5 × 2–4 mm; peduncle 1.5–5 mm; involucral bracts 3–3.5 × 0.8–1 mm, ovate-lanceolate, acuminate, denticulate distally, subglabrous, brown or purplish. Florets white or pinkish; female *c*.4 mm; male 2–3 mm. Achene 2.5–5 × 0.7–1.2 mm, ellipsoid, densely white-lanate. $2n = c.36$. *Flowers November to February*. (Fig. 213).

Mud, sand and pebbles by coast, estuaries and saline lakes; 0–20 m.

N. Isla Grande. (Map 25f).
Argentina north to *c*.44°S.lat.
E. Chile north to *c*. 51°S.lat.

Tribe HELIANTHEAE Inner florets tubular, the outer usually with ligules; outer florets female or sterile, the inner hermaphrodite or functionally male. Stigmatic branches truncate.

17. *Madia* Mol.

Annual herbs. Leaves opposite to alternate. Capitula in terminal racemose or paniculate inflorescence; receptacle convex, with marginal bracts and hairs in the centre; involucral bracts in 1 series. Florets dimorphic; outer ligulate, female; inner tubular, 5-lobed, hermaphrodite, forming a disc. Pappus absent.

1. *M. sativa* Mol. 1782,
M. viscosa Cav. 1795, *Icon. Descr.* **3**: 50.

Stems 10–50 cm, erect, simple, hairy below, glandular-pubescent above. Lower leaves in a rosette, 30–100 × 1.5–8 mm, linear, acute, entire, hairy and ± glandular, sessile; cauline leaves similar, the lower opposite, the upper alternate. Capitula 9–12 × 7–10 mm; peduncles 6–35 mm, up to 80 mm in fruit, glandular-pubescent; involucral bracts 7–9 × 2–3 mm, lanceolate, acute, glandular-pubescent; receptacular bracts 4–5 mm, oblanceolate, scarious. Florets yellow; outer with ligule 2–3.5 mm, oblong, 3-dentate at apex; inner c. 3 mm, green or purplish. Achene *c*.5 × 1.5 mm, oblong-obovoid, somewhat compressed and curved, smooth, black. *Flowers January to March*. (Fig. 206).

Sands and gravels, open grassland, disturbed ground; 0–45 m.

Isla Grande, except the extreme west, and Isla Navarino, mainly in coastal areas. (Map 25g).
Argentina, particularly the west, north to *c*.35°S.lat.
Chile north to *c*.33°S.lat.

Helianthus annuus L. 1753, *Sp. Pl.* 904, a robust annual up to 80 cm, with ovate leaves and yellow, terminal, solitary, more or less nodding capitula 8–10 cm in diameter, having a disc and prominent ligulate marginal florets, is grown as a garden ornamental and occasionally appears as a casual in disturbed ground near habitations. It is introduced from Europe, though a native of North America.

Tribe ANTHEMIDEAE Florets tubular or outer with ligules; outer florets usually female or sterile, the inner hermaphrodite or functionally male. Stigmatic branches truncate.

18. *Achillea* L.

Perennial herbs. Leaves alternate, 2– to 3– pinnatisect. Capitula in terminal corymbs; receptacle flat or slightly convex, with bracts; involucral bracts in several imbricate series. Florets dimorphic; outer ligulate, female; inner tubular, 5-lobed, hermaphrodite, forming a disc. Pappus absent.

1. *A. millefolium L. 1753, *Sp. Pl.* 899.

Plant aromatic, with far creeping stolons; stems 15–59 cm, erect, simple, furrowed, ± lanate. Leaves linear- to oblong-lanceolate in outline, ± pubescent, the primary segments linear- to oblong-lanceolate, the ultimate segments *c*.2–3 mm, linear to subulate, acute; lower leaves with lamina 5.5–16(–37) × 1.3–2.8(–3.7) cm, the petiole 0.7–3(–7) cm; upper leaves 2.4–2.7 × 0.6–1 cm, sessile. Capitula 5–8 mm in diameter; involucral bracts 2.7–4 × 1–1.3 mm, elliptical, obtuse, ciliolulate, lanate, keeled, with pale brown to blackish scarious margin; receptacular bracts 3.2–4 × 0.3–1 mm, linear, scarious. Outer florets white or pinkish, the tube 1.5–1.8 mm, the

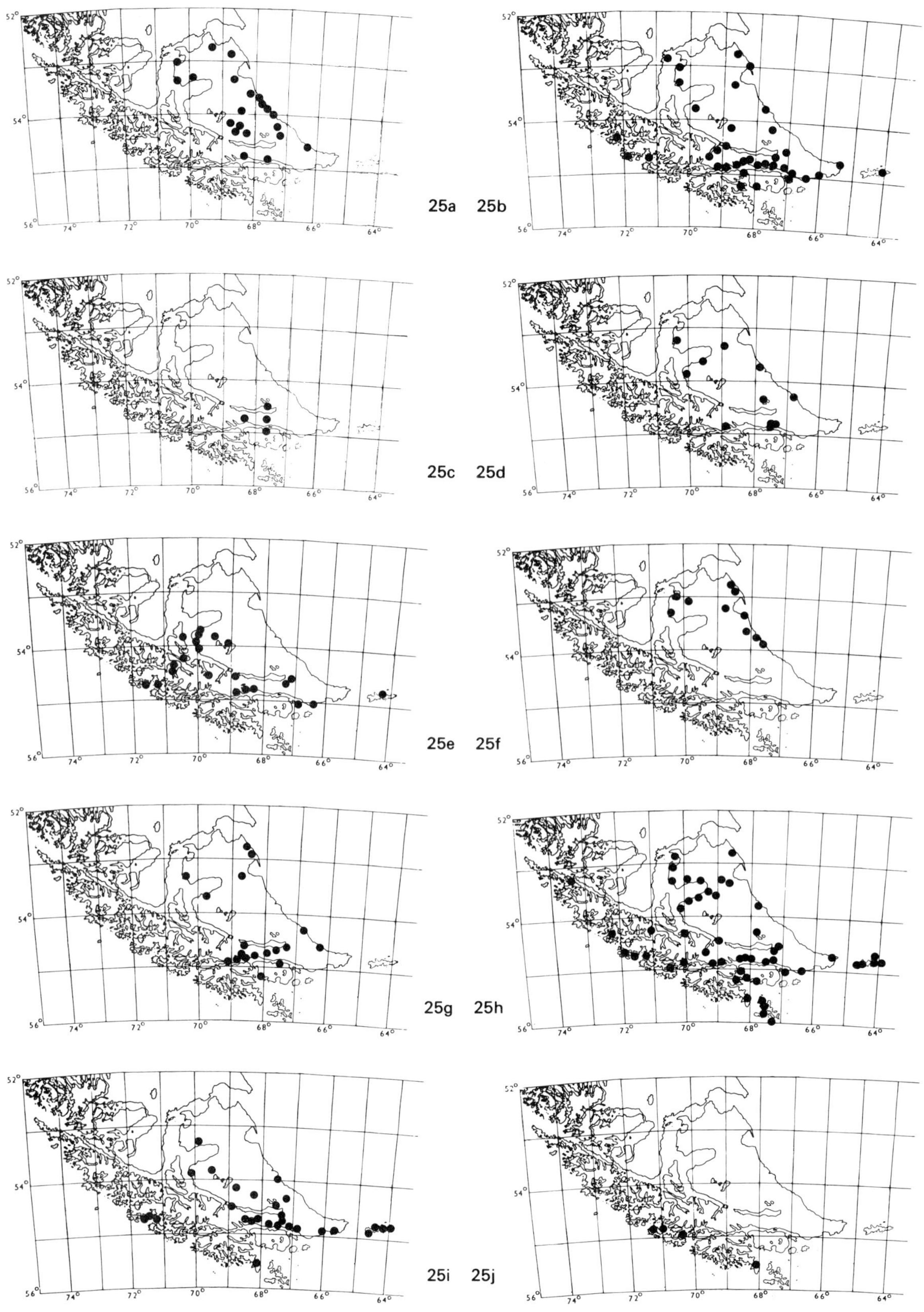

Map 25. a, *Antennaria chilensis*. b, *Gamochaeta spiciformis*. c, *Gamochaeta nivalis*. d, *Gamochaeta americana*. e, *Adenocaulon chilense*. f, *Eriachaenium magellanicum*. g, *Madia sativa*. h, *Cotula scariosa*. i, *Abrotanella emarginata*. j, *Abrotanella submarginata*.

ligule 1.2–3.6 × 1.1–3.1 mm, obovate to ± suborbicular, 3-dentate at apex; inner florets *c.*2.5 mm, white or cream. Achene *c.*2 × 0.8 mm, oblong to oblanceolate, truncate, compressed, somewhat winged, glabrous, shiny, greyish. *Flowers January to April.*

Open sand and soil; 0–50 m.

Much of Isla Grande except the extreme west and east; Isla Navarino.
Introduced from Europe and widely naturalized in temperate regions.

19. *Anthemis* L.

Annual herbs. Leaves alternate, 2– to 3– pinnatisect. Capitula solitary, terminal; receptacle convex, with bracts distally; involucral bracts in several imbricate series. Florets dimorphic; outer ligulate, sterile, sometimes absent; inner tubular, 5-lobed, hermaphrodite. Pappus absent.

*1. ***A. cotula*** L. 1753, *Sp. Pl.* 894.

Plant foetid; stems 20–60 cm, erect, branched, sparsely hairy. Leaves 1.5–5 × 1–2 cm, ovate to ovate-oblong in outline, the ultimate segments up to 2.5 × 0.4 mm, linear, acute, sparsely hairy. Capitula 12–25 mm in diameter, with peduncle up to 4.5 cm; involucral bracts 2.5–3 × 1–1.4 mm, ovate to ovate-oblong, obtuse to acute, ± glabrous, with green central stripe and pale brown scarious margin; receptacular bracts 1.5–1.8 mm, linear, subulate. Outer florets white, the ligule 5–8 × 2.4–2.7 mm, ± oblong-elliptical; inner florets yellow. Achene 1.3–2 mm, obovoid, with 8–10 tuberculate ribs. *Flowers December to May.*

Disturbed open ground by roads and habitations; 0–40 m.

Occasional on N, E. & S. coasts of Isla Grande. Introduced from Europe and widely naturalized in temperate regions.

20. *Matricaria* L.

Annual herbs. Leaves alternate, 2– to 3– pinnatisect. Capitula solitary, or in lax corymbose cyme, terminal; receptacle conical, without bracts; involucral bracts in 2 series. Florets dimorphic; outer ligulate, female; inner tubular, 5-lobed, hermaphrodite, forming a disc. Pappus a short membranous rim.

*1. ***M. perforata*** Mérat 1812, *Nouv. Fl. Env. Paris* 332.
M. inodora L. 1755, *Fl. Suec.*, Ed. 2, 297, nom illeg. – Dusén, 1900; Skottsberg, 1906; Cabrera, 1971.

Stems 15–58 cm, ascending to erect, simple or corymbosely branched above, glabrous, striate. Leaves 2.3–7.5 × 1.5–3.5 cm, oblanceolate in outline, glabrous, the ultimate segments filiform, apiculate. Capitula 30–45 mm in diameter, with peduncle up to 11 cm; involucral bracts 5–5.2 × 1.1–1.4 mm, oblong to triangular, obtuse, with narrow, brownish scarious margin. Outer florets white, the ligule 11–15 × 2.4–3.3 mm, elliptic-oblong; inner florets yellow. Achene 2–3 × 0.7–0.9 mm, obovoid, prominently 3-ribbed, brown. *Flowers December to February.*

Coastal sands and shingle, disturbed soil; 0–120 m.

Locally in N. E., C. & S. Isla Grande.
Introduced from Europe and widely naturalized in temperate regions.

21. *Chamomilla* S. F. Gray

Glabrous, annual herbs. Leaves alternate, 2– to 3– pinnatisect. Capitula solitary; receptacle conical, hollow, without bracts; involucral bracts in 2–3 series. Florets monomorphic, tubular, 4-lobed, hermaphrodite. Pappus a short, membranous rim.

*1. ***C. suaveolens*** (Pursh) Rydb. 1916, *N. Amer. Fl.* **34**: 232.
Matricaria matricarioides (Less.) Porter 1894, *Mem. Torrey Bot. Club* **5**: 341 – Cabrera, 1971.

Plant aromatic; stems 4.5–32(–55) cm, ascending to erect, usually branched, especially above. Leaves 1.2–5.5 × 0.5–1.3 cm, oblong in outline, the ultimate segments linear to elliptic-linear, acute, aristate. Capitula 5.5–10 × 6–9.5 mm, conical, with peduncle 2–10 mm; involucral bracts 2–4 × 1.5–2 mm, ovate to oblong or obovate, obtuse, the midvein green to brown, with scarious margin. Florets 1.4–1.6 mm, campanulate distally, greenish yellow. Achene 1.5–2(–2.2) × *c.*0.5 mm, cylindrical, inconspicuously 3– to 4– ribbed, pale brown. *Flowers December to May.*

Coastal sands and gravels, disturbed soil by roads and habitations; 0–100 m.

Isla Grande, except extreme west and east, mainly in coastal areas.
Introduced from Europe; a native of Asia widely naturalized in temperate regions.

208 209
210 211

Fig. 208. *Chiliotrichum diffusum*. Fig. 209. *Gamochaeta americana*. Fig. 210. *Senecio candidans*. Fig. 211. *Senecio tricuspidatus*.
(Sydney Parkinson)

22. *Tanacetum* L.

Perennial herbs. Leaves alternate, 1– to 2– pinnatisect. Capitula in corymb; receptacle convex, without bracts; involucral bracts in 3 series. Florets dimorphic or monomorphic; outer ligulate to tubular, 3– or 5– dentate, female or hermaphrodite; inner tubular, 5-dentate, hermaphrodite. Pappus a short, membranous rim.

a. Ligules yellow or absent; capitula not more than 10 mm in diameter **1. *vulgare***
a. Ligules white; capitula at least 15 mm in diameter **2. *parthenium***

★**1. *T. vulgare*** L. 1753, *Sp. Pl.* 844.

Plant aromatic, rhizomatous; stems up to 67(–93) cm, erect, angled, branched above, subglabrous, reddish. Leaves 12–14 × 4–10 cm, oblong in outline, 2-pinnatisect, glandular, dark green, the lower petiolate, the upper sessile, semi-amplexicaul; primary lobes up to 5.5 × 2 cm, lanceolate-ovate; ultimate segments up to 1.7 × 0.7 cm, lanceolate, dentate or with dentate lobes. Capitula 4–7 × 7–10 mm, discoid, numerous in dense, compound corymb; involucral bracts 2.7–3.5 × 1–1.8 mm, ovate-lanceolate, acute, pale green with scarious margin, the outer shorter and obtuse. Florets *c.*2 mm, yellow; outer 3-dentate, or with short ligule, female. Achene 1.2–1.8 mm, cylindrical-obovoid, 5-ribbed, glabrous, with sparse, sessile, transparent glands. *Flowers January to March (to May).*

Open sand and soil, disturbed areas near habitations; 0–50 m.

Local in parts of Isla Grande.
Introduced from Europe for cultivation as an ornamental and pot-herb, locally naturalized or persisting in abandoned gardens.

★**2. *T. parthenium*** (L.) Sch. Bip. 1844, *Tanacet.* 55.
Chrysanthemum parthenium (L.) Bernh. 1800, *Syst. Verz. Erfurt.* 145. – Cabrera, 1971.

Plant aromatic; stems 20–40 cm, erect, branched above, ridged, ± hairy. Lower leaves 4.5–6.8 × 1.7–3.5 cm, ovate in outline, 1– to 2– pinnatisect, with petiole 1.8–3 cm; primary lobes 1.5–2.2 × *c.*1.2 cm, ovate, pinnatisect to crenate or entire; upper leaves smaller, less divided. Capitula 15–20 mm in diameter, in lax corymbs; involucral bracts 2.5–3.5 × 0.6–0.7 mm, lanceolate to oblong, acute, lanate, green with scarious margin. Outer florets white, with ligule 4.5–7.5 × 2–2.8(–3.7) mm, elliptic-oblong; inner florets yellow. Achene 1.2–1.5 mm, cylindrical to narrowly ovoid, 5– to 8– ribbed, glabrous. *Flowers February to March.*

Coastal sands and disturbed soil near habitations; 0–20 m.

C. S. coast of Isla Grande.
Introduced from Europe for cultivation as an ornamental, locally naturalized.

23. *Leucanthemum* Miller

Perennial herbs. Leaves alternate, simple. Capitula solitary; receptacle convex, without bracts; involucral bracts in 2–3 series. Florets dimorphic; outer ligulate, female; inner tubular, 5-dentate, hermaphrodite, forming a disc. Pappus a short, membranous rim, on achenes of outer florets only.

★**1. *L. vulgare*** Lam. 1779. *Fl. Fr.* **2**: 137.
Chrysanthemum vulgare L. 1753, *Sp. Pl.* 888 – Ruiz-Leal, 1954; Cabrera, 1971.

Plant with stout, ± woody oblique stock bearing non-flowering leaf-rosettes and flowering stems; stems 10–46 cm, erect, simple or branched, ± ridged, hairy. Basal leaves 3–4(–8.5) × 0.5–1.5 cm, spathulate, obtuse, narrowed to petiole-like base, crenate-dentate to ± lobed, glabrous or sparsely hairy above; cauline leaves smaller, the upper oblong. Capitula 25–45 mm in diameter; involucral bracts 1.6–6 × 1.1–1.8 mm, lanceolate, obtuse, irregularly serrate-crenate, green with brownish scarious margin. Outer florets white, with ligule 12–20 × 3–5 mm, oblanceolate to elliptical; inner florets yellow. Achene 1.2–1.7 mm, cylindrical, dark brown, with *c.*10 paler ribs, glabrous. *Flowers January to May.*

Open, disturbed grassland; 0–30 m.

Local in parts of Isla Grande and N. side of Isla Navarino.
Introduced from Europe, probably with grass seed, widely naturalized in temperate regions.

24. *Cotula* L.

Dioecious, perennial herbs. Leaves alternate, pinnatifid. Capitula solitary, terminal and axillary; receptacle ± convex, without bracts; involucral bracts in 2–3 imbricate series. Florets monomorphic, tubular-infundibuliform, 4-lobed, yellow-green. Pappus absent.

Literature: D. G. Lloyd, *N. Z. Jour. Bot.* **10**: 277–372(1972).

1. *C. scariosa* (Cass.) Franchet 1889, *Miss. Sci. Cap Horn* **5**: 344.

C. acaenoides (Hooker & Arn.) Alboff 1902, *Anal. Mus. La Plata Bot.* **1**: ix – Cabrera, 1939.

Leptinella scariosa Cass. 1822, *Bull. Soc. Philom. Paris* **1822**: 127 – Hooker, 1847.

Stems up to 15 cm, prostrate or ascending distally, branched, rooted at lower nodes, sparsely hairy. Leaves with lamina 12–38 × 5–13(–27) mm, oblong to oblong-lanceolate in outline, sparsely appressed-hairy; lobes 7–11 pairs, 2.5–4 × 1.5–2 mm, ovate to ovate-lanceolate, acute, entire or serrate; petiole 5–25(–47) mm, sparsely hairy, sheathing at base. Capitula 5–11 mm in diameter; peduncle 1.9–7.5(–9.5) cm; involucral bracts 3–4.5 × 1.8–3 mm, ovate to ovate-oblong, obtuse, entire to fimbrio-dentate, sparsely hairy, purplish distally, scarious on margin. Female florets 1–1.2 mm, the stigma bifid; male florets 2–3 mm, the stigma undivided. Achene 1.5–2 mm, cylindrical – oblanceolate, compressed, glabrous, yellow-brown. $2n = 260$. *Flowers November to February*. (Fig. 205).

Moist ground by streams and marshes, wet coastal sands and gravels; 0–50 m.

Almost throughout Fuegia. (Map 25h).
W. Argentina north to *c.*41°S.lat.
Chile north to *c.*40°S.lat.
Falkland Islands (Islas Malvinas).
Islas Diego Ramirez.

25. *Abrotanella* Cass.

Perennial, moss-like herbs, often cushion-forming. Leaves alternate, simple. Capitula small, solitary, terminal; receptacle ± convex, without bracts; involucral bracts few, in 1–2 series. Florets tubular, indistinctly dimorphic; outer 2–4, 3-dentate, female; inner 2–8, 4-lobed, hermaphrodite or functionally male. Pappus a short, dentate or aristate, membranous rim.

a. Leaves bifid and with conspicuous scarious margin at apex; achene without pappus **1. *emarginata***

a. Leaves obtuse to retuse and with or without a narrow scarious margin at apex; achene with pappus:

b. Leaves linear, truncate or retuse and with narrow scarious margin at apex; achene ± hairy at base **2. *submarginata***

b. Leaves linear-spathulate, obtuse and without scarious margin at apex; achene glabrous **3. *linearifolia***

1. *A. emarginata* (Cass. ex Gaudich.) Cass. 1825, *Dict. Sci. Nat.* **36**: 27.

Oligosporus emarginatus Cass. ex Gaudich. 1825, *Ann. Sci. Nat.* **5**:104.

Forming small, usually compact cushions or mats; stems 2–4 cm, ascending to erect, branched, densely clothed with leaf-remains below. Leaves 2–5 × 1–1.5 mm, oblong, truncate and bifid at apex, sheathing at base, glabrous, with conspicuous scarious margin, particularly at apex, green, usually purple on sheath, coriaceous. Capitula sessile, among upper leaves; involucral bracts *c.*2.5 × 0.8 mm, oblong-lanceolate, truncate, ± bifid at apex, purplish. Florets 2–3 mm, dark purplish-brown distally, the tube pale yellow in male and narrower, pale green to yellow, in female. Achene *c.*1 mm, oblong-ellipsoid, ± 4-angled, glabrous, pale brown; pappus absent. $2n = 18.^{\star}$ *Flowers October to January*. (Fig. 215).

Empetrum heath, *Bolax* communities, Magellanic moorland, feldmark; (100–)250–1000 m.

C. & S. Fuegia. (Map 25i).
Chile north to *c.*50°S.lat.
Falkland Islands (Islas Malvinas).

2. *A. submarginata* A. Gray 1862, *Proc. Amer. Acad. Arts* **5**: 137.

Similar to *A. emarginata* but of somewhat laxer habit; leaves truncate or slightly retuse and with narrow scarious margin at apex; achene ± hairy at base and with pappus. *Flowers November to January*.

Moist rocks and pebbles near high water mark; 0–10 m.

Southernmost Fuegia. (Map 25j).
Chile north to 51°30′S.lat.

3. *A. linearifolia* A. Gray 1862, *Proc. Amer. Acad. Arts* **5**: 137.

Plant with slender rhizomes bearing ascending to erect stems up to 2 cm. Leaves 4–5 × *c.*1 mm, linear-spathulate, obtuse, ± sheathing at base, entire, glabrous. Capitula terminal, shortly pedunculate; involucral bracts 1–1.2 × 0.5–0.7 mm, oblong-elliptical, obtuse, grey brown, scarious. Florets 1.5–2 mm, the inner hermaphrodite. Achene *c.*1 mm, cylindrical, glabrous; pappus with few teeth or aristae. *Flowers December to February*. (Fig. 212).

Wet places in feldmark, mosses and grass along streams; 500–1000 m.

S. Fuegia. (Map 26a).
Chile north to 51°S.lat.

26. *Artemisia* L.

Perennial herbs, often woody at base, or small shrubs. Leaves alternate, 1– to 3– pinnatisect. Capitula small, usually pendent, in panicles or solitary and axillary to form a spike-like inflorescence; receptacle convex, without bracts, sometimes hairy; involucral bracts in few series. Florets indistinctly dimorphic; outer filiform, 3-dentate, female; inner tubular, 5-dentate, hermaphrodite. Pappus absent.

a. Capitula solitary in axils of upper leaves to form a long, spike-like inflorescence **1. magellanica**
a. Capitula in a dense panicle **2. *absinthium***

1. *A. magellanica* Sch. Bip. 1855, *Flora (Regensb.)* **38**: 116.

Stems 5–30 cm, ascending to erect, pubescent to glabrous, leaves 4–6 × 1–2.5 cm, elliptical to elliptic-oblanceolate in outline, 2– to 3– pinnatisect; ultimate segments short, linear-lanceolate; petiole about as long as lamina. Capitula 2–5 × 2–9 mm, sessile in axils of upper leaves to form a long, spike-like inflorescence; involucral bracts 4–5 × 1.8–2 mm, oblong-obovate, obtuse, irregularly denticulate at apex, with wide scarious margin; receptacle glabrous. Florets 1.2–2 mm, pale yellow, purplish distally. Achene *c.*1.2 × 0.5 mm, obovoid, glabrous, brown. $2n = 22$. *Flowers January to March.* (Fig. 217).

Sands and gravels, usually coastal; 0–50 m.

N. E. Isla Grande. (Map 26b).
W. Argentina north to *c.*35°S.lat.
Chile on north shore of Estrecho de Magallanes.

2. *A. absinthium L. 1753, *Sp. Pl.* 848.

Stems 50–60 cm, woody at base, erect, densely silver sericeous. Leaves 2.5–5 × 2–3.5 cm, obovate-cuneiform in outline, 1– to 2– pinnatisect, with petiole 1–2 cm; ultimate segments *c.*1.5–3 mm wide, lanceolate to linear-oblong, usually obtuse, densely sericeous. Capitula 2–3 × 3–5 mm, pendent, in a panicle; involucral bracts 2.5–3 × *c.*2 mm, obtuse, the outer oblong, the inner ovate, greenish with a wide scarious margin; receptacle hairy. Florets 1–1.2 mm, pale yellow, purplish distally. Achene *c.* 1.5 mm, obovoid, glabrous, brownish. *Flowers April.*

Distributed and degraded grassland; *c.*300 m.

Local in C. Isla Grande.
Introduced from Europe, probably for cultivation as a condiment, and locally adventive.

A. abrotanum L. 1753, *Sp. Pl.* 845, widely cultivated in Europe for ornament and flavouring, is known in and near some estancia gardens in N. E. Isla Grande. It is a strongly aromatic shrub up to *c.* 35 cm, with 1– to 3– pinnatisect leaves having filiform lobes, greyish-pubescent beneath, and a panicle of globose capitula with glabrous receptacle and yellow florets.

Tribe SENECIONEAE Florets tubular or outer with ligules; outer florets usually female, the inner hermaphrodite or functionally male. Stigmatic branches truncate.

27. *Senecio* L.

Perennial, rarely annual, herbs, sometimes woody at base, or shrubs. Leaves alternate, entire to variously divided. Capitula solitary or in corymbose cymes; receptacle flat, without bracts; involucral bracts in 1 series, often with some shorter supplementary bracts (calyculus) outside. Florets monomorphic or dimorphic; outer female, with ligules (capitula radiate) or short and filiform; inner hermaphrodite, tubular, 5-dentate or -lobed. Pappus of various series of strigulose hairs.

A genus in which several species show considerable, and often intergrading, variation in Fuegia. Certainly some of the variation is related to environmental differences, but interspecific hybridization is also possible. Because of these factors plants showing affinities with combinations of species **4–7, 9–10, 12** and **14**, can be very difficult to identify with confidence.

Literature: A. L. Cabrera, *Lilloa* **15**: 27–501(1949); *in* M. N. Correa (Ed.) *Fl. Patag.* **7**: 176–275(1971).

a. Capitula in corymbose cymes:
 b. Capitula radiate:
  c. Leaves orbicular or reniform to flabellate or spathulate:
   d. Leaves at least 20 × 20 mm, densely white-tomentose beneath **17. *websteri***
   d. Leaves not more than 20 × 4 mm, glabrous **19. *philippii***
  c. Leaves ovate to oblong-ovate:
   e. Leaves lobed, glabrous or sparsely pubescent beneath **16. *acanthifolius***
   e. Leaves dentate, glabrous above, densely white-tomentose beneath **18. *smithii***

b. Capitula discoid:
f. Leaves at least 4 cm wide **2. *candidans***
f. Leaves not more than 3 cm wide:
g. Leaves coarsely pinnatifid **13. *vulgaris***
g. Leave entire to dentate:
h. Leaves glabrous or somewhat glandular-pubescent:
i. Leaves entire or indistinctly dentate **15. *arnotti***
i. Leaves usually with 1–3 prominent teeth on each side **10. *tricuspidatus***
h. Leaves densely tomentose, at least beneath;
j. Leaves linear to linear-lanceolate; involucre not more than 8 mm in diameter **14. *patagonicus***
j. Leaves linear-oblanceolate to oblanceolate-spathulate; involucre at least 10 mm in diameter:
k. Leaves obtuse, entire or with 1–2 wide teeth distally on each side **7. *leucomallus***
k. Leaves acute, usually with 1–3 prominent teeth on each side **10. *tricuspidatus***

a. Capitula solitary on ends of sometimes scape-like branches:
l. Capitula radiate:
m. Ligules white; stems scape-like **12. *trifurcatus***
m. Ligules yellow; stems not scape-like:
n. Leaves deeply 3-fid, with linear segments **8. *eightsii***
n. Leaves entire or with 2–3 teeth on each side, obovate– to oblanceolate– spathulate **3. *darwinii***
l. Capitula discoid:
o. Leaves glabrous or glandular-pubescent:
p. Herb; leaves irregularly incised or dentate, the lobes often dentate; marginal florets female **11. *kingii***
p. Shrub; leaves entire or with 1–3 pairs of teeth; marginal florets hermaphrodite:
q. Leaves 0.5–2.5 mm wide, glandular-pubescent; involucre 6–8 × 8–9 mm **9. *miser***
q. Leaves 2–5 mm wide, glabrous to somewhat glandular; involucre 8–10 × 10–11 mm **10. *tricuspidatus***
o. Leaves tomentose to pubescent, at least beneath;
r. Leaves linear-lanceolate, with persistent sheathing base; peduncle at least (5–) 8 cm, scape-like **1. *magellanicus***
r. Leaves oblanceolate to oblanceolate-spathulate, without persistent sheathing base; peduncle not more than 4 cm, not scape-like:
s. Involucre not more than 7 × 8 mm; leaves not more than 2 mm wide:
t. Leaves acute; marginal florets hermaphrodite **4. laseguei**
t. Leaves obtuse; marginal florets female **5.*humifusus***
s. Involucre at least 9 × 10 mm; leaves at least 3 mm wide:
u. At least upper leaves ± deeply 3– fid; bracts (calyculus) at base of capitulum $\frac{1}{2}$–$\frac{2}{3}$ as long as involucral bracts **6. *alloeophyllus***
u. Leaves entire; or with 1–2 wide teeth on each side distally; bracts of calyculus not more than $\frac{1}{2}$ as long as involucral bracts **7. *leucomallus***

1. *S. magellanicus* Hooker & Arn. 1841, *Jour. Bot. (Hooker)* 3: 343.
Culcitium magellanicum (Hooker & Arn.) Hombron & Jacquinot 1845, Voy. Pôle Sud. Bot. tab. 11 – Hooker, 1847; Dusén, 1900; Skottsberg, 1916.

Perennial herb with oblique rhizome; stems (5–)8–18 cm, ascending to erect, branched, covered with leaf-remains towards base, densely leafy above and with reduced, distant leaves to give long scape-like apex, sericeous. Leaves 20–120 × 1–4 mm, linear-lanceolate, acute, attenuate at base and then expanded into basal sheath, entire, somewhat revolute, densely sericeous. Capitula 10–22 mm in diameter, discoid, solitary and terminal on scape-like stem – apices; involucral bracts 6–9 × 1–1.5 mm, linear-lanceolate, acute, densely velutinous; calyculine bracts as long as involucre, linear, acute, densely sericeous. Florets monomorphic, yellow, 6–8 mm, tubular, hermaphrodite. Achene 3.5–4 mm, cylindrical, ribbed, glabrous; pappus *c.*7 mm, white $2n = 80$. *Flowers December to March.*

Grassland, open *Empetrum* heath and *Chiliotrichum* scrub, feldmark; absent from woodland; 0–1100 m.

Much of Fuegia except the extreme west, south and east. (Map 26c).
Argentina north to *c.*50°S.lat.
E. Chile north to *c.*51°S.lat.

2. *S. candidans* DC. 1838, *Prodr.* 6: 412.
S. candicans auct., non Wall. – Franchet, 1889; Skottsberg, 1916.

Perennial herb with stout, horizontal to oblique rhizome; stems up to 1 m, erect, branched near base, densely white lanate. Basal leaves with lamina

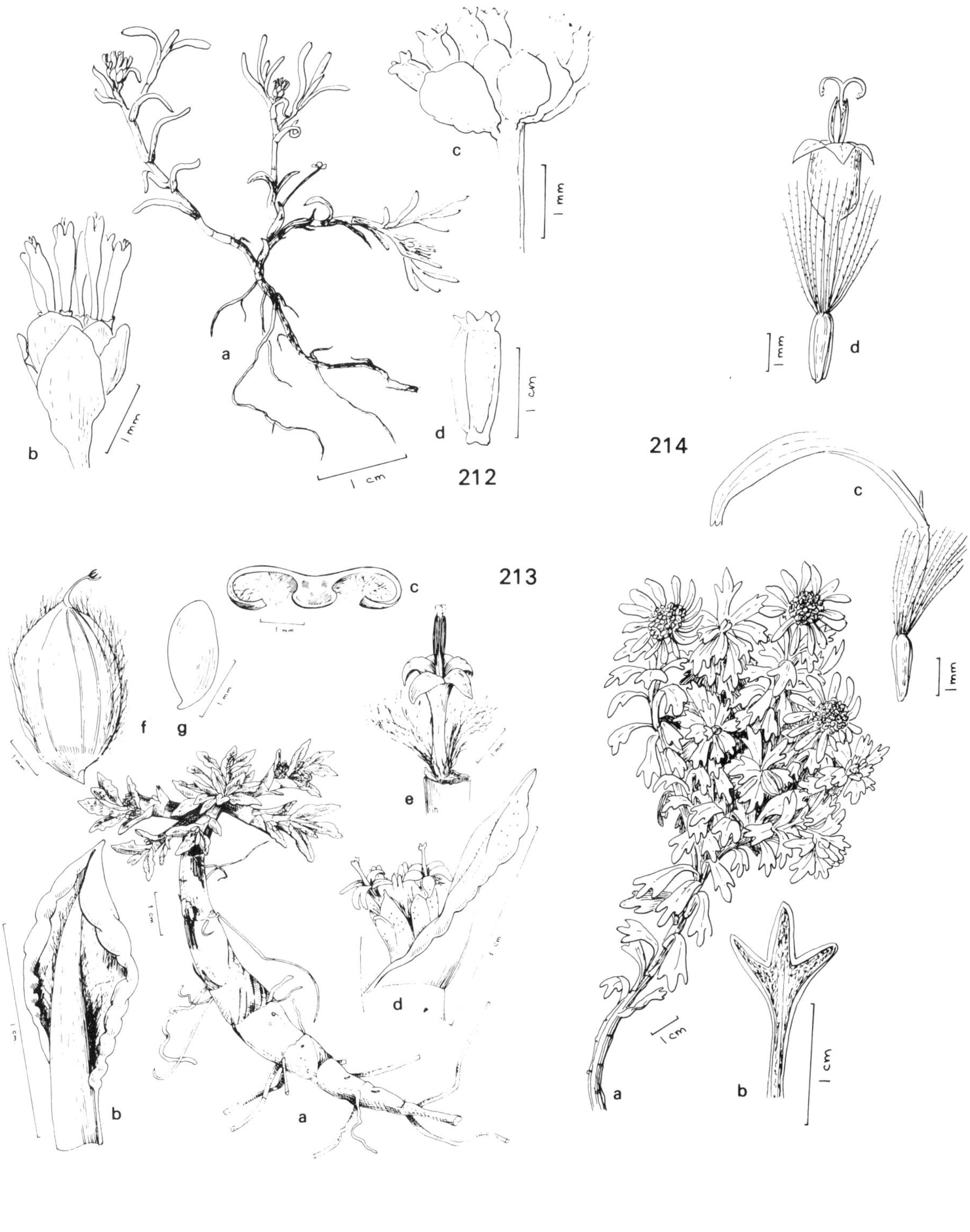

Fig. 212. *Abrotanella linearifolia*. a, Plant. b, Flowering capitulum. c, Fruiting achene. d, Achene. Scale = 1 cm (a, d) or 1 mm (b–c). Fig. 213. *Eriachaenium magellanicum*. a, Plant. b, Leaf. c, Leaf-section. d, Capitulum. e, Floret. f, Achene. g, Seed. Scale = 1 cm (a–b, d) or 1 mm (c, e–g). Fig. 214. *Senecio eightsii*. a, Plant. b, Leaf. c, Marginal floret. d, Disc floret. Scale = 1 cm (a–b) or 1mm (c–d).

(R. N. P. Goodall)

40–150 × 20–95 mm, ovate to ± suborbicular, obtuse, cuneate or truncate at base, crenate-serrate, rarely entire, densely white – lanate; petiole 5–12 cm, sheathing at base; upper leaves smaller, ± lanceolate, entire, sessile. Capitula 15–25 mm in diameter, discoid, in terminal or axillary corymbose cymes, the peduncles 1–6 cm, white-lanate, with few linear-lanceolate bracts; involucral bracts 6–10 × 1.5–3 mm, ovate to lanceolate-oblong, acute to acuminate, entire to serrate, green, with wide pale scarious margin, densely lanate towards base; calyculine bracts short. Florets monomorphic, yellow, 6–8 mm, tubular, hermaphrodite. Achene 7–8 mm, oblong, obtuse, indistinctly 8- to 10-ribbed, glabrous; pappus 7–8 mm, white. $2n = 40$. *Flowers November to January*. (Fig. 210).

Coastal sands and gravels; 0–10 m.

Coasts of Fuegia west to 70°30′W. long., except the extreme south. (Map 26d).
E. Argentina north to *c.*41°30′S.lat.
Chile north to *c.*37°20′S.
Falkland Islands (Islas Malvinas).

3. *S. darwinii* Hooker & Arn. 1841, *Jour. Bot. (Hooker)* **3**; 333.

Perennial herb, woody at base; stems 8–25 cm, procumbent to ascending, densely white lanate. Leaves 13–30(–45) × 4–10 mm, oblanceolate- to obovate-spathulate, obtuse, long attenuate at base, with 2–3 teeth on each side, rarely subentire, revolute, sparsely hairy to subglabrous above, densely white lanate beneath. Capitula 15–35 mm in diameter, radiate, solitary and terminal on branches; involucral bracts 8–12 × 0.8–1.2 mm, linear, acute to acuminate, entire, subglabrous, green, black distally; calyculine bracts short, linear, sparsely lanate. Florets dimorphic, yellow; marginal female, with ligule 10–12 × 2–5 mm, oblong-elliptical, the tube 4–5 mm; inner 8–9 mm, tubular, hermaphrodite, forming a disc. Achene 3–4 mm, cylindrical, ribbed, glabrous; pappus 7–8 mm, white. *Flowers December to March.*

Wet sand and open soil between rocks; 400–1000 m.

S. Fuegia. (Map 26e).
Chile north to 50°59′S.lat.

S. argyreus Phil. 1894, *Anal. Univ. Chile* **88**: 261, which occurs in W. Argentina and E. Chile between *c.*40° and 51°S.lat., has been recorded once from near Lago Fagnano in S. C. Isla Grande (Skottsberg, 1926). It is similar to *S. darwinii* but may be distinguished by its narrower (3–4 mm wide), entire leaves which have dense branched sericeous hairs above and beneath, and densely sericeous involucral bracts.

4. *S. laseguei* Hombron & Jacquinot 1846, *Voy. Pôle Sud. Bot.* tab. 13–D.
S. philippii auct., non Sch. Bip. ex Wedd. – Dusén, 1900.

Perennial herb; stems 3–11 cm, procumbent to ascending, glabrous and ± leafless below, white-tomentose and densely leafy above. Leaves 5–11 × 1–2 mm, oblanceolate, acute, cuneate at base, entire or 3-dentate distally, tomentose to subglabrous. Capitula 7–12 mm in diameter, discoid, solitary and terminal on branches; involucral bracts 5–6 × 1–1.5 mm, linear-lanceolate, acute, entire, tomentose to subglabrous, green with pale scarious margin; calyculine bracts short, linear, tomentose. Florets monomorphic, yellow, 5.5–7 mm, tubular, hermaphrodite. Achene *c.*2 mm, cylindrical, ribbed, glabrous to pubescent; pappus 4–7 mm, white. $2n = 40$. *Flowers December to January.*

Empetrum heath, open *Chiliotrichum* scrub; 10–200 m.

C. & N. Isla Grande. (Map 26f).
E. Argentina north to *c.*52°S.lat.
E. Chile north to 50°40′S.lat.

5. *S. humifusus* (Hooker f.) Cabrera 1969, *Bol. Soc. Argent. Bot.* **11**: 281.
Melalema humifusa Hooker f. 1847, *Fl. Antarct.* **1(2)**: 311.

Perennial herbs; stems 3–10 cm, procumbent to ascending or erect, ± densely tomentose, densely leafy throughout. Leaves 8–12 × 2–5 mm, oblanceolate-spathulate, obtuse, somewhat attenuate at base, entire or the lower 3-dentate, densely silver-tomentose. Capitula 8–15 mm in diameter, discoid, solitary and terminal on branches; involucral bracts 7–9 × 1–1.5 mm, oblong to ± lanceolate, acute to subobtuse, entire or denticulate distally, tomentose, green, blackish distally; calyculine bracts numerous, linear, densely tomentose, the inner as long as involucre. Florets indistinctly dimorphic, yellow; outer 2–2.5 mm, tubular, female; inner 4.5–6 mm, tubular, hermaphrodite. Achene 2.5–3.5 mm, cylindrical, ribbed, glabrous; pappus 4–5 mm, white. $2n = 120$. *Flowers December to February.*

Feldmark, screes, montane *Empetrum* heath; 500–1100 m.

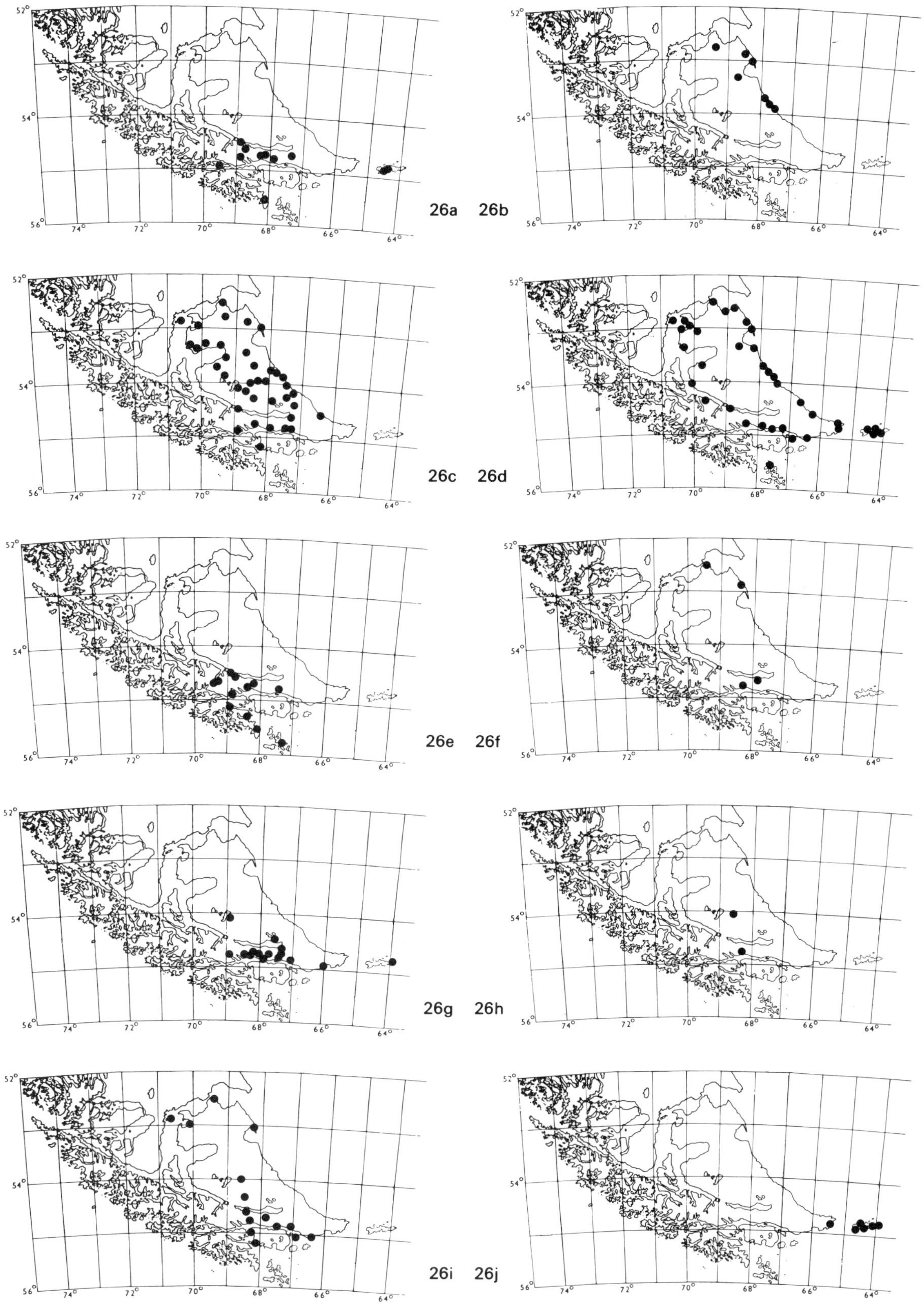

Map 26. a, *Abrotanella linearifolia*. b, *Artemesia magallanica*. c, *Senecio magellanicus*. d, *Senecio candidans*. e, *Senecio darwinii*. f, *Senecio laseguei*. g, *Senecio humifusus*. h, *Senecio alloeophyllus*. i, *Senecio leucomallus*. j, *Senecio eightsii*.

S. Isla Grande. (Map 26g).
Endemic.

6. *S. alloeophyllus* O. Hoffm. 1900, in Dusén, *Svenska Exped. Magellansl.* **3(5)**: 105.

Perennial herb, woody at base; stems 8–15(–20) cm, ascending, sparsely tomentose to subglabrous, densely leafy throughout. Leaves 12–25(–30) × 3–10 mm, oblanceolate, acute, attenuate at base, ± deeply 3-fid, the lower sometimes entire, revolute, densely white-tomentose, sometimes subglabrous above. Capitula 10–20 mm in diameter, discoid, solitary and terminal on branches; involucral bracts 9–11 × 0.5–2 mm, lanceolate, acuminate, entire, sparsely lanate, green with pale scarious margins, blackish and with brown hairs at apex; calyculine bracts up to $\frac{1}{2}$–$\frac{2}{3}$ as long as involucre, linear, ± lanate to subglabrous. Florets monomorphic, yellow, 6–7 mm, tubular, hermaphrodite. Achene 4–5 mm, cylindrical, ribbed, glabrous; pappus 6–7 mm, white. $2n = c.80$. *Flowers (October–) December to March.*

Loose shale and soil screes, cliffs, *Empetrum* heath, *Bolax* communities, usually in feldmark; (0–)250–1100 m.

S. Isla Grande. (Map 26h).
W. Argentina north to *c.*50°S.lat.
E. Chile north to *c.*51°S.lat.

Sometimes vegetatively similar to *S. humifusus*, with which it often occurs, but distinguished by the hermaphrodite outer florets.

7. *S. leucomallus* A. Gray 1862, *Proc. Amer. Acad. Arts* **5**: 141.
S. hyadesii Franchet 1889, *Miss. Sci. Cap Horn*, 348.
S. micropifolius auct., non DC. – Alboff, 1896.

Perennial herb, woody at base, or low shrub; stems up to 17 cm, ascending to erect, branched, densely white-tomentose and leafy throughout. Leaves 20–35 × 3–10 mm, oblanceolate-spathulate, obtuse, attenuate at base, entire or with 1–2 wide teeth on each side, often revolute, densely white-tomentose. Capitula 10–18 mm in diameter, discoid, solitary and terminal on branches or 2–4 in lax cyme with peduncles 1–4 cm; involucral bracts 8–10 × 1.5–2 mm, linear-oblong, acuminate, entire, densely tomentose, green with scarious margin and blackish at apex; calyculine bracts up to $\frac{1}{2}$ as long as involucre, linear, tomentose. Florets monomorphic, yellow, 6–7 mm, tubular, hermaphrodite. Achene *c.* 4 mm, cylindrical, ribbed, glabrous; pappus 6–8 mm, white. *Flowers November to January.*

Montane rocks; 350–500 m.

S. Fuegia. (Map 26i).
W. Argentina north to *c.*50°S.lat.
E. Chile north to 50°56′S.lat.

8. *S. eightsii* Hooker & Arn. 1841, *Jour Bot. (Hooker)* **3**: 332.

Dwarf shrub; stems 7–28 cm, procumbent to ascending, much branched, glabrous, but young branches white-lanate, leafy distally. Leaves deeply 3-fid, glabrous above, white tomentose beneath; basal part 5–18 × 0.5–1 mm, linear; segments 2–7 × 0.5 mm, linear, subobtuse, revolute, Capitula 16–20 mm in diameter, radiate, solitary and terminal on branches, with peduncle up to 2 cm; involucral bracts 6–8 × 1–1.2 mm, linear-oblanceolate, acuminate, entire, pale green with whitish scarious margin, blackish distally, glabrous to ± lanate; calyculine bracts *c.*$\frac{1}{2}$ as long as involucre, linear, ± lanate. Florets dimorphic, yellow; marginal female, with ligule 6–7 × 1.2–2 mm, oblong, the tube *c.*5 mm; inner 5.5–7 mm, tubular, hermaphrodite, forming a disc. Achene *c.*3.5 mm, cylindrical, ribbed, glabrous; pappus *c.*5 mm, white. *Flowers October to December.* (Fig. 214).

Open peat, rocks, wet grassland, *Chiliotrichum* scrub, dwarf shrub heath; 0–460 m.

Easternmost Isla Grande; Isla de los Estados. (Map 26j).
Endemic.

9. *S. miser* Hooker f. 1847, *Fl. Antarct.* **1(2)**: 314.
S. fueginus Phil. 1894, *Anal. Univ. Chile* **88**: 26.

Perennial herb, woody at base; stems 6–25 cm, ascending to erect, much branched, ± densely glandular-pubescent, leafy almost throughout. Leaves 8–32 × 0.5–3 mm, linear to linear-spathulate, acute, attenuate at base, entire or with 1–3 pairs of short teeth or lobes, ± revolute, glandular-pubescent and ± viscid. Capitula 9–12 mm in diameter, discoid, solitary and terminal on branches; involucral bracts 5–8 × 0.7–1 mm, linear-lanceolate, acute, ± glandular-pubescent, green, purplish distally; calyculine bracts short, linear. Florets monomorphic, yellow, 5–6 mm, tubular, hermaphrodite. Achene *c.*3 mm, cylindrical-oblanceolate, ribbed, glabrous; pappus 4–5 mm, white. *Flowers December to March.*

Coastal sands and gravels, open soil in grassland and *Empetrum* heath; 0–300 m.

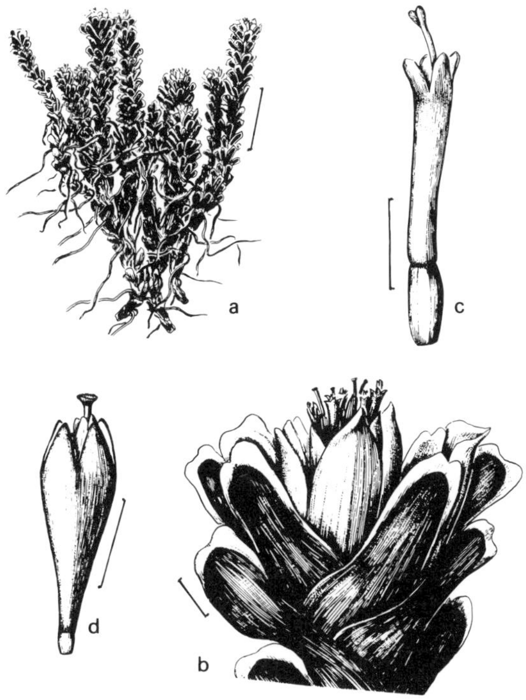

215

216

217

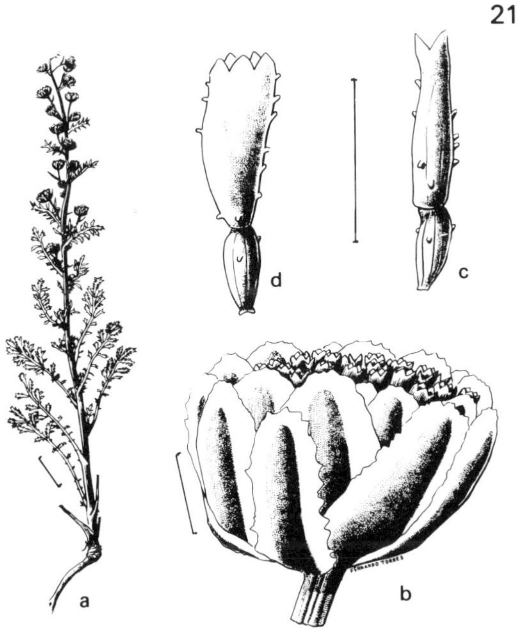

218

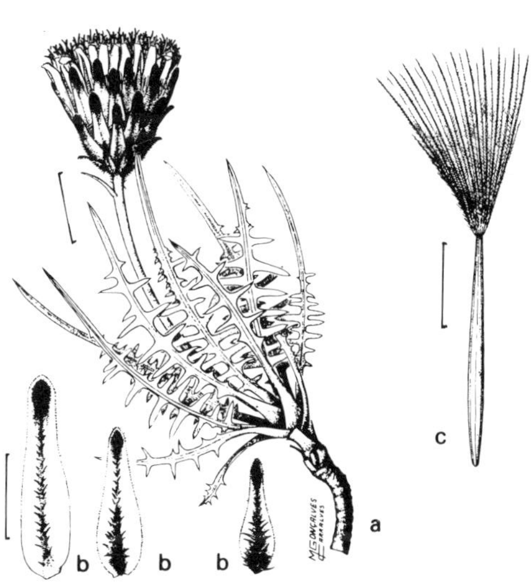

Fig. 215. *Abrotanella emarginata*. a, Plant. b, Capitulum. c, Female flower. d, Male flower. Scale = 1 cm (a) or 1 mm (b–d). Fig. 216. *Macrachaenium gracile*. a, Plant. b, Capitulum. c, Marginal floret. d, Disc floret. Scale = 1 cm (a), 5 mm (b) or 1 mm (c–d). Fig. 217. *Artemisia magellanica*. a, Plant. b, Capitulum. c, Marginal floret. d, Disc floret. Scale = 1 cm (a) or 1 mm (b–d). Fig. 218. *Hypochoeris incana*. a, Plant. b, Involucral bracts. c, Achene. Scale = 1 cm (a–b) or 5 mm (c)
(Flora Patagónica)

N. & C. S. Isla Grande. (Map 27a).
Argentina north to *c.*44°S.lat.
E. Chile north to 50°42′S.lat.

May be confused with narrower leaved variants of *S. tricuspidatus* but usually separable by its smaller, more numerous capitula and more densely glandular-pubescent leaves, which are often viscid.

10. *S. tricuspidatus* Hooker & Arn. 1841, *Journ. Bot. (Hooker)* **3**: 347.
S. longipes Hooker f. 1847, *Fl. Antarct.* **1(2)**: 314.
S. dumosus Phil 1894, *Anal. Univ. Chile* **88**: 24.
S. nordenskjöldii O. Hoffm. 1900, in Dusén, *Svenska Exped. Magellansl.* **3(5)**: 107.

Dwarf shrub up to 36 cm, the branches ascending to erect, glabrous to glandular-hairy. Leaves 15–45 × 2–8 mm, linear-oblanceolate, acute, attenuate at base, usually with 1–3 pairs of prominent lateral teeth, ± revolute, glabrous to glandular-pubescent. Capitula 10–20 mm in diameter, discoid, solitary and terminal on branches or 2–7 in lax cyme with peduncles 4–5 cm; involucral bracts 8–11 × 1.5–2 mm, lanceolate, acuminate, entire, glabrous to glandular-pubescent, green, with pale scarious margin and small black mark at tip; calyculine bracts short, linear. Florets monomorphic, yellow, 7.5–9 mm, tubular, hermaphrodite. Achene 4–5 mm, cylindrical, ribbed, glabrous; pappus 6–7.5 mm, white. $2n = 40, 80, 120$. *Flowers October to January.*

Open deciduous forest, *Chiliotrichum* scrub, dwarf-shrub heath, grassland, coastal sands; 0–500 m.

Almost throughout Isla Grande; Isla de los Estados. (Map 27b).
W. Argentina north to 45°30′S.
E. Chile north to 50°49′S.lat.

Very variable in pubescence and number of capitula. Densely pubescent plants have been ascribed to var. *dumosus* (Phil.) Cabrera, while those with cymes of 5 or more capitula have been termed var. *dentatus* (Sch. Bip.) Cabrera. The value of recognizing such varieties is doubtful, in view of the rather continuous pattern of variation shown by these characters in Fuegian material and the lack of correlation with the chromosome races. Possibly hybridization with *S. miser* and/or *S. patagonicus* has resulted in some of this variation, as well, perhaps, as explaining the specimens which can blur the species' delimitations.

11. *S. kingii* Hooker f. 1847, *Fl. Antarct.* **1(2)**: 314.
S. kurtzii Alboff 1896, *Rev. Mus. La Plata Bot.* **7**: 379.
S. paradoxus Alboff ex F. Kurtz 1900, *Bol. Acad. Nac. Ci.* **16**: 235.

Glabrous, perennial herb, with branched, oblique rhizome; stems 6–14 cm, ascending to suberect, densely leafy at base, scape-like above, Leaves 13–25 × 3–7 mm, linear-spathulate in outline, acute, attenuate at base, deeply and irregularly lobed or dentate, the larger lobes serrate-dentate, somewhat fleshy, pale green, Capitula 8–18 mm in diameter, discoid, solitary and terminal on scape-like branches; involucral bracts 8–11 × 1–1.5 mm, linear-oblong, acute; calyculine bracts short, linear. Florets indistinctly dimorphic, yellow; outer 7–8 mm, tubular, female; inner 8–9 mm, tubular, hermaphrodite. Achene 5–6 mm, cylindrical, ribbed, glabrous; pappus 6.5–8 mm, white. *Flowers November to January.*

Open sands and soil screes; 0–900 m.

Almost throughout Isla Grande, except the extreme west and east, usually in coastal areas. (Map 27c).
Argentina north to *c.*41°S.lat.
Chile north to 50°41′S.lat.

Plants with less deeply divided leaves have been recognized as var. *paradoxus* (Alboff) Cabrera; since they occur with typical forms the value of this variety must be doubtful.

12. *S. trifurcatus* (Forster f.) Less. 1832, *Syn. Gen. Comp.* 392
Tussilago trifurcata Forster f. 1789, *Comment. Gotting.* **9**: 38.

Perennial herb, with slender, oblique rhizome and sparsely leafy stolong. Stems 4–25 cm, erect, glabrous or somewhat papillate above, with leaves in basal cluster and sparse, linear bracts above. Leaves 15–35 × 3–8 mm, oblanceolate-spathulate, obtuse, long attenuate in sheathing petiole-like base, conspicuously 3– or 5– lobed distally, glabrous. Capitula 18–30 mm in diameter, radiate, solitary and terminal on scape-like stems; involucral bracts 5–7 × 1–1.5 mm, oblong, acute to acuminate, glabrous, green; calyculine bracts absent. Florets dimorphic; marginal female, white, with ligule 7–12 × 2.5–4.5 mm, oblong, the tube *c.*6 mm; inner 6–8 mm, tubular, hermaphrodite, yellow, forming a disc. Achene 2.5–4 mm, cylindrical, ribbed, glabrous; pappus 7–8 mm, white. $2n = 40$. *Flowers December to March.* (Fig. 219).

Moist coastal sands and cliffs, *Sphagnum – Astelia* bogs, wet areas in feldmark and at tree-line; 0–800 m.

S. Fuegia. (Map 27d).

W. Argentina north to 38°40′S.lat.
Chile north to *c.*38°30′S.lat.

13. *S. vulgaris L. 1753, *Sp. Pl.* 867.

Annual or overwintering herb; stems 8–45 cm, ascending to erect, simple or branched, rather succulent, glabrous or ± hairy. Lower leaves with lamina 25–70 × 10–18 mm, lanceolate to obovate in outline, pinnatifid, with distant, oblong, obtuse, irregularly serrate-dentate lobes, glabrous or ± floccose, the petiole 15–20 mm; upper leaves similar but smaller, sessile, semi-amplexicaul, with basal auricles. Capitula 8–10 × 3–5 mm, discoid, in dense terminal and axillary cymes; involucral bracts 7–8 × *c.*0.5 mm, linear, acute, glabrous, green, black-tipped; calyculine bracts *c.*¼ as long as and darker than involucral bracts. Florets monomorphic, yellow, 3–5 mm, tubular, hermaphrodite. Achene 1.5–2.6 mm, ± cylindrical, ribbed, densely appressed – hairy on ribs, reddish brown; pappus 3–7 mm, white. *Flowers November to April.*

Disturbed and open ground by roads and habitations, beaches; 0–100 m.

Isla Grande, except the extreme west.
Introduced from Europe and widely naturalized in temperate regions.

14. *S. patagonicus* Hooker & Arn. 1841, *Jour. Bot. (Hooker)* **3**: 344.

S. danyausii Hombron & Jacquinot 1846, *Voy. Pôle Sud, Atlas*, t. 13A. – Alboff, 1896.
S. exilis Hombron & Jacquinot 1846, *Voy. Pôle Sud, Atlas*, t. 13C.
S. floccidus Hombron & Jacquinot 1846, *Voy. Pôle Sud, Atlas*, t. 12.
S. andersonii Hooker f. 1847, *Fl. Antarct.* **1(2)**: 312 – Alboff, 1896.
S. micropifolius auct., non DC. – Franchet, 1889; Spegazzini, 1896; Alboff, 1896.

Perennial herb, woody at base; stems 10–50 cm, ascending to erect, branched, white-tomentose to laxly lanate, leafy. Leaves 10–40 × 1.5–5 mm, linear to linear-lanceolate, acute to subobtuse, attenuate at base, entire and revolute, or with 1–2 pairs of large marginal teeth, lanate to subglabrous above, densely tomentose beneath. Capitula 5–10 mm in diameter, discoid, in cymes, the peduncles 1–1.5 cm; involucral bracts 6–7.5 × 1–2 mm, linear-oblong, acute, lanate to subglabrous, green or brownish, darker distally; calyculine bracts up to ½ as long as involucral bracts, linear. Florets monomorphic, yellow, often reddish distally, 6–8 mm, tubular, hermaphrodite. Achene 2–3 mm, cylindrical, ± curved, ribbed, glabrous to sparsely pubescent; pappus 4–5 mm, white. $2n = 80$. *Flowers January to March.*

Coastal sands and rocks, grassland, *Lepidophyllum*, *Chiliotrichum* and *Berberis buxifolia* scrub; 0–100 (–300) m.

Isla Grande except the extreme west. (Map 27e).
W. Argentina north to *c.*39°S.lat.
Chile north to 45°30′S.lat.

Very variable in density of pubescence, leaf-size and indentation, and in number of capitula. Some of this variation has been formally recognized – var. *alyssoides* (Sch. Bip.) Cabrera, with small, densely pubescent leaves and 2–6 capitula; var. *andersonii* (Hooker f.) Cabrera, with dentate leaves; and var. *lobatifolius* (Hombron & Jacquinot) Cabrera, with smaller dentate leaves. Such varieties only deal with some of the many combinations of variables and seem of doubtful value. Some populations appear to grade towards *S. tricuspidatus*.

15. *S. arnottii* Hooker f. 1847, *Fl. Antarct.* **1(2)**: 314.

Glabrous perennial herb, woody at base; stems 20–40 cm, ascending to erect, densely leafy above. Leaves 25–40 × 3–6 mm, oblanceolate, acute to subobtuse, attenuate at base, entire and revolute or occasionally sparsely dentate. Capitula *c.*10 mm in diameter, discoid, in cymes, the peduncles 1.5–6 cm; involucral bracts 8–9 × 1–1.5 mm, linear-lanceolate, acuminate; calyculine bracts short, linear. Florets monomorphic, yellow, *c.*7 mm, tubular, hermaphrodite. Achene *c.*3 mm, cylindrical, glabrous; pappus 5–6 mm, white. *Flowers December to March.*

S. Isla Grande. (Map 27f).
W. Argentina north to *c.*39°S.lat.
E. Chile along N. side of Estrecho de Magallanes.

Very similar to glabrous, entire-leaved forms of *S. tricuspidatus*, and possibly conspecific.

S. subpanduratus O. Hoffm. 1900, in Dusén, *Svenska Exped. Magellansl.* **3(5)**: 108, described from N.E. Isla Grande (Cabo Domingo), has not been rediscovered despite intensive searching. It differs from *S. arnottii* in having irregularly dentate lower leaves, semi-amplexicaul upper leaves, and pubescent achenes.

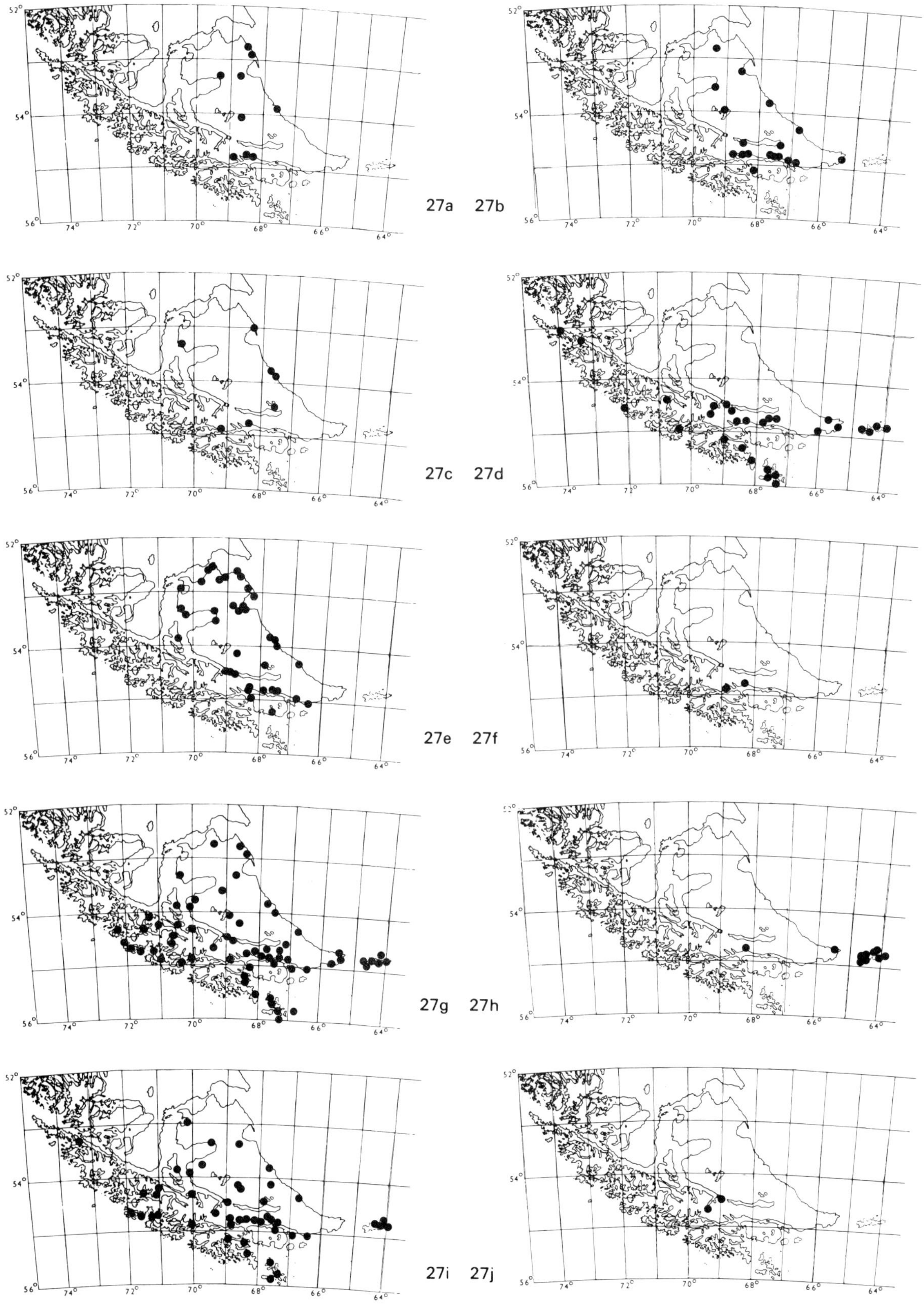

Map 27. a, *Senecio miser*. b, *Senecio tricuspidatus*. c, *Senecio kingii*. d, *Senecio trifurcatus*. e, *Senecio patagonicus*. f, *Senecio arnotti*. g, *Senecio acanthifolius*. h, *Senecio websteri*. i, *Senecio smithii*. j, *Senecio philippii*.

16. *S. acanthifolius* Hombron & Jacquinot 1846, *Voy. Pôle Sud., Atlas,* t. 11.

S. auriculatus Alboff 1896, *Rev. Mus. La Plata* **7**: 382.

S. ombrophyllus Skottsb. 1906, *Wiss. Ergebn. Schwed. Südpolarexped.* **4(4)**: 34.

Perennial herb, with short, stout rhizome; stems 9–90 cm, erect, glabrous. Lower leaves with lamina (5–)8–30 × 4–10 cm, ovate, obtuse, cordate at base, lobed, the lobes usually dentate or lobulate, glabrous, or sparsely pubescent beneath; petiole 2–10 cm, with wide, lobed or undulate wings, sheathing at base; upper leaves smaller, oblong-oblanceolate, acute, semi-amplexicaul, lobed, sessile. Capitula 20–30 mm in diameter, radiate, in terminal cymes, the peduncles up to 50 mm; involucral bracts 8–10 × 1–2.5 mm, oblong-lanceolate, acute, glabrous, green with pale scarious margin; calyculine bracts absent. Florets dimorphic; marginal female, white, with ligule 7–10 × 2–5 mm, oblong, the tube *c.*6 mm; inner 6–7 mm, tubular, hermaphrodite, yellow, forming a disc. Achene 4–5 mm, cylindrical, ribbed, dark brown, glabrous or sparsely lanate; pappus 5–6 mm, white. $2n = 40, 80$. *Flowers December to April.* (Fig. 223).

Moist or wet areas on coastal cliffs, grassland, scrub, *Nothofagus* forest, cushion plant communities, feldmark; 0–800 m.

Throughout Fuegia except the extreme west. (Map 27g).
W. Argentina north to *c.*40°S.lat.
Chile north to *c.*38°S.lat.

12. *S. websteri* Hooker f. 1847, *Fl. Antarct.* **1(2)**: 317

Perennial herb; stems 10–30 cm, ascending, glabrous below, tomentose above, leafy throughout. Leaves with lamina 15–30 × 15–50 mm, orbicular to flabellate or reniform, obtuse, subentire to irregularly crenate, glabrous above, densely white-tomentose beneath; petiole 15–35 mm. Capitula 15–25 mm in diameter, radiate, in terminal and axillary cymes, the peduncles 2–25 mm; involucral bracts 6–8 × 1.5–3 mm, linear-lanceolate, acute, glabrous to lanuginose, green with paler scarious margin; calyculine bracts short, linear. Florets dimorphic; marginal female, yellow, with ligule 6.5–8 × 2–3 mm, elliptical, the tube 3.5–4 mm; inner 4–5 mm, tubular, hermaphrodite, yellow to orange, forming a disc. Achene 1.5–2 mm, cylindrical, ribbed, glabrous; pappus 3.5–4.5 mm, white. $2n = c.40$. *Flowers October to November.* (Fig. 220).

Wet coastal rocks and cliffs, moist peat in *Poa flabellata* association; 0–100 m.

Isla de los Estados; local on S. coast of Isla Grande; Isla Hoste. (Map 27h).
Endemic.

18. *S. smithii* DC. 1838, *Prodr.* **6**: 412.

Perennial herb, with short, robust rhizome; stems 25–120 cm, erect, simple, leafy, sparsely lanuginose to subglabrous. Basal leaves with lamina 8–25(–40) × 5–14(–20) cm, oblong-ovate, acute, ± truncate to cordate at base, serrate-dentate, subglabrous to sparsely lanate above, more densely white-lanate beneath; petiole 5–20 cm; upper leaves smaller, ovate-lanceolate, acute to acuminate, sheathing or semi-amplexicaul, sessile. Capitula 30–50 mm in diameter, radiate, in terminal cyme, the peduncles 1–10 cm, lanuginose; involucral bracts 9–14 × 1–2 mm, linear-lanceolate, acuminate, lanuginose to subglabrous, green or brownish with pale scarious margin; calyculine bracts linear, the inner almost as long as involucral bracts. Florets dimorphic; marginal female, white, with ligule 10–18 × 3–6 mm, elliptical, the tube 4–6 mm; inner 6.5–7.5 mm, tubular, hermaphrodite, yellow, forming a disc. Achene 4.5–6 mm, cylindrical, ribbed, glabrous, grey-brown; pappus 6–7 mm, white. $2n = 40$. *Flowers December to April.*

Moist or wet areas in grassland, *Marsippospermum* communities, coastal forest, especially where streams cross pebbly beaches; 0–140(–500) m.

Almost throughout Fuegia. (Map 27i).
W. Argentina north to *c.* 41°S.lat.
Chile north to *c.*43°S.lat.

19. *S. philippii* Sch. Bip. ex Wedd. 1856, *Chloris Andina* **1**: 134.

Glabrous, perennial herb, woody at base; stems 4–20 cm, ascending to suberect, simple or sparingly branched, densely leafy at base, sparingly so above. Leaves 5–25 × 3–8 mm, oblanceolate-spathulate, obtuse to subacute, attenuate at base, with 2(3) pairs of dentate-serrations distally, Capitula 16–20 mm in diameter, radiate, 2–3 in lax terminal cyme, the peduncles 1.5–5.5 cm; involucral bracts 7–9 × 1–2 mm, linear-oblong, acute, entire, dark green with pale scarious margin; calyculine bracts small, few. Florets dimorphic, yellow; marginal female, with ligule 7–9 × 2–3 mm, oblong , the tube 3–4 mm; inner 6–8 mm, tubular, hermaphrodite, forming a disc. Achene 2.5–3.5 mm, cylindrical, glabrous or sparsely puberulent; pappus 5–7 mm, white. *Flowers January to March.*

Sandy soils with good drainage, and rock-crevices, by streams and on glacial moraines; 250–500 m.

C. S. Isla Grande. (Map 27j).
W. Argentina and Chile north to *c.*41°S.lat.

Tribe CARDUEAE Florets without ligules, though marginal florets often enlarged; florets usually hermaphrodite. Style thickened or hairy below the branches.

28. ***Carduus*** L.

Perennial herbs with spiny-winged stems. Leaves alternate, pinnatisect, spinose-dentate. Capitula depressed globose, solitary; receptacle flat or convex, densely setose; involucral bracts in 4–8 imbricate series. Florets monomorphic, tubular, 5-lobed with 1 lobe larger than others, hermaphrodite. Pappus of several series of scabrid hairs united at base, unequal, the inner longest.

*1. ***C. nutans*** L. 1753, *Sp. Pl.* 821.

Stem 32–100 cm, erect, simple or branched above, lanate, wings up to 10 mm. Leaves 3–15 × 1–5 cm, elliptical to lanceolate, acute, the basal attenuate to petiole-like base, the upper decurrent, sparsely lanate, the lobes triangular. Capitula 40–70 mm in diameter, ± pendent; involucral bracts 14–17 × (1.2–)2–4 mm, linear-triangular to lanceolate and narrowing abruptly to oblong base, acuminate, with apical spine, green or purplish, the outer ± strongly reflexed, the inner erect. Florets reddish-purple; tube *c.*13 mm, widening at apex; lobes linear, 6–7 mm, the longest *c.*8 mm. Achene 3–5 × *c.*1.5 mm, cylindrical, somewhat curved and compressed, glabrous, pale brown with paler transverse markings; pappus 13–18 mm, white. *Flowers February to March.*

Open coastal soils, disturbed ground near habitations and cultivated areas; 0–300 m.

Locally in N, C. & S. Isla Grande.
Introduced from Europe and widely naturalized in temperate regions.

29. ***Cirsium*** Miller

Spiny, biennial or perennial herbs. Leaves alternate, pinnatisect. Capitula solitary or in panicles or corymbs; receptacle flat, with bristles; involucral bracts in several imbricate series. Florets monomorphic, tubular, 5-lobed, hermaphrodite or unisexual. Pappus of several series of plumose hairs united at base.

a. Stems winged; leaves with rigid, pungent bristles above; florets hermaphrodite **1. *vulgare***
a. Stems not winged; leaves without pungent bristles above; florets unisexual **2. *arvense***

*1. ***C. vulgare*** (Savi) Ten. 1835–1838, *Fl. Nap.* **5**: 209.

Biennial; stems 30–150 cm, erect, branched above, with interrupted spiny wings, sparsely arachnoid-hairy. Leaves 13–23 × 4–9 cm, obovate-oblanceolate, acute, attenuate to petiole-like base, with rigid, pungent bristles above, arachnoid-lanate beneath, the lobes linear– to elliptic– lanceolate or narrowly triangular, with weak or strong apical spine 2–10 mm. Capitula 30–50 × 20–35 mm, ovoid-oblong, solitary or in small panicle or corymb, the peduncles short or long; involucral bracts 14–18 × 1.5–2 mm, lanceolate, acuminate, with pungent apical spine, green, sparsely arachnoid-hairy to subglabrous, the outer recurved, the inner erect. Florets reddish-purple, hermaphrodite; tube 17–22 mm; lobes 6–7 mm, linear. Achene 3.5–4.2 mm, irregularly ovoid, compressed, glabrous, pale brown with black streaks; pappus 19–25 mm, white. *Flowers January to April.*

Coastal sands and cliffs, degraded grassland, disturbed ground near habitations and especially by roads; 0–60 m.

N. Isla Grande, locally on S. coast.
Introduced from Europe and widely naturalized in temperate regions.

This species has spread rapidly in N. Fuegia during the past 25 years with the extensive construction of roads.

*2. ***C. arvense*** (L.) Scop. 1772, *Fl. Carn.*, ed. 2, **2**: 126.

Dioecious or monoecious perennial; stems 15–72 cm, erect, branched, without wings, glabrous or sparsely arachnoid-hairy above. Lower leaves 7–11 × 1.8–3.5 cm, oblanceolate, subacute, narrowed and ± decurrent at base, without pungent bristles and glabrous to sparsely arachnoid-hairy above, glabrous to arachnoid-tomentose beneath, the lobes trian-

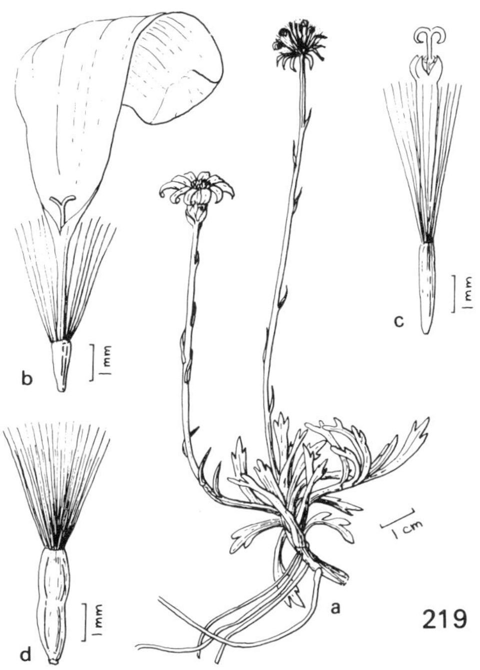

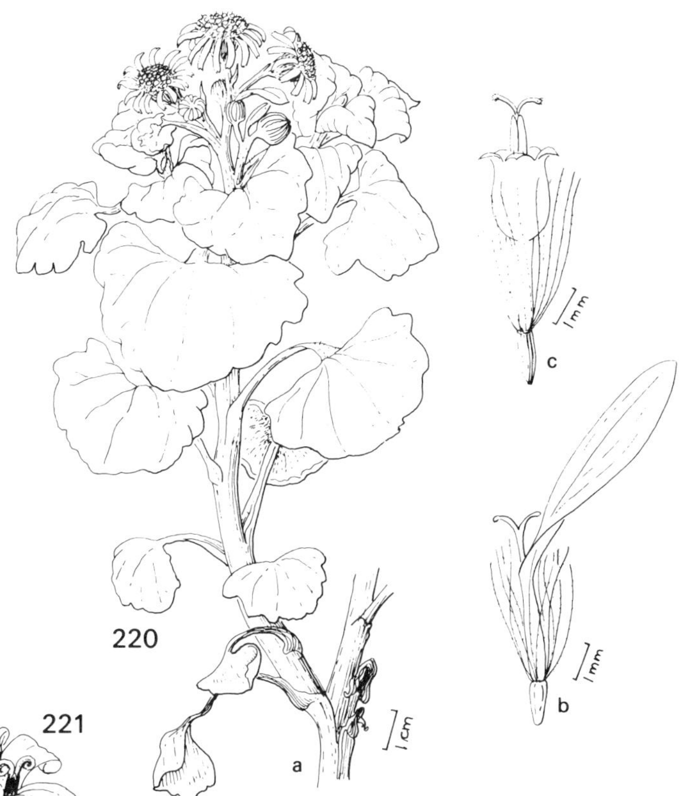

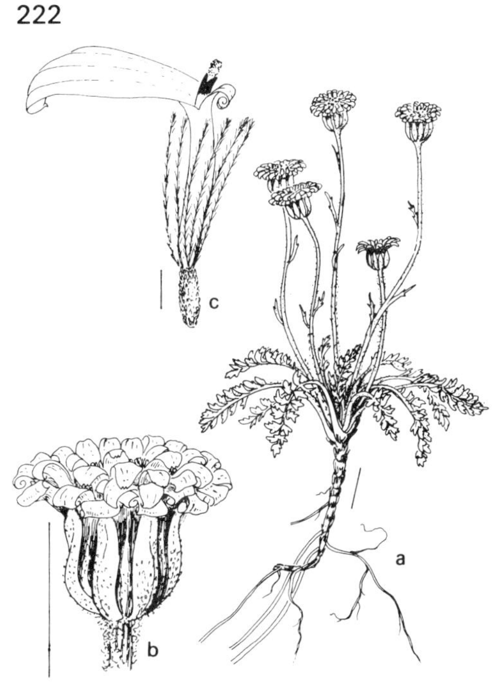

Fig. 219. *Senecio trifurcatus*. a, Plant. b, Marginal floret. c, Disc floret. d, Achene. Scale = 1 cm (a) or 1 mm (b–d). Fig. 220. *Senecio websteri*. a, Plant. b, Marginal floret. c, Disc floret. Scale = 1 cm (a) or 1 mm (b–c). Fig. 221. *Nassauvia pygmaea*. a, Plant. b, Capitulum. c, Leaf. d, Floret. e, Pappus-scale. Scale = 1 cm (a) or 1 mm (b–e). Fig. 222. *Leucheria purpurea*. a, Plant. b, Capitulum. c, Floret. Scale = 1 cm (a–b) or 1 mm (c).

(R. N. P. Goodall)

gular, with spiny apex and margins; upper leaves smaller, not decurrent, semi-amplexicaul. Capitula 10–17 × 15–25 mm, solitary or in small terminal corymbs, the peduncles short; involucral bracts glabrous to ± arachnoid-hairy, green or purplish distally; outer 4.5–6 × 1.5–1.8 mm, ovate, obtuse, with ± patent apical spine; inner *c.*13 × 1 mm, linear-lanceolate, acute, with erect apical spine. Florets pale purple to whitish, unisexual; male with tube *c.*9 mm, the lobes 4–5 mm; female with tube 10–14 mm, the lobes 2.7–3 mm. Achene 3.2–3.5 mm, cylindrical to obovoid, glabrous, pale brown. *Flowers March.*

Disturbed grassland in degraded scrub; 30–80 m.

Local in S. Isla Grande.
Introduced from Europe.

30. ***Silybum*** Adanson

Annual or biennial herbs. Leaves alternate, pinnatisect, white-veined, with strongly spiny margin. Capitula solitary, ovoid; receptacle flat, densely hairy; involucral bracts spiny, in several imbricate series. Florets monomorphic, tubular, 5-lobed, hermaphrodite. Pappus of several series of strigose hairs united at base.

*1. ***S. marianum*** (L.) Gaertner 1791, *Fruct. Sem. Pl.* **2**: 378.

Stem 30–40 cm, erect, simple or branched above, glabrous or slightly arachnoid-pubescent. Lower leaves 20–45 × 9–20 cm, oblong to lanceolate, obtuse, narrowed at base, deeply divided, the segments with few, ± rounded lobes, glabrous; upper leaves smaller, less deeply divided, auriculate-amplexicaul. Capitula 25–40 × 40–65 mm; involucral bracts 13–25 × 6–7 mm, suborbicular to ovate and spinose-ciliate at base, narrowing above and with apical spinose-ciliate appendage with stout yellowish spine, the spine patent in outer bracts. Florets reddish – purple; tube *c.*20 mm; lobes 6–7.5 mm. Achene 6–7 mm, oblong-ovoid, compressed, glabrous, dark brown, yellowish distally. *Flowers March.*

Weed of cultivated fields. *c.*20 m.

Locally on S. coast of Isla Grande.
Introduced from Europe and naturalized in temperate regions.

Tribe MUTISEAE. Florets tubular, bilabiate, hermaphrodite. Stigmatic branches thickened and hairy, or filiform with hairs only on the margins and at the apex.

31. ***Macrachaenium*** Hooker f.

Rhizomatous perennial herbs. Leaves alternate, deeply pinnatisect. Capitula solitary, terminal on scape-like upper part of stem; receptacle flat, without bracts; involucral bracts in 2–3 series. Florets monomorphic, tubular, 2-lipped, hermaphrodite. Pappus of 1 series of plumose hairs.

1. ***M. gracile*** Hooker f. 1847, *Fl. Antarct.* **1(2)**: 321.
M. foliosum Alboff 1896, *Rev. Mus. La Plata Bot.* **7**: 375.

Rhizome oblique; stems 20–50 cm, erect, simple, laxly tomentose and leafy below, densely tomentose and leafless above. Leaves with lamina 70–140 × 40–80 mm, elliptic-oblong in outline, glabrous or lanuginose above, ± densely greyish-lanate beneath; segments reaching almost to midvein, 15–30 mm wide, oblong to ovate, obtuse, entire or irregularly dentate to lobed; petiole 50–160 mm. Capitula 12–20 mm in diameter; involucral bracts 10–15 × 1.5–2 mm, linear-lanceolate, acute, lanate, green in centre, paler at margins. Florets white; tube 10–11 mm; lips 3–4 mm. Achene 8–10 mm, cylindrical, ± narrowed to apex, ribbed, glabrous, pale brown; pappus 8–10 mm, white. *Flowers December to February.* (Fig. 216).

Deciduous *Nothofagus* forest, rarely in damp shaded places outside forest; 0–300 m.

C. & S. Fuegia except the extreme south and west. (Map 28a).
W. Argentina north to *c.*41°S.lat.
Chile north to 47°12′S.lat.

Fuegian material belongs to var. *gracile* Further north, in W. Argentina, var. *radiatum* Cabrera, which has marginal florets with the external lip ligulate, becomes predominant.

32. ***Nassauvia*** Comm. ex Juss.

Perennial herbs, often woody at base, or dwarf shrubs. Leaves alternate, usually rigid and spiny. Capitula usually sessile, (4) 5-flowered, solitary or grouped into usually dense globose to ovoid inflorescence; receptacle flat, without bracts; involucral bracts in 2 series. Florets monomorphic, tubular, 2-lipped, hermaphrodite; outer lip somewhat larger, 3-dentate; inner lip 2-fid. Pappus of 3–15 caducous scales or numerous plumose bristles.

a. Capitula solitary or in terminal clusters, not in ovoid-globose spike:
  b. Achene velutinous; pappus of caducous scales **4. *fuegiana***
  b. Achene glabrous; pappus of plumose bristles:
    c. Plant densely sericeous-velutinous; leaves entire **6. darwinii**
    c. Plant glabrous or somewhat pubescent; leaves entire or spinose-dentate **7. *abbreviata***
a. Capitula in dense, terminal, ovoid or globose spike:
  d. Leaves densely velutinous or lanate, at least beneath:
    e. Leaves at least 10 mm wide; pappus of scales **3. *latissima***
    e. Leaves not more than 4 mm wide; pappus of plumose bristles **5. *lagascae***
  d. At least lower leaves glabrous, except on veins beneath, or subglabrous:
    f. Leaves dimorphic, the lower oblong-lanceolate, acute, the upper lanceolate, acuminate; spikes at least 20 mm in diameter **1. *magellanica***
    f. Leaves monomorphic, ovate-lanceolate; spikes not more than 13 mm in diameter **2. *pygmaea***

1. *N. magellanica* J. F. Gmelin 1792, *Syst. Nat.* ed. 13, *2*: 1281.
N. suaveolens Cass. 1825, *Dict. Sci. Nat.* **38**: 457 – Hooker, 1847; Franchet, 1889; Dusén, 1900; Wildeman, 1905.
N. heterophylla Alboff 1896, *Rev. Mus. La Plata Bot.* **7**: 373.
N. spinosa auct., non D. Don – Autran, 1905.

Perennial herb; stems 8–40 cm, procumbent to ascending, branched, densely covered with imbricate leaves, glabrous below, velutinous above. Leaves dimorphic, erecto-patent to ± recurved; lower 15–35 × 6–15 mm, lanceolate-oblong, acute, mucronate, with conspicuous, oblong basal sheath, with 4–5 deep dentate-serrations on each side, subglabrous; upper 10–30 × 4–10 mm, lanceolate, acuminate, mucronate, ± sheathing at base, entire or sparingly dentate-serrate, ± densely velutinous. Capitula numerous, grouped into dense, ± globose apical spike 20–30 mm in diameter; involucral bracts velutinous, the inner 5, 9–11 × 1.5–2 mm, lanceolate, acuminate, mucronate, the outer shorter, linear. Florets white, chocolate/vanilla-scented; tube 3–4.5 mm; outer lip 3–4.5 × 2–3 mm. Achene 3.5–4.5 mm, obovoid, ribbed, glabrous; pappus of 4–5, white, linear scales. $2n = 44$. *Flowers (October) December to March.* (Fig. 227, Pl. 8a).

Feldmark, rock and soil screes, rarely along river banks at low elevations; (200–)350–1100 m.

C. & S. Fuegia. (Map 28b).
W. Argentina north to *c.* 50°S.lat.
E. Chile north to 50°40′S.lat.

2. *N. pygmaea* (Cass.) Hooker f. 1847, *Fl. Antarct.* **1(2)**: 319.
N. pumila Poeppig ex Less. 1832, *Syn. Gen. Comp.* 399 – Spegazzini, 1896.
N. nordenskjoldii O. Hoffm. 1900, *Svenska Exped. Magellansl.* **3(5)**: 113.

Perennial herb; stems 2–15 cm, procumbent to ascending, branched, densely covered with imbricate leaves. Leaves 3–6 × 1–3 mm, ovate-lanceolate, acute; distal part ± thick, mucronate, with 1 pair of thick, mucronate teeth, recurved, glabrous or pubescent on midvein beneath; basal part longer and wider, oblong, membranous, dentate above, ciliate below, sheathing. Capitula numerous, grouped into dense, ± globose apical inflorescence 8–13 mm in diameter; involucral bracts glabrous, the inner 5, 4–5 × *c.* 1 mm, oblanceolate-oblong, acute, mucronate, purple-green with pale scarious margin, the outer shorter, linear. Florets white; tube *c.* 2 mm; outer lip 1.5–2 × *c.* 1 mm. Achene 1.5–2 mm, obovoid, ribbed, glabrous; pappus of 3–5, white, oblanceolate-linear scales. *Flowers December to March.* (Fig. 221).

Feldmark, rock and soil screes, rock-crevices; 450–1100 m.

C. & S. Isla Grande; Isla de los Estados. (Map 28c).
W. Argentina north to *c.* 40°S.lat.
E. Chile north to *c.* 50°S.Lat.

3. *N. latissima* Skottsb. 1916, *Kungl. Svenska Vetensk Akad. Handl.* **56(5)**: 330.
N. serpens auct., non D'Urv. – Skottsberg, 1906.

Perennial herbs; stems 8–30 cm, procumbent to ascending, branched, densely covered with imbricate leaves. Leaves 12–25 × 10–16 mm, broadly obovate-spathulate, obtuse, ± abruptly narrowed into basal sheath, serrate-dentate in distal part, ± densely velutinous, the teeth triangular, acute, ± mucronate. Capitula numerous, grouped into dense, ovoid apical spike 20–30 mm in diameter; involucral bracts velutinous distally, the inner 8–11 × 1–1.5 mm, oblong-lanceolate, acute. Florets white, sweet-scented; tube 3–3.5 mm; outer lip *c.* 2 × 1 mm. Achene 3–3.5 mm, obovoid, ribbed, glabrous; pappus of 4–5, white, linear scales. *Flowers December to February.* (Pl. 8b).

Feldmark, screes; 700–1000 m.

S. Isla Grande; Isla Navarino; Isla de los Estados. Endemic. (Map 28d).

4. *N. fuegiana* (Speg.) Cabrera 1931, *Physis* **10**: 285.

Strongylomopsis fuegiana Speg. 1899, *Com. Mus. Nac. Bs Aires* **1(4)**: 135.

S. struthonium auct., non Phil. – Spegazzini, 1896.

Nassauvia glomerulosa auct., non (Lag.) D. Don – Moore, 1974.

Dwarf shrub up to 10 cm; stems ascending to erect, much branched, densely covered with imbricate leaves. Leaves 6–15 × 0.5–1 mm, linear-subulate, distally recurved, mucronate, entire and thickened at margin, pubescent, with short, wide, entire, lanate basal sheath. Capitula 5–6.5 × 3–3.5 mm, solitary and terminal on short axillary branches; inner involucral bracts *c.*5, 6–7 × *c.*1 mm oblong-lanceolate, mucronate, scarious at margin, pubescent to glabrous, the outer 3–4, shorter, ovate-lanceolate, mucronate. Florets white; tube 2.5–3.5 mm; outer lip 2.5–3 × *c.*2 mm. Achene 1.5–2 mm, turbinate, velutinous; pappus of 3, ciliate, white, linear scales. *Flowers November.*

Open rocks and sands; 0–50 m.

N. Isla Grande. (Map 28e).
Argentina north to *c.*39°S.lat.

5. *N. lagascae* (D. Don) F. Meigen 1894, *Bot. Jahrb.* **18**: 445.

N. revoluta auct., non D. Don – Skottsberg, 1906.

Perennial herb; stems 4–8 cm, procumbent to ascending, sparingly branched, densely covered with imbricate leaves. Leaves 6–13 × 2–4 mm, oblong, acute, recurved and with 5–9 acute teeth towards apex, subglabrous above, densely lanate beneath. Capitula numerous, grouped into dense, globose apical spike 20–40 mm in diameter; involucral bracts 5–8 × 1.2–1.4 mm, the inner 5, elliptical, emarginate and mucronate, scarious at margin, glabrous or sparsely lanate distally, the outer 3–4, linear, acute. Florets white or pale purple; tube 2–3 mm; outer lip 1.5–2 × *c.*1 mm. Achene 2.5–3 mm, obovoid, glabrous; pappus of plumose bristles. *Flowers January to February.*

Open soil and rocks above timberline; 900–1100 m.

C. & S. Isla Grande. (Map 28f).
W. Argentina and E. Chile north to *c.*36°S.lat.

Fuegian material belongs to var. ***globosa*** Skottsb. 1916, *Svenska Vetensk Akad. Handl.* **56(5)**: 329, which extends north in W. Argentina to *c.*45°S.lat. and in E. Chile to *c.*51°S.lat. Var. *lagascae* differs in its smaller, oblong-lanceolate leaves, smaller inflorescence and shorter involucral bracts.

6. *N. darwinii* (Hooker & Arn.) O. Hoffm. & Dusén 1900, *Svenska Exped. Magellansl.* **3(5)**: 112.

Panargyrum darwinii Hooker & Arn. 1836, *Comp. Bot. Mag.* **2**: 43- Hooker, 1847; Spegazzini, 1896.

Perennial herb, woody at base; stems 3–15 cm, ascending to erect, branched at base, sericeous, densely covered with leaves towards base, more sparsely so above. Leaves 5–25 × 0.5–1.3 mm, linear, mucronate, entire, sericeous-velutinous, rigid. Capitula solitary or in small apical clusters; inner involucral bracts 5, 6.5–8 × 1–1.5 mm, oblong, acute to obtuse, with short apical spine, pale green with brownish midvein, often scarious at margin, lanate, the outer shorter, ovate-lanceolate, acute, lanate. Florets 4–5, white or pink; tube *c.*4 mm; outer lip 3–3.5 × 1–2 mm. Achene *c.*3 mm, ellipsoid, glabrous; pappus of 20–25 plumose bristles. $2n = 22$. *Flowers December to March.*

Coastal sands and rocks, open grassland, *Empetrum* heath; 0–60 m.

N. & C. Isla Grande, local in C.S. and adjacent islands of Canal Beagle. (Map 28g).
W. Argentina and E. Chile north to *c.*38°40′S.lat.

7. *N. abbreviata* (Hooker & Arn.) Dusén 1900, *Svenska Exped. Magellansl.* **3(5)**: 112.

Panagyrum abbreviatum Hooker & Arn. 1836, *Comp. Bot. Mag.* **2**: 43 – Hooker, 1847.

Nassauvia subspinosa (Phil.) Wildem. 1905, *Result. Voy. S. Y. Belgica* 177.

Similar in habit, size and inflorescence to *N. darwinii*, but distinguished by its usually wider (1–4 mm), lanceolate, glabrous to sparsely sericeous leaves, often with 1–2 pairs of spine-tipped marginal teeth, spine-tipped inner involucral bracts, and turbinate achene with more (*c.*50) plumose bristles. *Flowers November to March.*

Open soils and rocks, dry drassland; 0–800 m.

N. & C. Isla Grande. (Map 28h).
W. Argentina north to *c.*43°S.lat.
E. Chile north to *c.*51°S.lat.

Intermediate plants, perhaps hybrids, blur the distinctions between this species and *N. darwinii*.

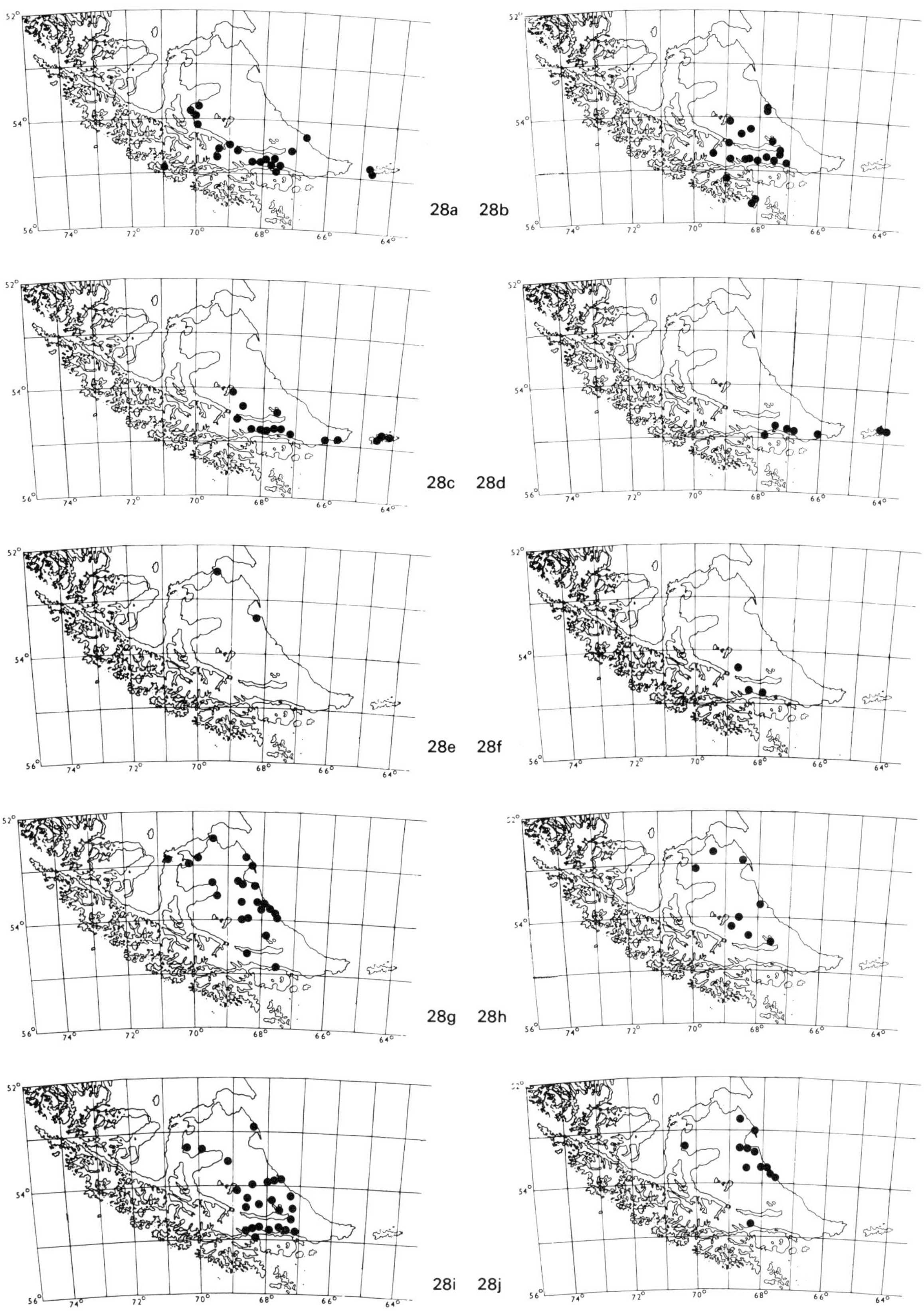

Map 28. a, *Macrachaenium gracile*. b, *Nassauvia magellanica*. c, *Nassauvia pygmaea*. d, *Nassauvia latissima*. e, *Nassauvia fuegiana*. f, *Nassauvia lagascae*. g, *Nassauvia darwinii*. h, *Nassauvia abbreviata*. i, *Leucheria hahnii*. j, *Leucheria purpurea*.

33. *Leucheria* Lag.

Perennial herbs. Leaves alternate, mostly in basal rosette, entire to pinnatifid. Capitula solitary, terminal; receptacle flat, without bracts; involucral bracts in 2–3 series. Florets indistinctly dimorphic, tubular, 2-lipped, hermaphrodite; inner lip 2-fid; outer lip 3-dentate, larger in marginal florets. Pappus of 1 series of plumose hairs, partly united at base.

Literature: J. V. Crisci, *Darwiniana* **20**: 9–126(1976).

a. Flowers dark red-purple, blackish when dry **2. *purpurea***

a. Flowers white or pink **1. *hahnii***

1. *L. hahnii* Franchet 1889, *Miss. Sci. Cap Horn* **5**: 349.

L. fuegina Phil. 1894, *Anal. Univ. Chile.* **87**: 98.
L. gracilis Alboff 1896, *Rev. Mus. La Plata Bot.* **7**: 375.
L. lanata Alboff 1896, *Rev. Mus. La Plata Bot.* **7**: 374.
L. patagonica Speg. 1897, *Rev. Fac. Agron. La Plata* **3**: 538 – Dusén, 1900.
L. candidissima auct., non D. Don – Spegazzini, 1896; Skottsberg, 1906.

Rhizome oblique to erect; stems 4–30 cm, ascending to erect, simple, white-lanate to, rarely, subglabrous. Basal leaves 30–150 × 3–20 mm, oblanceolate-spathulate, obtuse, long attenuate at base into sheathing petiole, entire to deeply pinnatifid, with lanceolate to oblong segments, densely white-lanate, rarely subglabrous, upper leaves absent or 1–2, small, linear, bract-like, usually white lanate. Capitula 18–30 mm in diameter; involucral bracts in 2 series, 9–11 × 1.5–2.2 mm, lanceolate, acute to subobtuse, entire, lanate and with some glandular hairs, to subglabrous, green or purplish. Florets white or pinkish; tube 4.5–6 mm; outer lip 7–9 × 2–3.5 mm in marginal florets, 2–5 × 1.5–2 mm in inner florets. Achene 1–2 mm, obovoid, sparsely glandular-pubescent, brown; pappus 5–6.5 mm, white. $2n = 40$. *Flowers November to March.*

Open sands and gravels, grassland, *Empetrum* heath, open *Chiliotrichum* and *Berberis* scrub, montane rocks, feldmark, screes; 0–1100 m.

Almost throughout Isla Grande except extreme west and east; S. shore of Canal Beagle. (Map 28i).
Argentina north to *c.*46°S.lat.
E. Chile north to *c.*51°S.lat.

An extremely variable species, especially in habit, indumentum and leaf-shape, within which numerous taxa have been recognized. There is some suggestion of ecotypic differentiation, since alpine populations comprise densely lanate plants of reduced stature and with relatively large capitula, thus differing significantly from plants lower on the mountain slopes; these differences are maintained in cultivation. However, some populations from dry lowland steppe in N. E. Fuegia closely resemble the alpine forms, perhaps as a result of exposure and available moisture. In view of these factors, formal infraspecific treatment seems inappropriate at this stage.

Cabrera (1971) erroneously reported *L. eriocephala* Speg. from Fuegia on the basis of a sparsely pubescent form of *L. hahnii* with very deeply pinnatifid leaves.

2. *L. purpurea* (Vahl) Hooker & Arn. 1836, *Comp. Bot. Mag.* **2**: 43.

Rhizome oblique to suberect; stems 5–18 cm, procumbent to ascending or erect, simple or once branched distally, white-lanate below and towards apex. Basal leaves with lamina 17–70 × 7–20 mm, oblanceolate, acute, with sheathing petiole 15–50 mm, deeply pinnatifid, with lanceolate to elliptical, lobed or dentate segments, glabrous to ± lanate; upper leaves 1–2, smaller, linear-lanceolate, acute, ± sessile, somewhat semi-amplexicaul at base, less divided to entire, glabrous or sparsely lanate. Capitula 13–25 mm in diameter; involucral bracts in 2–3 series, 6–9 × 1–2 mm, lanceolate, acute, entire or denticulate, ± white lanate and with some glandular hairs, to subglabrous, green to brownish or purplish, darker distally, paler and scarious at margin. Florets dark red-purple, becoming blackish on drying; tube 3–5 mm; outer lip 5–6 × 1.3–2 mm in marginal florets, 2–2.5 × 0.5–1 mm in inner florets. Achene 1.5–2.5 mm, cylindrical, densely glandular-pubescent, pale brown; pappus 5–6 mm, white. $2n = 40$. *Flowers November to February.* (Fig. 222).

Coastal sands, open grassland, rocks and pebbles; 0–70 m.

N. Isla Grande, locally on S. coast. (Map 28j).
Argentina north to *c.*49°S.lat.
E. Chile north to *c.*51°S.lat.

34. *Perezia* Lag.

Perennial herbs, sometimes woody at base. Leaves alternate, simple. Capitula solitary or 2–3 in cluster terminal on usually scape-like upper part of stem; receptacle flat, glabrous or with short hairs; involucral bracts in 3–9 series. Florets monomorphic or indistinctly dimorphic, hermaphrodite, 2-lipped;

inner lip 2–fid; outer lip 3-dentate, often larger in marginal florets. Pappus of several series of ± strigose hairs, united at base.

Literature: B. S. Vuilleumier, *Contr. Gray Herb.* **199**: 1–163(1970).

a. Leaves with strongly revolute margin bearing curved spines on rounded 'false-margin' **4. *recurvata***
a. Leaf-margin not revolute:
 b. Leaves entire **2. *lactucoides***
 b. Leaves pinnatifid:
  c. Leaves with linear segments, coriaceous **1. *pilifera***
  c. Leaves with ovate, obtuse segments, herbaceous **3. *magellanica***

1. *P. pilifera* (D. Don) Hooker & Arn. 1835, *Comp. Bot. Mag.* **1**: 34.
Clarionea pilifera D. Don 1832, *Philos. Mag.* **11**: 388.
C. lechleri Sch. Bip. 1855, *Flora (Regensb.)* **38**: 122.
Homoianthus humilis Phil. 1894, *Anal. Univ. Chile* **87**: 307.

Plant with stout, oblique stock; stems 2–5 cm, erect, simple, glabrous, often reddish. Leaves 12–35 × 3–8 mm, linear-oblong in outline, acute, sheathing at base, deeply pinnatifid, glabrous, coriaceous; segments linear, entire or lobed, with white apical spine. Capitula 12–20 mm in diameter, solitary, sessile among upper leaves or on short scape-like stem-apex; involucral bracts in 2–3 series, lanceolate, acute to mucronate, greenish to blue-black with pale scarious margin, rigid, the outer 7–12 × 1–2 mm, with long, soft, white marginal spines, subglabrous, the inner slightly larger, entire, glabrous. Florets blue to white, sweet-scented, the outer with tube 7–9 mm and outer lip 5–7 × 1–2 mm, oblong. Achene 3–4 mm, turbinate, densely velutinous; pappus 10–16 mm, white or pale brown. $2n = 16$. *Flowers December to February.*

Coastal sands and gravels, grassland, feldmark, absent from forests; 0–900 m.

Isla Grande except extreme west and east. (Map 29a).
W. Argentina north to *c.*29°30′S.lat.
Chile north to *c.*32°S.lat.

2. *P. lactucoides* (Vahl) Less. 1830, *Linnaea* **5**: 22.
Homoianthus magellanicus (Lam.) DC. 1838, Prodr. **7(1)**: 65 – Hooker, 1847; Alboff, 1896.

Plant with stems 6–53 cm, erect, simple, glabrous or sparsely pubescent above, green or reddish. Basal leaves with lamina 12–100 × 2–20 mm, lanceolate, acute, long cuneate at base, entire, glabrous, herbaceous, the petiole (10–)20–70 mm; upper cauline leaves similar but smaller, sessile, semi-amplexicaul. Capitula 15–30 mm in diameter, 1(2) on long, usually nodding, scape-like stem-apex; involucral bracts in 3–5 series, acute, glabrous or sparsely pubescent, the outer 4–10 × 1–4 mm, ovate, entire or serrulate, green, reddish or brownish with pale scarious margin, the inner somewhat longer, linear-lanceolate, entire. Florets white, rarely pinkish, the outer with tube 5–6 mm and outer lip 6–12 × 1.2–2 mm, lanceolate-oblong. Achene 6–7.5 mm, cylindrical-obovoid, sparsely velutinous, grey-brown; pappus 8–14 mm, pale brown. *Flowers December to February.* (Fig. 224).

Bogs, swamps, wet mosses by streams; 0–1000 m.

C. & S. Fuegia, except extreme west, south and east. (Map 29b).
W. Argentina north to *c.*50°S.lat.
Chile north to *c.*50°50′S.lat.

Fuegian material belongs to ssp. *lactucoides,* which has the above distribution. The larger ssp. *palustris* (Phil.) Vuilleumier, which has blue flowers, occurs further north *c.*40°–42°S.lat.

3. *P. magellanica* (L. f.) Lag. 1811, *Amen. Nat.* **1**: 31.
Percidium magellanicum L.f. 1781, *Suppl.* 376 – Forster, 1789.
Clarionea magellanica (L.f.) DC. 1838, *Prodr.* **7(1)**: 61 – Hooker, 1847.

Plant with stout, oblique stock; stems 3–20 cm, erect, simple, reddish glandular-pubescent. Basal leaves with lamina 9–40 × 3–12 mm, lanceolate-oblong in outline, acute to obtuse, attenuate to petiole, ± glandular-pubescent, deeply divided into 6–9 pairs of ovate, obtuse segments, the apical segment oblong to suborbicular, herbaceous, the petiole 5–35 mm; upper cauline leaves 0–4, up to 15 × 4 mm, lanceolate, acute, serrate. Capitula 20–35 mm in diameter, on long, scape-like stem-apex; involucral bracts in 3–4 series, acute, green or reddish, the outer 3–6 × 1–3 mm, ovate, entire or serrulate, glandular-pubescent, with narrow scarious margin, the inner much longer, lanceolate to oblong, entire. Florets white, the outer with tube 3–4.5 mm and outer lip 8–13 × 1.5–3.5 mm, oblong. Achene 3–5 mm, turbinate, densely velutinous; pappus 7–9 mm, pale brown, $2n = 24$. *Flowers November to February (to May).* (Fig. 225, Pl. 6d).

Bogs, swamps, wet grassland, wet rocks, feldmark; 0–800 m.

223
225

224
226

Fig. 223. *Senecio acanthifolius*. Fig. 224. *Perezia lactucoides*. Fig. 225. *Perezia magellanica*. Fig. 226. *Agoseris coronopifolia*.
(Sydney Parkinson)

Throughout Fuegia except N. & E. C. Isla Grande. (Map 29c).
W. Argentina north to *c*.49°20′S.lat.
Chile north to *c*.46°30′S.lat.

4. *P. recurvata* (Vahl) Less. 1830, *Linnaea* **5**: 21.
Homoianthus echinulatus Cass. 1825, *Dict. Sci. Nat.* **38**: 458 – Hooker, 1847; Alboff, 1896.

Plant forming mats or compact clumps; stems 3–45 cm, procumbent to erect, woody and branched at base, with dense leaf-remains below and imbricate leaves above, sparsely glandular-pubescent above, often reddish. Leaves 7–25 × 1–2 mm, linear to linear-lanceolate, spiny-apiculate, somewhat widened and sheathing at base, with ciliate, strongly revolute margins with curved spines on the rounded part, recurved, rigid, coriaceous, glabrous, dark glossy green. Capitula 12–32 mm in diameter, 1(–3), terminal on short, scape-like stem-apex; involucral bracts in 3–9 series, the outer 3–10 × 1–3 mm, ovate, mucronate, ciliate or spiny on margin, glabrous or sparsely pubescent, green, often reddish distally, scarious on margin, the inner longer, lanceolate. Florets blue or white, the outer with tube 8–11 mm and outer lip 6–11 × 1.5–2.5 mm, oblong. Achene 3–5 mm, ± turbinate, subglabrous to puberulent; pappus 7–15 mm, pale brown, rarely whitish. $2n = 24$. *Flowers November to March.*

Rocks and sands, *Empetrum* heath, river banks, dry open grassland; 0–450 m.

Widespread in Fuegia except the west and extreme east. (Map 29d).
Argentina north to *c*.30°S.lat., especially in the west.
Chile north to *c*.35°S.lat.
Falkland Islands (Islas Malvinas).

A variable species within which several taxa have been recognized. Vuilleumier (1970) has shown that there is a complex of continuous and partially discontinuous characters northwards. Cabrera (1971) has recognized several subspecies throughout the range; on this treatment Fuegian, S. Patagonian and Falkland Islands material belongs to ssp. *recurvata*, with the characters given above.

Tribe CICHORIEAE Florets with ligules, hermaphrodite. Stigmatic branches long, flattened, hairy beneath.

35. *Hypochoeris* L.

Perennial herbs. Leaves basal, simple to pinnatifid. Capitula solitary, terminal on simple or forked scape-like peduncles; receptacle flat, with bracts; involucral bracts in 3-numerous series. Florets monomorphic, ligulate, hermaphrodite; ligule 5-dentate at apex. Pappus of 1 series of plumose hairs, sometimes also with outer series of simple hairs.

Literature: A. L. Cabrera, *Bol. Soc. Argent. Bot.* **10**: 166–195(1963).

a. Leaves hispid; pappus-hairs in plumose inner and simple outer series **4. *radicata***
a. Leaves softly tomentose or lanate to glabrous; pappus of 1 series of plumose hairs:
  b. Flowers white or pink; involucral bracts obtuse; leaves usually tomentose to lanate **1. *incana***
  b. Flowers yellow; at least inner involucral bracts acute; leaves glabrous or almost so:
    c. Achene with long apical beak **2. *arenaria***
    c. Achene without apical beak **3. *patagonica***

1. *H. incana* —(Hooker & Arn.) Macloskie 1906, in Scott, *Rep. Princeton Univ. Exped. Patag.* **8(2)**: 899.
Achyrophorus coronopifolius Sch. Bip. 1845, *Nova Acta Acad. Leop. Carol.* **21**: 97 – Autran, 1905.
Leontodon coronopifolius (Sch. Bip.) Alboff 1896, *Rev. Mus. La Plata Bot.* **7**: 376.
Hypochoeris coronopifolia (Sch. Bip.) Franchet 1889, *Miss. Sci. Cap Horn* 350 – Spegazzini, 1896; Skottsberg, 1916.
H. tenuifolia auct., non (Hooker & Arn.) Griseb. – Hooker, 1847; Spegazzini, 1896.

Leaves 20–200 × 1–10 mm, linear to lanceolate, subacute, sheathing at base, entire to remotely dentate or deeply pinnatifid, the segments linear, entire to dentate, tomentose, sometimes lanuginose or subglabrous. Peduncle 5–300 mm, simple, with few linear bracts, ± tomentose. Capitula 15–35 mm in diameter; involucral bracts in numerous series, 10–20 × 1–2.5 mm, linear, obtuse, tomentose to glabrous, dark green, black at least distally. Florets white or pink, sweet-scented; ligule 7–10 × 2–3.5 mm. Achene 9–14 mm, fusiform, ± curved, ribbed, transversely rugose, narrowed to apex, without beak, pale brown; pappus-hairs 10–12 mm, in 1 series, plumose, white or brownish at base. $2n = 16$. *Flowers November to January*. (Fig. 218).

Coastal sands and gravels, open *Chiliotrichum*, *Berberis buxifolia* and *Pernettya mucronata* scrub, dry grassland and *Empetrum* heath; 0–150(–650) m.

Almost throughout Fuegia, except the extreme west, south and east. (Map 29e).
W. Argentina north to *c*.41°S.lat.
Chile north to *c*. 46°S.lat.

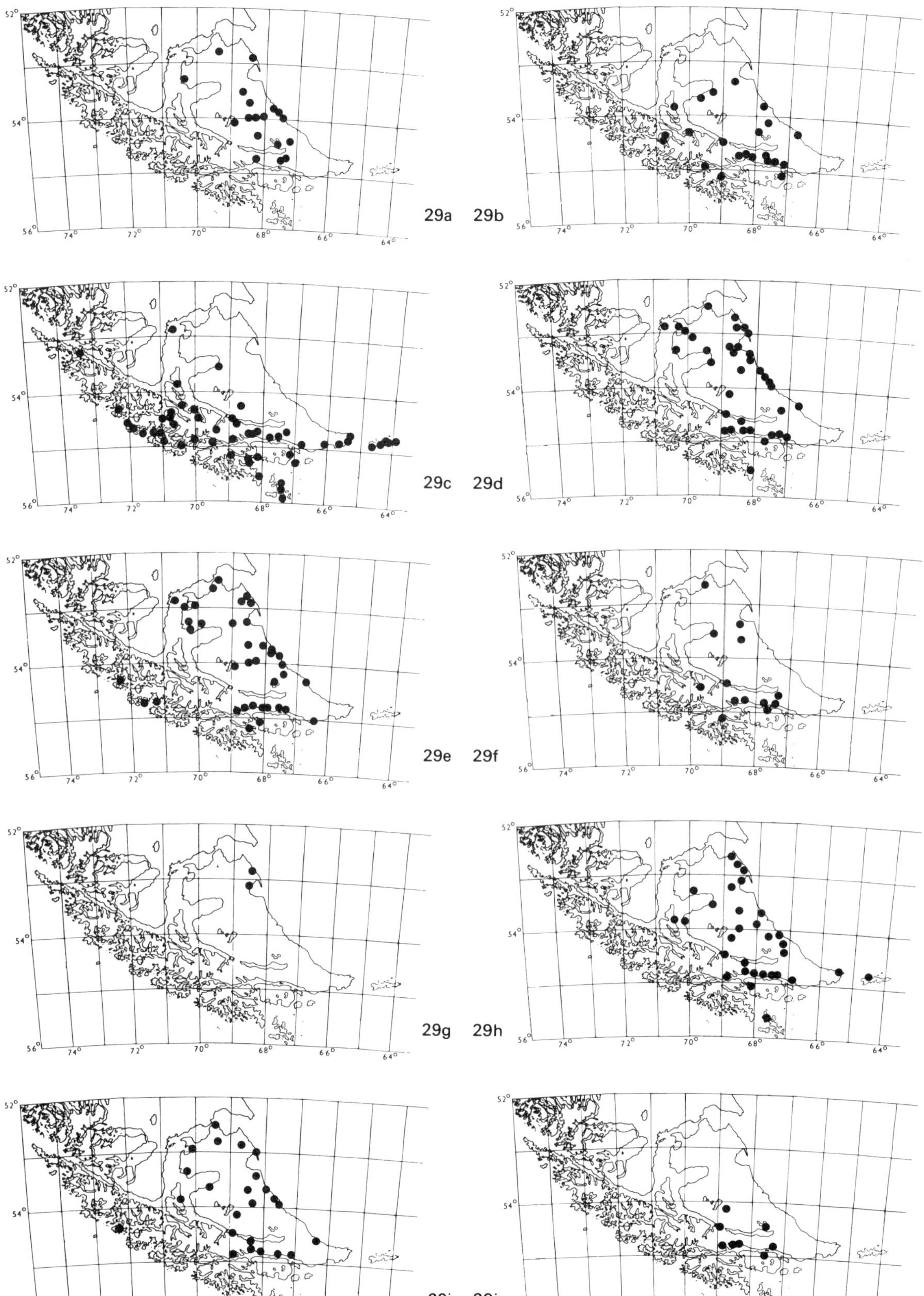

Map 29. a, *Perezia pilifera*. b, *Perezia lactucoides*. c, *Perezia magellanica*. d, *Perezia recurvata*. e, *Hypochoeris incana*. f, *Hypochoeris arenaria*. g, *Hypochoeris patagonica*. h, *Taraxacum gilliesii*. i, *Agoseris coronopifolium*. j, *Hieracium antarcticum*.

A very variable species within which several varieties have been recognized on the basis of leaf-shape and indumentum. However, since these features frequently vary within the same population, or even individual, formal recognition of infra-specific taxa seems premature.

2. *H. arenaria* Gaudich. 1825, *Ann. Sci. Nat.* **5**: 103.
Achyrophorus arenarius (Gaudich.) DC. 1838, *Prodr.* **7(1)**: 95 – Hooker, 1847.
Leontodon arenarius (Gaudich.) Alboff 1896, *Rev. Mus. La Plata Bot.* **7**: 376.
Hypochoeris ushuaiae Skottsb. 1906, *Wiss. Ergebn. Schwed. Südpolarexped.* **4(4)**: 37.
H. palustris auct., non (Phil.) Wildem. – Dusén, 1900.
H. tenerifolia auct., non (Remy) Dusén – Skottsberg, 1916.

Leaves (10–)30–150 × 5–17 mm, obovate to oblanceolate, obtuse to subacute, long attenuate into sheathing base, sinuate-dentate to lobed or irregularly and deeply pinnatifid, the segments linear to triangular-lanceolate, acute, glabrous or sparsely hairy on margins and veins beneath. Peduncle 10–230 mm, simple, rarely forked at base, with 0–4, ± linear, acute bracts, glabrous or hairy above. Capitula 10–20 mm in diameter; involucral bracts in 3–4 series, 5–25 × 1–2 mm, linear to linear-lanceolate, acute, or the outer subobtuse, coarsely hairy, dark brownish green to black with pale scarious margin. Florets yellow; ligule 4–6 × 1–1.5 mm. Achene 5–8 mm, narrowly pyriform, tapering to beak 2.5–4 mm, ribbed, muricate, dark brown; pappus-hairs 5–7 mm, in 1 series, plumose, white. $2n = 8$. *Flowers November to March.*

Open areas in most communities except forest; 0–800 m.

Isla Grande, except extreme west and east, and islands immediately south of Canal Beagle. (Map 29f).
W. Argentina and E. Chile north to *c.*38°S.lat.

A very variable species, particularly in stature, leaf-shape and size of capitula. Some of the variation is clearly due to environmental factors and no infra-specific categories can be usefully recognized at present.

3. *H. patagonica* Cabrera 1969, *Bol. Soc. Argent. Bot.* **11**: 289.

Similar to *H. arenaria* but leaves acute, glabrous; peduncle simple, glabrous; involucral bracts in 4–5 series, glabrous; achene without beak. *Flowers December to January.*

Coastal sands and saline grassland; 0–30 m.

N. E. Isla Grande. (Map 29g).
Argentina north to *c.* 45°30′S.lat.

4. *H. radicata L. 1753, *Sp. Pl.* 811.

Leaves 70–140 × 7–22(–40) mm, oblong-oblanceolate, obtuse, long attenuate into wide base, sinuate-dentate to -pinnatifid, hispid. Peduncles 16–53 cm, ascending or erect, simple or forked, with numerous ovate, acuminate, brownish bracts, usually glabrous. Capitula 15–35(–40) mm in diameter; involucral bracts in several series, 3–20 × 1.2–2 mm, lanceolate, acuminate, entire, glabrous or sparsely ciliolate, the outer hispid on midvein, brownish green, with scarious margin. Florets yellow; ligule 8–14 × 1.5–2.7 mm, linear-lanceolate. Achene 4.5–5(–7) mm, narrowly pyriform, tapering to beak 3–8 mm, or the outer without beak, ribbed, transversely rugose, muricate, dark reddish brown; pappus-hairs in 2 series, white, the inner 7.5–9 mm, plumose, the outer *c.*3 mm, scabrid. *Flowers December to March.*

Open ground in degraded grassland and forest, open scrub, coastal sands, disturbed areas near roads and habitations; 0–350 m.

Isla Grande, especially in coastal areas, except the extreme west and east.
Introduced from Europe and widely naturalized in temperate regions.

36. *Tragopogon* L.

Annual to perennial herbs. Leaves alternate, simple. Capitula solitary, on long terminal peduncles; receptacle flat, without bracts; involucral bracts in 1 series. Florets monomorphic, hermaphrodite, with 5-dentate ligule. Achene with apical beak; pappus of 1 series of plumose hairs.

1. *T. pratensis L. 1753, *Sp. Pl.* 789.

Stems 30–70 cm, erect, simple or sparingly branched above, glabrous or somewhat lanate when young. Basal leaves up to 240 × 6 mm, linear to linear-lanceolate, acuminate, wide and sheathing at base, entire, glabrous; cauline leaves smaller, similar but semi-amplexicaul. Capitula 20–40 mm in diameter, the peduncle enlarged just below capitulum; involucral bracts 22–32 × 4.5–8 mm, lanceolate to oblong-ovate, acuminate, glabrous, green, brown at base and blackish distally. Florets pale yellow; ligule

8–10 × 1.5–2 mm. Achene 11–14 mm, fusiform, ribbed, muricate, glabrous, yellow-brown, the beak *c.*2 mm; pappus 20–30 mm, white or brownish. *Flowers December to January.*

Coastal sands, disturbed grassland near habitations; 0–50 m.

N. E. Isla Grande.
Introduced from Europe and widely naturalized in temperate regions.

37. ***Taraxacum*** Weber

Perennial herbs with tap-root. Leaves in basal rosette, simple to pinnatifid. Capitula solitary, terminal on leafless scape; receptacle flat, without bracts; involucral bracts in several imbricate series. Florets monomorphic, ligulate, hermaphrodite; ligule 5-dentate. Achene with long beak; pappus of several series of strigose hairs.

a. Outer involucral bracts strongly reflexed **2. *officinale***
a. Outer involucral bracts appressed to spreading, not reflexed **1. *gilliesii***

1. *T. gilliesii* Hooker & Arn. 1835, *Comp. Bot. Mag.* **1**: 31.
T. magellanicum Comm. ex Sch. Bip. 1855, *Flora (Regensb.)* **38**: 122 – Skottsberg, 1916.
T. dens-leonis var. *laevigatum* Hooker f. 1847, *Fl. Antarct.* **1(2)**: 323.
T. laevigatum auct., non (Willd.) DC. – Spegazzini, 1896; Alboff, 1896; Dusén, 1900.

Leaves 25–100 × 4–12 mm, sublinear to oblong-obovate or oblanceolate, obtuse to subacute, tapering to base, distantly crenate to crenate-dentate or shallowly to deeply runcinate - pinnatifid, the lobes becoming smaller towards apex, glabrous or rarely sparsely white-hairy. Scapes 1 or more, 2–17 cm, erect, glabrous to sparsely hairy distally. Capitula 8–20 mm in diameter; involucral bracts 2–14 × 1.2–2.5 mm, ovate, acute to ± obtuse, glabrous but apex ciliate, dark green with wide pale scarious margin, often purplish distally, the outer appressed, later spreading. Florets yellow, the ligule 4–8 × 1–1.5 mm, linear-oblong, at least the outer green beneath. Achene 3–4 mm, cylindrical to obovoid, prominently ribbed, muricate at least distally, brownish, the beak 3–7 mm, slender; pappus about as long as achene and beak, white. $2n = 16$. *Flowers November to February.* (Fig. 231).

Coastal sands, open grassland, open scrub, among cushion plants and beside streams above timberline; 0–1100 m.

Almost throughout Fuegia except extreme west and south. (Map 29h).
W. Argentina north to *c.* 35°S.lat.
E. Chile north to *c.* 46°S.lat.
Falkland Islands (Islas Malvinas).
New Zealand.

2. *T. officinale Weber 1780, in Wiggers, *Prim. Fl. Holsat.* 56.

Similar to *T. gilliesii* but generally larger and more robust. Capitula 25–40 mm in diameter; outer involucral bracts linear to linear-lanceolate, green, without pale scarious margin at least distally, strongly reflexed. *Flowers November to March.*

Coastal sands, disturbed grassland and forest, open soil by habitations and roads; 0–100(–800) m.

Isla Grande, except the extreme west; Isla de los Estados; Isla Navarino.
Introduced from Europe and widely naturalized in open and modified habitats.

38. ***Sonchus*** L.

Annual herbs, sometimes overwintering. Leaves alternate, entire to pinnatifid. Capitula in irregular corymbose umbels; receptacle flat, without bracts; involucral bracts in several imbricate series. Florets monomorphic, ligulate, hermaphrodite, yellow; ligule 5-dentate. Achene without beak; pappus of 2 equal series of strigose hairs.

a. Cauline leaves with acute auricles; achene transversely rugose **1. *oleraceus***
a. Cauline leaves with rounded auricles; achene not transversely rugose **2. *asper***

1. *S. oleraceus L. 1753, *Sp. Pl.* 794.

Annual or overwintering; stems 17–70 cm, erect, branched above, glabrous or glandular-hairy above. Basal and lower cauline leaves 13–17 × 4–9 cm, usually ovate, attenuate at base, pinnatifid, glabrous, glaucous, with oblong, acute to subobtuse, dentate lobes, the terminal usually wider than the uppermost pair of lateral lobes; petiole up to 5 cm, winged, with acute, usually spreading auricles; upper leaves smaller, lanceolate, serrate, glabrous, glaucous. Capitula 20–25 mm in diameter; involucral bracts 4–12 × 1–1.8 mm, lanceolate, acute, glabrous or glandular-hairy, the outer shorter than the inner.

Florets with ligule 3–5 × c.1 mm, oblong, the outer purplish beneath. Achene 2.8–3.3 mm, oblanceolate, compressed, 3-ribbed on each face, transversely rugose, brown; pappus 8–10 mm, white. *Flowers November to January.*

Coastal sands and gravels, open disturbed ground by roads and habitations.

Locally on N. W. & S. coasts of Isla Grande.
Introduced from Europe and widely naturalized in temperate regions.

*2. ***S. asper*** (L.) Hill 1769, *Herb. Brit.* **1**: 47.

Similar to *S. oleraceus* but leaves less frequently pinnatifid, if so with the terminal lobe narrower than the uppermost pair of lateral lobes, the margin often crispate and spinose-ciliate, dark glossy green above and with rounded basal auricles appressed to the stem; achenes sometimes with narrow, ciliolate wings, not transversely rugose. *Flowers January to March.*

Open ground along coasts, degraded scrub, disturbed ground near habitations.

Coastal areas of N. E., C. N. W. & S. Isla Grande.
Introduced from Europe and widely naturalized in temperate regions.

39. *Agoseris* Rafin.

Perennial herbs. Leaves basal, entire to pinnatifid. Capitula solitary, terminal on leafless scape; receptacle flat, without bracts; involucral bracts in several imbricate series. Florets monomorphic, ligulate, hermaphrodite, yellow; ligule 5-dentate. Achene with long beak; pappus of several series of simple, strigulose hairs.

Literature: K. L. Chambers, *Quart. Rev. Biol.* **38(2)**: 124–140 (1963).

1. ***A. coronopifolium*** (D'Urv.) Chambers ex D. M. Moore 1968, *Brit. Antarct. Survey Sci. Rep.* **60**: 137.
   Macrorhynchus pumilus (Gaudich.) DC. 1838, *Prodr.* **7(1)**: 152 – Hooker, 1847; Franchet, 1889.
   M. pterocarpus auct., non Fischer & Meyen – Dusén, 1900; Autran, 1905.
   Troximon pumilum (Gaudich.) Wildem. 1905, *Result. Voy. S. Y. Belgica*, 182 – Skottsberg, 1916.
   T. poeppigii (DC.) Wildem. 1905, *Result. Voy. S. Y. Belgica* 182.
   T. glaucum auct., non (Nutt. ex Fraser) Pursh – Spegazzini, 1896.

Leaves 20–150 × 1–15 mm, oblanceolate to oblong or linear, acute to subobtuse, tapering to base, entire to ± deeply cut into linear to triangular-ovate, acute segments, glabrous to fimbrio-ciliate or hairy. Scape 1–30 cm, erect, usually ± appressed-hairy. Capitula 8–18 × 6–13 mm; involucral bracts 4–15 × 1–2 mm, linear to linear-lanceolate, acute, entire, white-ciliate, the outer ± lanuginose, dark green, with pale scarious margin. Florets with ligule 3–5 × 0.8–1 mm, the outer often purplish beneath. Achene 4–5 mm, narrowly pyriform, with beak 3.5–5 mm, ribbed, glabrous, brown; pappus 3.5–5 mm, white. $2n = 18$. Self-compatible, facultatively self-pollinated. *Flowers November to February.* (Fig. 226).

Coastal sands, grassland, open scrub and forest margins; 0–500 m.

Almost throughout Fuegia, except extreme west, south and east. (Map 29i).
W. Argentina north to c.40°S.lat.
Chile north to c.26°S.lat.
Falkland Islands (Islas Malvinas).

40. *Crepis* L.

Annual or biennial herbs. Leaves alternate, dentate to pinnatifid. Capitula in terminal corymbose cymes; receptacle flat, without bracts; involucral bracts in 2 series, the outer shorter. Florets monomorphic, ligulate, hermaphrodite, yellow; ligule 5-dentate. Achene without beak; pappus of several series of simple hairs.

*1. ***C. capillaris*** (L.) Wallr. 1841, *Linnaea* **14**: 657.

Stems 10–50 cm, ascending to erect, glabrous or hairy below. Basal and lower cauline leaves 30–100 × 3–15 mm, lanceolate to oblanceolate, obtuse to acute, narrowed at base, dentate to runcinate-pinnatifid, with distant, dentate lobes, glabrous or sparsely hairy; middle and upper cauline leaves lanceolate, acute, sagittate and amplexicaul at base, dentate. Capitula 6–13 mm in diameter, the pendunciles slender, glabrous to hairy; inner involucral bracts 5–8 × 1.2–1.3 mm, linear-lanceolate, acute, glandular hairy, green with pale scarious margin, the outer about ⅓ as long. Florets with ligule 4–6 × c. 1 mm, the outer often reddish beneath. Achene 2–2.8 mm, cylindrical, somewhat compressed, 10-ribbed, glabrous, pale brown; pappus 3–4 mm, white. *Flowers January to March.*

Disturbed ground near habitations, beside tracks and roads, coastal cliffs; 0–50 m.

W. & S. coasts of Isla Grande.
Introduced from Europe and widely naturalized in S. Chile and S. Argentina.

41. *Hieracium* L.

Perennial herbs, sometimes stoloniferous. Leaves alternate or in basal rosette, entire to dentate. Capitula ± campanulate, solitary and terminal or in corymbose cymes; receptacle flat, without bracts; involucral bracts in 2 or more irregular series. Flowers monomorphic, ligulate, hermaphrodite, yellow; ligule 5-dentate. Achene without beak, usually 10-ribbed; pappus of 1–2 series of simple hairs.

Literature: H. Sleumer, *Bot. Jahrb.* **77**: 85–148(1956).

a. Plant with stolons; achenes with ribs projecting to form a crenulate apex; pappus-hairs in 1 series:
 b. Scapes each with 1 capitulum **1. pilosella**
 b. Scapes each with 2–6 capitula **2. *flagellare***
a. Plant without stolons; achenes with ribs united distally into indistinct apical ring; pappus-hairs in 2 series:
 c. Leaves with dense, long, white or yellow hairs **5. *patagonicum***
 c. Leaves glabrous or with sparse short hairs, often with short glandular hairs at margin:
  d. Leaves ± obtuse, entire or inconspicuously sinuate-dentate; involucral bracts densely black-hairy, sometimes with a few stalked glands **4. *antarcticum***
  d. Leaves acute, serrate-dentate; involucral bracts with dense, black, glandular hairs **3. *murorum***

*1. *H. pilosella* L. 1753, *Sp. Pl.* 800.

Plant with long, slender, leafy stolons. Rosette-leaves 3–6 × 0.5–0.9 cm, oblanceolate, obtuse to subacute, entire or indistinctly denticulate, with sparse, long, simple hairs, usually also with stellate hairs beneath. Scapes up to 8 cm, each with 1 capitulum, with black glandular hairs and shorter, whitish, simple and stellate hairs. Capitula 8–16 mm in diameter; involucral bracts 6–10 × 0.5–1 mm, linear-lanceolate, acute, with dense, black glandular hairs. Florets with ligule *c.*3 × 0.8 mm, usually with red stripe beneath. Achene up to 2.5 mm, narrowly obconical, crenulate at apex; pappus-hairs in 1 series, *c.*1½ times as long as achene, whitish or pale yellowish. *Flowers December.*

Disturbed scrub/grassland, stabilized mine spoil-heaps; *c.*100 m.

N. W. Isla Grande (Altos del Bocquéron, Río del Oro).
Introduced from Europe, probably with grass-seed.

Fuegian material belongs to ssp. *euronotum* Naegeli & Peter.

*2. *H. flagellare* Willd. 1814, *Enum. Pl. Horti Berol., Suppl.* 54.

Plant with long, stout, leafy stolons. Rosette-leaves 3–5 × 0.7–1.5 cm, oblanceolate to spathulate, obtuse to subacute, entire to indistinctly undulate, with long, pale simple hairs, usually also with stellate hairs beneath. Scapes up to 20 cm, each with 2(–6) capitula, with black glandular and longer, paler, simple eglandular hairs. Capitula 12–20 mm in diameter; involucral bracts 6–10 × 0.5–1 mm, linear-lanceolate, acute, with dense, black glandular and longer simple hairs. Florets with ligule 3–4 × *c.*1.1 mm, usually with red stripe beneath. Achene *c.* 2 mm, narrowly obconical, crenulate at apex; pappus-hairs in 1 series, 1½–2½ times as long as achene, whitish. *Flowers January to February.*

Disturbed grassland in cleared *Nothofagus antarctica* forest; *c.*60 m.

S. Isla Grande (Ea Harberton).
Introduced from Europe, probably with grass-seed.

Fuegian material belongs to ssp. *flagellare*.

*3. *H. murorum* L. 1753, *Sp. Pl.* 802.

Stolons absent. Rosette-leaves with lamina 4–10.5 × 2–4 cm, oblong-lanceolate, acute, truncate at base, serrate-dentate, glabrous or with short, simple hairs above, more densely hairy beneath; petiole 1.5–7 cm, with long, pale, simple hairs. Scapes 20–60 cm, with 0–2 small leaves and 6–35 capitula in corymbose cyme, with sparse simple and stellate, sometimes glandular, hairs. Capitula 10–20 mm in diameter; involucral bracts 8–11 × 0.5–1 mm, linear-lanceolate, acute, with dense, black, glandular hairs. Florets with ligule 7–10 × 2–2.5 mm. Achene *c.*3 mm, cylindrical-obconical, with indistinct apical ring; pappus-hairs in 2 series, 2–3 times as long as achene, whitish. *Flowers December to January.*

Cleared *Nothofagus* forest, dry grassland, at tree-line in open soil with *Empetrum* and *Bolax*; 0–500 m.

S. Isla Grande (Ea Harberton).
Introduced from Europe, probably with grass-seed.

Reports of *H. rhomboides* (Sternstrom) K. Joh. by Moore and Goodall (1977) refer to material included here, since that species is described as having at least 2 cauline leaves.

4. *H. antarcticum* D'Urv. 1825, *Fl. Is. Mal.* 39.
H. philippii Alboff 1896, *Rev. Mus. La Plata Bot.* **7**: 377.

Stolons absent. Rosette-leaves with lamina 2–6 × 0.4–1.6 cm, elliptical to oblanceolate, ± obtuse, cuneate at base, entire or obscurely sinuate-dentate, glabrous or sparsely floccose-pubescent, usually with shortly stalked marginal glands; petiole $\frac{1}{2}$ to as long as lamina; cauline leaves smaller upwards. Scapes 3–21 cm, with (1) 2–6 capitula in corymbose cyme, with dense black, simple, glandular and eglandular hairs. Capitula 5–10(–15) mm in diameter; involucral bracts 4–10 × 1–1.5 mm, linear-lanceolate, acute to subobtuse, densely black-hairy, sometimes with a few stalked glands. Florets with ligule 2–3 × *c.*1 mm. Achene 3–3.5 mm, ± cylindrical, with indistinct apical ring; pappus-hairs in 2 series, *c.*1$\frac{1}{2}$ times as long as achene, pale yellowish brown. $2n = 10$. Self-compatible and self-pollinated. *Flowers December to March.* (Fig. 232).

Open, often rather dry, clay-sand soils by coast and rivers, screes, feldmark; 0–900 m.

C. & S. Isla Grande, N. Isla Navarino. (Map 29j).
W. Argentina north to *c.*37°S.lat.
E. Chile north to *c.*51°S.lat.
Falkland Islands (Islas Malvinas).

The recognition of 2 varieties, distinguished on stature and leaf-size (Sleumer, 1956), seems of dubious value.

5. *H. patagonicum* Hooker f. 1847, *Fl. Antarct.* **1(2)**: 324.
H. austroamericanum Dahlst. 1907, in Dusén, *Ark. Bot.* **7(2)**: 51 – Skottsberg, 1916.
H. chilense auct., non Less. – Alboff, 1896.

Stolons absent. Rosette-leaves 5–11 × 1–2.5 cm, oblong-oblanceolate, obtuse, narrowing to short, winged petiole, entire to obscurely sinuate-dentate, with dense, long, white or yellowish simple hairs; cauline leaves smaller upwards. Scapes up to 38 cm, with 6–20 or more capitula in corymbose cyme, with long, whitish simple hairs and with short, glandular hairs. Capitula 5–15 mm in diameter; involucral bracts 7–10 × 1–1.4 mm, lanceolate to oblong-lanceolate, acute, floccose towards base and with short, black, glandular hairs near mid-vein. Florets with ligule *c.*5 × 1.3 mm. Achene 3–3.5 mm, ± cylindrical, with indistinct apical ring; pappus-hairs *c.*1$\frac{1}{3}$ times as long as achene, pale brownish. $2n = 18$*. Self-compatible and self-pollinated. *Flowers January to February.*

Open scrub and grassland; 30–100 m.

C. & S. Isla Grande, Isla Navarino. (Map 30a).
W. Argentina north to *c.*42°40′S.lat.
E. Chile north to *c.*35°S.lat.
Falkland Islands (Islas Malvinas).

LXXV. JUNCAGINACEAE

Rhizomatous perennial herbs. Leaves obscurely alternate, basal, linear, with basal sheath. Flowers actinomorphic, or weakly zygomorphic, hermaphrodite or unisexual, in terminal ebracteate spike or raceme. Perianth-segments 4 or 6, in 2 series, free. Stamens 4 or 6, inserted at base of perianth-segments; anthers subsessile. Ovary superior, of 4 or 6 partly united carpels, each with 1 ovule; stigmas plumose. Fruit of indehiscent, 1-seeded follicles; seed without endosperm.

a. Leaves flat, folded; flowers unisexual; ovary with 4 fertile carpels **1. *Tetroncium***
a. Leaves cylindrical or semi-cylindrical; flowers hermaphrodite; ovary with 6 fertile carpels **2. *Triglochin***

1. *Tetroncium* Willd.

Dioecious, glabrous herbs. Leaves flat, folded, strongly keeled. Flowers deflexed, weakly zygomorphic. Perianth-segments 4. Stamens 4. Carpels 4, united at base. Fruit subulate.

1. *T. magellanicum* Willd. 1808, *Ges. Nat. Freunde Berl. Mag.* **2**: 17.

Rhizome ascending, with stems towards apex; stems 2.5–6 cm, ascending or erect, branched near base, densely covered with brown leaf-bases, with leaves towards apex. Leaves 17–110 × 1.5–3 mm, acute, rigid, distichous. Spike 1.3–3.5 cm; scape 2.5–18 cm, erect; pedicels *c.*0.5 mm, or absent. Perianth-segments 1.2–2 × 0.5–1.5 mm, the 2 upper larger, ovate to ovate-lanceolate, acute to subobtuse, entire or irregularly serrulate, reddish brown. Fruit 4–8 × *c.*1 mm, with 4 wide ribs which separate from the middle into 4, attenuate, distally curved horns formed by the persistent styles, smooth, reddish brown. *Flowers December to March.* (Fig. 258).

Sphagnum and *Astelia* bogs, Magellanic moorland; 0–140 m.

C. & S. Fuegia, except Isla de los Estados. (Map 30b).
W. Argentina north to *c.*40°S.lat.
Chile north to *c.*42°S.lat.

2. *Triglochin* L.

Glabrous herbs. Leaves cylindrical or semi-cylindrical; sheath with apical entire or 2-lobed ligule. Flowers actinomorphic, ± erect, hermaphrodite. Perianth-segments 6. Stamens 6. Carpels 6, ± fused to central axis (carpophore) from which they detach at maturity. Fruit oblong or ± linear.

a. Fertile carpels 6; carpophore without wings; fruit oblong **1. *concinna***
a. Fertile carpels 3; carpophore with wings; fruit linear **2. *palustris***

1. *T. concinna* Davy 1895, *Erythea* **3(7)**: 117.
T. maritimum auct., non L. – Alboff, 1896; Spegazzini, 1896; Skottsberg, 1916.
T. monanthos Speg. 1924, *Bol. Acad. Ci. Córdoba* **27**: 337.
T. striata auct., non Ruiz & Pavón – Dusén, 1900; Wildeman, 1905.

Rhizome ± horizontal. Leaves 1–16 cm; lamina 1–2 mm wide, semi-cylindrical, ± flattened towards apex, obtuse; sheath membranous, the ligule 2-lobed. Raceme 0.5–7 cm; scape 1–20 cm, erect; pedicels up to 3 mm or absent. Perianth-segments 1–1.5 × *c.*1 mm, the inner somewhat smaller, ovate, subobtuse, entire or crenulate, brownish. Fruit 3–4 × 2–2.5 mm, oblong, with 6 fertile follicles, brown, the carpophore without wings. $2n = 48$. *Flowers November to December.* (Fig. 228).

Wet coastal sands, muds and rock-crevices, often inundated by high tides, and inland saline lakes; 0–10 m.

N. & C. Fuegia, except the extreme west and east, but easily overlooked and probably under-recorded. (Map 30c).
W. & E. Argentina north to *c.*34°S.lat.
Chile north to *c.*35°S.lat.
North America.

2. *T. palustris* L. 1753, *Sp. Pl.* 338.

Rhizome ± horizontal, short, with bulbous stolons. Leaves 2.5–30 cm; lamina 1–3 mm wide, ± cylindrical, obtuse; sheath membranous, the ligule emarginate or 2-lobed. Raceme (1–) 7–20 cm; scape 8–37 cm, erect; pedicels up to 6 mm. Perianth-segments 1.5 × 1–1.5 mm, ovate-elliptical, subobtuse, entire, greenish or reddish. Fruit 7–10 × *c.*1 mm, linear, with 3 fertile follicles, brown, the carpophore with 3 wings. *Flowers December to February.*

Moist open ground by coast and inland waters; 0–40 m.

Almost throughout Fuegia, except the extreme west, south and east. (Map 30d).
Argentina north to *c.* 29°S.lat.
Chile north to *c.* 42°S.lat.
New Zealand. Eurasia. North America.

LXXVI. POTAMOGETONACEAE

Aquatic perennial herbs of fresh or brackish water. Leaves alternate, simple, with sheathing scale ('stipule') in leaf-axil. Flowers actinomorphic, hermaphrodite, in axillary spikes. Fruit a drupe-like achene, up to 4 from each flower; seed 1, without endosperm.

1. *Potamogeton* L.

Plant glabrous. Leaves often dimorphic, narrower and more delicate when submerged than when floating. Stamens sessile at base of perianth-segments. Fruit with fleshy exocarp and boney endocarp.

Literature: R. L. Pérez-Moreau, in M. N. Correa (Ed.) *Flora Patagónica* **2**: 18–21 (1969).

a. Leaves dimorphic, the submerged at least 5 mm wide **1. *linguatus***
a. Leaves monomorphic, all submerged, not more than 2 mm wide **2. *strictus***

1. *P. linguatus* Hagström 1901, in Dusén, *Öfvers. Kongl. Vet.-Akad. Förhandl.* **58(4)**: 259.
P. natans auct., non L. – Dusén, 1900.

Stems up to 20 cm, rather rigid, simple or branched, subterete. Leaves dimorphic, obtuse, entire; submerged leaves 40–60 × 5–15 mm, linear to linear-lanceolate, rather membranous, sessile or with petiole up to 30 mm; floating leaves 30–80 × 15–50 mm, elliptical to ovate, cuneate at base, coriaceous, often reddish, with petiole 4–8 cm; stipules 30–60 × 5–14 mm, lanceolate or ovate-lanceolate, obtuse, sheathing at base, persistent, free. Peduncle 3–15 cm, erect; spike 8–40 × 3–8 mm, cylindrical. Perianth-segments 2–3 × 1.5–2.5 mm, unguiculate. Achenes 2.5–3.5 × 1.5–2 mm, obovoid, compressed, irre-

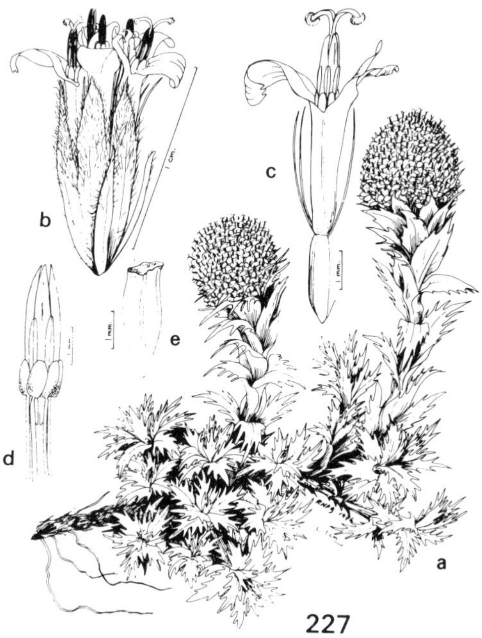

227

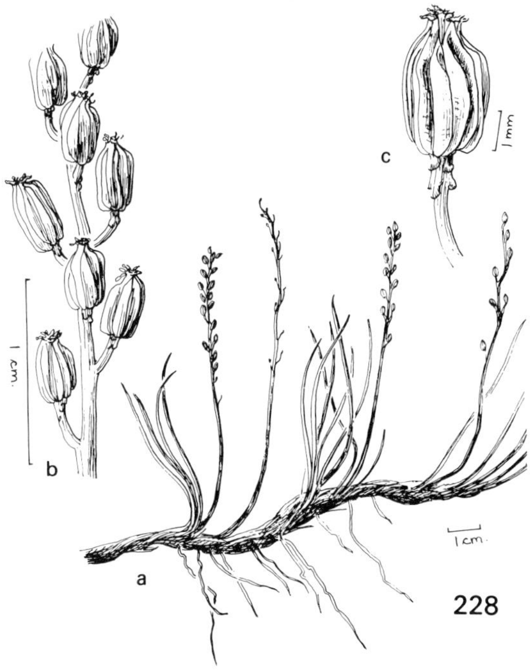

228

229

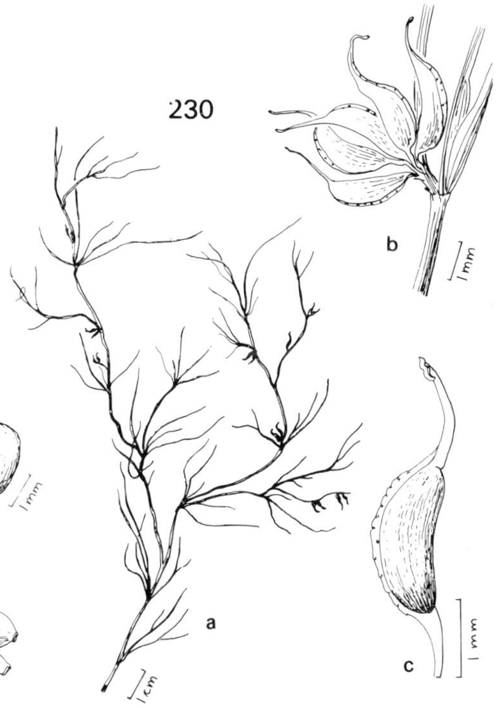

230

Fig. 227. *Nassauvia magellanica.* a, Plant. b, Capitulum. c, Floret. d, Staminal tube. e, Achene. Scale = 1 cm (a–b) or 1 mm (c–e). Fig. 228. *Triglochin concinna.* a, Plant. b, Infructescence. c, Fruit. Scale = 1 cm (a–b) or 1 mm (c). Fig. 229. *Potamogeton strictus.* a, Plant. b, Inflorescence. c, Flower. d, Infructescence (part). e, Fruit. Scale = 1 cm (a–b) or 1 mm (c–e). Fig. 230. *Zannichellia palustris.* a, Plant. b, Infructescence. c, Fruit. Scale = 1 cm (a) or 1 mm (b–c).

(R. N. P. Goodall)

gularly rugose, reddish brown. *Flowers January to February*.

Freshwater lakes and slow-moving streams; 0–80 m.

C. & S. Isla Grande, except the extreme west and east. (Map 30e).
W. Argentina and Chile north to *c*. 41°S.lat.
Falkland Islands (Islas Malvinas).

2. *P. strictus* PhiL. 1860, *Fl. Atacam*. 50.
P. pusillus auct., non L. – Spegazzini, 1896.

Stems up to 50 cm, simple or branched, terete. Leaves monomorphic, submerged, 30–150 × 1–2 mm, linear-filiform, obtuse, entire, membranous; stipules 5–20 × 1–2 mm, filiform, sheathing at base, united near middle to leaf-base. Peduncle 5–10 cm, erect; spike 15–40 × 3–5 mm, with flowers in 2–6 whorls. Perianth-segments *c*.1 × 1.5 mm, unguiculate. Achenes 2–2.5 × 1.5–2 mm, subobovoid, compressed, irregularly rugose. *Flowers December to January*. (Fig. 229).

Freshwater streams and lakes; 0–200 m.

N. & C. Isla Grande. (Map 30f).
W. Argentina north to *c*. 34°S.lat.
Chile north to *c*. 24°S.lat.

Fuegian material belongs to var. *magellanicus* Hagström, which has wider leaves and more completely united stipules than var. *strictus*, and occurs throughout much of the whole range of the species given above.

LXXVII. ZANNICHELLIACEAE

Monoecious, submerged, perennial aquatic herbs. Leaves alternate or opposite; stipules present. Flowers unisexual, in axillary cymes. Perianth absent. Stamen 1. Carpel 1; style simple; stigma peltate. Fruit an achene; seed 1, without endosperm.

1. *Zannichellia* L.

Plant glabrous. Leaves mostly opposite. Cyme of 1 male and 2–5 female flowers in a hyaline, deciduous, cup-shaped spathe.

1. *Z. palustris* L. 1753, *Sp. Pl.* 969.

Rhizome slender; stems 3–30 cm, filiform, much branched. Leaves 10–100 × 0.4–*c*.1 mm, filiform, acuminate, flat; stipule amplexicaul, tubular below, membranous. Male flower of 1 stamen on long, slender filament. Female flower of 1 carpel, with short, stout style, the stigma ± undulate and crenate. Achene 2–3 mm, oblong, ± dentate on dorsal margin, with persistent style 0.5–1.5 mm, subsessile or shortly stalked. *Flowers December to February*. (Fig. 230).

Shallow freshwater lakes and margins of slow-moving streams; 0–30 m.

E. C. Isla Grande. (Map 30g).
Almost throughout the temperate regions of the world.

LXXVIII. RUPPIACEAE

Perennial, submerged aquatic herbs. Leaves alternate or opposite, simple. Flowers hermaphrodite, ebracteate, in small pedunculate spikes. Perianth absent. Stamens 2, opposite, Carpels 4, free. Fruitlets achene-like; seed 1, without endosperm.

1. *Ruppia* L.

Leaves with shortly auriculate basal sheath. Spikes 2-flowered, subtended by 2 involucral leaves. Stamens sessile, with small dorsal appendage at base. Carpels with sessile, peltate stigma.

Literature: J. C. Gamarro, *Darwiniana* **14**: 575–608 (1968).

1. *R. filifolia* (Phil.) Skottsb. 1916, *Kungl. Svenska Vet. Akad. Handl.* **56(5)**: 171.

Stems up to 40 cm, slender, much branched, with leaves and roots at nodes. Leaves up to 20 cm, up to 0.5 mm wide, filiform, acute, entire but minutely denticulate towards apex, yellow-green; sheath 3–5 cm. Spike *c*.6 × 2 mm; peduncle short, elongating to 15–18 mm in fruit. Anthers *c*.2 × 2 mm, reniform, purplish-brown. Fruitlets 2–2.5 × 1.5 mm, obliquely ovoid, on stalk 1–2 mm. *Flowers December to January*. (Fig. 255).

Freshwater lakes and brackish pools and streams on coast; 0–80 m.

C. & S. Isla Grande except extreme west and east. (Map 30h).
Chile north to 51°28′S. and at *c*. 23°S.lat.
Falkland Islands (Islas Malvinas). ?Elsewhere.

Gamarro (1968) refers all South American material to either *R. cirrhosa* (Petagna) Grande or R. *maritima* L. On his key Fuegian material has the anther-size and -shape of the former but the narrower, acute

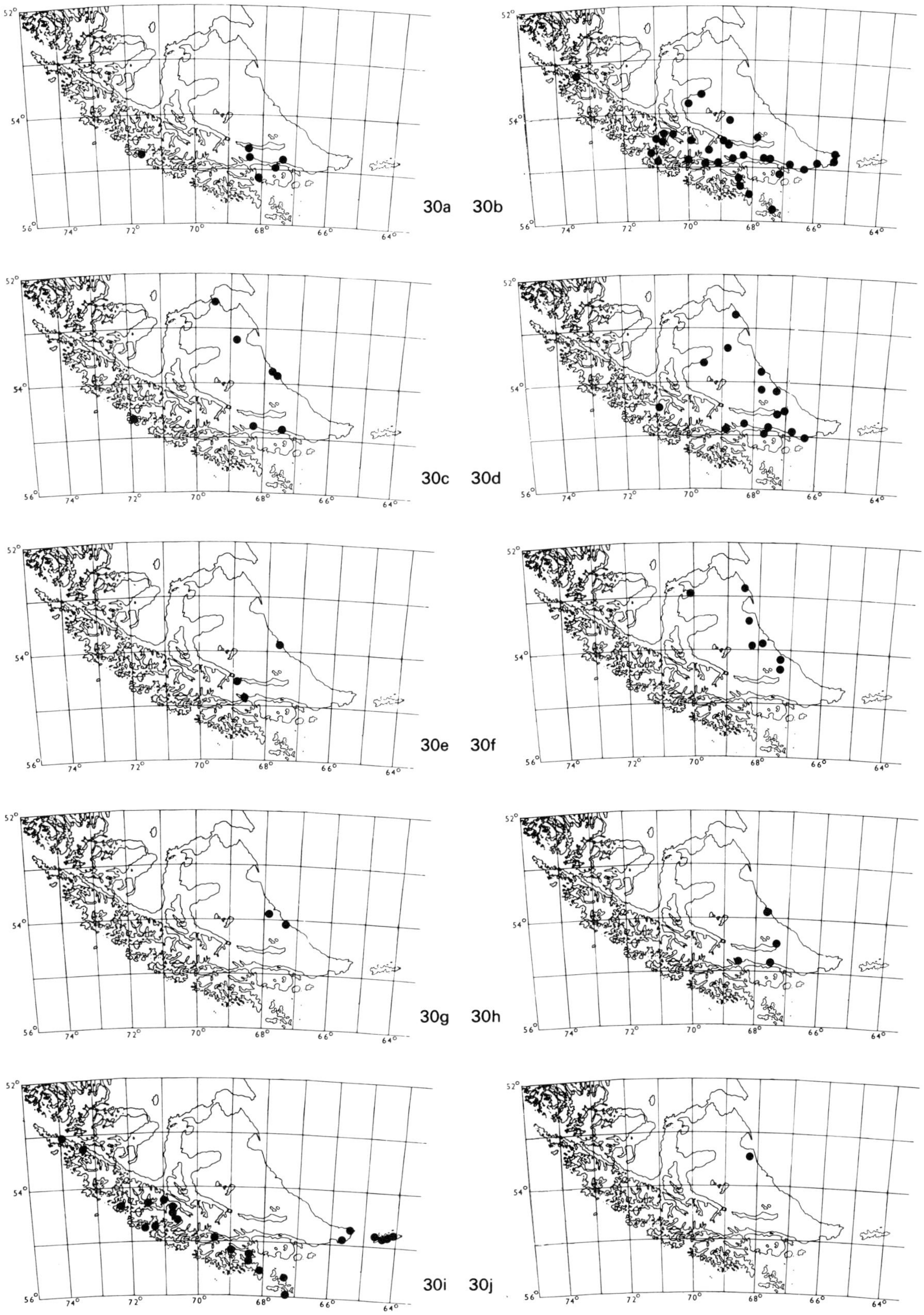

Map 30. a, *Hieracium patagonicum*. b, *Tetroncium magellanicum*. c, *Triglochin concinna*. d, *Triglochin palustris*. e, *Potamogeton linguatus*. f, *Potamogeton strictus*. g, *Zannichellia palustris*. h, *Ruppia filifolia*. i, *Gaimardia australis*. j, *Festuca arundinacea*.

leaves and shorter non-spiral fruiting peduncle of the latter species. Consequently, we adopt here the separate taxon until further work can clarify the variation in this difficult genus.

LXXIX. CENTROLEPIDACEAE

Perennial herbs. Leaves closely distichous. Flowers unisexual, in a pseudanthium composed of 0–2 male and 0 to several female flowers enclosed by 2 bracts. Perianth O. Stamen 1, with 1-celled anther. Carpel 1; style 1; stigma curved, papillose, Fruit with membranous pericarp; seed 1, with endosperm.

1. *Gaimardia* Gaudich.

Plant moss-like, densely tufted and cushion-forming. Leaves with wide membranous basal sheath having apical ligule. Pseudanthia with usually 2 male and 2 female flowers, usually laterally united. Fruit dehiscing loculicidally with 2 valves.

1. *G. australis* Gaudich. 1825, *Ann. Sci. Nat.* **5**: 100.

Stems 2–9 cm, erect or suberect, much branched, densely clothed in leaves which are green towards apices of stems. Leaves 4–9 × *c.*0.5 mm; lamina linear-subulate, acute to obtuse, strict, channelled above; sheath ± equalling lamina, with obtuse ligule 0.5–1 mm. Inflorescence usually enclosed by upper leaves. Lower bract *c.*4 × 1.5 mm, ovate, apiculate, green, with wide, pale scarious margin, the upper bract shorter and narrower. Anthers exserted, purple or yellow. Stigma reddish purple. Fruit *c.*4.5 × 0.9 mm, cylindrical to oblong-ovoid, compressed, smooth, golden brown, dark purplish brown at apex. $2n = 16^{\star}$. *Flowers November to January*. (Fig. 257).

Bogs with *Astelia, Donatia, Caltha* or *Empetrum*, wet rocks and screes, bryophyte carpets; 0–650 m.

S. & W. Fuegia. (Map 301).
Chile north to 42°15′S.lat.
Falkland Islands (Islas Malvinas).

LXXX. GRAMINEAE

Annual or perennial herbs, often with rhizomes or stolons. Stems of 2 sorts in perennials – flowering stems (culms) and sterile or vegetative shoots; in annuals all stems bear flowers; culms terete or compressed. Leaves alternate, 2-ranked, comprising:– sheath, encircling the stem, with free, overlapping or connate margins, sometimes with auricles at mouth; ligule, a small membranous flap of tissue or a row of hairs at junction of sheath and lamina; lamina linear to filiform, usually constricted below at junction with sheath. Inflorescence a panicle, raceme or spike, composed of 1-many spikelets; spikelet comprising 1-many distichous, ± imbricate, sessile florets on a slender, jointed axis (rhachilla), usually with 2 bracts (glumes) at base. Florets hermaphrodite or unisexual, with or without 2 small, delicate scales (lodicules) at base, usually enclosed between 2 bracts of which the lower (lemma) is coriaceous to membranous, sometimes with elongate base (callus), and the upper (palea) is usually thin and delicate but sometimes absent. Stamens 1–3. Ovary superior, 1-celled; styles usually 2, normally with plumose stigmas. Fruit (grain) a dry achene (caryopsis). (Fig. 283).

Literature: E. G. Nicora, in M. N. Correa (Ed.) *Flora Patagónica* **3**: 1–563 (1978).

a. At least some spikelets pseudoviviparous:
  b. Lemma rounded on back, the lower with short apical awn; leaf-sheath with auricles — **1. *Festuca***
  b. Lemma keeled, awnless; leaf-sheath without auricles — **3. Poa**
a. Spikelets sexual or sterile:
  c. Ligule a row of hairs:
    d. Leaf-sheath with auricles; lemma 7-veined, deeply 2-fid, with geniculate dorsal awn; anthers not more than 1.2 mm — **32. *Rytidosperma***
    d. Leaf-sheath without auricles; lemma 3-veined, truncate, with straight apical awn; anthers at least 2 mm — **31. *Cortaderia***
  c. Ligule membranous:
    e. Inflorescence a single terminal spike:
      f. Spikelets with back of lemma towards rhachis — **2. *Lolium***
      f. Spikelets with side of lemma towards rhachis:
        g. Spikelets in groups of 2–6 at each node of rhachis:
          h. Neither glumes nor lemma with awn — **9. *Leymus***
          h. Glumes and lemma awned:
            i. Spikelets 1-flowered, in groups of 3 at each node of rhachis, the central spikelet sessile, hermaphrodite, the lateral pedicellate, hermaphrodite, male or sterile; glumes reduced to long awns — **14. *Hordeum***
            i. Spikelets 2- to 4- flowered, in pairs at each node of rhachis, sessile, hermaphrodite; glumes lanceolate — **10. *Elymus***

g. Spikelets solitary at each node of rhachis:
j. Glumes subulate, with 0–1 veins **13. *Secale***
j. Glumes lanceolate to ovate, with 4 or more veins:
k. Glumes persistent; veins of lemma not confluent towards apex **12. *Triticum***
k. Glumes falling with florets when ripe; veins of lemma confluent towards apex **11. *Agropyron***
e. Inflorescence a lax to contracted, often dense and spike-like, panicle:
l. Spikelets with 1 hermaphrodite floret only:
m. Inflorescence a dense, spike-like panicle:
n. Ligule at least 8 mm; spikelets at least 12 mm **25. *Ammophila***
n. Ligule not more than 5 mm; spikelets not more than 7 mm;
o. Glumes glabrous, with rigid cilia on keel, with apical awn; lemma without awn **27. *Phleum***
o. Glumes with soft, appressed hairs on keel and sides, only rarely with short apical awn; lemma with dorsal or subapical awn **28. *Alopecurus***
m. Inflorescence a lax to contracted panicle:
p. Lemma with awn at least 30 mm, strongly indurate when mature **30. *Stipa***
p. Lemma awnless or with awn not more than 7 mm, not strongly indurate when mature:
q. Callus with hairs $\frac{1}{2}$ as long as to exceeding lemma **26. *Calamagrostis***
q. Callus glabrous or with hairs less than $\frac{1}{2}$ as long as lemma:
r. Rhachilla not prolonged beyond floret **24. *Agrostis***
r. Rhachilla prolonged beyond floret:
s. Culms not more than 40 cm; anthers not more than 0.4 mm; palea not more than 2.5 mm **19. *Deschampsia***
s. Culms at least 50 cm; anthers at least 1.8 mm; palea at least 4 mm **26. *Calamagrostis***
l. Spikelets with 2 or more florets, some sometimes sterile:
t. Spikelets with 1 fertile floret and 2 sterile florets below, sometimes reduced to a lemma:
u. Lemmas without awns; glumes equal, whitish with conspicuous green veins; plant not smelling of coumarin **29. *Phalaris***
u. Lower (sterile) lemmas awned; glumes unequal, without conspicuous green veins; plant smelling of coumarin **22. *Anthoxanthum***
t. Spikelets with 2 or more fertile florets, hermaphrodite or unisexual:
v. Lemma with a dorsal awn; glumes usually ± as long as or exceeding lowest floret:
w. Spikelets with 2 florets; upper floret male, the lemma awned; lower floret hermaphrodite, awnless **23. *Holcus***
w. Spikelets with 2 or more florets; at least lower floret with awned lemma; at least upper floret hermaphrodite:
x. Lower floret male, the lemma with long awn; upper floret hermaphrodite, the lemma with or without short awn **16. *Arrhenatherum***
x. All florets hermaphrodite, with similarly awned lemmas:
y. Spikelets with 3–4 florets; glumes at least 15 mm **15. *Avena***
y. Spikelets with 2–3 florets; longest glume not more than 9 mm:
z. Annual; rhachilla not or scarcely produced beyond upper floret; longest glume not more than 3.5 mm **20. *Aira***
z. Perennial; rhachilla distinctly produced beyond upper floret; glumes at least 4 mm:
aa. Spikelets with 2 florets; lemma rounded on back, obtuse, truncate or dentate; glumes subequal **19. *Deschampsia***
aa. Spikelets with 2–3 florets; lemma strongly keeled, subacute to deeply bifid; glumes unequal:
bb. Lemma deeply bifid, the lateral veins reaching the apex **18. *Trisetum***
bb. Lemma entire or shallowly 2-lobed at apex, the lateral veins not reaching apex **17. *Koeleria***
v. Lemma awnless or with apical awn, sometimes awned from or near apical sinus; glumes usually shorter than lowest floret:
cc. Spikelets strongly dimorphic, in clusters of 3–7 with laterally compressed fertile spikelet subtended by several sterile spikelets, borne on 1 side of axis **6. *Cynosurus***
cc. Spikelets not strongly dimorphic:
dd. Spikelets in dense 1-sided clusters

towards ends of the panicle-branches **5. *Dactylis***

dd. Spikelets not as above:

ee. Spikelets with 1 hermaphrodite or female floret and 2 larger male florets below **21. *Hierochloë***

ee. Spikelets not as above:

ff. Lemma with apical awn:

gg. Upper glume 5- to 7-veined; ovary with lobed hairy apical appendage; styles lateral below apex of ovary **8. *Bromus***

gg. Upper glume 3-veined; ovary glabrous or sparsely pubescent at apex; styles apical:

hh. Lemma rounded on back; leaf-sheaths with auricles; ligule not more than 3 mm **1. *Festuca***

hh. Lemma keeled; leaf-sheaths without auricles; ligule at least 4 mm **3. *Poa***

ff. Lemma without awn:

ii. Lower glume less than ½ as long as lemma, ovate to elliptical; lemma obtuse, 3-veined **7. *Catabrosa***

ii. Lower glume usually more than ½ as long as lemma, lanceolate to lanceolate-ovate; lemma acute to obtuse, 5-veined, or 3-veined and acute:

jj. Lemma strongly keeled **3. *Poa***

jj. Lemma rounded on back:

kk. Leaf-sheaths with auricles; lemma lanceolate to lanceolate-ovate, subobtuse to acute **1. *Festuca***

kk. Leaf-sheaths without auricles; lemma ovate to ovate-lanceolate, obtuse to subobtuse **4. *Puccinellia***

Tribe POEAE

Spikelets hermaphrodite, or unisexual, rarely some sterile, sometimes pseudo-viviparous, laterally compressed to terete, with (1)2– many florets. Glumes shorter than spikelet. Palea 2-keeled. Lodicules 2, entire or lobed. Stamens (1–)3. Ovary glabrous or hairy at apex.

1. *Festuca* L.

Caespitose or rhizomatous perennials. Ligule membranous, often ciliate; leaf-sheath with auricles. Inflorescence a lax to contracted, often spike-like, panicle. Spikelets subterete to compressed, with 3–8(–10) florets. Glumes usually keeled, awnless; upper 3-veined; lower 1(–3)-veined. Lemma 5-veined, rounded on back but ± keeled distally, usually with apical awn. Palea 2-veined, shortly 2-fid, ± as long as lemma. Lodicules acute or minutely lobed. Anthers 2–6 times as long as wide. Ovary usually glabrous, rarely sparsely pubescent at apex. Styles short, apical.

Literature: L. R. Parodi, *Rev. Argent. Agron.* **20**: 177–229(1953).

a. Lamina (1.5–)3–10 mm wide, flat or folded:

b. Rhizomes absent; auricles spreading, ciliate; lemma-awn 0 or 1–4 mm **1. *arundinacea***

b. Rhizomes present; auricles inconspicuous, glabrous; lemma-awn 0.2–1.5 mm **2. *purpurascens***

a. Lamina not more than 3.5 mm wide, setaceous, at least distally:

c. Glumes at least as long as adjacent lemmas **5. *contracta***

c. Glumes not more than ¾ as long as adjacent lemmas:

d. At least some spikelets pseudo-viviparous **10. monticola**

d. All spikelets sexual:

e. Plant with rhizomes:

f. Lemma 5–6 mm, glabrous or scabrid towards apex; anthers 2–3 mm **3. *rubra***

f. Lemma 7–10 mm, densely hairy; anthers 4–5 mm **4. *juncifolia***

e. Plant caespitose, without rhizomes:

g. Leaves at least 1 mm wide; auricles lobed, up to 3 mm in upper sheaths **6. *cirrosa***

g. Leaves usually less than 1 mm wide; auricles continuous with ligule, not more than 1 mm:

h. Lamina finely pointed at apex; anthers at least 2.5 mm **9. *gracillima***

h. Lamina obtuse; anthers not more than 2(–2.5) mm:

i. Lamina glabrous, or sparsely hairy above; anthers at least 1.8 mm **8. *ovina***

i. Lamina hairy above; anthers not more than 1.5 mm **7. *magellanica***

1. *F. arundinacea Schreber 1771, *Spicil. Fl. Lips.* 57.

Plant caespitose. Leaves 10–25 cm × 6–8 mm; lamina flat, acuminate, rather rigid, scabrid at least above; ligule up to 2 mm; sheath smooth, the auricles spreading, shortly ciliate. Culms 50–80 cm, erect, glabrous. Panicle 20–30 cm, lax; branches few, solitary or paired, scabrid. Spikelets 11–14 mm, with

4–6 florets. Upper glume 6–6.5 mm, lanceolate, glabrous, the lower 3–5 mm, narrower. Lemma 8–9.5 mm, ovate-lanceolate, scabridulous distally and on veins; awn 0 or 1–4 mm, scabridulous. Palea as long as lemma, scabrid on keels. Anthers 2.8–3.5 mm. *Flowers November to January*.

Disturbed ground by roadside.

N. E. Isla Grande (Ea Sara). (Map 30j).
Introduced from Europe for pasture-improvement.

2. *F. purpurascens* Banks & Sol. 1847, in Hooker f., *Fl. Antarct.* **1(2)**: 383.

Plant with long rhizomes. Leaves 15–50 cm × 1.5–10 mm; lamina flat or ± folded, acuminate, scabridulous distally; ligule 0.5–1 mm, truncate, ciliate; sheath smooth, glabrous, the auricles continuous with ligule. Culms 13–83 cm, erect, glabrous. Panicle 5–24 cm, lax; branches few, solitary or paired, scabrid distally. Spikelets 8–13 mm, with 3–6 florets, purplish brown. Upper glume 4.6–7 mm, ovate-lanceolate, glabrous; lower 2.5–5 mm, linear-lanceolate. Lemma 6–8.5 mm, lanceolate, glabrous or somewhat scabrid distally; awn 0.2–3 mm, from apical notch. Palea slightly shorter than lemma, scabrid on keels. Anthers 1.8–3 mm. Ovary pubescent at apex. *Flowers December to February*.

Moist places by streams, coastal rocks; 0–30 m.

Almost throughout Fuegia except the extreme west. (Map 31a).
W. Argentina and Chile north to *c.*39°S.lat.

3. *F. rubra L. 1753, *Sp. Pl.* 74.

Plant rhizomatous. Leaves 3–40 cm × 0.5–1 mm; lamina setaceous, obtuse to abruptly acute, ± hairy above; ligule very short, truncate, ± ciliate; sheath glabrous to ± hairy, the lower reddish, the auricles rounded. Culms 30–40 cm, slender, erect, glabrous. Panicle 4.5–8 cm, linear-ovoid, lax; branches solitary or paired, slightly scabrid. Spikelets 8–11 mm, with 5–7 florets, purplish green. Glumes acute, glabrous; upper 4.5–6 mm, ovate-lanceolate; lower 3.2–4 mm, linear-lanceolate. Lemma 5–6 mm, ovate-lanceolate, scabrid distally; awn 0.3–2 mm, scabrid. Palea as long as lemma, scabrid on keels. Anthers 2–3 mm.

Artificial grassland, open coastal sands.

Locally on W, E. & S. coasts of Isla Grande.
Introduced from Europe for pasture-improvement.

4. *F. juncifolia St-Amans 1821, *Fl. Agen.* 40.

Similar to *F. rubra* but lamina acuminate; lemma 7–10 mm, densely hairy; anthers 4–5 mm.

Coastal sands.

N.E. & C.W. coasts of Isla Grande.
Introduced from Europe and locally naturalized.

5. *F. contracta* T. Kirk 1895, *Trans. Proc. N.Z. Inst.* **27**: 353.

F. erecta D'Urv. 1825, *Fl. Is. Mal.* 31 – Hooker, 1847; Franchet, 1889; Kurtz, 1896; Spegazzini, 1896; Autran, 1905; Skottsberg, 1916; Parodi, 1953.
F. paludicola Roivainen 1954, *Ann. Bot. Soc. Zool. – Bot. Fenn. Vanamo* **28**: 200.

Plant densely caespitose. Leaves 3–11(–25) cm × 0.6–0.8 mm; lamina setaceous distally, acute, strongly convolute, ± hairy above; ligule *c.*0.5 mm, truncate, ciliolate; sheath glabrous, shiny, the auricles continuous with ligule. Culms 3–31 cm, erect, slightly scabrid. Panicle 3–10, ± spike-like, contracted; branches in pairs, densely hairy on ridges. Spikelets 8–13 mm, with 3–4 florets, greenish and violet. Glumes scabridulous distally; upper 4.4–8.5 mm, ovate, acute; lower 4–7 mm, rarely 3-veined. Lemma 5–7.5 mm, ovate-lanceolate, scabrid to hairy; awn 0.5–3 mm, scabrid. Palea slightly shorter than to as long as lemma, sparsely scabrid on keels. Anthers 0.5–1 mm. *Flowers November to January*.

Feldmark, screes and rocks by streams, *Sphagnum* bogs, *Marsippospermum* communities; 250–900 m.

S. Fuegia. (Map 31b).
Falkland Islands (Islas Malvinas); South Georgia; Iles de Kerguelen; Macquarie Island.

6. *F. cirrosa* (Speg.) Parodi 1953, *Rev. Argent. Agron.* **20**: 190.

F. erecta var. *cirrosa* Speg. 1896, *Anal. Mus. Nac. Bs. Aires* **5**: 95.
F. longidiurma Parodi 1953, *Rev. Argent. Agron.* **20**: 214 – Moore, 1974.

Plant caespitose. Leaves 3–30 cm × 0.8–2 mm; lamina rush-like, acute, strongly convolute, glabrous beneath, pubescent above, pale green; ligule 0.5–3 mm, truncate, rigid, slightly ciliate; sheath glabrous, yellow-brown, the lower with lobed auricles 0.5–1.5 mm, the upper with ± lanceolate auricles up to 3 mm. Culms 20–48 cm, glabrous. Panicle 10–14 cm, linear-oblong, contracted; branches paired, short, subglabrous. Spikelets 14–18 mm, with 4–6(–8) florets, green to purplish. Glumes lanceolate, acuminate, ciliolate, ± scabrid on keel; upper 7–10 mm, the lower 6–9 mm, narrower. Lemma 8–11 mm, ovate-lanceolate, densely pubescent; awn 1.5–4 mm. Palea shorter than lemma, scabrid on keels. Anthers 2–3.5 mm. *Flowers November to March.*

Coastal sands and rocks; 0–50 m.

S. & E. Fuegia; C. W. coast of Isla Grande. (Map 31c).
Endemic.

Readily distinguished from the apparently closely related *F. contracta* by its subglabrous panicle-branches, glumes shorter than the adjacent lemmas and longer anthers.

7. *F. magellanica* Lam. 1788, *Encycl. Méth. Bot.* **2**: 461.

F. ovina var. *magellanica* (Lam.) Hackel 1900, in Dusén, *Svenska Exped. Magellansl.* **3(5)**: 228.
F. pyrogea Speg. 1896, *Anal. Mus. Nac. Bs. Aires* **5**: 97 – Parodi, 1953; Nicora, 1978.

Plant densely caespitose. Leaves (0.3–)1–5(–13) cm × 0.6–1(–3) mm; lamina setaceous, obtuse, glabrous to densely pubescent beneath, densely hairy above; ligule very short, truncate, ciliate; sheath glabrous or pubescent, smooth, brownish at base, the auricles continuous with ligule. Culms 4–37 cm, slender, glabrous or scabrid to hairy. Panicle 1–9 cm, linear; branches usually solitary, ± scabrid to hairy. Spikelets 6–13 mm, with 3–8 florets, green and purplish. Glumes acute, ± ciliate distally, rarely scabridulous; upper (2.4–)3.5–5.2(–6.7) mm, ovate-lanceolate; lower (1.4–)2.2–4.5(–5.5) mm, linear-lanceolate. Lemma 3.5–6(–7.5) mm, ovate-lanceolate, glabrous or partly scabrid to densely pubescent; awn 0.5–4.2 mm, usually longer in upper than basal florets of spikelet, scabrid to pubescent. Palea subequalling lemma, scabrid on keels. Anthers 0.5–1.4 mm. *Flowers November to February.* (Fig. 233).

Coastal sands and gravels, open scrub, grassland, *Empetrum* heath, river banks, screes, feldmark; 0–1000 m.

Almost throughout Fuegia except extreme west. (Map 31d).
W. Argentina north to *c.*37°S.lat.
Chile north to *c.*45°S.lat.

F. pyrogea has been distinguished from *F. magellanica* by its hairy lemmas. However, the degree of hairiness is very variable and cannot be correlated with any other character; furthermore there is some evidence that it is at least partly related to habitat dryness, and the 2 species are therefore considered synonymous.

8. *F. ovina L. 1753, *Sp. Pl.* 73.

Similar to *F. magellanica* but leaves longer and finer; lamina glabrous or sometimes sparsely hairy above; culms longer; lemma only slightly scabrid and with awn shorter in all florets; anthers 1.8–2.5 mm.

Hay meadows and disturbed ground.

S. & N. E. coasts of Isla Grande (Ea Harberton, Ea Viamonte).
Introduced from Europe for pasture-improvement.

9. *F. gracillima* Hooker f. 1847, *Fl. Antarct.* **1(2)**: 383.

Plant densely caespitose. Leaves 4–20 cm × 0.5–0.8 mm; lamina setaceous-filiform, acuminate, glabrous beneath, scabrid above; ligule up to 0.5 mm, truncate, ciliate; sheath glabrous, pale yellow-brown, the auricles continuous with ligule. Culms 7–65 cm, glabrous to scabrid. Panicle 3–8 cm, linear; branches solitary, scabrid, few, patent to erect. Spikelets 10–22 mm, with 3–7 florets, yellow-brown. Glumes lanceolate, glabrous or ± scabrid distally; upper 5–8.5 mm; lower 3.7–6.5 mm. Lemma 6.7–10.5 mm, lanceolate, acuminate, slightly scabrid on keel distally; awn 1.5–3.5 mm, scabrid. Palea ± as long as lemma, linear-ovate, puberulent on keels. Anthers 2.5–5 mm. Ovary slightly hairy at apex. *Flowers November to February.*

Coastal sands and rocks, dry grassland in which it is often dominant, open *Chiliotrichum* scrub; 0–60 m.

Isla Grande except extreme west; N. Isla Navarino and islands of Canal Beagle; Isla de los Estados. (Map 31e).
Argentina and Chile north to *c.*50°40′S.lat.

10. *F. monticola* Phil. 1873, *Anal. Univ. Chile* **43**: 576.

F. subantarctica Parodi 1953, *Rev. Argent. Agron.* **20**: 225 – Moore, 1974; Moore and Doggett, 1976; Nicora, 1978.
F. commersonii Franchet 1889, *Mis. Sci. Cap Horn* 388.

Plant caespitose. Leaves 4.5–20 cm × 1.4–3.5 mm; lamina rush-like, subobtuse, convolute, somewhat acute, rigid, glabrous beneath, scabridulous above; ligule 0.5–1 mm, truncate, ciliolate; sheath glabrous,

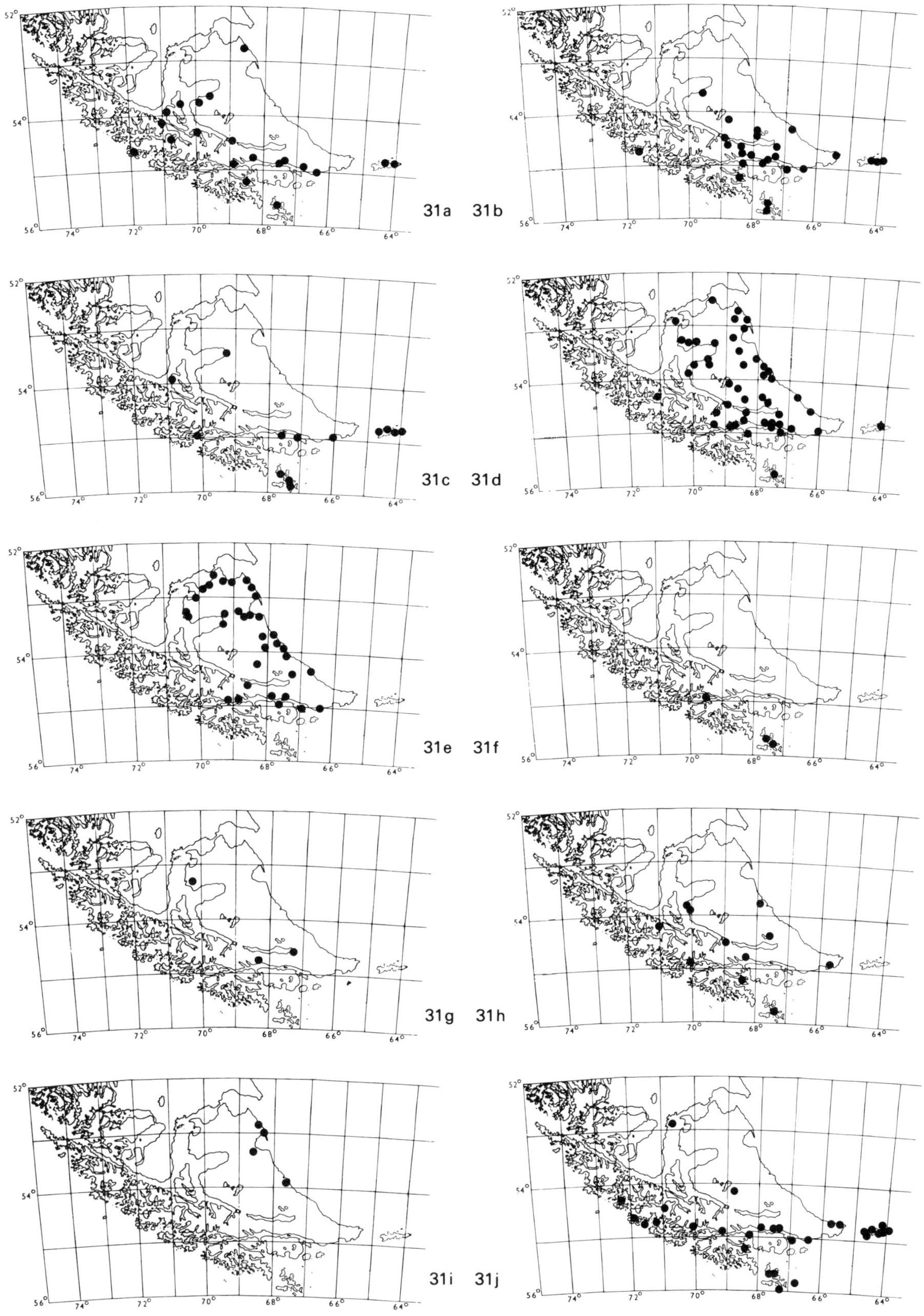

Map 31. a, *Festuca purpurascens*. b, *Festuca contracta*. c, *Festuca cirrosa*. d, *Festuca magellanica*. e, *Festuca gracillima*. f, *Festuca monticola*. g, *Poa scaberula*. h, *Poa yaganica*. i, *Poa atropidiformis*. j, *Poa darwiniana*.

brownish, the auricles continuous with ligule. Culms 38–60 cm, glabrous. Panicle 7–15 cm, lax; branches solitary or paired, slightly scabrid distally; spikelets mostly pseudo-viviparous. Sexual spikelets 8–11 mm, with 3–5 florets, brownish to purplish; glumes scabrid on keel, the upper 4.7–6.2 mm, ovate-lanceolate, the lower 4–5.5 mm, lanceolate; lemma 5–7 mm, ovate-lanceolate, densely pubescent, with awn *c.* 1 mm; palea ± as long as lemma, ovate-lanceolate, scabrid on keels; anthers 2.1–2.5 mm. Pseudo-viparous spikelets with 3(–5) florets, with glumes similar to those of sexual spikelets; lemma longer than in sexual spikelets, subglabrous; palea reduced; axis of spikelet produced into green, leafy shoot at apex to give vegetative progagule; anthers and ovary reduced.

Marsippospermym bogs, wet grassland with scattered *Nothofagus betuloides*, soil screes and open rocks; 100–500 m.

S. & ?W. Fuegia. (Map 31f).
W. Argentina and Chile north to *c.*40°S.lat.

Fuegian and southern continental material (north to *c.*51°S.lat.), which is partially or completely pseudo-viviparous, was separated as *F. subantarctica* from the more northerly sexual populations in which the plants have gibbous hairs on the upper surface of the leaves. Recent extensive collections in the S. Andes confirm the suggestion of Moore and Doggett (1976) that the southern material comprises an asexually reproducing facies of *F. monticola* in response to the more inclement conditions.

Superficially similar to some examples of the pseudo-viviparous *Poa alopecurus* ssp. *fuegiana* which, however, has hairs at the base of the florets, a lanceolate ligule more than 1 mm, and a keeled, awnless lemma.

2. *Lolium* L.

Annuals to perennials. Ligule membranous, obtuse; leaf-sheaths with or without auricles. Inflorescence a simple spike. Spikelets laterally compressed, alternating on opposite sides of axis, with back of lemma towards axis, with 6–18 florets. Glumes rounded on back, 4- to 7-veined, obtuse, the lower present only in apical spikelet. Lemma 5-veined, rounded on back. Palea acute. Lodicules lanceolate, acute. Anthers 3–5 times as long as wide. Styles short, apical.

a. Perennial, with non-flowering shoots at anthesis; leaves flat or folded when young; lemma usually awnless **2. *perenne***
a. Usually annual or biennial, without non-flowering shoots at anthesis; leaves convolute when young; lemma with subapical awn **1. *multiflorum***

*1. ***L. multiflorum*** Lam. 1779, *Fl. Fr.* **3**: 621.

Annual to short-lived perennial; non-flowering shoots absent at anthesis. Leaves 3.5–10 cm × 1.5–5 mm; lamina convolute when young, later flat, acuminate, glabrous beneath, usually scabridulous above; ligule 0.5–2 mm; sheath glabrous, usually with narrow, spreading auricles. Culms 36–90 cm, glabrous. Spike 9–30 cm, erect. Spikelets 9–25 mm, oblong, sessile, with 6–18 florets. Upper glume 3–8 mm, oblong, obtuse, glabrous, the lower similar but slightly smaller when present. Lemma 4.5–7 mm, oblong-lanceolate, glabrous, usually with straight, subapical awn 1–10 mm. Palea as long as lemma, the keels scabridulous. Anthers 3–4.5 mm. *Flowers January to May.*

Disturbed and open ground, especially near habitations.

Locally on N.E and S. coast of Isla Grande.
Introduced from Europe for pasture-improvement and widely naturalized in temperate regions.

*2. ***L. perenne*** L. 1753, *Sp. Pl.* 83.

Similar to *L. multiflorum* but perennial, with non-flowering shoots at anthesis; lamina flat or folded when young; sheath with or without inconspicuous auricles; spikelets with up to 10 florets; lemma usually awnless.

Disturbed and open ground, especially near habitations.

Locally on S. coast of Isla Grande.
Introduced from Europe for pasture-improvement and widely naturalized in temperate regions.

3. *Poa* L.

Annuals or perennials, caespitose or rhizomatous, often dioecious or gynodioecious. Ligule membranous; leaf-sheaths without auricles. Inflorescence a lax to contracted, often spike-like, panicle. Spikelets compressed, with (1)2–5(–9) florets. Glumes keeled, awnless; upper 3-veined; lower 1- to

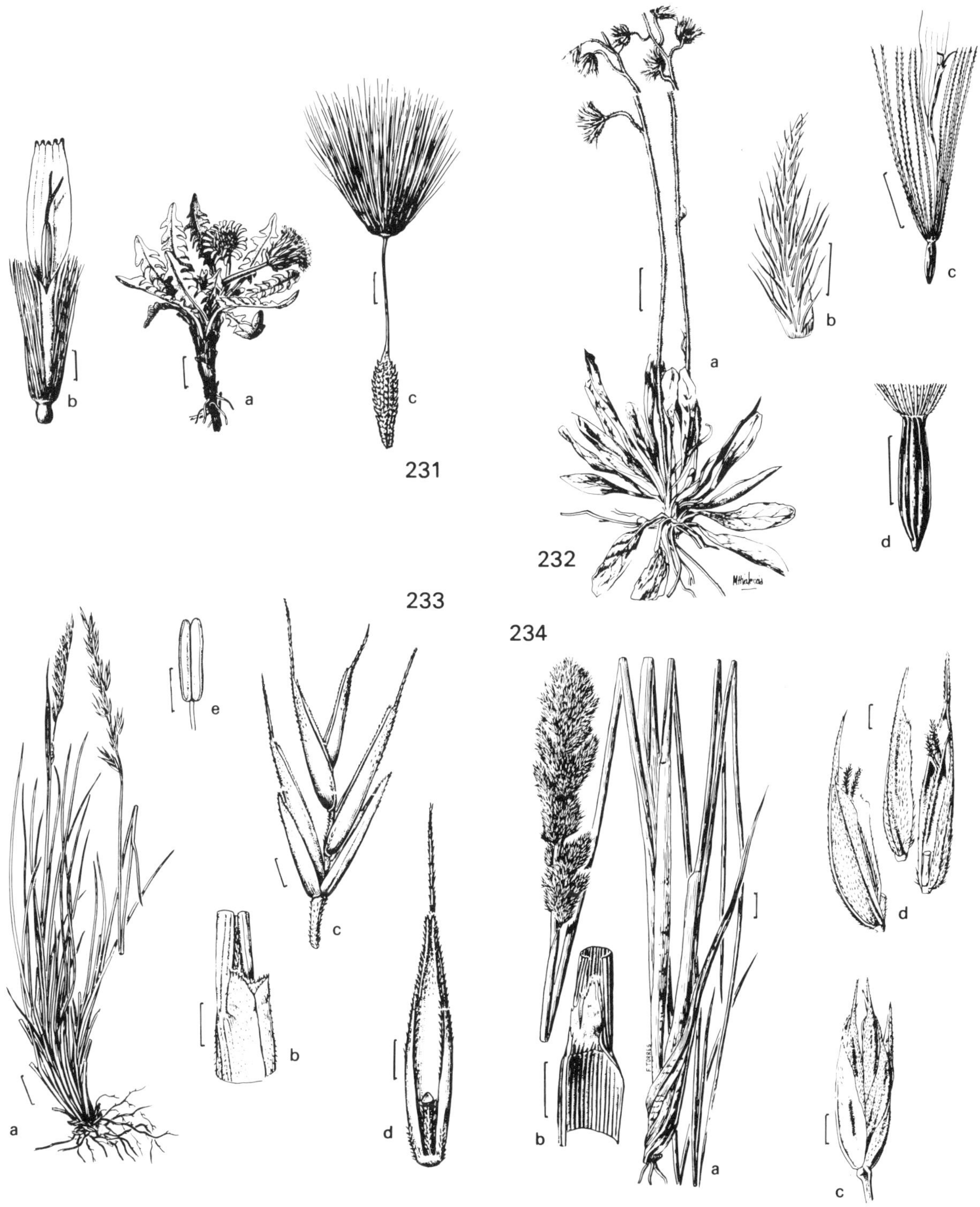

Fig. 231. *Taraxacum gilliesii*. a, Plant. b, Floret. c, Achene. Scale = 1 cm (a) or 1 mm (b–c). Fig. 232. *Hieracium antarcticum*. a, Plant. b, Involucral bract. c, Floret. d, Achene. Scale = 1 cm (a) or 1 mm (b–d). Fig. 233. *Festuca magellanica*. a, Plant. b, Ligule. c, Spikelet. d, Lemma. e, Anther. Scale = 1 cm (a), 1 mm (c–d) or 0.5 mm (b, e). Fig. 234. *Poa flabellata*. a, Plant. b, Ligule. c, Spikelet. d, Florets. Scale = 1 cm (a), 1 mm (c–d) or 0.5 mm (b).

(Flora Patagónica)

3-veined. Lemma 3- to 5-veined, keeled, rarely with apical awn. Palea 2-veined, bifid, shorter than to ± as long as lemma. Lodicules ovate, 2-lobed, somewhat swollen at base. Anthers 2–5 times as long as wide. Ovary glabrous. Styles short, apical.

A difficult genus in southern S. America. The monographic treatment of the Patagonian species by Nicora (1978) has clarified many of the problems posed by the Fuegian species but some groups still need extensive experimental study.

a. At least some spikelets pseudo-viviparous:
 b. Lemma in pseudo-viviparous spikelets 8–22 mm; leaves markedly pungent **10. *robusta***
 b. Lemma in pseudo-viviparous spikelets 5–8.5 mm; leaves non pungent **(11–12). *alopecurus*** group
a. Spikelets all sexual:
 c. Panicle lax, narrowly ovoid to pyramidal, often contracted when immature:
  d. Annual; keels of palea ciliate along whole length **1. *annua***
  d. Perennial; keels of palea scabrid or ciliate only in basal half:
   e. Ligule of uppermost leaf 5–10 mm **2. *trivialis***
   e. Ligule of uppermost leaf not more than 5 mm:
    f. Rhizomes absent; callus glabrous or sparsely lanate; lower glume 3-veined **4. *nemoralis***
    f. Rhizomes present; callus with few to many long, often woolly, hairs; lower glume usually 1-veined **(5–6) *pratensis*** group
 c. Panicle spike-like or coarctate:
  g. Lemma with apical awn:
   h. Lemma-awn more than 1 mm; lower glume not more than 5.5 mm, 1-veined; leaves not pungent; ligule at least 4 mm **9. *flabellata***
   h. Lemma-awn not more than 1 mm; lower glume at least 5.5 mm, 3-veined; leaves pungent; ligule not more than 3 mm **10. *robusta***
  g. Lemma awnless:
   i. Dioecious or gynodioecious; spikelets at least 4.5 mm; anthers at least 1.5 mm:
    j. Gynodioecious; female flowers with staminodes 1.5–3 mm **10. *robusta***
    j. Dioecious; female florets with staminodes not more than 0.5 mm **(11–12) *alopecurus*** group
   i. Plants with hermaphrodite florets; spikelets not more than 5 mm; anthers not more than 0.8 mm:
    k. Panicle not more than 2.5 cm; culms not more than 10 cm **8. *darwiniana***
    k. Panicle at least 5 cm; culms at least 10 cm:
     l. Culms glabrous or sparsely scabrid below panicle; leaves glabrous; lemma glabrous or scabridulous distally; glumes as long as or longer than lowest lemma **7. *atropidiformis***
     l. Culms and leaves scabridulous to scabrid; lemma with long hairs at base; glumes shorter than lowest lemma **3. *scaberula***

Subgen. POA
Plants with hermaphrodite flowers.

***1. P. annua** L. 1753, *Sp. Pl.* 68.

Laxly caespitose annual, erect to shortly creeping. Leaves 0.7–5.5 cm × 0.8–3.5 mm; lamina flat or folded, acute to subobtuse, ± flaccid, often transversely wrinkled, scabridulous, or glabrous beneath; ligule (0.6–)1–5 mm, obtuse, somewhat serrate; sheath keeled, glabrous. Culms 2–12(–23) cm, erect or ascending, slender, glabrous. Panicle 0.7–6.5 cm, ± pyramidal, lax; branches solitary or in pairs, patent or deflexed after flowering, glabrous. Spikelets 3–6.5 mm, with 2–6 florets. Glumes scabridulous on keel, purplish; upper 1.7–2.6 mm, ovate-lanceolate; lower 1.5–2.4 mm, lanceolate, 1-veined. Lemma 2.6–3.5 mm, ovate, usually sparsely hairy on veins. Palea 2.1–3 mm, ovate-lanceolate, ciliate along whole length of keels. Anthers 0.5–1 mm. *Flowers November to March.*

Coastal sands and rocks, open soil and gravels, by streams, roads and habitations; 0–200 m.

Scattered almost throughout Fuegia except extreme west.
Introduced from Europe and widely naturalized in temperate regions.

***2. P. trivialis** L. 1753, *Sp. Pl.* 67.

P. chrysantha Lindman ex Skottsb. 1916, *Kungl. Svenska Vet. Akad. Handle.* **16(5)**: 176 – Roivainen, 1954; Nicora, 1978.
P. breviculmis Pilger 1913, *Feddes Repert.* **11**: 307 – Skottsberg, 1916.
P. limicola Pilger, 1913, *Feddes Repert.* **12**: 308 – Skottsberg, 1916.

Laxly caespitose perennial, with short rhizomes. Leaves 3–32 cm × 0.5–6 mm; lamina flat or ± folded, acute to subobtuse, rather soft, scabridulous, or glabrous beneath; ligule of upper leaves 3.5–10 mm, acute; sheath glabrous to scabridulous. Culms 21–90 cm, erect or ascending from decumbent base, glabrous to scabridulous. Panicle 3.5–20 cm, ovoid to pyramidal, lax; branches arising 2–5 together, patent, scabrid. Spikelets 2–6 mm, with 2–4 florets. Glumes lanceolate, scabrid on keel; upper 1.7–6 mm; lower 1.2–5 mm, 1-veined. Lemma 2–5 mm, ovate to lanceolate, sparsely ciliate on keel, green or purplish. Palea 1.9–4 mm, lanceolate, scabridulous on keels. Anthers 1.2–2 mm. *Flowers November to January.*

Moist areas near streams and in swampy grassland; 0–100 m.

C. & S. Isla Grande, except extreme west and east. Introduced from Europe for pasture-improvement and widely naturalized in temperate regions.

P. breviculmis, described from Seno Almirantazgo, with which *P. limicola* and *P. chrysantha*, respectively from Rio Azopardo and Lago Fagnano, are synonymous, is based on larger specimens of *P. trivialis* having larger floral parts than normal.

3. ***P. scaberula*** Hooker f. 1847, *Fl. Antarct.* **1(2)**: 378.

Laxly caespitose perennial. Leaves 2–13 cm × 0.8–3 mm; lamina flat or ± folded, acute to subobtuse, scabrid or scabridulous; ligule 1–3 mm, acute to obtuse, scabridulous beneath; sheath scabridulous to scabrid. Culms 12–21(–40) cm, erect or ascending from decumbent base, slender, scabridulous to scabrid. Panicle 2–10.5 cm, coarctate, interrupted at base, dense; branches appressed, scabrid, with numerous spikelets on short pedicels. Spikelets 2–4.5 mm, with 2–4 florets. Glumes lanceolate, scabrid on keel; upper 1.6–3.5 mm; lower 1.5–2.5 mm, 1-veined. Lemma 2–2.6 mm, ovate-lanceolate, scabrid on keel above, ciliate below; callus lanuginose. Palea 1.5–1.8 mm, lanceolate, scabridulous on keels. Anthers 0.3–0.5 mm. *Flowers December to January.*

Moist grassland and open *Chiliotrichum* scrub; 100–250 m.

Locally in N. W. & C. S. Isla Grande. (Map 31g).
W. Argentina north to *c.*38°S.lat.
Chile on north side of Estrecho de Magallanes. Bolivia. Peru.

*4. ***P. nemoralis*** L. 1753, *Sp. Pl.* 69.

Laxly caespitose perennial. Leaves 7–14 cm × 1.3–2.2(–3) mm; lamina flat, acute, ± scabrid above, scabridulous to glabrous beneath; ligule up to 0.5(–1.1) mm, truncate, sometimes serrate; sheath terete, glabrous to scabridulous. Culms 18–49 cm, slender, glabrous. Panicle 3–12(–20) cm, elongate-ovoid, lax; branches arising 2–5 together, ascending, scabridulous, with spikelets distally. Spikelets 2.8–4(–6) mm, with (1)2–5 florets, green or reddish. Glumes lanceolate, scabrid on keel; upper 2.5–3(–3.5) mm; lower (2–) 2.3–3 mm, 3-veined. Lemma 2.9–3.6 mm, lanceolate-ovate, scabrid on keel above, ciliate below and on marginal veins; callus sparsely lanate. Palea 2.7–3 mm, lanceolate, scabridulous on keels. Anthers 1–1.5(–2) mm. *Flowers December to February.*

Moist grassland and open woodland; 0–100 m.

Coasts of N. E. & S. Isla Grande.
Introduced from Europe and locally naturalized in modified and improved pastures.

P. stenantha Trin. 1831, *Mém. Acad. Sci. Pétersb.*, Ser. 6, **1**: 376, described from N. E. Asia (Kamchatka), has been recorded from S. Isla Grande near Ushuaia (Kurtz, 1896; Nicora, 1978). The plants are similar to *P. nemoralis* but are distinguished by their longer, acute ligule (4–6 mm), larger spikelets (6–10 mm) and longer glumes (4–5.3 mm), lemma (5–5.8 mm) and anthers (2–2.5 mm). The status and affinities of these plants require further study.

(5–6). *P. pratensis* group
A highly variable group within which the frequent occurrence of apomixis gives rise to a series of at least partially true-breeding biotypes which make specific limits difficult to define. To what extent some members of the group in southern S. America are truly native and to what extent they are segregates from European *P. pratensis* introduced for pasture-improvement remains in doubt and detailed cytogenetical study is required.

*5. ***P. pratensis*** L. 1753, *Sp. Pl.* 67.
P. ianthoides Roivainen 1954, *Ann. Bot. Soc. Zool. – Bot. Fenn. Vanamo* **27**: 199.
P. cenisia auct., non All. – Skottsberg, 1906.

Laxly caespitose perennial with long rhizomes. Leaves 2–18 cm × 2–4.5 mm; lamina flat or folded, subobtuse, ± scabrid, or glabrous beneath; ligule very short, truncate, usually serrate in lower leaves, up to 2.5(–4) mm, obtuse, in upper leaves; sheath ± keeled, glabrous, sometimes scabridulous beneath. Culms 4–65 cm, erect or ascending, glabrous, rarely scabridulous. Panicle 2.5–13.5 cm, elongate-conical, usually lax; branches arising (2)3–5 together, patent, subglabrous to scabrid. Spikelets 3.5–6 mm, with

2–4 florets, purplish to yellow-brown. Glumes scabrid distally on keel, also sometimes on veins; upper 2.5–3.4(–5) mm, ovate-elliptical; lower 2.2–3.2(–4.1) mm, ovate, 1- or (3)-veined. Lemma 2.6–3.5(–4.5) mm, ovate, sparsely to densely ciliate on keel and veins, scabrid distally on keel; callus with few to many, long, ± lanate hairs. Palea (1.6–)2.2–3.4 mm, ovate, scabrid on keels. Anthers (0.8–)1–1.5(–1.9) mm. *Flowers October to March.*

Sands, clays, gravels and rocks by coast, rivers and lakes, moist grassland and open scrub, *Empetrum* heath, open *Embothrium* and deciduous forest; 0–250 m.

Isla Grande except extreme west; Isla de los Estados. Introduced from Europe and widely naturalized in temperate regions.

P. ianthoides, described from Ea Vicuña in W. C. Isla Grande, has larger than average lemma and palea but in all other characters it falls well within the range typical of *P. pratensis*, with which it is clearly synonymous.

P. oligeria Steudel 1854, *Syn. Pl. Glum.* **1**: 426, described from S. Chile (Punta Arenas), has been recorded from the S. coast of Isla Grande, as well as W. Argentina (*c.*45°S.lat.). It has been considered to differ from *P. pratensis* in its wider leaves (2–4(–5) mm), longer spikelets (5.5–6.5(–7) mm) and lemma (4–4.5 mm), but these suggest that it also comprises a larger than average facies of that species.

6. *P. yaganica* Speg. 1896, *Anal. Mus. Nac. Hist. Nat. Bs Aires* **5**: 90.

Similar to *P. pratensis* but leaf-lamina rigid, subacute; ligule 1–5 mm, lanceolate-triangular, scabridulous above; spikelets 4.5–8 mm, often with rhachilla produced *c.* 1 mm beyond upper floret; glumes slightly scabrid on keel distally, the upper 3–6.5 mm, the lower 2.4–6 mm; palea 3.2–5.2 mm, oblong; anthers up to 3 mm. *Flowers October to December.*

Raised parts of *Astelia* bog, moist areas by lakes, open *Nothofagus* and *Embothrium* forest; 0–20 m.

C. & S. Fuegia except extreme west and east. (Map 31h).
Endemic.

The above complex of characters leads me to recognize this as a distinct species within the *P. pratensis* group, though with some reservations. It may be pseudoviviparous and frequently produces only sterile florets in very compressed, membranous spikelets in the upper part of the panicle as can be clearly seen in the type specimen and as was noted by Spegazzini (1896); furthermore, the anthers are often poorly developed, as are the ovaries, indicating interference with normal sexual reproduction.

7. *P. atropidiformis* Hackel ex Dusén, 1900, *Svenska Exped. Magellansl.* **3(5)**: 224.

Caespitose perennial. Leaves 2–4 cm × *c.* 1 mm; lamina convolute, acute, rigid, often ± curved, glabrous, becoming glaucous; ligule 2–3 mm, ovate-lanceolate, subacute, often lacerate; sheath glabrous. Culms 10–20 cm, erect, glabrous or scabridulous distally. Panicle 5–8 cm, linear, rather dense; branches short, erect, densely scabrid, the lower in pairs. Spikelets 3.5–5 mm, with 2–4 florets, green to purplish. Glumes 3–4.5 mm, subequal, lanceolate, scabrid on keel, the lower 1-veined. Lemma 2.5–4 mm, ovate-oblong, scabrid on keel distally; callus glabrous. Palea as long as lemma, scabrid on keels. Anthers 0.4–0.8 mm. *Flowers December to January.*

Saline grassland, *Lepidophyllum* scrub; 0–30 m.

N. E. Isla Grande. (Map 31i).
Argentina north to *c.*50°S.lat.

8. *P. darwiniana* Parodi 1937, *Rev. Argent. Agron.* **4**: 243.
Triodia antarctica Hooker f. 1847, *Fl. Antarct.* **1(2)**: 380.

Densely caespitose perennial. Leaves 1–4 cm × 0.8–1.4 mm; lamina strongly convolute, acute, scabrid above, glabrous beneath; ligule 0.8–2.5 mm, triangular, serrulate; sheath glabrous. Culms 2–5 cm, sheathed to base of panicle. Panicle 1–1.5(–2.5) cm, spike-like, dense; branches in pairs, glabrous. Spikelets 2.5–3.5 mm, with 2–3 florets. Glumes acuminate, scabridulous on margin; upper 2.3–3 mm, ovate; lower 2–2.6 mm, lanceolate, 1-veined. Lemma 2.5–3 mm, ovate, strigose, especially on keel. Palea 2.2–2.6 mm, strigose, especially on keels. Anthers 0.3–0.5 mm. *Flowers October to January.*

Wet rocks, usually coastal; 0–30(–400)m.

Southernmost Fuegia; Isla de los Estrados. (Map 31j).
Endemic.

9. *P. flabellata* (Lam.) Raspail 1829, *Ann. Sci. Obs.* **2**: 86.
Dactylis caespitosa Forster f. 1789, *Comment. Gotting.* **9**: 12 – Hooker, 1847; Hariot, 1884.

Poa controversa Steudel 1854, *Syn. Pl. Glum.* **1**: 260 – Spegazzini, 1896.
P. forsteri Steudel 1854, *Syn. Pl. Glum.* **3**: 260.
P. caespitosa auct., non Poiret – Spegazzini, 1896.
Festuca flabellata Lam. 1788, *Encycl. Méth. Bot.* **2**: 462.

Densely caespitose perennial; stems aggregating to form erect stool up to 1 m high, up to *c.*1 m in diameter and crowned by the ± spreading leaves. Leaves 5–70 cm × 3–10 mm; lamina flat or ± folded, involute at margins, scabrid and striate above, glabrous beneath; ligule 4–10 mm, obtuse, usually lacerate, scabridulous; sheath ± keeled, scabrid to glabrous. Culms 10–82 cm, erect. Panicle 2.5–15.5 cm, spike-like, cylindrical, dense, yellowish brown; branches very short, clustered, glabrous. Spikelets 5–8 mm, with 2–4 florets. Glumes unequal, lanceolate, acute to acuminate, sparsely scabrid on keel distally; upper 3.9–6 mm; lower 3.4–5.5 mm, 1-veined. Lemma 3.3–6.5 mm, ovate-lanceolate, 3-dentate at apex, with scabrid awn 1.2–2.4 mm. Palea 2.6–3.6 mm, ciliolate-scabrid on keels. Anthers (0.8–)1.7–4 mm. *Flowers October to December*. (Fig. 234).

Coastal rocks and peats, often in coastal bogs with *Astelia, Caltha* and *Empetrum*; 0–100 m.

S. Fuegia. (Map 32a).
W. Chile on north side of Estrecho de Magallanes.
Falkland Islands (Islas Malvinas); South Georgia.
Islas Diego Ramirez.

Subgen. ANDINAE
Plants gynodioecious.

10. *P. robusta* Steudel 1854, *Syn. Pl. Glum.* **1**: 426.
Festuca arenaria Lam. 1791, *Tabl. Encycl. Méth. Bot.* **1**: 191 pro parte – Hooker, 1847; Franchet, 1889; Spegazzini, 1896; Skottsberg, 1916.

Stout rhizomatous perennial, sometimes pseudo-viviparous. Leaves 4–12(–18) cm × (1–)3–9 mm, rigid; lamina flat or folded, pungent, involute at margin, strigose above, scabridulous beneath, glaucous to yellowish; ligule 0.8–1(–3) mm, truncate, ciliate. Culms 17–80 cm, erect or ascending, smooth. Panicle 5–20 cm, spike-like, dense, often yellowish; branches in pairs, scabridulous. Sexual spikelets 6–12 mm, with 2–4 florets. Glumes unequal, lanceolate, acute to acuminate, scabrid on margin; upper 6.2–10 mm; lower 5.5–9.2 mm, 3-veined. Lemma 6–8.7 mm, ovate-lanceolate, scabridulous on keel distally, often with apical awn 0.5–1 mm. Palea 5.5–6.6 mm, scabrid on keels. Anthers 2–4 mm. Pseudo-viviparous spikelets with lemma 8–22 mm, at least the upper transformed into first leaf of propagule, often with a ligule, the florets with progressively reduceed palea, anthers and ovary upwards in spikelet. *Flowers October to February*. (Fig. 235).

Coastal rocks and sands, often immersed at high tide; 0–20 m.

C. & S. Fuegia, local in N. Isla Grande. (Map 32b).
Chile north to *c.*47°S.lat.
Falkland Islands (Islas Malvinas).

Plants with at least some spikelets pseudoviviparous occur in the extreme south and west of the species' range (Moore and Doggett, 1976).

Narrow-leaved variants from N. Isla Grande, in which the panicle is interrupted at the base and has fewer spikelets, have been placed in ***P. pungionifolia*** Speg. 1902, *Anal. Mus. Nac. Hist. Nat. Bs Aires* **7**: 199, which also occurs in W. Argentina north to *c.*49°30′S.lat. These plants occur in moist areas, often with *Caltha*, and may represent ecotypic forms of *P. robusta*, with which they intergrade in the 2 apparently distinguishing characters.

Subgen. DIOICOPOA
Plants dioecious.

(11–12) *P. alopecurus* group
A very variable group within which numerous taxa have been described. The conspicuous variation in, for example, leaf-width, spikelet-size and indumentum is frequently independent in the different characters, and such features also often differ in male and female plants. The occurrence of pseudo-vivipary adds a further dimension to the problem. Recent treatments have varied from major 'lumping' within the group (Moore and Doggett, 1976) to the recognition of at least the major variants as distinct species (Nicora, 1978). Here, I adopt a rather conservative treatment whilst attempting to indicate the more important morphological modes, but detailed field and experimental studies are now required to clarify the situation.

11. *P. alopecurus* (Gaudich.) Kunth 1829, *Révis. Gram.* 116.
Arundo alopecurus Gaudich. 1825, *Ann. Sci. Nat.* **5**: 100.
Festuca arenaria Lam. 1791, *Tabl. Encycl. Méth. Bot.* **1**: 191 pro parte – Hooker, 1847.
F. arundo Hooker f. 1847, *Fl. Antarct* **1(2)**: 382
Poa magellanica (Phil.) Speg. 1896, *Anal. Mus. Nac. Bs Aires* **5**: 91.

Densely caespitose. Leaves 4–25(–30) cm × 0.4–3.5(–5) mm; lamina flat or folded, often strongly convolute, acute, scabrid, or glabrous beneath, usually glaucous; ligule (1–)3–12 mm, acute, scabridulous on midline; sheath ± keeled, smooth, glabrous, yellow-brown to purplish. Culms 2.5–46 cm, erect, terete, glabrous. Panicle (1–)3–10 cm, cylindrical to obovoid, spike-like, the male usually smaller than female, yellow-brown, usually tinged purplish or greenish; branches arising 2(–4) together, ± scabrid. Spikelets 4.5–11 mm, with 2–5 florets. Glumes lanceolate to ovate-lanceolate, glabrous to scabrid, especially on keel; upper 4–8 mm in female, 3.3–6.5 mm in male; lower 1(3)-veined, 3.6–7.7 mm in female, 2.7–6 mm in male. Lemma ovate-lanceolate, acute, 3- to 5-veined, the female (3.7–)4.3–8.2 mm, scabrid to lanate on keel, the male (3.3–)3.9–7.6 mm, usually scabrid on keel; callus usually with lanate hairs 0.5–4 mm in female. Palea lanceolate, acute, (3–)3.5–5(–6.7) mm in female, scabrid or usually with hairs on keel, 3–5.5(–7) mm in male, scabrid or rarely with short hairs on keel. Anthers 1.5–3.5 mm, rudimentary in female florets. *Flowers November to March.* (Fig. 236).

Ssp. ***alopecurus***.
Sexual plants to which above description applies.

Coastal sands and gravels, open grassland, *Lepidophyllum* and *Chiliotrichum* scrub, *Bolax-Empetrum* heath, rock-crevices, screes, feldmark; 0–1000 m.

Isla Grande except extreme west; Isla de los Estados. (Map 32c).
Argentina north to *c.*51°S.lat.
Chile north to *c.*51°S.lat.
Falkland Islands (Islas Malvinas).

Ssp. ***fuegiana*** (Hooker f.) D. M. Moore & Doggett 1976, *Bull. Brit. Antarct. Surv.* **43**: 105.

Festuca fuegiana Hooker f. 1847, *Fl. Antarct.* **1(2)**: 380 – Spegazzini, 1896;
F. pogonantha Franchet 1889, *Miss. Sci. Cap Horn 387*.
Poa fuegiana (Hooker f.) Hackel ex Dusén 1900, *Svenska Exped. Magellansl.* **3(5)**: 225 – Dusén, 1900; Skottsberg, 1916.
P. poganantha (Franchet) Parodi 1953, *Rev. Argent. Agron.* **20(4)**: 180 – Nicora, 1978.

Panicle with some or all spikelets pseudo-viviparous. Lower spikelets usually with lower 1–2 florets bearing anthers or ovary which are often vestigial. Upper glume 5–9.2 mm, the lower 4–8.2 mm, 1-veined. Lemma 4.3–9 mm, becoming modified into first leaf of propagule; callus with lanate hairs 1.5–6 mm.

Wet coastal sands and rocks, wet areas by streams, bogs, wet areas in feldmark; 0–1100 m.

C. & S. Fuegia. (Map 32d).
W. Argentina north to *c.*41°S.lat.
Chile north to *c.*49°S.lat.

P. rigidifolia Steudel 1854, *Syn. Pl. Glum.* **1**: 260.

Festuca antarctica (D'Urv.) Kunth 1829, *Révis. Gram.* 132 – Hooker, 1847.
Poa antarctica (D'Urv.) St. Yves 1927, non (Hooker f.) Stapf 1899 – Franchet, 1889; Dusén, 1900.
P. commersonii Franchet 1889, *Miss. Sci. Cap Horn* 385.

Material from S. Isla Grande, southernmost Chile and the Falkland Islands (Islas Malvinas), with convolute leaves not more than 3 mm wide, spikelets 6–9.5 mm and glumes and lemma glabrous to ± scabrid distally have been separated as this species. Since these features vary independently and continuously in material of this group studied by me I can see little justification for maintaining the species distinct from *P. alopecurus*.

P. poecila Phil. 1873, *Anal. Univ. Chile* **43**: 573.

Aira spiciformis Steudel 1854, *Syn. Pl. Glum.* **1**: 424–Spegazzini, 1896.

Material from N. Isla Grande and southernmost Argentina has been separated as this species, distinguished by the convolute leaves less than 1 mm wide, the culms not more than 15 cm and the spikelets 5.5–7 mm. Its association with more xeric conditions within the region and the known variability of the above characters leads me, on available information, to consider it to represent ecotypic variants with *P. alopecurus*.

P. shuka (Speg.) Parodi 1953, *Rev. Argent. Agron.* **20 (4)**: 180.

Festuca shuka Speg. 1896, *Anal. Mus. Nac. Bs Aires* **5**: 95.

Described from Isla de los Estados and recorded from Isla Navarino, this species comprises plants from coastal rocks with glabrous, flat to convolute leaves up to 3 mm wide and glabrous glumes and lemma. Nicora (1978) notes that it may not be distinct from *P. rigidifolia* and I concur.

12. *P. patagonica* Phil. 1896, *Anal. Univ. Chile* **94**: 168.

P. bonariensis auct., non (Lam.) Kunth – Dusén, 1900.
P. lanuginosa auct., non Poiret – Dusén, 1900; Skottsberg, 1916.

Similar to *P. alopecurus* but rhizomatous; leaves up to 5 mm wide, ± convolute; upper ligules 6–17 mm; spikelets 6–10 mm, with 5–9 florets, the male shorter

235

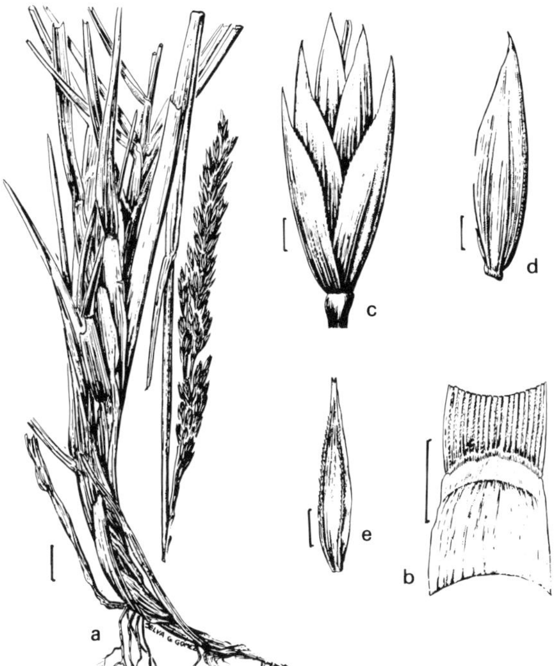

236

237

238

Fig. 235. *Poa robusta*. a, Plant. b, Ligule. c, Spikelet. d, Lemma. e, Palea. Scale = 1 cm (a), 5 mm (b) or 1 mm (c–e). Fig. 236. *Poa alopecurus*. a, Plant. b, Ligule. c, Spikelet. d, Floret. e, Palea. Scale = 1 cm (a) or 1 mm (b–e). Fig. 237. *Puccinellia magellanica*. a, Plant-base and inflorescence. b, Spikelet. c, Lower glume. d, Upper glume. e, Lemma. f, Palea. Scale = 1 cm (a) or 1 mm (b–f). Fig. 238. *Catabrosa aquatica*. a, Plant. b, Ligule. c, Spikelet. d, Glumes. e, Floret. Scale = 1 cm (a) or 1 mm (b–e).

(Flora Patagónica)

than the female; glumes ovate-lanceolate, acute to acuminate, scabrid on keel, 5–7.5 mm in female spikelets, 3.5–4.2 mm, in male spikelets, both with the lower somewhat shorter than the upper; lemma glabrous to hairy on keel in female florets, glabrous to shortly hairy on keel in male florets; callus glabrous, or with short hairs in female florets.

Coastal sands and dry, open grassland; 0–30 m.

N. E. Isla Grande.
W. Argentina north to *c.*42°S.lat.
E. Chile north to 50°40′S.lat.

In virtually all the characters noted above this species coincides well with *P. alopecurus*, with which it has been considered conspecific (e.g. Moore, 1974). However, its possession of rhizomes, which are not known in *P. alopecurus*, leads me to maintain it as a distinct species within this group until more is known about the prevalence and importance of rhizome-production in these austral dioecious Poas.

4. *Puccinellia* Parl.

Caespitose perennials. Ligule membranous; leaf-sheaths without auricles. Inflorescence a ± spreading panicle. Spikelets subterete, with 2–8 florets. Glumes unequal, somewhat keeled, awnless; upper 3- or 5-veined; lower 1(3)-veined. Lemma rounded on back, 5-veined, awnless. Palea 2-veined, bifid, ± equalling lemma. Lodicules lanceolate. Anthers 3–4 times as long as wide. Ovary glabrous. Stigmas sessile, apical.

a. Upper glume not more than 1.9 mm; lemma not more than 2 mm:
 b. Ligule at least 3 mm, acute; panicle-branches glabrous **2. skottsbergii**
 b. Ligule not more than 2.5 mm, ± obtuse; panicle-branches scabridulous **4. *pusilla***
a. Upper glume at least 2.3 mm; lemma at least 2.9 mm:
 c. Panicle-branches appressed at anthesis; upper glume at least 4 mm **1. *biflora***
 c. Panicle-branches erecto-patent at anthesis; upper glume not more than 4 mm **3. *magellanica***

1. *P. biflora* (Steudel) Parodi 1937, *Not. Mus. La Plata (Bot.)* **11(2)**: 14.
Festuca biflora Steudel 1854, *Syn. Pl. Glum.* **1**: 229
Glyceria leptostachys Speg. 1896, *Anal. Mus. Nac. Hist. Nat. Bs Aires* **5**: 94.
Atropis biflora (Steudel) St-Yves & Camus 1926, *Bull. Mus. Nat. Hist. Nat. (Paris)* **32**: 306 – Hauman and Parodi, 1929.

Plant with rhizomes. Leaves 2.5–6 cm × 0.3–1.5 mm; lamina convolute, subobtuse, glabrous; ligule 1.5–4 mm, lanceolate, acute; sheath striate, glabrous. Culms 8–35 cm, glabrous. Panicle 3–17 cm, linear; branches solitary or in pairs, appressed, glabrous, with few spikelets. Spikelets 6.5–15 mm, with 2–8 florets, the rhachilla produced beyond upper floret. Glumes ovate-lanceolate, obtuse; upper 4–5.5 mm, (3–) 5-veined; lower 3–5 mm, (1) 3-veined. Lemma 4–6 mm, ovate, obtuse to acute. Palea 3.8–5.5 mm, oblong, scabrid on keels. Anthers 0.6–1.2 mm. *Flowers December to February.*

Saline grassland by coast; 0–10 m.

N., N. W., N. E. and C. S. coasts of Isla Grande. (Map 32e).
S. Chile and S. Argentina on north shore of Estrecho de Magallanes.

2. *P. skottsbergii* (Pilger) Parodi 1937, *Not. Mus. La Plata (Bot.)* **11(2)**: 16
Atropis skottsbergii Pilger 1913, *Feddes Repert.* **12**: 305.

Leaves 8–15 cm × 1–1.2 mm; lamina convolute, obtuse, glabrous; ligule 3–5 mm, triangular, acute; sheath striate, glabrous. Culms 10–25 cm, erect, glabrous. Panicle 10–25 cm, conical to oblong; branches solitary or in pairs, patent or patent-deflexed, glabrous. Spikelets 4.5–5 mm, with 2–3 florets, the rhachilla produced *c.*1 mm beyond upper floret. Glumes ovate, obtuse, glabrous; upper 1–1.4 mm, 3-veined; lower 0.7–1 mm, 1-veined. Lemma 1.7–1.9 mm, ovate, subobtuse. Palea 1.6–1.8 mm, oblong, scabridulous on veins. Anthers 0.3–0.7 mm.

Saline grassland; 0–30 m.

N. E. Isla Grande. (Map 32f).
W. Argentina north to 45°S.lat.

3. *P. magellanica* (Hooker f.) Parodi 1937, *Not. Mus. La Plata (Bot.)* **11(2)**: 15.
Catabrosa magellanica Hooker f. 1847, *Fl. Antarct.* **1(2)**: 387.
Glyceria fuegiana Speg. 1896, *Anal. Mus. Nac. Bs Aires* **5**: 91.
G. antarctica Speg. 1896, *Anal. Mus. Nac. Bs Aires* **5**: 92.
Puccinellia antarctica (Speg.) Parodi 1937, *Not. Mus. La Plata (Bot.)* **11(2)**: 15.

Leaves 1–15 cm × 0.6–4 mm; lamina convolute, subacute to subobtuse, glabrous or scabridulous above and on margin; ligule 1–3.6 mm, ± lanceolate, obtuse; sheath striate, glabrous. Culms 3–33 cm, glabrous. Panicle 2.5–20 cm, oblong-ovoid; branches usually arising 2 or 3 together, erecto-patent, glabrous to scabridulous. Spikelets

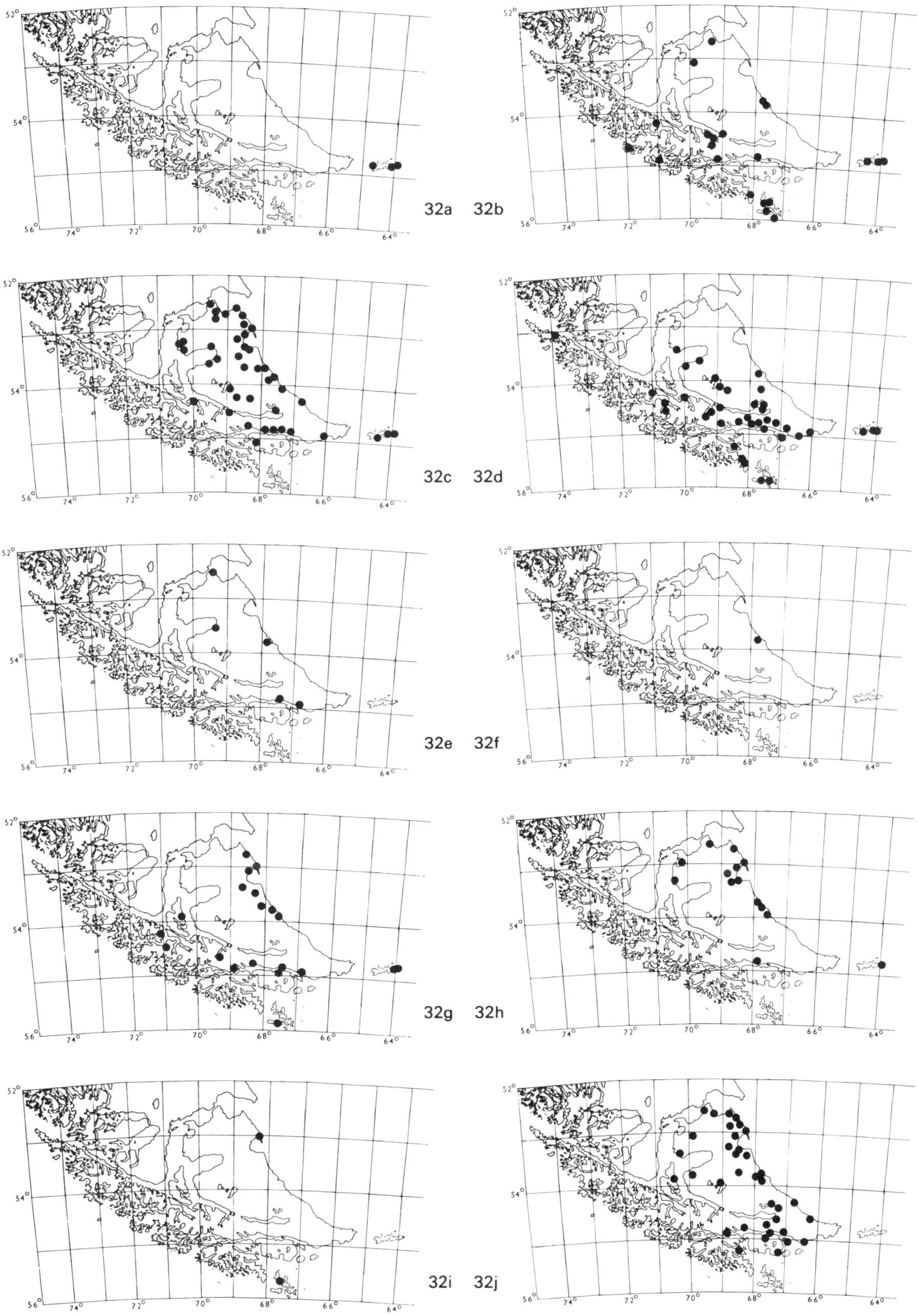

Map 32. a, *Poa flabellata*. b, *Poa robusta*. c, *Poa alopecurus* ssp. *alopecurus*. d, *Poa alopecurus* ssp. *fuegiana*. e, *Puccinellia biflora*. f, *Puccinellia skottsbergii*. g, *Puccinellia magellanica*. h, *Puccinellia pusilla*. i, *Catabrosa aquatica*. j, *Bromus unioloides*.

2.5–9(–15) mm, with 2–6 florets, the rhachilla produced beyond upper floret. Glumes ovate- to elliptic-lanceolate, obtuse to subacute; upper 2.3–4 mm, 3-veined; lower 1.4–3.1 mm, 1(3)-veined. Lemma 2.9–5 mm, lanceolate-ovate, subobtuse, shortly hairy or scabrid. Palea 2.5–4.3 mm, oblong- to lanceolate-ovate, ± scabrid on keels. Anthers 0.5–1 mm. *Flowers December to February*. (Fig. 237).

Coastal sands and rocks, saline grassland, *Lepidophyllum* scrub; 0–20 m.

N. E. Isla Grande; S. Fuegia. (Map 32g).
E. Argentina north to *c*.50°S.lat.
Chile on north shore of Estrecho de Magallanes.

P. laxa (Pilger) Parodi 1937, *Not. Mus. La Plata (Bot.)* **11(2)**: 15, described from Isla Dawson, is here included in this rather variable species. It comprises narrow leaved (0.8–1 mm) variants with shorter ligules (1–2 mm) and spikelets tending to be produced only towards the ends of the panicle-branches.

4. *P. pusilla* (Hackel) Parodi 1937, *Not. Mus. La Plata (Bot.)* **11(2)**: 15.
Atropis preslii Hackel ssp. *pusilla Hackel 1900, in Dusén, Svenska Exped. Magellansl.* **3(5)**: 227.
A. parviflora Hackel 1900, in Dusén, *Svenska Exped. Magellansl.* **3(5)**: 226 – Skottsberg, 1916.
Pucinellia parviflora (Hackel) Parodi 1937, *Not. Mus. La Plata (Bot.)* **11(2)**: 15 – Miante, in Nicora, 1978.

Leaves 0.5–4 cm × 0.4–1 mm; lamina convolute, obtuse, glabrous; ligule 0.5–2.5 mm, oblong-lanceolate, ± obtuse; sheath striate, glabrous. Culms 3–20 cm, glabrous. Panicle 1–12 cm, linear to ovoid, ± contracted to lax; branches solitary or in pairs, erect to patent, scabridulous, with 1–6 spikelets distally. Spikelets 1–6 mm, with 2–5 florets, the rhachilla shortly produced beyond upper floret. Glumes ovate-lanceolate, obtuse, glabrous; upper 1–1.9 mm, 3-veined; lower 0.4–1 mm, 1-veined. Lemma 1.5–2 mm, ovate, subobtuse, puberulent towards base. Palea 1–1.8 mm, ovate-lanceolate, sparsely scabrid on veins. Anthers 0.2–0.7 mm. *Flowers December to February*.

Coastal sands, open saline grassland, *Lepidophyllum* scrub; 0–20 m.

N. Isla Grande and locally on C. S. coast; Isla de los Estados. (Map 32h).
Argentina north to *c*.43°S.lat.
Falkland Islands (Islas Malvinas).

Plants included in *P. parviflora* tend to have wider leaves (0.8–1 mm), longer ligules (1.5–2.5 mm) and shorter upper glumes (1–1.4 mm), having an erose apex. However, such characters show a continuous pattern of variation with those possessed by plants from S. Fuegia and the Falkland Islands (Islas Malvinas) included in *P. pusilla* s. strict., and often the distinctions are obscured within a single individual. The generally smaller size ascribed to *P. pusilla* is probably an environmental response.

5. *Dactylis* L.

Glabrous, perennial herbs, with strongly compressed vegetative shoots. Ligule membranous; leaf-sheath without auricles. Inflorescence a compound panicle, the lowest branches usually long. Spikelets compressed, with 2–5 florets, in dense, 1-sided clusters towards ends of panicle-branches. Glumes keeled, awnless or with short apical awn; upper 3-veined; lower 1-veined. Lemma 5-veined, keeled, with short subapical awn. Palea 2-veined, 2-fid, slightly shorter than to as long as lemma. Lodicules linear, 2-lobed. Anthers 6–8 times as long as wide. Ovary glabrous. Styles very short, apical.

1. *D. glomerata L. 1753, *Sp. Pl.* 71.

Caespitose, often glaucous. Leaves 5–50 cm × 1–8 mm; lamina flat, folded when young, acute, usually scabrid; ligule 2–12 mm, ovate, serrate; sheath glabrous to scabridulous. Culms 18–135 cm, ± erect, scabridulous. Panicle 4–30 cm, oblong to ovoid; branches rigid, angular, ± scabrid, the lower often distant. Spikelets 5–8 mm. Glumes lanceolate to ovate, acuminate or with short awn, scabrid on keel; upper 3.5–6.5 mm; lower 5–5.5 mm. Lemma 3.7–6.5 mm, lanceolate to oblong, compressed, scabrid on veins and keel, with scabrid awn 0.2–2 mm. Palea scabrid on keels. Anthers 2–3.5 mm.

Coastal rocks and shingles, moist open scrub, forest clearings, disturbed ground near habitations; 0–50 m.

Coasts of Isla Grande except extreme west and east. Introduced from Europe for pasture-improvement and widely naturalized in temperate regions.

6. *Cynosurus* L.

Glabrous, caespitose perennials. Ligule obtuse; leaf-sheath without auricles. Inflorescence a 1-sided, spike-like panicle. Spikelets dimorphic, the fertile laterally compressed, with 2–5 florets, the sterile below and almost concealing the fertile, consisting of narrow, rigid glumes and lemmas only. Glumes subequal, keeled, 1-veined, acute. Lemma 5-veined, rounded on back, with short, apical awn. Palea emarginate or shortly 2-fid. Lodicules short, acutely 2-lobed. Anthers 5–7 times as long as wide. Styles short, apical.

1. *C. cristatus L. 1753, *Sp. Pl.* 72.

Leaves 4–27 cm × 0.7–2.5 mm; lamina flat, acuminate, scabridulous towards apex, usually minutely hairy above; ligule 0.6–1 mm; sheath terete, glabrous. Culms 15–60 cm. Panicle 5–6.5 cm, narrowly oblong, erect, dense, green or purplish, the branches short, scabrid. Fertile spikelets 3–5.5 mm, cuneate, sessile. Upper glume 3–4.5 mm, narrowly lanceolate, scabrid distally on veins, the lower somewhat smaller. Lemma 2.5–3.5 mm, ovate, obtuse, scabridulous distally; awn 0.7–1 mm. Palea ± as long as lemma, the keels scabridulous distally. Anthers 1–2 mm. Sterile spikelets 4–6 mm, the glumes and lemmas linear-lanceolate, shortly awned. *Flowers December to February.*

Disturbed ground and improved grasslands near settlements.

S. coast of Isla Grande (Ea Harberton).
Introduced from Europe in seed mixtures for pasture-improvement and naturalized in many temperate regions.

7. *Catabrosa* P. Beauv.

Stoloniferous perennials. Ligule obtuse; leaf-sheath without auricles. Inflorescence a lax panicle. Spikelets subterete, with (1)2(3) florets. Glumes unequal, obtuse, the upper 3-veined, the lower obscurely 1-veined. Lemma 3-veined, rounded on back. Lodicules ± lanceolate. Anthers *c.* 5 times as long as wide. Styles short, apical.

1. *C. aquatica* (L.) P. Beauv. 1812, *Ess. Agrost.* 97.

Leaves 4–14 cm × 2–10 mm; lamina folded when young, later flat, obtuse, smooth, soft; ligule 2–8 mm; sheath compressed, with free margins. Culms 5–75 cm, ascending to erect, succulent, glabrous. Panicle 5–30 cm, ovoid to oblong, the branches clustered, slender. Spikelets 3–5 mm, ovate to oblong, shortly pedicellate. Glumes purplish to whitish; upper 1.5–2.5 mm, oblong-ovate, membranous, smooth; lower 1–1.5 mm, ovate to elliptical, narrower. Lemma 2.5–3.5 mm, elliptic-oblong, truncate, glabrous or minutely hairy on veins. Palea ± as long as lemma, glabrous or scabridulous on keels. Anthers *c.*1.5 mm. *Flowers January to March.* (Fig. 238).

Shallow freshwater and wet sandy places.

N. E. Isla Grande (El Páramo), Isla Bayly. (Map 32i).
S. E. Argentina *c.*51°S.lat.
Eurasia; N. America.

Collected once in Isla Grande (Dusén, 1900). Very palatable to sheep and perhaps formerly more common. Recently discovered in Islas Wollaston (Pisano, pers. comm.).

Tribe BROMEAE
Spikelets hermaphrodite, laterally compressed to terete, with 1–many florets. Glumes shorter than spikelet. Palea 2-keeled. Lodicules 2, entire, united below. Stamens (2)3. Ovary with apical, fleshy, hairy appendage.

8. *Bromus* L.

Perennials, rarely annuals. Ligule membranous; leaf-sheath without auricles. Inflorescence a panicle. Spikelets somewhat compressed, with 3–12 florets. Glumes shorter than spikelet, awnless; upper (3–) 5- to 7-veined; lower 1- to 3(–5)-veined. Lemma (5–)7- to 11(–13)-veined, with straight subapical awn. Palea 2-veined, emarginate or 2-fid, ½ as long as to equalling lemma. Lodicules oblong, obtuse. Anthers 1–5 times as long as wide. Styles lateral below hairy apex of ovary.

Literature: J. C. Hernández, in M. N. Correa (Ed.) *Fl. Patag.* **3**: 77–93(1978).

a. Lemma strongly keeled:
  b. Lemma glabrous to scabridulous or sparsely pubescent distally, the awn not more than 8 mm — **1. *unioloides***
  b. Lemma pubescent over whole surface, the awn at least 8 mm — **2. *stamineus***
a. Lemma rounded or flat:
  c. Annual; upper glume 5- to 7-veined — **3. *hordeaceus***
  c. Perennial; upper glume 3- to 5-veined:
    d. Leaves at least 5 mm wide; ligule at least 4 mm — **6. *araucana***
    d. Leaves not more than 3 mm wide; ligule not more than 3.2 mm:
      e. Glumes and lemma densely hairy; palea and

lemma equal or subequal; upper glume obtuse **4. pellitus**

e. Glumes acute, glabrous or puberulent distally; lemma glabrous or puberulent on margins near base; palea $c.\frac{3}{4}$ as long as lemma **5. setifolius**

1. *B. unioloides* Humb., Bonpl. & Kunth 1815, *Nov. Gen. Sp.* **1**: 151.
B. willdenowii Kunth 1829, *Révis. Gram.* 134.
B. coloratus Steudel 1854, *Syn. Pl. Glum.* **1**: 429 – Franchet, 1889; Spegazzini, 1896.

Densely caespitose. Leaves 1.5–28 cm × 0.6–11.5 mm; lamina flat, often folded when young, subacute, glabrous to densely hairy, the lower more hairy than the cauline; ligule 1–4 mm, serrate; sheath glabrous on upper and hairy on lower leaves. Culms 6–107 cm, glabrous, often geniculate. Panicle 1.2–24 cm, lax; branches solitary, paired or in groups of 4, patent or nodding, often longer than spikelets. Spikelets 13–36 mm, with 3–8 florets. Glumes lanceolate, acute, keeled, usually scabridulous distally on keel; upper 7–13 mm, 5- to 7-veined; lower 5.8–11.5 mm 3- to 5-veined. Lemma 9.4–17 mm, lanceolate, shortly 2-fid, strongly keeled, glabrous to scabridulous or sparsely pubescent distally, (5–)9- to 11(–13)-veined; awn 0.2–8 mm, scabridulous. Palea 5–12 mm, $\frac{1}{2}$–$\frac{3}{4}$ as long as lemma, linear-lanceolate, acute, scabrid to ciliate on keels. Anthers 0.5–0.9(–3.8) mm, the shorter in the frequent cleistogamous florets. *Flowers December to March.* (Fig. 239).

Coastal sands and gravels. *Lepidophyllum* scrub, open deciduous forest, disturbed ground by roads and habitations; 0–60 m.

Almost throughout Fuegia except west, extreme south and extreme east. (Map 32j).
W. Argentina and Chile north to $c.$40°S.lat.
North along Andes to Ecuador.

Plants with smaller spikelets (up to 21 mm), glabrous glumes and longer lemma-awns (5–8 mm) have been distinguished as *B. coloratus*, having the distribution shown above. Such forms grow intermingled and intergraded completely with the rest of the material, with which they are consequently included in *B. unioloides*. Inbreeding, resulting from the frequent cleistogamy, probably accounts for the perpetuation of such biotypes.

2. *B. stamineus* Desv. 1856, in C. Gay, *Hist. Chile Bot.* **6**: 440.

Similar to *B. unioloides* but lemma pubescent over whole surface; lemma awn 8–12 mm.

Moist grassland; 0–30 m.

N. E. Isla Grande (Ea Viamonte). (Map 33a).
W. Argentina north to $c.$39°30′S.lat.

Perhaps a further variant of *B. unioloides*, with which it is readily confused.

3. *B. hordeaceus L. 1753, *Sp. Pl.* 77.
B. mollis L. 1762, *Sp. Pl.*, ed. 2, 112 – Nicora, 1978.

Annual. Leaves up to 20 cm × 2–7 mm; lamina flat, acuminate, soft, pubescent; ligule 0.5–2.5 mm, truncate, dentate, hairy; sheath densely hairy. Culms 5–80 cm, erect or ascending. Panicle 5–10 cm, oblong, dense, greyish green to purplish; branches clustered, pubescent, with 1–5 spikelets. Spikelets 5–20 mm, with 5–12 florets. Glumes acute, pubescent; upper 5–8 mm, ovate, 5- to 7-veined; lower somewhat shorter, lanceolate, 3- to 5-veined. Lemma 7–9 mm, ovate, obtuse, rounded on back, hairy; awn 5–10 mm, stout. Palea 6–9 mm, oblong-ovate, ciliate on veins. Anthers 0.2–2 mm.

Locally naturalized in disturbed ground by habitations.
Introduced from Eurasia.

4. *B. pellitus* Hackel 1900, in Dusén, *Svenska Exped. Magellansl.* **3(5)**: 230.

Caespitose. Leaves 1.5–6.5 cm × 0.8–2.5 mm; lamina flat or folded, subacute, densely hairy; ligule 0.3–1 mm, truncate, serrate, sparsely hairy; sheath densely hairy, brownish. Culms 9–27 cm, often geniculate, glabrous or densely pubescent distally. Panicle (0.2–)3–6.2 cm, dense, with 1–10 spikelets; branches short. Spikelets 13–17 mm, with 3–7 florets. Glumes densely hairy; upper 6.2–7.5 mm, ovate, obtuse, sometimes denticulate at apex, 3- to 5-veined; lower 5.5–6.7 mm, lanceolate, subobtuse, 3-veined. Lemma 8–9.8 mm, ovate, rounded on back, densely hairy, 7-veined; awn 3–4 mm. Palea 8–9 mm, ± as long as lemma, ciliate on veins. Anthers 0.9–1.5 mm. *Flowers November to March.*

Dry coastal grassland; 0–20 m.

N. E. Isla Grande (Río Grande). (Map 33b).
Argentina north to $c.$51°30′S.lat.
E. Chile north to $c.$51°S.lat.

5. *B. setifolius* C. Presl 1830, *Reliq. Haenk.* **1**: 261.
B. pictus Hooker f. 1847, *Fl. Antarct.* **1(2)**: 387 – Spegazzini, 1896.

Laxly caespitose. Leaves 1.5–12 cm × 0.7–3 mm; lamina flat or folded, subacute, densely hairy; ligule 0.7–3.2 mm, ovate, serrate, sparsely hairy; sheath densely hairy, brownish. Culms 15–46 cm, often geniculate, pubescent. Panicle 4–9.5 cm, dense, with 3–6 (–17) spikelets; branches solitary, paired or in groups of 4, up to 15 mm, densely pubescent. Spikelets 16–30 mm, with 3–8 florets. Glumes ovate-lanceolate, acute, glabrous or puberulent distally; upper 7–14 mm, (3–) 5-veined; lower 5.5–11 mm, (1–)3-veined. Lemma 9–14 mm, ovate, obtuse, rounded or flat on back, glabrous or puberulent on margins near base, 7- to 9-veined; awn 4–7 mm, scabrid. Palea 7–11 mm, *c.*$\frac{3}{4}$ as long as lemma, ciliate on veins. Anthers 1.1–1.5 mm. *Flowers December to February*.

Coastal sands and saline soils; 0–50 m.

N. W., E. and C. S. coasts of Isla Grande; Isla Izabel. (Map 33c).
W. Argentina north to *c.*39°42′S.lat.
E. Chile north to *c.*50°42′S.lat.

6. *B. araucana* Phil. 1896, *Anal. Univ. Chile* **94**: 343.

Caespitose. Leaves 15–30 cm × 5–7 mm; lamina flat, acute, sparsely hairy; ligule 4–5 mm, serrate, hairy; sheath hairy. Culms up to 90 cm, subglabrous. Panicle 15–22 cm, ± lax; branches solitary or paired, erecto-patent, usually longer than spikelets. Spikelets 17–20 mm, with 4–6 florets. Glumes linear-lanceolate, subacute, ± scabrid distally; upper 9–11 mm, 3-veined; lower 6–7 mm, 1-veined. Lemma 13–18 mm, lanceolate, rounded on back, pubescent, sometimes with long hairs at base, 7-veined; awn 4–5 mm. Palea slightly shorter than lemma, scabrid on keels. Anthers 0.8–1.1 mm.

Open deciduous woodland; 0–20 m.

S. Isla Grande (Ea Harberton). (Map 33d).
W. Argentina north to *c.*40°S.lat.
E. Chile north to *c.*38°S.lat.

Tribe TRITICEAE
Spikelets hermaphrodite or some male or sterile, with side of lemma towards rhachis, solitary or in groups of 2–6, sessile or subsessile on alternate sides of rhachis, with 1–numerous florets. Glumes $\frac{1}{2}$ as long to almost as long as spikelet. Palea 2-keeled. Lodicules 2, entire or ± dentate, ciliate. Stamens 3. Ovary hairy at apex.

9. *Leymus* Hochst.

Rhizomatous perennials. Ligule membranous; leaf-sheath with auricles. Inflorescence a spike. Spikelets in pairs, with 2–6 florets. Glumes equal, keeled, 3- to 5-veined, awnless. Lemma rounded on back, 7-veined, awnless. Palea shorter than lemma. Lodicules lanceolate, acute. Anthers several times as long as wide. Stigmas sessile, apical.

1. *L. arenarius (L.) Hochst. 1848, *Flora (Regensb.,)* **31**: 118.
Elymus arenarius L. 1753, *Sp. Pl.* 83 – Nicora, 1978.

Plant caespitose, glaucous, with long rhizomes. Leaves 6–36(–60) cm × 3–13 mm; lamina flat or folded, acute, rigid, with involute margins, scabridulous above; ligule up to 1 mm, truncate; sheath glabrous, with narrow spreading auricles. Culms 24–95 cm, stout, glabrous, rarely sparsely hairy. Spike 10–29 cm, stout, dense, sometimes interrupted at base. Spikelets 15–26 mm, sessile, the rhachilla hairy. Glumes 17–23 mm, linear-lanceolate, acute, glabrous or ciliate on keel distally. Lemma 14.5–20 mm, smaller in upper florets, lanceolate, acute, densely hairy, coriaceous. Palea 12–15 mm, ciliate on keels. Anthers 5–8.5 mm. *Flowers December to March*.

Coastal sands and gravels; 0–40 m.

N. & C. Isla Grande.
Introduced from Europe for coast-stabilization and widely naturalized in temperate regions.

10. *Elymus* L.

Similar to *Leymus* but spikelets with 2–4 florets; glumes 2- to 5-veined, with apical awn; lemma 5-veined, with apical awn.

1. *Elymus agropyroides* C. Presl 1830, *Reliq. Haenk.* **1**: 265.
E. antarcticus Hooker f. 1847, *Fl. Antarct.* **1(2)**: 388 – Spegazzini, 1896; Kurtz, 1896; Moore, 1974.
E. albowianus F. Kurtz 1896, in Alboff, *Rev. Mus. La Plata Bot.* **7**: 401 – Dusén, 1900; Skottsberg, 1916.

Laxly caespitose, with short rhizomes. Leaves 3.5–30 cm × 1.5–7 mm; lamina flat or folded, acuminate, with strongly involute margins, glabrous or scabridulous above, scabridulous distally beneath; ligule up to 1.2 mm, truncate; sheath glabrous, hairy when young, shiny, with obscure auricles. Culms 21–100 cm, glabrous. Spike 4.5–12 cm, dense, green to purple or yellow-brown. Spikelets 11–26 mm, sessile; rhachilla scabridulous. Glumes 6.5–14 mm, the upper slightly the larger, lanceolate, acute, ± 2-

lobed, scabridulous distally; awn 1–3 mm. Lemma 7–12 mm, smaller in upper florets, ovate-lanceolate, scabrid distally; awn 4–14 mm, scabrid. Palea about as long as lemma, scabridulous on keels. Anthers 1.6–2 mm, $2n = 28$. *Flowers December to April.* (Fig. 240).

Coastal sands and rocks, moist open areas by rivers and streams; 0–150 m.

Much of Isla Grande and islands immediately to south. (Map 33e).
W. Argentina north to *c*.40°S.lat.
Chile north to *c*. 44°S.lat.

E. antarcticus has been separated on the basis of its longer awns, but since awn-length shows continuous variation in Fuegian material, and there is often considerable variation within a single spike, it must be included with *E. agropyroides*.

11. *Agropyron* Gaertner

Tough, often rhizomatous perennials. Ligule membranous; leaf-sheath with auricles. Inflorescence a spike. Spikelets solitary, with 2–7(–9) florets. Glumes $\frac{1}{2}$–$\frac{3}{4}$ as long as spikelet, rounded on back, 1- to 7-veined, falling with florets when ripe. Lemma 3- to 7-veined. Palea emarginate or 2-fid, slightly shorter than lemma. Lodicules ovate, acute. Anthers 3–5 times as long as wide. Style short, subapical.

As noted by Nicora (1978) this is a difficult genus, especially in S. Patagonia and Tierra del Fuego. This account probably oversimplifies the treatment of variation shown by Fuegian *Agropyron*, but detailed biosystematic studies are needed to resolve the situation further.

Literature: L. R. Parodi, *Rev. Mus. La Plata, Bot.* **3**: 1–63(1940).

a. Glumes less than 0.5 mm wide **5. *elymoides***
a. Glumes more than 0.5 mm wide:
  b. Spikelets not imbricate, the rhachis with internodes at least as long as spikelets:
    c. Anthers less than 3.5 mm; glumes asymmetrical **4. *fuegianum***
    c. Anthers at least 3.5 mm; glumes symmetrical:
      d. Glumes and lemma acuminate or awned **1. *repens***
      d. Glumes and lemma truncate **2. *elongatum***
  b. Spikelets imbricate, the rhachis with internodes shorter than spikelets:
    e. Lemma, excluding awn, at least 12 mm **3. *pubiflorum***
    e. Lemma, excluding awn, at least 12 mm:
      f. Glumes scabrid-pubescent at least towards base and on veins **3. *pubiflorum***
      f. Glumes glabrous or slightly scabrid on veins **4. *fuegianum***

1. *A. repens (L.) P. Beauv. 1812, *Ess. Agrost.* 102.

Plant with long rhizomes. Leaves 6–30 cm × 3–10 mm; lamina flat, acuminate, glabrous to sparsely or densely hairy above, scabrid beneath; ligule less than 1 mm, truncate; sheath glabrous or the lower hairy, with short auricles. Culms 20–80 cm, glabrous. Spike 5–20 cm, the rhachis glabrous or scabridulous, with internodes longer than spikelets. Spikelets 8–17 mm, with 3–6 florets. Glumes 5–15 mm, subequal, lanceolate, mucronate or with awn up to 4 mm, (3–)5- to 7-veined, scabrid on veins distally. Lemma 6–11 mm, lanceolate-oblong, acuminate, keeled distally, glabrous, sometimes with awn up to 7 mm. Anthers 3.5–5 mm.

Disturbed and open areas, especially near habitations.

N. E. and C. S. coasts of Isla Grande.
Introduced from Europe and widely naturalized in temperate regions.

2. *A. elongatum (Host) P. Beauv. 1812, *Ess. Agrost.* 102.
A. × laxum auct., non (Fries) Almq. – Moore and Goodall, 1977.

Plant laxly caespitose, with rhizomes. Leaves 5–7 cm × 3–4.5 mm; lamina flat or folded, acuminate, involute, sparsely hairy above, yellowish, glabrous beneath; ligule up to 0.5 mm, truncate; sheath glabrous, yellowish to brown, with narrow, spreading auricles. Culms 65–68 cm, glabrous, contracted at nodes. Spike 13–19 cm, sometimes curved, with internodes longer than spikelets. Spikelets 16–19 mm, with 6–7 florets, yellowish – glaucous. Glumes 7–8.5 mm, subequal, oblong-ovate, truncate, prominently 6- to 7-veined, glabrous. Lemma 10–11 mm, oblong-ovate, obtuse, sometimes denticulate at apex, sometimes with awn up to 1 mm, 5-veined, glabrous or scabridulous distally. Palea entire or obscurely 2-fid, ciliolate on keels. Anthers 3–4.5 mm. *Flowers February to May.*

Coastal sands and gravels; 0–10 m.

N. E. Isla Grande (Río Grande).
Introduced from Europe and locally naturalized in open, disturbed habitats.

239

240

241

242

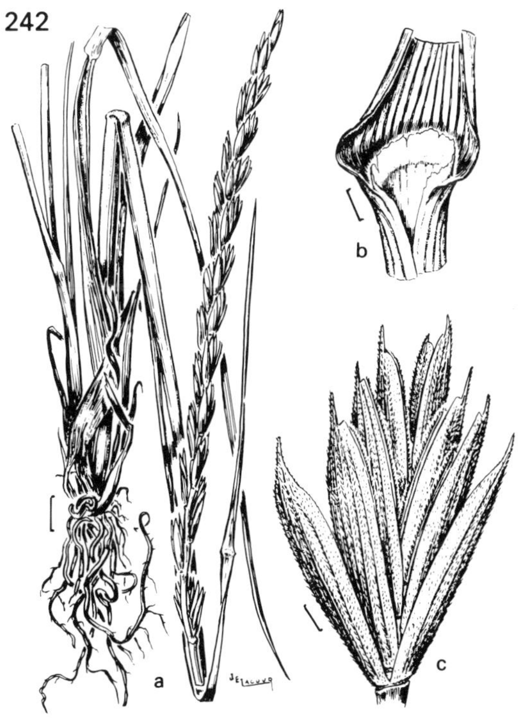

Fig. 239. *Bromus unioloides*. a, Plant. b, Ligule. c, Spikelet. d, Lemma. Scale = 1 cm (a), 5 mm (c) or 1 mm (b, d). Fig. 240. *Elymus agropyroides*. a, Plant. b, Ligule. c, Spikelet. d, Upper florets. e, Lower floret. f, Glumes. Scale = 1 cm (a), 2 mm (c) or 1 mm (b, d–f). Fig. 241. *Hordeum comosum*. a, Plant. b, Ligule. c, Lateral spikelet. d, Central spikelet. Scale = 1 cm (a), 5 mm (c) or 1 mm (b, d). Fig. 242. *Agropyron pubiflorum*. a, Plant. b, Ligule. c, Spikelet. Scale = 1 cm (a) or 1 mm (b–c). (Flora Patagónica)

3. *A. pubiflorum* (Steudel) Parodi 1940, *Rev. Mus. La Plata, Bot.* **3**: 36.

Triticum pubiflorum Steudel 1854, *Syn. Pl. Glum.* **1**: 429.
T. repens var. *magellanicum* Desv. 1853, in C. Gay, *Hist. Chile Bot.* **6**: 452.
T. magellanicum (Desv.) Speg. 1896, *Anal. Mus. Nac. Bs Aires* **5**: 98.
T. glaucum auct., non Desf. – D'Urville, 1825.
T. repens auct., non L. – Hooker, 1847; Franchet 1889.
Agropyron magellanicum (Desv.) Hackel 1900, in Dusén, *Svenska Exped. Magellansl.* **3(5)**: 231 – Skottsberg, 1916; Parodi, 1940; Moore, 1974; Nicora, 1978.
A. antarcticum Parodi 1940, *Rev. Mus. La Plata Bot.* **3**: 48 – Moore, 1974; Nicora, 1978.

Laxly caespitose, usually rhizomatous. Leaves 2.5–35 cm × (1–)2–7 mm; lamina subconvolute to folded or flat, acute, glabrous or scabridulous above; ligule 0.3–2 mm, truncate; sheath glabrous or scabridulous. Culms 14–100 cm, glabrous, terete to ± compressed. Spike 4–19 cm, dense, greenish or purplish, with spikelets longer than internodes, the rhachis glabrous to scabrid on margins. Spikelets 7–25 mm, with 2–7(–9) florets. Glumes 4.5–16 mm, equal or the upper slightly longer, asymmetrically lanceolate to ± oblanceolate, obtuse to acute, some shortly 2-fid, awnless or with up to 5 mm, with wide scarious margins, glabrous or scabrid – pubescent towards base and on veins, 3- to 6-veined. Lemma 6–17 mm, lanceolate to ± oblanceolate, acuminate or shortly 2-fid, scabrid to pubescent, at least towards base, 5- to 7-veined; awn 1–10 mm. Palea ovate-oblong to linear-lanceolate, acute, sparsely pubescent, or scabrid on keels. Anthers 1–3 mm. $2n = 28$. *Flowers December to March.* (Fig. 242).

Coastal sands and shingles, saline grassland, rocks and screes, stream-margins; 0–100 m.

Almost throughout Fuegia except extreme west. (Map 33f).
W. Argentina and Chile north to *c.*50°S.lat.
Falkland Islands (Islas Malvinas).

A rather variable species, especially in leaf-width, pubescence and length of glume – and lemma-awns. *A. magellanicum* was separated from *A. pubiflorum* by the absence of rhizomes in the latter. Since this character is not associated with others, and appears more to be related to habitat-differences, I cannot maintain this separation on present evidence.

A. antarcticum comprises 1 or 2 specimens with glume-awns (4–5 mm) and lemma-awns (5–10 mm) longer than elsewhere in the complex. Since awn-development is variable, even within a single plant, it seems difficult to justify the specific separation of this small amount of material, which is thus included in *A. pubiflorum*, although with hesitation.

4. *A. fuegianum* (Speg.) F. Kurtz 1896, in Alboff, *Rev. Mus. La Plata Bot.* **7**: 47.

Triticum fuegianum Speg. 1896, *Anal. Mus. Nac. Bs Aires* **5**: 99.

Caespitose, shortly rhizomatous. Leaves 3.5–21.5 cm × 1–4.8 mm; lamina flat to ± convolute, acute, scabrid to pubescent above; ligule 0.3–1.3 mm, truncate; sheath glabrous. Culms 27–70 cm, glabrous. Spike 7.2–15.5 cm, with spikelets longer or shorter than internodes, the rhachis compressed, scabridulous on margins. Spikelets 5.5–15.5 mm, with 3–5 florets. Glumes 7–10 mm, equal or the upper slightly longer, asymmetrically lanceolate-oblanceolate, acute, sometimes ± 2-fid, awnless or with scabrid awn up to 1.5 mm, with wide scarious margins, glabrous or scabridulous on veins, 3- to 5-veined. Lemma 8.5–12 mm, lanceolate, acuminate or slightly 2-fid, with scabrid awn 0.5–4 mm, scabrid-pubescent, 5- to 7-veined. Palea ovate-oblong, scabrid on keels. Anthers 1.5–2.8 mm. *Flowers December to February.*

Coastal sands, dry grassland, open soil and screes; 0–500 m.

Much of Isla Grande and islands immediately to west and south. (Map 33g).
W. Argentina and Chile north to *c.*50°S.lat.

A. patagonicum (Speg.) Parodi 1940, *Rev. Mus. La Plata Bot.* **3**: 23 (*Triticum fuegianum* var. *patagonicum* Speg. 1897, *Rev. Fac. Agron. La Plata* **3**: 588) comprises individuals with lax spikes, but since it differs in no other significant way from *A. fuegianum* I consider the species to be synonymous.

A. fuegianum and *A. pubiflorum* are frequently very difficult to separate and the occurrence of many apparent intermediates suggests that interspecific hybridization is common, as elsewhere in the genus.

5. *A. elymoides* Hackel ex Dusén 1900, *Svenska Exped. Magellansl.* **3(5)**: 232.

Cockaynea elymoides (Hackel ex Dusén) Zotov 1943, *Trans. Proc. Roy. Soc. N.Z.* **73**: 233.

Caespitose. Leaves 5–18 cm × 2–3 mm; lamina flat or folded, acute, pubescent; ligule 0.5–1 mm, denticulate; sheath glabrous. Culms 30–60 cm, geniculate or erect. Spike 5–9 cm, slender, with densely imbricate spikelets; rhachis compressed, glabrous. Spikelets 14–15 mm, with 2–3 florets. Glumes 5–6 × up to 0.5 mm, linear, equal,

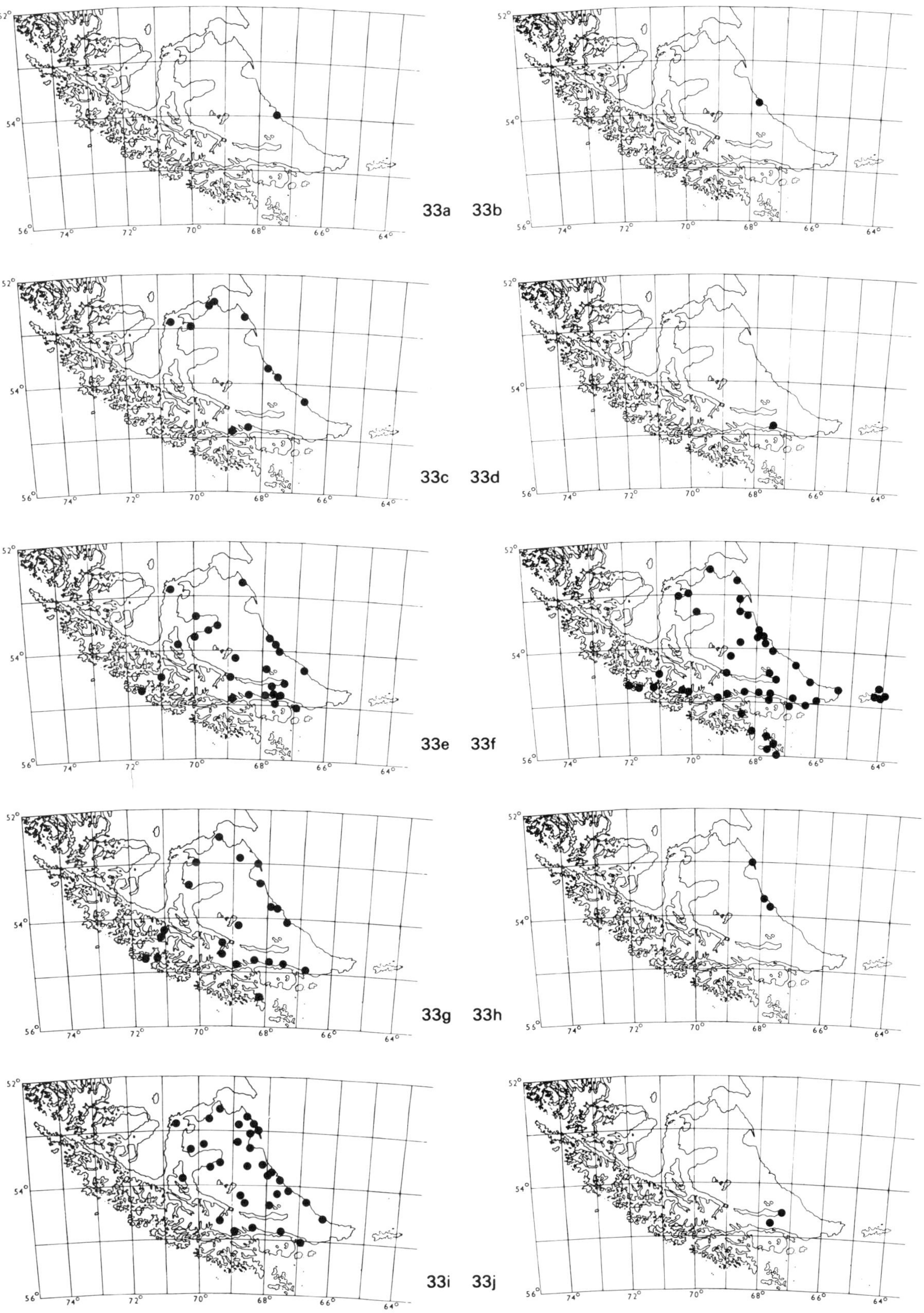

Map 33. a, *Bromus stamineus*. b, *Bromus pellitus*. c, *Bromus setifolius*. d, *Bromus araucana*. e, *Elymus agropyroides*. f, *Agropyron pubiflorum*. g, *Agropyron fuegianum*. h, *Agropyron elymoides*. i, *Hordeum comosum*. j, *Koeleria fueguina*.

pubescent, 1- to 3-veined; awn 5–6 mm, Lemma 8–9 mm, linear-lanceolate, acuminate or shortly 2-fid, pubescent, 5-veined; awn 2–5 mm. Palea ciliolate on nerves. Anthers 2–3 mm.

Coastal sands and rocks.

N. E. Isla Grande (Río Grande). (Map 33h).
S. Chile (*c.*52°40′S.lat.).

Probably a sterile intergeneric hybrid between *Agropyron* and *Hordeum* (Parodi, 1940; Hunziker and Itria, 1980), possibly *A. fuegianum* and *H. comosum* (Melderis, pers. com.).

12. *Triticum* L.

Annuals. Ligule membranous; leaf-sheath with auricles. Inflorescence a spike. Spikelets solitary, with 5 florets. Glumes $\frac{1}{2}$–$\frac{2}{3}$ as long as spikelet, keeled, 5- to 13-veined, awned. Lemma 7-veined, awned. Palea as long as lemma. Lodicules linear-ovate, truncate, ciliate. Anthers *c.* 5 times as long as wide. Styles very short, apical.

*1. ***T. aestivum*** L. 1753, *Sp. Pl.* 85.

Laxly caespitose. Leaves 2.5–13 cm × 2–9 mm; lamina flat, acute, involute at margins, scabridulous beneath; ligule 1–2 mm, truncate; sheath glabrous, with amplexicaul auricles. Culms 80–90 cm, glabrous. Spike 7–8 cm, ± dense, ± 4-angled in cross-section. Spikelets 10.3–15.5 mm, closely appressed to rhachis. Glumes 7.5–8 mm, ovate, acute to slightly 2-fid, the keel asymmetrical, scabrid; awn 4.5–6 mm, scabrid. Lemma 9–10.5 mm, ovate, rounded on back, glabrous; awn 40–55 mm, scabrid. Palea ovate, scabridulous on keels distally. Anthers 1.5–2 mm.

Disturbed open ground near habitations.

S. Isla Grande (Ushuaia).
Introduced from Europe and cultivated in continental Argentina and Chile. In Fuegia locally cultivated for forage and also brought in with animal foodstuffs.

13. *Secale* L.

Annuals. Ligule membranous; leaf-sheaths with auricles. Inflorescence a spike. Spikelets solitary, with 2 florets. Glumes $\frac{1}{2}$–$\frac{3}{4}$ as long as spikelet, keeled, 1-veined. Lemma keeled, 5-veined, awned. Palea ± equal to lemma. Anthers *c.*10 times as long as wide. Styles short, apical.

*1. ***S. cereale*** L. 1753, *Sp. Pl.* 84.

Caespitose. Leaves 9–80 cm × 1.5–4 mm; lamina flat, acute, puberulent; ligule *c.*0.5 mm, truncate, denticulate; sheath glabrous, with short auricles. Culms 56–75 cm, glabrous, pubescent near spike, with prominent nodes. Spike 5–7.5 cm, dense. Spikelets 12–17 mm (excluding awns), imbricate, greenish. Glumes 4.5–10 × up to 1 mm, linear-subulate, acuminate or with very short awn, densely scabrid on keel. Lemma 10–14 mm, lanceolate, densely scabrid on keel, with scabrid, apical awn 10–35 mm. Palea ciliate on keels. Anthers *c.*8 mm.

Disturbed, open ground by coast.

N. E. Isla Grande (Río Grande).
Introduced from Europe for local cultivation as forage crop and sometimes naturalized.

14. *Hordeum* L.

Annuals or perennials. Ligule membranous; leaf-sheath with or without auricles. Inflorescence a spike, compressed. Spikelets with 1 floret, in groups of 3 (triplets) at each node of rhachis, the central spikelet sessile or shortly pedicellate, hermaphrodite, the lateral pedicellate, usually sterile, sometimes male. Glumes rounded on back, linear-lanceolate or subulate, 1-veined. Lemma 5-veined, rounded on back, usually with apical awn. Palea subequal to lemma, narrowly ovate. Lodicules lanceolate, fimbriolate. Anthers 4–6 times as long as wide. Stigmas sessile.

A difficult genus in which morphological and cytological variation, together with hybridization, make the present treatment highly tentative.

a. Annual; lateral spikelets of triplet at least as long as the central; glumes of central spikelet ciliate **1. *murinum***
a. Perennial; lateral spikelets of triplet shorter than the central; glumes of central spikelet glabrous to pubescent, not ciliate:
  b. Triplets at least 18 mm; glumes of central spikelet subulate or awn-like; lemma of central spikelet with awn at least 8 mm, longer than lemma **2. *comosum***
  b. Triplets not more than 13 mm; glumes of central spikelet linear to linear-lanceolate; lemma of central spikelet mucronate or with awn up to 2.5 mm, shorter than lemma **3. *secalinum***

1. *H. murinum L. 1753, *Sp. Pl.* 85.
H. glaucum Steudel 1854, *Syn. Pl. Glum* **1**: 325 – Nicora, 1978.

Annual. Leaves 5–12 cm × 2–7 mm; lamina flat, acute, flaccid, ± pubescent to shortly hairy; ligule 0.7–1.5 mm, truncate; sheath ± pubescent to hairy, ± inflated above. Culms 5–50 cm, ascending, subglabrous to hairy. Spike 2–7(–10) cm, oblong, strongly compressed. Triplets 20–80 mm; lateral spikelets as long as or longer than the central. Central spikelet with floret on pedicel 1–1.2 mm; glumes 12–40 mm including awn, lanceolate, ciliate; lemma 6–8 mm, lanceolate, scabrid above, with awn 8–25 mm; palea obtuse, ± hairy; anthers 0.2–1.8 mm. Lateral spikelets with pedicels 1–1.2 mm; lower glume subulate; upper glume as in central spikelet; lemms filiform, longer than in central spikelet; palea up to ½ as long as in central spikelet. *Flowers November to February.*

Disturbed ground and modified grassland.

Locally on S. coast of Isla Grande.
Introduced from Europe and widely naturalized in temperate regions.

Two of the 3 subspecies recognized in Europe have been recorded from Tierra del Fuego. Ssp. *glaucum* (Steudel) Tzvelev has glaucous leaves, the central spikelet with anthers 0.2–0.5 mm and the rhachilla of the lateral spikelets with a stout, brownish prolongation. Ssp. *leporinum* (Link) Arcangeli has green leaves, the central spikelet with anthers 0.7–1.4 mm and the rhachilla of the lateral spikelets with a slender, greenish prolongation.

2. *H. comosum* C. Presl 1830, *Reliq. Haenk.* **1**: 327.
H. jubatum auct., non L. – Dusén, 1900; Autran, 1905.

Perennial. Leaves 1–15 cm × 0.4–5 mm; lamina flat or somewhat convolute, acuminate, scabridulous to ± puberulent above, pubescent beneath; ligule up to 0.8 mm, truncate; sheath glabrous to pubescent, brownish to purplish, with inconspicuous auricles. Culms 7–45 cm, glabrous. Spike 3–11 cm, cylindrical-obovoid. Triplets 18–40(–56) mm; lateral spikelets shorter than the central. Central spikelet with glumes 20–45(–50) mm including awn, subulate or awn-like, scabrid to pubescent; lemma 6.5–15 mm, ovate-lanceolate, scabrid distally, with awn 8–40 mm; palea obtuse or shortly 2-fid, scabridulous distally on keels; anthers 0.6–4.5 mm. Lateral spikelets with pedicels 0.5–1.5 mm; glumes 13–44 mm, awn-like; lemma 2–17 mm, subulate or awn-like, scabrid; palea usually vestigial. *Flowers December to February.* (Fig. 241).

Coastal sands, dry grassland, open clay soils; 0–150 m.

Isla Grande, except extreme west and east; Isla Dawson; islands of Canal Beagle and E. Estrecho de Magallanes. (Map 33i).
W. Argentina and E. Chile north to *c.*33°S.lat. Falkland Islands (Islas Malvinas) (? introduced).

A rather variable species within which other species have been recognized in Fuegian material here included in *H. comosum.*

H. pubiflorum Hooker f. 1847, *Fl. Antarct.* **1(2)**: 388, has been distinguished by its muticous or shortly aristate lemmas and hairy glumes in the lateral spikelets. In Fuegia, the lateral spikelets are usually sterile, with awn-like lemmas, and the indumentum is very variable in the complex, both characters apparently related to the habitat; in saline grassland and by rivers plants have less pubescent glumes and better developed lateral spikelets.

H. lechleri (Steudel) E. Schenk 1907, *Bot. Jahrb.* **40**: 108, has been applied to Fuegian plants with longer than average triplets and lateral spikelets often male and with more shortly awned lemmas. I can find no correlated features or discontinuities in the two quantitative characters, whilst the influence of habitat on the development of the lateral spikelets was mentioned above. In Fuegia the specimens with male lateral spikelets all occur on or near the east coast of Isla Grande.

3. *H. secalinum Schreber 1771, *Spicil. Fl. Lips.* 148.
H. andicola auct., non Griseb. – Dusén, 1900.

Laxly caespitose perennial. Leaves 0.5–10 cm × 0.2–2 mm; lamina flat, subacute to obtuse, involute, scabrid to puberulent above; ligule up to 0.5 mm, truncate, serrate; sheath scabrid to hairy below, the inconspicuous auricles continuous with ligule. Culms 9–40 cm, glabrous. Spike 2.5–5 cm, cylindrical, dense, the rhachis readily breaking up at maturity. Triplets 9.5–13 mm; lateral spikelets shorter than the central. Central spikelet with glumes 7–13 mm, linear to linear-lanceolate, scabrid; lemma 8–11.5 mm, linear-lanceolate, mucronate or with awn up to 2.5 mm, scabrid distally; palea linear-lanceolate, mucronate, scabridulous distally on keels; anthers 0.9–1.3 mm. Lateral spikelets with pedicels 0.7–1 mm; glumes 5.7–10 mm, bristle-like, scabrid; lemma 2.4–5 mm, linear, convolute, scabrid. *Flowers November to February.*

Coastal sands and gravels, margins of rivers and lakes, open soil and rocks; 0–50 m.

N. Isla Grande, locally on C. S. coast, and Isla Navarino.
Introduced from Europe and widely naturalized in temperate regions.

Fuegian material included here has been placed in ***H. chilense*** Roemer & Schultes var. ***magellanicum*** Parodi & Nicora 1977, *Hickenia* **1**: 61. However, although the plants are somewhat smaller than average for *H. secalinum* they fit closely to much European material and there seems no reason for specific separation of the material found in Tierra del Fuego.

Tribe AVENEAE
Spikelets hermaphrodite, somewhat laterally compressed, with 1-many florets. Glumes often longer than spikelet. Palea 2-keeled. Lodicules (0)2(3), entire or 2-dentate. Stamens 3. Ovary glabrous to hairy all over.

15. *Avena* L.

Annuals. Ligule membranous; leaf-sheath without auricles. Inflorescence a panicle. Spikelets with 3–4 florets, pendent; rhachilla produced beyond upper floret, ± silky. Glumes subequal, as long as spikelet, awnless; lower 7- to 9-veined; upper 9-veined. Lemma rounded on back, 7-veined, with subapical awn. Palea slightly shorter than lemma, coriaceous. Lodicules lanceolate, acute. Anthers several times longer than wide. Ovary pubescent; styles short, lateral.

*1. ***A. sativa*** L. 1753, *Sp. Pl.* 79.

Leaves 2.5–15 cm × 1.7–5.2 mm; lamina flat, subacute, glabrous to scabridulous beneath; ligule 1.2–2.5 mm, truncate, often torn; sheath scabridulous, brownish. Culm solitary, 7–75 cm, glabrous. Panicle 4–13 cm, lax, open and conical or ± contracted and 1-sided; branches long, scabridulous; spikelets solitary. Glumes 15–36 mm, lanceolate, acuminate, glabrous, scarious, persistent. Lemma 14–19 mm, ovate-lanceolate, 2-fid, involute, scabridulous to glabrous; awn 14–28 mm, scabridulous, straight or bent, absent in upper florets. Palea lanceolate-ovate, ciliolate on keels. Anthers 1.9–3.5 mm.

Coastal sands and disturbed ground; 0–30 m.

N. E and C. S. coasts of Isla Grande.
Introduced and locally cultivated for forage, sometimes escaping and persisting.

A. byzantina C. Koch 1848, *Linnaea* **21**: 392 is recorded as an adventive on the south coast of Isla Grande (Ushuaia). It is distinguished by the rhachilla-segments breaking at their base and falling attached to the upper floret; in *A. sativa* the segments break at their apex and fall attached to the lower floret.

16. *Arrhenatherum* P. Beauv.

Laxly caespitose perennials. Ligule membranous; leaf-sheath without auricles. Inflorescence a panicle. Spikelets with 2(–4) florets; lower floret usually male, the upper hermaphrodite. Glumes unequal, awnless; lower $\frac{1}{2}$–$\frac{3}{4}$ as long as spikelet, 1-veined; upper ± equal to spikelet, 3-veined. Lemma 7-veined, the lower with dorsal awn, the upper awnless or with subapical awn. Palea ± equal to lemma. Anthers 9–10 times as long as wide. Ovary hairy; styles short, lateral.

*1. ***A. elatius*** (L.) P. Beauv. ex J. & C. Presl 1819, *Fl. Čechica* 17.

Leaves 1.5–26 cm × 1.7–6 mm, flat, acuminate, scabridulous to sparsely hairy; ligule 0.3–2.5 mm, truncate, serrate; sheath rounded, glabrous or laxly hairy. Culms 60–125 cm, ± ascending to erect, glabrous. Panicle 8.5–19 cm, lanceolate-ovoid, erect or nodding, lax to rather dense, green or purplish; branches clustered, scabrid. Spikelets 7–10 mm, oblong. Glumes acuminate, persistent, scabridulous; upper 6.5–9.5 mm, narrowly ovate; lower 4–8 mm, lanceolate. Lemma 6–9 mm, linear- to oblong-ovate, acute, laxly hairy, in lower floret with geniculate awn 8–12.5 mm from basal half, in upper floret with subapical awn 0–4 mm. Palea membranous, with ciliate keels. Anthers 3.6–4.5 mm. *Flowers December to February*.

Coastal gravels, open grassland; 0–50 m.

Local on east and south coasts of Isla Grande.
Introduced from Europe for forage.

Fuegian plants, with the culms having swollen, often hairy, basal nodes, belong to ssp. *bulbosum* (Willd.) Schübler & Martens.

17. *Koeleria* Pers.

Rhizomatous perennials. Ligule membranous; leaf-sheath without auricles. Inflorescence a contracted panicle. Spikelets with 2–3 florets. Glumes subequal, keeled, awnless, ± equal to spikelet; upper 3-veined; lower 1-veined. Lemma 5-veined, the lateral veins not reaching apex, strongly keeled, with dorsal awn. Palea shorter than lemma. Lodicules 2-lobed. Anthers *c*.3 times as long as wide. Ovary glabrous; styles short, apical.

1. ***K. fueguina*** Calderón 1978, in Correa, *Fl. Patag.* **3**: 63.

Leaves 6–18 cm × 1–2 mm; lamina flat or ± convolute, acuminate, pubescent; ligule 0.5–2 mm, obtuse, ciliate; sheath striate, hairy, the lower subglabrous. Culms 16–45 cm, often geniculate near base, pubescent above. Panicle 3.5–10 cm, lanceolate-cylindrical, spike-like, dense, interrupted at base; rhachis, branches and pedicels pubescent. Spikelets 5–7 mm, yellow-brown. Glumes 4–7 mm, lanceolate, the upper slightly larger, scabrid on keel distally. Lemma 4–6 mm, lanceolate, subacute to shallowly 2-fid, scabrid; awn 1.5–4 mm, straight or ± curved. Palea 3–5 mm, 2-fid at apex. Anthers 0.5–0.8. *Flowers November to January*. (Fig. 243).

Moist grassland and bogs; 0–150 m.

C. and S. Isla Grande. (Map 33j).
W. Argentina and E. Chile north to *c*.51°30′S.lat.

18. *Trisetum* Pers.

Caespitose perennials. Ligule membranous; leaf-sheath without auricles. Inflorescence a panicle. Spikelets with 2–3 florets. Glumes unequal, keeled, awnless, shorter to longer than spikelet; upper 3-veined; lower 1-veined. Lemma strongly keeled, obscurely 5-veined, deeply 2-fid, with geniculate dorsal awn. Palea about equal to lemma. Lodicules minutely 2-lobed. Anthers 2–4 times as long as wide. Ovary glabrous or hairy at apex; styles very short, apical.

Literature: E. Hultén, *Svenska Bot. Tidskr.* **53**: 203–228(1959).

a. Panicle compact, spike-like; culm densely pubescent **2. *spicatum***
a. Panicle lax, ± nodding; culm glabrous or sparsely hairy above **1. *cernuum***

1. ***T. cernuum*** Trin. 1831, *Mém. Acad. Sci. Pétersb.* Ser. 6, **1**: 61.
Avena leptostachys Hooker f. 1847, *Fl. Antarct.* **1(2)**: 378 – Skottsberg, 1906.

Leaves 12–25 cm × 4.5–10 mm; lamina flat, acuminate, sparsely hairy above; ligule 3–5 mm, oblong, ciliate; sheath glabrous or sparsely hairy. Culms 30–70 cm, glabrous or sparsely hairy above. Panicle 11–20 cm, lax, ± nodding, with few spikelets; branches clustered, up to 5.5 cm, scabrid. Lower glume 2–4 mm, linear-lanceolate, acuminate, scabridulous distally and on margin; upper glume 4–5.7 mm, ovate-lanceolate, acute, glabrous, sparsely ciliate distally. Lemma of lowest floret 4.5–7.5 mm, lanceolate, acute, scabridulous; awn 9–16 mm; callus hairy. Palea oblong-lanceolate, scabrid on keels. Anthers 1–1.4 mm, Rhachilla hairy, produced beyond 2nd floret, the upper floret much reduced. *Flowers January to March*.

Open deciduous forest; 200–500 m.

S. Isla Grande. (Map 34a).
W. Argentina north to *c*.40°S.lat.
Chile north to *c*.47°S.lat.
North America.

2. ***T. spicatum*** (L.) K. Richter 1890, *Pl. Eur.* **1**: 59.
T. subspicatum (L.) P. Beauv. 1812, *Ess. Agrost.* 88 – Franchet, 1889; Kurtz, 1896; Spegazzini, 1896.
T. variabile auct., ? non Desv. – Kurtz, 1896; Spegazzini, 1896; Autran, 1905.
T. phleoides (D'Urv.) Kunth 1829, *Révis. Gram.* 101 – Roivainen, 1954; Nicora, 1978.
T. fuegianum Gand. 1913, *Bull. Soc. Bot. Fr.* **60**: 28.
T. antarcticum auct., non (Forster) Trin. – Spegazzini, 1896.

Leaves 0.5–22 cm × 0.8–4 mm; lamina flat or folded, acuminate, involute, glabrous to hairy beneath, appressed-hairy above; ligule 0.5–2 mm, truncate, hairy; sheath usually pubescent. Culms (1.5–)7–50 cm, densely appressed-hairy. Panicle 0.8–11 cm, spike-like, dense; branches mostly paired, short, hairy. Spikelets 4.5–8 mm, with 2(3) florets, the lower largest; rhachilla hairy, produced 1–2 mm beyond upper floret. Glumes lanceolate, acute, scabrid on keels distally; lower (2.6–)3.5–7 mm; upper 4–8.2 mm. Lemma of lower floret (3–) 4.2–6 mm, lanceolate, acute, scabridulous; awn (1.8–) 3–6.5 mm, geniculate, scabrid. Palea 3.4–5.2 mm, oblong-lanceolate, scabrid on keels. Anthers 0.6–1.2 mm. *Flowers December to March*. (Fig. 244).

Coastal sands and rocks, *Sphagnum* bogs, *Empetrum* heath, *Chiliotrichum* scrub, open deciduous forest, feldmark; 0–1000 m.

Throughout Fuegia except extreme west. (Map 34b).
W. Argentina and E. Chile north to *c*.39°S.lat.;

north along Andes to Colombia.
Falkland Islands (Islas Malvinas).
North America. Eurasia.
New Zealand; S. E. Australia.

T. phleoides has been separated from *T. spicatum* by its unequal, larger, acute glumes exceeding the florets, and both species have been recorded from Fuegia (Nicora, 1978). Such variation in European material has been recognized at the subspecific level but in Tierra del Fuego it shows no geographical pattern and varietal status at most seems appropriate.

19. *Deschampsia* P. Beauv.

Caespitose perennials, rarely with rhizomes. Ligule membranous; leaf-sheath without auricles. Inflorescence a panicle. Spikelets with (1) 2 florets; rhachilla silky, often produced beyond upper floret. Glumes enclosing florets, keeled, awnless; lower 1-veined; upper 3-veined. Lemma rounded on back, 5-veined, with straight or geniculate dorsal awn. Palea ± equal to lemma, 2-fid. Lodicules linear-lanceolate, acute. Anthers 1–5 or more times as long as wide. Ovary glabrous; styles distinct, lateral.

Literature: L. R. Parodi, *Darwiniana* **8(4)**: 415–475(1949).

a. Lamina at least 1.4 mm wide, flat:
 b. Lamina puberulent; interfloral rhachilla *c.*1 mm; anthers up to 1.1 mm **1. *atropurpurea***
 b. Lamina glabrous; interfloral rhachilla 1.5–2.5 mm; anthers at least 1.3 mm **2. *kingii***
a. Lamina not more than 1 mm wide, folded to setaceous:
 c. Ligule not more than 2 mm; interfloral rhachilla not more than 0.8 mm **5. *flexuosa***
 c. Ligule at least 4 mm; interfloral rhachilla at least 1 mm:
  d. Lamina at least 0.6 mm wide, folded or convolute **2. *kingii***
  d. Lamina not more than 0.5(–1) mm wide, setaceous or filiform:
   e. Panicle open; lemma awn ± straight **3. *antarctica***
   e. Panicle contracted; lemma awn spirally twisted towards base **4. *parvula***

1. *D. atropurpurea* (Wahlenb.) Scheele 1844, *Flora (Regensb.)* **27**: 56.
Aira magellanica Hooker f. 1847, *Fl. Antarct.* **1(2)**: 376.
A. atropurpurea Wahlenb. 1812, *Fl. Lappon.* 37 – Skottsberg, 1916.

Leaves 4–9 cm × 1.4–6 mm; lamina flat, acute, puberulent; ligule 2–5 mm, obtuse to truncate, somewhat erose, puberulent; sheath terete, striate, puberulent above. Culms up to 48 cm, erect, glabrous. Panicle 6.5–9 cm, lax, purple; branches paired at base, slender, flexuous, scabridulous. Spikelets with 2 florets; interfloral rhachilla *c.*1 mm, hairy. Glumes 4.8–6.1 mm, subequal, lanceolate, acute, ciliate on keel and veins. Lemma 2.5–3 mm, broadly elliptical, denticulate and hairy at apex; awn 1.5–3 mm, scabridulous, in distal $\frac{1}{3}$ of lemma; callus hairy. Palea 2.2–2.6 mm, oblong, obtuse, puberulent. Anthers 0.8–1.1 mm. *Flowers February to March.*

Montane wet grassland, feldmark; 550–1100 m.

S. Isla Grande. (Map 34c).
W. Argentina north to *c.*41°S.lat.
Chile north to *c.*53°S.lat.
North America.
N. Eurasia.

2. *D. kingii* (Hooker f.) Desv. 1853, in C. Gay, *Hist. Chile Bot.* **6**: 335.
Aira kingii Hooker f. 1847, *Fl. Antarct.* **1(2)**: 376 – Hariot, 1884.
Trisetum dozei Franchet 1889, *Miss. Sci. Cap Horn* 384.

Leaves 2–45 cm × 0.6–3.5 mm; lamina flat or folded, acuminate, pungent, glabrous, smooth, shiny; ligule 2–6 mm, the upper up to 9.5 mm, acuminate; sheath terete, glabrous, smooth. Culms up to 11.5 cm, erect, rigid, glabrous. Panicle 11–30 cm, lax, greenish purple; branches 1–many at base, often flexuous, glabrous to scabridulous. Spikelets with 2(3) florets; interfloral rhachilla 1.5–2.5 mm, hairy, the prolongation *c.*2.5 mm. Glumes lanceolate, acute, glabrous, hyaline distally; lower 5–8.5 mm; upper 5.7–8.8 mm. Lemma 3–4 mm, oblong-lanceolate, irregularly 4-dentate at apex, glabrous; awn 1.5–3.8 mm, scabridulous, at or about middle of lemma; callus with hairs at least $\frac{1}{2}$ as long as lemma. Palea equal to lemma, linear-oblong, 2-fid, scabrid on keels. Anthers 1.3–1.6 mm. *Flowers December to April.*

Wet sands and gravels by sea and rivers, boggy areas; 0–300 m.

C. & S. Fuegia. (Map 34d).
W. Argentina north to *c.*50°S.
Chile north to *c.*51°30′S.lat.

3. *D. antarctica* Desv. 1853, in C. Gay, *Hist. Chile Bot.* **6**: 338.
Aira antarctica Hooker, non Forster f. – Hooker, 1847;
A. aciphylla Franchet 1889, *Miss. Sci. Cap Horn* 384.
Deschampsia aciphylla (Franchet) Speg. 1896, *Anal. Mus. Nac. Bs Aires* **5**: 89 – Kurtz 1896; Autran, 1905.
D. tenella Phil. 1896, *Anal. Univ. Chile* **94**: 25 – Autran, 1905.
D. elegantula (Steudel) Parodi 1949, *Darwiniana* **8(4)**: 452.

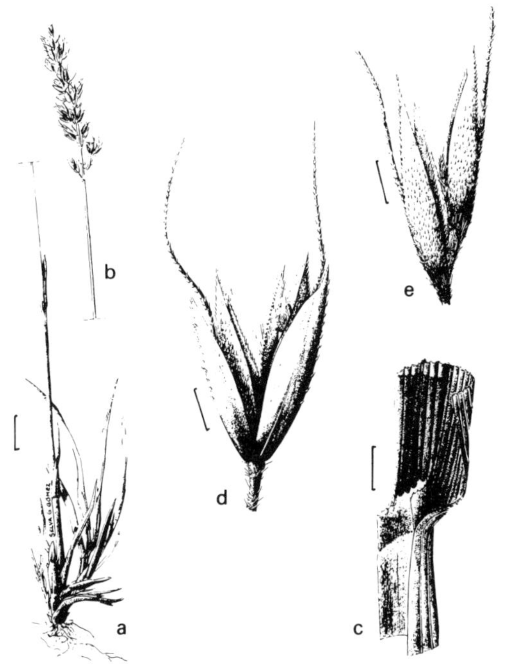

243 244

245 246

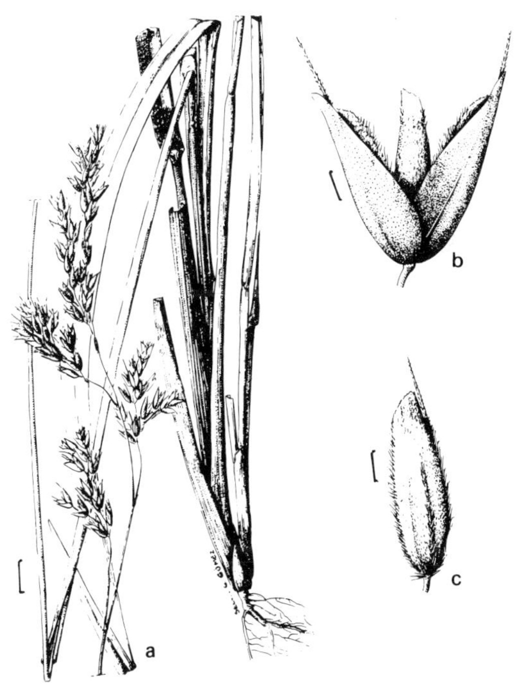

Fig. 243. *Koeleria fueguina*. a, Plant. b, Spikelet. c, Floret. d, Anther. Scale = 1 cm (a, d) or 1 mm (b–c). Fig. 244. *Trisetum spicatum*. a, Base of plant. b, Inflorescence. c, Ligule. d, Spikelet. e, Florets. Scale = 1 cm (a) or 1 mm (b–e). Fig. 245. *Deschampsia antarctica*. a, Plant. b, Inflorescence. c, Spikelet. d, Florets. e, Terminal floret and rhachilla. Scale = 1 cm (a) or 1 mm (b–e). Fig. 246. *Hierochloe redolens*. a, plant. b, Spikelet. c, Floret. Scale = 1 cm (a) or 1 mm (b–c).

(Flora Patagónica)

Leaves 1.2–15 cm × (0.1–)0.2 – *c.*1 mm; lamina setaceous, convolute, acute, glabrous; ligule 2–9 mm, acuminate; sheath terete, striate, glabrous. Culms 4–38 cm, ascending to erect, slender, glabrous. Panicle 6–15 cm, lax, spreading at maturity, green tinged with purple; lower branches arising 2–4 together, filiform, scabridulous. Spikelets with 2 florets; interfloral rhachilla 1–1.5 mm, hairy. Glumes lanceolate, acuminate, scabrid on veins; lower 3.8–6.7 mm; upper 4.3–7.7 mm. Lemma 1.8–3.3 mm, lanceolate, irregularly 4-dentate at apex, glabrous or scabridulous distally; awn 1.5–6.5 mm, straight or ± curved, scabridulous, in lower $\frac{1}{2}$–$\frac{1}{3}$ of lemma; callus with hairs up to $\frac{1}{3}$(–$\frac{2}{3}$) as long as lemma. Palea 2–2.5 mm, oblong, scabrid on veins. Anthers 0.25–0.8 mm. Self-compatible, often cleistogamous. *Flowers December to February*. (Fig. 245).

Coastal sands, moist areas by streams, swampy areas, *Lepidophyllum* scrub; 0–50 m.

Almost throughout Fuegia except extreme west, south and east. (Map 34e).
W. Argentina north to *c.*34°10′S.lat.
Chile north to *c.*51°S.lat.
Falkland Islands (Islas Malvinas), South Georgia, Iles de Kerguelen, South Orkney Is., South Shetland Is, South Sandwich Is, Antarctic Peninsula.

D. patula (Phil.) Pilger ex Skottsb. 1916, *Kungl. Svenska Vetensk Akad. Handl.* **56(5)**: 175 has been separated because of its shorter awns and lemma, longer callus-hairs and anthers more than 0.5 mm. These characters are found in plants which occur sporadically with the more usual forms and the recognition at most at the varietal level (Parodi, 1949) seems appropriate.

4. *D. parvula* (Hooker f.) Desv. 1853, in C. Gay, *Hist. Chile Bot.* **6**: 338.
Aira parvula Hooker f. 1847, *Fl. Antarct.* **1(2)**: 377 – Franchet, 1889;
Trisetum parvulum (Hooker f.) Speg. 1896, *Anal. Mus. Nac. Bs Aires* **5**: 89.

Leaves 1.5–6 cm × 0.1–0.6 mm; lamina filiform, pungent, convolute, glabrous; ligule (3.5–) 6–10(–14) mm, acuminate; sheath terete, glabrous. Culms 6–24 cm, ascending to erect, glabrous or scabrid above. Panicle 1.8–6 cm, contracted, green or purplish; branches short, ascending, scabrid. Spikelets with (1)2 florets; interfloral rhachilla 1.3–1.7 mm, hairy. Glumes 4.2–8.2 mm, subequal, lanceolate, acuminate, scabrid on keel. Lemma 1.8–3.4 mm, lanceolate, 4-dentate at apex, glabrous; awn 3.5–6.5 mm, geniculate, spirally twisted in basal part, scabrid, in lower $\frac{1}{3}$ of lemma; callus with hairs $\frac{1}{4}$–$\frac{1}{3}$ as long as lemma. Palea ± equal to lemma, oblong, scabrid on keels. Anthers 0.7–1 mm. *Flowers December to March.*

Feldmark and rock-crevices; 600–1000 m.

C. & S. Fuegia. (Map 34f).
W. Argentina north to 50°30′S.lat.
Chile north to 51°50′S.lat.
Falkland Islands (Islas Malvinas).

5. *D. flexuosa* (L.) Trin. 1836, *Bull. Sci. Acad. Imp. Sci. Pétersb.* **1**: 66.
Aira flexuosa L. 1753, *Sp. Pl.* 65 – Hooker, 1847; Franchet, 1889.
Deschampsia vestita (Steudel) Hauman & Parodi 1929, *Physis* **9**: 337.
D. discolor auct., non (Thuill.) P. Beauv. – Kurtz, 1896.

Plant sometimes with slender rhizomes. Leaves (2.5–)4–12(–25) cm × 0.2–0.6 mm; lamina setaceous, convolute, subobtuse, glabrous; ligule 0.5–2 mm, the upper up to 4 mm, obtuse or truncate; sheath terete, striate, glabrous. Culms 20–60 cm, erect or ± geniculate at base, rigid, glabrous. Panicle 4.5–12.5 cm, conical, lax, purplish to brown or silvery; branches paired at base, flexuous, scabrid; pedicels filiform. Spikelets with 2 florets; interfloral rhachilla 0.5–0.8 mm, silky. Glumes ovate-lanceolate, acuminate, glabrous; lower 3.5–5.5 mm; upper (4.5–)5–6.5 mm. Lemma 3.5–5.5 mm, ovate-oblong to lanceolate, obtuse to denticulate at apex; awn (3–)4.5–8 mm, straight or curved, sometimes twisted near base, near base of lemma; callus with hairs *c.*1 mm. Palea ± equal to lemma, oblong-lanceolate, scabrid on keels. Anthers 2–2.8 mm. *Flowers December to March.*

Bogs, *Empetrum* heath, river margins, rock crevices, clearings and margins of *Nothofagus* forest; 0–500 m.

Almost throughout Fuegia except extreme west and south. (Map 34g).
W. Argentina and E. Chile north to *c.*42°30′S.lat.
Falkland Islands (Islas Malvinas).
North America, Eurasia.

20. *Aira* L.

Slender, usually glabrous annuals. Ligule membranous; leaf-sheath without auricles. Inflorescence a panicle. Spikelets with 2 florets; rhachilla not produced beyond upper floret. Glumes equal, as

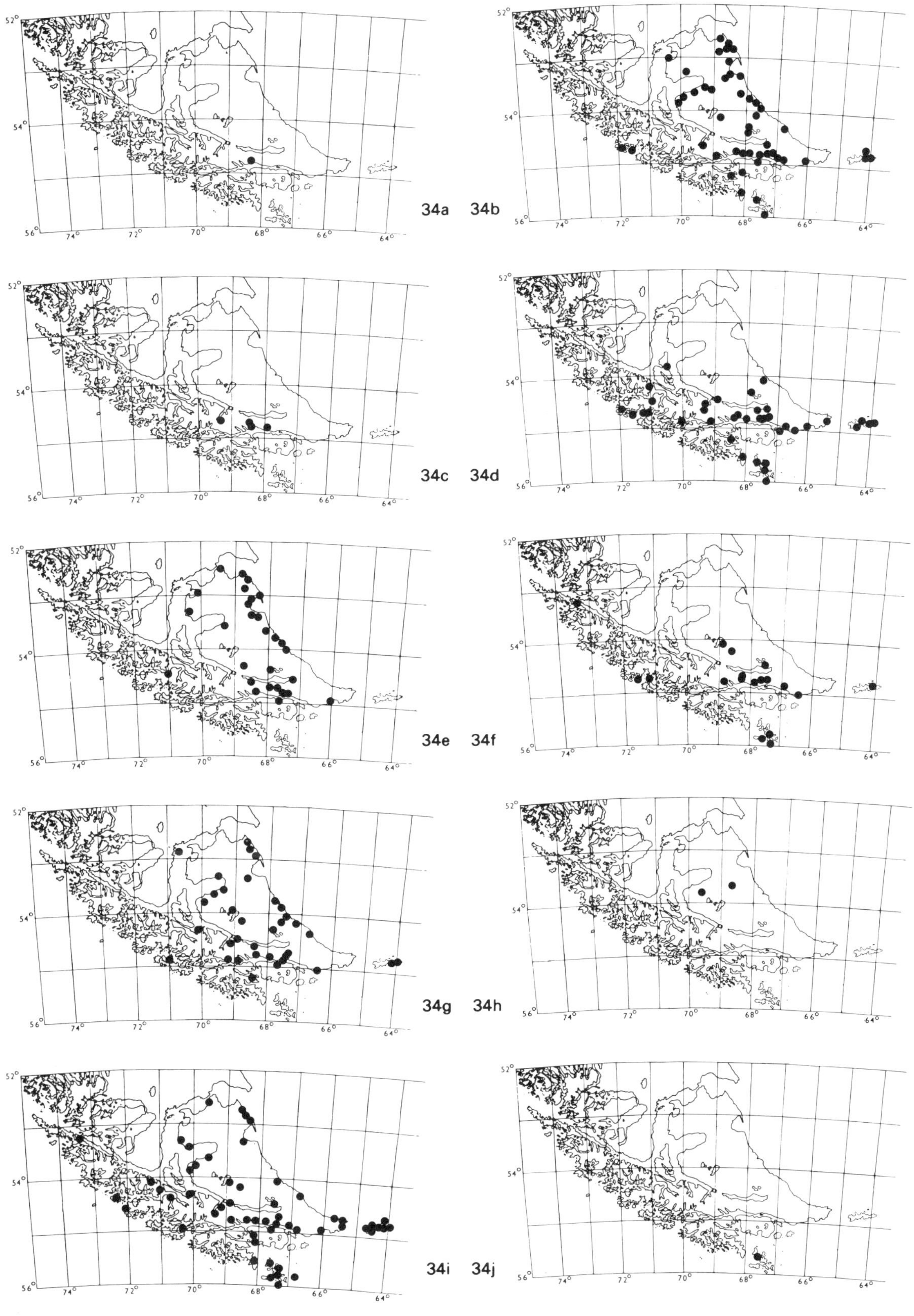

Map 34. a, *Trisetum cernuum*. b, *Trisetum spicatum*. c, *Deschampsia atropurpurea*. d, *Deschampsia kingii*. e, *Deschampsia antarctica*. f, *Deschampsia parvula*. g, *Deschampsia flexuosa*. h, *Hierochloe pusilla*. i, *Hierochloe redolens*. j, *Agrostis uliginosa*.

long as or longer than spikelet, ± keeled, 3-veined, awnless. Lemma rounded on back, obscurely 5-veined, with twisted, geniculate, dorsal awn in basal $\frac{1}{2}$. Palea shorter than lemma, obtuse or 2-fid. Lodicules ovate-lanceolate, acute, Anthers scarcely longer than wide. Ovary glabrous; stigmas subsessile, subapical.

a. Panicle narrow, dense, with short, erect branches; leaf-sheath smooth **1. *praecox***
a. Panicle wide, lax, with long, erecto-patent branches; leaf-sheath scabridulous **2. *caryophyllea***

1. *A. praecox L. 1753, *Sp. Pl.* 65.

Leaves 0.3–3 cm × 0.3–0.8 mm; lamina convolute, obtuse, smooth or scabrid, often reddish; ligule 0.8–3.3 mm, lanceolate-oblong, subacute; sheath smooth, often reddish. Culms 1.5–10.5 cm, slender, glabrous. Panicle 0.2–2.5 × 0.2–0.6 cm, narrow, dense; branches in groups of 2–3, scarcely longer than spikelets, erect. Spikelets 2–3.5 mm (excluding awns), crowded. Glumes ovate, acute, scabrid on keel. Lemma 2.3–3 mm, lanceolate, scabridulous distally; awn 3–4 mm; callus shortly hairy. Palea 2–2.6 mm, ovate. Anthers 0.2–0.3 mm. *Flowers November to February.*

Coastal sands, open soils in grassland and forest-margins; 0–70 m.

N. W. & S. Isla Grande, Isla Navarino, Isla Hoste.
Introduced from Europe and widely naturalized in temperate regions.

2. *A. caryophyllea L. 1753, *Sp. Pl.* 66.

Leaves 0.8–1 cm × 0.4–0.5 mm; lamina convolute, obtuse, glabrous; ligule 1–2.5 mm, lanceolate-oblong, subacute; sheath scabridulous, often reddish. Culms 1.2–9.5 cm, slender, glabrous. Panicle 1.3–3.2 × 1.4–2.5 cm, lax; branches in groups of 2–3, much longer than spikelets, erecto-patent. Spikelets 2.1–3 mm (excluding awns), clustered at ends of branches. Glumes ovate, acute, scabrid on keel. Lemma 2–2.3 mm, lanceolate, scabridulous distally; awn 3–4 mm; callus shortly hairy. Palea 1.5–1.8 mm, ovate, Anthers 0.3–0.6 mm. *Flowers December to January.*

Coastal sandy humus; 0–20 m.

Local on C. S. coast of Isla Grande.
Introduced from Europe and widely naturalized in temperate regions.

21. *Hierochloë* R. Br.

Rhizomatous perennials. Ligule membranous; leaf-sheath without auricles. Inflorescence a panicle. Spikelets with 3 florets, the uppermost hermaphrodite or female, the 2 lower male, larger. Glumes ± equal to spikelet, keeled, awnless, 3-veined. Lemma keeled, 5-veined, muticous or awned in male florets. Palea shorter than lemma, 2-fid. Lodicules ovate-lanceolate, shortly 2-lobed. Anthers 6–8 times as long as wide. Ovary glabrous; styles long, apical.

Literature: L. R. Parodi, *Rev. Mus. La Plata, Bot.* **3(14)**: 183–212(1941); V. D. Zotov, *N. Z. Jour. Bot.* **11**: 561–579(1973); M. E. de Paula de Brooks, *Darwiniana* **19**: 442–457(1975); D. M. Moore and M. C. Doggett, *Bol. Soc. Argent. Bot.* **16**: 390–396(1975).

a. Culms more than 10 cm; male florets with awned lemma **2. *redolens***
a. Culms less than 10 cm; male florets without awns **1. *pusilla***

1. *H. pusilla* Hackel 1907, in Dusén, *Ark. Bot.* **7(2)**: 4.

Plant caespitose, with long rhizomes. Leaves 0.4–4 cm × 2–3 mm; lamina flat or folded, subobtuse, sparsely hairy beneath, scabridulous above; ligule up to 0.5 mm, truncate, ciliate; sheath glabrous. Culms 2–7 cm, puberulent. Panicle 0.8–1.3 cm, contracted, dense, yellow-brown or purplish; branches solitary or paired, very short, with 1–2 spikelets, glabrous. Spikelets 4–5 mm. Glumes broadly ovate, ± obtuse, membranous. Uppermost floret with lemma 3–3.5 mm, ovate, glabrous, brownish; palea 2.8–3.2 mm, linear; anthers *c.*0.6 mm. Lower florets with lemma 4–5 mm, ovate, muticous, scabridulous on keel; palea *c.*3.5 mm, linear-ovate; anthers 2–2.2 mm. *Flowers December.*

Open grassland and rock outcrops; 100–150 m.

N. C. Isla Grande. (Map 34h).
W. Argentina north to *c.*48°S.lat.
E. Chile north to *c.*50°40′S.lat.

2. *H. redolens* (Vahl) Roemer & Schultes 1817, *Syst. Veg.* **2**: 514.
Holcus redolens Vahl 1791, *Symb. Bot.* **2**: 102.
Hierochloë arenaria Steudel 1854, *Syn. Pl. Glum.* **1**: 416 – Zotov, 1973.
H. magellanica Hooker f. 1847, *Fl. Antarct.* **1(2)**: 375 – Franchet, 1889; Hariot, 1894; Skottsberg, 1916.
H. antarctica auct., non (Labill.) R. Br. – Dusén, 1900.
H. moorei De Paula 1974, *Bol. Soc. Argent. Bot.* **15(4)**: 393.
H. sorianoi De Paula 1974, *Bol. Soc. Argent. Bot.* **15(4)**: 396.

Plant laxly caespitose, with short rhizomes. Leaves 2–54 cm × 2–13 mm; lamina flat, acuminate, convolute distally, glabrous beneath, scabridulous above; ligule 1–9 mm, triangular-ovate, truncate; sheath glabrous, brownish. Culms 10–120 cm, glabrous. Panicle 3.5–20 cm, lax, yellowish green to purplish; branches few, glabrous or scabridulous and with spikelets distally. Spikelets 6–14 mm. Glumes lanceolate, acute, membranous; upper 5.3–14 mm; lower 4–13 mm. Uppermost floret with lemma 3–5.5 mm, ovate, ciliate, scabrid on keel, brownish yellow, the awn up to 2 mm, dorsal; palea 2.8–4.5 mm, lanceolate; anthers 0.2–0.8 mm. Lower florets with lemma 5–9 mm, ovate, glabrous to scabrid, ciliate on margins and keel, with scabrid awn 1.3–4.5 mm, arising from distal $\frac{1}{3}(-\frac{2}{3})$; callus hairy; palea 4–5 mm, lanceolate-ovate; anthers (1.7–)2.6–3.5(–4.5) mm. *Flowers October to February*. (Fig. 246).

Damp or wet ground, usually with some water-movement, from coastal zone to feldmark, including open forest; 0–800 m.

Throughout Fuegia. (Map 34i).
W. Argentina north to *c*.40°S.lat.
Chile north to *c*.30°S.lat.
Ecuador.
Falkland Islands (Islas Malvinas).
New Zealand, S. E. Australia.
? New Guinea.

A variable species, especially in stature, laxness of panicle, glume-size and point of insertion of the lemma-awn. However, the continuous and uncorrelated pattern of variation shown by these characters suggests that attempts to recognize formal taxa within it are premature (Moore and Doggett, 1975).

22. *Anthoxanthum* L.

Caespitose perennials, smelling strongly of coumarin. Ligule membranous; leaf-sheath with auricles. Inflorescence a spike-like panicle. Spikelets with 3 florets, the uppermost hermaphrodite, the 2 lower sterile, larger. Glumes very unequal, keeled; lower 1-veined; upper 3-veined, acute or mucronate. Lemma rounded on back, 5- to 7-veined and awnless in fertile florets; 3-veined, with dorsal awn, in sterile florets. Palea 1-veined, absent in sterile florets. Lodicules absent. Stamens 2; anthers 5–7 times as long as wide. Ovary glabrous; styles long, subapical.

1. *A. odoratum L. 1753, *Sp. Pl.* 28.

Leaves 1.3–23 cm × 0.6–4 mm; lamina flat, acuminate, glabrous to sparsely white-hairy; ligule 1.5–3 mm, obtuse; sheath terete, glabrous to sparsely hairy, pubescent above. Culms 31–55 cm, glabrous. Panicle 2–5 cm, cylindrical or ovoid-cylindrical, dense, green or purplish brown; branches very short, hairy. Spikelets lanceolate. Glumes sparsely hairy; upper 5.6–7.6 mm, ovate-lanceolate, mucronate, enclosing florets; lower 2.8–4 mm, ovate, acute. Fertile floret with lemma 1.7–2.3 mm, suborbicular, glabrous, shiny; palea 1.5–2 mm; anthers 3–5 mm. Sterile florets with lemma 2.3–3.5 mm, oblong, emarginate, with brown silky hairs towards base, the lower with straight awn about as long as lemma, arising in distal half, the upper with geniculate awn, twisted below and 2–3 times as long as lemma, arising near base. *Flowers November to January*.

Open woods and moist grassland; 0–100 m.

S. Isla Grande.
Introduced from Europe for pasture improvement and widely naturalized in temperate regions.

23. *Holcus* L.

Softly pubescent, caespitose perennials. Ligule membranous; leaf-sheath without auricles. Inflorescence a panicle. Spikelets with 2(3) florets, the lower floret hermaphrodite, the upper 1(2) male; rhachilla shortly produced beyond upper floret, glabrous. Glumes longer than spikelet, strongly keeled; lower 1-veined; upper 3-veined, with short apical awn. Lemma strongly keeled, obscurely 3- to 5-veined, the upper with dorsal awn. Palea 2-veined, 3-lobulate at apex, equal to lemma. Lodicules lanceolate, acuminate. Anthers 3–4 times as long as wide. Ovary glabrous; stigmas sessile, apical.

1. *H. lanatus L. 1753, *Sp. Pl.* 1048.

Leaves 0.6–14 cm × 0.6–6 mm; lamina flat, acuminate, rarely subglabrous, greyish-green; ligule 0.5–3.2 mm, truncate, ± dentate; sheath white-hairy. Culms 5–75 cm, erect or geniculate at base, sometimes subglabrous. Panicle 1–10.5 cm, lanceolate to cylindrical or ovoid, dense to rather lax, erect or nodding, whitish or pale green to pinkish or purple; branches hairy. Spikelets oblong to elliptical. Glumes 3–6 mm, equal or the upper slightly longer, strigose on keel and veins, puberulent elsewhere; lower linear-lanceolate; upper ovate to elliptical, with apical awn up to 1 mm. Lower lemma 1.7–2.6 mm, ovate, obtuse, shiny; upper 1.5–2.6 mm, lanceolate, with recurved, hook-like, dorsal awn 0.5–2 mm, from distal third. Anthers 1.2–2.2 mm. *Flowers December to March*.

Coastal sands and gravels, disturbed grassland and cleared forest; 0–100 m.

C. & S. Isla Grande.
Introduced from Europe for pasture improvement and widely naturalized.

24. *Agrostis* L.

Caespitose perennials, sometimes rhizomatous. Ligule membranous; leaf-sheath without auricles. Inflorescence a panicle. Spikelets with 1 floret; rhachilla not produced beyond floret. Glumes subequal to unequal, longer than floret, usually 1-veined, awnless. Lemma rounded on back, 5-veined, often awned; callus glabrous or with hairs less than ½ as long as lemma. Palea equalling or, usually, much shorter than lemma. Lodicules lanceolate, acute. Anthers 2–4 times as long as wide. Ovary glabrous; styles short, apical.

Literature: K.-G. Widén, *Fl. Fenn.* **5**: 1–209(1971); E. G. Nicora, *In*: M. N. Correa (Ed.) *Fl. Patag.* **3**: 369–394(1978).

a. Palea at least ½ as long as lemma:
 b. Ligule not more than 1(–2) mm, wider than long; lower glumes not more than 2 mm; anthers at least 0.9 mm **1. *capillaris***
 b. Ligule at least (1.5–)2.5 mm, longer than wide; lower glumes at least 2.1 mm; anthers not more than 0.7 mm:
  c. Panicle lax, the branches with spikelets distally; lemma with dorsal awn at least 3.5 mm **3. *uliginosa***
  c. Panicle contracted, the branches with spikelets from base; lemma with or without very short dorsal awn **2. *stolonifera***
a. Palea not more than ½ as long as lemma:
 d. Lamina convolute:
  e. Panicle dense; branches not more than 1 cm, convergent, not verticillate **8. meyenii**
  e. Panicle ± lax; branches up to 6 cm, divergent at anthesis, semi-verticillate **9. *flavidula***
 d. Lamina flat or folded, at least in basal part:
  f. Lemma-awn at least 2 mm; lower glume at least 3.5 mm **4. *magellanica***
  f. Lemma-awn not more than 1.5 mm; lower glume not more than 3.5 mm:
   g. Panicle-branches dichotomous, at angle of more than 90°to each other; anthers at least 0.9 mm **5. *inconspicua***
   g. Panicle-branches not dichotomous, at an angle of less than 90° to each other; anthers not more than 0.9 mm:
    h. Callus glabrous; lemma 1.1–1.9 mm; anthers at least 0.8 mm **6. *leptotricha***
    h. Callus hairy; lemma 1.8–2.6 mm; anthers not more than 0.7 mm:
     i. Glumes, at least the upper, glabrous; panicle-branches up to 4.5 cm, glabrous or slightly scabrid distally; lemma 2.1–2.6 mm **10. *pyrogea***
     i. Glumes scabrid on keel; panicle-branches up to 11 cm; scabrid; lemma 1.8–2.1 mm **7. *fuegiana***

1. *A. capillaris L. 1753, *Sp. Pl.* 62.
A. tenuis Sibth. 1794, *Fl. Oxon.* 36 – Nicora, 1978.

Rhizomatous. Leaves 2–14 cm × 1–4 mm; lamina flat or slightly convolute, acuminate, glabrous to scabridulous; ligule 0.5–1 mm, truncate, scabridulous; sheath smooth to ± scabrid. Culms 20–60 cm, glabrous or sparsely scabridulous. Panicle 4.5–10.5 cm, ovoid, open, rarely contracted; branches patent at anthesis, glabrous or scabridulous distally. Glumes ovate-lanceolate, acute, sparsely scabrid on keel, 1-veined; upper 1.3–1.9 mm; lower 1.4–2 mm. Lemma 1.2–1.7 mm, ovate, awnless; callus glabrous. Palea 0.7–1.2 mm, ovate, 2-fid. Anthers 0.9–1.3 mm. *Flowers January to March.*

Moist sands, gravels and grassland; 0–30 m.

Locally in C. & S. Isla Grande.
Introduced from Europe and widely naturalized in temperate regions.

2. *A. stolonifera L. 1753, *Sp. Pl.* 62.

Stoloniferous, sometimes with short rhizomes. Leaves 4–18.5 cm × 2–3.5 mm; lamina flat, acuminate, scabrid distally; ligule 2.5–4 mm, oblong, truncate, lacerate; sheath striate, glabrous. Culms 24–53 cm, ascending, glabrous. Panicle 5.5–9 cm, oblong, ovoid, contracted, yellow-brown; branches patent at anthesis, clustered, scabrid, with spikelets throughout. Glumes ovate-lanceolate, acute, scabrid, especially on keel, 1-veined; upper 2.3–3 mm; lower 2–2.8 mm. Lemma 1.3–2.2 mm, ovate, awnless or with very short dorsal awn; callus glabrous. Palea 1–1.5 mm, ovate-oblong, obtuse. Anthers 0.5–0.7 mm. *Flowers ? January to March.*

Wet places by streams; 30–50 m.

N. C. Isla Grande, local.
Introduced from Europe and occasionally naturalized in temperate regions.

3. *A. uliginosa* Phil. 1865, *Anal. Univ. Chile* **27**: 323.

A. paucinodis Hackel 1885, *Bot. Jahrb.* **6**: 242 – Skottsberg, 1916; Moore, 1974.

Leaves 16–32 cm × 1.6–5.2 mm; lamina flat or folded when young, acuminate, sparsely scabridulous at margin and apex; ligule 4.5–8.5 mm, oblong-ovate, lacerate with age; sheath striate, glabrous. Culms 60–112 cm, scabridulous distally. Panicle 11–23 cm, conical, lax; branches in whorls, scabrid, with spikelets distally. Glumes 4–5 mm, narrowly lanceolate, acute, scabrid on keel distally. 1-veined. Lemma 1.9–2.2 mm, oblong, truncate, with geniculate dorsal awn 3.5–4.5 mm in distal $\frac{1}{2}$–$\frac{1}{3}$; callus hairy. Palea 1–1.4 mm, ovate, obtuse. Anthers 0.6–0.7 mm. *Flowers February to March.*

Moist grassland; 0–200 m.

S. Isla Grande; Isla de los Estados. (Map 34j).
W. Argentina north to *c.*39°S.lat.
Chile north to *c.*40°S.lat.

4. *A. magellanica* Lam. 1791, *Tabl. Encycl. Méth. Bot.* **1**: 160.

A. antarctica Hooker f. 1847, *Fl. Antarct.* **1(2)**: 374.
A. cognata Steudel 1854, *Syn. Pl. Glum.* **1**: 421 – Skottsberg, 1916.
A. araucana Phil. 1896, *Anal. Univ. Chile* **94**: 14 – Moore, 1974.

Sometimes with short rhizomes. Leaves 1–16 cm × 0.5–5 mm; lamina flat or folded, at least towards base, acuminate, glabrous to scabridulous; ligule 1.5–7 mm, oblong, obtuse, serrate, often scabridulous; sheath striate, glabrous. Culms 10–90 cm, glabrous or scabridulous; sheath striate, glabrous. Culms 10–90 cm, glabrous or scabridulous distally. Panicle (2.5–)5–15 cm, cylindrical – conical, lax to ± compact, often interrupted towards base; branches clustered, scabrid-pubescent, the lower with spikelets distally. Glumes lanceolate, acute, scabrid distally on keel. 1-veined; upper 3.2–4.5 mm; lower 3.5–5 mm. Lemma 1.5–2.3 mm, oblong-ovate, acute, with geniculate, dorsal awn 2–4.8 mm at about middle; callus glabrous or with sparse basal hairs. Palea 0.7–1 mm, suborbicular, glabrous, or absent. Anthers 0.5–0.8 mm. *Flowers December to March.* (Fig. 247).

Sphagnum bogs, stream-margins, wet cliffs, feldmark, marshes; 0–800 m.

S. Fuegia. (Map 35a).
W. Argentina north to *c.*40°S.lat.
Chile north to *c.*38°S.lat.
Falkland Islands (Islas Malvinas).

5. *A. inconspicua* G. Kunze 1845, *Mém. Acad. Sci. Pétersb.* Ser 6, **41**: 331.

A. airiformis Steudel 1854, *Syn. Pl. Glum.* **1**: 172.
A. airoides Franchet 1889, *Miss. Sci. Cap Horn* 382 – Spegazzini, 1896.

Leaves 0.5–10 cm × 0.4–3.1 mm; lamina flat or folded, subobtuse, involute at margins, ± scabrid; ligule 2.8–4 mm, oblong-lanceolate, often lacerate, scabrid; sheath scabridulous. Culms 15–60 cm, scabridulous. Panicle 3.5–15 cm, lax; branches dichotomous, at an angle of more than 90° to each other, ± scabridulous, the spikelets solitary, remote. Glumes lanceolate, acute, scabrid on keel distally; upper 2.1–3.1 mm, 3-veined; lower 2–3.2 mm, 1-veined. Lemma 1.6–1.8 mm, ovate, denticulate at apex, ± pubescent, with dorsal awn 0–1 mm at about middle; callus with sparse basal hairs. Palea 0.1–0.3 mm. Anthers 0.9–1.1 mm. *Flowers January to February.*

Wet coastal sands and bogs; 0–30 m.

S. coast Isla Grande (Bahía Slogett); Isla Izabel. (Map 35b).
W. Argentina north to *c.*40°S.lat.
Chile north to *c.*38°S.lat.

6. *A. leptotricha* Desv. 1853, in C. Gay, *Hist. Chile Bot.* **6**: 316.

A. lechleri Steudel 1854, *Syn. Pl. Glum* **1**: 422 – Moore, 1974.

Rhizomes often present. Leaves 7.5–60 cm × 0.4–2.5 mm; lamina flat, acuminate, the margins involute, scabrid above, scabridulous beneath; ligule 2.5–6 mm, lanceolate, acuminate, lacerate, scabrid; sheath glabrous. Culms 75–90 cm, glabrous. Panicle 7.5–13 cm, lax; branches in ± whorls of 6 at base of panicle, up to 2.5 cm, scabrid. Spikelets purplish, rather remote. Glumes lanceolate, acute, scabrid on keel; upper 2–2.6 mm, 3-veined; lower 2.1–2.8 mm, 1-veined. Lemma 1.4–1.9 mm, ovate, scabridulous, with scabrid, dorsal awn 0.5–1.5 mm in basal half; callus ± glabrous. Palea not more than 0.2 mm. Anthers 0.8–1 mm. *Flowers January to February.*

Open *Nothofagus pumilio* forest; 0–40 m.

S. coast Isla Grande. (Map 35c).
W. Argentina and Chile north to *c.*40°S.lat.

7. *A. fuegiana* Hackel 1900, in Dusén, *Svenska Exped. Magellansl.* **3(5)**: 220.

Similar to *A. leptotricha* but without rhizomes; leaves up to 5 mm wide; panicle-branches up to 11 cm; lemma 1.8–2.1 mm, awnless or with awn up to 0.6 mm; callus with short hairs; anthers 0.4–0.7 mm. *Flowers January to March.*

Open deciduous forest; 0–80 m.

C. Isla Grande (Seno Almirantazgo, Río Azopardo). (Map 35d).
W. Argentina and Chile north to *c.*51°25′S.lat.

Perhaps this should be included in *A. leptotricha*, as suggested by Nicora (1978), but the available material, including the type, shows sufficient differences to warrant further study.

8. *A. meyenii* Trin. 1841, *Agrost.* 66.
A. canina auct., non L. 1753, *Sp. Pl.* 62 – Franchet, 1889; Spegazzini, 1896; Moore, 1974.
A. falklandica Hooker f. 1847, *Fl. Antarct.* **1(2)**: 373.

Leaves 1–5 cm × 0.2–1.5 mm; lamina convolute, acuminate to subobtuse, ± scabrid; ligule 0.4–2 mm, obtuse to subacute, fimbrio-ciliate, scabrid; sheath glabrous to scabridulous. Culms 4–14 cm, erect or ascending, glabrous to scabridulous. Panicle 1.5–5 cm, narrowly lanceolate-cylindrical, dense; branches up to 1 cm, ascending, convergent, sparsely scabrid. Glumes lanceolate, acute, scabrid on keel; upper 2.3–2.8 mm, 3-veined; lower 2.7–2.9 mm, 1-veined. Lemma 1.7–2.1 mm, ovate, obtuse, awnless or with dorsal awn up to 2.3 mm about middle; callus with a few hairs. Palea up to 0.4 mm. Anthers 0.6–0.9 mm. *Flowers January to March.*

Rock crevices, screes, sands and gravels; 0–800 m.

C. & S. Isla Grande. (Map 35e).
W. Argentina north to *c.*38°S.lat.
E. Chile north to *c.*35°S.lat.
Falkland Islands (Islas Malvinas).

Fuegian and Falkland Islands (Islas Malvinas) plants have been hitherto included in the very variable European *A. canina*. However, recent treatments have recognized several species within the complex and, since our plants are smaller, with more compact panicles and shorter ligules than in the typical European form, together with geographical distribution, I follow Nicora (1978) in recognizing it as a distinct species.

9. *A. flavidula* Steudel 1854, *Syn. Pl. Glum.* **1**: 421.
A. aberrans Steudel 1854, *Syn. Pl. Glum.* **1**: 422 – Skottsberg, 1916.
A. brachyathera Steudel 1854, *Syn. Pl. Glum.* **1**: 422 – Kurtz, 1896; Spegazzini, 1896.
A. kufuim Speg. 1896, *Anal Mus. Nac. Bs Aires* **5**: 82.
A. exarata Trin. var. *angustifolia* Hackel 1900, in Dusén, *Svenska Exped. Magellansl.* **3(5)**: 219.

Leaves 1–11 cm × 0.25–2 mm; lamina convolute, subacute, scabridulous; ligule 0.5–2(–3) mm, oblong, obtuse, ± serrate; sheath striate, glabrous. Culms 5–38 cm, ascending to erect, glabrous. Panicle 2–15 cm, ± lax, purplish to yellowish; branches up to 6 cm, divergent at anthesis, semi-verticillate, scabridulous to glabrous. Glumes subequal, 1.9–3.4 mm, lanceolate, acute, scabrid on keel, 1-veined. Lemma 1.4–2.2 mm, ovate, obtuse, scabridulous on veins, with dorsal awn 0.1–2.5 mm, about middle; callus with few short hairs. Palea 0.2–0.5 mm. Anthers 0.5–0.8 mm. *Flowers December to March.*

Sands and gravels, *Empetrum and Bolax* heath, open grassland, saline flats; 0–600 m.

Throughout Isla Grande except extreme east. (Map 35f).
W. Argentina north to *c.*45°S.lat.
Chile north to *c.*50°50′S.lat.

The minor differences in the degree of contraction and colouring of the panicle, and the development of the lemma-awn, used to distinguish the species here reduced to synonymy, are not correlated and form parts of a continuum.

10. *A. pyrogea* Speg. 1896, *Anal. Mus. Nac. Bs Aires* **5**: 83.

Rhizomatous. Leaves 1–9 cm × 1–3 mm; lamina flat or folded, ± acute, ± rigid, scabrid; ligule 2.5–4 mm, oblong-lanceolate, acute, scabridulous; sheath glabrous. Culms 30–42 cm, glabrous. Panicle 8.2–10.5 cm, lax, conical; branches up to 4.5 cm, erecto-patent, glabrous or slightly scabrid distally, with spikelets distally. Glumes subequal, 2.3–3.5 mm, ovate, truncate, glabrous, 1-veined. Lemma 2.1–2.6 mm, ovate, truncate, with subapical awn 0.2–0.5 mm; callus with short hairs. Palea 0.3–0.5 mm. Anthers 0.7–1 mm. *Flowers January to February.*

Moist sands and bogs; 0–40 m.

Locally on W. & S. coasts of Isla Grande. (Map 35g).
W. Argentina north to *c.*38°S.lat
Chile north to *c.*47°S.lat.

Superficially similar to *A. flavidula* but with flat leaves and larger glumes and lemmas.

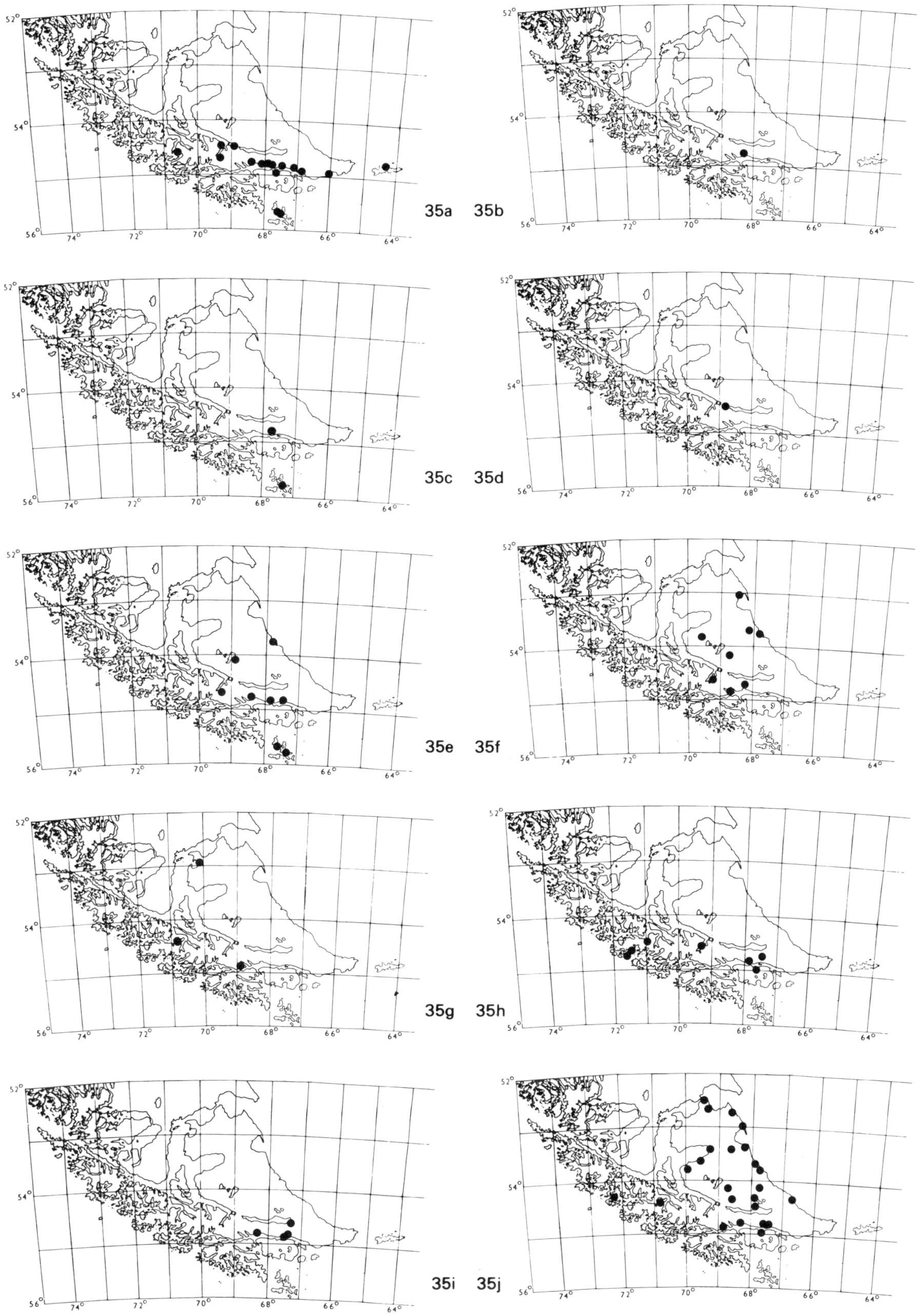

Map 35. a, *Agrostis magellanica*. b, *Agrostis inconspicua*. c, *Agrostis leptotricha*. d, *Agrostis fuegiana*. e, *Agrostis meyenii*. f, *Agrostis flavidula*. g, *Agrostis pyrogea*. h, *Calamagrostis suka*. i, *Calamagrostis antoniana*. j, *Calamagrostis stricta*.

25. *Ammophila* Host

Robust, subglabrous, rhizomatous perennials. Ligule membranous; leaf-sheath without auricles. Inflorescence a dense panicle. Spikelets strongly laterally compressed, with 1 floret; rhachilla produced beyond floret, silky. Glumes equal to or exceeding floret; lower 1-veined; upper 3-veined. Lemma 3- to 5-veined, shortly 2-fid, with short subapical awn. Palea ± equal to lemma. Anthers *c.*10 times as long as wide. Ovary glabrous; styles short, apical.

1. *A. arenaria (L.) Link 1827, *Hort. Berol.* **1**: 105.

Rhizomes far-creeping, stout. Leaves 9–44 cm × 2–4.5 mm; lamina convolute, rigid, pungent, ribbed and puberulent above, smooth and glabrous beneath; ligule 8–26 mm, deeply 2-fid; sheath terete, ribbed. Culms 58–65 cm, ascending to erect, terete, rigid. Panicle 8–16 cm, lanceolate-cylindrical, whitish; branches erect, appressed, scabrid. Spikelets imbricate. Glumes ± equal, 12–15 mm, linear-lanceolate, acuminate, scabrid on keel. Lemma 10–12 mm, lanceolate, the scabrid awn 0.3–0.5 mm; callus with white hairs up to ⅓ as long as lemma. Palea lanceolate, scabrid. Anthers 6–6.2 mm. *Flowers December to February.*

Coastal sands; 0–30 m.

Locally on N. W. & N. E. coasts of Isla Grande. Introduced from Europe for stabilizing sand dunes.

26. *Calamagrostis* Adanson

Caespitose, usually rhizomatous, perennials. Ligule membranous; leaf-sheath without auricles. Inflorescence a lax to contracted, often spike-like, panicle. Spikelets with 1 floret, rarely also with a rudimentary floret. Glumes longer than floret; lower 1-veined; upper 3-veined. Lemma rounded on back, 5-veined, with dorsal awn; callus with ring of hairs longer to shorter than lemma. Palea ⅔ as long as to equalling lemma, 2-veined. Lodicules linear-ovate, subacute. Anthers 2–6 times as long as wide. Styles short, apical.

The Fuegian and Patagonian species have been placed in the genus *Deyeuxia* Clarion by various authors, most recently by Rúgolo de Agrasar (1978), distinguished from *Calamagrostis* by the length of the callus-hairs relative to the lemma and by the prolongation of the rhachilla. Since these differences do not seem satisfactory in a world context I take here, with diffidence, a conservative view, including the Fuegian species in *Calamogrostis* sect. *Deyeuxia* (Clarion) P. Beauv.

Literature: Z. E. Rúgolo de Agrasar, *Darwiniana* **21**: 417–453 (1978).

a. Culms not more than 10 cm; leaves not more than 1 mm wide; anthers not more than 0.8 mm **4. *erythrostachya***
a. Culms at least 17 cm; leaves usually more than 1 mm wide; anthers at least 0.8 mm:
 b. Glumes not more than 5 mm; lower branches of panicle appressed to axis **3. *stricta***
 b. Glumes at least 5 mm; lower branches of panicle somewhat diverging:
  c. Callus-hairs much shorter than lemma; lemma-awn at least 4.5 mm; rhachilla produced beyond floret **2. *antoniana***
  c. Callus-hairs as long as or longer than lemma; lemma-awn not more than 2.5 mm; rhachilla not produced **1. *suka***

1. *C. suka* Speg. 1896, *Anal. Mus. Nac. Bs Aires* **5**: 86.
Deyeuxia suka (Speg.) Parodi 1953, *Rev. Argent. Agron.* **20(1)**: 14 – Rúgolo 1978.

Rhizomes ascending. Leaves 4–17 cm × 1–3.5 mm; lamina flat or folded, subacute, involute at margin, glabrous beneath, scabridulous above; ligule 1–3.7 mm, truncate, serrate; sheath glabrous. Culms 39–67 cm, glabrous. Panicle 6.5–9.5 × 1.5–2.5 cm, somewhat contracted; lower branches somewhat diverging, glabrous to scabridulous distally. Spikelets brownish to purplish; rhachilla not produced beyond floret. Glumes lanceolate, acute, scabridulous distally and on keel and margin; lower 5–6.2 mm; upper 5.3–6.3 mm. Lemma 3–3.3 mm, lanceolate-ovate, truncate, denticulate at apex, scabridulous to scabrid, especially on keel and veins; awn 2–2.4 mm, scabrid, arising 0.5–1.1 mm from lemma-apex; callus with hairs 3.5–4.7 mm. Palea 2.8–3 mm. Anthers 0.8–1.8 mm. *Flowers January to February (? March).*

Sphagnum bogs and other moist areas; 0–500 m.

S. Isla Grande and immediately adjacent islands. (Map 35h).
W. Argentina north to *c.*49°20′S.lat.
Chile north to 51°24′S.lat.

2. *C. antoniana* (Griseb.) D. M. Moore, ***comb. nov.***
Agrostis antoniana Griseb. 1879, *Symb. Fl. Argent.* 293.

Calamagrostis gusindei Pilger ex Skottsb. 1926, *Acta Horti Gothob.* **2**: 36 – Moore, 1974.
Deyeuxia gusindei (Pilger ex Skottsb.) Parodi 1953, *Rev. Argent. Agron.* **20**: 14.
D. antoniana (Griseb.) Parodi, 1953, *Rev. Argent. Agron.* **20**: 14 – Rúgolo, 1978.

Rhizomes ascending to prostrate, short. Leaves 5–35 cm × 0.8–1.5 mm; lamina convolute, acuminate, glabrous to scabridulous; ligule 3–10 mm, lanceolate, acute; sheath glabrous. Culms 50–85 cm, scabrid distally. Panicle 10–15.5 × 1–2.5 cm, somewhat contracted; lower branches somewhat diverging, scabrid, especially distally. Spikelets green to purplish; rhachilla produced 1.2–2 mm, with hairs up to 5 mm. Glumes lanceolate, acute, scabridulous on keel and margin; lower 5.4–7.5 mm; upper 5.7–8 mm. Lemma 5.2–5.8 mm, lanceolate, truncate, denticulate at apex, densely scabridulous; awn 4.8–5.5 mm, scabrid, arising 3–4.2 mm from lemma-apex; callus with hairs 0.7–1 mm. Palea 4–4.8 mm. Anthers 1.8–2.2 mm. *Flowers February to March.*

Open grassland and rocky areas near coast; 0–60 m.

C. S. Isla Grande and islands of Canal Beagle. (Map 35i).
W. Argentina north to *c.*46°S.lat. and at *c.*24°S.lat.
E. Chile north to 51°S.lat.
Peru.

3. ***C. stricta*** (Timm) Koeler 1802, *Descr. Gram.* 105.
C. neglecta auct., non (Ehrh.) P. Beauv. –
C. poaeoides Steudel 1854, *Syn. Pl. Glum.* **1**: 423 – Spegazzini, 1896; Roivainen, 1954.
C. magellanica Phil. 1896, *Anal. Univ. Chile* **94**: 20 – Martín, 1949.
C. fuegiana Speg. 1896, *Anal. Mus. Nac. Bs Aires* **5**: 85.
Deyeuxia poaeoides (Steudel) Rúgolo 1978, *Darwiniana* **21**: 439.
? *D. freticola* Speg. 1896, *Anal. Mus. Nac. Bs Aires* **5**: 87.

Rhizomes prostrate, long. Leaves 3.5–37 cm × 0.8–4 mm; lamina flat to convolute, acute to acuminate, glabrous to scabridulous beneath, ± scabrid above; ligule 0.1–2 mm, truncate, serrate at apex; sheath glabrous. Culms 17–74 cm, ± scabrid distally. Panicle 4.5–22 × 0.6–2.2 cm, spike-like, dense; branches ± appressed to axis, scabrid. Spikelets yellow-brown to purplish; rhachilla scarcely produced, rarely with vestigial 2nd floret. Glumes lanceolate, acute, sparsely scabridulous distally on veins; lower 3.4–4.5 mm; upper 3.2–4.7 mm. Lemma 2.5–4.1 mm, ovate, truncate, denticulate at apex, scabrid, especially distally; awn 1.7–3 mm, scabrid, arising 1.4–2.1 mm from lemma-apex; callus with hairs 1.5–2.5 mm. Palea 1.7–2.8 mm. Anthers (1–)1.5–2 mm. *Flowers December to March.* (Fig. 248).

Moist sands, gravels and soils by coast, streams and lakes, *Sphagnum* bogs, wet, open *Chiliotrichum* scrub; 0–50 m.

Isla Grande and islands immediately to west and south. (Map 35j).
W. Argentina north to *c.*42°S.lat.
Chile north to *c.*51°S.lat.
North America. N. Eurasia.

4. ***C. erythrostachya*** (Desv.) Hackel ex Stuckert 1906, *Anal Mus. Nac. Bs Aires* **13**: 476.
Deyeuxia erythrostachya Desv. 1853, in C. Gay, *Hist. Chile Bot.* **6**: 323 – Parodi, 1953; Rúgolo 1975, 1978.

Densely caespitose, with slender rhizomes. Leaves 0.6–5.5 cm × 0.5–1 mm; lamina setaceous, folded, ± acute, usually curved, scabridulous on margin; ligule 1.5–7.5 mm, lanceolate, glabrous; sheath glabrous. Culms 3.5–9 cm, glabrous. Panicle 2–6 × 0.6–2.2 cm, spike-like, dense; branches appressed to axis, sparsely scabrid distally. Spikelets grey-green or purplish; rhachilla shortly produced, with hairs 2–3 mm. Glumes lanceolate, acute, scabridulous distally on midvein; lower 3.7–4.9 mm; upper 3.9–5.3 mm. Lemma 2.2–2.5 mm, ovate, truncate, serrate at apex, glabrous; awn 3.7–5 mm, scabrid, arising 2–2.2 mm from lemma-apex; callus with hairs 2.7–2.8 mm. Palea 1.6–1.9 mm. Anthers 0.5–0.8 mm. *Flowers February to March.*

Montane grassland, rock and soil screes, feldmark; 600–1000 m.

S. Isla Grande. (Map 36a).
W. Argentina and E. Chile north to *c.*36°S.lat.

27. *Phleum* L.

Caespitose perennials, sometimes rhizomatous. Ligule membranous; leaf-sheath without auricles. Inflorescence a dense, spike-like panicle. Spikelets with 1 floret. Glumes longer than floret, glabrous but with rigid, whitish cilia on keel, with scabrid, apical awn, 3-veined. Lemma rounded on back, obscurely 5-veined, awnless. Palea equalling lemma, 2-veined, slightly keeled. Lodicules irregularly obovate, acute to truncate. Anthers 4–5 times as long as wide. Ovary glabrous; styles long, apical.

a. Panicle cylindrical, usually at least 4 times as long as wide; awns 1–2 mm **2. *pratense***

a. Panicle broadly cylindrical to ovoid, usually not more than 4 times as long as wide; awns 1–2.7 mm **1. *alpinum***

1. *P. alpinum* L. 1753, *Sp. Pl.* 59.

P. commutatum Gaudin 1808, *Alpina (Winterthur)* **3**: 4 – Martín, 1949; Nordenskjöld, 1945; Nicora, 1978.

Laxly caespitose, often with short rhizomes. Leaves 0.3–10(–21) cm × 1–6 mm; lamina flat or folded, acute, the margin involute distally, scabridulous; ligule 0.7–2(–2.7) mm, truncate; sheath glabrous, the upper somewhat inflated. Culms 4–80 cm, glabrous. Panicle 1–3.5 × 0.6–1.1 cm, broadly cylindrical to ovoid, green or purplish; branches short, erect. Glumes 4–6.5 mm, subequal, narrowly oblong, the veins sparsely scabridulous distally; awn 1–2.7 mm, straight or slightly curved. Lemma 1.7–2.6 mm, elliptic-oblong, obtuse, puberulent distally on veins. Palea oblong-ovate, puberulent on keels. Anthers 1–1.6 mm. $2n = 28$★. *Flowers December to April.* (Fig. 249).

Moist rocks, clays and sands, open *Nothofagus antarctica* forest and *Chiliotrichum* scrub, montane grassland, feldmark; 0–700 m.

Isla Grande, except extreme west; Isla Navarino. (Map 36b).
W. Argentina and E. Chile north to *c.*32°S.lat. South Georgia. North America. Eurasia.

Fuegian material belongs to ssp. *alpinum*. The diploid ssp. *rhaeticum* Humphries is restricted to the mountains of C. & S. Europe (Humphries, 1978).

★2. *P. pratense* L. 1753, *Sp. Pl.* 59.

Similar to *P. alpinum* but ligule 1–4 mm, truncate; panicle 1.8–6 × 0.6–0.9 cm, cylindrical, green, often purplish; glumes 3.5–5.5 mm, the awn 1–2 mm; lemma 1.9–3 mm; anthers 1–1.7 mm. *Flowers November to February.*

Coastal sands and gravels, disturbed open *Chiliotrichum* scrub and deciduous forest, roadsides; 0–100 m.

N. E. and C. S. Isla Grande.
Introduced from Europe for pasture-improvement and widely naturalized in temperate regions.

28. *Alopecurus* L.

Subglabrous perennials, rarely annuals. Ligule membranous; leaf-sheath without auricles. Inflorescence a dense, spike-like panicle. Spikelets with 1 floret. Glumes as long as or slightly exceeding floret, ± equal, united at least at base, with soft appressed hairs on keel and sides, with or without short apical awn, 3-veined. Lemma keeled, (4) 5-veined, with dorsal or subapical awn, with margins united towards base. Palea absent. Lodicules absent. Anthers 4–6 times as long as wide. Ovary glabrous; styles long, apical, united at base.

Literature: L. R. Parodi, *Rev. Fac. Agron. Vet. Bs Aires* **7**: 345–369(1931).

a. Spikelets at least 3 mm; glumes acute:

b. Panicle not more than 4 times as long as wide; lemma obtuse **4. *magellanicus***

b. Panicle at least 6 times as long as wide; lemma acute **1. *pratensis***

a. Spikelets not more than 3 mm; glumes obtuse:

c. Anthers orange at maturity, twice as long as wide; lemma obtuse, the margins united for basal $\frac{1}{3}$–$\frac{1}{2}$, the awn not or slightly exserted beyond glumes **3. *aequalis***

c. Anthers brown at maturity, 3–4 times as long as wide; lemma truncate, the margins free, the awn distinctly exserted **2. *geniculatus***

★1. *A. pratensis* L. 1753, *Sp. Pl.* 60.

Caespitose. Leaves 2–30 cm × 1.5–8 mm; lamina flat, acuminate, ± scabridulous, sometimes glabrous beneath; ligule 1–2.5 mm, truncate; sheath glabrous. Culms 27–100 cm, erect or ascending, glabrous. Panicle 3–12 × 0.5–1 cm, cylindrical to ovoid, green or purplish. Glumes 4.7–6.5 mm, lanceolate, acute, strongly keeled, the margins united in basal $\frac{1}{3}$–$\frac{1}{2}$, awnless. Lemma 4.7–6 mm, ovate, acute, sparsely ciliate distally on veins and keel, the margins united for basal $\frac{1}{3}$; awn exserted for 3–5 mm, geniculate, scabrid. Anthers 2.5–3 mm. *Flowers December to January.*

Improved pastures and meadows near habitations; 0–50 m.

C. E. & C. S. coasts of Isla Grande.
Introduced from Europe for pasture-improvement and locally naturalized or adventive.

★2. *A. geniculatus* L. 1753, *Sp. Pl.* 60.

Plant with long stolons. Leaves 1.2–12.5 cm × 0.8–4 mm; lamina flat, acute, scabridulous; ligule 1–5 mm, truncate; sheath ± glabrous, slightly inflated. Culms 17–56 cm, decumbent, rooted at lower nodes, glabrous. Panicle 1.5–7 × 0.3–0.7 cm, narrowly

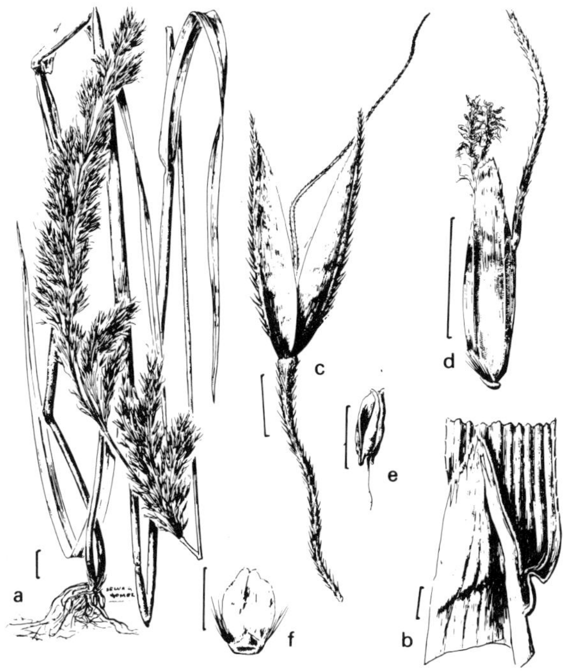

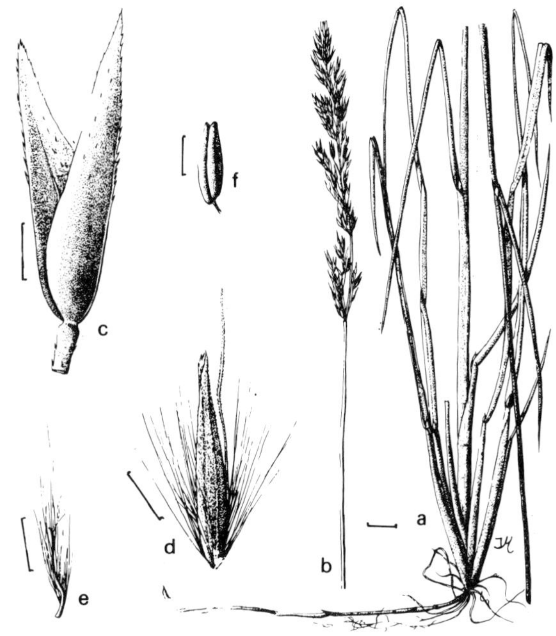

247 248

249 250

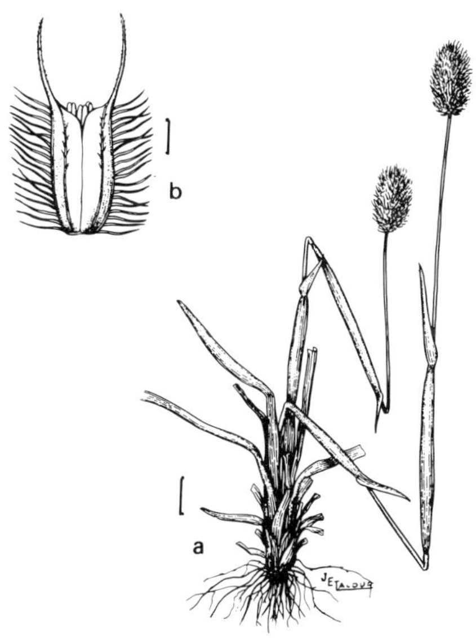

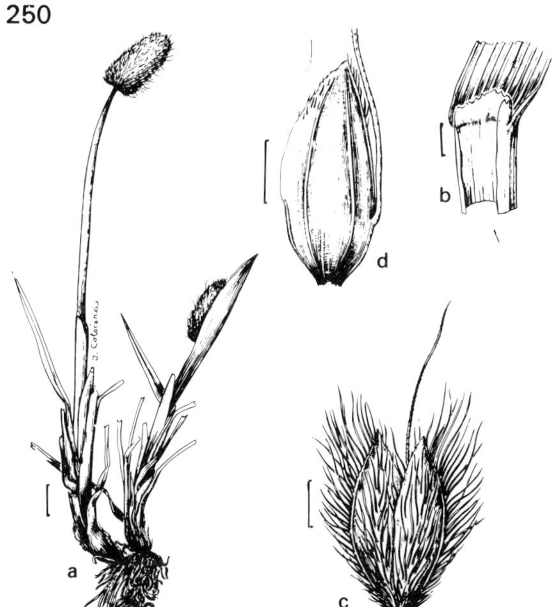

Fig. 247. *Agrostis magellanica*. a, Plant. b, Ligule. c, Spikelet. d, Floret. e, Anther. f, Palea. Scale = 1 cm (a), 1 mm (b–d) or 0.5 mm (e–f). Fig. 248. *Calamagrostis stricta*. a, Plant-base. b, Inflorescence. c, Glumes. d, Floret. e, Rhachilla. f, Anther. Scale = 1 cm (a) or 1 mm (b–f). Fig. 249. *Phleum alpinum*. a, Plant. b, Spikelet. Scale = 1 cm (a) or 1 mm (b). Fig. 250. *Alopecurus magellanicus*. a, Plant. b, Ligule. c, Spikelet. d, Floret. Scale = 1 cm (a) or 1 mm (b–d).

(Flora Patagónica)

cylindrical, pale to greyish green, Glumes 2.4–2.7 mm, oblong, obtuse, keeled, the margins united at base, awnless. Lemma 2.4–2.7 mm, oblong-ovate, truncate, the margins free; awn exserted for 2–3.5 mm, geniculate, scabrid distally. Anthers 1–2 mm, 3–4 times as long as wide, brown at maturity. *Flowers December to February.*

Wet open valleys in *Nothofagus pumilio* forest; 0–50 m.

C. S. Isla Grande.
Introduced from Europe and locally naturalized.

*3. ***A. aequalis*** Sobol. 1799, *Fl. Petrop.* 16.
A. fulvus Sm. f. *violacea* Hackel 1900, in Dusén, *Svenska Exped. Magellansl.* **3(5)**: 218.

Similar to *A. geniculatus* but annual or short-lived perennial; glumes 2–2.5 mm, the margins united for basal $\frac{1}{5}$–$\frac{1}{4}$; lemma 2–2.8 mm, obtuse, the margins united for basal $\frac{1}{3}$–$\frac{1}{2}$, the awn not or only slightly exserted; anthers 0.7–1 mm, twice as long as wide, orange at maturity. *Flowers January to February.*

Swampy ground and margins of lakes and rivers subject to periodic flooding, sometimes floating; 0–20 m.

C. E. coast of Isla Grande. (Map 36c).
Native of Europe probably introduced but its status uncertain.

4. ***A. magellanicus*** Lam. 1791, *Tabl. Encycl. Méth. Bot.* **1**; 168.
A. antarcticus Vahl 1791, *Symb. Bot.* **2**: 18 – Skottsberg, 1906, 1916; Martín, 1949.
A. alpinus auct., non Sm. – Hooker, 1847; Hariot, 1884; Alboff, 1889; Spegazzini, 1896; Kurtz, 1896; Dusén, 1900; Autran, 1905;

Caespitose. Leaves (2.5–)4–29 cm × 1–4.5(–5.5) mm; lamina flat to somewhat convolute, acuminate, scabrid at least beneath, glaucous; ligule 1–3.8 mm, acuminate to truncate or dentate; sheath glabrous. Culms 10–70(–90) cm, erect, glabrous. Panicle (1–) 1.5–4.3 × 0.4–1.3 cm, cylindrical to ovoid, silver purple. Glumes 3–6 mm, ovate-lanceolate, acute or with short awn, keeled, the margins united only at base. Lemma (2.3–)3–4.5(–6.8) mm, ovate, obtuse, united for basal $\frac{1}{2}$; awn (2–)4–9 mm, usually well exserted, geniculate, ± scabrid, sometimes absent. Anthers 1.8–3.2 mm. *Flowers November to March.* (Fig. 250).

Margins of rivers and lakes, swamps and bogs, wet *Bolax – Empetrum* heath, feldmark; 0–1000 m.

Throughout Fuegia except much of the west. (Map 36d).
W. Argentina and E. Chile north to *c.*33°S.lat. Falkland Islands (Islas Malvinas). South Georgia.

Tribe PHALARIDEAE

Spikelets hermaphrodite, strongly laterally compressed, with 1 fertile floret and 2 sterile below. Glumes longer than spikelet. Lodicules 2. Stamens 3. Ovary glabrous.

29. *Phalaris* L.

Annuals or perennials. Ligule membranous; leaf-sheath without auricles. Inflorescence a dense, spike-like panicle. Spikelets with the 2 sterile florets reduced to lemmas. Glumes equal, whitish with 3 conspicuous green veins, awnless. Fertile lemma obscurely 5-veined, the sterile 3-veined, both awnless. Palea ± equal to lemma, 2-keeled. Lodicules ovate, irregularly serrate. Anthers *c.*7 times as long as wide. Styles long, apical.

*1. ***P. canariensis*** L. 1753, *Sp. Pl.* 54.

Caespitose annual. Leaves 2.5–10 cm × 2–6.5 mm; lamina flat, acuminate, scabridulous; ligule 2–6.5 mm, obtuse; sheath glabrous or scabridulous, the upper inflated. Culms 10–50 cm, glabrous. Panicle 1.3–2.6 × 0.3–2 cm, ovoid-oblong; branches very short. Glumes 6.5–8 mm, oblanceolate, acute, with green wing on keel in distal $\frac{1}{2}$, scabridulous. Sterile lemmas 3–3.8 mm, ovate-lanceolate, acute; fertile lemma 4.5–5.2 mm, elliptic-ovate, densely hairy. Palea hairy on keels. Anthers 3–4.2 mm. *Flowers February to March.*

Disturbed and open coastal sands and gravels; 0–10 m.

Locally on C. S. (Ushuaia) and N. E. (Río Grande) coasts of Isla Grande.
Native of S. Europe; introduced, probably with animal foods, and occasionally adventive in disturbed areas near habitations.

Tribe STIPEAE

Spikelets hermaphrodite, terete, with 1 floret. Glumes longer than floret, 0- to 3-veined. Lodicules 2–3. Stamens 3. Ovary glabrous.

30. *Stipa* L.

Caespitose perennials. Ligule membranous; leaf

sheath with or without auricles. Inflorescence a lax to contracted panicle. Glumes subequal, awnless. Lemma rounded on back, 5-veined, with long, twisted or geniculate, apical awn. Palea ± equal to lemma, not keeled. Lodicules 2–3, linear-lanceolate. Anthers 5–10 times as long as wide. Styles short, apical.

Literature: L. R. Parodi, *Rev. Argent. Agron.* **27**: 65–106(1960); O. R. Matthei, *Gayana (Bot.)* **13**: 1–137(1965); F. A. Roig, in M. N. Correa (Ed.) *Fl. Patag.* **3**: 288–330(1978).

a. Panicle contracted; glumes at least 15 mm; lemma-awn geniculate, with white pappus at base, scabrid distally **1. *chrysophylla***
a. Panicle lax; glumes not more than 4.5 mm; lemma-awn twisted or spiralled, scabrid **2. *rariflora***

1. *S. chrysophylla* Desv. 1853, in C. Gay, *Hist. Chile Bot.* **6**: 278.
S. humilis auct., non Cav. – Moore, 1974.

Plant with short stolons. Leaves 2–5 cm × 0.6–0.8 mm; lamina convolute, acuminate, glabrous; ligule 0.2(–3) mm, truncate; sheath subglabrous, the short, ciliate, semi-amplexicaul auricles continuous with ligule. Culms 7–14 cm, glabrous. Panicle 3–3.5 × *c.*1 cm, contracted, white or purplish; branches 5–10 mm, scabrid. Glumes 15.5–18 mm, linear-lanceolate, acuminate, glabrous. Lemma 6–9 mm, linear-lanceolate, covered with hairs *c.*3 mm, tapering to geniculate awn 25–40 mm; basal 10–16 mm of awn twisted, with white hairs 4–5 mm, the distal part of awn scabridulous. Palea ciliate. Lodicules 3, the lateral larger. Anthers 1.6–3 mm. *Flowers December to January.*

Coastal sands; 0–20 m.

N. & N. E. coasts of Isla Grande. (Map 36e).
W. Argentina north to *c.*40°S.lat.
E. Chile north to *c.*29°S.lat.

2. *S. rariflora* (Hooker f.) Bentham 1881, *Bot. Jour. Linn. Soc. London* **19**: 81.
Muhlenbergia rariflora Hooker f. 1847, *Fl. Antarct.* **1(2)**: 371–Kurtz, 1896;
Orthachne rariflora (Hooker f.) Hughes 1923, *Kew Bull.* **8**: 301 – Roig, 1978.

Plant with short rhizomes. Leaves (2–)3–12 cm × 0.3–0.8 mm; lamina convolute or ± flat, pungent, glabrous; ligule 0.3–1(–1.5) mm, truncate; sheath striate, without auricles. Culms up to 45 cm, glabrous. Panicle 3–10 cm, very lax, spreading, purple; branches up to 20 mm, flexuous, scabrid. Glumes 2.7–4.5 mm, lanceolate, acuminate, glabrous. Lemma 3.6–4.7 mm, linear-lanceolate, scabrid at base; awn 30–50 mm, twisted or spiralled when dry, scabrid throughout. Palea scabrid distally. Lodicules 2. Anthers 1.1–1.7 mm. *Flowers January to March.* (Fig.251).

Wet rocks, margins of rivers and streams, bogs, screes, feldmark; 30–1100 m.

S., C. & W. Fuegia. (Map 36f).
W. Argentina north to *c.*40°S.lat.
Chile north to *c.*46°S.lat.

Tribe CORTADERIEAE

Dioecious or gynodioecious. Spikelets laterally compressed, with 2–many florets. Palea 2-veined. Lodicules 2, ciliate. Stamens 3. Ovary glabrous.

31. *Cortaderia* Stapf

Caespitose, gynodioecious perennials. Ligule a line of hairs. Inflorescence a contracted, plume-like panicle. Spikelets with 4–6 florets, the upper 1–2 sterile and often reduced. Glumes ± equal to spikelet, keeled, 1-veined, with straight, apical awn. Lemma keeled, 3-veined, truncate, with long apical awn, with long, white dorsal hairs. Palea bifid. Lodicules oblong, obtuse. Anthers 5–6 times as long as wide, much shorter or absent in female florets. Styles short, subapical.

Literature: H. J. Conert, *Die Systematik und Anatomie der Arundineae*. Weinheim, Cramer (1961).

1. *C. pilosa* (D' Urv.) Hackel 1899, in F. Kurtz, *Bol. Acad. Nac. Ci.* **16**: 253.
Arundo pilosa D'Urv. 1825, *Fl. Is. Mal.* 33 – Hooker, 1847; Franchet, 1889; Kurtz, 1896.

Leaves 9–25(–60) cm × 0.6–2(–3.5) mm; lamina plicate or convolute, rarely flat, acuminate, scabrid on veins beneath, ciliate at base; ligule-hairs 0.3–2.5 mm, white; sheath rounded, glabrous or appressed-hairy. Culms 30–60 cm, erect, terete or somewhat compressed, smooth, glabrous or ± strigose at nodes. Panicle 4.5–7 cm, lanceolate- to oblong-ovoid, dense, often interrupted at base, purplish to pale brown; branches ascending, rigid, scabridulous. Spikelets 8–12 (–16) mm, obovate to obovate-cuneiform. Glumes subequal, 9.5–13 mm, linear-lanceolate, glabrous; awn *c.*$\frac{1}{5}$ as long as glume, scabrid. Lemma 4.5–8 mm, ovate-lanceolate, with hyaline, ciliolate margin, the dorsal hairs 3.5–5.5 mm; awn 2.8–8.5 mm, scabrid. Palea 4.2–6.5 mm, ovate-lanceolate, scabrid on veins. Anthers 2–3.5 mm, $2n = 36^{\star}$. *Flowers January to March.* (Fig. 252).

251

252

253

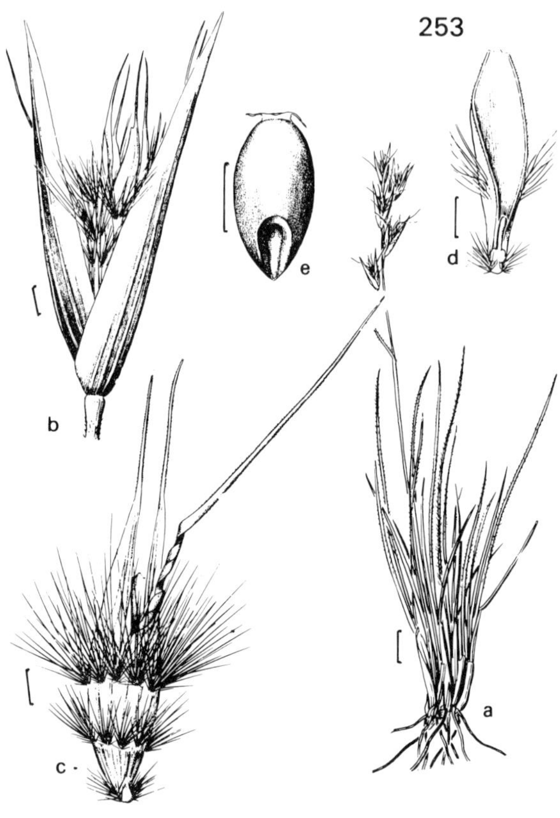

Fig. 251. *Stipa rariflora*. a, Plant. b, Ligule. c, Spikelet. d, Floret. Scale = 1 cm (a) or 1 mm (b–d). Fig. 252. *Cortaderia pilosa*. a, Plant-base. b, Inflorescence. c, Spikelet. d, Floret. e, Anther. f, Ovary and staminodes. Scale = 1 cm (a–b) or 1 mm (c–f). Fig. 253. *Rytidosperma virescens*. a, Plant. b, Spikelet. c, Floret. d, Palea. e, Caryopsis. Scale = 1 cm (a) or 1 mm (b–e).
(Flora Patagónica)

Moist, open areas, including raised parts of *Sphagnum* and *Astelia* bogs, locally dominant; 0–900 m.

S. Fuegia, except extreme east. (Map 36g).
W. Argentina north to *c.*40°S.lat.
Chile north to *c.*41°S.lat.
Falkland Islands (Islas Malvinas).

Fuegian material belongs to var. *pilosa*, which is of smaller stature but has larger spikelets and glumes than var. *minima* (Connert) Nicora, to which material from the northern part of the species range has been ascribed.

Tribe DANTHONIEAE

Spikelets hermaphrodite, somewhat laterally compressed, with 2–10 florets. Palea 2-keeled. Lodicules 2, comose on 1 side of apex. Stamens 3. Ovary glabrous.

32. *Rytidosperma* Steudel

Caespitose perennials. Ligule a line of hairs; leaf-sheath with auricles. Inflorescence a contracted panicle. Spikelets with 3–6 florets, the upper sterile. Glumes enclosing florets, keeled, 4- to 9-veined, acute. Lemma 7- to 10-veined, deeply bifid at apex, with twisted geniculate dorsal awn, with 2 rows of tufted dorsal cilia. Palea 3-denticulate. Lodicules square to oblong. Anthers 2–3 times as long as wide. Styles long, lateral.

1. ***R. virescens*** (Desv.) Nicora 1973, *Darwiniana* **18**: 91.
   Danthonia virescens Desv. 1853, in C. Gay, *Hist. Chile Bot.* **6**: 363.
   D. picta Nees & Meyen var. *patagonica* Speg. 1902, *An. Mus. Nac. Hist. Bs Aires* **7**: 193.

Leaves 0.5–6 cm × 0.8–1.2 mm; lamina convolute, pungent, hairy beneath, scabridulous to glabrous above; ligule-hairs up to 3 mm, brownish; sheath rounded, ± glabrous. Culms 4–13 cm, erect, smooth, glabrous. Panicle 2–3.5 cm, with 1–7 spikelets on short, hairy branches. Glumes subequal, 12–16 mm, broadly lanceolate, glabrous, greenish to purple with whitish margin. Lemma 9–16 mm, ovate-lanceolate, divided to *c.*½-way, purplish, the cilia 4–5 mm; awn 9–14 mm; callus ciliolate. Palea 4–5 mm, lanceolate, ciliolate on veins. Anthers 0.9–1.2 mm. *Flowers December to January*. (Fig. 253).

Open grassland and rocky areas; 0–150 m.

N. E. to C. Isla Grande. (Map 36h).
W. Argentina north to *c.*40°S.lat.
E. Chile 39–37°S.lat.

Fuegian material has been assigned to 2 of the 3 varieties recognized within this species on the basis of the size, colour and hairiness of the glumes; such formal recognition of this variation seems of doubtful merit.

LXXXI. JUNCACEAE

Perennial, rarely annual herbs, caespitose or with creeping rhizome. Leaves alternate or basal, with basal sheath; lamina sometimes absent; stipules absent. Flowers actinomorphic, hermaphrodite, solitary or in terminal cymes. Perianth-segments in 2 whorls of 3, free, scarious. Stamens 3 or 6. Ovary superior, 1- or 3-celled; style 1; stigmas 3. Fruit a 3-valved loculicidal capsule; seeds 3 or numerous, with endosperm.

Literature: M. Barros, *Darwiniana* **10**: 279–460(1953).

a. Flowers 2 or more:
  b. Leaves glabrous; capsule with numerous seeds — **3. *Juncus***
  b. Leaves at least sparsely hairy; capsule with 3 seeds — **4. *Luzula***
a. Flowers solitary:
  c. Leaves transversely septate — **3. *Juncus***
  c. Leaves not transversely septate:
    d. Leaves 1 per stem; lamina terete; bracts shorter than perianth; fruit at least 3 times as long as wide — **2. *Marsippospermum***
    d. Leaves 2 or more per stem; lamina channelled; upper bract much exceeding perianth; fruit not more than 1½ times as long as wide — **1. *Rostkovia***

1. *Rostkovia* Desv.

Caespitose, often rhizomatous, glabrous, perennial herbs. Leaf-lamina long and grass-like, channelled, without internal septa. Flowers solitary and terminal on scape-like stem, subtended by two bracts, the lower bract much exceeding perianth. Stamens 6, inserted on perianth-segments. Ovary 1-celled; style long; stigmas filiform. Capsule 1-celled; seeds numerous.

1. ***R. magellanica*** (Lam.) Hooker f. 1847, *Fl. Antarct.* **1(2)**: 81.
   Juncus magellanicus Lam. 1789, *Encycl. Méth. Bot.* **3**: 266 – Franchet, 1889.

Rhizome, if present, slender, bearing clusters of stems. Stems 7–30 cm, *c.*0.8 mm in diameter, erect,

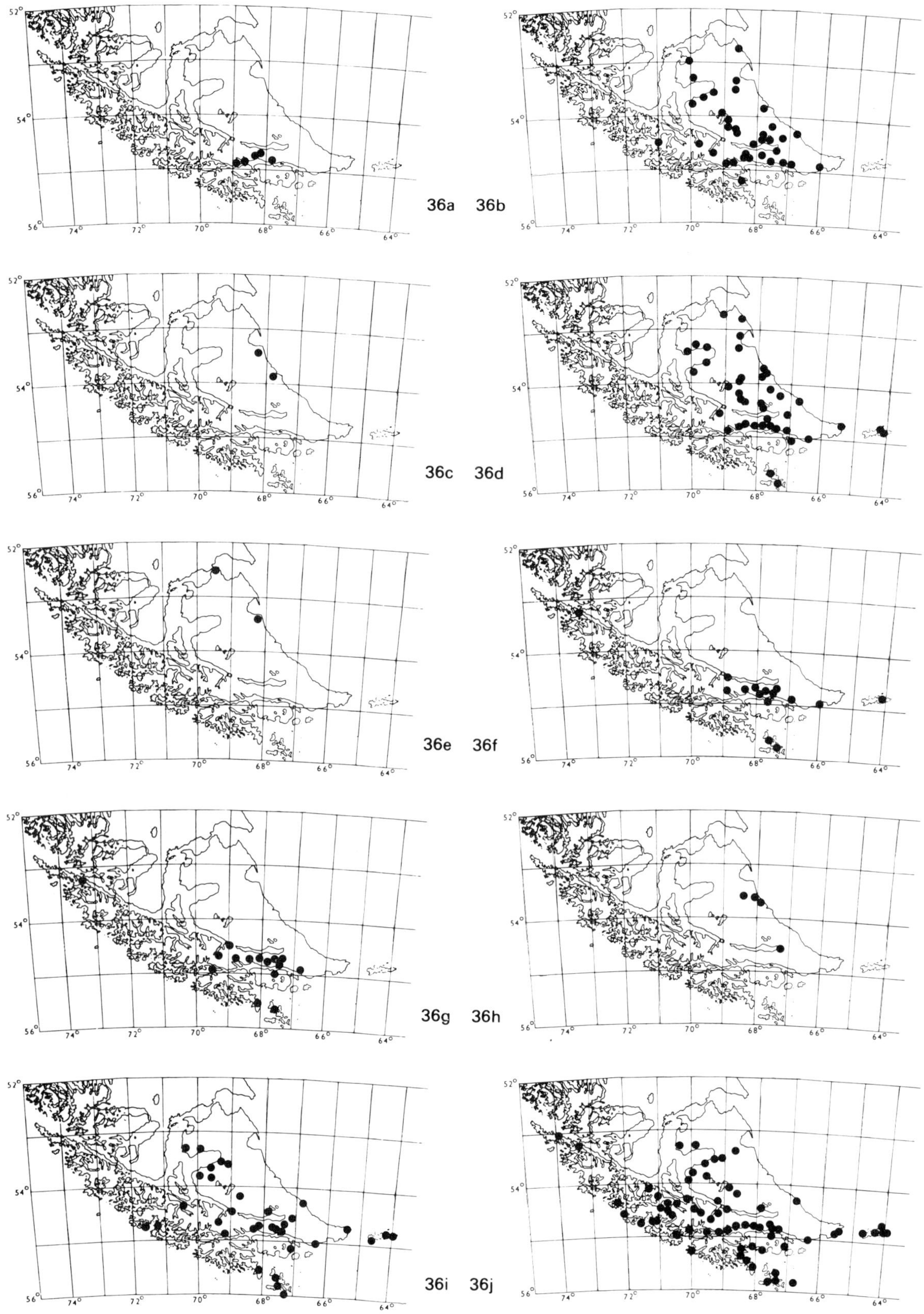

Map 36. a, *Calamagrostis erythrostachya*. b, *Phleum alpinum*. c, *Alopecurus aequalis*. d, *Alopecurus magellanicus*. e, *Stipa chrysophylla*. f, *Stipa rariflora*. g, *Cortaderia pilosa*. h, *Rytidosperma virescens*. i, *Rostkovia magellanica*. j, *Marsippospermum grandiflorum*.

branched near base, ± terete, faintly striate. Leaves 6–35 cm, *c.*1 mm wide; lamina linear, acute, channelled, ± erect, green or brownish green, rudimentary in outermost leaves; sheath yellowish brown, with obtuse auricles about as long as wide. Lower bract 18–40 mm, ovate-lanceolate near base, subulate above; upper bract about equalling perianth. Perianth-segments 7–8.5 × 1.2–2 mm, oblong to oblong-lanceolate, acute or acuminate, reddish brown, with paler scarious margins, shiny, persistent; anthers longer than filaments; stigmas held above anthers, exserted. Capsule 6–10 × 4.5–7 mm, equalling or slightly exceeding perianth, cylindrical- to ovoid-globose, abruptly narrowing to short beak, blackish to reddish brown, shiny; seeds 1.5–1.7 mm, ellipsoid-obovoid, with short acute appendage, yellow-brown, shiny. *Flowers October to November (December).* (Fig. 257).

Wet sands and rocks, wet grassland, bogs, Magellanic moorland, river margins; 0–800 m.

Almost throughout Fuegia except the extreme north. (Map 36i).
W. Argentina north to *c.*48°S.lat.
Chile north to *c.*46°40′S.lat.
Falkland Islands (Islas Malvinas), South Georgia, Campbell Island, Auckland Islands, New Zealand (S. Island).

2. *Marsippospermum* Desv.

Similar to *Rostkovia* but with all except one leaf per stem reduced to basal sheath; lamina terete and stem-like; bracts usually 3, shorter than perianth; capsule much longer than wide, partially 3-celled.

a. Leaf solitary, at least 13 cm; capsule at least 15 × 4 mm **1. *grandiflorum***
a. Leaves several, not more than 10 cm; capsule not more than 9 × 2.5 mm **2. *reichei***

1. *M. grandiflorum* (L.f.) Hooker f. 1843, *Icon. Pl.* **6**, t. 533.
Rostkovia grandiflora (L.f.) Hooker f. 1847, *Fl. Antarct.* **1(1)**: 82.
Juncus grandiflorus L.f. 1781, *Suppl.* 209 – Forster, 1789; Franchet, 1889.

Rhizome up to 5 mm in diameter, creeping, often branched, bearing clusters of stems at intervals. Stems 16–30(–50) cm, *c.*1–1.2 mm in diameter, erect, terete. Leaf solitary, 13–45 cm; lamina terete, stem-like, acute, shallowly channelled near base; sheath pale greenish, with brownish, obtuse auricles about as long as wide; remaining leaves with pale greenish to reddish brown sheath and short, filiform, rudimentary lamina. Bracts (2–)3(–4), 2–5 × 2–3 mm, ovate, acute to obtuse, reddish brown, scarious. Perianth-segments 16–36 × 2–2.5 mm, the outer whorl longer than the inner, linear-lanceolate, long-acuminate, reddish brown, rather shiny, persistent. Anthers much longer than filaments. Capsule 15–20 × 4–5 mm, shorter than outer perianth-segments, cylindrical-ovoid, subtrigonous, gradually tapering to short beak, dark reddish brown, shiny; seeds *c.*3.5 mm, fusiform, acute at ends, pale yellowish, rather shiny. $2n = c.50$.* (Figs. 254, 260).

Wet sands, open moist peaty grassland, river margins, bogs, Magellanic moorland, often locally dominant; 0–350 m.

Almost throughout Fuegia. (Map 36j).
W. Argentina north to *c.*37°S.lat.
Chile north to *c.*47°S.lat.
Falkland Islands (Islas Malvinas).

2. *M. reichei* Buchenau 1901, *Ber. Deutsch. Bot. Ges.* **19**: 159.

Caespitose. Stems 3–6 cm, *c.*0.5 mm in diameter, erect, terete. Leaves several, 3–10 cm; lamina subterete, acute, shallowly channelled near base; sheath pale brownish, with short, obtuse auricles; some leaves with rudimentary lamina. Bract usually solitary, 1–2 × 1–2 mm, oblong-ovate, obtuse, yellowish, scarious. Perianth-segments 6–10 × 0.5–1.5 mm, the outer whorl longer than the inner, linear-lanceolate, long-acuminate, yellowish brown. Anthers much longer than filaments. Capsule 4–9 × 1.5–2.5 mm, shorter than outer perianth-segments, cylindrical-ovoid, gradually tapering to short beak, yellowish brown, shiny; seeds 1.5–2 mm, ovoid, yellowish brown.

Wet rocks and screes; *c.*400 m.

Southernmost Fuegia (Islas Wollaston). (Map 37a).
W. Argentina north to *c.*37°S.lat.
Chile north to 50°30′S.lat.

3. *Juncus* L.

Glabrous, perennial, rarely annual herbs, caespitose or with creeping rhizomes. Leaf-sheath open, bearing auricles; lamina long and grass-like, terete or channelled, often with internal septa, sometimes rudimentary. Inflorescence a cluster of terminal or

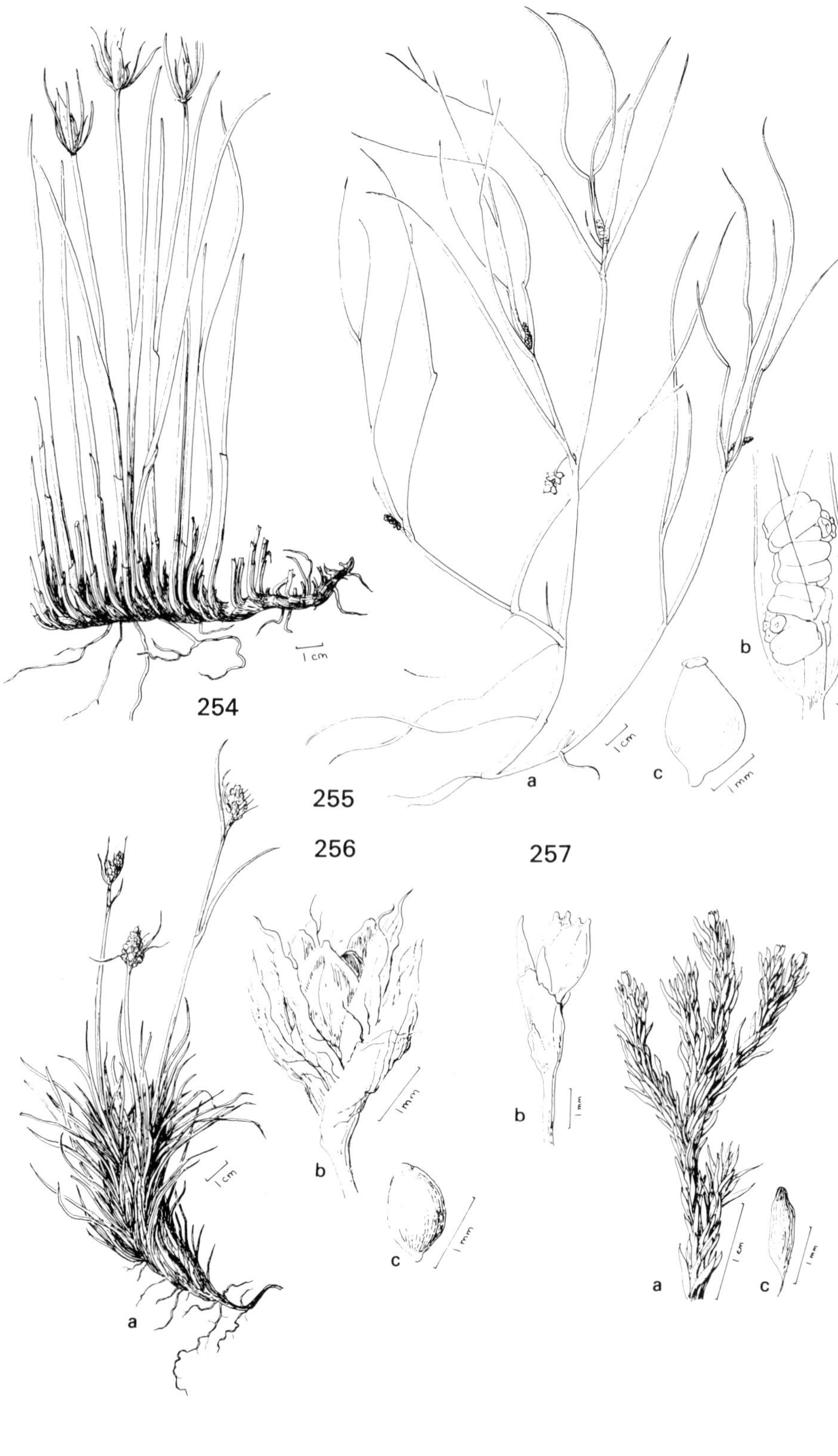

Fig. 254. *Marsippospermum grandiflorum*. Scale = 1 cm. Fig. 255. *Ruppia filifolius*. a, Plant. b, Inflorescence. c, Fruitlet. Scale = 1 cm (a) or 1 mm (b–c). Fig. 256. *Luzula alopecurus*. a, Plant. b, Fruit. c, Seed. Scale = 1 cm (a) or 1 mm (b–c). Fig. 257. *Gaimardia australis*. a, Plant. b, Pseudanthium. c, Seed. Scale = 1 cm (a) or 1 mm (b–c).

(R. N. P. Goodall)

apparently lateral cymes, subtended by two bracts. Flowers few to many, rarely solitary, often subtended by bracteoles. Perianth-segments greenish or brownish. Stamens 3 or 6, free. Ovary 1- or 3-celled; style ± long to almost absent; stigmas filiform. Capsule 1- or 3-celled; seeds many, sometimes with appendage.

Literature: A. Lourteig, *CNFRA - Biologie* **23**: 33–49(1968).

a. Leaf-lamina cylindrical, often ± compressed, with transverse septa; sheath with 2 apical auricles; flowers not subtended by bracteoles **3–5. *scheuchzerioides*** group
a. Leaf-lamina flat or cylindrical, without transverse septa; sheath without auricles; flowers subtended by 2 bracteoles:
 b. Annual; leaf-lamina flat; inflorescence terminal **1. *bufonius***
 b. Perennial; leaf-lamina cylindrical or absent; inflorescence apparently lateral **2. *balticus***

1. *J. bufonius L. 1753, *Sp. Pl.* 328.

Annual; stems 3–30 cm, ascending to erect, often ± recurved. Leaves 1–10 cm × 0.5–2 mm; lamina linear, acute, flat, the margin often ± convolute; sheath without auricles. Inflorescence much branched, forming upper $\frac{1}{3}$–$\frac{1}{2}$ or more of plant; flowers sessile, solitary, rarely in clusters; lower bracts leaf-like, the upper resembling bracteoles; bracteoles 3–6 × 2–3 mm, broadly ovate, obtuse, scarious. Perianth-segments 4.5–8 mm; inner lanceolate, acute to apiculate, scarious; outer longer than inner, ovate, acute to subulate, greenish with narrow scarious margin. Stamens (3)6; anther as long as filament. Capsule 3–5 mm, subellipsoid to ovoid-trigonous, 3-celled, brown, shiny; seeds 0.35–0.55 mm, obovoid, shortly apiculate, reticulate, reddish brown. *Flowers January to March.*

Coastal sands and sandy soil, disturbed ground especially near habitations, open moist ground above timberline; 0–800 m.

S. Isla Grande.
Introduced, probably from Europe, but widely naturalized throughout much of the world.

2. *J. balticus* Willd. 1809, *Ges. Naturf. Freunde Berlin Mag.* **3**: 298.

Glaucous perennial; rhizome horizontal, long; stems 10–25 cm, erect, rigid, with several brownish leaf-sheaths towards base, the upper 1(2) of which has a cylindrical, channelled lamina shorter than the stem. Inflorescence 5- to 20- flowered, in upper $\frac{1}{4}$ of apparent stem, appearing lateral, the lower bract forming an apparent prolongation of the stem. Perianth-segments 3.2–5 mm, subequal, ovate-lanceolate, acute, brownish. Stamens 6; anther 3 times as long as filament. Capsule 3–4.5 mm, ovoid-trigonous, shortly mucronate, ± 3-celled, brown, shiny; seeds 0.6–1 mm, ovoid, transversely striate, brownish grey. *Flowers December to February.*

Coastal and estuarine sands; 0–10 m.

N.E. Isla Grande. (Map 37b).
W. Argentina and Chile north to *c.*30°S.lat.
North America, Europe.

Fuegian material belongs to var. ***mexicanus*** (Willd.) O. Kuntze 1898, *Revis. Gen.* **3**: 320, which has the above distribution and is distinguished from other varieties by the presence of leafy upper sheaths and the longer anthers. The species extends to Europe and Australasia.

3–5. *J. scheuchzerioides* group.

The species included here show considerable variation in response to the differences in climate and habitat which they encounter throughout their ranges. Many features which distinguish them in other areas prove unsatisfactory in S. Patagonia and Tierra del Fuego. For this reason, the commonest Fuegian species, *J. scheuchzerioides*, is described fully while only the most reliable differential characters are given for the other two species.

3. *J. scheuchzerioides* Gaudich. 1825, *Ann. Sci. Nat.* **5**: 100.

J. inconspicuus D'Urv. 1825, *Fl. Is. Mal.* 34 – Buchenau, 1906; Barros, 1953, 1969.
J. pusillus auct., non Buchenau – Barros, 1953, 1969.

Perennial; rhizome long-spreading, branched, purplish, bearing tufts of leaves. Stems (1–)2–18(–20) cm., ascending to erect, simple or branched, leafy, often with scales towards base. Leaves 1.5–14 cm., usually exceeding flowering stems; lamina linear, terete or ± compressed, with obvious septa, somewhat flattened and acute at apex, often shallowly channelled near base, short or absent in basal leaves; sheath conspicuous, produced above into semicircular or oblong, obtuse, scarious auricles which are usually wider than long. Inflorescence (1–)2- to 6(–7)-flowered; outer bract 6–17 × 1–2 mm, narrowly ovate-acicular, 1–1$\frac{1}{2}$ times as long as inflorescence. Flowers subsessile or on pedicels up to 2 mm. Perianth-segments 3–5 × 1–1.5 mm, ovate-

258

259

260

261

Fig. 258. *Tetroncium magellanicum.* Fig. 259. *Rostkovia magellanica.* Fig. 260. *Marsippospermum grandiflorum.* Fig. 261. *Juncus scheuchzerioides.*

(Sydney Parkinson)

lanceolate, acuminate, the outer somewhat keeled, reddish brown, greenish on midvein, with narrow, white, hyaline margin in basal two-thirds. Stamens 6 or less; anthers ± as long as filaments. Capsule 3–4 × 1.3–1.8 mm, equalling or slightly exceeding perianth, cylindrical-ovoid, trigonous, mucronate, ± 1-celled, dark shiny brown; seeds 0.5–0.7 × 0.3–0.4 mm., oblong-obovoid, somewhat trigonous, pale brown. $2n = 40^{\star}$. Self-compatible; self-pollinated, sometimes cleistogamous. *Flowers October to February*. (Fig. 261).

Moist areas in most non-forested communities; 0–500 m.

Almost throughout Fuegia. (Map 37c).
W. Argentina north to *c*.34°30′S.lat.
Chile north to *c*.28°30′S.lat.
Falkland Islands (Islas Malvinas); South Georgia; Marion Island; Iles Crozet; Iles Kerguelen; Macquarie Island; Campbell Islands; Auckland Islands; Antipodes Island.

Depauperate specimens with the inflorescence reduced to a solitary flower have been referred to var. *inconspicuus* (D'Urv.) Hooker f. 1847, *Fl. Antarct.* **1(1)**: 80.

4. *J. stipulatus* Nees & Meyen 1843, in Meyen, *Nova Acta Acad. Leop. – Carol.* Ser. 1, **13** Suppl. **1**: 126.

Distinguished from *J. scheuchzerioides* by: auricles at apex of leaf-sheath subacute; perianth-segments shortly aristate-mucronate; anthers shorter than filaments; capsule oblong, mucronate; seeds 0.3–0.57 × 0.2–0.35 mm, with several denticulate longitudinal ridges.

Moist areas in open habitats; 0–200 m.

S. side Isla Grande. (Map 37d).
W. Argentina north to *c*.22°S.lat.
Chile north to *c*.30°S.lat.
Bolivia, Ecuador, Colombia.

5. *J. chilensis* C. Gay 1853, *Hist. Chile Bot.* **6**: 146.
J. depauperatus Phil. 1860, *Fl. Atacam.* 53 – Buchenau, 1906; Barros, 1953, 1969.

Distinguished from *J. scheuchzerioides* by: auricles at apex of leaf-sheath shallow, obtuse; outer perianth-segments lanceolate, acute, the inner oblong, obtuse; anthers shorter than filaments; capsule subglobose-trigonous; seeds 0.37–0.5 × 0.18–0.3 mm, with 2 shallowly denticulate longitudinal ridges, finely tuberculate.

Moist areas in open habitats; 0–50 mm.

S. side Isla Grande. (Map 37e).
W. Argentina north to *c*.22°S.lat.
Chile north to *c*.30°S.lat.
Bolivia, Peru.

4. *Luzula* DC.

Caespitose, sometimes stoloniferous, perennial herbs, grass-like. Leaves basal, sometimes also alternate and cauline, with closed sheathing base lacking auricles; lamina flat or channelled, with long marginal hairs. Flowers actinomorphic, hermaphrodite, subtended by bracteoles, in few- to many-flowered cymes subtended by 1–4 bracts, sometimes condensed into a head. Stamens (3) 6, free. Ovary 1-celled; style long to very short; stigmas filiform. Capsule 1-celled; seeds 3, usually with conspicuous appendage.

a. Leaves acuminate; at least outer perianth-segments prominently laciniate; anthers not longer than filaments; style very short **1. *alopecurus***
a. Leaves obtuse or subobtuse to subacute; perianth-segments not prominently laciniate; anthers longer than filaments; style long:
  b. Inflorescence a lax, somewhat umbellate, drooping panicle with 1 sessile and 3–6 stipitate, 3- to 12- flowered clusters **3. *campestris***
  b. Inflorescence spike-like, sometimes interrupted towards base **2. *chilensis***

1. *L. alopecurus* Desv. 1808, *Jour. Bot. Rédigé* **1**: 159.
L. antarctica Hooker f., 1847, *Fl. Antarct.* **1(2)**: 550 – Barros, 1953, 1969.

Root-stock short, erect or suberect; stems (6–)10–30(–40) cm, terete, glabrous. Basal leaves 5.5–20(–25) cm × 2–6 mm, acuminate, with dense white marginal hairs; cauline leaves usually 1–2, 3–5 cm. Inflorescence 15–24 mm, ovoid to oblong, compact, terminal, sometimes with a smaller cluster below, subtended by 1–4 bracts; lower bracts 1.2–3 cm, equalling or exceeding inflorescence, leaf-like, the upper smaller, membranous, laciniate; bracteoles 3–6 mm, ovate-oblong to linear-lanceolate, long-acuminate, deeply laciniate, ciliate, whitish membranous. Outer perianth-segments 3–5 mm., linear-lanceolate, fimbrio-ciliate, dark reddish brown, with pale scarious margin, the inner similar but shorter,

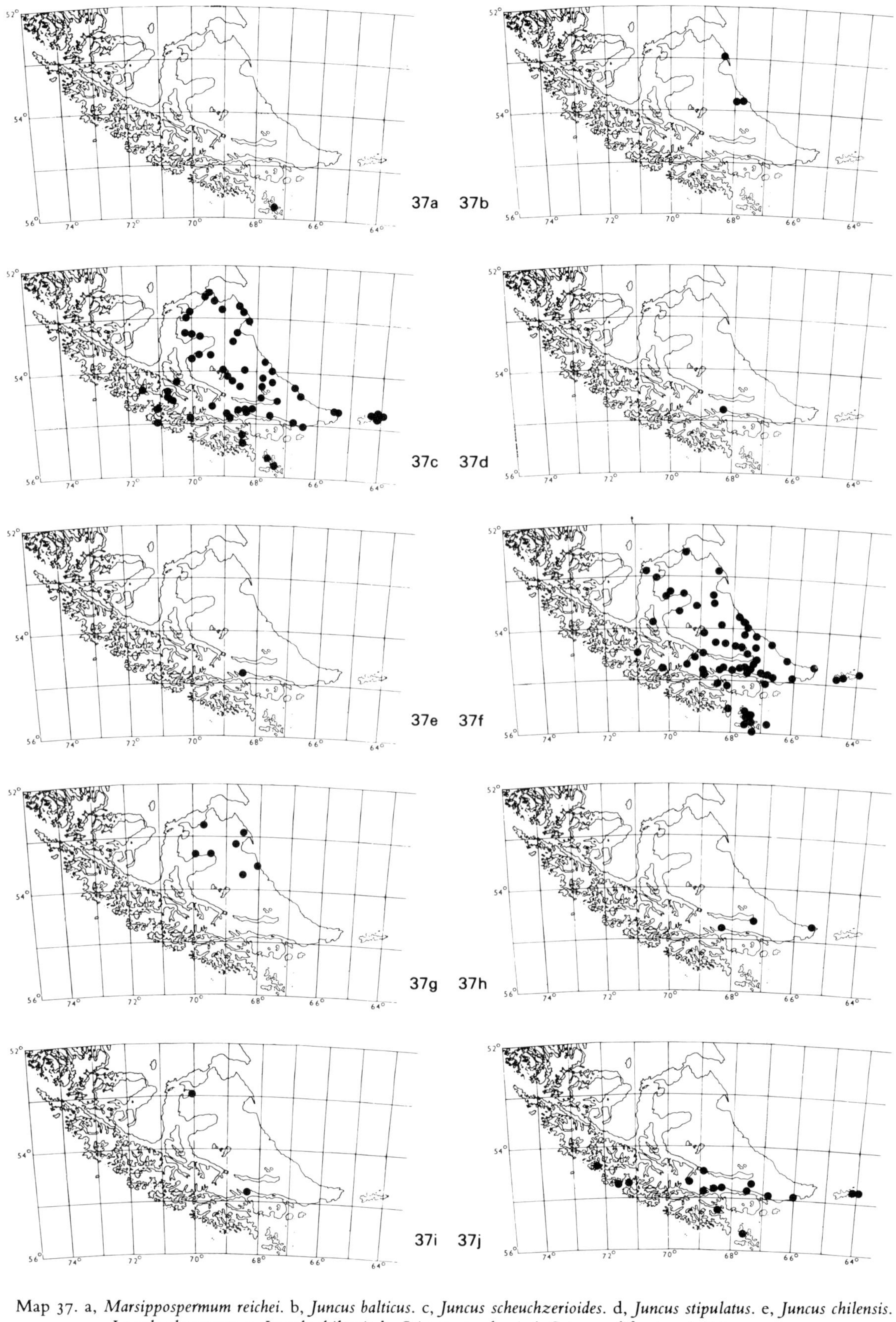

Map 37. a, *Marsippospermum reichei*. b, *Juncus balticus*. c, *Juncus scheuchzerioides*. d, *Juncus stipulatus*. e, *Juncus chilensis*. f, *Luzula alopecurus*. g, *Luzula chilensis*. h, *Scirpus nevadensis*. i, *Scirpus californicus*. j, *Scirpus cernuus*.

not ciliate. Anthers equalling to *c.* half as long as filaments; style very short. Capsule 1.2–2 mm, ovoid-trigonous, reddish to dark brown, glabrous, shiny; seeds 0.5–1.2 mm, cylindrical, with pale, narrow, wing-like aril, dark brown, yellowish at base, with short, white coma. $2n = 24$.* Self-compatible and probably usually self-pollinated. *Flowers November to February.* (Fig. 256).

Open grassland, cliffs, screes, rocky areas, feldmark, *Chiliotrichum* scrub, *Bolax-Empetrum* heath; 0–1100 m.

Almost throughout Fuegia. (Map 37f).
W. Argentina north to *c.*50°S.lat.
Chile north to *c.*51°S.lat.
Falkland Islands (Islas Malvinas).

L. pumila Hooker f. 1864, *Handb. N.Z. Fl.* 293, a New Zealand species, has been reported once from feldmark above Puerto Cook on Isla de los Estados (Spegazzini, 1896; Moore, 1974). It is a compact, ± cushion-forming species, with stems 2–6(–8) cm, obtuse, subglabrous leaves 1–4 cm × 0.7–1.5 mm, small compact inflorescence of 1–3 crowded clusters of flowers having acute, dark brown perianth-segments with a narrow to wide pale scarious margin, and a dark brown to almost black capsule. In all these features it approaches some forms of *L. alopecurus* from exposed, extreme habitats and the occurrence of *L. pumila* in Fuegia must remain in doubt until it can be confirmed by more abundant material than that now available.

2. *L. chilensis* Nees & Meyen 1841, in Kunth, *Enum. Pl.* **3**: 312.
L. racemosa auct., non Desv. – Dusén, 1900.

Rootstock short, oblique; stems 4–18 cm, terete, glabrous. Basal leaves 3–10 cm × 1–4 mm, subacute to subobtuse, with dense white marginal hairs at least when young; cauline leaves (0)1–3, 2.5–6 cm. Inflorescence 8–15 mm, ± ovoid, compact, terminal, sometimes interrupted at base, subtended by 1–3 bracts; lower bracts 1–4 cm, shorter than to exceeding inflorescence, leaf-like, the upper smaller, membranous, ciliate to shortly laciniate; bracteoles 2.5–4 mm, ± lanceolate, acute, entire or shortly laciniate, white, hyaline. Outer perianth-segments 2.5–4 mm, lanceolate, acuminate, often aristate, reddish brown with wide scarious margin, the inner similar but shorter. Anthers longer than filaments; style short. Capsule 2–2.5 mm, globose-trigonous, blackish to reddish-brown, glabrous, shiny; seeds 1.5–1.8 mm, ellipsoid, with pale, narrow, wing-like aril, reddish-brown, with short yellowish coma. *Flowers November to December.*

Open sandy soils, coastal scrub and grassland, *Empetrum* heath; 0–50 m.

N. Isla Grande. (Map 37g).
W. Argentina north to *c.*33°S.lat.
Chile north to *c.*30°30′S.lat.

3. *L. campestris (L.) DC. 1805, in Lam. & DC., *Fl. Fr.*, ed. 3, **3**: 161.

Root-stock short, erect, sometimes with short stolons; stems 2–15(–25) cm., terete, glabrous. Leaves 1.5–10 cm × 2–4 mm, obtuse and usually somewhat swollen at apex, sparsely to rather densely ciliate, especially towards base; cauline leaves 2–4, 1.5–9 cm. Inflorescence 8–18 mm, a loose, conical to somewhat umbellate, often drooping panicle, with 1 sessile and 3–6 lateral, stipitate, 3- to 12-flowered clusters, subtended by 2(–3) bracts; bracts 7–20 mm, shorter than to somewhat exceeding inflorescence, leaf-like; bracteoles 1.5–3 mm, ovate, acute, sparsely laciniate especially towards base, whitish membranous. Perianth-segments 3–4 mm, ovate, lanceolate, long-apiculate, reddish brown, with pale scarious margins, subequal or the inner somewhat shorter. Anthers 2–6 times as long as filaments; style long and filiform. Capsule 2–2.8 mm, cylindrical to obovoid, trigonous, dark reddish brown, glabrous, shiny; seeds 1.2–1.5 mm, cylindrical-globose, with small apical wing, dark reddish brown, with yellowish wing-like basal aril up to half the total length. $2n = 12$. *Flowers October to December.*

Disturbed grassland and open pebbles and sands; 0–250 m.

S. coast of Isla Grande.
Introduced from Europe, probably with grass seed, and locally naturalized in areas disturbed by human activity.

Fuegian material belongs to the Eurasian ssp. *campestris*.

LXXXII. CYPERACEAE

Perennial, rarely annual, often rhizomatous, herbs. Leaves usually linear, some or all often reduced to sheaths; stipules absent. Flowers hermaphrodite or unisexual, subtended by a bracteole (glume) and arranged in spikelets; spikelets in a capitate cluster, a spike or in a simple or compound umbel or panicle. Perianth of bristles, scales or hairs, sometimes absent. Stamens 3(–1). Ovary superior, 1-celled; style 1; stigmas 2–3. Fruit a nut, sometimes included in thin, rather hard envelope (utricle); seed with endosperm.

Literature: M. Barros, *Anal Mus. Nac. Bs Aires* **34**: 427–496(1928); **37**: 133–263(1935).

a. Flowers all unisexual; nut enclosed in a utricle:
 b. Rhachilla produced into a hooked bristle (rhacheole); spikes simple **6. *Uncinia***
 b. Rhacheole absent or, if present, not hooked; spikes simple or compound **7. *Carex***
a. Flowers both unisexual and hermaphrodite; nut not enclosed in a utricle:
 c. Perianth of 6 plumose setae **5. *Carpha***
 c. Perianth absent, or of 3–6 non-plumose setae or scales:
  d. Inflorescence not subtended by bracts; leaf-sheaths without lamina **2. *Eleocharis***
  d. Inflorescence subtended by a ± leaf-like bract; leaf-sheaths with lamina, which is sometimes rudimentary:
   e. Inflorescence of 1 terminal, 1-flowered spikelet and up to 2 lateral sterile spikelets; perianth of 6 scales **4. *Oreobolus***
   e. Inflorescence of 1 or more, at least 2-flowered spikelets; perianth absent or of 3–6 smooth or strigose setae:
    f. Glumes spirally arranged, all with flowers or the lower 1(2) sterile in each spikelet; rhachilla not flexuous; the internodes ± equal **1. *Scirpus***
    f. Glumes distichous, the lower 3–4 in each spikelet sterile; rhachilla flexuous, short between sterile nodes and long between fertile nodes **3. *Schoenus***

1. *Scirpus* L.

Glabrous perennials, rarely annuals, tufted or rhizomatous; stems terete or 3-angled. Leaves sometimes with rudimentary lamina. Inflorescence of 1-many spikelets, in apparently lateral capitate cluster, with a bract appearing as a prolongation of the stem. Spikelets 1– to many– flowered; glumes spirally arranged, all with flowers or the lowest 1(2) sterile. Flowers hermaphrodite; perianth of 0 or 3–5 strigose setae; stamens 2–3; stigmas 2–3.

a. Stems up to 300 cm, 3-angled; leaf-sheaths without lamina or with rudimentary lamina **2. *californicus***
a. Stems not more than 40 cm, terete; leaves with lamina:
 b. Annual; perianth absent; stamens 2; stigmas 3 **3. *cernuus***
 b. Perennial; perianth of 6 unequal strigose setae; stamens 3; stigmas 2 **1. *nevadensis***

1. *S. nevadensis* S. Watson 1871, *Bot. King's Exped.* 360.

Rhizome robust, horizontal; stems 15–40 cm, terete. Leaves shorter than stem, semi-cylindrical, obtuse, channelled, scabrid on margin. Bract much exceeding inflorescence. Spikelets 3–8, 8–15 mm, ovoid-ellipsoid, many-flowered. Glumes 3.5–4.5 mm, oblong-ovate, obtuse or emarginate, ciliolate distally, reddish brown with yellow-green midrib. Perianth of 3–5 unequal setae. Stamens 3. Stigmas 2. Nut 1.5–1.8 mm, obovoid, ± flattened on one side, smooth, greenish to reddish brown. *Flowers November to December.*

Moist areas at lake margins; 0–200 m.

S.C. Isla Grande (Lago Fagnano). (Map 37h).
W. Argentina north to *c.*29°S.lat.
Chile north to 51°S.lat.
North America.

Fuegian material belongs to var. *remireoides* (Griseb.) Beetle.

2. *S. californicus* (C. A. Meyer) Steudel 1841, *Nomencl. Bot.*, ed. 2, **2**: 538.

S. riparius C. Presl. 1828, *Reliq. Haenk.* **1**: 193 – Dusén, 1900; Skottsberg, 1916; Barros, 1935.
Schoenoplectus riparius (C. Presl.) Palla 1888, *Bot. Jahrb.* **10**: 299.

Rhizome robust, creeping. Stems up to 300 cm, acutely 3-angled, erect, dark green; sheaths thin, membranous, the upper usually with a short lamina. Inflorescence 2–5 cm, with branches 0.5–5 cm. Lower bract *c.*$\frac{3}{4}$–1$\frac{1}{2}$ times as long as mature inflorescence. Spikelets many, 5–10 mm, ovoid, reddish brown; pedicels up to *c.*6 mm. Glumes 4–4.5 mm, oblong- to suborbicular-ovate, somewhat emarginate, with terminal mucro up to 1.7 mm, ciliate, hyaline, brownish, with green midrib. Perianth of (2) 3 setae. Stamens 3. Stigmas 2. Nut 2–2.5 mm, obovoid, compressed, shiny, reddish brown. *Flowers December to March.*

In shallow water at lake margins; 0–50 m.

N.W. Isla Grande. (Map 37i).
North along Andes to *c.*10°S.lat.
Falkland Islands (Islas Malvinas).
N. America.

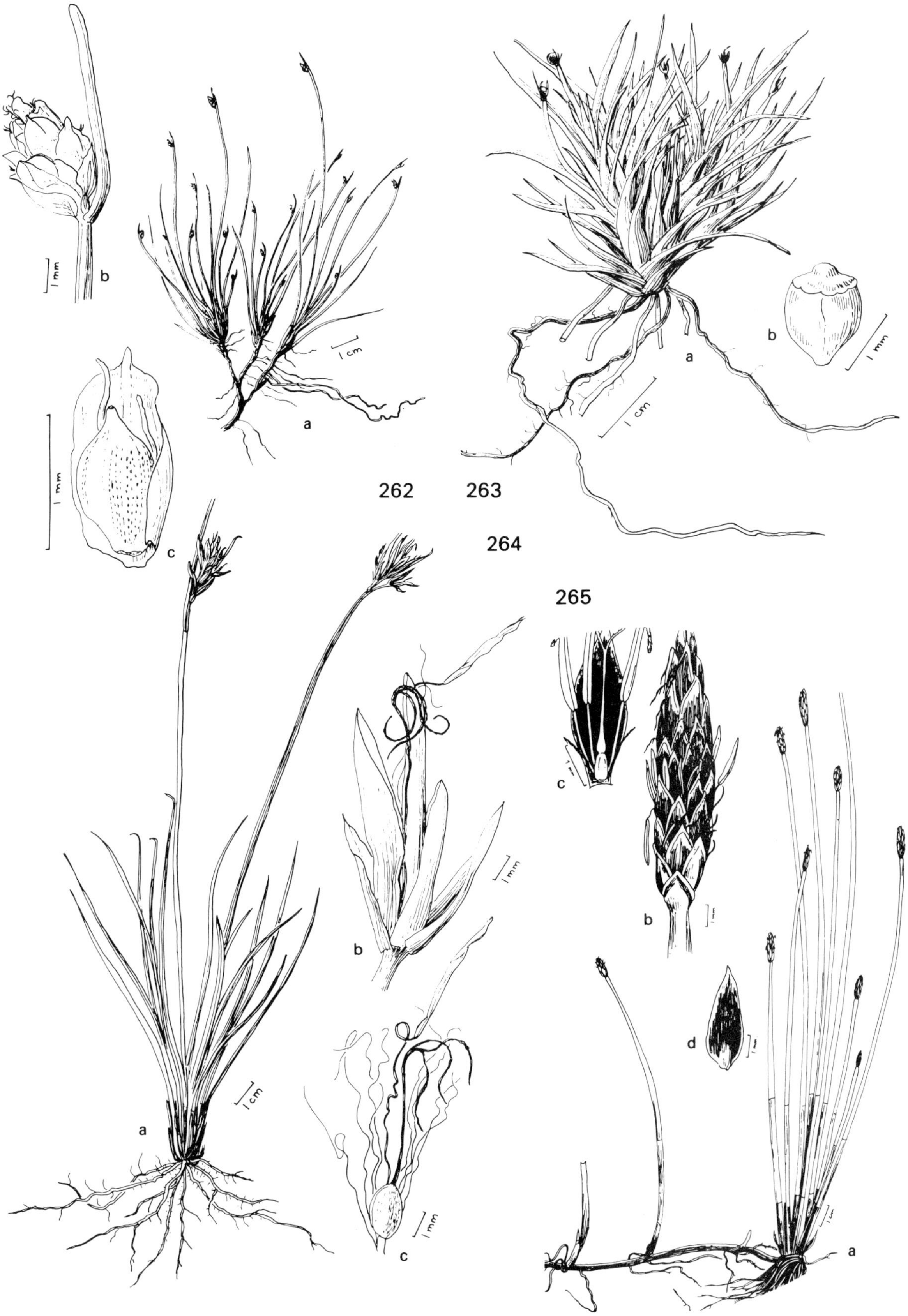

Fig. 262. *Scirpus cernuus*. a, Plant. b, Spikelet. c, Flower. Scale = 1 cm (a) or 1 mm (b–c). Fig. 263. *Oreobolus obtusangulus*. a, Plant. b, Nut. Scale = 1 cm (a) or 1 mm (b). Fig. 264. *Schoenus antarcticus*. a, Plant. b, Spikelet, c, Flower. Scale = 1 cm (a) or 1 mm (b–c). Fig. 265. *Eleocharis melanostachys*. a, Plant. b, Spikelet. c, Flower. d, Glume. Scale = 1 cm (a) or 1 mm (b–d). (R. N. P. Goodall)

3. *S. cernuus* Vahl 1805, *Enum. Pl.* **2**: 245.
S. brevis D'Urv. 1825, *Fl. Is. Mal.* 29.
Isolepis pygmaea Kunth 1837, *Enum.* **2**: 191 – Hooker, 1847; Franchet, 1889.
I. magellanica Gaudich. 1826, *Freyc. Voy.* 414.

Annual; stems 2–13 cm, terete, filiform. Leaves 1–2, usually shorter than stem. Lower bract shorter or slightly longer, occasionally much longer, than inflorescence. Spikelets 1–3, 2–5 mm., ovoid. Glumes 1.3–1.8 mm, broadly ovate, mucronate, with hyaline margins, greenish or reddish brown with green midrib or green spot on either side of midrib. Perianth absent. Stamens 2. Stigmas 3. Nut *c.*1 mm, broadly trigonous-obovoid or subglobose, smooth, greyish or reddish brown, rather dull. $2n = 48$★. *Flowers October to March.* (Fig. 262).

Moist places in most open communities; 0–500 m.

Almost throughout Fuegia, uncommon in N. Isla Grande. (Map 37j).
Virtually cosmopolitan.

2. *Eleocharis* R. Br.

Glabrous perennials; stems terete. Sheaths leafless. Spikelet solitary, terminal, many-flowered, the lowest glume usually sterile and differing in shape from the others. Flowers hermaphrodite; perianth of 3 or 6 scabrid setae, shorter than or slightly exceeding nut; stamens 3; stigmas 2–3; style with a swollen persistent base.

a. Spikelet at least 6 mm; stigmas 2 **1. *melanostachys***
a. Spikelet not more than 6 mm; stigmas 3 **2. *albibracteata***

1. *E. melanostachys* (D'Urv.) C. B. Clarke 1901, *Bot. Jahrb.* **30** Beibl. 68: 20.
E. palustris auct., non (L.) Roemer & Schultes – Hooker, 1847.

Rhizome stout, producing clusters or dense stands of stems. Stems (4–) 9–45 cm, 1.5–3 mm in diameter, usually reddish at base; sheaths obtuse to subacute, green to yellowish brown. Spikelet 6–17 mm, cylindrical to cylindrical-obconical, dark reddish or brownish purple. Lowest 1(2) glumes sterile, up to ½ as long as spikelet but usually much less. Glumes ovate, obtuse to subacute, reddish purple, darker or almost black towards apex, with green midrib, with hyaline margin, the lowest often green and purplish only at margins. Perianth-setae usually 3. Stigmas 2. Nut *c.*1.5–1.8 mm, obovoid, somewhat compressed, yellowish, with short, conical, terminal beak. $2n = 24$.★ *Flowers December to February.* (Fig. 265).

Shallow pools, streams and lake margins; 0–20 m.

Locally on S. coast of Isla Grande. (Map 38a).
W. Argentina north to *c.*39°S.lat.
Chile north to *c.*33°S.lat.
Falkland Islands (Islas Malvinas).

2. *E. albibracteata* Nees & Meyen ex Kunth 1837, *Enum. Pl.* **2**: 143.

Rhizome slender, producing clusters of stems. Stems 2–10 cm, up to *c.*0.5 mm in diameter, reddish at base; sheaths obtuse, purplish to reddish. Spikelet 2–6 mm, ovoid, reddish brown to blackish. Lowest glume *c.*½ as long as spikelet. Glumes ovate, obtuse, dark reddish brown with white hyaline margin. Perianth-setae 6. Stigmas 3. Nut 1.2–1.35 mm, obovoid, angled, whitish or yellowish, with short, conical terminal beak. *Flowers December to January.*

Lake and stream margins, wet bryophyte communities; 0–80 m.

N. & C. Isla Grande. (Map 38b).
W. Argentina and E. Chile north to *c.*26°S.lat.
Peru? Falkland Islands (Islas Malvinas).

3. *Schoenus* L.

Rhizomatous perennials; stems terete. Leaves with linear lamina. Inflorescence of 12–20 pedicellate spikelets, paniculate; bracts equalling or exceeding panicle. Spikelets with 1–2 fertile flowers; glumes distichous, the lower 3–4 empty; rhachilla flexuous, short between sterile nodes and long betweeen fertile nodes. Flowers hermaphrodite; perianth of 6 smooth or scabrid setae; stamens 3; stigmas 3.

a. Perianth-setae scabrid, ± twice as long as nut **2. *antarcticus***
a. Perianth-setae smooth, several times longer than nut **1. *andinus***

1. *S. andinus* (Phil.) Pfeiffer 1927, *Feddes Repert.* **23**: 348.
Elynanthus sodalium Hariot ex Franchet 1889, *Miss. Sci. Cap Horn* 375.
Schoenus sodalium (Hariot ex Franchet) Phil. 1896, *Anal. Univ. Chile* **93**: 484 – Alboff, 1896.

Rhizome up to *c.*30 cm, producing clusters of stems and leaves. Stems 14–50 cm, yellowish green, ± arcuate. Leaves 10–30 cm × 1–2 mm; lamina flat below, ± terete and channelled above, acute; sheath reddish brown to blackish, with paler margin. Panicle 3–5 cm, contracted; lower bract 5–15 cm. Spikelets 9–14 mm; pedicels 3–7 mm; glumes lanceolate, acute, dark reddish brown to blackish. Perianth-setae several times as long as nut, smooth. Nut 1.5–2 mm, obovoid-ellipsoid, 3-ribbed, reddish brown, shiny. *Flowers November to January.*

Sphagnum swamps, *Roskovia* meadows, boggy ground, stream-margins; 0–600 m.

Almost throughout Fuegia, but only occasional in N. Isla Grande. (Map 38c).
W. Argentina north to *c.*39°S.lat.
Chile north to *c.*46°S.lat.

2. *S. antarcticus* (Hooker f.) Dusén 1900, *Svenska Exped. Magellansl.* **3(5)**: 213.
Elynanthus antarcticus (Hooker f.) Franchet 1889, *Miss. Sci. Cap Horn* 374 – Spegazzini, 1896.
Chaetospora antarctica Hooker f. 1847, *Fl. Antarct.* **1(2)**: 361.

Similar to *S. andinus* but distinguished by the scabrid perianth-setae *c.*2 times as long as nut and usually with paler yellowish to reddish brown glumes. *Flowers November to January.* (Fig. 264).

Boggy ground, Magellanic moorland; 0–50 m.

S. & W. Fuegia. (Map 38d).
W. Chile north to *c.*45°S.lat.

Readily confused with *S. andinus* and most previous reports from Isla Grande probably refer to that species.

4. *Oreobolus* R. Br.

Densely tufted, glabrous, perennial herbs, forming low, compact cushions. Leaves densely imbricate, subulate, with conspicuous sheaths. Inflorescence of terminal, 1-flowered spikelet and 1–2 lateral, sterile spikelets. Glumes (2–)3–4, deciduous. Flowers hermaphrodite; perianth of 6 persistent scales in 2 whorls ; stamens 3; style with swollen base; stigmas 3.

1. *O. obtusangulus* Gaudich. 1825, *Ann. Sci. Nat.* **5**: 99.

Cushions *c.*10–25 cm in diameter. Stems 5–10 cm, ascending or erect, branched, densely leafy. Leaf-lamina 1–3 cm, acicular-subulate, long-acuminate, channelled above, rigid, scabrid on margin, glossy green, often brownish towards apex; sheath 0.5–1 cm, *c.*3 times as wide as lamina, ciliate-scabrid on margin, reddish brown, shiny. Culm axillary, not exceeding uppermost leaves, compressed, rectangular distally, smooth or slightly scabrid on angles. Lower glume 4.5–8 mm, acicular, keeled, scabrid on margin and keel, pale greenish or yellowish brown, reddish brown at margin. Perianth-scales 1.4–1.6 mm, ovate-triangular, acute to acuminate, sparsely ciliate, reddish brown. Nut 1–1.5 mm, pyriform, with raised margin at apex, smooth or with few narrow ribs, dark purplish. $2n = 48$.* *Flowers November to January (February).* (Fig. 263).

Astelia and *Sphagnum* bogs, *Rostkovia* communities, wet areas beside streams, Magellanic moorland; 0–300(–500) m.

C. & S. Fuegia. (Map 38e).
W. Argentina north to *c.*36°30′S.lat.
Chile north to *c.*43°S.lat.
Falkland Islands (Islas Malvinas).
Ecuador – Colombia.

5. *Carpha* Banks & Sol. ex R. Br.

Tufted, glabrous, grass-like perennial herbs. Inflorescence a compact terminal corymb of 1– to 2–flowered spikelets. Glumes 3–6, distichous, the lower smaller, empty. Flowers hermaphrodite; perianth of 6 plumose setae; stamens 3; style-base persistent; stigmas 3.

1. *C. alpina* R. Br. 1810, *Prodr. Fl. Nov. Holl.*, Ed. 1, 230.
C. schoenoides Banks & Sol. ex Hooker f. 1847, *Fl. Antarct.* **1(2): 362.**

Rhizome short; stems 10–40 cm, leafy at base, ribbed, ± arcuate, rigid. Leaves 5–20 cm × 0.5–1.5 mm., flat or ± channelled above, obtuse; sheath wide, brown, shiny, serrulate distally. Inflorescence with 2–3 bracteate groups of spikelets; lower bract 4.5–8 cm, leaf-like. Spikelets 7–12 mm; glumes 4–7 mm, lanceolate, ± obtuse, yellowish brown. Nut 2–3 mm, oblong-ellipsoid, trigonous, with long beak, dark brown, shorter than perianth-setae. *Flowers January to February.* (Fig. 266).

Sphagnum swamps, *Rostkovia* and *Marsippospermum* communities, stream margins and boggy ground; 0–900 m.

Much of C. & S. Fuegia except extreme east. (Map 38f).
Chile north to *c.*40°S.lat.
New Zealand, Australia, New Guinea.

South American material belongs to var. *schoenoides* (Banks & Sol. ex Hooker f.) Kük. 1939, *Feddes Repert.* **47**: 113, distinguished by its minutely scabrid style-base

6. *Uncinia* Pers.

Tufted, glabrous, perennial herbs. Leaf-lamina with ligule at base. Stems solid, usually leafy and ± triquetrous. Inflorescence a solitary, terminal spike, of many 1-flowered spikelets. Flowers unisexual, each subtended by a glume. Male flowers with 3 stamens; perianth 0. Female flowers surrounded by a compressed, urceolate sac (utricle) having terminal beak from which project the 3 stigmas and a long, rigid, exserted bristle (rhacheole) with hooked tip, which arises from the base of the ovary within the utricle. Nut trigonous, compressed, enclosed in utricle.

a. Inflorescence not more than 12 mm; leaves not more than 0.5 mm wide **1. *kingii***
a. Inflorescence at least 30 mm; leaves at least 3.5 mm wide **2. *lechleriana***

1. *U. kingii* Boott 1847, in Hooker f., *Fl. Antarct* **1(2)**: 370.

Rhizome slender, stoloniferous. Leaves 2–8 cm × *c.*0.5 mm, filiform, channelled, acuminate, the margin scabrid distally; ligule very short, acute. Culms 5–15 cm, erect to ascending. Spike 8–10(–12) mm, obconical, the male part ill-defined. Female glumes 4.5–6 × 1–1.5 mm, linear-lanceolate, reddish brown with green midvein. Utricles 2–7, (4.5–)6–8 mm, linear-subulate, dark reddish brown; rhacheole exserted 3–3.5 mm. Male glumes 2, *c.*4 mm, linear-lanceolate, acute, involute, reddish brown. *Flowers November to February.*

Wet areas, especially streams and seepages, usually in feldmark and high montane grassland; (50–)450–900 m.

C. & S. Fuegia. (Map 38g).
Chile north to 50°59′S.lat.

2. *U. lechleriana* Steudel 1855, *Syn, Pl. Glum.* **2**: 244.

U. triquetra Kük. 1900, *Bot. Centr.* **82**: 97 – Autran, 1905.
U. brevicaulis auct., non (Thouars) Kunth – Wildeman, 1905.
U. macloviana auct., non Gaudich. – Spegazzini, 1896.

Rhizome ± stout, rhizomatous. Leaves up to 30 cm × 3.5–7 mm, flat, acuminate, the margin smooth or scabrid. Culms 15–52(–66) cm, erect. Spike (33–)40–75 mm, cylindrical to cylindrical-clavate, the male part 5–8 mm. Female glumes (6–)6.5–7.7(–10) × 2–2.5 mm, oblong to ovate, obtuse to subacute, white-green, reddish brown at margin. Utricles numerous, slightly exceeding glumes, ellipsoid, stipitate, prominently 2-ribbed, ciliate above; rhacheole *c.*2 times as long as utricle. Male glumes *c.*3–3.5 mm, ovate, obtuse, green, brownish at margin, white at apex. *Flowers December to February.* (Fig. 268).

Wet areas by streams, *Sphagnum* and *Marsippospermum* bogs, wet peats in *Drimys* and *Nothofagus* forest; 0–800 m.

C. & S. Fuegia, except extreme west. (Map 38h).
W. Argentina north to *c.*40°S.lat.
Chile north to 51°30′S.

The most widely distributed form of this species has dense spikes 5–8 mm wide and is referable to var. *lechleriana*. A series of populations most frequent in the open deciduous *Nothofagus* forests and associated grassland, having much laxer spikes (3–)4–5 mm wide, which apparently have affinities with ***U. tenuis*** Poeppig ex Kunth 1837, *Enum. Pl.* **2**: 525, have been placed in var. ***triquetra*** (Kük.) Kük. 1909 in Engler, *Pflanzenreich* **38(IV 20)**: 58. The status of these taxa requires further investigation.

U. macrolepis Decne 1853, in D'Urv., *Voy. Pole Sud* **2**: 13 (*U. sinclairii* auct., non Boott – Dusén, 1900), reported once from S.C. Isla Grande (Río Azopardo) by Skottsberg (1916), is a poorly known species said to differ from *U. lechleriana* by its shorter spike (10–20 mm), culms shorter than the leaves and distally hispid utricles. Further material is required to confirm the presence of this species in Fuegia and indeed its status in relation to *U. lechleriana* and the closely allied *U. brevicaulis* D'Urv of the Falkland Islands (Islas Malvinas) and Tristan da Cunha.

7. *Carex* L.

Perennial, rhizomatous herbs. Stems solid, usually leafy, usually 3-angled. Leaves linear, ± keeled or involute, sometimes flat, the base usually sheathing; sheaths entire; ligule present, all except apical 0.5–1 mm adnate to lamina. Inflorescence of one or more spikes, each of which may be subtended by a bract. Flowers unisexual, in 1-flowered spikelets, each subtended by a glume. Male flowers with 2–3 stamens; perianth 0; female flowers as in *Uncinia* but usually without basal bristle (rhacheole) and this, if

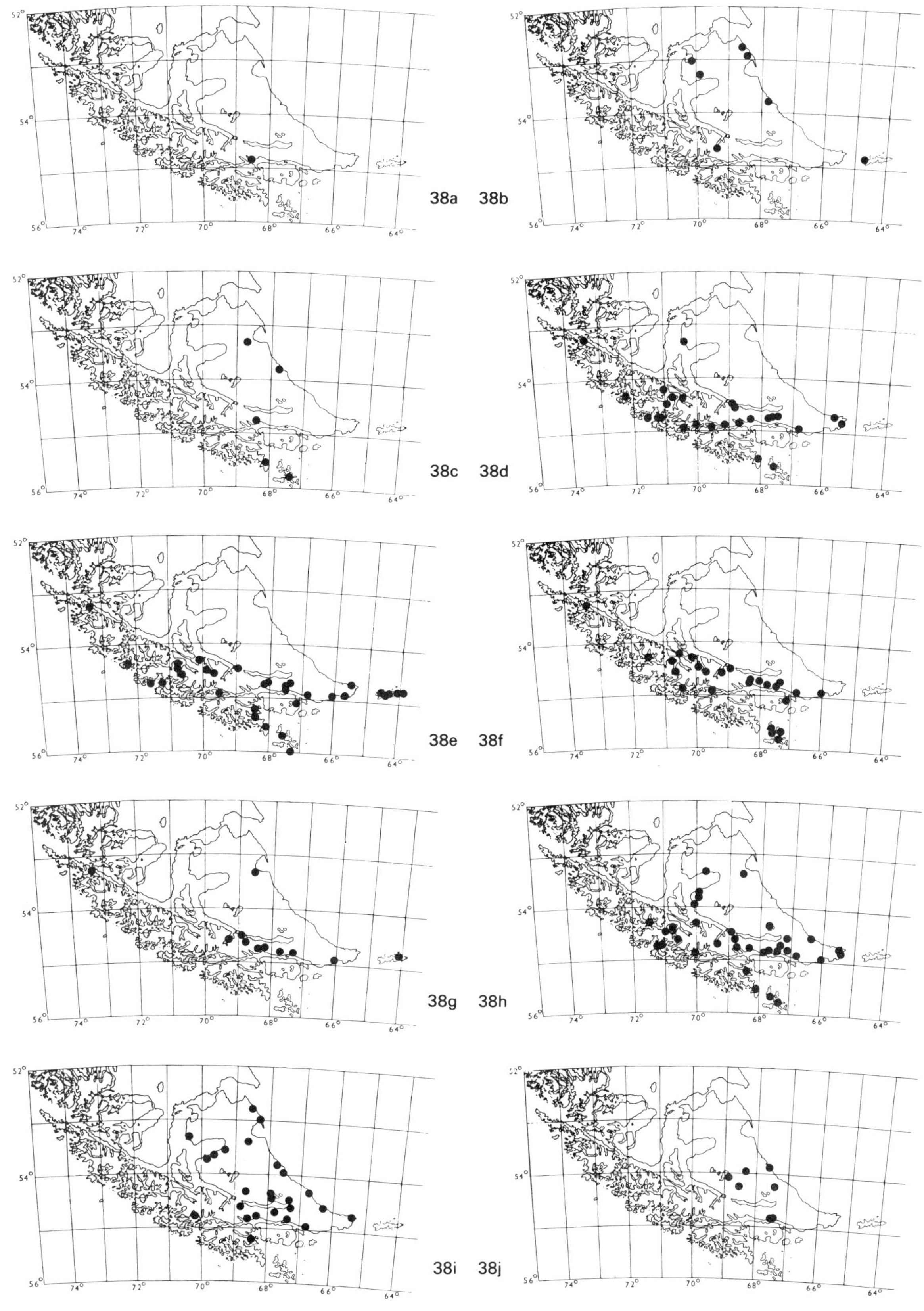

Map 38. a, *Eleocharis melanostachys*. b, *Eleocharis albibracteata*. c, *Schoenus andinus*. d, *Schoenus antarcticus*. e, *Oreobolus obtusangulus*. f, *Carpha alpina*. g, *Uncinia kingii*. h, *Uncinia lechleriana*. i, *Carex fuscula*. j, *Carex acaulis*.

present, not hooked; ovary trigonous and stigmas 3, or ovary biconvex and stigmas 2. Nut trigonous or biconvex, enclosed in utricle.

The distribution of male and female flowers in the inflorescence is of taxonomic value in this genus. Usually the terminal spike and often some of the upper lateral spikes are male and the rest female (spp. 1–8, 10) although the female spikes often have some male flowers distally and the male spike may have female flowers at the base. Other species have male and female flowers in the same spike, the male flowers being either basal (spp. 9, 13–14) or distal (spp. 15–19) in the spike.

a. Spike solitary, terminal; leaves ± filiform, at least near base, less than 2 mm wide:
 b. Fruit deflexed when ripe; basal bristle (rhacheole) present, exserted from utricle with stigmas **15. *microglochin***
 b. Fruit not deflexed when ripe; rhacheole absent:
  c. Stigmas 3 **18. *vallis-pulchrae***
  c. Stigmas 2: **19. *capitata***
   d. Spike without subtending bract
   d. Spike subtended by bract:
    e. Culms at least 6 cm; lamina ± filiform, channelled; bracts much longer than spike **16. *caduca***
    e. Culms not more than 6 cm; lamina semi-cylindrical at base, flat distally; bracts glume-like, scarcely longer than spike **17. *sorianoi***
a. Spikes at least 2; leaves flat or keeled, at least 2 mm wide:
 f. Spikes all ± similar in appearance, sessile, the terminal not entirely male:
  g. Spikes pale brown or whitish, not crowded into a head **11. *curta***
  g. Spikes reddish to dark brown, crowded into a head, which is sometimes interrupted at base:
   h. Glume longer than utricle **14. *gayana***
   h. Glume shorter than utricle:
    i. Spikes with male spikelets at apex; utricle without serrate wing **12. *maritima***
    i. Spikes with female spikelets at apex; utricle with serrate wing on margin **13. *macloviana***
 f. Spikes dissimilar in appearance, sessile or pedunculate, the terminal usually entirely male:
  j. Utricle pubescent **5. *aematorryncha***
  j. Utricle glabrous:
   k. Utricle at least 14 mm **10. *macrosolen***
   k. Utricle not more than 8 mm:
    l. Spikes sessile or with peduncles not more than 4 mm:
     m. All spikes bisexual, with female spikelets at apex **9. *atropicta***
     m. Spikes unisexual:
      n. Stigmas usually 2; female glume with midvein at most shortly excurrent; fruit with 3–5 prominent, raised lateral veins **2. *acaulis***
      n. Stigmas 3; female glume with prominent apical mucro; fruit smooth or with very indistinct lateral veins **1. *fuscula***
    l. At least lower spikes with peduncles at least 5 mm:
     o. Lower bract shorter than inflorescence; female glumes long acuminate; spikes usually distinctly nodding **6. *magellanica***
     o. Lower bract usually longer than inflorescence; female glumes not long acuminate; spikes erect or ± nodding:
      p. Stigmas 3; utricle at least 4 mm:
       q. Terminal male spike at least 35 mm; female glumes with serrate excurrent midrib at least 3 mm **4. *trifida***
       q. Terminal male spike not more than 30 mm; female glumes without excurrent midrib **3. *banksii***
      p. Stigmas 2; utricle not more than 4 mm:
       r. Terminal male spike not more than 20 mm; leaves not more than 4 mm wide; female glume obtuse **7. *decidua***
       r. Terminal male spike at least 30 mm; leaves at least 5 mm wide; female glume acute, often mucronate **8. *darwinii***

Subgenus CAREX

Spikes several, unbranched, often pedunculate, dissimilar in appearance, normally unisexual. Peduncles subtended by often sheathing bracts. Stigmas 2 or 3.

1. *C. fuscula* D'Urv. 1825, *Fl. Is. Mal.* 28.
C. indecora Kunth 1837, *Enum. Pl.* **2**: 448 – Hooker, 1847.

Rhizome rather short. Leaves 4.5–25 cm × 2–6 mm; lamina flat or ± keeled, scabrid on margins and midvein beneath, often yellowish, sometimes rather glaucous; ligule 1.8–4.5 mm, obtuse to truncate; upper sheaths yellowish brown, the lower dark brown. Culms (1–) 5–25(–35) cm, smooth. Bracts leaf-like, longer than their spikes, at least the lower exceeding inflorescence. Male spike 1.5–12 mm, cylindrical to subclavate, subsessile or with erect peduncle up to 4 mm; male glumes 2.5–7 mm, ovate to oblong-ovate, obtuse or acute, mucronate, purplish brown, hyaline. Female spikes 2–4, 8–14 mm, cylindrical or cylindrical-ovoid, on suberect peduncle 2–5 mm, or subsessile, the lower often remote; female glumes 2–3 mm, oblong to oblong-ovate, obtuse or acute to emarginate, prominently mucronate, ciliate on mucro and midvein, green or purplish brown with green midvein. Stigmas 3. Fruit 2.5–3.5 mm, ovoid, tapering to beak, smooth, golden-brown, with prominent marginal often ± winged vein, the lateral veins indistinct; beak *c.*1 mm, shortly bifid, rather winged, scabrid. $2n = c.50$★. *Flowers October to December.*

Moist grassland and *Chiliotrichum* scrub, stream and lake margins; 0–500 m.

Almost throughout Fuegia except the extreme west, south and east. (Map 38i).
W. Argentina north to *c.*27°S.lat.
Chile north to *c.*35°S.lat.
Falkland Islands (Islas Malvinas).

2. *C. acaulis* D'Urv. 1825, *Fl. Is. Mal.* 29.

Similar to *C. fuscula* but rhizomes usually much longer; culms up to 3 cm, much shorter than leaves; male glumes 2–2.5 mm, ovate, acute or subacute, serrate and scabridulous towards apex, white and scarious with green midvein; female glumes similar but up to 5.5 mm, ovate or oblong-ovate, rarely with midvein shortly excurrent; stigmas 2(3); fruit 2–3 mm, ovoid, tapering to beak, with prominent often ± winged marginal veins, with 3–5 prominent veins on each side; beak up to 3 mm, smooth or sparsely scabrid at tip. $2n = 50$.★ *Flowers October to December.* (Fig. 270).

Wet sands and gravels by sea and lakes, wet grassland; 0–250 m.

C. & S. Isla Grande except the extreme west and east. (Map 38j).
W. Argentina north to *c.*40°S.lat.
Falkland Islands (Islas Malvinas).

3. *C. banksii* Boott 1846, *Trans. Linn. Soc. London* **20**: 119.
C. trifida auct., non Cav. – Franchet, 1889.
C. trifida var. *franchetii* F. Kurtz 1896, *Rev. Mus. La Plata Bot.* **7(2)**: 394.

Rhizome stoloniferous. Leaves 7–30 cm × 3–8 mm; lamina flat or ± keeled, ± scabrid on margins; ligule 2.5–3.5 mm, obtuse; sheaths brown. Culms (4–) 15–46 cm, smooth. At least lower bracts leaf-like, exceeding their spikes. Male spike 15–30 mm, clavate, subsessile or on erect to nodding peduncle up to 20 mm, sometimes female at apex; female spikes 2–4, 15–30 mm, cylindrical-ovoid, on suberect to ± nodding peduncles up to 30 mm; glumes 5–7 mm, oblong-lanceolate, prominently mucronate, dark purple-brown, with yellowish midvein. Stigmas 3. Fruit 4–7 mm, elliptical to suborbicular, compressed, glabrous, whitish to pale brown; beak *c.*1 mm, smooth, purplish. *Flowers November to March.*

Sphagnum swamps, *Astelia* bogs, wet places by streams, lakes and seepage areas; 0–900 m.

S. Fuegia, except the extreme west, occasional further north. (Map 39a).
W. Argentina north to *c.*39°S.lat.
Chile north to *c.*40°S.lat.

4. *C. trifida* Cav. 1799, *Icon. Descr.* **5**: 41.

Similar to *C. banksii* but terminal spike always entirely male, 35–50 mm; glumes 4–5 mm, ovate to oblong, deeply 2-fid, reddish brown, with whitish midvein excurrent into scabrid awn 3–8 mm; fruit cylindrical to obovoid.

Wet coastal rocks; 0–10 m.

Southernmost Fuegia (Isla Otter, Islas Wollaston). (Map 39b).
W. Chile north to 43°35′S.lat.
Falkland Islands (Islas Malvinas), Macquarie Island, Auckland Islands, Campbell Island, Snares Island, New Zealand (South Island).

5. *C. aematorryncha* Desv. 1853, in C. Gay, *Hist. Chile Bot.* **6**: 224.
C. filiformis auct., non L. – Spegazzini, 1896.

Rhizome rather slender. Leaves 13–60 cm × 4–6 mm; lamina flat or ± plicate towards base, scabrid on margin and midvein beneath; ligule *c.*0.5 mm, subacute; upper sheaths brownish green, the lower purplish or dark brown. Culms 20–80 cm, trigonous, scabrid on angles. Lower bracts leaf-like, exceeding their spike, usually exceeding inflorescence; bracts of terminal spikes setaceous, longer or shorter than spikes, brown. Male spikes 2–3, 17–35 mm, cylindrical, often ± trigonous, the lower sometimes reduced; peduncles 2–9 mm, scabrid; male glumes 4.5–6.5 mm, oblong to lanceolate-oblong, acute to obtuse and emarginate, scarious, reddish brown, with paler or greenish midrib excurrent into scabrid awn *c.*1–2 mm. Female spikes 2–3, 20–30 mm, cylindrical, remote, the uppermost often with male flowers distally; peduncles 2–14

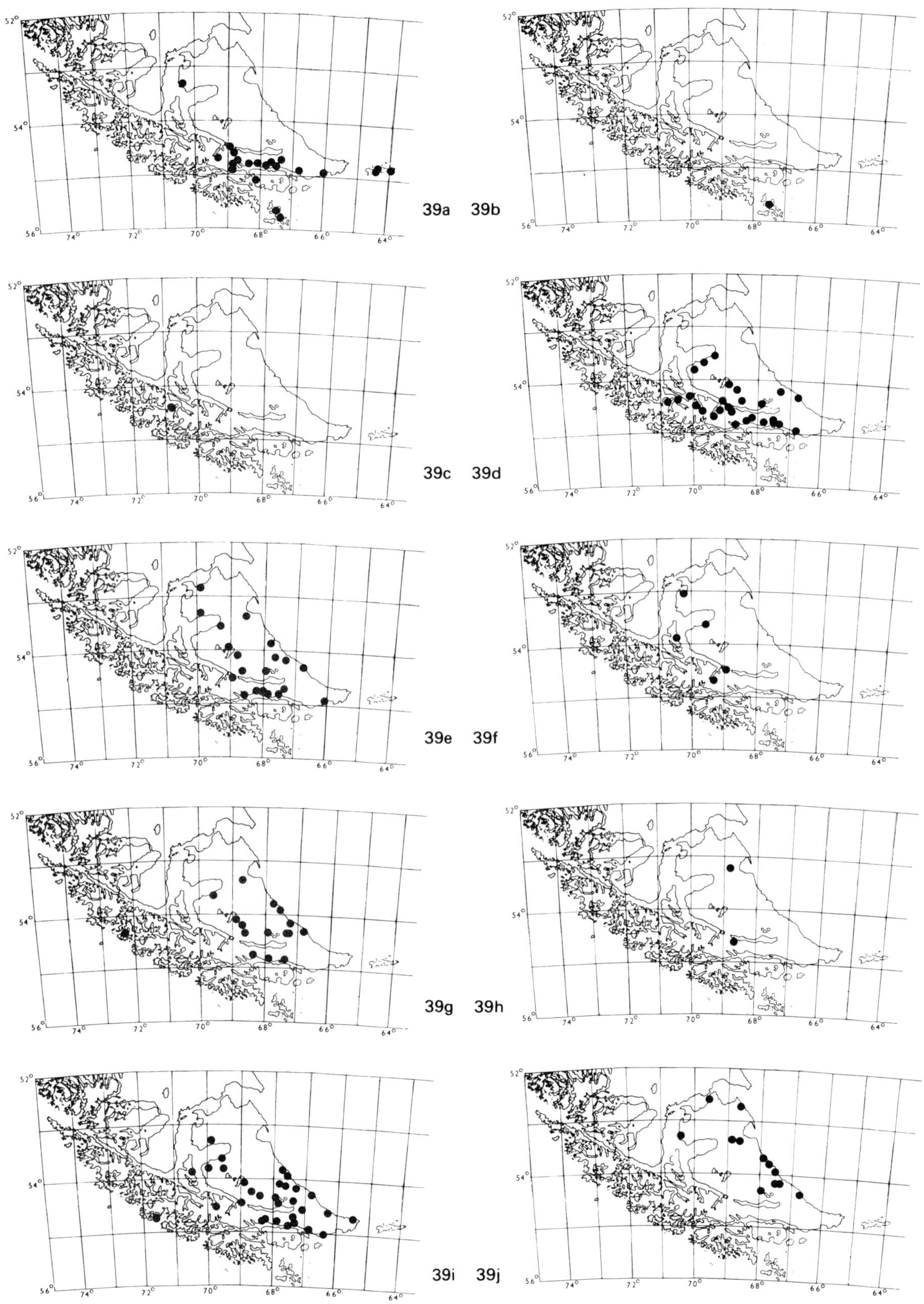

Map 39. a, *Carex banksii*. b, *Carex trifida*. c, *Carex aematorryncha*. d, *Carex magellanica*. e, *Carex decidua*. f, *Carex darwinii*. g, *Carex atropicta*. h, *Carex macrosolen*. i, *Carex curta*. j, *Carex maritima*.

mm, erect or slightly nodding, scabrid; female glumes similar to male but with awn up to 5 mm. Stigmas 3. Fruit 3–5 mm, ovoid, somewhat trigonous, densely golden- to reddish brown-pubescent; beak up to 1 mm, shortly bifid, stout, ciliate, purplish.

Wet areas by streams; 0–20 m.

S.W. Isla Grande (Fjordo Sarmiento). (Map 39c).
W. Argentina north to *c*.44°S.lat.
Chile north to *c*.36°40′S.lat.
Falkland Islands (Islas Malvinas).

6. *C. magellanica* Lam. 1792, *Encycl. Méth. Bot.* **3**: 385.

Rhizome short. Leaves (6.5–)10–30 cm × 2–3 mm; lamina flat, scabrid on margin; ligule 4–5 mm, acute; sheaths purplish to pale brown. Culms 10–30 cm, trigonous, smooth, somewhat nodding. Bracts leaf-like, the lowest exceeding inflorescence; peduncles 15–30 mm, slender, smooth, usually nodding. Male spike 1, 10–20 × 2–3 mm, cylindrical, with some female flowers; male glumes *c*.4 mm, lanceolate, acute or subacute, reddish brown. Female spikes 2–3, 6–15 × 4–8 mm, ovoid to oblong-ovoid, nodding; female glumes ovate-lanceolate, long-acuminate, smooth, purplish brown, caducous. Stigmas 3. Fruit *c*.3 mm, much wider than glume, ovoid or subglobose, trigonous, compressed, obtuse, with very short emarginate beak, minutely papillose, brownish green. $2n = 58$. *Flowers December to February.*

Sphagnum swamps, wet grassland by rivers and lakes, boggy areas in forest clearings; 0–500 m.

C. & S. Isla Grande. (Map 39d).
W. Argentina north to *c*.40°S.lat.
Chile north to *c*.50°S.lat.
Falkland Islands (Islas Malvinas).
North America, Eurasia.

Southern Hemisphere material belongs to ssp. *magellanica*, which differs from the circumboreal ssp. *irrigua* (Wahlenb.) Hultén in having more female flowers in the terminal spike, and longer and wider female glumes (Moore and Chater, 1971).

7. *C. decidua* Boott 1846, *Trans. Linn. Soc. London* **20**: 119.

Rhizome rather slender. Leaves 15–35 cm × 2–3 mm; lamina flat or somewhat keeled, scabrid on margins and midvein beneath; ligule 3–5 mm, obtuse; sheaths pale green to yellow-brown. Culms 15–30 cm, trigonous, scabrid on angles distally. Bracts leaf-like, the 2 lower exceeding inflorescence. Male spike 1, 14–20 × 2.5–3 mm, oblanceolate-cylindrical, erect, often with some female flowers; male glumes 3.5–4 mm, oblong to ovate-oblong, obtuse, purplish brown, with green midvein. Female spikes 3–6, 12–22 × 4–6 mm, cylindrical, erect on short smooth peduncles, the lowest often remote; female glumes 3–3.5 mm, oblong-ovate, obtuse, brownish purple to blackish with green midvein. Stigmas 2. Fruit 2.5–3.5 mm, larger than glumes, oblong-ovate, compressed, abruptly narrowed to beak, minutely papillose, green; beak short, stout, truncate or slightly emarginate. *Flowers November to January.*

Sphagnum and *Astelia* bogs, wet areas especially by streams and lakes; 0–1000 m.

Almost throughout Isla Grande except the extreme west. (Map 39e).
W. Argentina north to *c*.40°S.lat.
Chile north to *c*.33°S.lat.
Falkland Islands (Islas Malvinas).

8. *C. darwinii* Boott 1845, *Proc. Linn. Soc. London* **1**: 261.

Rhizome rather stout. Leaves 21–80 cm × 5–11 mm; lamina flat; ligule *c.* 2 mm, obtuse; sheaths green to pale brown. Culms 20–70 cm, trigonous, ± winged on angles. Bracts leaf-like, at least the lowest exceeding inflorescence. Male spikes 2–5, 15–30 × 3–6 mm, ± cylindrical, erect or somewhat nodding; male glumes 4–6 mm, lanceolate, obtuse, purplish brown, with green midvein. Female spikes 5–8, 16–45 × 5–9 mm, cylindrical, erect to ± nodding on peduncles up to 50 mm, the upper sometimes male distally; female glumes 3–3.5 mm, ovate, acute, often mucronate, brown, with wide green midvein. Stigmas 2. Utricle 3–4 mm, ovoid-ellipsoid, smooth, yellowish; beak short, stout. *Flowers January to February.*

Wet grassland and stream-sides; 0–30 m.

W. Isla Grande; Isla Dawson. (Map 39f).
W. Argentina north to *c*.40°S.lat.
Chile north to *c*.41°S.lat.

9. *C. atropicta* Steudel 1855, *Syn, Pl. Glum.* **2**: 204.

Rhizome short, stout. Leaves 7–40 cm × 2–4 mm;

lamina flat; ligule 2–3 mm, obtuse to subacute; sheaths purplish brown. Culms 12–60 cm, trigonous. Bracts leaf-like, exceeding inflorescence. Spikes 2–3, bisexual with female spikelets at apex, the terminal 10–13 × 5–8 mm, cylindrical to ellipsoid, sessile, the lateral smaller and with peduncles up to 4 mm; female glumes 5–5.5 mm, ovate-elliptical, obtuse, dark purplish brown, yellow along midvein, whitish at margin, the male similar. Stigmas 2. Utricle 4–6 mm, oblong-obovate, compressed, winged and ciliate distally, yellowish at base, dark purplish above; beak short, stout, emarginate, scabrid. *Flowers December to March.*

Grassland, open scrub, river gravels; 0–350 m.

C. & S. Isla Grande, and some islands further west. (Map 39g).
W. Argentina north to *c.*33°S.lat.
Chile north to *c.*50°S.lat.

10. ***C. macrosolen*** Steudel 1855, *Syn. Pl. Glum.* **2**: 210.

Rhizome stout. Leaves 2–5 cm × 2–4 mm; lamina linear-triangular, long acuminate, flat, scabrid on margin; ligule *c.*1 mm, obtuse; sheaths brown. Culms very short. Male spike 1, 8–20 × 2–3 mm, cylindrical, with short, setiform bract, on erect, rigid peduncle 10–40 mm; male glumes 4–5 mm, lanceolate, dark purplish brown with greenish midvein. Female spikes 2–3, 15–20 mm, clavate, subsessile at base of male peduncle; female glumes 7–9 mm, similar to the male. Stigmas 2. Utricle 14–18 mm, lanceolate-subulate, yellowish green; beak long, ± scabrid. *Flowers November to December.*

Wet open grassland; 0–50(?–250) m.

N. & S.C. Isla Grande. (Map 39h).
W. Argentina north to *c.*50°S.lat.
Chile along north shore of Estrecho de Magallanes.

Subgenus VIGNEA (P.Beauv.) Kük.

Spikes several, often branched, sessile, all ± similar, usually with male and female flowers, rarely the upper or lower unisexual, subtended by usually small bracts. Stigmas 2.

11. ***C. curta*** Gooden. 1794, *Trans. Linn. Soc. London* **2**: 145.
C. canescens auct., non L. – Kurtz, 1896; Dusén, 1900; Skottsberg, 1916; Barros 1935, 1969.
C. similis D'Urv. 1825, *Fl. Is. Mal.* 28 – Spegazzini, 1896.

Rhizomes short. Leaves 10–30 cm × 2–3 mm; lamina almost flat, scabridulous on margin and midvein beneath, pale green; ligule 2.5–3 mm, obtuse to subacute; lower sheaths leafless, brown. Culms 12–25 cm, triquetrous, scabrid on angles. Bracts usually shorter than spike, resembling the glumes but with scabrid midvein, the lower with prominent scabrid seta. Inflorescence 9–30 mm; spikes (2–)4–8(–10), 5–9 mm, ovoid, ± remote. Glumes 2–2.5 mm, oblong-obovate to sub-orbicular, cuspidate, hyaline, pale brownish, usually with green midvein. Utricle 2–2.5 mm, ovoid, plano-convex, gradually narrowed to beak, pale yellowish green; beak *c.*0.5 mm, emarginate, sparsely scabrid, brownish. $2n = 56$★. Self-compatible. *Flowers November to January.*

Wet open grassland, bogs, often with *Sphagnum*, river and lake margins; 0–500 m.

Widespread in Fuegia except the extreme west, south and east. (Map 39i).
W. Argentina north to *c.*35°S.lat.
E. Chile north to *c.*50°S.lat.
Falkland Islands (Islas Malvinas).
S.E. Australia, New Guinea. N. Eurasia, North America.

South American material belongs to var. ***robustior*** (Blytt ex Andersson) D. M. Moore & Chater 1971, *Bot. Not.* **124**: 326, which has wider leaves and utricles, and a generally longer utricle than circum-boreal material to which the type belongs.

12. ***C. maritima*** Gunn. 1772, *Fl. Norv.* **2**: 131.
C. incurva Lightf. 1777, *Fl.Scot.* **2**: 544 – Dusén, 1900; Skottsberg, 1916; Barros, 1935, 1969.

Rhizome long, slender. Leaves 2–20 cm × up to 1(–2) mm; lamina plicate or with involute margins; ligule *c.*1 mm, truncate; sheaths pale brown. Culms 2–18 cm, trigonous, smooth, often curved. Bracts absent. Inflorescence 5–18 mm, ovoid, with 3–8 appressed spikes. Glumes 3–4 mm, orbicular-ovate, obtuse to acute, usually mucronulate, reddish to dark brown, with scarious margin. Utricle 3–5 mm, ovoid, brown to blackish; beak *c.*0.5 mm, smooth or sparsely scabrid. *Flowers November to January.*

Open sands and gravels; 0–50 m.

N. & C. Isla Grande. (Map 39j).
W. Argentina north to *c.*33°S.lat.
E. Chile north to *c.*28°S.lat.
N. Eurasia, North America.

13. *C. macloviana* D'Urv. 1825, *Fl. Is. Mal.* 28.

C. festiva Dewey 1836, *Amer. Jour. Sci. Arts* **29**: 246 – Franchet, 1889; Spegazzini, 1896.
C. propinqua Nees & Meyen 1834, in Meyen, *Reise Erde* **1**: 116 – Kurtz, 1896.
C. ovalis auct., non Gooden. – Hooker, 1847.

Rhizomes short. Leaves 3.5–10 cm × 2–3 mm; lamina flat or somewhat keeled near base, scabrid on margin and midvein beneath, bright green; ligule 1.5–2.5 mm, subacute to obtuse; lower sheaths leafless, brown. Culms 7–30 cm, trigonous, scabrid on angles distally. Bracts shorter than spike, resembling glumes but usually with short scabrid seta. Inflorescence 10–25 mm, ± ovoid, spikes 4–9, congested, with female spikelets distally. Male glumes 3–3.5 mm, ovate, acute to shortly acuminate or subobtuse, hyaline, brownish, with pale, often green, midvein and paler margins; female glumes similar but often ovate-lanceolate. Utricle 3–4.5 mm, longer and wider than glumes, elliptic-ovate, narrowed to beak, plano-convex, with narrow wing, ciliate and serrulate on margins distally, light brown; beak 1–2.5 mm, bifid, scabrid. *Flowers November to January.*

Open grassland, lake and stream margins, boggy areas; 0–500 m.

Throughout Isla Grande except the extreme west. (Map 40a).
W. Argentina north to *c.*27°S.lat.
Chile north to *c.*37°S.lat.
Bolivia. Falkland Islands (Islas Malvinas).
North America, arctic-alpine Eurasia.

14. *C. gayana* Desv. 1853, in C. Gay, *Hist. Chile Bot.* **6**: 224.

Similar to *C. macloviana* but spikes 6–*c.*30, usually unisexual, rarely the female with male spikelets distally; utricle 2–2.4 mm, shorter than glumes, orbicular-oblong, with beak *c.*0.2 mm. *Flowers November to February.*

Wet grassland, lake and river margins; 0–250 m.

Throughout Isla Grande except extreme west and east. (Map 40b).
W. Argentina and Chile north to *c.*33°S.lat.
W. North America.

Subgenus PRIMOCAREX Kük.
Spike solitary, terminal, unisexual or bisexual. Bracts present or absent. Stigmas 2 or 3. Rhacheole often present.

15. *C. microglochin* Wahlenb. 1803, *Kungl. Svenska Vet. – Akad. Handl.* **24**: 140.

C. camptoglochin Krecz. 1937, *Not. Syst. (Leningrad)* **7**: 34 – Kukkonen, 1970.
C. incurva auct., non Lightf. – Dusén, 1900.
Uncinia microglochin (Wahlenb.) Sprengel 1826, *Syst. Veg.* **3**: 830 – Franchet, 1889.

Rhizome short, slender. Leaves 3–9 cm × *c.*0.5 mm; lamina flat to filiform, shallowly channelled above, ± obtuse, rigid, smooth; ligule *c.*1 mm, subacute to truncate; lower sheaths leafless or almost so, brown. Culms 7–15 cm, subterete, channelled towards base, ridged distally. Spike 6–8 × *c.* 4–5 mm, cylindrical to subovoid, 3- to 6-flowered, bisexual, male distally. Male glumes *c.*3 mm, lanceolate, subacute to obtuse, reddish brown, with pale midrib and pale hyaline margins; female glumes rather shorter, oblong-ovate, obtuse, reddish brown, with pale midrib, with hyaline margins and apex, caducous. Stigmas 2. Utricle 4–6 mm, narrowly conical and rather curved, obscurely ribbed, brownish yellow, with stout, smooth, brown rhacheole projecting for up to 2 mm beyond the top of the utricle; pedicel short and deflexed when fruit ripe. $2n = 58$. *Flowers November to January.*

Sphagnum bogs, *Rostkovia* communities, wet areas by streams and lakes; 0–600 m.

C. & S. Fuegia. (Map 40c).
W. Argentina north to *c.*30°S.lat.
Chile north to 49°S.lat.
Falkland Islands (Islas Malvinas).
North America, N. Eurasia.

Southern American material belongs to ssp. *fuegina* Kük. 1899, *Bot. Jahrb.* **27**: 546, which differs most conspicuously from ssp. *microglochin* in its larger utricle and longer pedicels (Moore and Chater, 1971).

16. *C. caduca* Boott 1867, *Illustr.* **4**: 157.

C. ortegae Phil. 1896, *Anal. Univ. Chile* **93**: 488.
C. trichodes auct., ? non Steudel – Alboff, 1896.

Rhizome short. Leaves 3–20 cm × *c.*1 mm; lamina rather filiform, channelled, curved, rigid, scabridulous on margins; ligule *c.*0.5 mm, rounded; lower sheaths leafless, brown or reddish brown. Culms 6–28 cm, terete below, subtrigonous distally, shallowly ridged. Spike 6–10 × 5–7 mm, ovoid-globose, 10– to 16– flowered, usually male distally, subtended and exceeded by leaf-like bract. Male glumes 3–3.5 mm, ovate-lanceolate, subacute to obtuse, shortly apiculate, hyaline, reddish brown

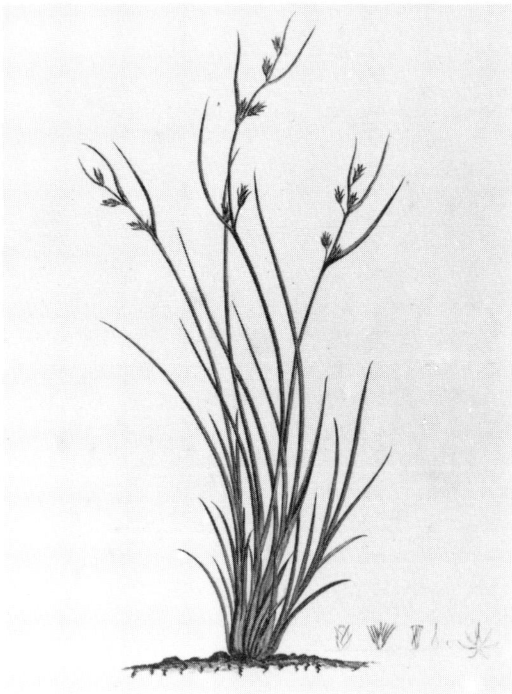

266 267

Fig. 266. *Carpha alpina*. Fig. 267. *Astelia pumila*.

(Sydney Parkinson)

with paler midvein; female glumes often caducous, similar to male but usually wider and with prominent, pale, scabrid, terminal awn. Stigmas 2. Utricle 3–3.5 mm, larger than glumes, ovoid, tapering to beak, plano-convex, smooth, greenish brown; beak emarginate, smooth or slightly scabrid; pedicels *c.*1 mm., erect. *Flowers December to March.*

Wet rock-crevices, stream-margins, bogs; (0–)100–800 m.

C. & S. Fuegia except the extreme west, south and east. (Map 40d).
W. Argentina at *c.*49°S.lat.
Chile north to 50°55′S.lat.
Falkland Islands (Islas Malvinas).

17. *C. sorianoi* Barros 1949, *Darwiniana* **9**: 166.

Similar to *C. caduca* but leaves up to 1.5 mm wide, semi-cylindrical at base, flat distally; culms not more than 6 cm; spike 4- to 6- flowered, scarcely exceeded by ± glume-like lower bract; utricle ellipsoid, hispid-scabrid on margin. *Flowers December to January.*

Coastal sands and saline soils; 0–20 m.

N.E. Isla Grande. (Map 40e).
W. Argentina north to *c.*46°S.lat.

18. *C. vallis-pulchrae* Phil. 1896, *Anal. Univ. Chile* **93**: 487.

Similar to *C. caduca* but culms and leaves not exceeding 5 cm; leaf-margin smooth; awn of basal glumes not scabrid; stigmas 3; utricle *c.*2 mm, the beak never scabrid. *Flowers November to January.*

N. Isla Grande, also S. coast. (Map 40f).
W. Argentina north to *c.*52°30′S.lat.
E. Chile *c.* 33–36°S.lat.
Falkland Islands (Islas Malvinas).

19. *C. capitata* L. 1759, *Syst. Nat.* **2**: 1261.
C. antarctogena Roivainen 1954, *Ann. Bot. Soc. Zool. Bot. Fenn. Vanamo* **28(2)**: 197.

Plant densely caespitose. Leaves 5–14 cm × *c.*0.5 mm, setaceous; basal sheaths purplish brown; ligule *c.*0.2 mm, obtuse. Culms 10–40 cm, trigonous, scabrid above. Bract absent. Spike 4–9 mm, ovoid to subglobose, many-flowered, dense, male distally. Female glumes 1.5–2.5 mm, ovate, obtuse, yellowish to reddish brown, with scarious margin; male glumes similar but narrower. Stigmas 2. Utricle

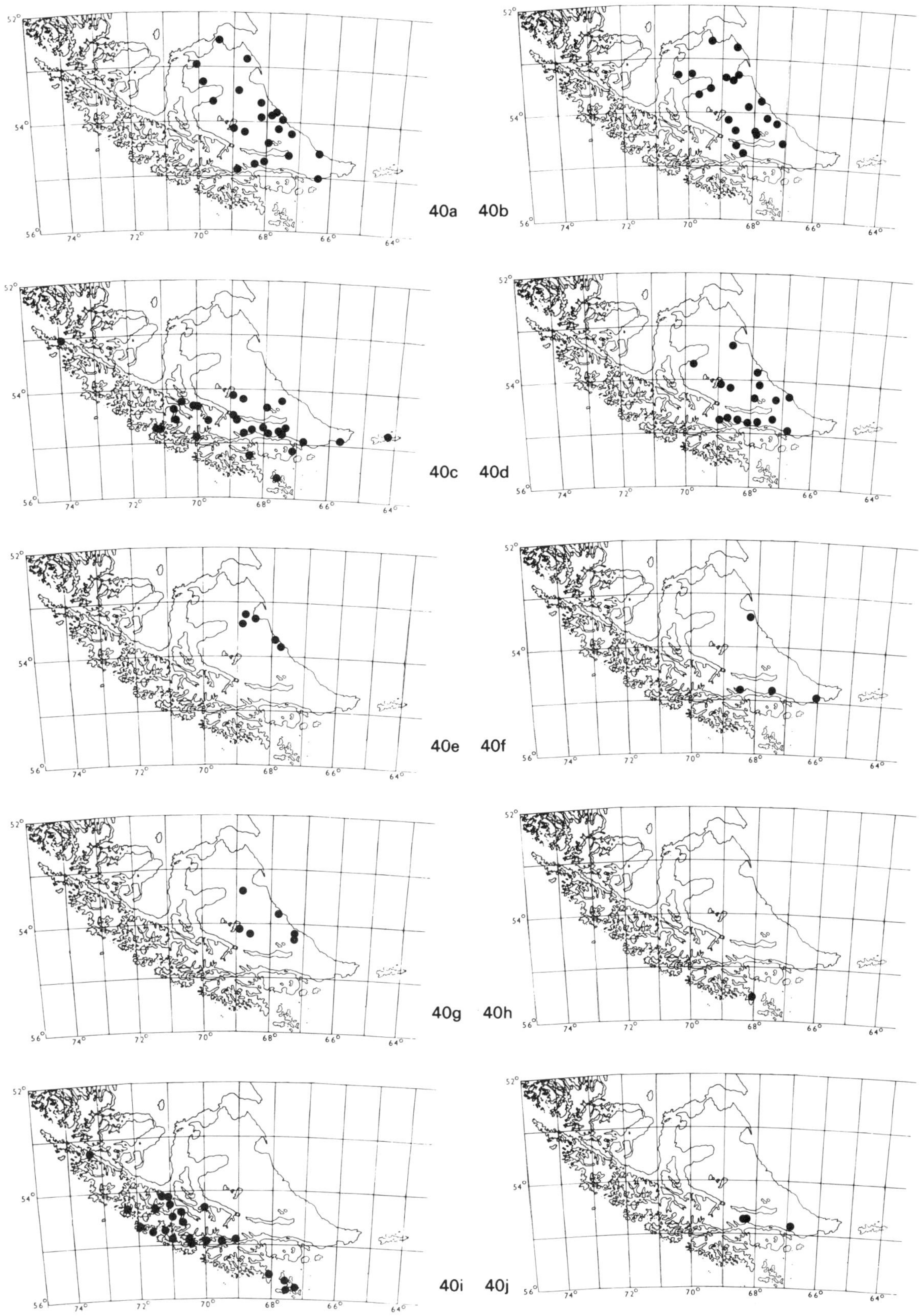

Map 40. a, *Carex macloviana*. b, *Carex gayana*. c, *Carex microglochin*. d, *Carex caduca*. e, *Carex sorianoi*. f, *Carex vallis-pulchrae*. g, *Carex capitata*. h, *Spirodela punctata*. i, *Tapeinia pumila*. j, *Tapeinia obscura*.

2–3.5 mm, ovoid, abruptly narrowing to beak, plano-convex, greenish, often brownish distally, smooth; beak *c.*0.5 mm, smooth or scabrid. *Flowers December to January.*

Wet grassland; 0–150 m.

N. & C. Isla Grande. (Map 40g).
W. Argentina north to *c.*40°S.lat.
E. Chile north to *c.*50°S.lat.
North America; N. Eurasia.

LXXXIII LEMNACEAE

Small aquatic perennial herbs. Stems and leaves not distinct, the plant constituting an undifferentiated frond, with or without simple roots. Inflorescence minute, consisting of 2 stamens and 1 ovary borne in a sheath on dorsal surface of frond. Ovary 1-celled; style short, simple. Fruit a utricle; seeds 1–4, without endosperm.

Literature: C. den Hartog and F. van der Plas, *Blumea* **18**: 355–368(1970).

1. ***Spirodela*** Schleiden

Fronds floating, separate or 2–5 connected, flat to inflated, asymmetrical, with dorsal and ventral scales, veined; roots present. Ovary globose. Fruit ± globose.

1. ***S. punctata*** (C. A. Meyer) C. H. Thompson 1898, *Ann. Rep. Missouri Bot. Gard.* **9**: 28.
Lemna punctata C. A. Meyer 1818, *Prim. Fl. Esseq.* 262.

Fronds 1–1.6 × 2–3.1 mm, elliptical to obovate-oblong, obtuse, rounded at base, thick, indistinctly 3-nerved, convex or flat above, flat beneath, both surfaces with irregular brown spots. Roots 1–3, up to 6 cm. Stamens with long filaments. Style short; stigma capitate. Fruit *c.*1–1.2 mm in diameter; seeds longitudinally ribbed.

Stationary fresh water.

S. Fuegia (Bahía Orange, Isla Hoste). (Map 40h).
?Guyana.

LXXXIV IRIDACEAE

Rhizomatous, perennial herbs. Leaves equitant, ensiform to linear. Flowers actinomorphic, hermaphrodite, 1-several in terminal inflorescence. Perianth-segments in 2 whorls of 3, free or united at base, petaloid, usually with hypanthial tube. Stamens 3, free or the filaments united in a tube or column. Ovary inferior, 3-celled; style 3-lobed, the stigmatic branches simple or divided. Fruit a capsule; seeds numerous, with endosperm.

a. Flowers solitary; ovules and seeds grouped in lower part of cells **1. *Tapeinia***
a. Flowers 2 or more in cymes; ovules and seeds distributed throughout cells:
b. Perianth not more than 8 mm, the segments free, yellow with brown veins **2. *Sisyrinchium***
b. Perianth at least 16 mm, the segments united into conspicuous basal tube, white with purple veins to orange or reddish **3. *Phaiophleps***

1. ***Tapeinia*** Juss.

Rhizome oblique to vertical, short. Leaves distichous, equitant at least at base. Flowers solitary, subtended by 2-valved, herbaceous spathe. Perianth-segments united at base, the inner somewhat narrower than the outer. Filaments united into column for at least $\frac{1}{4}$ of length. Style – branches, entire. Ovules and seeds grouped in lower part of cells.

Literature: D. M. Moore *Bot. Not.* **124**: 82–86(1971).

a. Leaves green, with white basal sheath at least 4 mm wide; perianth at least 8 mm; filaments united almost to apex **2. *obscura***
a. Leaves glaucous, with brownish basal sheath up to 2.5 mm wide; perianth not more than 6 mm; filaments united in basal $\frac{1}{4}$ **1. *pumila***

1. ***T. pumila*** (Forster f.) Baillon 1894, *Hist. Pl.* **13**: 149.
Ixia pumila Forster f. 1789, *Comment. Gotting*, ser. 2, **9**: 21.
I. magellanica Lam. 1789, *Encycl. Méth. Bot.* **3**: 334.
Tapeinia magellanica (Lam.) J. F. Gmelin 1791, *Syst. Veg.* **2**: 108 – Hooker, 1847; Alboff, 1889; Spegazzini, 1896; Skottsberg, 1916.

Rhizome short, oblique to vertical, with branched, often ± woody, short stem above. Leaves 7–27 × 0.5–1.3 mm, ensiform, rigid, glabrous, glaucous, with brown, scarious basal sheath up to 2.5 mm wide. Peduncle 4–7 mm, up to 14 mm in fruit. Perianth-segments 4–6 × *c.*1.5 mm, oblanceolate, obtuse, white, united for basal $\frac{1}{6}$. Filaments united in basal $\frac{1}{6}$–$\frac{1}{4}$. Capsule 3.5–5 × 3–4.5 mm, obovoid, glabrous, brown; seeds 1.8–2 mm, oblong, brownish, rugulose. *Flowers November to January.* (Fig. 269).

Magellanic moorland, bogs, wet coastal rocks, wet feldmark; 0–400 m.

S. & W. Fuegia. (Map 40i).
Chile north to *c.*40°S.lat.

2. ***T. obscura*** (Cav.) D. M. Moore 1971, *Bot. Not.* **124**: 85.
Galaxia obscura Cav. 1788, *Monad. Class. Diss. Dec.* 341, t. 189 fig. 4.
Ona obscura (Cav.) Ravenna 1972, *Anal. Mus. Hist. Nat. Valparaiso* **5**: 99.

Rhizome very short, vertical, with simple or sparingly branched short stem above. Leaves 18–120 × 0.8–1 mm, linear-subulate, acuminate, glabrous, with pale scarious basal sheath 8–18 × 4–9 mm. Peduncle up to 10 mm, up to 90 mm in fruit. Perianth-segments 8–14 × 2.5–4.5 mm, oblanceolate, obtuse to truncate, shallowly emarginate and mucronulate, white or very pale yellowish, united for basal $\frac{1}{7}$–$\frac{1}{6}$. Filaments united almost to apex. Capsule 5–8 × 4–6 mm, oblong-ovoid, glabrous, brown; seeds 1.6–2 mm, globose to ovoid, somewhat compressed, rugulose, brown. *Flowers December to January*. (Fig. 271).

Wet areas, usually having good water movement, above timberline, with *Caltha*, grasses, byrophytes; 625–1000 m.

S. Fuegia, probably under-recorded. (Map 40j).
Chile north to *c.*49°S.lat.

2. *Sisyrinchium* L.

Rhizome oblique to vertical. Leaves distichous, equitant. Flowers in umbellate cymes of 2– several, each subtended by ± equally 2-valved, herbaceous spathe. Perianth-segments free, the inner somewhat narrower than the outer. Filaments united ± to apex, inserted at top of very short hypanthial tube. Style branches simple. Ovules and seeds distributed throughout cells.

Literature: P. F. Ravenna, in M. N. Correa (ed.) *Fl. Patag.* **2**: 172–182(1969).

1. ***S. patagonicum*** Phil. ex Baker 1892, *Handb. Irid* 126.
S. chilense auct., non Hooker – Dusén, 1900; Skottsberg, 1926.
S. graminifolium auct., non Lindley – Spegazzini, 1896; Dusén, 1900.
S. iridifolium auct., non Humb., Bonpl. & Kunth – Spegazzini, 1896.
S. junceum auct., non E. Meyer ex C. Presl – Alboff, 1896.
S. maculatum auct., ?non Hooker – Spegazzini, 1896.

Basal leaves 30–160 × 1–4 mm, linear-ensiform, ascending to erect, often ± glaucous. Stems 8–28 cm, with wide wings, erect, often ± arcuate, bearing 1–3 small cauline leaves which become bract-like above. Cymes 1–3, 2– to 5– flowered on peduncles up to 70 mm; pedicels up to 20 mm. Perianth-segments 6–8 × 1.5–2.5 mm, oblanceolate, obtuse, apiculate, yellow with brown veins. Capsule 5–6 × 4–5 mm, subglobose to oblong-ellipsoid or obovoid, glabrous to ± pubescent, pale brown; seeds *c.*1 mm, turbinate, angled, rugulose, black. $2n = 64$. Self-compatible, facultatively self-pollinated. *Flowers November to February*. (Fig. 272).

Grassland, dry *Empetrum* heath, open *Chiliotrichum* and *Berberis buxifolia* scrub, coastal sands; 0–100 m.

Almost throughout Isla Grande; islands south of Canal Beagle to *c.*55°S.lat. (Map 41a).
W. Argentina north to *c.*38°S.lat.
E. Chile north to *c.*50°30′S.lat.

A difficult genus which still needs considerable study. The present treatment of Fuegian material is provisional and, although not conspicuously variable, its relationship to the entities present elsewhere in southern South America remains in doubt.

3. *Phaiophleps* Rafin.

Rhizome very short, vertical. Leaves mostly basal. Flowers in 1(–3) terminal cymes enclosed by equally 2-valved spathe. Perianth-segments united into conspicuous basal tube. Filaments ± completely united. Style-branches simple, short.

Literature: D. M. Moore, S. C. Arroyo and M. C. Doggett, *Bot. Jour. Linn. Soc. London* **84**: 103–113(1982).

1. ***P. biflora*** (Thunb.) R. C. Foster 1939, *Contr. Gray Herb.* **127**: 43.
Symphyostemon biflorus (Thunb.) Dusén 1900, *Svenska Exped. Magellansl.* **3(5)**: 203 – Skottsberg, 1916.
S. narcissoides (Cav.) Miers ex Klatt 1881–82, *Linnaea* **31**:569 – Hooker, 1847; Spegazzini, 1896.
Solenomelus biflorus (Thunb.) Baker 1877, *Bot. Jour. Linn. Soc. London* **16**: 121 – Franchet, 1889.

Basal leaves 1–5, 40–220 × 1–2.5 mm, elliptical in cross-section, striate, including basal sheath 10–70 mm; flowering stems 9–70 cm, striate, with 1–3 cauline leaves 30–170 mm. Cymes 2– to 7– flowered, the spathe-valves 20–40 mm; peduncles 0–110 mm; pedicels 15–55 mm. Perianth-tube 7–20 mm, ± cylindrical to infundibuliform; perianth-lobes 9.5–18 × 5–7.5 mm, oblong-ovate, subacute to ± obtuse and apiculate. Staminal column 7.5–13 mm. Capsule 6–9 mm, globose to obovoid, glabrous, green-brown; seeds 2–2.2 mm, irregularly angular, reddish brown. *Flowers November to January*.

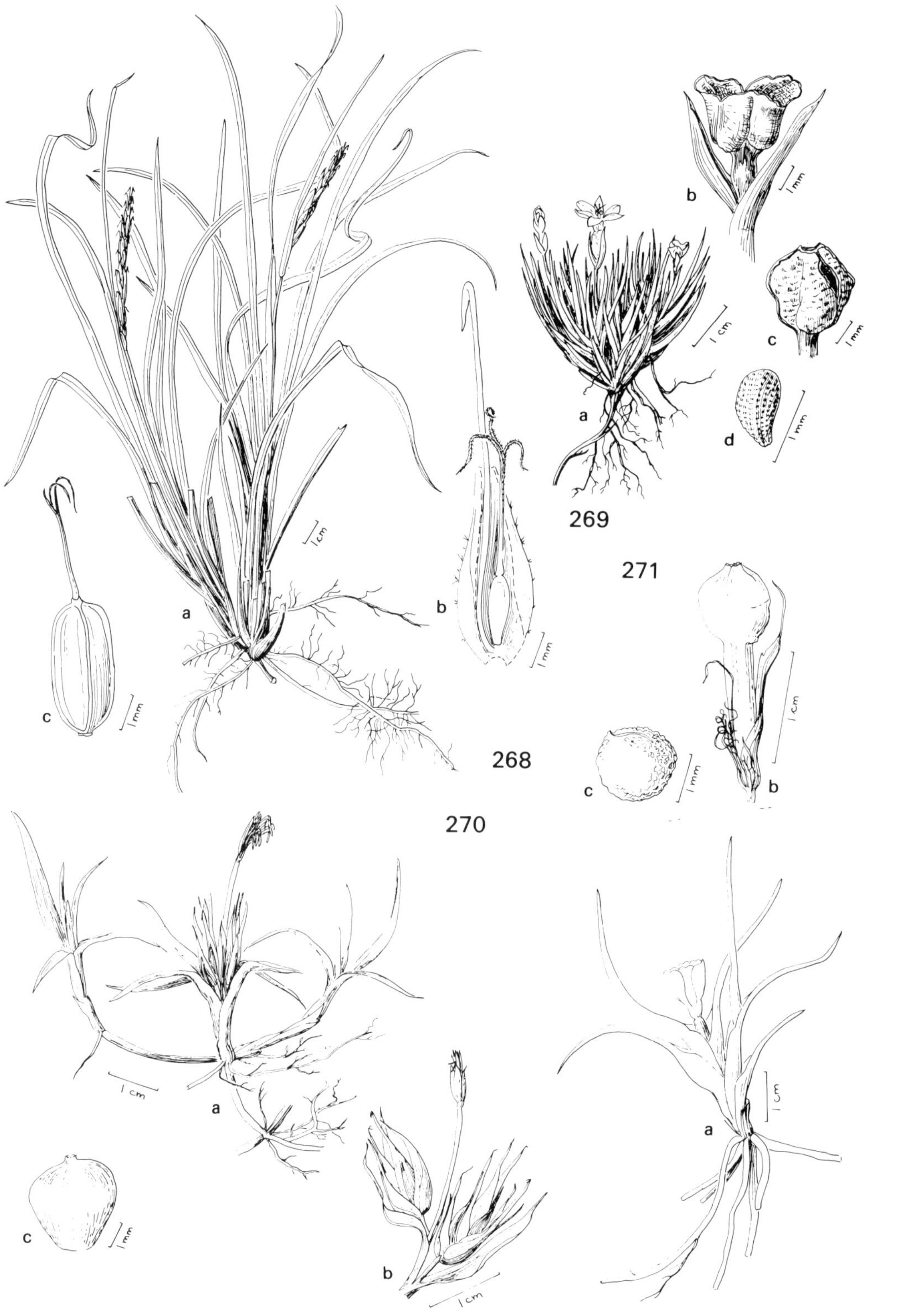

Fig. 268. *Uncinia lechleriana*. a, Plant. b, Female flower. c, Fruit. Scale = 1 cm (a) or 1 mm (b–c). Fig. 269. *Tapeinia pumila*. a, Plant. b, Dehisced capsule. c, Capsule. d, Seed. Scale = 1 cm (a) or 1 mm (b–d). Fig. 270. *Carex acaulis*. a, Plant. b, Inflorescence.c, Nut. Scale = 1 cm (a–b) or 1 mm (c). Fig. 271. *Tapeinia obscura*. a, Plant. b, Capsule. c, Seed. Scale = 1 cm (a–b) or 1 mm (c).

(R. N. P. Goodall)

Grassland, *Empetrum* heath, *Chiliotrichum, Berberis buxifolia* and *Pernettya* scrub, open soils and rock pavements, feldmark; 0–700 m.

Almost throughout Isla Grande and adjacent islands to south.
W. Argentina north to *c.*46°S.lat.
Chile north to *c.*50°40′S.lat.

Two rather distinctive sets of populations are found in this species. In Fuegia they differ in a number of morphological features and in habitat and have been recognized as different species (Dusén, 1900; Moore and Goodall, 1974). However, further north, in prov. Santa Cruz, Argentina, morphological and ecological intermediates occur and consequently the two entities have been recognized as altitudinal subspecies.

a. Ssp. ***biflora***: Ridges on stems and leaves usually straight, smooth. Peduncles (20–)30–110 mm. Perianth white or cream with purple veins; tube usually infundibuliform. $2n = 18$. (Fig. 273).

Grassland and open scrub; 0–150 m.

Throughout the range of the species. (Map 41b).

b. Ssp ***lyckholmi*** (Dusén) D. M. Moore 1982, *Bot. Jour. Linn. Soc. London* **84**: 112.
Symphyostemon lyckholmii Dusén 1900, *Svenska Exped. Magellansl.* **3(5)**: 201).

Ridges on stems and leaves usually undulate, tuberculate. Peduncles 0–33 mm. Perianth deep burgundy red or reddish brown to orange or yellowish orange; tube usually cylindrical. (Fig. 274).

Open stony soils and rock pavements; 150–700 m.

C. & C.N. Isla Grande. (Map 41c).
W. Argentina north to *c.*48°50′S.lat.
E. Chile north to *c.*50°38′S.lat.

LXXXV LILIACEAE

Perennial herbs, usually with swollen rhizomes or bulbs, or shrubs. Leaves alternate or whorled to basal, simple; stipules absent. Flowers actinomorphic, hermaphrodite or unisexual. Perianth-segments in 2 whorls of 3, petaloid. Stamens 6. Ovary superior, 3-celled; style 1 or absent; stigmatic lobes 3. Fruit a berry or capsule; seeds with endosperm.

a. Plant with prominent aerial stems; leaves alternate:
 b. Perianth-segments subequal, up to 16 mm, white. **2. *Luzuriaga***
 b. Inner perianth-segments much longer than the outer, at least 40 mm, usually pink **1. *Philesia***
a. Plant without prominent aerial stems; leaves basal or in imbricate whorls:
 c. Plant dioecious, forming compact carpets, with swollen rhizomes; leaves rigid, linear- to triangular-lanceolate **3. *Astelia***
 c. Plant with hermaphrodite flowers, not carpet-forming, with bulbs; leaves linear or fistular, flaccid:
  d. Perianth white or pinkish; flowers in ± globose inflorescence; leaves fistular **5. *Allium***
  d. Perianth dark purple-brown and green; flowers in lax umbel-like cyme; leaves linear **4. *Tristagma***

1. *Philesia* Comm. ex Juss.

Evergreen shrub. Leaves alternate, distichous, coriaceous. Flowers hermaphrodite, solitary, pendent. Perianth-segments free, the outer much shorter than the inner. Filaments united for *c.*$\frac{2}{3}$ length. Fruit a berry; seeds numerous.

1. *P. magellanica* J. F. Gmelin 1792, *Syst. Nat.*, ed. 13, **2**: 1012.
P. buxifolia Lam. 1819, *Tabl. Encycl. Méth. Bot.* **2** 387 – Hooker, 1847; Franchet, 1889; Dusén, 1900.

Stems up to 3 m, 2–4 mm in diameter, creeping and with adventitious roots below, suberect above, much branched, often scrambling up tree-trunks. Leaves with lamina 10–35 × 2–6 mm, oblong, obtuse, mucronate, cuneate at base, entire, revolute at margin, glaucous beneath, dark shiny green above; petiole 2–5 mm, ± twisted. Peduncle short, with up to 8 small brownish bracts. Perianth campanulate, pink, rarely white; inner segments 40–50 × 12–28 mm, oblanceolate-spathulate, obtuse, the outer 14–18 × 4–6 mm, oblong-lanceolate, obtuse, often greenish. Style 32–42 mm, filiform, somewhat exceeding stamens, exserted; stigma capitate, shallowly 3-lobed. Berry 12–14 × 9–11 mm, subglobose to ellipsoid or ovoid, apiculate, green-brownish; seeds *c.*3 mm, ovoid, brown. $2n = 12$. *Flowers November to February*. (Fig. 275, Pl. 8c).

Drimys, Nothofagus betuloides and *Pilgerodendron* forest; 0–100 m.

S. & W. Fuegia. (Map 41d).

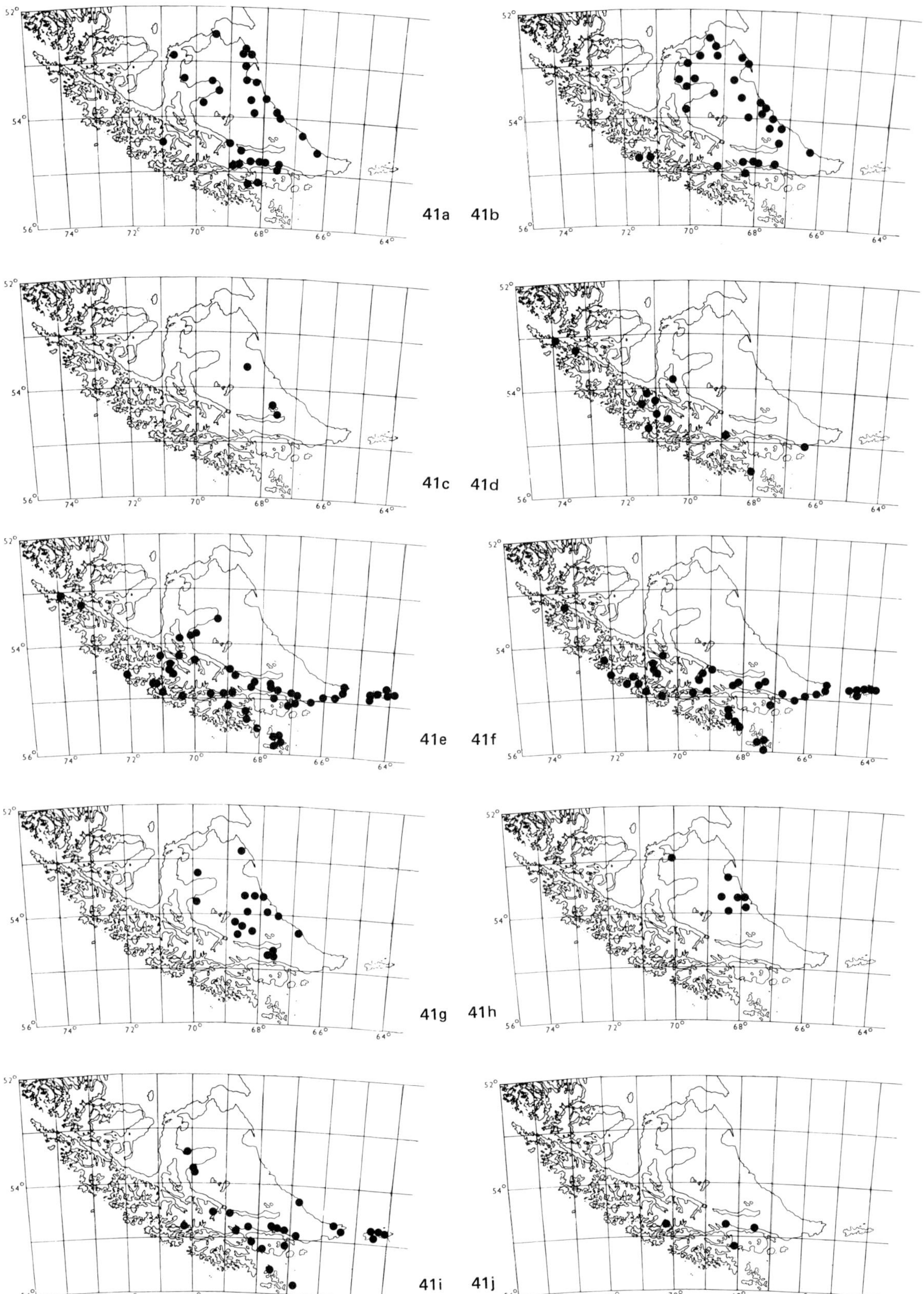

Map 41. a, *Sisyrinchium patagonicum*. b, *Phaiophleps biflora* ssp. *biflora*. c, *Phaiophleps biflora* ssp. *lyckholmi*. d, *Philesia magellanica*. e, *Luzuriaga marginata*. f, *Astelia pumila*. g, *Tristagma nivale*. h, *Alstroemeria patagonica*. i, *Codonorchis lessonii*. j, *Gavilea australis*.

W. Argentina at *c.*49–50°S.lat.
Chile north to *c.*42°S.lat.

2. *Luzuriaga* Ruiz & Pavón

Prostrate perennial herbs, woody at base. Leaves alternate, distichous. Flowers hermaphrodite, solitary (rarely 2–3), terminal on axillary bracteate peduncles. Perianth-segments subequal, free. Filaments free. Fruit a berry; seeds few.

1. *L. marginata* (Banks & Sol. ex Gaertner) Bentham & Hooker f. 1883, *Gen. Pl.* **3(2)**: 768.
Enargea marginata Banks & Sol. ex Gaertner 1788, *Fruct. Sem. Pl.* **1**: 283 – Spegazzini, 1896; Skottsberg, 1916.
Callixene marginata (Banks & Sol. ex Gaertner) Lam. 1819, *Tabl. Encycl. Méth. Bot.* **2**: 387 – Hooker, 1847; Franchet, 1889; Alboff, 1896; Dusén, 1900.

Stems up to 2 m, *c.*2–3 mm in diameter, creeping, branched, woody, rooted at lower nodes, glabrous, bearing small, triangular, brown scales below terminal leafy part. Leaves 6–22 × 2–8 mm, oblong or elliptical to elliptic-lanceolate, acuminate, apiculate, cuneate at base, entire, rigid, glabrous, setulose-spinulose on margin, pale green and shiny above, duller and paler beneath; petiole 0.5–3 mm. Peduncle *c.*4 mm, glabrous, with 2–4 bracts clustered at base. Flowers sweetly scented. Perianth shallowly campanulate, white; segments 10–16 × 5–9 mm, oblong to oblong-lanceolate, obtuse, often with 2 green spots at base. Style 3–5 mm, filiform, somewhat exceeding stamens; stigma capitate, shallowly 3-lobed. Berry 7–10 mm in diameter, globose, apiculate, dark purple; seeds *c.*2.5 mm, oblong-ovoid, slightly compressed. $2n = 20$*. *Flowers January to March.* (Fig. 277).

Forest floor, wet grassland and scrub, moist rock-crevices and stream-sides; 0–400 m.

Almost throughout Fuegia except N. & E. Isla Grande. (Map 41e).
W. Argentina north to *c.*48°S.lat.
Chile north to *c.*50°50′S.lat.
Falkland Islands (Islas Malvinas).

3. *Astelia* Banks & Sol. ex R.Br.

Dioecious, perennial herbs. Leaves in several closely imbricate whorls. Flowers bracteate, solitary or few in racemose panicle terminal on axillary peduncle. Perianth-segments subequal, shortly united at base. Filaments free. Male flowers with aborted ovary, the female with staminodes. Stigma shortly 3-lobed. Fruit a berry; seeds numerous.

1. *A. pumila* (Forster f.) Gaudich. 1825, *Ann. Sci. Nat.* **5**: 101.
Melanthium pumilum Forster f. 1789, *Comment. Gotting.*, Ser. 2, **9**: 30.

Plant forming compact carpets up to several square metres in area. Stems up to 5 cm, branched, with a few long silky hairs and leaf-scars below, densely leafy distally, reddish brown. Leaves 10–23 × 3–7 mm, linear- to triangular-lanceolate, acute, entire, rigid, thick, fimbriate on margins and midrib beneath, dark shiny green above, paler and duller beneath, with imbricate somewhat sheathing base which is densely covered with long silvery white linear scales at base. Peduncle 10–16 mm, erect, axillary, simple or branched from about middle, with minute triangular or linear-triangular, silvery, scarious scales. Flowers 1–3, terminal, subtended by greenish silvery bract *c.*5 × 1 mm. Perianth-segments 3–5 × 1–1.4 mm., ± oblong, obtuse, creamy white. Stamens about equalling perianth-segments, the staminodes *c.* half as long, green. Style slightly exceeding perianth-segments. Fruit *c.*7 mm., ovoid to sublgobose. $2n = 64$. *Flowers November to February.* (Fig. 267).

Wet areas, often dominant or co-dominant to form conspicuous patches; 0–600 m.

C. & S. Fuegia. (Map 41f).
Chile north to *c.*40°S.lat.
Falkland Islands (Islas Malvinas).

4. *Tristagma* Poeppig

Slender herbs with tunicate bulb. Leaves basal. Flowers hermaphrodite, (1–)2–6 in umbel-like cyme, enclosed by 2-valved spathe terminal on leafless scape. Perianth-segments subequal, united into tube at base. Stamens in 2 whorls, those opposite the outer perianth-segments inserted in the perianth-tube above those opposite the inner segments. Stigma capitate, obscurely 3-lobed. Fruit a loculicidal capsule; seeds numerous.

1. *T. nivale* Poeppig 1833, *Fragm. Syn. Pl. Phan.* 9.
T. australe Neger 1900, in Dusén, *Svenska Exped. Magellansl.* **3(5)**: 207 – Skottsberg, 1916.

Bulb 12–32 × 6–16 mm, oblong-ovoid. Leaves 3–8, 30–300 × 1–4 mm, linear, subobtuse, entire, often falcate or spiral distally. Scapes 4–23 cm, slender, erect; spathe-valves 12–22 mm, linear-lanceolate, acuminate, ± united at base, scarious, white to purplish; pedicels 5–25 mm, slender, ± nodding. Perianth dark purple-brown and green; tube 8–12

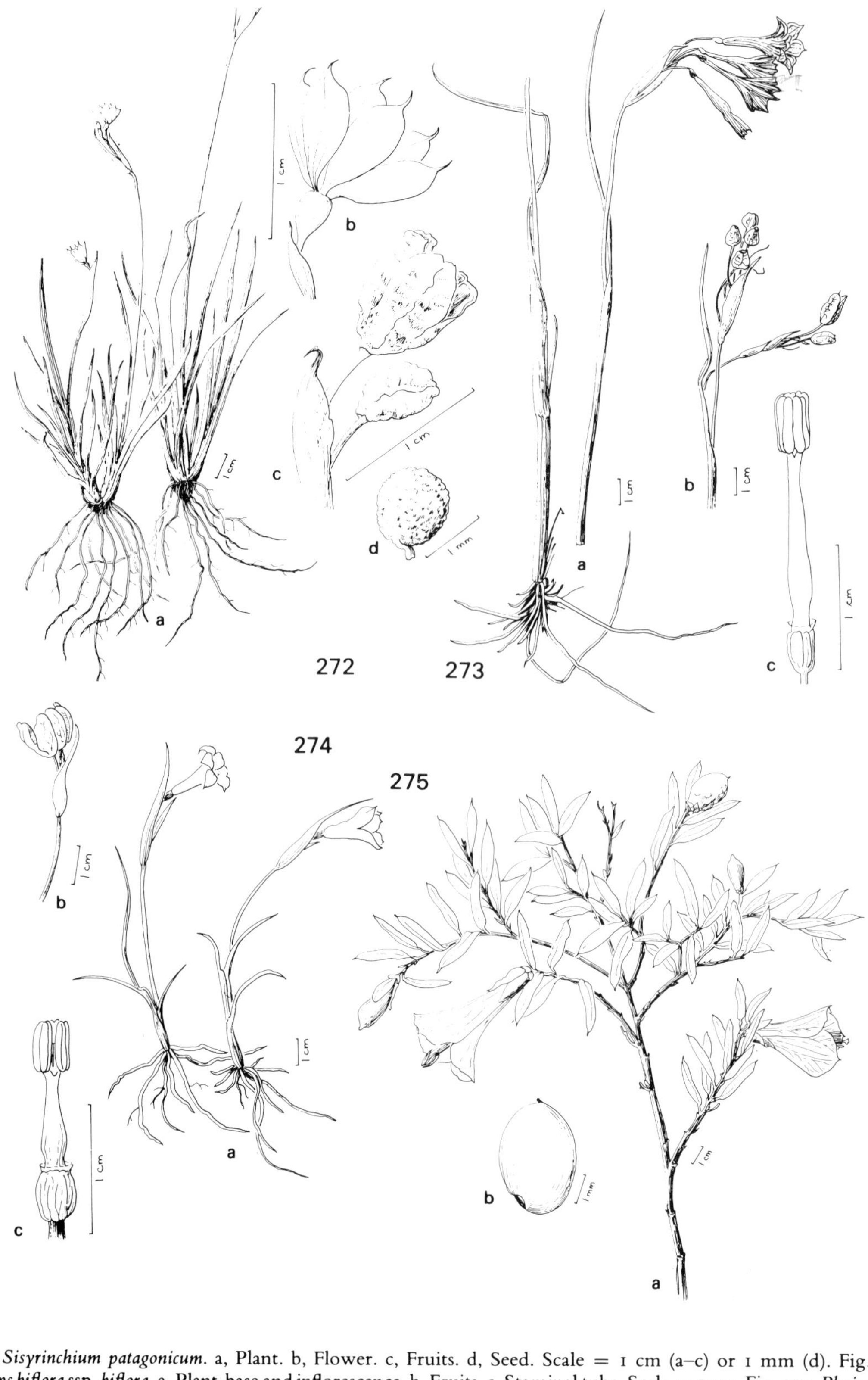

Fig. 272. *Sisyrinchium patagonicum*. a, Plant. b, Flower. c, Fruits. d, Seed. Scale = 1 cm (a–c) or 1 mm (d). Fig. 273. *Phaiophleps biflora* ssp. *biflora*. a, Plant-base and inflorescence. b, Fruits. c, Staminal tube. Scale = 1 cm. Fig. 274. *Phaiophleps biflora* ssp. *lyckholmi*. a, Plants. b, Fruit. c, Staminal tube. Scale = 1 cm. Fig. 275. *Philesia magellanica*. a, Branch with flowers and fruit. b, Seed. Scale = 1 cm (a) or 1 mm (b).

(R. N. P. Goodall)

mm, cylindrical to ellipsoid; segments 3–10 × 0.5–1.2 mm, linear, obtuse, ± revolute. Upper filaments 0.8–1.4 mm, inserted *c*.3.5 mm from base of perianth-tube, the lower filaments somewhat longer, inserted 1–2 mm from base. Style short. Capsule 7–10 × 6–7 mm, subglobose to ovoid, glabrous, brown; seeds 2–2.5 mm, oblong-ellipsoid, angular, black. *Flowers November to January*. (Fig. 276).

Open sands, rock and soil screes, less commonly in grassland; 0–800 m.

Much of Isla Grande except the extreme west and east. (Map 41g).
W. Argentina north to *c*.40°S.lat.
E. Chile north to *c*.37°S.lat.

5. *Allium* L.

Bulbous herbs, smelling of onion; bulbs tunicate, clustered. Leaves basal, fistular. Flowers hermaphrodite, 8–30 in umbel enclosed by spathe terminal on leafless scape. Perianth-segments subequal, free or slightly united at base. Stamens free, inserted at base of perianth-segments. Stigma capitate. Fruit a loculicidal capsule; seeds few.

1. *A. schoenoprasum L. 1753, *Sp. Pl.* 301.
Bulbs 5–10 mm in diameter, narrowly ovoid. Leaves 1–2, up to 280 × 1–4 mm, cylindrical, sheathing in lower ⅓. Scape up to 37 cm, erect; spathe up to 13 mm, with 2–3 ovate, purplish lobes; umbel 13–25 mm in diameter, globose-ovoid; pedicels 2–15 mm. Perianth-segments 7–15 × 2.5–4 mm, ± lanceolate, acute to acuminate, purplish, rarely white. Capsule *c*. 4 mm in diameter, subglobose to ovoid; seeds angular, black. *Flowers December (? later)*.

Disturbed and open moist ground; 0–120 m.

Locally in S. Isla Grande.
Introduced from Europe and cultivated for its edible leaves (chives); persisting in abandoned gardens and locally naturalized near habitations.

LXXXVI AMARYLLIDACEAE

Perennial herbs with slender rhizomes and tuberous roots. Leaves in rosettes, simple; stipules absent. Flowers zygomorphic, hermaphrodite. Perianth-segments in 2 whorls of 3, petaloid. Stamens 6. Ovary inferior, 3-celled; style 1; stigma 3-fid. Fruit a capsule; seeds with endosperm.

1. *Alstroemeria* L.

Flowers 1(2), terminal. Perianth-segments free, the inner somewhat longer than the outer and with the upper more prominent. Filaments curved downwards. Capsule loculicidal.

1. *A. patagonica* Phil. 1894, *Anal. Univ. Chile* **84**: 160.
A. pygmaea auct., non Herbert – Spegazzini, 1896; Dusén, 1900.

Stems 5–40 mm, with white, linear scales below and rosette of leaves at apex. Leaves 10–20 × 1–3 mm, oblanceolate, obtuse, crispate-undulate, glabrous. Peduncle *c*.5 mm. Perianth-segments golden yellow; inner 16–19 × 5–6.5 mm, oblanceolate, acuminate, apiculate, the upper with prominent red spots in lower ⅔; outer 15–20 × 2–5 mm, linear-oblanceolate to oblanceolate, acute. Stamens and style 9–11 mm. Capsule 9–15 × 10–12 mm, globose to globose-cylindrical, pale brown; seeds 1.5–2 mm in diameter, globose, rugose, brown. $2n = 16$. *Flowers November to January*. (Fig. 278).

Coastal sands, open dry grassland; 0–20 m.

N. & E.C. Isla Grande. (Map 41h).
Argentina north to *c*.39°S.lat., especially in the west.
E. Chile north to *c*.50°S.lat.

LXXXVII ORCHIDACEAE

Rhizomatous perennial herbs. Leaves spirally arranged or in a whorl, simple; stipules absent. Flowers hermaphrodite, zygomorphic. Perianth-segments in 2 whorls of 3, petaloid; median inner segment (*labellum*) usually larger than and of different shape from the others, usually directed downwards owing to the ovary or pedicel twisting through 180°. Anthers and stigma borne on a *column* formed from fused filaments and style; stamen 1, with sessile 2-celled anther at summit of column. Ovary inferior, 1-celled; stigmas 3, 1–2 fertile, the other forming a ± beak-like process (*rostellum*) between anther and fertile stigma(s). Fruit a capsule, opening by 3 longitudinal slits; seeds numerous, minute, without endosperm.

a. Leaves in a whorl at or above middle of stem; flower solitary, terminal **1. *Codonorchis***
a. Leaves spirally arranged up stem; flowers in a terminal spike:
 b. Labellum distinctly 3-lobed; column not more than 7 mm **2. *Gavilea***
 b. Labellum entire or obscurely 3-lobed; column at least 12 mm **3. *Chloraea***

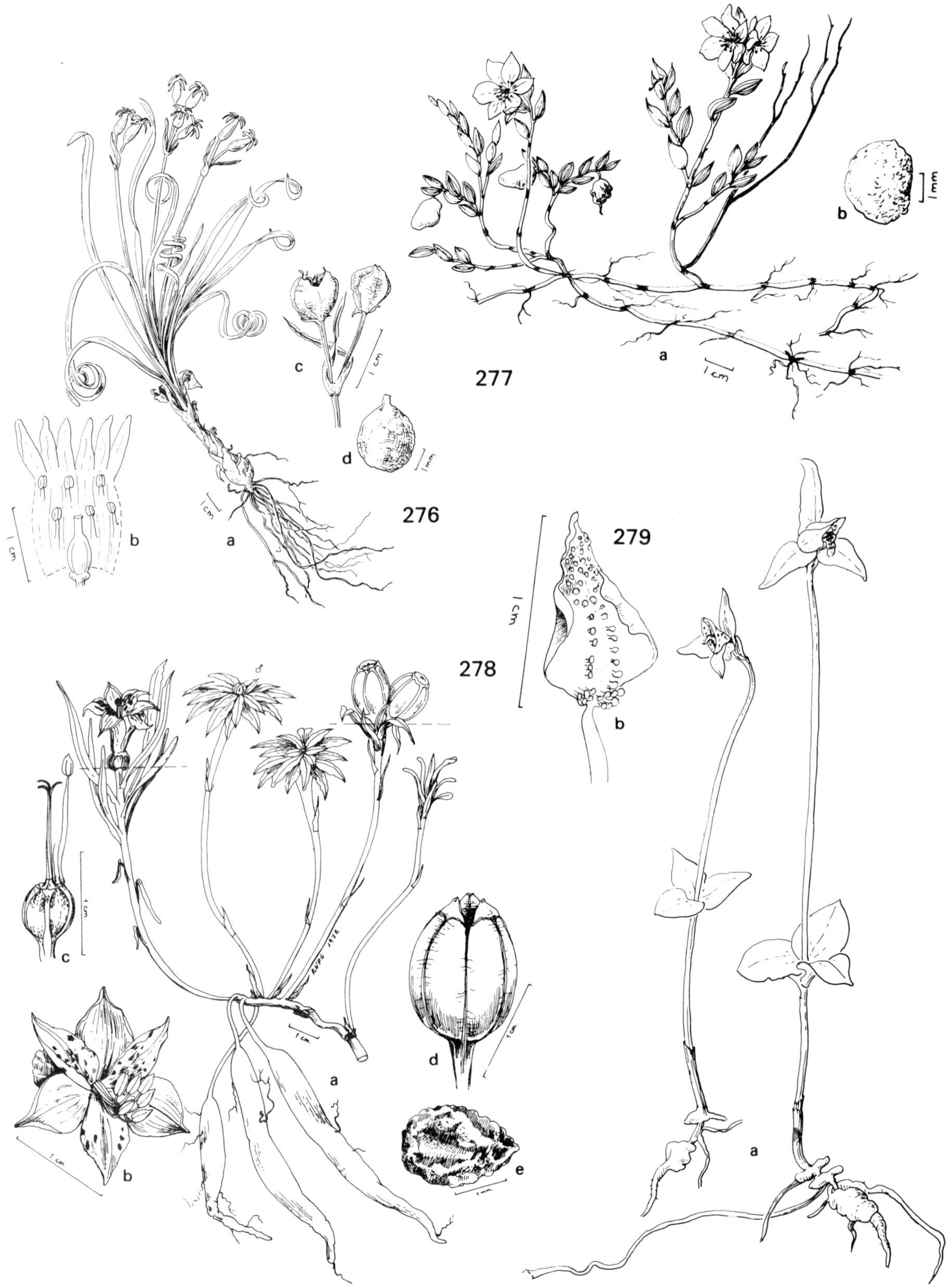

Fig. 276. *Tristagma nivale*. a, Plant. b, Opened perianth. c, Capsules. d, Seed. Scale = 1 cm (a–c) or 1 mm (d). Fig. 277. *Luzuriaga marginata*. a, Plant. b, Seed. Scale = 1 cm (a) or 1 mm (b). Fig. 278. *Alstroemeria patagonica*. a, Plant. b, Flower. c, Gynoecium and stamen. d, Capsule. e, Seed. Scale = 1 cm (a–d) or 1 mm (e). Fig. 279. *Codonorchis lessonii*. a, Plants. b, Labellum. Scale = 1 cm.

(R. N. P. Goodall)

1. *Codonorchis* Lindley

Plant glabrous, with tuberous rhizome. Leaves in a whorl of (2)3(4) at or above middle of stem, entire. Flower solitary, terminal, subtended by 1 bract. Outer perianth-segments equal, free, spreading, the inner much shorter, erect; labellum entire, free. Column long; fertile stigmas 2.

1. *C. lessonii* (D'Urv.) Lindley 1840, *Gen. Spec. Orchid.* 411.

Calopogon lessonii (D'Urv.) Brongn. 1829, *Voy. Coquille*, 188.
Epipactis lessonii D'Urv. 1825, *Fl. Is. Mal.*, 36.

Stem 6–35 cm, erect., or somewhat nodding in fruit, fleshy, bearing pinkish, membranous scale-leaves in basal half. Leaves (9–)18–25 × (6–)12–25 mm, broadly ovate, subacute, cuneate at base, shiny green, sessile or subsessile. Bract 7–10 × 2–6 mm, ovate, acute, equalling ovary. Outer perianth-segments 8–22 × 3.5–8 mm, ovate-lanceolate, acute, white, often with greenish mid-veins, and purplish towards base. Lateral inner perianth-segments 6–17 × 3–11 mm, ovate-lanceolate, acute or obtuse, of similar colour to outer segments but usually with conspicuous purplish blotches. Labellum 5–16 × 3–10 mm, ovate-lanceolate, subacute or subacuminate to obtuse, white, often purplish towards margin, with prominent, greenish stipitate glands distally and on each side of central vein towards base. Column 9–13 mm. Capsule 12–14×8–10 mm, cylindrical to cylindrical-ellipsoid, dark purplish brown. *Flowers November to March.* (Fig. 279).

Nothofagus forest, open moist scrub of *Chiliotrichum*, *Berberis buxifolia* and *Nothofagus antarctica*, rarely in moist grassland among rocks; 0–350 m.

C. & S. Fuegia except the west. (Map 411).
W. Argentina north to *c.*40°S.lat.
Chile north to *c.*39°S.lat.
Falkland Islands (Islas Malvinas).

2. *Gavilea* Poeppig

Plant glabrous, with short rhizome and fleshy roots. Leaves spirally arranged, with overlapping bases, entire. Flowers subtended by leaf-like bracts, in terminal spike. Outer perianth-segments free, membranous, spreading, the upper ± concave, the inner lateral similar to the outer; labellum distinctly 3-lobed. Column short; fertile stigma 1.

Literature: C. J. F. Skottsberg, *Acta Horti Gothob.* **1**: 211–224(1924). M. N. Correa, *Bol. Soc. Argent. Bot.* **6**: 73–86(1956); *Fl. Patag.* **2**: 191–199(1969).

a. Labellum wider than long; lateral outer perianth-segments acuminate, not thickened **3. *littoralis***
a. Labellum as long as or longer than wide; lateral outer perianth-segments ± abruptly narrowed and thickened distally:
b. Flowers white or cream with green veins; lateral inner perianth-segments papillose on veins towards base; labellum at least 8 mm **1. *australis***
b. Flowers yellow with green papillae and apices to outer perianth-segments; lateral inner perianth-segments without papillae on veins; labellum not more than 6 mm **2. *lutea***

1. *G. australis* (Skottsb.) Correa 1956, *Bol. Soc. Argent. Bot.* **6**: 77.

Asarca australis Skottsb. 1913, Kungl. Svenska Vet. Akad. Handl. **50(3)**: 22 – Skottsberg, 1924.

Stems 20–50 cm, sometimes leafy only in basal half. Basal leaves in rosette, 5–7 × 2.5–3 cm, ovate to ovate-oblong, acute, tapering to wide, sheathing base; cauline leaves narrower. Lower bracts 22–27 × 3–5 mm, linear-lanceolate, decreasing upwards. Inflorescence 5–10 cm; flowers 3–10, white or cream with green veins. Upper outer perianth-segment 12–14 × 6–7 mm, ovate, acuminate; lateral segments 15–16 × 3.5–7 mm, obliquely lanceolate, abruptly contracting at apex to thickened tube *c.*5 mm. Lateral inner perianth-segments 9–10 × 4–6 mm, ovate, obtuse, sparsely papillose on veins in basal half. Labellum 8–11 × 8–11 mm, as long as or slightly longer than wide, subcordate; lateral lobes suborbicular, with lacerate outer margin; median lobe 3.5–4 mm wide, oblong-ovate, subacute, with thickened and glandular-lacerate margin, papillose in basal half to two-thirds. Column 5–7 mm. Capsule ovoid-oblong. *Flowers November to January.* (Fig. 280).

Moist grassland and open scrub; 0–60 m.

S. Isla Grande; Isla Navarino. (Map 41j).
Chile along N. shore of Estrecho de Magallanes.
Falkland Islands (Islas Malvinas).

2. *G. lutea* (Pers.) Correa 1956, *Bol. Soc. Argent. Bot.* **6**: 78.

Asarca lutea (Pers.) Skottsb. 1916, *Kungl. Svenska Vet. Akad. Handl.* **56(5)**: 199.
Chloraea commersonii Brongn. 1829/34, *Voy. Coquille* t. 44B – Franchet, 1889; Alboff, 1896; Spegazzini, 1896; Autran, 1905.

Stem 20–60 cm. Basal leaves in rosette, 6–17 × 2–5

Fig. 280. *Gavilea australis*. a, Plant-base. b, Inflorescence. c, Flower. d, Labellum and column. Scale = 1 cm (a–c) or 1 mm (d).

(R. N. P. Goodall)

cm, ovate-lanceolate, obtuse, tapering to wide sheathing base; upper leaves smaller, becoming bract-like. Inflorescence 6–9 cm; flowers 7–18, bright yellow with green papillae, subtended by acuminate, lanceolate bracts 15–30 × 4–11 mm. Upper outer perianth-segment 11–15 × 3–6 mm, lanceolate, acute; lateral segments 14–16 × 3–5 mm, lanceolate, gradually narrowed to green thickened tube 3–4 mm. Lateral inner perianth-segments 11–14 × 2.5–5 mm, ovate, subobtuse, without papillae. Labellum 4–7 × 3–4 mm, longer than wide, ± ovate in outline; lateral lobes ± oblong, entire; median lobe 1.5–2 × 1–1.5 mm, oblong-triangular, obtuse, with glandular-lacerate margin and sparse papillae on veins. Column *c*.3.5 mm. Capsule ovoid-oblong. *Flowers November to February*. (Fig. 282).

Evergreen forest and scrub; 0–60 m.

C. & S. Isla Grande; Isla de los Estados; Isla Navarino. (Map 42a).
W. Argentina and E. Chile north to *c*.37°S.lat.

3. *G. littoralis* (Phil.) Correa 1969, *Flora Patag.* **2**: 195.

Asarca littoralis (Phil.) Reiche 1910, *Anal. Mus. Nac. Chile Bot.* **18**: 16 – Skottsberg, 1924.
A. macroptera Kraenzlin 1903, *Orchid. Gen. Spec.* **2**: 38 – Skottsberg, 1916.
A. odoratissima auct., non Poeppig & Endl. – Hooker, 1847.
Gavilea macroptera (Kraenzlin) Correa 1956, *Bol. Soc. Argent. Bot.* **6**: 82.
Chloraea magellanica auct., non Hooker f. *pro parte* – Alboff, 1896.

Stem 25–50 cm. Leaves 10–20 × 3–4 cm, oblong-lanceolate, subobtuse, tapering to wide sheathing base. Inflorescence 12–15 cm; flowers 10–14, yellow with green veins, subtended by lanceolate, acute bracts 10–22 × 2–6 mm. Upper outer perianth-segment 10–14 × 4–6 mm, oblong to lanceolate, acuminate; lateral segments 11–15 × 4–6 mm, obliquely oblong-lanceolate, rather abruptly acuminate. Upper inner perianth-segments 9–11 × 3–5.5 mm, oblong, obtuse or usually retuse. Labellum 8–9 × (6–)10–11 mm, usually wider than long, broadly ovate; lateral lobes ± oblong, obtuse, entire; median lobe 2–3 mm wide, oblong to lanceolate, obtuse, obscurely sinuate, densely papillose. Column *c*.5 mm. Capsule ovoid. *December to February*.

S. Isla Grande. (Map 42b).
Chile north to *c*.40°S.lat.
Falkland Islands (Islas Malvinas).

3. *Chloraea* Lindley

Like *Gavilea* but labellum entire or only obscurely 3-lobed; column long.

Literature: M. N. Correa, *Darwiniana* **15**: 374–500(1969).

1. *C. magellanica* Hooker f. 1847, *Fl. Antarct.* **1(2)**: 350.

C. bouganvilleana Franchet 1889, *Miss. Sci. Cap Horn* 366.

Stem 11–40 cm, erect, fleshy. Leaves 7–20 × 1–4 cm, lanceolate to ovate-lanceolate, acute, sheathing at base, entire, dark shiny green, the upper often pinkish. Inflorescence lax, with up to 5 bracteate flowers; bracts 25–35 × 7–12 mm, ovate, acute to acuminate, green, often whitish with green veins distally. Outer perianth-segments 25–35 × 5–11 mm, oblong to oblong-lanceolate, acute to subobtuse, often slightly cucullate at apex, white to pale green with darker green veins, membranous. Lateral inner perianth-segments 18–25 × 8–12 mm, elliptical, otherwise like outer segments. Labellum 15–20 × 8–11 mm, ovate, entire or obscurely 3-lobed, subobtuse, with irregularly lacerate margins in basal half, somewhat cucullate and entire distally, covered with greenish papillae. Column 12–16 mm. Capsule 20–25 × 7–10 mm, cylindrical-ovoid. *Flowers December to February*. (Fig. 281, Pl. 8d).

Moist grassland, open *Chiliotrichum* and *Pernettya mucronata* scrub; 0–250 m.

Much of Isla Grande except extreme west and east. (Map 42c).
W. Argentina north to *c*.40°S.lat.
E. Chile north to *c*.39°S.lat.

Fig. 281. *Chloraea magellanica.* a, Plant. b, Labellum. Scale = 1 cm

(R. N. P. Goodall)

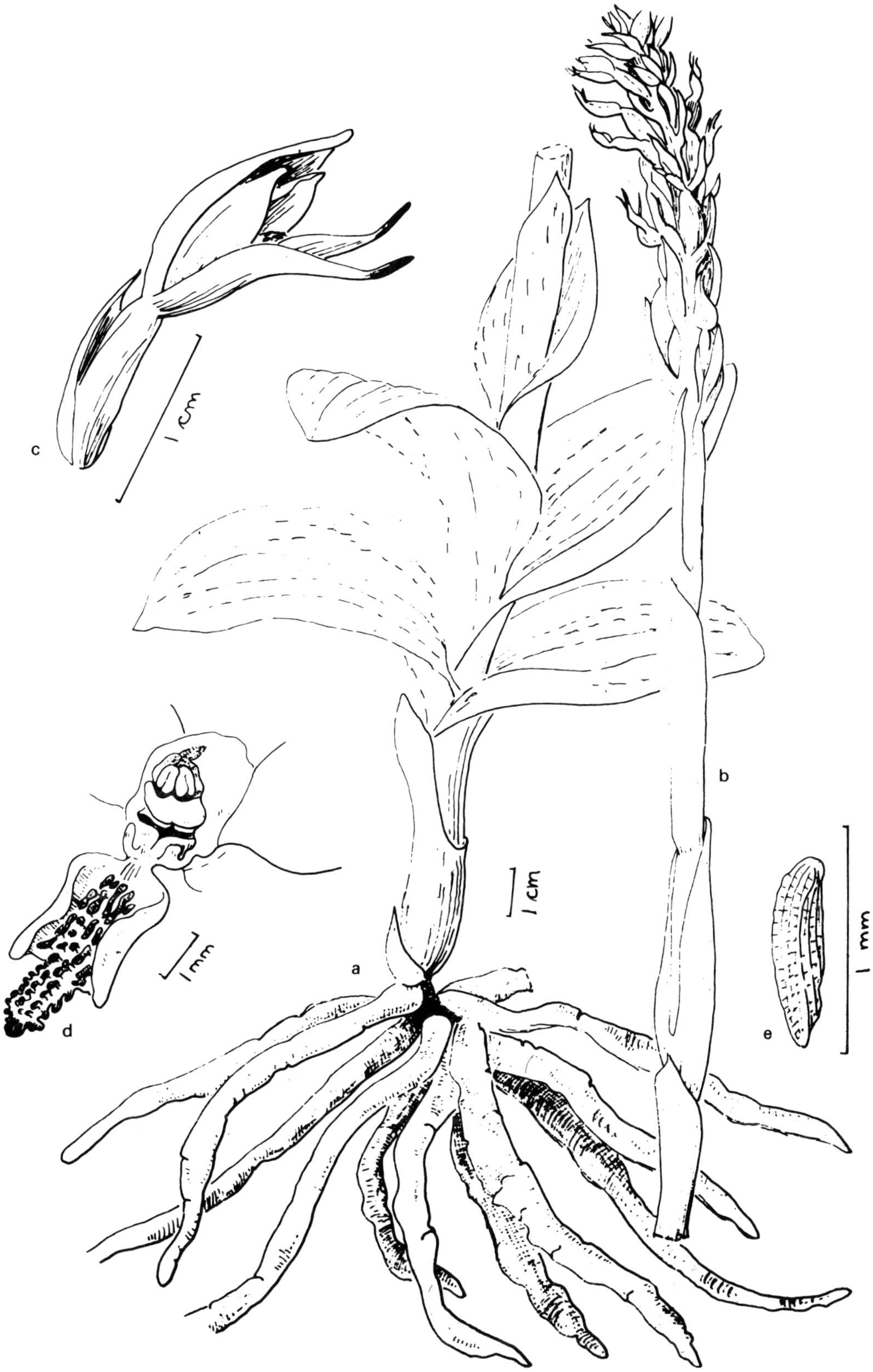

Fig. 282. *Gavilea lutea.* a, Plant-base. b, Inflorescence. c, Flower. d, Labellum and column. e, Seed. Scale = 1 cm (a–c) or 1 mm (d–e).

(R. N. P. Goodall)

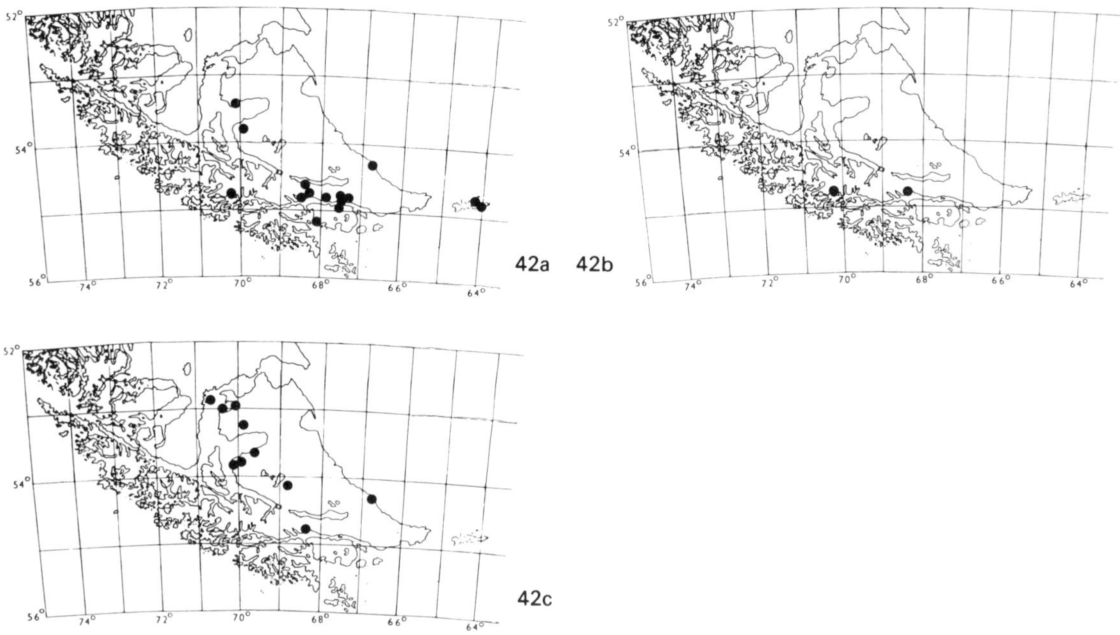

Map 42. a, *Gavilea lutea*. b, *Gavilea littoralis*. c, *Chloraea magellanica*.

D. Glossary (including abbreviations)

achene: a small dry indehiscent one-seeded fruit.
acicular: stiff and pointed, like a needle.
actinomorphic: radially symmetrical.
acuminate: tapering to a fine point, the sides ± concave (Fig. 284c).
acute: sharply pointed, making an angle of less than 90° (Fig. 284b).
adnate: united to an organ of a different kind, as stipules to petiole.
adventitious: arising irregularly and in an unusual position, as roots from stems.
alien: a species not native to the area, usually introduced directly or indirectly by Man.
alternate: placed singly along an axis, not in opposite pairs.
amplexicaul: clasping the stem by the base.
annual: completing the life cycle in 1 year.
anthesis: period during which flower is open.
apiculate: terminating by a short slender ± flexible point (Fig. 284n).
appressed: closely and flatly pressed against a surface.
arcuate: curved or arched.
ascending: directed somewhat obliquely or indirectly upwards.
auct.: *auctorum*, of author(s).
auricle: rounded lobe or appendage; at base of leaf-blade, or at summit of leaf-sheath, in many grasses.
awn: a stiff or bristle-like projection, often from tip or back of an organ.

berry: a fleshy fruit containing several to many seeds.

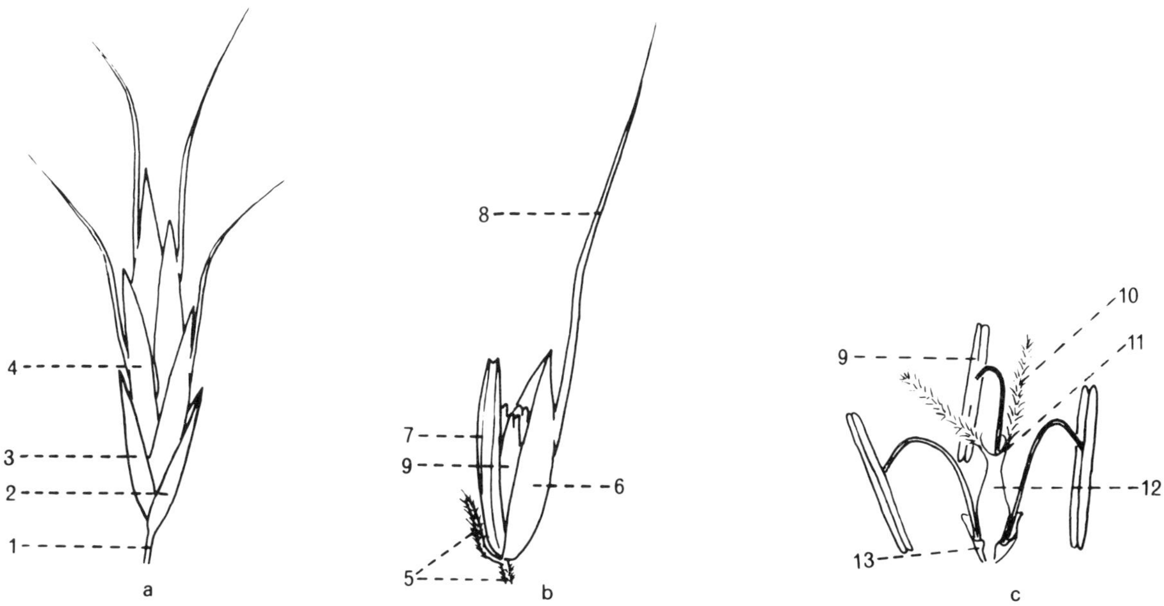

Fig. 283 Structure of a grass spikelet. a, Spikelet. b, Floret. c, Floret with palea and lemma removed. 1. pedicel; 2. outer glume; 3. inner glume; 4. floret; 5. rhachilla; 6. palea; 7. lemma; 8. awn; 9. anther; 10. stigma; 11. style; 12. ovary; 13. lodicule.

bi-: a prefix meaning two, e.g. bilabiate: with two lips.
biennial: living 2 years and flowering only in the second.
biternate: of leaves in which the common petiole bears three secondary petioles, each with three leaflets.
bract: a modified, often much-reduced, leaf subtending a flower, usually associated with an inflorescence.
bracteole: a secondary bract borne on a pedicel instead of subtending it.

c.: *circa*, about, approximately.
caducous: falling off at an early stage.
caespitose: growing in ± dense tufts.
campanulate: bell-shaped.
capitate: head-like.
capitulum: a dense head-like inflorescence of numerous, usually sessile flowers, as for example most Compositae.
capsule: a dry dehiscent fruit formed by the union of two or more carpels.
carpophore: a prolongation of the floral axis between the carpels, as in Umbelliferae.
cauline: pertaining or belonging to the stem, especially the upper aerial part.
ciliate: fringed with hairs along the margin; dimin. **ciliolate**.
circinate: coiled, with the apex innermost, as in young fern fronds.
circumscissile: dehiscing as if cut circularly around, the top valve coming off like a lid.
clavate: club-shaped, thickened towards the apex.
claw: the narrow stalk-like base of a petal or sepal.
cleistogamous: of flowers that remain closed and self-fertilised.
cochleariform: spoon-shaped.
column: the combination of stamens and styles into a solid central body in orchids.
commissure: the face by which two carpels cohere, as in Umbelliferae.
comose: bearing a tuft or tufts of hairs.
compound: composed of several ± similar parts; cf. simple.
compressed: flattened, especially laterally.
connate: joined together, usually secondarily and of similar parts.
connective: the tissue connecting the two lobes of an anther.
connivent: converging together, usually of organs with bases separated and apices approaching one another.
convolute: rolled together longitudinally.
cordate: heart-shaped; also of bases with two equal rounded lobes projecting beyond point of attachment of petiole. (Fig. 284c).
coriaceous: of somewhat leathery texture, tough.
crenate: with shallow rounded teeth, the sinus acute (Fig. 284m); dimin. **crenulate**.
crispate: with the margin crumpled or thrown into close waves.

cucullate: hood-shaped, hood-like.
culm: stem of grasses bearing leaves and inflorescences.
cuneate: gradually and evenly narrowed to point of attachment, applied to leaf-bases (Fig. 284e).
cuneiform: wedge-shaped, applied to total shape of leaf, petal or sepal.
cupule: see p. 81.
cuspidate: tapering gradually into a rigid point (Fig. 284o).
cyathium: see p. 162.
cymbiform: boat-shaped, especially of a leaf with the edges inrolled, giving the appearance of a row-boat.
cyme: a broad ± flat-topped inflorescence with the oldest flowers in the centre.

decurrent: prolonged below the point of insertion.
decussate: of leaves that are opposite with successive pairs at right-angles to each other so forming four rows.
deflexed: bent sharply downwards.
dentate: with sharp teeth perpendicular to the margin (Fig. 284k); dimin. **denticulate**.
diadelphous: of stamens whose filaments are united to give two groups.
dichasium: a cyme in which the branches are opposite and of ± equal length.
dichotomous: forked into two equal branches.
didymous: of two similar parts attached by a short length of their inner surface.
dioecious: having unisexual flowers with male and female on separate plants.
disc: a flattened often ± fleshy development of the receptacle.
distal: towards the free, as opposed to the attached, or proximal, end of an organ.
distant: not close together.
distichous: conspicuously two-ranked, in two rows.
dorsal: relating to the back, the surface away from the axis.
drupe: a fleshy, one-seeded indehiscent fruit, with the seed enclosed in a stony endocarp.
drupelet: one small drupe of an aggregate fruit, as in a raspberry.

Ea: Estancia.
echinate: covered with numerous rigid hairs or straight prickles.
ed.: edition.
ellipsoid: of a solid object elliptical in section or outline.
elliptical: in the shape of an ellipse, widest at the middle and more than twice as long as wide.
emarginate: with a wide shallow notch at the apex (Fig. 284a).
endocarp: the inner layer of the fruit wall, sometimes stony, lying next to the seed.
endosperm: the nutritive tissue outside the embryo in a seed.
entire: with a continuous margin completely lacking indentations.
erect: upright in relation to the ground.
excurrent: projecting beyond the apex, as the midrib of an apiculate leaf.

f.: *filius*, son (following author name).
falcate: sickle-shaped.
-fid: divided nearly to the base into a particular number of segments, e.g. trifid (or 3-fid) (Fig. 284i).
fig.: figure.
filiform: thread-like, long and very slender.
fimbriate: fringed, the hairs longer and coarser than in ciliate.
fistulose: hollow and cylindrical.
floccose: bearing tufts of woolly hairs, usually easily removed.
-foliolate: of compound leaves having two or more leaflets borne from the same point on the petiole, as 3-foliolate (Fig. 285j).
follicle: a dry, dehiscent fruit formed from one carpel and dehiscing along one side.
foveate: pitted; dimin. **foveolate**.
funiculus: the stalk attaching the ovule or seed to the placenta or ovary wall.
fusiform: spindle-shaped.

geniculate: with a knee-like bend.
glabrous: without hairs of any sort.
glabrescent: becoming glabrous.
glaucous: of a bluish green colour.
glume: the chaffy bract(s) subtending the spikelet of Cyperaceae and Gramineae (Fig. 283).

gynodioecious: having female and hermaphrodite flowers on separate plants.

hastate: shaped like an arrowhead with the basal lobes pointed or narrow and abruptly spreading at a wide angle (Fig. 284h).
hemiparasite: a plant attached to and deriving part of its nourishment from another living plant.
herb: a non-woody vascular plant lacking persistent above-ground stem and naturally dying to the ground at the end of the growing season.
herbaceous: having the texture or colour of a foliage leaf.
hermaphrodite: of a flower having both functional stamens and carpels.
heterogamous: of a capitulum bearing two kinds of flowers, as in Compositae.
heterostylous: with style length relative to other floral parts not the same in all plants of a species.
hispid: bearing stiff ± bristle-like hairs.
homogamous: of a capitulum with flowers all alike.
hyaline: thin and transluscent.
hypanthium: a cup-shaped enlargement of the receptacle on which perianth and stamens are inserted.
hypogynous: in which perianth and stamens arise below the ovary.

I: Isla.
imbricate: overlapping, like roof-tiles.
indusium: outgrowth of tissue ± covering the sorus in some ferns.
inferior ovary: ovary with the perianth situated on top.
inflexed: turned sharply inwards.
inflorescence: flower-cluster.
internode: the part of the axis between two nodes.
involucre: a cluster of bracts subtending a flower or inflorescence.
involute: rolled inwards or to upper side.

keeled: with a central dorsal ridge.

labellum: the posterior inner perianth-segment of orchids, usually enlarged.
lacerate: irregularly divided by deep incisions.
lamina: the expanded part or blade of a leaf.
lanate: clothed with woolly hairs, usually intertwined.
lanceolate: lance-shaped, several times longer than wide with greatest width below the middle (Fig. 284d).
latex: milky juice.
lemma: the lower of the two chaffy bracts immediately enclosing the floret in grasses (Fig. 283).
ligule: a strap-shaped body (hence *ligulate*), as the limb of the ray floret in Compositae; the thin scarious projection from the top of the leaf-sheath in grasses and sedges.
limb: the expanded part of a sepal or petal.
linear: very narrow with parallel margins (Fig. 284b).
lobed: partly divided into a number of segments.
loculicidal: splitting longitudinally down the middle of each cell of the ovary.
loculus: a compartment or cavity of an organ, e.g. anther, ovary or fruit.
lodicule: a small scale outside the stamens in a grass flower (Fig. 283).
lyrate: pinnatifid, with the terminal lobe considerably larger than the others.

marginate: distinctly margined.
membranous: thin and ± pliable.
mericarp: a half-fruit in the Umbelliferae.
monadelphous: of stamens whose filaments are united to give a cylinder or column.
monocarpic: of a plant which flowers and fruits once before dying.
monoecious: having unisexual flowers borne on same plant.
Mte(s): Monte(s).
mucro: a short sharp tip or excurrent midrib, hence **mucronate** (Fig. 284p); dimin. **mucronulate**.

nectary: an organ which secretes nectar, hence **nectariferous**.
node: a place on a stem marked by the attachment of one or more leaves.
nut: an indehiscent one-seeded fruit with hard woody wall.

ob-: signifies inversion, e.g. an obovate leaf is attached at narrow end (Fig. 284e).
obdiplostemonous: with the stamens in two whorls, the outer opposite, the inner alternating with, the petals.
oblong: longer than wide with parallel margins (Fig. 284a).
obsolete: not obvious, poorly developed.
obtuse: blunt, making an angle greater than 90° (Fig. 284d).
opposite: of a pair of organs arising at the same level on opposite sides of the stem.
orbicular: rounded in outline (Fig. 284f).
ovate: egg-shaped, attached by the wider end (Fig. 284c).
ovoid: of a solid body ovate in outline or longitudinal section.

palea: the inner bract of a grass floret (Fig. 283).
palmate: consisting of more than three leaflets arising from the same point.
panicle: a branched inflorescence, strictly a branched racemose inflorescence.
papillose: covered with minute, pimple-like processes.
pappus: a persistent calyx of special form crowning achene in Compositae (p. 219).
patent: spreading ± at right-angles to axis.
pedicel: the stalk of an individual flower in a compound inflorescence.
peduncle: the stalk of a solitary flower or main stalk of a compound inflorescence.
peltate: shield-like, with the stalk attached well inside the margin (Fig. 284f).
pendent: hanging downwards.
penin: península.
perennial: with a life-span of more than 2 years.

perianth: the sepals and petals considered together, especially when they are not clearly different.
pericarp: the wall of the fruit.
perigynium: the inflated sac-like organ surrounding the style in Cyperaceae.
persistent: remaining attached, not falling off.
petiole: the stalk of a leaf; dimin. **petiolule**.
phyllode: a green dilated petiole resembling a leaf.
pilose: bearing soft shaggy hairs.
pinna: a primary division of a pinnate leaf; dimin. **pinnule**.
pinnate: when simple leaflets are arranged on each side of a common petiole (Fig. 103); in all examples in this Flora a terminal leaflet or tendril is present and the term is strictly *imparipinnate*, as opposed to *paripinnate* in their absence; **2-pinnate** if leaflets are themselves pinnate.
pinnatifid: of leaves pinnately cut but the divisions not reaching midrib.
pinnatisect: like pinnatifid but with some of lower divisions almost or quite reaching midrib.
pl.: plate.
plicate: folded lengthwise into pleats.
plumose: feather-like.
pod: a superior, l-celled fruit usually dehiscent into two valves, having the seeds attached along the central line of junction of the two carpels, as in a pea.
pollinium: the coherent mass of pollen, as in orchids.
procumbent: lying ± flat along the ground but not rooting.
pro parte: in part.
protandrous: of an hermaphrodite flower in which the anthers mature before the stigma.
protogynous: of an hermaphrodite flower in which the stigma matures before the anthers.
Prov.: Provincia.
pseudoviviparous: with the flowers proliferating vegetatively and not forming seeds.
Pta: Punta.
Pto: Puerto.
pubescent: clad in short soft hairs; dimin. **puberulent**.

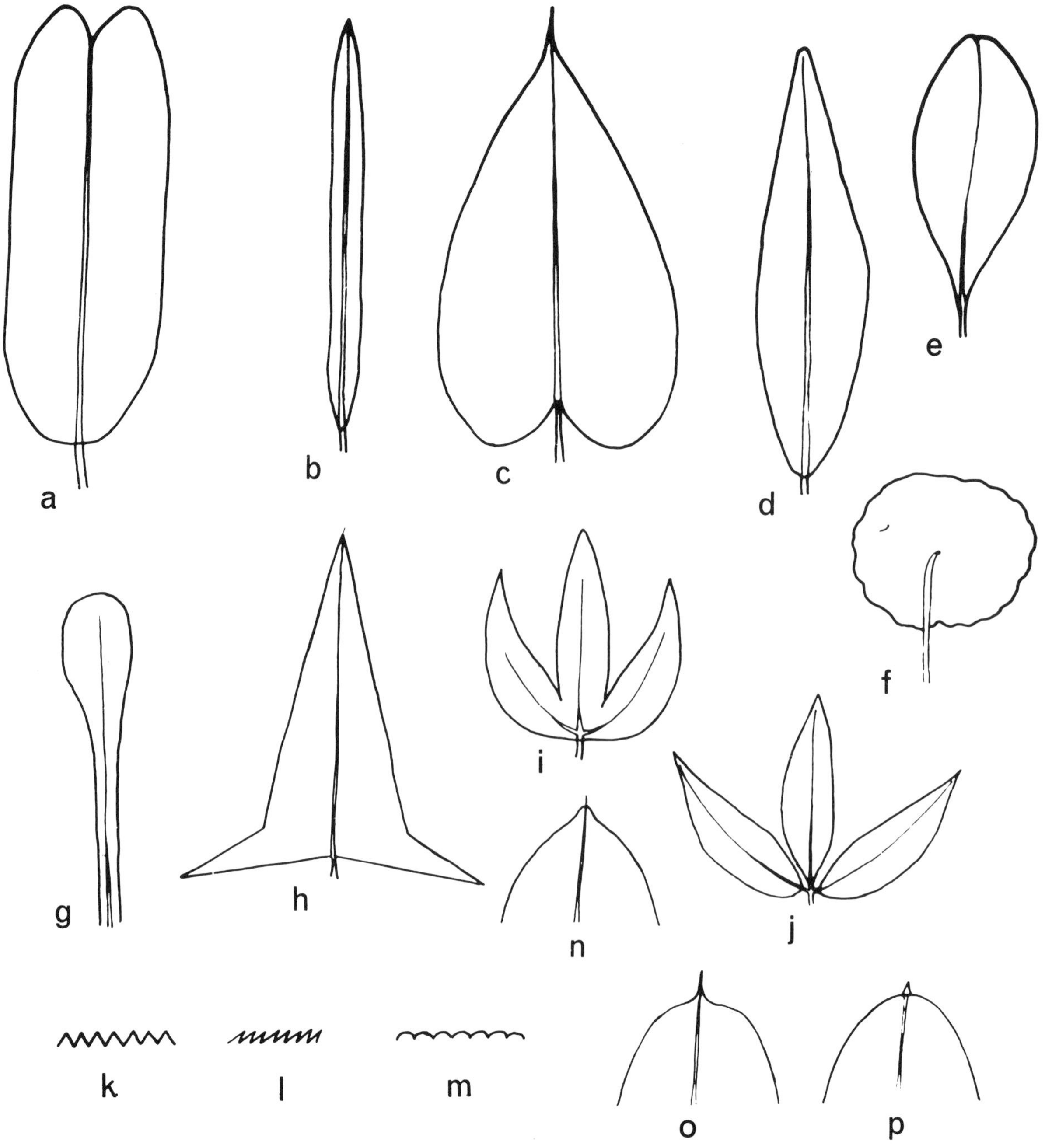

Fig. 284. Diagrams illustrating the use of some terms employed in describing the outline, apex, base and margin of leaves. a, oblong, base truncate, apex emarginate; b, linear, base cuneate, apex acute; c, ovate, base cordate, apex acuminate; d, lanceolate, base cordate, apex obtuse; e, obovate, base cuneate, apex retuse; f, orbicular and peltate; g, spathulate; h, hastate; i, trifid; j, trifoliolate; k, dentate margin; l, serrate margin; m, crenate margin; n, apiculate apex; o, cuspidate apex; p, mucronate apex.

punctate: with dot-like markings.
pungent: ending in a stiff sharp point.
pyriform: pear-shaped.

raceme: an unbranched, ± elongate inflorescence with stalked flowers, those at the base the oldest.
receptacle: the ± expanded apex of the stalk on which the flower or flower-head is borne.
reflexed: bent sharply backwards.
remote: not close together.
reniform: kidney-shaped.
reticulate: in the form of a network.
retuse: with the apex rounded and having a shallow notch (Fig. 284e).
revolute: rolled outwards or to the lower side.
rhachilla: the axis of grass spikelet that bears the florets.
rhizome: an underground stem lasting more than one growing season.
rhomboid: ovate, with sides angular rather than rounded towards middle.
rostellum: the slender extension from the upper edge of the stigma in orchids.
rotate: wheel-shaped, of a corolla with flat, circular limb at right-angles to short tube.
ruderal: a plant growing in waste places.
rugose: wrinkled; dimin. **rugulose**.
runcinate: pinnately and rather sharply lobed with the lobes directed backwards.

saccate: bag-shaped.
scabrid: with very short bristly hairs; rough to the touch.
scabridulous: slightly rough to the touch.
scape: a leafless ± elongate peduncle arising from the crown with or without scales or bracts.
scapigerous: bearing scapes.
scarious: very thin, dry and ± translucent.
self-compatible: of a plant capable of being fertilised by its own pollen.
sens. lat.: *sensu lato*, in a wide sense.
septate: divided by partitions or cross-walls.
septicidal: splitting longitudinally between the cells of the ovary.
ser.: series.
sericeous: clothed with simple ± appressed hairs.
serrate: sharply toothed with the teeth pointing forward (Fig. 284l); dimin. **serrulate**.
sessile: without a stalk.
setaceous: bristle-like.
sheath: the tubular and ± tubular basal part of the leaf that encloses the stem, as in Gramineae and Cyperaceae.
shrub: a low woody plant.
silicula: capsule of Cruciferae, less than 3 times as long as wide.
siliqua: capsule of Cruciferae, more than 3 times as long as wide.
simple: not divided into several separate parts, not compound, unbranched.
sinuate: with a wavy margin.
sorus: a cluster of sporangia in ferns.
sp.: species; plural **spp**.
spathulate: paddle-shaped (Fig. 284g).
spicule: a diminutive, fine, often fleshy, erect point.
spike: an unbranched ± elongate inflorescence with sessile flowers, those at the base oldest.
spikelet: the ultimate unit of the inflorescence in Gramineae, consisting of two glumes and one or more florets (Fig. 283), and in Cyperaceae.
sporangium: a sac containing spores.
sporophyll: a ± modified leaf-like structure bearing sporangia.
ssp.: subspecies; plural **sspp**.
stamen: the male organ of the flower, bearing pollen in terminal *anther* which is usually attached by a stalk-like *filament*.
stellate: star-shaped.
stipitate: having a stalk or a stalk-like base.
stipule: one of a pair of scale-like or leaf-like appendages at the base of a petiole.
stock: the tissue at the junction of the root and stem.
stolon: an overground creeping stem rooted at nodes and of short duration.
striate: with fine grooves.
strict: upright, straight and ± rigid.
strigose: with sharp, stiff, ± appressed hairs and bristles; dimin. **strigulose**.
style: the contracted part of the pistil between the ovary and the stigma.
stylopodium: the enlarged basal part of a style, as in Umbelliferae.
sub-: prefix meaning somewhat, slightly or not quite.
subulate: awl-shaped; very narrow, tapering to a sharp apex from a rather wide base.
suffruticose: rather woody at base with herbaceous shoots.
superior ovary: an ovary with the perianth inserted below it.

terete: circular in cross-section.
tetragonous: four-angled.
tomentose: with a dense ± matted covering of soft ± appressed hairs.
trigonous: of a solid body triangular in cross-section with the angles obtuse.
triquetrous: of a solid body triangular in cross-section with the angles acute.
truncate: ending abruptly as though cut squarely across (Fig. 284a).
tuberculate: covered with small wart-like swellings.

umbel: a flat or convex inflorescence in which the pedicels arise from a common point like the rays of an umbrella.
umbel, compound: an umbel with each ray itself bearing an umbel.
undulate: waved in a plane at right-angles to the surface.
unisexual: of one sex only.
urceolate: pitcher-like, hollow, swollen in lower part and contracted at the mouth.

var.: variety.
verrucose: covered with small wart-like excrescences.
villous: with long, silky, straight hairs.
viscid: very sticky.

whorl: an arrangement of three or more parts or organs at the same level round an axis.

zygomorphic: bilaterally symmetrical.

VI. References

Aellen, P., 1929. Beitrag zur Systematik der *Chenopodium*-Arten Amerikas. *Feddes Repert.* **26**: 119–160.

Aellen, P. and Just, T., 1943. Key and synopsis of the American species of the genus *Chenopodium* L. *Amer. Midl. Nat.* **30**: 47–76.

Ahrendt, L. W. A., 1961. *Berberis* and *Mahonia*. A taxonomic revision. *Bot. Jour. Linn. Soc. London* **57**: 1–410.

Alboff, N., 1896a. Contributions à la flore de la Terre de Feu. I. Observations sur la végétation du canal de Beagle. *Rev. Mus. La Plata* **7**: 277–308.

Alboff, N., 1896b. Contributions à la Flore de la Terre de Feu. II. Enumération des plantes du canal de Beagle et de quelques autres endroits de la Terre de Feu. *Rev. Mus. La Plata* **7**: 355–392.

Alboff, N., 1902. Flore raisonée de la Terre de Feu. *Anal. Mus. La Plata, Bot.* **1**: 1–85 + XXIII.

Andrews-Speed, C. P., 1979. General Account. *Rep. Cambridge Exped. Tierra del Fuego.* pp. 1–9. Cambridge.

Anon., 1977. *The Times Atlas. Comprehensive Edition.* Ed. 5. Times Books, London.

Arctowski, H., 1901. Voyage d'exploration dans la région des canaux de la Terre de Feu. *Bull. Soc. Belge Geog.* **25**: 33–62.

Arroyo, S., 1975. *Lebetanthus*, género austroamericano de *Epacridaceae* y sus diferencias con el género *Prionotes* de Tasmania. *Darwiniana* **19**: 312–330.

Arroyo, S., 1980. The genus *Benthamiella* (Solanaceae). *Bot. Notiser* **133**: 67–76.

Aubert de la Rue, E., 1959. Quelques observations faites aux isles Diego Ramírez (Chili). *Bull. Mus. Hist. Nat. (Paris)* 2e Serie **31(4)**: 387–391.

Auer, V., 1956. The Pleistocene of Fuego-Patagonia. I: The ice and interglacial ages. *Ann. Acad. Sci. Fenn., Ser. AIII*, **45**: 1–226.

Auer, V., 1958. The Pleistocene of Fuego-Patagonia. II. The history of the flora and vegetation. *Ann. Acad. Sci. Fenn., Ser. AIII*, **50**: 1–239.

Auer, V., 1959. The Pleistocene of Fuego-Patagonia. III: Shoreline displacements. *Ann. Acad. Sci. Fenn., Ser. AIII*, **60**: 1–247.

Auer, V., 1960. The Quaternary history of Fuego-Patagonia. *Proc. Roy. Soc. London, Ser. B*, **152**: 507–516.

Auer, V., 1965. The Pleistocene of Fuego-Patagonia. IV: Bog profiles. *Ann. Acad. Sci. Fenn., Ser. AIII*, **80**: 1–160.

Auer, V., 1970. The Pleistocene of Fuego-Patagonia. V: Quaternary problems of southern South America. *Ann. Acad. Sci. Fenn., Ser. AIII*, **100**: 1–194.

Autran, Eugène, 1905. Énumération des plantes récoltées par Miles Stuart Pennington pendant son premier voyage á la Terre de Feu en 1903. *Rev. Univ. Bs Aires*, **4(No.18)**: 287–305.

Bacigalupo, N. M., 1979. El género *Callitriche* en la flora argentina. *Darwiniana* **22**: 377–396.

Barclay, W. S., 1926. *The Land of Magellan.* Methuen, London.

Barros, M., 1928. Ciperáceas Argentinas I. Género *Heleocharis* R. Br. *Anal. Mus. Nac. Hist. Nat. Bs Aires* **34**: 427–496.

Barros, M., 1935. Ciperáceas Argentinas. II. Géneros *Kyllingia, Scirpus, Carex. Anal. Mus. Nac. Hist. Nat. Bs Aires* **37**: 133–263.

Barros, M., 1953. Las Juncáceas de la Argentina, Chile y Uruguay. *Darwiniana* **10**: 219–430.

Beaglehole, J. C., 1961. *The Journal of Captain James Cook. The Voyage of the Resolution and Adventure, 1772–1775.* University Press, Cambridge.

Beaglehole, J. C., 1962. *The Endeavour Journals of Joseph Banks, 1768–1771. Vols. I and II.* Angus and Robertson, Sydney.

Beauvoir, J. M., 1915. *Los Shelknam, Indígenas de la Tierra del Fuego.* Inst. Salesiano, Buenos Aires.

Belza, J. E., 1973. 1886, el primer año de gobierno territorial. *Karukinka* **4**.

Bonarelli, G., 1917. Tierra del Fuego y sus turberas. *Anal. Min. Agric. Nación, Secc. Geol., Min., Minería* **12(3)**: 1–119.

Borsini, O. E., 1966. Valerianáceas de Chile. *Lilloa* **32**: 375–476.

Bove, G., 1883. Informes preliminares de la Expedición Austral Argentina. Dept. Agricultura, Buenos Aires.

Brackenridge, W. D., 1854. *United States Exploring Expedition.* Vol. XVI. *Botany. Cryptogamia.* Sherman, Philadelphia.

Brand, A., 1931. Borraginaceae–

Borraginoideae–Cryptantheae. *Pflanzenreich* **97(IV.252)**: 1–236.
Bridges, E. L., 1948. *The Uttermost Part of the Earth.* Hodder and Stoughton, London.
Bridges, T., 1933. *Yamana – English, a Dictionary of the Speech of Tierra del Fuego.* St. Gabriel, Mödling.
Buchinger, M. P., 1956. Sinopsis preliminar de las especies argentina del género *Polygonum* L. *Bol. Soc. Argent. Bot.* **6**: 98–106.
Burkart, A., 1935. Las especies de *Lathyrus* de la República Argentina. *Rev. Fac. Agron. Vet. Bs Aries* **8**: 41–127.
Burkart, A., 1964. Contribución al estudio del género *Adesmia* (Leguminosae). V. *Darwiniana* **13**: 9–66.
Burkart, A., 1967. Sinopsis del género sudamericano de Leguminosas *Adesmia* DC. *Darwiniana* **14**: 463–568.
Cabrera, A. L., 1945. Sinopsis del género *Lepidophyllum* (Compositae). *Bol. Soc. Argent. Bot.* **1**: 48–58.
Cabrera, A. L., 1949. El género *Senecio* en Chile. *Lilloa* **15**: 27–501.
Cabrera, A. L., 1954. Las especies del género *Nardophyllum*. *Not. Mus. La Plata, Bot.* **17**: 55–56.
Cabrera, A. L., 1957. Una nueva especie del género *Antennaria* (Compositae). *Not. Mus. La Plata Bot.* **19**: 73–79.
Cabrera, A. L., 1961. Observaciones sobre las Inuleae – Gnaphalineae (Compositae) de América del Sur. *Bol. Soc. Argent. Bot.* **9**: 359–386.
Cabrera, A. L., 1963. Estudios sobre el género *Hypochoeris*. *Bol. Soc. Argent. Bot.* **10**: 166–195.
Cabrera, A. L., 1966. The genus *Lagenophora* (Compositae). *Blumea* **14**: 285–307.
Cabrera, A. L., 1971a. *Nardophyllum*. *In*: Correa, M. N. (Ed) *Fl. Patag.* **7**: 34–37.
Cabrera, A. L., 1971b. Fitogeografía de la República Argentina. *Bol. Soc. Argent. Bot.* **14**: 1–42.
Carolin, R. C., 1963. J. R. and J. G. A. Forster and their collections. *Proc. Linn. Soc. New South Wales* **88(2)**: 108–111.
Casper, S. J., 1966. Monographie der Gattung *Pinguicula* L. *Biblioth. Bot. (Stuttgart)* **127–128**: 1–209.
Castellanos, A., 1945. Las exploraciones botánicas en la epoca de la independencia, 1810–1853. *Holmbergia* **4(8)**: 3–14.
Chambers, K. L., 1963. Amphitropical species pairs in *Microseris* and *Agoseris* (Compositae: Cichorieae). *Quart. Rev. Biol.* **38**: 124–140.
Chamorro, C., 1936. *Bajo el Cielo Austral.* Illustración, Santiago de Chile.
Chapman, W., 1964. *The Loneliest Continent.* N.Y. Graphic Soc., Greenwich, Conn.
Clausen, R. T., 1948. The *Ophioglossum* of the Falkland Islands and the relationships and distribution of *Ophioglossum crotalophoroides*. *Amer. Fern. Jour.* **38**: 176–186.
Coiazzi, A., 1914. Los indios del archipiélago fueguino. *Rev. Chil. Hist. Geog.* **9**: 288–352.
Conert, H. J., 1961. *Die Systematik und Anatomie der Arundineae.* Cramer, Weinheim.
Constance, L. and Shan, R. H., 1948. The genus *Osmorhiza* (Umbelliferae). *Univ. Calif. Publs. Bot.* **23**: 111–156.
Cooke, C. D. K., 1966. A monographic study of *Ranunculus* subgenus *Batrachium* (DC.) A. Gray. *Mitt. Bot. Staatssam. Munchen* **6**: 47–237.
Correa, M. N., 1956. Las especies argentinas del género *Gavilea*. *Bol. Soc. Argent. Bot.* **6**: 73–86.
Correa, M. N., 1966. Las Frankeniaceae argentinas. *Darwiniana* **14**: 68–94.
Correa, M. N., 1969. *Chloraea*, género sudamericano de Orchidaceae. *Darwiniana* **15**: 374–500.
Correa, M. N. (Ed.) 1969. *Flora Patagónica.* II. INTA, Buenos Aires.
Correa, M. N. (Ed.) 1971. *Flora Patagónica.* VII. INTA, Buenos Aires.
Correa, M. N. (Ed.) 1978. *Flora Patagónica.* III. INTA, Buenos Aires.
Correa Luna, H., 1964. El Parque Nacional Tierra del Fuego. *Anal. Parques Nac.,* **10**: 61–72.
Cowan, J. M. 1954. Some information on the Menzies and Jack collection in the Herbarium, Royal Botanic Garden, Edinburgh. *Notes R. Bot. Gard. Edinb.* **21**: 219–227.
Cozzo, D. 1949. El bosque de Bahía Aguirre en Tierra del Fuego (Argentina). *Rev. Argent. Agron.* **16**: 193–198.
Crisci, J. V., 1976. Revisión del género *Leucheria* (Compositae: Mutisieae). *Darwiniana* **20**: 9–126.
Crow, G. E., 1975. Vegetation of Isla de los Estados, Argentina. *Antarct. Jour. U.S.* **10**: 81–85.
Cunningham, R. O., 1871. *The Natural History of the Strait of Magellan.* Edmonston and Douglas, Edinburgh.
Dabbene, R., 1904. Viaje a la Tierra del Fuego y a la Isla de los Estados. *Bol. Inst. Geogr. Argent.* **21**: 3–78.
Dainelli, G., 1922. Piante della Terra del Fuoco. *Scrit. Vari. Geog. Geol. (Firenze)* **1922**: 273–278.
Dalziel, I. W. D., de Wit, M. J. and Palmer, K. F., 1974. Fossil marginal basin in the southern Andes. *Nature (London)* **250**: 291–294.
Darwin, C., 1839. *Narrative of the Surveying Voyages of His Majesty's Ships 'Adventure' and 'Beagle' between the years 1826 and 1836. Vol. 3. Journal and Remarks, 1832–36.* Colburn, London.
Davidson, J. F., 1950. The genus *Polemonium* (Tournef.) L. *Univ. Calif. Publs. Bot.* **23(5)**: 209–282.
Dawson, G., 1944. Las Santaláceas Argentinas. *Rev. Mus. La Plata Bot., n. ser.* **6**: 5–80.
Den Hartog, C. and Van der Plas, F., 1970. A synopsis of the Lemnaceae. *Blumea* **18**: 355–368.
de Paula de Brooks, M. E., 1975. Las especies del género *Hierochloë* (Gramineae) de Argentina y Chile. *Darwiniana* **19**: 442–457.
Diaz, C., Avilez, C. and Roberts, C., 1960. Los grandes grupos de suelos de la provincia de Magallanes. *Agric. Téc. (Chile)* **19/20**: 227–308.
Diem, J., 1958. Observaciones sobre *Polystichum mohrioides* var. *plicatum* (Poepp.) C. Christensen. *Bol. Soc. Argent. Bot.* **7**: 94–98.
Diem, J. and Lichenstein, J. S. de, 1959. Las Himenofiláceas del área argentino–chilena del sur.

Darwiniana **11**: 611–761.
Dimitri, M. J. (Ed.) 1972. *La Región de los Bosques Andino–Patagónicos.* INTA, Buenos Aires.
Dollenz, O., 1978. Estado de la flora vascular en Puerto Espora, Tierra del Fuego, contaminada por el petróleo del B/T 'Metula'. II. Reconocimiento de la entrada de mar suroeste. *Anal. Inst. Patag. Punta Arenas (Chile)* **9**: 133–139.
Dollenz, O., 1980. Estudios fitosociológicos en el archipielago Cabo de Hornos. I. Relevamientos en caleta Lientur, isla Wollaston y surgidero Romanche, isla Bayly. *Anal. Inst. Patag. Punta Arenas (Chile)* **11**: 225–238.
Dudley, T. R., 1981. Taxonomic and nomenclatural notes on the Flora of Isla de los Estados (Staten Island), Tierra del Fuego, Argentina. *Rhodora* **83**: 477–519.
D'Urville, J. S. C. D., 1825. *Flore des Iles Malouines.* Paris.
Dusén, P., 1897. Über die Vegetation der Feuerlandischer Inselgruppe. *Bot. Jahrb.* **24**: 179–196.
Dusén, P., 1900. Die Gefässpflanzen der Magellansländer. *Svensk Exped. Magellansl.* **3(5)**: 77–266.
Dusén, P., 1905. Die Pflanzenvereine der Magellansländer nebst einem Beitrage zur Ökologie der Magellanischen Vegetation. *Svensk Exped. Magellansl.* **3(10)**: 351–521.
Epling, C., 1938. Las Labiadas de la Argentina, Paraguay y Uruguay. *Rev. Mus. La Plata, Bot.* **2(8)**: 89–178.
Etcheveherre, P. H., 1972. Los suelos de la región Andino–Patagónica. In: M. J. Dimitri (Ed.) *La Región de los Bosques Andino–Patagónicos.* INTA, Buenos Aires pp.83–95.
Fitzroy, R. (Ed.), 1839. *Narrative of the Surveying Voyages of HMS Adventure and Beagle between the years 1826 and 1836.* 3 vols. London.
Forster, G., 1777. *A Voyage Round the World in his Brittanic Majesty's Sloop Resolution, Commanded by Capt. James Cook during the years 1772, 1773, 1774 and 1775.* London.
Forster, G., 1787. *Fasciculus Plantarum Magellanicum.* Göttingen.
Forster, G., 1789. De plantis magellanicis et atlanticis commentationes. *Comment. Gott.* **9**: 13–74.
Franchet, A. R., 1889. *Mission Scientifique du Cap Horn, 1882–1883.* Vol. 5. *Phanérogamie.* Paris.
Fuchs, H. P., 1962. Nomenklatur, Taxonomie und Systematik der Gattung *Isoetes* L. in geschichtlicher Betrachtung. *Nova Hedwig. Beih.* **3**: 1–160.
Gallardo, C. R., 1910. *Los Onas–Tierra del Fuego.* Cabaut, Buenos Aires.
Gamerro, J. C., 1968. Observaciones sobre la biología floral y morfología de la Potamogetonácea *Ruppia cirrhosa* (Petag.) Grande (= *R. spiralis* L. ex Dum.). *Darwiniana* **14**: 575–603.
Gerlache, A. de, 1902. *15 Mois dans L'Antarctique. Le Voyage de la Belgica.* Bulens, Brussels.
Godley, E. J., 1960. The botany of southern Chile in relation to New Zealand and the Subantarctic. *Proc. Roy. Soc. London, Ser. B,* **152**: 457–475.
Godley, E. J., 1963. Royal Society expedition to southern Chile: botanist's itinerary. *N.Z. Jour. Bot.* **1**: 316–324.
Godley, E. J., 1964. Contributions to the Plant Geography of Southern Chile. *Rev. Univ.* **48**: 31–39.
Godley, E. J., 1965. Botany of the Southern Zone, exploration to 1843. *Tuatara* **13(3)**: 140–181.
Godley, E. J., 1970. Botany of the Southern Zone, exploration 1847–1891. *Tuatara* **18(2)**: 49–93.
Gómez-Sosa, E., 1979. Las especies sudamericanas del género *Astragalus* (Leguminosae). I. Las especies patagónicas argentinas. *Darwiniana* **22**: 313–376.
Goodall, R. N. P., 1976. *Tierra del Fuego, Argentina (Map).* Shanamaiim, Buenos Aires.
Goodall, R. N. P., 1979. *Tierra del Fuego* Ed. 3. Shanamaiim, Buenos Aires.
Goodspeed, T. H., 1941. *Plant Hunters in the Andes.* Farrar and Rinehart, New York.
Gray, A., 1854. *United States Exploring Expedition.* Vol. XV. *Botany. Phanerogamia* Pt. 1. Sherman, Philadelphia.
Grondona, E. M., 1948. Las especies argentinas del género *Polygala. Darwiniana* **8**: 279–405.
Grondona, E. M., 1964. Las especies argentinas del género *Acaena* (Rosaceae). *Darwiniana* **13**: 208–342.
Groves. E. W., 1962. Notes on the botanical specimens collected by Banks and Solander on Cook's First Voyage, together with an itinerary of landing localities. *Jour. Soc. Bibl. Nat. Hist.* **4**: 57–62.
Guiñazu, J. R., 1934. Los depósitos de turba de Tierra del Fuego, su extensión y posibles usos. *Min. Agric. Nac. Minas Geol.* **103**: 1–24.
Gunckel, L. H., 1971. Las primeras plantas herborizadas en Chile en 1690. *Anal. Inst. Patag. Punta Arenas (Chile)* **2**: 134–141.
Gusinde, M., 1937. *Die Feuerland Indianer.* 3 Vols. St. Gabriel, Mödling.
Hara, H., 1957. Synopsis of the genus *Chrysosplenium* L. (Saxifragaceae). *Jour. Fac. Sci. Univ. Tokyo* III, **7**: 1–90.
Hariot, P., 1884. Liste des plantes vasculaires observées dans le detroit de Magellan et a la Terre de Feu. *Bull. Soc. Bot. Fr.* **1884**: 151–164.
Hariot, P., 1900. Liste des phanérogames et des cryptogames vasculaires récoltées a la Terre de Feu par MM. Willems et Rousson. *Journ. Bot.* **14**: 148–153.
Hauman, L. 1916. Note préliminaire sur les *Hordeum* spontanés de la flore argentina. *An. Mus. Nac. Hist. Nat. Bs Aires* **27**: 263–316.
Heckard, L. R., 1960. Taxonomic studies in the *Phacelia magellanica* polyploid complex. *Univ. Calif. Publs. Bot.* **32**: 1–126.
Heckard, L. R., 1963. Amphitropical relationships in the herbaceous flora of the Pacific coast of North and South America: The Hydrophyllaceae. *Quart. Rev. Biol.* **38**: 117–123.
Hicken, C. M., 1921. Las Himenofiláceas argentino-chilenas y los "continentes pacíficos". *Rev. Chile Hist. Nat.* **25**: 253–262.
Hitchcock, L. C., 1945. The South American species of *Lepidium. Lilloa* **11**: 75–134.
Holdgate, M. W., 1961. Vegetation and soils in the south Chilean islands. *Jour. Ecol.* **49**: 559–580.
Holmberg, E. A., 1906. Viaje al interior de la Tierra del

Fuego. *An. Min. Agr. Argent.* **1**: 1–95.
Holmgren, P. K. and Keuken, W., 1974. *Index Herbariorum*. Pt. 1. The herbaria of the world, ed. 5. *Reg. Veg.* **92**: 1–397.
Hooker, J. D., 1847. *The Botany of the Antarctic Voyage. I. Flora Antarctica. Part II, Botany of Fuegia, The Falklands, Kerguelen's Land, etc.* Reeve, London.
Hueck, K. and Seibert, P., 1972. Vegetationskarte von Südamerika. *Veget. Monog. Grossräume* **2a**: 1–69.
Hultén, E., 1959. The *Trisetum spicatum* complex. *Trisetum spicatum* (L.) Richt., an arctic-montane species with world-wide range. *Svenska Bot. Tidskr.* **53**: 203–228.
Humphries, C. J., 1978. Notes on the genus *Phleum* L. *In*: Heywood, V. H. (Ed.), Flora Europaea Notulae Systematicae ad Floram Europaeam spectantes. *Bot. Jour. Linn. Soc.* **76**: 337–340.
Humphrey, P. S., Bridges, D., Reynolds, P. W. and Peterson, R. T., 1970. *Birds of Isla Grande (Tierra del Fuego)*. Smithsonian, Washington.
Hunziker, A. T., 1967. Estudios sobre Solanaceae. IV. *Kurtziana* **4**: 131–138.
Hunziker, J. H., 1949. Sinopsis de las especies argentinas del género *Ephedra*. *Lilloa* **17**: 147–174.
Hunziker, J. H. and Itria, C. D., 1980. Hibridación natural entre *Agropyron* y *Hordeum* en Tierra del Fuego. *Bol. Soc. Argent. Bot.* **19**: 91–102.
Jerez, M. and Arancibia, M., 1972. Trazado de isoyetas del sector centro oriental de la provincia de Magallanes. *Monogr. Inst. Patag., Punta Arenas (Chile)* **4**: 1–28.
Johnston, I. M., 1927. A revision of the South American Boraginoideae. *Contr. Gray Herb.* **78**: 1–118.
Johnston, I. M., 1947. *Astragalus* in Argentina, Bolivia and Chile. *Jour. Arn. Arb.* **28**: 336–409.
King, P. P., 1839. *Narrative of the Surveying Voyages of His Majesty's Ships 'Adventure' and 'Beagle' between the years 1826 and 1836. Vol. 1. Proceedings of the First Expedition*, 1826–30. Colburn, London.
Knuth, R., 1930. Oxalidaceae. *Pflanzenreich* **95(IV.130)**: 1–481.
Kranck, E. H., 1932. Geological investigations in the cordillera of Tierra del Fuego. *Acta Soc. Geogr. Fenn.* **4(2)**: 1–231.
Kubitzki, K., 1966. Untersuchungen über den Blütenbau von *Oreobolus* R. Br. *Bot. Jahrb.* **85**: 80–87.
Kunkel, G., 1966. Notes on the Gleicheniaceae of southern Chile. *Nova Hedwigia* **11**: 507–511.
Kurtz, F., 1896. Cyperaceae et Gramineae, in Alboff, N., Contributions á la flore de la Terre du Feu. II. *Rev. Mus. La Plata* **7(2)**: 393–402.
Laming-Emperaire, A., Lavalle, D. and Humbert, R., 1972. Le site de Marazzi en Terre de Feu. *Objets et Mondes* **12**: 225–244.
Lenzi, J. H., 1967. *Tierra del Fuego, Antartida e Islas del Atlantico Sur*. Progreso, Buenos Aires.
Lista, R., 1887. *Viaje al País de los Onas. Tierra del Fuego. Nuñez, Buenos Aires.*
Lista, R., 1896. Plantas patagónicas. Anal. Soc. Ci. Argent. **42**: 321–385.
Lloyd, D. G., 1972. A revision of the New Zealand, Subantarctic and South American species of *Cotula*, Section *Leptinella*. *N.Z. Jour. Bot.* **10**: 277–372.
Looser, G., 1934. Sobre Gleichenia Bibreae y sobre las Gleicheniáceas chilenas en general. *Rev. Univ.* **19**: 761–775.
Looser, G., 1944. Sinopsis de los *Asplenium* (Filices) de Chile. *Lilloa* **10**: 233–264.
Looser, G., 1961. Los Pteridófitos o helechos de Chile. I. *Rev. Univ.* **46**: 213–262.
Looser, G., 1962. Los Pteridófitos o helechos de Chile. II. Gleicheniacea. *Rev. Univ.* **47**: 17–31.
Lothrop, S. K., 1928. *The Indians of Tierra del Fuego*. Heye Foundation, New York.
Lourteig, A., 1942. Primulaceae argentinae. *Lilloa* **8**: 231–269.
Lourteig, A., 1947. Primulaceae argentinae addenda. *Darwiniana* **7**: 556–559.
Lourteig, A., 1952. Ranunculáceas de Sud América templada. *Darwiniana* **9**: 397–608.
Lourteig, A., 1967. *Primula farinosa* L. y *Primula magellanica* Lehmann. *Bol. Soc. Argent. Bot.* **11**: 140–145.
Lourteig, A., 1968. Révision de *Juncus* subgenus *Septati* Buch. *Com. Nat. Fr. Rech. Antarct., Biol.* **23**: 33–49.
Lourteig, A. and O'Donnell, C. A., 1942. Acalypheae Argentinae (Euphorbiaceae). *Lilloa* **8**: 273–333.
Luti, R., unpub. Fieldbooks, in Herbarium, Univ. Córdoba.
Macloskie, G., 1903–1906. Flowering plants. *Princeton Univ. Exped. Patagonia* **8(1–9)**: pp. 981.
Madsen, H. B., Nielsen, E. S. and Ødum, S., 1980. The Danish scientific expedition to Patagonia and Tierra del Fuego, 1978–1979. *Geog. Tidsskr.* **80**: 1–28.
Marchionatto, J., 1940. Las especies de *Cyttaria* y *Cyttariella* en la Argentina. *Darwiniana* **4**: 9–32.
Marett, R. R. and Penniman, T. K. (Eds.) 1931. *Spencer's Last Journey*. Clarendon, Oxford.
Martinez-Crovetto, R., 1967. Catálogo preliminar de las Caryophyllaceae de la Argentina y del Uruguay. *Bonplandia* **3**: 187–264.
Martinez-Crovetto, R., 1968. Estudios Etnobotánicos. IV – Nombres de plantas y su utilidad, según los indios Onas de Tierra del Fuego. *Fac. Agron. Vet. Corr. Feb. No.* 3.
Martial, L.-F., 1888. *Mission Scientifique du Cap Horn. Vol. I. Histoire du Voyage*. Gauthier-Villars, Paris.
Martín, R. R., 1949. Consideraciones generales sobre la flora actual de las turberas de Tierra del Fuego. *Rev. Argent. Agron.* **16**: 149–155.
Martinić, M., 1976. La expansión económica de Punta Arenas sobre los territorios argentinos de la Patagonia y Tierra del Fuego, 1885–1925. *Anal. Inst. Pat., Punta Arenas (Chile)* **7**: 5–42.
Martinić, M., 1977. *Historia del Estrecho de Magallanes*. Andres Bello, Santiago de Chile.
Mathias, M. E. and Constance, L., 1955. The genus *Oreomyrrhis* (Umbelliferae). *Univ. Calif. Publs. Bot.* **27**: 347–416.
Mathias, M. E. and Constance, L., 1971. A first revision of

Huanaca (Umbelliferae-Hydrocotyloideae). *Kurtziana* **6**: 7–23.

Matthei, O. R., 1965. Estudio crítico de las gramíneas del género *Stipa* en Chile. *Gayana (Bot.)* **13**: 1–137.

Melchior, H., 1964. A. Engler's *Syllabus der Pflanzenfamilien. I.* Borntraeger, Berlin.

Melchior, H. and Werdermann, E., 1954. A. Engler's *Syllabus der Pflanzenfamilien. II.* Borntraeger, Berlin.

Mexia, Y., unpub. Field books, in Bancroft Library, Univ. Calif., Berkeley.

Middleton, R. M., 1909. The First Fuegian Collection. *Jour. Bot.* **47**: 207–212.

Milbraed, J., 1908. Stylidiaceae. *Pflanzenreich* **35(IV.278)**: 19–22.

Molina, A. M., 1978. El género *Gunnera* en la Argentina y Uruguay (Gunneraceae). *Darwiniana* **21**: 473–489.

Moore, D. M., 1963. The subspecies of *Montia fontana* L. *Bot. Notiser* **116**: 16–30.

Moore, D. M., 1967. Chromosome numbers of Falkland Islands angiosperms. *Bull. Brit. Antarct. Surv.* **14**: 69–82.

Moore, D. M., 1968. The vascular flora of the Falkland Islands. *Brit. Antarct. Surv. Sci. Rep.* **60**: 1–202.

Moore, D. M., 1969. *Saxifragodes*, a new genus of Saxifragaceae from Tierra del Fuego. *Bot. Notiser* **122**: 322–329.

Moore, D. M., 1970. Studies in *Colobanthus quitensis* (Kunth) Bartl. and *Deschampsia antarctica* Desv. II. Taxonomy, distribution and relationships. *Bull. Brit. Antarct. Surv.* **23**: 63–80.

Moore, D. M., 1970b. Additions to the vascular flora of Tierra del Fuego. *Bol. Soc. Argent. Bot.* **13**: 1–9.

Moore, D. M., 1971. *Tapeinia* Comm. ex Juss. (Iridaceae) and the rediscovery of *Galaxia obscura* Cav. *Bot. Notiser* **124**: 82–86.

Moore, D. M., 1972. Connections between cool temperate floras, with particular reference to southern South America. *In*: D. H. Valentine (ed.) *Taxonomy, Phytogeography* and *Evolution*. pp.115–138. Academic Press, London.

Moore, D. M., 1974. Catálogo de las plantas vasculares nativas de Tierra del Fuego. *Anal. Inst. Pat., Punta Arenas (Chile)* **5**: 105–121.

Moore, D. M., 1975. The alpine flora of Tierra del Fuego. *Anal. Inst. Bot. Cavanilles* **32**: 419–440.

Moore, D. M., 1978. Post-glacial vegetation in the South Patagonian territory of the giant ground sloth, *Mylodon*. *Bot. Jour. Linn. Soc. London* **77**: 177–202.

Moore, D. M., 1979. Southern oceanic wet-heathlands (including Magellanic moorland). In: R. L. Specht (Ed.) *Heathlands and Related Shrublands of the World* pp.489–497. Elsevier, Amsterdam.

Moore, D. M., 1979b. Vascular plant collections and observations. *Rep. Cambridge Exped. Tierra del Fuego*, 54–59. Cambridge.

Moore, D. M., 1981. Chromosome numbers of Fuegian angiosperms. *Bol. Soc. Brot.*, sér. 2, **53**: 995–1012.

Moore, D. M., Arroyo, S. C. and Doggett, M. C., 1982. The status of *Phaiophleps lyckholmi* (Dusén) R. C. Foster (Iridaceae). *Bot. Jour. Linn. Soc. London* **84**: 103–113.

Moore, D. M. and Chater, A. O., 1971. Studies on bipolar species. I. *Carex* L. *Bot. Notiser* **124**: 317–334.

Moore, D. M. and Doggett, M. C., 1975. *Hierochloë redolens* (Vahl.) Roem. and Schult. in Tierra del Fuego and adjacent regions. *Bol. Soc. Argent. Bot.* **16**: 390–396.

Moore, D. M. and Goodall, R. N. P., 1973. Interspecific hybridisation in Fuegian *Caltha* L. *Bol. Soc. Argent. Bot.* **15**: 72–76.

Moore, D. M. and Goodall, R. N. P., 1974. Further additions to the native vascular flora of Tierra del Fuego. *Bot. Notiser* **127**: 38–43.

Moore, D. M. and Goodall, R. N. P., 1977. La flora adventícia de Tierra del Fuego. *Anal. Inst. Patag., Punta Arenas (Chile)* **8**: 263–274.

Moore, D. M. and Scotter, B., 1976. *Vicia* L. in Tierra del Fuego. *Darwiniana* **20**: 371–386.

Moore, D. M. and Yates, B., 1974. *Armeria* in South America. *Bot. Notiser* **127**: 183–192.

Muñoz, C., 1960. *Las Especies de Plantas Descritas por R. A. Philippi en el Siglo XIX.* Universidad de Chile, Santiago.

Navas, B. E., 1961. El género *Urtica* en Chile. *Bol. Soc. Argent. Bot.* **9**: 395–413.

Nicora, E., 1978. *Poa*. *In*: Correa, M. N. (Ed.) *Flora Patagonica* **3**: 138–207.

Ojeda, R., 1966. Breve ensayo sobre el clima y la dinámica de Punta Arenas. (Mimeo), FACH.

Ødum, S., Hjerting, J. P. and Søegaard, B., 1977. The Nordic arboretum expedition to southernmost Argentina and adjacent Chile, 1975. *Royal Vet.-Agric. Univ., Hørsholm Arboretum.*

Oliver, S. P., 1909. *The Life of Philibert Commerson.* Murray, London.

Orchard, A. E., 1981. A revision of South American *Myriophyllum* (Haloragaceae) and its repercussions on some Australian and North American species. *Brunonia* **4**: 27–65.

Orsi, M. C., 1976. Sinopsis de las especies argentinas del género *Berberis* (Berberidaceae). *Bol. Soc. Argent. Bot.* **17**: 127–149.

Parodi, L. R., 1931. Revisión de las gramíneas austroamericanas del género *Alopecurus*. *Rev. Fac. Agron. Vet. Bs Aires* **7**: 345–369.

Parodi, L. R., 1940. Estudio crítico de las gramíneas austro-americanas del género *Agropyron*. *Rev. Mus. La Plata, Bot.* **3**: 1–63.

Parodi, L. R., 1941. Las gramíneas sudamericanas del género *Hierochloë*. *Rev. Mus. La Plata, Bot.* **3(14)**: 183–212.

Parodi, L. R., 1949. Las gramíneas sudamericanas del género *Deschampsia*. *Darwiniana* **8**: 415–475.

Parodi, L. R., 1953. Las especies de *Festuca* de la Patagonia. *Rev. Argent. Agron.* **20**: 177–229.

Parodi, L. R., 1960. Las especies de *Stipa* del subgénero *Pappostipa* de la Argentina y Chile. *Rev. Argent. Agron.* **27**: 65–106.

Parodi, L. R., 1961. Ciento cinquenta años de botánica en la República Argentina. *Bol. Soc. Argent. Bot.* **9**: 1–68.

Parris, B. S. and Given, D. R., 1976. A taxonomic revision

of the genus *Grammitis* Sw. (Grammitidaceae: Filicales) in New Zealand. *N.Z. Jour. Bot.* **14**: 85–111.
Paz, F. M., 1886. Territorios australes: Expedición a la bahía de San Sebastián. *Bol. Inst. Geogr. Argent.* **7**: 217–219.
Pérez-Moreau, R. A., 1937. Las umbelíferas argentinas del género *Lilaeopsis*. *Lilloa* **1**: 283–306.
Pérez-Moreau, R. A., 1938. Revisión de las *Hydrocotyle* argentinas. *Lilloa* **2**: 413–463.
Pichi Sermolli, R. E. G. and Bizzarri, M. P., 1978. The botanical collections (Pteridophyta and Spermatophyta) of the AMF Mares-G.R.S.T.S. Expedition to Patagonia, Tierra del Fuego and Antarctica. *Webbia* **32(2)**: 455–534.
Pilger, R., 1928. Die Gattung *Plantago* in Zentral- und Sud-Amerika. *Bot. Jahrb.* **62**: 1–113.
Pisano V., E., 1971. Comunidades vegetales del área del Fiordo Parry, Tierra del Fuego. *Anal. Inst. Patag. Punta Arenas (Chile)* **2**: 93–133.
Pisano V., E., 1972a. Comunidades vegetales del área de Bahía Morris, Isla Capitan Aracena, Tierra del Fuego. *Anal. Inst. Patag. Punta Arenas (Chile)* **3**: 103–130.
Pisano V., E., 1972b. Observaciones fito-ecológicas en las Islas Diego Ramirez. *Anal. Inst. Patag. Punta Arenas (Chile)* **3**: 161–169.
Pisano V., E., 1973a. Fitogeografía de la Península de Brunswick, Magallanes. I. *Anal. Inst. Patag. Punta Arenas, Chile* **4**: 141–206.
Pisano V., E., 1973b. La vida en los parques nacionales de Magallanes. *Monog. Inst. Patag. Punta Areans (Chile)* **6**: 1–64.
Pisano V., E., 1974. Estudio ecológico de la región continental sur del área andino–patagónica. II. Contribución a la fitogeografía de la zona del Parque Nacional 'Torres del Paine'. *Anal. Inst. Patag. Punta Arenas (Chile)* **5**: 59–104.
Pisano V., E., 1976a. Adiciones a la flora vascular nativa de Tierra del Fuego. *Anal. Inst. Patag. Punta Arenas (Chile)* **7**: 155–157.
Pisano V., E., 1976b. Contaminación por petroleo del B/T 'Metula' en vegetación fanerogámica litoral. *Anal. Inst. Patag. Punta Arenas (Chile)* **7**: 139–153.
Pisano V., E., 1977a. Fitogeografía de Fuego–Patagonia Chilena I. – Comunidades vegetales entre las latitudes 52° y 56°S. *Anal. Inst. Patag. Punta Arenas (Chile)* **8**: 121–250.
Pisano V., E., 1977b. Contribución de Enrique Ibar Sierra al conocimiento de la naturaleza de la Patagonia Oriental Austral. *Anal. Inst. Patag. Punta Arenas (Chile)* **8**: 95–119.
Pisano V., E., 1980a. Catálogo de la flora vascular del archipiélago del Cabo de Hornos. *Anas. Inst. Patag. Punta Arenas (Chile)* **11**: 151–189.
Pisano V., E., 1980b. Distribución y características de la vegetación del archipiélago del Cabo de Hornos. *Anal. Inst. Patag. Punta Arenas (Chile)* **11**: 191–224.
Pontiroli, A., 1963. Flora Argentina. Calyceraceae. *Rev. Mus. La Plata, N. ser.* **9**: 175–241.
Quinn, C. J., 1982. Taxonomy of *Dacrydium* Sol. ex Lamb. emend. de Laub. (Podocarpaceae). *Austral. Jour. Bot.*, **30**: 311–320.
Räsänen, V., 1932. Zur Kenntnis der Flechtenflora Feuerlands. *Ann. Bot. Soc. Zoo. Bot. Fenn. Vanamo* **2(1)**: 1–65.
Ratter, J. A., 1973. Cytogenetic studies in *Spergularia*: VII, cryptic speciation in *S. media* (L.) C. Presl and *S. marina* (L.) Griseb. *Not. Roy. Bot. Gard. Edinb.* **32**: 291–296.
Raven, P. H., 1963. Amphitropical relationships in the floras of North and South America. *Quart. Rev. Biol.* **38**: 151–177.
Rechinger, K. H., 1934. Die süd- und zentralamerikanischen Arten der Gattung *Rumex*. *Ark. Bot.* **26A**: 1–58.
Ricardi, M., 1963. Rehabilitación del género *Oreopolus* Schlecht. *Gayana (Bot.)* **6**: 3–16.
Roivainen, H., 1933. Contribuciones a la flora de Isla Elizabeth, Río de las Minas y Puerto San Isidor de Prov. de Magallanes, de Puerto Barroso de Prov. de Chiloe y de los alrededores de Termas de Chillan de Prov. de Ñuble, Chile. *Ann. Bot. Soc. Zoo. Bot. Fenn. Vanamo* **4(8)**: 1–22.
Roivainen, H., 1954. Studien über die Mooren Feuerlands. *Ann. Bot. Soc. Zoo. Bot. Fenn. Vanamo* **28(2)**: 1–205.
Rousson, H. and Willems, P., 1891. Mission scientifique a la Terre de Feu. *Compt. Rend. Soc. Geog.* **1891**: 176–183.
Rousson, H. and Willems, P., 1892. La Terre de Feu et ses habitants. *Compt. Rend. Ass. Fr. Annee Sci.* **2**: 961–965.
Rúgolo de Agrasar, Z. E., 1978. Las especies australes del género *Deyeuxia* (Gramineae) de la Argentina y Chile. *Darwiniana* **21**: 417–453.
Ruiz-Leal, A., 1954. Flora de las inmediaciones de Ushuaia (Tierra del Fuego). *Rev. Fac. Cienc. Agrar. Mendoza* **4**: 1–34.
Samuelsson, G., 1923. Revision des südamerikanischen *Epilobium* – Arten. *Svensk Bot. Tidskr.* **17(2)**: 241–296.
Savatier, P. A. L., 1880. Contributions a la géographie médicale. Station navale de l'Océan Pacifique. *Arch. Médecin Navale* **33**: 5–35.
Schulz, O. E., 1927. Cruciferae – *Draba* et *Erophila*. *Pflanzenreich* **89(IV.105)**: 1–396.
Schwaar, J., 1981. Amphi-arktische Pflanzengesellschaften in Feuerland. *Phytocoenologia* **9(4)**: 547–572.
Skottsberg, C. J. F., 1905. Feuerlandische Blüten. *Wiss. Ergebn. Schwed. Südpolar-Exped.* **4(2)**: 1–75.
Skottsberg, C. J. F., 1906. Zur Flora des Feuerlandes. *Wiss. Ergebn. Schwed. Südpolar-Exped.* **4(4)**: 1–39.
Skottsberg, C. J. F., 1909. Pflanzenphysiognomische Beobachtungen aus dem Feuerlande. *Wiss. Ergebn. Schwed. Südpolar-Exped.* **4(9)**: 1–63.
Skottsberg, C. J. F., 1911. *The Wilds of Patagonia*. Arnold, London.
Skottsberg, C. J. F., 1916. Vegetationsverhältnisse langs der Cordillera de los Andes S. von 40°S. Br. *K. Svensk. Vetensk Akad. Handl.* **56(5)**: 1–366.
Skottsberg, C. J. F., 1924. Bemerkungen zu einigen *Chloraea* und *Asarca* Arten. *Acta Horti Gothoburg* **1**:

211–224.
Skottsberg, C. J. F., 1926. Eine kleine Pflanzenliste aus dem Feuerlande. *Acta Horti Gothoburg.* **2**: 27–37.
Skottsberg, C. J. F., 1958. Myzodendraceae. *Pflanzenreich* **62(IV.68)**: 1–16.
Sleumer, H., 1935. Revision der Gattung *Pernettya* Gaudich. *Notizbl. Bot. Gart. Mus. Berlin* **12**: 626–645.
Sleumer, H., 1951. Las Ericáceas Argentinas. *Lilloa* **25**: 515–556.
Sleumer, H., 1956. Die Hieracien Argentiniens. *Bot. Jahrb.* **77**: 85–148.
Sleumer, H., 1968. Die Gattung *Escallonia* (Saxifragaceae). *Verh. K. Ned. Akad. Wet. Afd. Natuurk.* **58**: 1–146.
Smit, P. G., 1973. A revision of *Caltha* (Ranunculaceae). *Blumea* **21**: 119–150.
Smith, A. C., 1943. The American species of *Drimys*. *Jour. Arnold Arbor.* **24**: 1–33.
Smith, R. I. L., 1973. The occurrence of *Empetrum rubrum* Vahl ex Willd. on South Georgia. *Bull. Brit. Antarct. Surv.* **33/34**: 173–174.
Solbrig, O. T., 1962. The South American species of *Erigeron*. *Contr. Gray Herb.* **191**: 1–79.
Solbrig, O. T., 1966. The South American species of *Gutierrezia*. *Contr. Gray Herb.* **197**: 3–42.
Soraru, S. B., 1972. Revisión de las *Urticaceae* argentinas. *Darwiniana* **17**: 246–325.
Soriano, A., 1942. Las especies del género *Suaeda* en la República Argentina. *Rev. Argent. Agron.* **9**: 343–353.
Soriano, A., 1947. Las Quenopodiáceas de la tribu Salicornieae en la República Argentina. *Rev. Argent. Agron.* **14**: 148–172.
Soriano, A., 1948. Las exploraciones botánicas en la Patagonia Argentina. *Ciencia Investig.* **4**: 443–453.
Soriano, A., 1948b. El género *Benthamiella* (Solanaceae). *Darwiniana* **8**: 233–262.
Sota, E. R. de la, 1960. Polypodiaceae y Grammitidaceae argentinas. *Opera Lilloana* **5**: 1–229.
Sota, E. R. de la, 1966. Consideraciones sobre las especies australes de *Grammitis*, anatomía de *Grammitis magellanica* Desv. y su forma *nana* (Grammitidaceae-Filices). *Darwiniana* **14**: 42–54.
Sota, E. R. de la, 1970. Notas sobre las especies austrosudamericanas del género *Blechnum* L. (Blechnaceae-Pteridophyta) I. *Bol. Soc. Argent. Bot.* **13**: 129–139.
Sparre, B., 1949. Systematical and nomenclaturistical studies in the genus *Viola*. I. An attempt at a revision of the section *Chilenium*. *Lilloa* **17**: 377–416.
Sparre, B., 1950. Systematical and nomenclatorial studies on the genus *Viola*. II. Further annotations on the section *Chilenium*. *Lilloa* **23**: 347–364.
Spegazzini, C., 1883. Informe VIII. Relación botánica. *In*: Bove, G. (Ed.) *Expedición Austral. Argentina: Informes Preliminares.* Dept. Nac. Agric., Buenos Aires.
Spegazzini, C., 1896. Plantae per Fuegiam a Carolo Spegazzini anno 1882 collectae. *Anal. Mus. Nac. Bs Aires. 2 ser.* **5**: 39–103.
Spegazzini, C., 1924. Relación de un paseo hasta el Cabo de Hoorn. *Bol. Acad. Ci. Córdoba* **27**: 321–404.
Stearn, W. T., 1969. A Royal Society appointment with Venus in 1769: The Voyage of Cook and Banks in the Endeavour in 1768–1771 and its botanical Results. *Not. Rec. Royal Soc. London* **24(1)**: 64–90.
Stebbins, G. L., 1974. *Flowering Plants: Evolution above the Species Level.* Arnold, London.
Suárez de Cullen, D. C. A., 1953. Las especies argentinas del género *Calandrinia*. *Bol. Soc. Argent. Bot.* **5**: 1–29.
Subils, R., 1977. Las especies de *Euphorbia* de la República Argentina. *Kurtziana* **10**: 83–248.
Townsend, C. H., 1910. A naturalist in the Straits of Magellan. *Pop. Sci. Monthly*, **July**: 1–16.
Vuilleumier, B. S., 1970. The systematics and evolution of *Perezia* sect. *Perezia* (Compositae). *Contr. Gray Herb.* **199**: 1–163.
Walton, D. W. H., 1975. Taxonomic notes on South American species of the genus *Acaena* L. (Rosaceae). *Darwiniana* **19**: 500–509.
Weberling, F. and Uhlarz, H., 1977. Morphologische, anatomische und palynologische Untersuchungen an der Gattung *Aretiastrum* (Valerianaceae). *Pl. Syst. Evol.* **127**: 217–242.
Webster, W. H. B., 1834. *Narrative of a Voyage to the Southern Atlantic in the years of 1828, 29, 30, performed on H.M. Sloop 'Chanticlear' under the command of the late Capt. Henry Foster, FRS.* London.
Widén, K.-G., 1971. The genus *Agrostis* in eastern Fennoscandia. Taxonomy and distribution. *Fl. Fenn.* **5**: 1–209.
Wildeman, É. de, 1905. *Voyage du S.Y. 'Belgica'. Botanique.* Buschmann, Anvers.
Wilkes, C., 1845. *Narrative of the United States Exploring Expedition during the years 1838, 1839, 1840, 1841, 1842.* Sherman, Philadephia.
Wimmer, F. E., 1955. Campanulaceae-Lobelioideae. *Pflanzenreich* **107(IV.276b)**: 1–813.
Young, S. B., 1970. Vascular plant investigations in Magellanic Chile, September–November 1969. *Antarct. Jour. U.S.* **5**: 119–120.
Young, S. B., 1973. Subantarctic rain forest of Magellanic Chile: distribution, composition, and age and growth rate studies of common forest trees. *Antarct. Res. Ser. U.S.* **20**: 307–322.
Zardini, E. M., 1971. Las especies del género *Sedum* (Crassulaceae) espontáneas en la República Argentina. *Bol. Soc. Argent. Bot.* **14**: 95–106.
Zotov, V. D., 1973. *Hierochloë* R. Br. (Gramineae) in New Zealand. *N.Z. Jour. Bot.* **11**: 561–579.

VII. Indian Names and Uses of Plants

When Europeans settled in Tierra del Fuego just over a hundred years ago the area was the home of four distinct tribes of Indians: the Haush (Aush), Ona (Shelknam), Yahgans (Yamana) and Alacaluf. Now the Fuegian Indians are extinct or virtually so, and the few survivors have lost practically all of the old customs and languages. Living, as they did, in close contact with the environment the Indians had considerable natural lore on the plants and animals of Tierra del Fuego (e.g. Bridges, 1948). Although much of this information must now be permanently lost to students of the area valuable data are provided in the literature, notably by Bridges (1933, 1948), Gusinde (1937), Lothrop (1928) and Martinez-Crovetto (1968). These, together with other relatively reliable sources (see below), have been used to compile the list of Indian names and uses for Fuegian plants given in this section; only the Ona and Yahgans, the two most numerous tribes, have been considered.

There must inevitably be errors in any compilation of plant names based upon records for which there seem to be few botanical voucher specimens, in many cases assembled by people who were not primarily botanists. Some of the more obviously erroneous records have been omitted here, such as an Ona name for *Bolax caespitosa* which does not occur in the territory formerly occupied by that tribe (Martinez-Crovetto, 1968), but others undoubtedly remain. A number of the names attributed to several species are clearly the generalised terms, such as 'daisy', 'margarita', 'grass' and 'clover', commonly applied by most peoples to plants of little social or economic significance. Thus, the Ona used the word álcha (álchai, hálcha) for a number of small herbs, and their similar generalised use of koóshpa (= flower), se (= ear) and the picturesque term kéjke t'al (= plover's nest) can be understood in this context; the significance of t'ol (young guanaco), followed by harr(e) (hide) or kóche (leg), is much less obvious. The Yahgans' use of the suffix amaaim (= berry) is clearly related to the economic attributes of the species to which it is applied. In contrast, taxa such as *Apium australe, Berberis buxifolia, B. ilicifolia, Chiliotrichum diffusum, Empetrum rubrum, Nothofagus, Osmorhiza*, and *Ribes*, important in providing food and materials for the Indians, have several names, associated with the various utilisable parts of the plants, while the equally important *Pernettya pumila* even has Yahgan names for different habits, depending upon whether it grows in dry (kupöna) or wet (shanamaim) conditions.

Since the indians had no written records, various phonetic systems have been employed by authors. No attempt has been made here to correlate such systems and the spellings follow those of the source references. It may be noted, however, that š, č and ö represent, respectively, the sounds sh, ch and up. The sources are listed in order of apparent accuracy, dependent upon the author's knowledge of and contact with the Indians, although this is clearly a highly subjective decision. In the following lists Yahgan names are indicated by an asterisk, as are the uses which follow these names given for each species. Both taxa and Indian names are listed alphabetically. The sequence of authorities is as follows: 1. Bridges (1933); 2. Bridges, G. G., unpub., Indian names noted by his daughter, C. M. Bridges de Goodall; 3. Bridges (1948); 3a. T. L. Bridges (pers. comm.); 3b. R. N. Prosser de Goodall, (pers. com.); 4. Lothrop (1928); 5. Gallardo (1910); 6. Beauvoir (1905); 7. Gusinde (1937); 8. Martinez-Crovetto (1968); 9. Coiazzi (1914):

LIST OF TAXA

Acaena spp.
hálcha (8), kéjke t'al (8), tâpl (8)

A. lucida
kéjke t'al (8)

Andenocaulon chilense
álcha (8), álchai (8), hálcha (8)
Adesmia lotoides
kiárksh (8)
Rhizomes eaten (8)
Anemone multifida
bul* (1)
Antennaria chilensis
koóshpa (8)
Arenaria serpens
chepl (8), chipl (8), chispl (8), hálcha (8)
Apium australe
alcho (6), hušun* (1, 7), kiél (i) (8), kugy-in* (1), laem-uska* (1), shon (6).
Leaves and roots eaten (8). *Used for pads to make canoes watertight (1).
Arjona patagonica
téen (8)
Prized for eating (8)
Armeria maritima
t'ol kóche (8), yöšama* (1), yöxama* (7).
*Roots eaten (1).
Artemesia magellanica
chárr (i) (8).
Aster vahlii
kóoshpa (8).
Azorella filamentosa
tésh(ue)n (8)
A. lycopodioides
tésh (8), tésh(ue)n (8)
A. selago
tésh(ue)n (8), tísh(ue)n (8)
A. trifurcata
tésh(ue)n (8), tísh(ue)n (8)
A. spp. Roots of all *Azorella* eaten raw or warmed over coals (8).

Baccharis patagonica
kóoshpa se (8)
Berberis buxifolia
humërzamaim* (7), kor (6, 9), kôr (8), me'ch (8), meech (5), metz (4), mich (8), miích (8), tjohorr (6), ummöš* (1), umm-öšaaicix* (1), umm-öšamaiim* (1), umöš* (1), umöš-amaiim* (1), umöš-auacix* (1), úmosh*(4), umush*(2), umushamaim* (3).
Wood used for arrows (3, 4, 8) and fire-tongs (4). Berries eaten after arrival of Europeans (5). *Wood for fire-tongs (4).
B. empetrifolia
mich kan (8)
Fruit eaten (8)
B. ilicifolia
celia* (1, 7), cilia-maiim* (1), celiamaim* (7), celiummöš* (1), cháukel (6, 8), chelia* (2), chélia* (4), cilërz* (7), huma* (7), mech (3a), meech (5), mich (8), umm–a* (1), yaan (2, 3a), yahn (4).
Preferred wood for arrows (3, 4, 8).
Bolax gummifera
tanna-wöla* (1), tésh(ue)n (8), tísh(ue)n (8).
Root eaten raw or warmed over coals (8).
Boopis australis
íshta (8)
Roots eaten roasted over coals (8).

Calceolaria uniflora
haiaien (2)
Caltha sagittata
châl (8), shin (8).
Cardamine glacialis
álcha (8), álchai (8), hálcha (8)
Carex curta
hoshl (8), hushl (8), séssur (8), shésh-hu(e)rr (8)
Cerastium spp.
huruf* (1)
C. arvense
álcha (8), cho(u) (8), haarn(e) (8), kálun ját (8), koóshpa (8)
Chiliotrichum diffusum
góorrh (4), gorhi (2), kó-or (8), kóor (8), kors (6), korsch (6, 9), korschpa (9), kpeen (6), sherr (6), yeya* (1), yéya* (4), yiya* (2).
Wood used for arrows (4). Plant used in tattooing (8).
Flowers rubbed over eyes to clear vision.
Colobanthus subulatus
t'ol (8)
Coronopus didymus
káuon (8)
Cotula scariosa
kéjke t'al (8)
Cystopteris fragilis
cilaaia-gömaka* (1), kéjke t'al (8), kelpin (6)

Descurainia antarctica (? + *Capsella bursa–pastoris*)
hálcha(i) (8), táaiiu (8), táaiu (8), tai (8, 10), tâiiu (8), táiu (8), tay (5), thai (6).
Seeds ground to make bread (5, 9).
Drimys winteri
aikuš* (1), chôl (8), choól (8), kalchenika (6), school (5), ukúshta* (4), ukušta* (1), ushcuta* (2), uškutta* (1).
Wood used for harpoons because it is heavy and sinks (8). Infusion of bark used against dandruff and diseases such as scurvy (8). *Wood used as spear-shafts (3). *Bark ground for pepper (3). *Small twigs used to line and protect bark of canoes (1).

Embothrium coccineum
maku* (7), mögu* (1), mögu-luš* (1), mugu* (2), mu'gu* (4).
Empetrum rubrum
asana* (7), fruta-wáshj (8), kol(e) (8), schal (6), sebisa* (1), sebusa* (1), sépice* (5), sepisa* (1, 3), sépisa* (2), sépisamaiim* (2), sepisamaim* (7), shalu (2), wáshege (5).
*Berries eaten (3).
Erigeron spp.
cho(u) (8), harr âl (8), harrn(e) (8).

Fuchsia magellanica
yeya-rax* (1).

Galium spp.
huruf* (1)
G. aparine
álcha (8)
Gentianella magellanica
poótrl (8)
Geranium patagonicum
kilpl orr (8)
Geum magellanicum
shipi (8), téuten (8)
Gunnera magellanica
o(u)ltá waásh (8), paiyo-l* (1), se (8), tuxalamaim* (7)

Hebe elliptica
kökaila★ (1), wieng-u★ (1)
Hypochoeris incana
álbi (8), sóol (8)
Toasted roots eaten (8).

Jaborosa magellanica
parr (8)

Lebetanthus myrsinites
chauter (6)
Luzula alopecurus
hoshl (8), hushl (8), sésuer (8), shésh-hu(e)rr (8)
Luzuriaga marginata
amann★ (2)

Madia sativa
hoshl (8), hushl (8)
Marsippospermum grandiflorum
kartay (6), mápi★ (4), mapi★ (7), möpi★ (1), taáiiu (8), taíiu (8), táiu (8), táyu (5), thai (6), thayii (6).
Used for making baskets, ankle- and wrist-bands (4, 5, 8). ★Used for baskets (1, 3, 4, 7).
Maytenus magellanica
aiaku★ (1), aíaku★ (4), aiakus★ (7), aiyakus★ (7), háaiko (8), haikó (6, 8), háiko (8), iacu★ (3, 3b), iacu-liush★ (3b), jaku★ (1), kaáiko(u) (8), káiko (8), liuš★ (1), tailó (6).
Wood used for arrow-shafts, bows, harpoons and scraper-handles (8). ★Used for arrows (1).
Myriophyllum quitense
kaar (8)
Misodendrum spp.
haniskstauwöla★ (1), téno (8), tenoká (8), ténokán (8).
Body rubbed with plant to ease muscular pains (8).

Nanodea muscosa
aultal-kashkal (6), cilaaia-'amaaim★ (1)
Nassauvia darwinii
cho(u) (8), harrn(e) (8)
Nothofagus antarctica
char(n) (4), c'tern★ (2), cucharrn (5), kechárn (8), kicharrn (3), kökauiya★ (1), kötölli-ran★ (1), kötöran★ (1), kucharn (6), kutúran★ (4), showínshi (8), wín (8), wínchi (8), wínshi (8).
Used for making bows (3, 4, 5, 8).
N. betuloides
iéñú (8), kienñú(n) (8), kieñú (8), kíniu (8), kiñiú (8), mögatöx★ (1), öšpötöx★ (1), serzci★ (7), shuschi★ (1), shushi★ (3), šöpaia-tösiluwa★ (1), šöšči (1), šošči-sin★ (1), šöš-paia★ (1), súschi★ (4), sushpeya★ (3a).
Wood used for harpoons (8). Bark-torches for hunting (8). ★Bark for canoes and buckets (3, 4, 7, 9).
N. pumilio
comink (2), cuálchinke (5), hámo★ (4), hanis★ (1, 3, 7), hanislush★ (2), kawaltchénk (4), kualchink (3, 8), kaulchínke (8), walchin (6), walcim (9)
Sap and scrapings eaten in spring (5, 6, 8, 9). Used for making torches (8). ★Sap and scrapings eaten in spring (3).

Oreomyrrhis hookeri
kufy-in★ (1), seltái (8)
Root edible
Osmorhiza chilensis
álcha (8), álchai (8), auwön-im★ (1), auwön-imapi★ (1), awanim★ (7), hálcha(i) (8), lux★ (1), owwunim★ (1), pulöšči★ (1)
★Roots edible and considered a delicacy (1, 2, 3, 7)
Oxalis enneaphylla
al haístikin (8)

Perezia recurvata
t'ol harr (8)
Pernettya mucronata
amai-a-inguš★ (1), amaiisgurz★ (7), auttal-kaskál (9), chals (6), goosh★ (3), gus★ (4), gush★ (2, 5), scheuk (10), shal (8), uashaj (2), wáshege (5), yah (n) (4).
P. pumila
küpona★ (1, 2), sanamaim★ (7), shal (8), shanamaim★ (1, 2, 3, 5) wasch (9), wáshege (5)
Fruit eaten (5, 8, 9). ★Fruit eaten (1, 2, 3, 7)
Phacelia secunda
âl (8), kástjen (8), se (8)
Phaiophleps biflorus
t'ol harre (8), t'ol kóche (8)
Pilgerodendron uvifera
lipaix★(7), shios (6)
★Valued for spear-shafts
Plantago barbata
sésuer (8), shéshen-hu(e)rr (8)
P. maritima
karr (8)
Poa flabellata
garuga★ (3b)
Polygonum aviculare
káuon (8)
Primula magellanica
cho(u)j(n) (8)

Ranunculus spp.
kiél(i) (8), wökat-an-amaiim★ (1)
Ribes magellanicum
hupus★ (7), hupusamaim★ (7), jaujton (6), schal (6), schiteren (9), shéthen (8), shetrr (8), shétrrhen (8), shíterhén (4), shiterjen (2), shitrr (8), upuš-amaim★ (1), upush★ (2), upúsh★ (4), upushamaiim★ (2).
Wood used for arrow-shafts (4, 8). Fruit eaten (8). Infusion of leaves and bark taken for stomach-ache (8)
Rostkovia magellanica
taáiiu (8), taáiu (8), tai (8), taíiu (8), táiu (8)
Rubus geoides
belacimaim★ (3), belaka-aachij★ (2), belak-aacix★ (1), belakamaiim★ (2), belak-'amaiim★ (1), palaxaacix★ (7), schal (6), schinamos awat (9), waásh-shal (8).
Fruit eaten (8)
Rumex acetosella
óte(n) (8), shésh-hu(e)rr (8)
R. maritimus
se (8)

Senecio spp.
kufy-in★ (1), laem-uska★ (1)
S. acanthifolius
koóshpa (8), se (8)
S. candidans
masa★ (2), mufkuwa★ (1), se kôj (8)
S. laseguei
kpen (8)
S. magellanicus
harr âl (8), poótrl (8)
S. miser
kpen (8)

S. smithii
koóshpa se (8), se (8)
S. tricuspidatus
kpen (8)
Scirpus californicus
tâíiu (8)
Silene magellanica
poól (8)
Sisyrinchium patagonicum
kóóshpa (8)
Stellaria spp.
álchai (8)

Taraxacum spp.
hui-tai (3b), oité (10), oitén (6), wetie (2), weyeta* (2), wi-tay (3b), wiya-ta* (1), wiyata* (7)
Whole plant eaten (8)
T. gilliesii
aitá (8), oí'ta (8), oitá (8), oitái (8)
T. officinale
oí'ta (8), ôitá (8), pitá (8)

Urtica magellanica
mohopeli (6), pasaan-göla* (1), tâtl (8)

Valeriana carnosa
kuí (8)
Veronica peregrina
tâpl (8)
Vicia bijuga
êl (8)

LIST OF INDIAN NAMES

aiaku*, áiaku*, aiakus* Maytenus magellanica
aikuš* Drimys winteri
aiyakus* Maytenus magellanica
aitá Taraxacum gilliesii
âl Phacelia secunda
álbi Hypochoeris incana
álcha Andenocaulon chilense
Cardamine glacialis
Cerastium arvense
Galium aparine
Osmorhiza chilensis
álchai Adenocaulon chilense
Cardamine glacialis
Stellaria debilis
Osmorhiza chilense
alcho Apium australe
al haístikin Oxalis enneaphylla
amai-a-inguš* Pernettya mucronata
amaiisgurz* Pernettya mucronata (fruit)
amann* Luzuriaga marginata
asana* Empetrum tubrum
aultal-kashkal Nanodea muscosa
auttal-kaskâl Pernettya mucronata (fruit)
awanim* Osmorhiza chilensis
auwön-im* Osmorhiza chilensis
auwön-im* Osmorhiza chilensis
auwön-imapi* Osmorhiza chilensis (seeds)

belacamaim*, belakamaiim* Rubus geoides
belak-'amaiim*
belaka-aachij* Rubus geoides
belak-aacix* Rubus geoides (Flowers)
bul* Anemone multifida

celia* Berberis ilicifolia
celiamiam*, celia-maiim* Berberis ilicifolia (berries)
celiummöš* Berberis ilicifolia
châl Caltha sagittata
chals Pernettya mucronata
char(n) Nothofagus antarctica
chárr(i) Artemesia magellanica
cháukel Berberis ilicifolia
chauter Lebetanthus myrsinites
chelia*, chélia* Berberis ilicifolia
chepl, chipl, chispl Arenaria serpens
cho(u) Cerastium arvense
Erigeron spp.
Nassauvia darwinii
cho(u)j(n) Primula magellanica
choól (chôl) Drimys winteri
cilaaia-'amaaim* Nanodea muscosa
cilaaia-gömaka* Cystopteris fragilis
cilërz* Berberis ilicifolia
cí'tá Taraxacum gilliesii
comink Nothofagus pumilio
c'tern* N. antarctica
cuálchinke N. pumilio
cucharrn N. antarctica
c(u)ltá waásh Gunnera magellanica

êl Vicia bijuga

fruta-wáshj Empetrum rubrum

garuga* Poa flabellata
góorh Chiliotrichum diffusum
goosh* Pernettya mucronata
gorhj Chiliotrichum diffusum
gus(h)* Pernettya mucronata

hahâien Calceolaria uniflora
haiaien Calceolaria uniflora
haikó, háaiko, háiko Maytenus magellanica
hálcha Acaena spp.
A. pinnatifida
Adenocaulon chilense
Arenaria serpens
Descurainia antarctica
Cardamine glacialis
Osmorhiza chilensis
hámo* Nothofagus pumilio
hanis* Nothofagus pumilio
haniskstawöla Misodendrum spp.
hanislush* Nothofagus pumilio
harrn(e) Cerastium arvense
Erigeron spp.
Nassauvia darwinii
harr âl Erigeron spp.
Senecio magellanicus
hoshl Carex curta
Luzula alopecurus
Madia sativa
hui-tai Taraxacum spp.
huma* Berberis ilicifolia
humërzamaim* B. buxifolia (ripe fruit)

hupus*, hupusamaim* Ribes magellanicum
huruf* Cerastium spp.
hushl Carex curta
Luzula alopecurus
Madia sativa
husun* Apium australe (fruit)

iacu* Maytenus magellanica
iacu-liush* Maytenus magellanica
iéñu Nothofagus betuloides
íshta Boopis australis

jaku Maytenus magellanica
jaujton Ribes magellanicum

kaáiko(u), káiko Maytenus magellanica
kaar Myriophyllum quitense
kalchenika Drimys winteri
kálun ját Cerastium arvense
karr Plantago maritima
kartay Marsippospermum grandiflorum
kástjen Phacelia secunda
káuon Coronopus pidymus
Polygonum aviculare
kawaltchénk Nothofagus pumilio
kechárn N. antarctica
kéjke t'al Acaena spp.
A. lucida
Cotula scariosa
Cystopteris fragilis
kelpin Cystopteris fragilis
kiárksh Adesmia lotoides
kicharrn Nothofagus antarctica
kiél(i) Apium australe
Ranunculus spp.
kieñú, kienñú(n) Nothofagus betuloides
kilpl orr Geranium patagonicum
kíniu, kiñiú Nothofagus betuloides
kökaila* Hebe elliptica
kökauiya* Nothofagus antarctica (saplings)
kol(e) Empetrum rubrum
ko'or, kóor Chiliotrichum diffusum
koóshpa Antennaria chilensis
Aster vahlii
Cerastium arvense
Sisyrinchium patagonicum
koóshpa se Boopis australis
Senecio acanthifolius
S. smithii
kôr Berberis buxifolia (fruit)
kors, korsch, korschpa Chiliotrichum diffusum
kötölli-ran* Nothofagus antarctica
kötöran* Nothofagus antarctica
kpeen Chiliotrichum, diffusum
kpen Senecio laseguei
S. miser
S. tricuspidatus
kualchínk(e) Nothofagus pumilio
kucharn N. antarctica
kufy-in* Apium australe (stems)
Osmorhiza chilensis
Senecio smithii
kuí Valeriana carnosa
kupöna* Pernettya pumila (growing on dry ground)
kutúran* Nothofagus antarctica

laem-uska* Apium australe (flowers)
Senecio magellanicus
lipaix* Pilgerodendron uvifera
liuš* Maytenus magellanica (foliage)
lux* Osmorhiza chilensis

maku* Embothrium coccineum
mápi*, mapi* Rostkovia magellanica
masa* Senecio candidans
mech Berberis ilicifolia
meech, me'ch, miích Berberis buxifolia
metch, metq Berberis buxifolia
mich Berberis buxifolia B. ilicifolia
mich kan B. empetrifolia
mögat-öx* Nothofagus betuloides (sapling)
mögu*, mögu-luš* Embothrium coccineum
mohopeli Urtica magellanica
möpi* Marsippospermum grandiflorum
mufkuwa* Senecio candidans
mugu*, mu'gu* Embothrium coccineum

oitá, oitái Taraxacum gilliesii
ói'tá, ôitá T. officinale
oité, oiten Taraxacum spp.
öspötöx Nothofagus betuloides
óte(n) Rumex acetosella
owwunim* Osmorhiza chilensis

paiyo–l* Gunnera magellanica
palaxaacix* Rubus geoides
parr Jaborosa magellanica
pasaan-göla* Urtica magellanica
pitá Taraxacum officinale
poól Silene magellanica
poótrl Gentianella magellanica
Senecio magellanicus
pulöšči* Osmorhiza chilensis (young root)

sanamaim* Pernettya pumila
schal Empetrum rubrum
Ribes magellanicum (berry)
Rubus geoides
scheuk Pernettya mucronata
schinamose awatal Rubus geoides
schiteren Ribes magellanicum
school Drimys winteri
se Gunnera magellanica
Phacelia secunda
Rumex maritimus
Senecio smithii
sebisa*, sebusa* Empetrum rubrum
se kôj Senecio candidans
seltái Oreomyrrhis hookeri
sépice*, sepisa*, sépisa Empetrum rubrum
sepisamaim*, sépisamaiim Empetrum rubrum (berries)
sërzci Nothofagus betuloides
séseur Carex curta
sésuer Luzula alopecurus
Plantago barbata
shal Pernettya mueronata
P. pumila
shalu Empetrum rubrum
shanamaim* Pernettya pumila (growing in swamps)
sherr Chiliotrichum diffusum
shésh-hu(e)rr Carex curta

Luzula alopecurus
Plantago barbata
Rumex acetosella
shéthen Ribes magellanicum
shétrr(hen) Ribes magellanicum
shin Caltha sagittata
shios Pilgerodendron uvifera
shipi Geum magellanicum
shíterhen, shiterjen Ribes magellanicum
shitrr Ribes magellanicum
shon Apium australe
showínshi Nothofagus antarctica
shuschi, shushchi* N. betuloides
sóol Hypochoeris incana
šöpaia-tösiluwa* Nothofagus betuloides (diseased leaves)
šöšči Nothofagous betuloides
šošči–sin* Nothofagus betuloides
šöš–paia* Nothofagus betuloides (foliage)
súschi* Nothofagus betuloides
sushpaya* Nothofagus betuloides

tai, taáiu, taáiiu Descurainia antarctica
Rostkovia magellanica
tâiiu Descurainia antarctica
Scirpus californicus
taikó Maytenus magellanica
táiu Rostkovia magellanica
tana-wöla* Berberis ilicifolia
tâpl Acaena spp.
Veronica peregrina
tâtl Urtica magellanica
tay Capsella bursa-pastoris
táyu Marsippospermum grandiflorum
téen Arjona patagonica
téno, tenoká, ténokán Misodendrum spp.
tesh Azorella lycopodioides
tésh(ue)n Bolax gummifera
Azorella filamentosa
téuten Geum magellanicum
thai Capsella bursa-pastoris
thayii Marsippospermum grandiflorum
tísh(ue)n Bolax gummifera
Azorella selago
A. trifurcata
tjohorr Berberis buxifolia
t'ol Colobanthus subulatus
t'ol harr Perezia recurvata
Phaiophleps biflorus
t'ol kóche Ameria maritima
Phaiophleps biflorus
tuxalamaim* Gunnera magellanica (fruit)

uashaj Pernettya mucronata
ukúshta*, ukušta* Drimys winteri
umm-a* Berberis ilicifolia
umm-öšaaicix* Berberis buxifolia (flowers)
umm-öšamaiim* Berberis buxifolia (berries)
umöš*, ummöš*, úmosh* Berberis buxifolia
umöš–amaiim* Berberis buxifolia (berries)
umöš–auačix* Berberis buxifolia (flowers)
umush*, umushamaim* Berberis buxifolia
upush*, upúsh* Ribes magellanicus
upushamaiim*, upuš-amaim* Ribes magellanicus (berries)
ushcuta*, uskutta* Drimys winteri

waásh shal Rubus geoides
walchin, walcim Nothofagus pumilio
wasch Pernettya pumila
wáshege Empetrum rubrum
Pernettya mucronata
P. pumila
wásje Empetrum rubrum
wetie Taraxacum spp.
weyeta* Taraxacum spp.
wieng-u* Hebe elliptica
wín, wínchi, wínshi Nothofagus antarctica
wi-tay, wiya-ta*, wiyata* Taraxacum spp.
wokat-an-amaiim* Ranunculus minutiflorus

yaan, yahn Berberis ilicifolia
yéya*, yiya* Chiliotrichum diffusum
yeya-raz* Fuchsia magellanica
yöšama*, yöxama* Armeria maritima

VII. Spanish and English Names

Almost all the common names of plants in Tierra del Fuego are those used in mainland Chile and Argentina north of the Estrecho de Magallanes. In most cases the same species are involved but in a number of instances this is not so; thus, for example, on the mainland the names Mata negra and Mata verde refer to, respectively, *Verbena tridens* and *Chiliotrichum diffusum*, while in Tierra del Fuego they are used for, respectively, *Chiliotrichum diffusum* and *Lepidophyllum cupressiform*. Furthermore, as in most parts of the world, groups of similar species, or even genera, are known by a single common name; for example, most species of *Acaena* are known as Abrojo, while almost all members of the Compositae with daisy-like flowers will be usually called Margarita.

The few English names employed for Fuegian plants are mostly those used in the Falkland Islands (Islas Malvinas), with which some of the earlier settlers had connections. Other names are those commonly applied to the same or similar groups in Europe or North America. It is possible to coin further names but these have not generally been included in the list below since they are not widely used by the local population. For the most extensive list of possible Spanish and English common names the work of Goodall (1979) should be consulted.

Abrojo *Acaena* spp.
Abrojo común *Acaena magellanica*
Achicoria *Taraxacum gilliesii*
Agropiro *Agropyron* spp.
Almond flower *Luzuriaga marginata*
Amor seco *Acaena pinnatifida*
Anémone *Anemone multifida*
Apio nativo *Apium australe*
Arvejilla *Lathyrus* spp.

Balsam bog *Bolax gummifera*
Bedstraw *Galium* spp.
Beech *Nothofagus* spp.
Berro nativo *cardamine glacialis*
Botón de oro *Ranunculus peduncularis*
Buttercup *Ranunculus* spp.
Butterwort *Pinguicula antarctica*

Cadillo *Acaena pinnatifida*
Calafate *Berberis buxifolia*
Calafatillo *Berberis empetrifolia*
Campanilla *Phaiophleps biflorus*
Canchalagua *Gentianella magellanica*
Canelo *Drimys winteri*
Cebada patagónica *Hordeum comosum*
Cebadilla *Bromus* spp.
Centella *Anemone multifida*
Cerastio *Cerastium arvense*
Chacai (Chacay) *Discaria chacaye*
Chaura *Pernettya mucronata*
Chilco, Chilcón *Fuchsia magellanica*
Chillum *Nertera depressa*
Chique *Pernettya mucronata*
Christmas bush *Baccharis magellanica*
Ciprés *Pilgerodendron uvifera*
Ciruelillo *Embothrium coccineum*
Coicopihue *Philesia magellanica*
Coihue *Nothofagus betuloides*
Coirón *Festuca gracillima*
Coirón amargo *Stipa* spp.
Comida de Culebras *Nertera depressa*
Cortadera *Cortaderia pilosa*
Couch grass *Agropyron* spp.
Cress *Cardamine glacialis*

Dandelion *Taraxacum* spp.
Diddle-dee *Empetrum rubrum*
Dock *Rumex* spp.
Dog orchid *Codonorchis lessonii*

Edelweiss fueguino *Perezia magellanica*
Escorzonera *Hypochoeris arenaria*
Estrella de la vega *Arjona pusilla*

Fachine *Chiliotrichum diffusum*
Firebush *Embothrium coccineum*
Flor de la cascada *Ourisia ruelloides*
Fox berry *Nanodea muscosa*

Frutilla del diablo *Gunnera magellanica*
Frutilla de Magallanes *Rubus geoides*

Geranio *Geranium* spp.
Guindo *Nothofagus betuloides*

Hair grass *Deschampsia* spp.
Hierba de sapo *Myriophyllum quitense*

Junco *Scirpus californicus*
Junco canasto *Marsippospermum grandiflorum*
Junco común *Rostkovia magellanica*

Lady's slipper *Calceolaria* spp.
Lahuán *Pilgerodendron uvifera*
Land Tussac *Festuca contracta*
Leña dura *Maytenus magellanica*
Lenga *Nothofagus pumilio*
Lengua de gato *Galium aparine*
Llalante *Geum magellanicum*
Llareta *Bolax gummifera*

Madi *Madia sativa*
Maillico *Caltha sagittata*
Maitén enano *Maytenus magellanica*
Malvina berry *Myrteola nummularia*
Margarita *Aster vahlii*
Mata negra *Chiliotrichum diffusum*
Mata verde *Lepidophyllum cupressiforme*
Mealy primrose *Primula magellanica*
Medallita *Lebetanthus myrsinites*
Melosa *Madia sativa*
Michay *Berberis ilicifolia*
Monte alto *Nothofagus pumilio*
Monte bajo *Nothofagus antarctica*
Mosaiquillo *Baccharis magellanica*
Murta *Pernettya mucronata*
Mutilla *Empetrum rubrum, Pernettya pumila*

Nancú-lahuén *Valeriana carnosa*
Ñire *Nothofagus antarctica*
Notro *Embothrium coccineum*

Ojo de agua *Oxalis enneaphylla*
Orquídea amarilla *Gavilea lutea*
Ortiga caballuna *Urtica magellanica*

Palomita *Codonorchis lessonii*
Parrilla *Ribes magellanicum*
Pega-pega *Galium antarcticum*
Pichoa *Euphorbia collina*
Pickwood *Maytenus magellanica*
Plantain *Plantago* spp.
Primavera *Primula magellanica*
Primula *Primula magellanica*

Quilineja *Luzuriaga marginata*

Rainberry *Rubus geoides*
Ratonera *Hierochloë redolens*
Romaza común *Rumex magellanicus*
Romerillo *Baccharis patagonica*

Sauco del diablo *Pseudopanax laetevirens*
Sea cabbage *Senecio candidans*
Siempreviva *Armeria maritima*
Streaked maiden *Phaiophleps biflorus*
Sundew *Drosera uniflora*

Taique *Desfontainia spinosa*
Té malvinero *Myrteola nummularia*
Té pampa *Satureja darwinii*
Tepú *Tepualia stipularis*
Topa-topa *Calceolaria biflora*
Triguillo *Elymus antarcticus*
Tussac grass *Poa flabellata*

Violet *Viola* spp.
Violeta amarilla *Viola maculata, magellanica, reichei*
Violeta blanca *Viola commersonii*
Violeta de la montaña *Viola tridentata*

Waterfall plant *Ourisia ruelloides*
Wild celery *Apium australe*
Wild currant *Ribes magellanicum*
Winter's bark *Drimys winteri*
Woodrush *Luzula* spp.

Zanahoria salvaje *Daucus montanus*
Zapatilla de la Virgen *Calceolaria uniflora*

IX. Abbreviations of Author's Names

Aellen P. Aellen (1896–1973)
Aiton W. Aiton (1731–1793)
Alboff N. M. Alboff (1866–1897)
All. C. Allioni (1728–1804)
Alm. S. O. I. Almquist (1844–1923)
Alston A. H. G. Alston (1902–1958)
Andersson N. J. Andersson (1821–1880)
Arcangeli G. Arcangeli (1840–1921)
Ard. P. Arduino (1728–1805)
Arn. G. A. W. Arnott (1799–1868)
Baillon H. E. Baillon (1827–1895)
Baker J. G. Baker (1834–1920)
Balbis G. B. Balbis (1765–1831)
Banks J. Banks (1743–1820)
Barnéoud F. M. Barnéoud (b. 1821)
Barros M. Barros (1880–1973)
Bartl. F. G. Bartling (1798–1875)
Baumg. J. C. G. Baumgarten (1765–1843)
Beauv., P. A. M. F. J. Palisot de Beauvois (1752–1820)
Beetle A. A. Beetle (b. 1913)
Bentham G. Bentham (1800–1884)
Berg O. C. Berg (1815–1886)
Bernh. J. J. Bernhardi (1774–1850)
Bert. C. G. Bertero (1789–1831)
Bitter F. A. G. Bitter (1873–1927)
Blytt M. N. Blytt (1789–1862)
Blume C. L. von Blume (1796–1862)
Bocquet G. Bocquet (b. 1927)
Boiss. P. E. Boissier (1810–1885)
Bonpl. A. J. A. Bonpland (1773–1858)
Boott F. Boott (1792–1863)
Boreau A. Boreau (1803–1875)
Bory J. B. G. M. Bory de Saint-Vincent (1778–1846)
Br., N. E. N. E. Brown (1849–1934)
Br., R. R. Brown (1773–1858)
Brack. W. D. Brackenridge (1810–1893)
Briquet J. I. Briquet (1870–1931)
Britton N. L. Britton (1859–1934)
Brongn. A. T. Brongniart (1801–1876)
Buchenau F. G. P. Buchenau (1831–1906)
Burkart A. E. Burkart (1906–1975)
Cabrera A. L. Cabrera (b. 1908)
Calderón C. E. Calderón (b. 1931)
Campd. F. Campderá (1793–1862)
Camus, A. A. Camus (1879–1965)
Carmichael D. Carmichael (1772–1827)
Cass. A. H. G. Cassini (1781–1832)
Cav. A. J. Cavanilles (1745–1804)
Chaix D. Chaix (1703–1799)
Cham. L. A. von Chamisso (1781–1838)
Chambers K. L. Chambers (b. 1929)
Chater A. O. Chater (b. 1933)
Ching R. Ching (b. 1899)
Chevall. F. F. Chevallier (1796–1840)
Christ H. Christ (1833–1933)
C. Chr. C. F. A. Christensen (1872–1942)
Clarke, C. B. C. B. Clarke (1821–1906)
Clos D. Clos (1821–1908)
Colla L. A. Colla (1766–1848)
Comm. P. Commerson (1727–1773)
Conert H. J. Conert (b. 1929)
Const. L. Constance (b. 1909)
Correa M. N. Correa (b. 1914)
Coville F. V. Coville (1867–1937)
Cronq. A. J. Cronquist (b. 1919)
Dahlst. H. G. A. Dahlstedt (1856–1934)
Davy J. B. Davy (1870–1940)
DC. A. P. de Candolle (1778–1841)
Dcne J. Decaisne (1807–1882)
De Paula M.E. de Paula de Brooks (b. 1940)
Desf. R. L. Desfontaines (1750–1833)
Desr. L. A. J. Desrousseaux (1753–1838)
Desv. A. N. Desvaux (1784–1856)
Dewey C. Dewey (1784–1867)
Diem J. Diem (b. 1899)
Doell J. C. Doell (1808–1885)
Doggett M. C. Doggett (b. 1951)
Domin K. Domin (1882–1953)
Don, D. D. Don (1799–1841)
Don, G. G. Don (1764–1814)
Douglas D. Douglas (1798–1834)
Duby J. E. Duby (1798–1885)
Dudley, T. T. R. Dudley (b. 1936)
Dumort. B. C. J. Dumortier (1797–1878)
D'Urv. J. S. C. D. D'Urville (1790–1842)
Dusén P. K. H. Dusén (1855–1926)
Ehr. J. F. Ehrhart (1742–1795)
Endl. S. L. Endlicher (1804–1849)
Engelm. G. Engelmann (1809–1884)
Engler H. G. A. Engler (1844–1930)
Fabris H. A. Fabris (1924–1976)
Fassett N. C. Fassett (1900–1954)
Fenzl E. Fenzl (1808–1879)
Fischer F. E. L. von Fischer (1782–1854)

Florin C. R. Florin (1894–1965)
Forrest G. Forrest (1873–1932)
Forskål P. Forskål (1732–1763)
Forster J. R. Forster (1729–1798)
Forster f. J. G. A. Forster (1754–1794)
Foster, R. C. R. C. Foster (b. 1904)
Fourn., E. E. P. N. Fournier (1834–1884)
Franchet A. R. Franchet (1834–1900)
Fraser J. Fraser (1750–1811)
Fries E. M. Fries (1794–1878)
Gaertner J. Gaertner (1732–1791)
Gaertner, C. F. C. F. von Gaertner (1772–1850)
Gand. M. Gandoger (1850–1926)
Gaudich. C. Gaudichaud-Beaupré (1789–1854)
Gaudin J. F. A. T. G. P. Gaudin (1766–1833)
Gay, C. C. Gay (1800–1873)
Gilg E. F. Gilg (1867–1933)
Gillies J. Gillies (1792–1834)
Gmelin, C. C. C. C. Gmelin (1762–1837)
Gmelin, J. F. J. F. Gmelin (1748–1804)
Gooden S. Goodenough (1743–1827)
Graebner K. O. P. P. Graebner (1871–1933)
Grande L. Grande (1878–1965)
Gray, A. A. Gray (1810–1888)
Greene E. L. Greene (1843–1915)
Grev. R. K. Greville (1794–1866)
Griseb. A. H. R. Grisebach (1814–1879)
Gunn. J. E. Gunnerus (1718–1773)
Hackel E. Hackel (1850–1926)
Haenke T. Haenke (1761–1816)
Hagström J. O. Hagström (1860–1922)
Hariot P. A. Hariot (1854–1917)
Hauman L. Hauman (1880–1965)
Hausskn. H. K. Haussknecht (1838–1903)
Hegelm. C. F. Hegelmaier (1834–1906)
Herbert W. Herbert (1778–1847)
Hicken C. M. Hicken (1875–1933)
Hieron. G. H. E. Hieronymus (1846–1921)
Hill J. Hill (1716–1775)
Hochst. C. F. Hochstetter (1787–1860)
Hoffm. G. F. Hoffmann (1761–1826)
Hoffm., O. O. Hoffmann (1853–1909)
Hombron J. B. Hombron (1800–1852)
Hooker W. J. Hooker (1785–1865)
Hooker, f. J. D. Hooker (1817–1911)
Host N. T. Host (1761–1834)
Hughes D. K. Hughes (1899–1932)
Hultén E. O. G. Hultén (b. 1894)
Humb. F. H. A. von Humboldt (1769–1859)
Humphries C. J. Humphries (b. 1947)
Jackson B. D. Jackson (1846–1927)
Jacq. N. J. von Jacquin (1727–1817)
Jacq. f. J. F. von Jacquin (1766–1839)
Jacquinot H. Jacquinot (1814–1887)
Jalas J. Jalas (b. 1920)
Johnston, I. M. I. M. Johnston (1898–1960)
Juss. A. L. de Jussieu (1748–1836)
Kaulf. G. F. Kaulfuss (1786–1830)
Kerner A. J. Kerner von Marilaun (1831–1898)
Kirk, T. T. Kirk (1828–1898)
Klatt F. W. Klatt (1825–1897)
Knuth, R. R. G. P. Knuth (1874–1957)
Koch W. D. J. Koch (1771–1849)
Koch, C. C. H. E. Koch (1809–1879)
Koeler G. L. Koeler (1765–1807)
Koern. F. A. Koernicke (1828–1908)
Kraenzlin F. W. L. Kraenzlin (1847–1934)
Krasser F. Krasser (1863–1922)
Krecz. V. I. Kreczetowicz (1901–1942)
Kuhn M. F. A. Kuhn (1842–1894)
Kük. G. Kükenthal (1864–1955)
Kunth C. S. Kunth (1788–1850)
Kuntze, O. K. E. O. Kuntze (1843–1907)
Kunze, G. G. Kunze (1793–1851)
Kurtz, F. F. Kurtz (1854–1920)
L. C. von Linné (1707–1778)
L. f. C. von Linné (1741–1783)
Labill. J. J. H. de Labillardière (1755–1834)
Lag. M. Lagasca y Segura (1776–1839)
Lam. J. B. A. P. Monnet de la Marck (1744–1829)
Lambert, A. N. A. B. Lambert (1761–1842)
Laterrade J. F. Laterrade (1784–1858)
Le Gall N. J. M. le Gall (1787–*c.* 1860)
Lehm J. G. C. Lehmann (1792–1860)
Less. C. F. Lessing (1810–1862)
L'Hérit. C. L. L'Héritier de Brutelle (1746–1800)
Lichtenstein J. S. de Lichtenstein (b. 1902)
Lightf. J. Lightfoot (1735–1788)
Lindley J. Lindley (1799–1865)
Lindman C. A. M. Lindman (1856–1928)
Link J. H. F. Link (1767–1851)
Looser G. Looser (1898–1981)
Lorentz P. G. Lorentz (1835–1881)
Lourteig A. Lourteig (b. 1913)
Macloskie G. Macloskie (1832–1920)
Mart. C. F. P. von Martius (1794–1868)
Martens G. M. von Martens (1788–1872)
Mathias M. E. Mathias (b. 1906)
Medicus F. C. Medicus (1736–1808)
Menz. A. Menzies (1754–1842)
Mérat F. V. Mérat (1780–1851)
Mett. G. H. Mettenius (1823–1866)
Meyen F. J. F. Meyen (1804–1840)
Meyer, C. A. C. A. von Meyer (1795–1855)
Meyer, E. E. H. F. Meyer (1791–1858)
Michx A. Michaux (1746–1802)
Miers J. Miers (1789–1879)
Miller P. Miller (1691–1771)
Mirbel C. F. B. Mirbel (1776–1854)
Mol. J. I. Molina (1740–1829)
Moore, D. M. D. M. Moore (b. 1933)
Moore, T. T. Moore (1821–1887)
Moq. C. H. B. A. Moquin-Tandot (1804–1863)
Moris G. G. Moris (1796–1869)
Müller Arg. J. Müller of Aargau (1828–1896)
Murray J. A. Murray (1740–1791)
Naegeli C. W. von Naegeli (1817–1891)
Naudin C. V. Naudin (1815–1899)
Nees C. G. D. Nees von Esenbeck (1776–1858)
Neger F. W. Neger (1868–1923)
Nicora E. G. Nicora (b. 1912)
Niederl. G. Niederlein (1858–1924)
Nutt. T. Nuttall (1786–1859)
Oersted A. S. Oersted (1816–1872)
Palla E. Palla (1864–1922)
Parodi L. R. Parodi (1895–1966)
Pavón J. Pavón (1750–1844)
Pennell F. W. Pennell (1886–1952)
Pérez-Moreau R. A. Pérez-Moreau (1905–1975)
Pers. C. H. Persoon (*c.*1762–1836)

Petagna V. Petagna (1734–1810)
Peter G. A. Peter (1853–1937)
Pfeiffer H. Pfeiffer (b. 1896)
Phil. R. A. Philippi (1808–1904)
Pichi Serm. R. E. G. Pichi-Sermolli (b. 1912)
Pilger R. K. F. Pilger (1876–1953)
Poeppig E. F. Poeppig (1798–1868)
Poiret J. L. M. Poiret (1755–1834)
Pollich J. A. Pollich (1740–1780)
Porter T. C. Porter (1822–1901)
Prantl K. A. E. Prantl (1849–1893)
Presl. C. C. B. Presl (1794–1852)
Presl, J. J. S. Presl (1791–1849)
Pursh F. T. Pursh (1774–1820)
Rafin. C. S. Rafinesque-Schmaltz (1783–1840)
Raspail F. V. Raspail (1794–1878)
Ravenna P. F. Ravenna (b. 1938)
Regel E. A. von Regel (1815–1892)
Reiche C. F. Reiche (1860–1929)
Remy E. J. Remy (1826–1893)
Ricardi M. Ricardi S. (b. 1921)
Richard, A. A. Richard (1794–1852)
Richter, K. K. Ritcher (1855–1891)
Roemer J. J. Roemer (1763–1819)
Rohrbach P. Rohrbach (1847–1871)
Roivainen H. Roivainen (b. 1900)
Rúgolo Z. Rúgolo de Agrasar (b. 1940)
Ruiz H. Ruiz López (1754–1815)
Rydb. P. A. Rydberg. (1860–1931)
Savi G. Savi (1769–1844)
Sch. Bip. K. H. Schultz (1805–1867)
Scheele G. H. A. Scheele (1808–1864)
Schenk, E. E. Schenk (1880–1965)
Schlecht. D. F. L. von Schlechtendal (1794–1866)
Schrank F. von Paula von Schrank (1747–1835)
Schreber J. C. D. Schreber (1739–1810)
Schübler G. Schübler (1787–1834)
Schultes J. A. Schultes (1773–1831)
Schulz, O. E. O. E. Schulz (1874–1936)
Scop. G. A. Scopoli (1723–1788)
Seem. B. C. Seemann (1825–1871)
Seringe N. C. Seringe (1776–1858)
Sibth. J. Sibthorp (1758–1796)
Skottsb. C. J. E. Skottsberg (1880–1963)
Sm. J. E. Smith (1759–1828)
Sm., W. W. W. Smith (1875–1956)
Sobol. G. F. Sobolewski (1741–1807)
Sol. D. C. Solander (1733–1782)
Soriano A. Soriano (b. 1919)
Sota, de la E. de la Sota (b. 1933)
Speg. C. Spegazzini (1858–1926)
Sprengel K. P. J. Sprengel (1766–1833)
St.-Amans J. F. B. de Saint-Amans (1748–1831)
Stapf O. Stapf (1857–1933)
Steudel E. G. von Steudel (1783–1856)
St.-Yves A. St.-Yves (1855–1933)
Swartz O. P. Swartz (1760–1818)
Tausch I. F. Tauch (1793–1848)
Ten. M. Tenore (1780–1861)
Thellung A. Thellung (1881–1928)
Thompson C. H. Thompson (1870–1931)
Thouars L. M. A. Aubert du Petit-Thouars (1758–1831)
Thuill. J. L. Thuillier (1757–1822)
Thunb. C. P. Thunberg (1743–1828)
Timm J. C. Timm (1734–1805)
Tindale M. D. Tindale (contemp.)
Tort R. Tortosa (b. 1946)
Trin. K. B. von Trinius (1778–1844)
Urban I. Urban (1848–1931)
Vahl M. H. Vahl (1749–1804)
Van den Bosch R. B. van den Bosch (1810–1862)
Van Tiegh. P. E. L. van Tieghem (1839–1914)
Vent. E. P. Ventenat (1757–1808)
Vill. D. Villars (1745–1814)
Volkens G. L. A. Volkens (1855–1917)
Wahlenb. G. Wahlenberg (1780–1851)
Wall. N. Wallich (1786–1854)
Wallr. K. F. W. Wallroth (1792–1857)
Walpers W. G. Walpers (1816–1853)
Walter T. Walter (1740–1789)
Walton D. W. H. Walton (b. 1945)
Watson, S. S. Watson (1826–1892)
Webb P. B. Webb (1793–1854)
Weber G. H. Weber (1752–1828)
Weddell H. A. Weddell (1819–1877)
Wildeman E. A. J. de Wildeman (1866–1947)
Willd. C. L. Willdenow (1765–1812)
Yates B. Yates (b. 1948)
Zotov V. D. Zotov (1908–1977)

X. Abbreviations of Books and Periodicals cited

Acta Horti Gothob. *Acta Horti gothoburgensis (gotoburgensis)./Meddelanden från Göteborgs botaniska Trädgård.* Göteborg. 1924→.

Agrost. Trunius, C. B. von *Agrostideae.* II. *Callo rotundo (Agrostea) Petropolis, typ. Academiae.* Leipzig. 1841.

Alpina (Winterthur). *Alpina.* Winterthur. **1–4**, 1806–1809.

Amer. Jour. Sci. Arts *American Journal of Science and Arts.* New Haven, Connecticut, 1818→.

Amen. Nat. Lagasca y Segura, M. *Amenidades naturales de las Españas.* **1(1)–1(2).** Madrid. 1811–1821.

Analecta Pterid. Kunz, G. *Analecta pteridographica.* Leipzig. 1837.

Anal. Mus. Nac. Chile, Bot. *Anales del Museo Nacional de Historia Natural. Botánica.* Santiago de Chile.

Anal. Mis. Hist. Nat. Valparaiso. *Anales del Museo de Historia Natural de Valparaiso.* Valparaiso.

Anal. Mus. Nac. Bs Aires. *Anales del Museo nacional de historia natural de Buenos Aires.* Titled *Anales del Museo nacional de Buenos Aires* 1864–1931, *Anales del Museo argentino de Ciencias Naturales "Bernadino Rivadavia"* 1931–1947. Buenos Aires.

Anal. Soc. Ci. Argent. *Anales de la Sociedad Científica Argentina. Buenos Aires. 1876→*

Anal. Univ. Chile. *Anales de la Universidad de Chile.* Valparaiso and Santiago. **1–50**, 1843–1922. Ser. 2, **1–8**, 1923–1930. Ser. 3, **1**, → 1931 →.

Ann. Bot. Soc. Zool.–Bot. Fenn. Vanamo. *Annales botanici Societatis zoologicae-botanicae fennicae 'Vanamo'.* Helsinki. **1–35**, 1931–1964.

Ann. Nat. Hist. *The Annals and Magazine of natural History, including Zoology, Botany and Geology London.* **6–20**, 1841–1842. Ser. 2, **1–20**, 1843–1857. Ser. 3, **1–20**, 1858–1867. Many later series.

Ann. Rep. Missouri Bot. Gard. *Annual Report. Missouri Botanical Garden.* St. Louis. Missouri. **1–23**, 1890–1912.

Ann. Sci. Nat. *Annales des Sciences naturelles.* Paris. Ser. 1, **1–30**, 1824–1833. Ser. 2, **1–20**, 1834–1843. Ser. 3, **1–20**, 1844–1853. Ser. 4, **1–20**, 1854–1863. Ser. 5, **1–20**, 1864–1874. Ser. 6, **1–20**, 1875–1885. Ser. 7, **1–20**, 1885–1895. Ser. 8, **1–20**, 1895–1904. Ser. 9, **1–20**, 1905–1917. Ser. 10, **1–20**, 1919–1938. Ser. 11, **1–20**, 1939–1959. Ser. 12, **1** →, 1960 →. (Ser. 2–10, *Botanique.* Ser. 11 →. *Botanique et Biologie végétale*).

Ann. Sci. Obs. *Annales des Sciences d'Observation.* Paris, **1–4**, 1829–1830.

Ark. Bot. *Arkiv för Botanik.* Uppsala. Ser. 1, **1–33**, 1903–1949. Ser. 2. **1** →, 1949 →.

Ber. Deutsch. Bot. Ges. *Berichte der deutschen botanischen Gesellschaft.* Berlin, 1833 → (**59–61**, 1941–1944 at Jena; **62** →, 1949 →, at Stuttgart).

Biblioth. Bot. (Stuttgart). *Bibliotheca botanica. Abhandlungen aus dem Gesammtgebiete der Botanik.* Cassel. 1886 →. (**29** →, 1894 →, at Stuttgart).

Bol. Acad. Ci. Córdoba. *Boletín de la Academia Nacional de Ciencias en Córdoba.* Córdoba, Argentina. **1** →. 1874 →.

Bol. Soc. Argent. Bot. *Boletin de la Sociedad Argentina de Botánica.* La Plata and Buenos Aires. 1945 →.

Bot. Centr. *Botanisches Centralblatt.* Jena & Dresden. 1880 →.

Bot. Gaz. *Botanical Gazette,* Hanover, Indiana. **2** →, 1876 →. (**22** →, 1896 →, at Chicago, Illinois).

Bot. Jahrb. *Botanische Jahrbücher für Systematik, Pflanzengeschichte und Pflanzengeographie.* Leipzig. 1880 →. (**69** →, 1938 →, at Stuttgart).

Bot. Jour. Linn. Soc. Cf. *Jour. Linn. Soc. Loncon (Bot.).*

Bot. King's Exped.

Bot. King's Exped. Watson, S., and others. *Botany. Report of United States Geological Exploration of the Fortieth Parallel.* Clarence King, Geologist-in-Charge. Vol. **5**, pp. 1–525. Washington. 1871.

Bot. Mag. *The botanical Magazine or Curtis' botanical Magazine.* London 1793 →.

Bot. Misc. Hooker, W. J. *Botanical Miscellany.* **1–2, Suppl.** London. 1829–1833.

Bot. Not. *Botaniska Notiser.* Lund. 1839–1980.

Bot. Zeit. *Botanische Zeitung.* Berlin. **1–68**, 1843–1910 (**14–68**, 1856–1910, at Leipzig).

Bull. Brit. Antarct. Survey. *British Antarctic Survey Bulletin.* London. 1963 →.

Bull. Mus. Nat. Hist. Nat. (Paris). *Bulletin du Muséum d'Histoire naturelle.* Paris. **1–12**, 1895–1906; titled *Bulletin du Muséum national d'Histoire naturelle,* **13** →, 1907.

Bull. Sci. Acad. Imp. Sci. Pétersb. *Bulletin scientifique publié par l'Académie impériale des Sciences de Saint-Pétersbourg.* Saint-Pétersbourg & Leipzig. **1–10**, 1836–1842.

Bull. Soc. Bot. Fr. *Bulletin de la Société botanique de France.* Paris. 1854 →.

Bull. Soc. Philom. Paris *Nouveau Bulletin des Sciences de la Société Philomatique de Paris.* Paris. 1807–25.

Candollea *Candollea. Organe du Conservatoire et du Jardin botaniques de la Ville de Genève.* Genève. 1922.

Char. Gen. Pl. Forster, J. R. & J. G. A. *Characteres Generum Plantarum, quas in Itinere ad Insulas Maris australis collegerunt, descripserunt, delinearunt ... Joannes Reinoldus Forster ... et*

Georgius Forster. Ed. 1. Londini. 1775. Ed. 2. Londini. 1776.

Chloris Andina Weddell, H. A. *Chloris andina. Essai d'une flore de la région alpine des cordillères de l'Amérique du sud*. Paris. 1855–61.

Collect. Bot. von Jacquin, N. J. *Collectanea ad Botanicam, Chemiam et Historiam naturalem spectantia*. Vindobonae. 1787–1797. (**1** in 1787; **2** in 1789; **3** in1790; **4** in 1791; **5** (*Supplementum*) in 1797.

Coll. Mém. Candolle, A. P. de. *Collection de mémoires*. **1–10**. Paris and Strasburg. 1828–1838.

Comment. Gotting. *Commentationes Societatis Regiae Scientiarum Gottingensis*. Ser. 2, **1–16**. Gottingen. 1779–1808.

Com. Mus. Nac. Bs Aires *Comunicaciones del Museo Nacional, Buenos Aires* (later titled *Communicaciones del Museo de Historia Natural*). Buenos Aires. 1898–1901.

Comp. Bot. Mag. Hooker, W. J. *Companion to the botanical magazine*. **1–2**. London. 1835–37.

Comp. Fl. Ital. Arcangeli, G. *Compendio della Flora italiana, ossia Manuale per la Determinazione delle Piante che trovansi selvatiche od inselvatichite nell'Italia e nelle Isole adiacenti*. Ed. 1. *Torino*. 1882. Ed. 2. Torino & Roma. 1894.

Contr. Gray Herb. *Contributions from the Gray Herbarium of Harvard University*. Cambridge, Massachusetts. Nov. Ser., **1** →, 1891 →.

Contr. U.S. Nat. Herb. *Contributions from the United States National Herbarium*. Washington. 1890 →.

Cyclop. Rees, A. *The cyclopaedia; or universal dictionary of arts, sciences and literature*. London. 1802–20.

Darwiniana. *Darwiniana*. Buenos Aires. 1922 →.

Del. Sem. Horto Hamburg. Lehmann, J. G. C. *Delectus seminum quae in horto hamburgensium botanico*. Hamburg. 1849–1852.

Demonstr. Pl. Linne, C. von. *Demonstrationes Plantarum in Horto upsaliensi 1753* ... Uppsala. 1753.

Denkschr. Schweiz. Naturf. Ges. *Denkschriften der schweizerischen naturforschenden Gesellschaft/Mémoires de la Société helvétique de Sciences naturelles*. Zürich. 1920 →.

Descr. Gram. Koeler, G. L. *Descriptio Graminum in Gallia et Germania tam sponte nascentium quam humana Industria copiosius provenientium*. Frankfurt. 1802.

Descr. Pl. Cavanilles, A. J. *Descripción des las plantas* ... Madrid. 1801–1802.

Deutschl. Fl. Sturm, J. *Deutschlands Flora*. Nürnberg. 1798–1855.

Deutschl. Fl. Taschen. Hoffmann, G. F. *Deutschlands Flora, oder botanisches Taschenbuch für das Jahr 1791*. Ed. 1. Erlangen. 1791. *Kryptogamie ... für das Jahr 1795*. Erlangen. 1795. Ed. 2 ... *für das Jahr 1800*. Erlangen. 1800.

Dict. Sci. Nat. *Dictionnaire des sciences naturelles*. **1–60**. Paris 1816–30.

Eclogae Pl. Rar. Jacquin, J. F. *Eclogae graminum rariorum*. Vienna. 1813–1844.

Encycl. Méth. Bot. Monnet de la Marck, J. B. A. P. *Encyclopédie méthodique botanique*. Paris. **1–13**. 1783–1817. (**5–13** by J. L. M. Poiret).

Enum. Kunth, C. S. *Enumeratio plantarum omnium hucusque cognitarum*. **1–5**. Stuttgart and Tubingen. 1833–50.

Enum. Fil. Kaulfuss, G. F. *Enumeratio filicum*, quas in itinere circa terram legit clar. Adalbertus de Chamisso adjectis ... Leipzig. 1824.

Enum. Pl. Vahl, M. H. *Enumeratio plantarum vel ab aliis, vel ab ipso observatarum, cum earum differentiis specificis, synonymis selectis et descriptionibus succinctis*. **1–2.** Köbenhavn. 1804–05.

Enum. Pl. Horti. Berol. Willdenow, C. L. *Enumeratio plantarum Horti regii botanici Berolinensis continens descriptiones omnium vegetabilium in Horto dicto cultorum*. Berlinl. 1809.

Enum. Stirp. Transs. Baumgarten, J. C. G. *Enumeratio stirpium magno Transsilvaniae principatui praeprimis indigenarum*. Vienna. 1816.

Erythea. *Erythea. A Journal of Botany*. Berkeley, California. 1893 →.

Ess. Agrost. Palisot de Beauvois, A. M. F. J. *Essai d'une nouvelle agrostographie*. Paris. 1812.

Exped. Río Negro. Lorentz, P. G. & Niederlein, G. *Enumeración sistemática de las plantas colectas durante la expedición. Informe official* ... *de la expedición al Río Negro (Patagonia) realizada ... 1879, bajo las ordenes del Grl Julio A. Roca*. pp. 171–294. Buenos Aires. 1881

Feddes Repert. *Repertorium novarum Specierum Regni vegetabilis*. Berlin. **1–8**: 1905–1910; titled *Repertorium Specierum novarum Regni vegetabilis*, **8–51**: 1910–1942; titled *Feddes Repertorium Specierum novarum Regni vegetabilis*, **52** →, 1943 →.

Festschr. Ascherson —Urban, I. and Graebner, K. O. P. P. *Fetschrift zur Feier des siebzigsten Geburtstages des ... Dr. P. Ascherson*. Leipzig. 1904.

Fil. Afr. Kuhn, M. *Filices Africanae... additamentis braunianis ... mettenianis adaucta, accedunt filices deckenianae et Peter siaiae*. Leipzig. 1868.

Fil. Lechl. Mettenius, G. *Filices Lechlerianae, Chilenses ac Peruanae*. **1–2**. Leipzig. 1856–59.

Fl. Agen. de Saint-Amans, J. F. B. *Flore agenaise*. Agen. 1821.

Fl. Amer. Sept. Pursh, F. T. *Flora Americae septentrionalis*. Ed. 1. **1–2**. London. 1814. Ed. 2. **1–2**. London. 1816.

Fl. Antarct. Hooker, J. D. *The botany of the Antarctic voyage*. **1**. *Flora Antarctica*. Pt. **1** *Botany of Lord Auckland's Group and Campbell's Island*. Pt. **2**. *Botany of Fuegia, the Falklands, Kerguelen's Land, etc*. London. 1847.

Fl. Atacam. Philippi, R. A. *Florula atacamensis*. Halle. 1860.

Fl. Belg. Dumortier, B. C. J. *Florula belgica, Operis majoris Prodromus. Staminecia*. Tornaci Nerviorum. 1827.

Fl. Bordel. Laterrade, J.-F. *Flore bordelaise*. Bordeaux. 1811.

Fl. Brit. Smith, J. E. *Flora Britannica*. **1–3**. London. 1800–04.

Fl. Carn. Scopoli, J. A. *Flora Carniolica, exhibens plantas Carnoliae indigenas*. 2nd ed., **1–2**. Vienna. 1772.

Fl. Carol. Walter, T. *Flora Carolianiana, secundum systema vegetabilium perillustrus Linnaei digesta* ... London. 1788.

Fl. Čechica Presl, J. S. & C. B. *Flora čechica*. Prague. 1819.

Fl. Centr. Fr. Boreau, A. *Flore du centre de la France*. Ed. 1, **1–2**, 1840. Ed. 2. **1–2**, 1849. Ed. 3. **1–2**, 1857. Angers.

Fl. Chile Reiche, C. F. *Flora de Chile*. 1–6. Santiago. 1894–1911.

Fl. Fr. Monnet de la Marck, J. B. A. P. *Flore française*. Ed. 1. **1–3**. Paris. 1779. Ed. 2. **1–3**. Paris. 1795. Ed. 3. by J. B. A. P. Monnet de la Marck & A. P. de Candolle. **1–5**. Paris. 1805–1815.

Fl. Gén. Env. Paris Chevallier, F. F. *Flore générale des Environs de Paris*. Ed. 1. **1–2**. Paris. 1826–1827. Ed. 2. Paris. 1836.

Fl. Ins. Austr. Prodr. Forster, J. G. A. *Florulae insularum Australium prodromus*. Göttingen. 1786.

Fl. Is. Mal. D'Urville, J. S. C. D. *Flore des Îles Malouines*. Paris. 1825. (Preprint of article published in *Mémoires de la Société Linneenne de Paris*, **4**, 573–621, 1826).

Fl. Lapp. Wahlenberg, G. *Flora lapponica*. Berolini. 1812.

Fl. Nap. Tenore, M. *Flora napolitana*. **1–5**. Naples. 1811–1838.

Fl. Norv. Gunnerus, J. E. *Flora norvegica*. **1–2**. Nidrosiae & Hafniae. 1766–1772

Fl. Oxon. Sibthorp, J. *Flora Oxoniensis*. Oxford. 1794.

Fl. Patag. Correa, M. N. (Ed.) *Flora Patagónica* pts. **2**, **3**, **7**.

Buenos Aires. 1969–1978.

Fl. Pedem. Allioni, C. *Flora pedemontana, sive Enumeratio methodica Stirpium indigenarum Pedemontii.* **1–3**. Turin. 1785.

Fl. Peruv. Ruiz Lopez, H. & J. Pavón. *Flora Peruviana* et *Chilensis, sive descriptiones et icones plantarum Peruvianarum et Chilensium.* **1–4**. Madrid. 1798–1802.

Fl. Petrop. Sobolewski, G. *Flora petropolitana.* Petropoli. 1799.

Fl. Scot. Lightfoot, J. *Flora scotica.* **1–2**. London. 1777.

Fl. Suec. Linne, C. von. *Flora Suecica, exhibens plantas per regnum Sueciae crescentes, systematice cum differentiis specierum.* Ed. 2. Stockholm. 1755.

Fl. Tristan d'Acugna. Petit-Thouars, L. M. A. du. *Mélanges de botanique et de voyages. V. Esquisse de la flore de l'Isle de Tristan d'Acugna.* Paris. 1808.

Flora (Regensb.). *Flora, oder allgemeine botanische Zeitung.* Regensburg. 1818 →. (**72–95**, Marburg; **96** →, Jena).

Fragm. Syn. Pl. Poeppig, E. F. *Fragmentarum Synopseos plantarum phanerogamarum chilensis.* Leipzig. 1833.

Freyc. Voy. Freycinet, L. de. *Voyage autour du monde, entrepris par order du roi, exécuté sur les corvettes de S. M. l'Uranie et la Physicienne, 1817–20. Botanique* par M. Gaudichaud-Beaupré. Paris. 1826.

Fruct. Sem. Pl. Gaertner, J. *De fructibus et seminibus plantarum.* **1–3**. Stuttgart, Tübingen and Leipzig. 1788–1807. (**3** by C. F. Gaertner).

Gard. Dict. Miller, P. *The gardeners dictionary.* 8th Ed. London. 1768.

Gayana, Bot. *Gayana, Botánica.* **1** →, Concepción. 1961 →.

Gen. Pl. Bentham, G. and J. D. Hooker. *Genera plantarum ad exemplaria imprimis in herbariis Kewensibus servata definita.* London. 1862–83.

Gen. Pl. Ord. Endlicher, S. L. *Genera Plantarum secundum Ordines naturales disposita.* Vienna 1836–1841. **Suppl. 1–5** 1842–1850.

Gen. Sp. Gent. Grisebach, A. H. R. *Genera et species Gentianearum* ... Stuttgart and Tubingen. 1839.

Gen. Syst. Don, G. *A general system of Gardening and Botany,* **1–4**. London. 1831–1838.

Gen. Spec. Orchid. Lindley, J. *The genera and species of orchidaceous plants.* London. 1830–40.

Ges. Nat. Freunde Berl. Mag. *Der Gesellschaft naturforschender Freunde zu Berlin, Magazin für die neuesten Entdeckungen in der gesammten Naturkunde.* **1–8**. Berlin. 1807–18.

Handb. Irid. Baker, J. G. *Handbook of the Irideae.* London, New York. 1892.

Handb. N. Z. Fl. Hooker, J. D. *Handbook of the New Zealand Flora.* Pts. 1–2. London. 1864–1867.

Herb. Brit. Hill, J. *Herbarium britannicum* ... **1–2**. London. 1769–1770.

Hickenia *Hickenia, boletín del Darwinión.* San Isidro. 1976 →.

Hist. Chile Bot. Gay, C. *Historia física y política del Chile* ... *Botánica.* **1–8**. Paris, Santiago. 1845–1854.

Hist. Pl. Baillon, H. E., *Histoire des Plantes,* **1–13**. Paris. 1866–1895.

Hist. Pl. Dauph. Villars, D. *Histoire des Plantes de Dauphiné.* **1–3**. Grenoble. 1786–1789.

Hist. Pl. Palat. Pollich, J. A. *Historia plantarum in Palatinatu electorali sponte nascentium incepta* ... **1–3**. Mannheim. 1776–1777.

Hort. Berol. Link, J. H. F. *Hortus regius botanicus berolinensis, descriptus.* **1–2**. Berlin. 1827–1833.

Hort. Kew. Aiton, W. *Hortus Kewensis, or a catalogue of the plants cultivated in the Royal Botanic Gardens at Kew.* London. Ed. 1, **1–3**, 1789; Ed. 2, **1–5**, 1810–13.

Hymenoph. Presl, C. B. *Hymenophyllaceae.* Prague. 1844.

Icon. Descr. Cavanilles, A. J. *Icones et descriptiones plantarum, quae aut sponte in Hispania crescunt, aut in hortis hospitantur.* **1–6**. Madrid. 1791–1801.

Icon. Fil. Hooker, W. J. and R. K. Greville. *Icones filicum* ... **1–2**. London. 1828–1831.

Icon. Pl. Hooker, W. J. *Icones plantarum* ... London. 1837–1854.

Illustr. Boott, F. *Illustrations of the genus Carex.* **1–4**. London. 1858–1867.

Ind. Fil. Moore, T. *Index Filicum* **1**. *Synopsis of the genera.* 1857. **2**. *Enumeration of the species.* 1857–1862.

Ind. Kew. Jackson, B. D. *Index Kewensis.* **1–15**. London. 1893–1974.

Jour. Bot. (Hooker) *The Journal of Botany, being a second Series of the Botanical Miscellany* ... *by W. J. Hooker.* London. **1–4**, 1834–1842.

Jour. Bot. (London). *The Journal of Botany, British and foreign.* London. **1–80**, 1863–1942.

Jour. Bot. Appl. *Journal de Botanique, appliquée à l'Agriculture, à la Pharmacie, à la Médecine et aux Arts.* Paris. **1–4**. 1813–1816.

Jour. Bot. Rédigé. Desvaux, A. N. and others, ed. *Journal de Botanique rédigé par une Société de Botanistes.* **1–2**. Paris. 1808–1809.

Jour. Hist. Nat. Paris. *Journal d'Histoire naturelle.* Paris **1–2**, 1792.

Jour. Linn. Soc. London (Bot.) *The Journal of the Proceedings of the Linnean Society. Botany.* London. **1–7**, 1856–1864; *Botanical Journal of the Linnean Society,* **62** →, 1969 →.

Kew Bull. *Kew Bulletin,* London. **1** →, 1946. →.

Kungl. Svenska Vet. Akad. Handl. *Kungliga svenska Vetenskaps-Akadamiens Handlingar,* Nov. ser. **26–63**, 1855–1923. Ser. 3, **1–25**, 1924–1948. Ser. 4, **1** →, 1951 →.

Lilloa. *Lilloa.* Tucumán. 1937 →

Linnaea *Linnaea. Ein Journal für die Botanik in ihrem ganzen Umfange.* Berlin. **1–43**, 1826–1882. (**9–34**, 1834–1866, at Halle.

Mantissa Linné, C. von. *Mantissa Plantarum.* Stockholm. 1767.

Med. Repos. *The Medical Repositary.* New York. Ser. 1, **1–15**. 1797–1812. Nov. ser., **1–15**, 1812–1826.

Mém. Acad. Sci. Pétersb. *Mémoires de l'Academie impériale des Sciences de St. Pétersbourg (Classe des Sciences Physiques et Mathematiques).* St. Pétersbourg. Ser. 5, **1–11**, 1809–1830, Ser. 6, **1–10**, 1831–1859. Ser. 7, **1–42**, 1859–1897.

Mém. Acad. Sci. (Turin). *Mémoires de l'Academie royale des Sciences.* Turin. Ser. 1, **1–22**, 1784–1816.

Mém. Soc. Linn. Paris. *Mémoires de la Société linnéenne de Paris.* Paris. **1–6**. 1822–1828.

Mém. Soc. Hist. Nat. Paris. *Mémoires de la Société d'Histoire naturelle de Paris.* Paris **1799**, 1799; **1–5**, 1823–1834.

Mem. Torrey Bot. Club *Memoirs of the Torrey botanical club.* New York. 1889 →. (**18–19**, 1931–1941, at Menasha, Wisconsin; **20**, 1943–1954, at Lancaster, Pennsylvania; **21** →, 1958 →, at Durham, North Carolina).

Miss. Sci. Cap Horn Franchet, A. R. *Phanérogamie.* (In *Mission scientifique du Cap Horn,* 1882–1883. **5**, 313–400). Paris. 1889.

Monad. Class. Diss. Dec. Cavanilles, A. J. *Monadelphiae classis dissertationes decem.* Madrid. 1785–1790.

Monogr. Epilob. Haussknecht, C. *Monographie der Gattung Epilobium.* Jena. 1884.

Monogr. Primul. Lehmann, J. G. C. *Monographia generis Primularum.* Leipzig. 1817.

Monogr. Rumex Campderá, F. *Monographie des Rumex, précédée de quelques vues générales sur la famille des Polygonées.* Paris. 1819.

N. Amer. Fl. Rydberg. P. A. *et al. North American Flora.* Ser. 1, **1–34**. New York. 1905–1949.

Natürl. Pflanzenfam. Engler, H. G. A. and Prantl, K. A. E., *Die natürlichen Pflanzenfamilien nebst ihren Gattungen und wichtigeren Arten, insbesondere den Nutzpflanzen.* Ed. 1, **1–4**. Leipzig. 1887–1915. Ed. 2, **13–21**. Leipzig and Berlin. 1925–1953.

Ned. Kruidh. Arch. *Nederlandsch Kruidkundig Archief.* Leiden. 1846–1951.

Neues Jour. Bot. Schrader, H. A. *Neues Journal für die Botanik.* **1–4**. Erfurt. 1806–1810.

Nomencl. Bot. Steudel, E. G. von. *Nomenclator botanicus.* **1–2**. Ed. 1. Stuttgart & Tubingen. 1821. Ed. 2. **1–2**. Stuttgart & Tubingen. 1840–1841.

Not. Mus. La Plata (Bot.) *Notas del Museo de La Plata. Botánica.* Buenos Aires. **1–6**. 1935–1941. La Plata **7** →, 1942 →.

Not. Syst. (Leningrad) *Notulae systematicae ex Herbario Instituti botanici Academiae Scientiarum URSS,* **7–8(3)**. 1937–1938.

Notas Helechos Chile Looser, G. *Notas sobre Helechos,* 2. Santiago de Chile. 1936.

Notes Roy. Bot. Gard. Edinb. *Notes from the Royal Botanic Garden, Edinburgh.* Edinburgh. 1900. →.

Nova Acta Acad. Leop.-Carol. *Nova Acta (Leopoldina) Academiae Caesareae Leopoldino-Carolinae germanicae Natural curiosorum.* Norimbergae 1757–1982. Nov. Ser. 1934 →.

Nov. Gen. Sp. Humboldt, F. H. A. von, Bonpland, A. J. A. and C. S. Kunth. *Nova genera et species plantarum quas in peregrinatione orbis novi collegerunt, descripserunt, partim adrumbraverunt Amatus Bonpland et Alexander de Humboldt ...* **1–7**, Paris. 1816–25.

Nouv. Fl. Env. Paris Mérat, F. V. *Nouvelle Flore des Environs* de Paris. Ed. 1. Paris. 1812. Ed. 2. **1–2**. Paris. 1821. Ed. 3. **1–2**. Paris. 1831–1834. Ed. 4. **1–2**. Paris. 1836. Ed. 5. **1–2**. Bruxelles. 1837–1838. Ed. 6. Bruxelles. 1841.

N. Z. Jour. Bot. *New Zealand Journal of Botany.* Wellington. 1963 →.

Öfvers. Kongl. Vet.-Akad. Förhandl. *Öfversigt af kongl. Vetenskaps-Akademiens Förhandlingar.* Stockholm. **1–63**, 1844–1921.

Opera Lilloana *Opera Lilloana.* Tucumán. 1957 →.

Orchid. Gen. Spec. Kraenzlin, F. W. L. *Orchidacearum genera et species exposuit.* **1–2**. Berlin. 1897–1904.

Österr. Bot. Zeitschr. *Österreichische botanische Zeitschrift.* Vienna. 1858–1974. Titled **Pl. Syst. Evol.** *Plant Systematics and Evolution.* **123** →, 1975 →.

Parodiana. *Parodiana.* Buenos Aires. 1981 →.

Pflanzengatt. Medicus, F. C. *Pflanzengattungen nach dem Inbegriffe sämmtlicher Fruktifikationstheile gebildet und nach dem Sexual-Pflanzenregister geordnet.* Mannheim. 1792.

Pflanzenreich Engler, H. G. A., ed. *Das Pflanzenreich, Regni Vegetabilis Conspectus.* **1–107**. Leipzig and Berlin. 1900–53.

Philos. Mag. *The Philosophical Magazine.* London. **1** →, 1798 →.

Physis *Physis.* Buenos Aires 1 →, 1912 →.

Pittonia. *Pittonia. A series of Papers relating to Botany and Botanists.* **1–5**. Berkeley. 1887–1905.

Pl. Eur. Richter, K. *Plantea Europeae, enumeratio systematica et synonymica plantarum phanerogamicarum in Europa sponte crescentium vel mere inquilinarum.* **1–2**. Leipzig. 1890–1903.

Pl. Icon. Ined. Smith, J. E. *Plantarum icones hactenus ineditae, plerumque ad plantas in Herbario Linneano conservatas delineatae.* London. 1789–91.

Prim. Fl. Esseq. Meyer, C. A. *Primitiae florae essequeboensis ...* Gottingen. 1818.

Prim. Fl. Holsat. Wiggers, F. H. *Primitiaue florace holsaticae. Kiel.* 1780.

Pl. Syst. Evol. See *Österr. Bot. Zeitschr.*

Proc. Amer. Acad. Arts *Proceedings of the American Academy of Arts and Sciences.* Boston. 1846 →.

Proc. Linn. Soc. London *Proceedings of the Linnean Society of London.* London. 1838 →.

Prodr. Candolle, A. P. de. *Prodromus systematis naturalis regni vegetabilis, sive enumeratio contracta ordinum, generum, specierumque plantarum ...* **1–17**. Paris. 1824–74.

Prodr. Aethéog. Beauvois, A. M. F. J. Palisot de. *Prodrome des cinquième et sixième familles de l'Aethéogamie. Les Mousses, les Lycopodes.* Paris. 1804–05.

Prodr. Stirp. Gotting. Murray, J. A. *Prodromus designationis stirpium Gottingensium.* Göttingen. 1770.

Reg. Veg. Syst. Nat. Candolle, A. P. de. *Regni vegetabilis systema naturale, sive ordines, genera et species plantarum.* **1–2**. Paris. 1817–21.

Reise Erde Meyen. F. J. F. *Reise um die Erde ....* **1–4**. Berlin. 1834–1843.

Reliq. Haenk. Presl, C. B. *Reliquiae Haenkeanae seu descriptiones et icones plantarum quas ... collegit T. Haenke.* **1–2**. Prague 1830–35.

Rep. Princeton Univ. Exped. Patag. *Reports of the Princeton University Expeditions to Patagonia 1896–1903. Flora Patagonica, Flowering Plants,* by G. Macloskie **8**. pts. **1–9**. 1903–1906. *Revision of Flora Patagonica,* by G. Macloskie and P. Dusén **8. Suppl**. 1914.

Résult. Voy. S.Y. Belgica Wildeman, E. de. *Les Phanérogames des Terres Magellaniques. Résultats du Voyage de la Belgica, Rapports Scientifiques, Botanique.* Anvers. 1905.

Rev. Argent. Argon. *Revista Argentina de Agronomia.* **1** →, Buenos Aires. 1934 →.

Rev. Fac. Agron. La Plata. *Revista de la Facultad de Agronomia. Universidad Nacional de la Plata.* **1–6**. 1895–1904. Ser. 2. **1** →, 1905 →.

Revis. Gen. Kuntze, K.E.O. *Revisio Generum Plantarum vascularium omnium.* **1–3**. Leipzig. 1891–1898.

Révis. Gram. Kunth. C. S. *Révision des Graminées publiées dans les Nova Genera et Species Plantarum de Humboldt et Bonpland.* Paris. 1829–1834.

Rev. Mus. La Plata Bo. *Revista del Museo de La Plata. Sección Botánica.* La Plata 1936/38 →.

Rhodora *Rhodora. Journal of the New England botanical Club.* Boston, Massachusetts. 1899 →.

Sinensia *Sinensia. Contributions from the Metropolitan Museum of Natural History* Academia sinica. Nanking. **1** →, 1929 →. Later at Shanghai.

Spicil. Fl. Lips. von Schreber, J. C. D. *Spicilegium Florae lipsicae.* Leipzig. 1771.

Spicil. Fl. Rumel. Grisebach, A. H. R. *Spicilegium Florae rumelicae et bithynicae.* **1–2**. Brunswick. 1843–1845.

Sp. Fil. Hooker, W. J. *Species filicum ...* **1–5**. London. 1846–1864.

Sp. Pl. Linné, C. von. *Species plantarum, exhibentes plantas rite* *cognitas, ad genera relatas, cum differentiis specificis, nominibus trivialibus* ... Ed. 1. Stockholm. 1753; Ed. 2. 1762–63; Ed. 3. **1–5**. Berlin. 1797–1810. Ed. 4, by C. L. Willdenow.

Spec. Umbellif. Sprengel, C. *Species Umbelliferarum minus cognitae*. Halle. 1818.

Sunyatsenia *Sunyatsenia. Journal of the Botanical Institute, Sun Yatsen University*. **1–6**. Canton. 1930/34–1946.

Suppl. Linné, C. von. *Supplementum plantarum systematis vegetabilium editionis XII, generum plantarum editionis VI, et specierum plantarum editionis II*. Brunswick. 1781.

Svensk Bot. Tidskr. *Svensk botanisk Tidskrift*. Stockholm. 1907 →. (**16** →, 1922 →, at Uppsala).

Svenska Exped. Magellansl. *Svenska Expeditionen till Magellansländerna*. **3(5)**. *Die Gefässpflanzen der Magellansländer* by P. Dusén. Stockholm. 1900.

Symb. Bot. Vahl, M. H. *Symbolae botanicae, sive plantarum, tam earum, quas in itinere imprimis orientali, collegit Petrus Forskal*. **1–3**. Stockholm. 1790–94.

Symb. Fl. Argent. Grisebach. A. H. R. *Symbolae ad floram argentinam*. Göttingen. 1879.

Syn. Conif. Endlicher, S. L. *Synopsis coniferarum*. Sankt-Gallen. 1847.

Syn. Fil. Hooker, W. J. and Baker, J. G. *Synopsis filicum, or a synopsis of all known ferns etc*. London. Ed. 2. 1874.

Syn. Fil. Lycopod. Swartz, O. P. *Synopsis filicum, earum general et species systematica complectens. Adjectis Lycopodineis et descriptionibus novarum et rariorum specierum*. Kiel. 1806.

Syn. Gen. Comp. Lessing, C. F. *Synopsis generum Compositarum* ... Berlin. 1832.

Syn. Pl. Persoon, C. H. *Synopsis plantarum, seu enchiridium botanicum, complectens enumerationem systematicam specierum hucusque cognitarum*. **1–2**. Paris and Tübingen. 1805–07.

Syn. Pl. Glum. Steudel, E. G. von. *Synopsis plantarum glumacearum*. **1–2**. Stuttgart. 1853–55.

Syst. Bemerk. Grisebach, A. H. R. *Systematische Bemerkungen* ... Göttingen. 1854. (Preprint of article published in *Abhandlungen der Königlichen Gesellschaft Wissenschaften zu* Göttingen **6**: 89–138. 1856.)

Syst. Nat. Linné, C. von. *Systemae Naturae*. Ed. 11, **1–2**. Leipzig. 1762. Ed. 12, **2–3**. Stockholm. 1767–1768. Ed. 13, by J. F. Gmelin, **2**, Leipzig, 1791–1792. Ed. 15 by Roemer, J. J. and Schultes, J. A., **1–7**, Stuttgart, 1817–1830. Ed. 16 by Springel, C. P. J., **1–5**, Göttingen, 1824–1828.

Syst. Veg. *Systema vegetabilium*. Ed. 14 by J. A. Murray. Göttingen and Gotha. 1774. Ed. 15 by J. J. Roemer and J. A. Schultes, Stuttgart. 1817–30. Ed. 16 by C. P. J. Sprengel. Göttingen. 1824–28.

Syst. Veg. Peruv. Ruiz, H. and Pavón, J. A. *Systema vegetabilium florae peruvianae et chilensis*. **1**. Madrid. 1798.

Syst. Verz. Erfurt. Bernhardi, J. J. *Systematisches Verzeichniss der Pflanzen, welche in der Gegend um Erfurt gefunden werden*. Erfurt. 1800.

Tent. Pteridogr. Presl, C. B. *Tentamen pteridographiae, seu genera filicacearum praesertim juxta venarum decursum et distributionem exposita*. Prague. 1836.

Tabl. Encycl. Méth. Bot. Monnet de la Marck, J. B. A. P. *Tableau encyclopédique et méthodique des trois Règnes de la Nature. Botanique*. **1–3**. Paris. 1791–1823.

Tanacet. Schultz, C. H. *Ueber die Tanaceteen, mit besonderer Berücksichtigung der deutschen Arten*. Neustadt. 1844.

Trans. Linn. Soc. London *Transactions of the Linnean Society*. London. Ser. 1, **1–30**. 1791–1825. Ser. 2, *Botany*, **1** →, 1875 →. From 1939 *Botany* and *Zoology* are combined.

Trans. Proc. N. Z. Inst. *Transactions and Proceedings of the New Zealand Institute*. Wellington, **1–63**, 1869–1934.

Trans. Proc. Roy. Soc. N. Z. *Transactions and Proceedings of the Royal Society of New Zealand*. Wellington. 1934–1952.

Univ. Calif. Publ. Bot. *University of California Publications in Botany*. Berkeley, California. 1902 →.

U. S. Expl. Exped. *The Botany of the United States Exploring Expedition during the Years 1838–42 under the Command of C. Wilkes*. Vol. 15. *Phanerogamia* by A. Gray. Philadelphia. 1854/56.

Veg. Siber.-Mong. Front. Printz, H. *The Vegetation of the Siberian–Mongolian Frontiers*. Trondhjem. 1921.

Veg. Syst. Hill, J. *The vegetable system*. London. **1–26**. 1759–75.

Verh. Bot. Ver. Brandenb. *Verhandlungen des botanischen Vereins für die Provinz Brandenburg und die angrenzenden Länder*. Berlin. **1** →, 1859 →.

Voy. Coquille. Brongniart, A. T. *Botanique (Phanérogamie) du voyage autour du monde ... sur la corvette la Coquille*. Paris. 1829–34.

Voy. Pôle Sud. Dumont d'Urville, J. S. C. *Voyage au Pôle sud et dans l'Océanie sur les corvettes l'Astrolabe et la Zelée* ... Paris. *Botanique* **2** *Plantes Vasculaires*, by J. Decaisne. 1853. *Atlas* by J. B. Hombron and C. Jacquinot, pts. **1–13**. 1843–1853.

Wiss. Ergebn. Schwed. Südpolarexped. *Wissenschaftliche Ergebnisse der Schwedische Südpolar-Expedition 1901–1903*. **1–5**. Stockholm. 1905–1909.

XI. Index to Scientific Names of Fuegian Vascular Plants

Synonyms are shown in italics

52
53°
54°
55°
Isla Desolación
Bahía Felipe
Bahía Gente Grande
R. del Oro
Punta Arenas
Porvenir
Altos del Boq
BAHÍA INU
Peninsula de Brunswick
Isla Santa Inés
ESTRECHO DE MAGALLANES
Isla Dawson
CANAL WHITESIDE
Isla Capitán Aracena
Isla Clarence
R. Grande
Canal Cockburn
Peninsula Brecknock
Cordillera
PACIFIC OCEAN
Isla G
Islas Londonderry
75°
74°
73°
72°
71°
70°